C0-AXQ-783

Conservation Principles
and the
Structure of Engineering
Fifth Edition

Charles J. Glover
Department of Chemical Engineering
Texas A&M University

Kevin M. Lunsford
Bryan Research and Engineering
Bryan, Texas

John A. Fleming
Department of Engineering
Texas A & M University

McGraw–Hill, Inc.
College Custom Series

New York St. Louis San Francisco Auckland Bogotá
Caracas Lisbon London Madrid Mexico Milan Montreal
New Delhi Paris San Juan Singapore Sydney Tokyo Toronto

McGraw-Hill's College Custom Series consists of products that are produced from camera-ready copy. Peer review, class testing, and accuracy are primarily the responsibility of the author(s).

Conservation Principles and the Structure of Engineering

Copyright © 1996, 1994, 1993, 1992, 1991 by McGraw-Hill, Inc. All rights reserved. Printed in the United States of America. Except as permitted under the United States Copyright Act of 1976, no part of this publication may be reproduced or distributed in any form or by any means, or stored in a data base retrieval system, without prior written permission of the publisher.

Pages F17-19, F21-F36 from *Handbook of Chemistry & Physics*, Sixty-Fourth Edition by Chemical Rubber Company. Copyright © 1983 by CRC Press, Inc. Reprinted with permission from CRC Press, Boca Raton, Florida.

3 4 5 6 7 8 9 0 DOC DOC 9 0 9 8 7

ISBN 0-07-024259-3

Editor: Lorna Adams
Cover Design: Maggie Lytle
Printer/Binder: R. R. Donnelley & Sons Company

Dedicated to the memory of

Carl A. Erdman

for his dedication and leadership

to engineering education

This book is one of a series of four textbooks which are developed for use in a complete sophomore-level engineering science core in all university undergraduate engineering departmental curricula. The new engineering science core is being produced with support from the National Science Foundation's Office of Undergraduate Science, Engineering, and Mathematics and the College of Engineering at Texas A&M University.

This curriculum is a set of "first courses" in the engineering disciplines, cutting across the traditional disciplinary boundaries. As such, it seeks to provide students with an understanding of the handful of truly fundamental scientific principles which govern our world and the role they play in determining the structure of engineering, problem solving, and design. The presentation of these concepts and structure in a unified and deliberate way early in the curriculum provides a framework for effective assimilation of more advanced material in the subsequent courses in each engineering discipline. Providing a cohesive view of engineering early in the curriculum as opposed to discipline-based, limited-scope, and calculation-intensive courses, is the most important objective of this series. By introducing a single framework, we believe that students can absorb and appreciate their own disciplines more fully.

There are, however, other issues which provide us with further incentive for a broader approach. In recent years, there has been considerable growth in technologies which span the traditional disciplinary boundaries and therefore increase the need for all engineers to have a solid and broad engineering science foundation. Also, the international competitive picture has highlighted the need for improvements in our ability to invent and design technological products. It is our belief that understanding the fundamental engineering science base is an absolutely essential element in the design process.

The new curriculum consists of four courses, designed to be presented in a sequence of two in the fall semester and two in the spring. The course titles at Texas A&M are as follows:

1. Conservation Principles in Engineering
2. Properties of Matter
3. Modeling and Behavior of Engineering Systems
4. Conservation Principles for Continuous Media

It should be noted that the first two courses are designed as corequisites, to be taken concurrently. Material in each of these courses is required for a complete understanding of the other, although the Properties of Matter course is much more able to stand alone. The order of the other two courses is also important, although it is not essential that they be taken at the same time. Each of the courses builds upon the foundations which have been laid in the previous courses. The first two courses require a solid freshman-year foundation in physics (mechanics), calculus I, and chemistry.

The major theme for the new engineering science core deals with instilling an understanding of the foundational conservation concepts (and related concepts, including the second law of thermodynamics), the structure which they provide to engineering science, analysis, and design, and their application to the traditional engineering disciplinary domains (solid and fluid mechanics, electrical science, thermodynamics, and heat transfer). The first course, Conservation Principles in Engineering, presents the foundation and structure for all engineering science through development of the conservation principles for lumped or macroscopic systems and applications in each of the above mentioned domains. This approach emphasizes the fundamental role of conservation in understanding and predicting the behavior of simple engineering systems.

The second course, Properties of Matter, parallels the first with a science-based development of the properties of solids, liquids, and gases which is essential for the application of conservation principles to the elements of engineering systems. These include mechanical properties, electrical and magnetic properties, optical properties, and thermal properties.

The third course, Modeling and Behavior of Engineering Systems, extends the application of conservation principles to more complex systems in all of the engineering science domains, including multi-domain systems. The focus is on predicting the response of systems through the application of engineering judgment to reduce the complexity of systems and develop sound engineering problem-solving methodology.

The final course in the new sequence, Conservation Principles for Continuous Media, further develops the general theme by examining the application of conservation principles to continuous media for which internal changes are important to understanding system behavior. This course deals with areas such as solid and fluid mechanics, heat transfer, and introduces the need for understanding the use of partial derivative information for the solution of many engineering problems. The inclusion of numerical tools for the solution of realistic problems is an important part of this course.

We have not intended for this curriculum to provide a one-for-one replacement for existing courses. Rather, we started with a clean sheet of paper and attempted to define what the engineering science core should be. As with any major curriculum modifications, other changes to present departmental curricula are inevitable. Thus, it will be necessary to make some adjustments to upper level courses in order to reap the full benefit of this new approach.

One further note about design: we believe that the use of realistic design examples for lectures, problems, and projects is a most effective pedagogical tool for developing understanding of the fundamentals of engineering science. Accordingly, in parallel with the development of this textbook series, a Design Resource Handbook is being developed to assist instructors in the use of design information with this objective in mind.

David G. Jansson
Carl A. Erdman
Texas A& M University
August, 1990

AUTHORS' PREFACE

Conservation Principles and the Structure of Engineering is divided into seven parts. Part I provides a perspective or framework for the rest of the text, and in fact for much of the engineering curriculum. Chapter 1 is an introduction to engineering problems and the way that engineers solve problems. It also provides more detail on the purposes of the conservation principles course and on the purposes of future courses. Chapter 2 continues with this perspective by presenting the overall picture of engineering analysis and the importance of the conservation concept to this picture.

Part II presents the fundamental laws upon which engineering and science are based. Chapters 3 through 8 address the concepts and equations representing the laws of mass, charge, linear momentum, angular momentum, and energy conservation, plus the second law of thermodynamics. The concepts are presented and illustrated by fairly simple problems. By the end of Chapter 8 the fundamental equations governing a great many areas of engineering including statics and dynamics, fluid mechanics, and heat transfer have been developed.

Parts III through VI cover applications of these principles to a number of traditional engineering areas. This material presents a great many more examples and applications of the laws presented in Part II. The examples are from a wide range of disciplines including rigid body mechanics, fluid mechanics, electrical circuits, and thermodynamics. In addition to providing the student with experiences in applying the physical laws, such problems also illustrate how mathematical tools and techniques are applied to meaningful engineering problems. Throughout, an effort is made to provide a clear separation between those aspects which are physics and those which are solution techniques.

This text focuses on conservation concepts: what is meant by a conservation law and an accounting statement, how we count both conserved and non-conserved properties, which properties are conserved and which are not, and why these laws are so important to our achieving an understanding of engineering and science. Consequently, in the early chapters the calculation of the various properties which appear in the conservation laws and accounting equations are de-emphasized. Then, in Parts III through VI, as material properties and their calculation become familiar to the student through a companion properties of matter course, problems which include such calculations are included.

Finally, Section VII provides a summary and perspective. Included is, first, a tabular summary of the principle laws covered in this text, and second, an analogous tabular presentation of the laws written for a differential element of a continuum. Their derivation, application, and solution are well beyond the scope of this text, but their inclusion further emphasizes that such equations for solving detailed solid and fluid mechanics problems are nothing more than the same conservation and accounting equations in another form.

In a sense, an engineering curriculum may be viewed as a giant and complex jigsaw puzzle. Each course is a piece of the larger puzzle and the parts of a course are themselves pieces of a smaller puzzle. It is our belief that by seeing a coherent picture of the puzzle early in the curriculum, students can assemble the pieces more easily and with better results.

Needless to say, this book would not have happened without the contributions of a great many people whom we gratefully acknowledge. Dave Jansson and Carl Erdman saw a need for a different approach to the engineering core material. A multidisciplinary team which included Louis Everett, Harry Jones, Mike Rabins, Cyrus Ostowari, Tom Pollock and Vic

Willson wrestled with the philosophy and content of these courses. Additionally, the members of the external advisory board provided their time and many helpful suggestions.

Dorothy Jordan performed exceptional service in transcribing first-draft tapes of the text, working through many tedious rewrites, and numerous other details required to prepare the final draft. Chuck Gazaway, Tracy Ingram, Louis Malarcher, Jennifer Nottingham, and Jeff Moore were lifesavers in drafting many of the figures at the last minute. Thanks also go to Christine Sanchez and Lillie Lewis for making last minute corrections to the copy.

The students of the inaugural teaching of this course in the fall of 1989 risked the unknown and helped us greatly with feedback on materials: E. J. Beam, B. S. Beardsley, C. L. Booker, B. J. Bunker, C. L. Castrianni, D. E. Dascher, F. M. Davis, J. D. Deakins, E. J. Deal, D. L. Duren, D. R. Faulkner, K. R. Faulkner, J. R. Fowler II, C. S. Gallun, M. E. Grein, J. P. King, D. E. May, R. G. McDonald, W. C. McDonald, J. K. Miller, K. D. Perkins, P. E. Poirot, M. K. Poor, B. A. Pratt, L. A. Prince, M. L. Pusey, N. M. Rendon, C. P. Rhodes, C. E. Rodriguez-Cayro, J. T. Rogers, B. A. Ryza, D. E. Sanders, C. M. Schuster, C. L. Shipman, T. T. Shippy, E. H. Simonson, B. P. Smaistrla, R. G. Spitzer, J. C. Thomas, M. E. Thompson, K. A. Tilley, M. K. Usey, N. J. Wyman, and G. D. Younglove.

And, of course, we thank the National Science Foundation, Office of Undergraduate Science, Engineering, and Mathematics for supporting a fresh look at engineering education.

Finally, we especially appreciate the love, understanding, and patience of our families for putting up with us for the last year and a half. Surely this was the most difficult task of all! Thank you Diane, James, and Matthew (CJG), Ron and Joyce (KML), and Jo-Ann (JAF).

Charles J. Glover
Kevin M. Lunsford
John A. Fleming
August, 1990

TABLE OF CONTENTS

Dedication   iii

Preface   iv

I PERSPECTIVES   1

1   Introduction   3

2   The Big Picture   7

2.1   Introduction   7
2.2   The Accounting Concept   8
2.3   Conservation Laws vs. Accounting Statements   13
2.4   Conservation Laws and the Structure of Engineering   20
2.5   Review   23

II THE FUNDAMENTAL LAWS   33

3   The Conservation of Mass   35

3.1   Introduction   35
3.2   Multicomponent Systems   38
3.3   Rate Equation   42
3.4   Multicomponent Rate Expressions   44
3.5   Multiple Inlets and Outlets   48

3.6	Steady State	48
3.7	Multiple Unit Processes	50
3.8	Recycle and Bypass	52
3.9	Density in Conservation of Mass	55
3.10	Chemical Reactions	59
3.11	Closed and Open Systems	67
3.12	Review	67
3.13	Notation	68
3.14	Suggested Reading	68

| 4 | **The Conservation of Charge** | 79 |

4.1	Introduction	79
4.2	Accounting of Positive or Negative Charge	79
4.3	Conservation of Net Charge	81
4.4	Rate Equation	89
4.5	Steady State–Kirchoff's Current Law	94
4.6	Review	97
4.7	Notation	98
4.8	Suggested Reading	98

| 5 | **The Conservation of Linear Momentum** | 105 |

5.1	Introduction	105
5.2	Conservation of Linear Momentum Equations	107
5.3	Linear Momentum and Newton's Laws of Motion	114
5.4	Steady State	116
5.5	Other Special Cases	116
5.6	Reference Frame	124
5.7	Review	125
5.8	Notation	125
5.9	Suggested Reading	126

| 6 | **The Conservation of Angular Momentum** | 135 |

6.1	Introduction	135
6.2	Conservation of Angular Momentum Equations	137
6.3	Steady State	140
6.4	Angular Momentum of the System	142
6.5	Rigid Body Angular Momentum Conservation	155
6.6	The Equation of Rotational Motion	165
6.7	Linear Momentum Versus Angular Momentum	165
6.8	Review	166
6.9	Notation	167
6.10	Suggested Reading	168

7	The Conservation of Energy	175
7.1	Introduction	175
7.2	Energy Possesed by Mass	177
7.3	Energy Possesed by Charge	183
7.4	Energy in Transition	185
7.5	The Conservation of Energy Equations	187
7.6	Steady State	193
7.7	Energy Accounting Equations	197
7.8	Review	218
7.9	Notation	219
7.10	Suggested Reading	221

8	The Second Law of Thermodynamics	231
8.1	Introduction	231
8.2	Entropy Accounting Equations	234
8.3	Other Concepts Related to Entropy and the Second Law	242
8.4	Review	244
8.5	Notation	244
8.6	Suggested Reading	245

III RIGID BODY MECHANICS — 255

9	Rigid Body Statics	257
9.1	Introduction	257
9.2	Particle Versus Rigid Body	257
9.3	Applicability of the Basic Equations	258
9.4	Particle Statics	259
9.5	Rigid Body Statics	262
9.6	Analysis of Structures and Machines	277
9.7	Rigid-Body Statics Problem-Solving Tools	299
9.8	Review	299
9.9	Notation	300
9.10	Suggested Reading	310

10	Rigid Body Dynamics	313
10.1	Introduction	313
10.2	Kinematics	314
10.3	Particle Dynamics	338
10.4	Rigid Body Dynamics	352
10.5	Rigid-Body Dynamics Problem-Solving Tools	372
10.6	Review	373
10.7	Notation	374

10.8	Suggested Reading	375

IV MACROSCOPIC FLUID MECHANICS — 383

11   Fluid Statics — 385

11.1	Introduction	385
11.2	Dependence of Pressure on Position in a Static Fluid	386
11.3	Forces on Submerged Areas	392
11.4	Buoyancy	404
11.5	Thin-Walled Pressure Vessels	407
11.6	Pressure Drop Across Static Curved Fluid Interfaces	410
11.7	Fluid Statics Problem-Solving Tools	413
11.8	Review	414
11.9	Notation	414
11.10	Suggested Reading	416

12   Fluid Dynamics — 423

12.1	Introduction	423
12.2	Review of the Basic Equations	424
12.3	Flow in Pipes and Other Conduits	433
12.4	Forces Associated with Fluid Systems	459
12.5	Fluid Dynamics Problem-Solving Tools	467
12.6	Review	467
12.7	Notation	468
12.8	Suggested Reading	470

V ELECTRICAL CIRCUITS — 475

13   Resistive Electrical Circuit Analysis — 477

13.1	Introduction	477
13.2	Equivalent Resistance	480
13.3	Source Transformation	489
13.4	The Superposition Principle	491
13.5	Nodal Anaylsis	493
13.6	Mesh Anaylsis	497
13.7	Thevenin Equivalent Circuit	500
13.8	Maximum Power Transfer	502
13.9	Resistive Circuit Problem-Solving Tools	505
13.10	Review	510
13.11	Notation	507

13.12	Suggested Reading	507

14	Inductance and Capacitance	513
14.1	Introduction	513
14.2	Review of the Basic Equations	513
14.3	First Order Response	519
14.4	Second Order Response	527
14.5	Inductive, Capacitive Circuit Problem-Solving Tools	529
14.6	Review	530
14.7	Notation	531
14.8	Suggested Reading	531

VI THERMODYNAMICS

535

15	Calculation of Thermal Properties of Materials	537
15.1	Introduction	537
15.2	Applicability of the Fundamental Laws	538
15.3	Thermodynamic Properties of Homogeneous Materials of Constant Composition	540
15.4	Volumetric Properties of Materials	572
15.5	Thermodynamic Problem Solving Tools and Strategy	580
15.6	Thermodynamics Problems and Special Forms of the Property Relationships	585
15.7	Phase Transitions	603
15.8	Enthalpy Changes Associated with Chemical Reactions	609
15.9	Review	615
15.10	Notation	618
15.11	Suggested Reading	619

16	More on Entropy and the Second Law of Thermodynamics	629
16.1	Introduction	629
16.2	Analysis of Cyclic Processes	630
16.3	Physical Observations Relating to Temperature and the Second Law of Thermodynamics	634
16.4	Availability	640
16.5	Irreversibility	643
16.6	Review	643
16.7	Notation	644
16.8	Suggested Reading	645

17	Power and Refrigeration Cycles	649
17.1	Introduction	649
17.2	Power Cycles	650
17.3	Refrigeration Cycles	670
17.4	Review	676
17.5	Notation	677
17.6	Suggested Reading	678

VII PERSPECTIVES AGAIN — 683

Appendix A: Dimensions and Units — 687

Appendix B: Vectors — 697

Appendix C: Engineering Data — 717

Index — 747

CONSERVATION PRINCIPLES
AND THE
STRUCTURE OF ENGINEERING

PART

I

PERSPECTIVES

CHAPTER
ONE
INTRODUCTION

As we assess the last fifty, seventy-five or 100 years, we have to be amazed at the great strides which have been made, both in terms of science, our knowledge of the world around us, and in technology, our ability to use that knowledge to change the manner in which we live.

Power generation cycles in the form of internal combustion engines used for transportation and for conversion to electric power have transformed our transportation systems and given us a ready supply of energy with which to perform an almost unimaginable array of tasks. In ground transportation, development of the steam engine led to intercontinental rail transportation. And then the internal combustion engine produced the automobile. And, of course, at the turn of the century the use of lightweight internal combustion engines and a highly efficient aircraft structure led to the first controlled man flight in 1903, marking the beginning of a revolution in travel.

In electronics, the invention of vacuum tubes led to sophisticated feedback circuit design and the development of signal processing, TV, radio, logic circuits and the first digital computers. The discovery of the ability to use semiconductor devices for the same functions led to miniaturization, first using transistors and then integrated circuits, to the point that today we have whole computers on a single chip. The use of these electronic devices has enabled us to expand our society in a great many ways and to change the way we do business and the way we live.

In the chemical industry, the development and manufacture of new chemicals and new materials also has contributed to our technological revolution. With the commercial introduction of nylon by DuPont in 1939, a whole new generation of materials, polymers, was introduced. Then, during World War II with the loss of our source of natural rubber in Indonesia, our country was faced with conducting a war without rubber. Our chemical industry developed the expertise and technology for the production of rubber and then designed and built the plants required to manufacture synthetic rubber to maintain our war effort. The design of new polymers and the design of processes for producing these polymers has led to their use in aircraft, spacecraft, everyday appliances, devices in our offices and homes, body implants, eyeglasses, membranes for separating or filtering materials, and on and on. The list is virtually endless.

Engineering advances in medicine also seem miraculous. The kidney dialysis machine is a lifesaving device for those with kidney malfunctions. Biotechnology is just beginning a major expansion in science and technology. Although in existence for centuries (fermentation processes and more recently the production of penicillin), many new processes are being developed for producing materials from biochemical reactions.

The point of these examples, and of course, many others could be listed, is to begin a discussion of what it takes to accomplish results such as these. What does it take to bring about the major scientific and engineering accomplishments? What does it take to implement new science in new devices and machines for use in our everyday lives? What does it take to understand the many different kinds of processes which are involved when we discuss the range of activities from electronics, to space exploration, to computers, to chemical reactions and production of all the synthetic materials which we use? What does it take to understand the biochemical processes which are now being developed? The science and mathematics courses which you have had to date have begun to develop your understanding. This course and your later engineering courses will further contribute to that development.

Throughout, you must keep in mind that there are really two levels of understanding which must be addressed. First is a *qualitative* level of understanding at which we are concerned with physical concepts which govern the various processes. This qualitative understanding has its roots in our understanding of physical, chemical, electrical, and biological phenomena. Features such as the flow of current from high potential to low potential, the flow of a fluid from a high pressure to a low pressure, the tendency of compounds or elements to react to form new compounds, the knowledge that an unsupported object will fall to the ground are all part of our experiences and lead us to understand or suspect the way certain things are going to happen.

However, this kind of qualitative understanding of phenomena is grossly inadequate from the viewpoint of designing practical devices or processes. Obviously, understanding from a qualitative, physical viewpoint is extremely important for the purpose of conceiving processes or devices, but from the standpoint of actually making them work, it is not enough. The second level of our understanding must be *quantitative*. That is, we have to know not just that current flows from high potential to low potential, but how much will flow in a given situation. When we allow an object to move through a fluid as in a free fall, we would like to know not just that it will fall but how fast. Concerning chemical reactions, we would like to know not just what reactions will occur but also how quickly they will occur and in what proportions the reactants are needed for the reaction.

It is this ability to quantify which is of paramount importance to scientists and engineers in being able to devise practical processes. Without an ability to quantify, we are not able to assess two different qualitative judgments about a process; which one is going to be more important and therefore, which one is going to dominate the design of the process. The quantitative level of understanding allows us to put all of the various qualitative ideas into perspective. Without this perspective, we cannot proceed.

It is our purpose to provide you with the big picture of engineering analysis and design from both the qualitative (conceptual) and quantitative (mathematical) levels. You also will begin to fill in many of the details of this picture and learn how to approach and solve a large number of engineering problems. Additionally, the foundation will be laid for solving a great many more problems which, because of their complexity, you are not yet ready to solve, but for which you are able to understand the foundations. This process will be an on-going and iterative process throughout your academic and professional career. Finally, we hope to provide considerable experience at analyzing a wide variety of problems from a number of engineering and scientific disciplines. By addressing this variety of problems, and by the manner in which they are approached, it is hoped that you will develop an understanding of work done in a great many engineering areas from a viewpoint which transcends the limits of any one particular discipline. Additionally, you will see a similarity in the manner in which problems are solved. It is a unified picture of analysis which is the overriding objective of this course and which you should keep well in mind throughout your academic career.

Your future engineering courses will provide more details of specific areas of engineering, science or design. Frequently, attention to the details will necessitate less attention to the big picture and to the underlying principles which are the foundation of our engineering analysis. Mathematical methods, procedures, numerical analysis, and jargon all are details which make up the trees in the forest. Without obtaining a good understanding of these details of a problem, you will not be able to proceed to a solution. The advanced courses will provide you with a depth of knowledge which is absolutely essential to obtaining a good understanding of engineering practice and application and also to broaden the base of your knowledge to the point necessary for creative solutions to tomorrow's problems. However, throughout these courses, while becoming immersed in these details, you must not forget the big picture, the forest which is created by these many trees. In every field of endeavor and in every application, you must be aware of the underlying principles which give rise to the governing equations. Only by keeping in mind these underlying principles can you truly develop a physical feeling and understanding for the problems at hand.

The big picture of engineering and science analysis will be outlined in more detail in the next chapter. At the heart of this picture is a handful of fundamental principles which govern the physical behavior and provide the need for other relations such as material properties. Interwoven within these laws and properties is a mathematical framework which is necessary as a means of expression and communication. Then, with this engineering, science, and mathematics foundation established, the challenge is to become skilled at applying it to specific situations and at solving the mathematical problem which arises.

As students, you should focus throughout your academic career on two distinct objectives. First, you must learn how to do things, what are the techniques, what are the procedures, what are the engineering rules of thumb, etc. Second, although not second in importance, is to understand why things are done in this way. Concentrating on understanding will give you a depth to your perspective of learning how to do things and it will enable you to learn and remember procedures much better. Keep in mind that it often is easy to become excessively wrapped up in the techniques of problem solving to the extent of overlooking the underlying foundations and physics of a problem. Only by focusing on why things are done the way they are can we develop sufficient understanding, physical feeling, and intuition for problems to be able to attack new and perhaps seemingly unrelated problems in creative ways.

CHAPTER
TWO
THE BIG PICTURE

2.1 INTRODUCTION

Conservation is the most important concept in engineering and science. A handful of conservation principles, together with the second law of thermodynamics, provide the governing laws for virtually all physical behavior. These laws, together with material properties and certain "given" conditions fully establish or dictate the destiny of the process at hand in a physical sense. With suitable mathematical descriptions of these laws, methods for calculating the material properties, and initial and/or boundary conditions, we can (at least in principle) describe the behavior of the process mathematically. Solutions of this mathematical problem provide a means of calculating answers to questions about its behavior, questions which may be concerned with designing a new device or process or with understanding or predicting the behavior of a given (existing) device or process. A clear understanding of conservation enables one to attack a wide variety of problems in a wide variety of disciplines with confidence and creativity.

The governing conservation laws are *mass, electrical charge, linear momentum, angular momentum, and energy*. These laws tell us that whatever processes we may devise or try to understand, they must obey these principles. The *second law of thermodynamics* places a further constraint on processes in that it says that not all of the processes that we can conceive which conserve these quantities are viable processe. Other processes, though they may fully satisfy conservation concepts, cannot exist. For example, a cup of hot coffee placed in a cold room will get colder rather than hotter even though the reverse process would not violate any of the conservation principles.

The analysis picture to be outlined in this chapter and followed throughout this text is not new or unique. In fact, the picture is presented and followed *because* it is not new and has withstood the test of time and experience; it is followed by engineers and scientists in their design, analysis, trouble-shooting, and research. What is new, however, is to emphasize conservation as a unifying concept for understanding the foundations of a wide variety of disciplines and the behavior of seemingly different phenomena and to do so up front in the engineering curriculum. By doing so, we hope to provide you with a well-defined framework upon which to build your understanding and application of a great many concepts, problems, processes, and phenomena. It is hoped that such a framework, well understood in the beginning, will help you to absorb information, and to develop a perspective of many disciplines that will spark your interest, creativity, and excitement in engineering and the world around you.

In this chapter, we present an overall picture of the conservation analysis approach. First, we discuss the procedure of accounting for things that can be counted and the elements that are required in order to adequately define and perform such an accounting. Second, we define the special concept of conservation in terms of this accounting concept and the conservation laws which provide the basis of engineering analysis. Third, we look at an overall scheme for analyzing engineering problems based upon using these fundamental laws. Finally, we consider several forms in which conservation statements may be made, for example, for finite or differential regions of space and for finite or differential time periods, the latter being a rate equation.

2.2 THE ACCOUNTING CONCEPT

2.2.1 Introduction to Accounting

The accounting concept must be stated and understood in very precise and complete terms before it can be applied. In order to quantify an accounting of any property, you must establish three things. First, you must recognize exactly which property you are counting: mass, momentum, charge, etc. Second, you must define a region of the universe within which the property is to be counted (called the "system"). Third, you must define the exact time period over which the accounting is to be made. Each of these elements is considered in more detail below.

First, *define the property to be counted*. This could be any of the properties or entities which can be counted. We define *extensive property* to be any property which can be counted. The extensive properties which may be counted are total mass, individual species mass (in a multicomponent system), electrical charge, linear momentum, angular momentum, total energy and entropy. Additionally, mechanical energy and thermal energy, subsets of total energy, may be counted separately. In the above list, total mass conservation is not independent of the species accountings; if there are n species, then there are n independent mass accounting statements to be chosen from the n species statements plus the 1 total mass conservation law ($n + 1$ mass equations, total). Note that the same property must be counted in each term within a single equation; you cannot include oranges in an accounting of apples (although both would be included in an accounting of *fruit*) and likewise in an accounting of mass, all quantities must be mass; in an accounting of energy, all quantities must be energy, etc.

Second, *define a system*. This means that you must divide the universe into two parts by defining a closed surface. The system is "inside" this boundary and the surroundings are "outside." This region of space may be quite large and contain within it a great many processes or activities, or it may be very small, even of differential size, and contain a single phase material undergoing no changes or processes. The shape, size, and location of the boundary used to define the system is entirely up to you and will be dictated by the information which you desire to learn. Defining the boundary says nothing whatsoever about what is happening within the system.

Third, *define a time period*. All elements of the accounting statement must be evaluated over the same time period.

Figure 2-1 is a schematic of these steps in performing the accounting process. The property to be counted, the system boundary, and the time period are defined by you as is convenient for solving the problem at hand and provide the context for the accounting statements. *Without that context, an accounting statement has no meaning.*

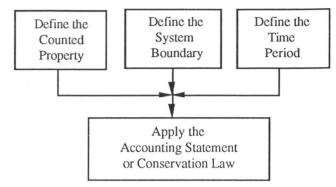

Figure 2-1: Steps to Accounting

Furthermore, a pictorial representation of a system and surroundings is given in Figure 2-2. The arrows depict an extensive property entering and leaving the system by crossing the system boundary. All quantities must be evaluated during the same time period (also called the "basis").

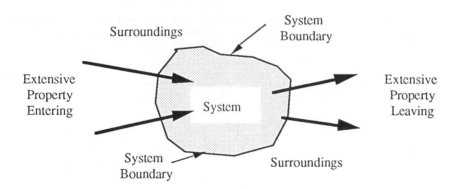

Figure 2-2: System and Surroundings

With these three elements understood, you are ready to make a verbal statement of accounting. In this statement you must be very explicit about the role or importance of each of these elements. The essence of the accounting statement is that the amount of the property being counted which is inside the system changes because of input, output, generation, and consumption. The general statement of accounting for a extensive property (E.P.) is:

$$\left\{ \begin{array}{c} \text{Amount of} \\ \text{Accumulation of} \\ \text{E.P. within system} \\ \text{during time period} \end{array} \right\} = \left\{ \begin{array}{c} \text{Amount of} \\ \text{E.P. entering system} \\ \text{during time period} \\ \text{(Input)} \end{array} \right\} - \left\{ \begin{array}{c} \text{Amount of} \\ \text{E.P. leaving system} \\ \text{during time period} \\ \text{(Output)} \end{array} \right\}$$

$$+ \left\{ \begin{array}{c} \text{Amount of} \\ \text{E.P. generated} \\ \text{within system} \\ \text{during time period} \end{array} \right\} - \left\{ \begin{array}{c} \text{Amount of} \\ \text{E.P. consumed} \\ \text{within system} \\ \text{during time period} \end{array} \right\}$$

$$(2-1)$$

Where the accumulation is defined by the following equation:

$$\left\{ \begin{array}{c} \text{Accumulation} \\ \text{of E.P. within} \\ \text{system during} \\ \text{time period} \end{array} \right\} \equiv \left\{ \begin{array}{c} \text{Amount of E.P.} \\ \text{contained within} \\ \text{system at the end} \\ \text{of the time period} \end{array} \right\} - \left\{ \begin{array}{c} \text{Amount of E.P.} \\ \text{contained within} \\ \text{system at the beginning} \\ \text{of the time period} \end{array} \right\} \qquad (2-2)$$

Input and *output* refer only to the extensive property crossing the system boundary and then only during the time period of interest. *Generation* and *consumption* refer only to the system itself (inside the system boundaries). Generation outside the system is totally irrelevant to the system and accounting statement which is currently of interest. *Accumulation*, as shown above, is the difference between the amount of the extensive property contained within the system at the end of the time period and that contained within the system at the beginning of the time period. This is a definition of accumulation. It does not refer to the amount contained elsewhere or at other times.

Example 2-1: An Accounting of Dollars. To illustrate the accounting concept quantitatively, consider the process of analyzing your bank account upon receiving a monthly statement. In this case the entity being counted is dollars. The time period is the length of time since the last statement. The system is your bank account. That is, imagine that a boundary defines your checking account and that dollars cross the system boundary when entering or leaving the account. Also, dollars exist in the account. Deposits and interest fall in the category of dollars entering the account during the time period. Checks written since the last statement and service charges are dollars that left the account. Do not be concerned with any transactions that go on outside of your account (at least from the view point of analyzing this one particular account). Also, do not be concerned with interest or service charges except those which affect this account, and do not be concerned with deposits, withdrawals or checks made to any other account.

Table 2-1: Bank Account Statement of Activity (Dollars)

Date	Descr.	$ In account at end of time period	–	$ In account at beginning of time period	=	$ Added during time period	–	$ Removed during time period
		$(\$_{sys})_{end}$	–	$(\$_{sys})_{beg}$	=	$(\$)_{in}$	–	$(\$)_{out}$
7/31	Previous Balalance			500				
8/1	Pay Check					1000		
8/3	Apartment							200
8/4	Television							10
8/10	Utility							60
8/14	Phone							40
8/28	Service Charge							10
8/28	Interest					5		
9/1	Books							75
9/2	Food							50
9/3	Ending Balance	1060						
9/3	Sum	$ 1060	–	$ 500	=	$ 1005	–	$ 445

Table 2-1 shows a listing of transactions which occurred during the month for your account. The transactions are summarized in rows by date and in columns by category (as related to the above accounting statement) of the transaction. At the bottom of the table is the sum of each column indicating total input to the account during the time period, total withdrawals from the account during the time period, and the beginning and ending dollars in the account. Note that generation and consumption of dollars in the account are zero and therefore this category is omitted from the table. (See question 10 at the end of the chapter).

This last line total for the account illustrates the manner in which the ending dollar amount is calculated for the account and corresponds exactly to that described above in our general accounting statement, that is,

$$
\left\{\begin{array}{c} \$ \text{ in account} \\ \text{at end of} \\ \text{time period} \end{array}\right\} - \left\{\begin{array}{c} \$ \text{ in account} \\ \text{at beginning} \\ \text{of time period} \end{array}\right\} =
$$

$$
\left\{\begin{array}{c} \$ \text{ entering account} \\ \text{during time period} \end{array}\right\} - \left\{\begin{array}{c} \$ \text{ leaving account} \\ \text{during time period} \end{array}\right\} + \left\{\begin{array}{c} \$ \text{ generated} \\ \text{within account} \\ \text{during time period} \end{array}\right\} - \left\{\begin{array}{c} \$ \text{ consumed} \\ \text{within account} \\ \text{during time period} \end{array}\right\}
$$

(2 – 3)

Substituting the numerical values yields $1060 − $500 = $1005 − $445 + $0 − $0 and hence $560 = $560.

One further point should be made about this example. When you review your monthly bank statement there are really two time periods which are important. The first one is that discussed in the table above and is that used by the bank. This time period is the time from the previous statement to the closing of the current statement. All transactions and activity which occurred during that time period and which has actually passed through the account by the bank are included. The second time period is that which you consider when you reconcile your checkbook with the bank statement. In this case, the time period is that from the end of the previous statement to the time that you are reconciling your records. Consequently, all checks, deposits, etc., which occurred since the closing of the current monthly statement to the present time are included as well. Furthermore, any transactions which occurred before the beginning of the statement time period and which are still outstanding must also be included. These transactions may have cleared the bank by the time you are reconciling your account. When you have completed this reconciliation, see Table 2-2, you will have verified your own ledger against the bank's figures, and you will have the current account dollar amount should all outstanding items clear.

Table 2-2: Bank Account Reconciliation (Dollars)

Date	Descr.	$ In account at end of time period	–	$ In account at beginning of time period	=	$ Added during time period	–	$ Removed during time period
		$(\$_{sys})_{end}$	–	$(\$_{sys})_{beg}$	=	$(\$)_{in}$	–	$(\$)_{out}$
7/31	Previous Balance			500				
8/1	Pay Check					1000		
8/3	Apartment							200
8/4	Television							10
8/10	Utility							60
8/14	Phone							40
8/28	Service Charge							10
8/28	Interest					5		
9/1	Books							75
9/2	Food							50
9/3	Bank State.	$ 1075	–	$ 500	=	$ 1020	–	$ 445
9/5	Deposit					500		
9/6	Bill							65
9/10	Today	$ 1510	–	$ 500	=	$ 1520	–	$ 510

It should be emphasized again that this exact type of accounting will be followed throughout your engineering courses. It may not always be stated in this way or this explicitly, but you must understand that this is always what is done. Systems, the property being counted, and time periods will change so that the equation used in one case may bear little resemblance to that in another, but you must keep this kind of physical picture in mind.

The importance of how a system can be defined and the consequences of that system definition are described in the following example.

Example 2-2: An Accounting of Water. For illustration, consider a bucket which is half full of water containing 7 pounds mass (lb_m). (1 lb_m-read, "pound-mass"-is defined to be that amount of mass that weighs 1 lb_f-read "pound force"-when subjected to a gravitational force per mass (g) of 32.174 (ft / s^2). See Appendix A.) During some time period, lets say 1 hour, we add an additional 5 (lb_m) of water. Determine the amount of water contained in the bucket after the additional water was added. (The obvious answer is 12 pounds mass; however, this example is helpful to understanding the accounting concept.)

For the first case we define the system boundary to be the volume enclosed by the bucket. The extensive property is water, and the time period is 1 hour. Figure 2-3 gives a simple schematic of the process.

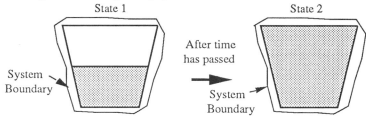

Figure 2-3: Accounting of Water System 1

The verbal statement of the accounting of water is given below and recognizes that the amount of water in the bucket (system) changes because of water added:

$$\left\{\begin{array}{c} \text{Water in system} \\ \text{at end of} \\ \text{time period} \end{array}\right\} - \left\{\begin{array}{c} \text{Water in system} \\ \text{at beginning} \\ \text{of time period} \end{array}\right\} = \left\{\begin{array}{c} \text{Water entering system} \\ \text{during time period} \end{array}\right\} - \left\{\begin{array}{c} \text{Water leaving system} \\ \text{during time period} \end{array}\right\}$$
$$+ \left\{\begin{array}{c} \text{Water generated} \\ \text{within system} \\ \text{during time period} \end{array}\right\} - \left\{\begin{array}{c} \text{Water consumed} \\ \text{within system} \\ \text{during time period} \end{array}\right\} \qquad (2-4)$$

For this system, there is no water leaving the system. Also, no water is being generated or consumed in the system during the time period. Therefore, the equation reduces to:

$$\left\{\begin{array}{c} \text{Water in system} \\ \text{at end of} \\ \text{time period} \end{array}\right\} - \left\{\begin{array}{c} \text{Water in system} \\ \text{at beginning} \\ \text{of time period} \end{array}\right\} = \left\{\begin{array}{c} \text{Water entering system} \\ \text{during time period} \\ \text{as input} \end{array}\right\} \qquad (2-5)$$

Table 2-3: Accounting of Water (Mass)

Water in System at end (m_W)end		Water in System at beginning (m_W)beg		Water Added during time period (m_W)in		Water Removed during time period (m_W)out		Water Generated time period (m_W)gen		Water Consumed time period (m_W)con
(lb_m)	$-$	(lb_m)	$=$	(lb_m)	$-$	(lb_m)	$+$	(lb_m)	$-$	(lb_m)
?	$-$	7	$=$	5	$-$	0	$+$	0	$-$	0

This equation is presented in tabular form (as was done for the bank account problem) in Table 2-3. The row of numbers in the table provides the equation to be solved:

$$? \ (lb_m) - 7 \ (lb_m) = 5 \ (lb_m)$$

The final amount of water in the system (bucket) at the end of the time period is 12 (lb_m).

Now, define a different system and apply the accounting statement. This new system is the volume defined by the water that was in the bucket at the begining of the time period. Figure 2-4 shows this system along with the process.

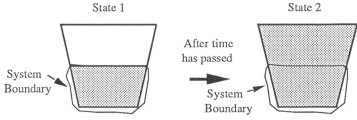

Figure 2-4: Accounting of Water System 2

The verbal accounting statement is given below for this system.

$$\left\{ \begin{array}{c} \text{Water in system} \\ \text{at end of} \\ \text{time period} \end{array} \right\} - \left\{ \begin{array}{c} \text{Water in system} \\ \text{at beginning} \\ \text{of time period} \end{array} \right\} = \left\{ \begin{array}{c} \text{Water entering system} \\ \text{during time period} \end{array} \right\} - \left\{ \begin{array}{c} \text{Water leaving system} \\ \text{during time period} \end{array} \right\}$$

$$+ \left\{ \begin{array}{c} \text{Water generated} \\ \text{within system} \\ \text{during time period} \end{array} \right\} - \left\{ \begin{array}{c} \text{Water consumed} \\ \text{within system} \\ \text{during time period} \end{array} \right\} \qquad (2-4)$$

For this system, there is no consumption or generation and (net) water entering the system is zero (why?). Therefore:

$$\left\{ \begin{array}{c} \text{Water in system} \\ \text{at end of} \\ \text{time period} \end{array} \right\} - \left\{ \begin{array}{c} \text{Water in system} \\ \text{at beginning} \\ \text{of time period} \end{array} \right\} = 0 \qquad (2-6)$$

There is no accumulation of water. Also, the fact that the amount of water in the bucket increased is irrelevant.

Although this problem was very trivial, the point here is that we are free to define any system for any problem. However, once that system is defined, the accounting statement applies strictly to that system. In this example, the second system that was defined did not enable us to calculate the amount of water contained in the bucket. For some problems, however, different systems can provide additional useful information. Remember, the system definition can be as creative as your imagination.

2.3 CONSERVATION LAWS VS. ACCOUNTING STATEMENTS

In the previous discussion, the concept of writing an accounting statement for extensive properties was presented. This concept involves defining a system, a property to be counted, and a time period. Then we account for the property by including input and output terms, generation and consumption terms, and changes in the amount of the property contained within the system. We can make this kind of statement for any extensive property, that is, for anything we can count. However, a distinction needs to be made between conservation laws and accounting statements.

Some properties are special in that not only can they be counted, but also, they are conserved.

> By conserved we mean that the property is neither created (generated) nor destroyed (consumed).

For example, we are familiar with the concepts of conservation of mass and conservation of energy meaning that mass and energy are neither created nor destroyed.

To expand upon this idea, consider the system and surroundings together. If we have an accounting statement for a conserved property for the defined system along with an accounting statement for the same property for the surroundings, then these two statements add together to give an accounting statement for the system and surroundings combined as shown in Table 2-4. Here all the input and output terms for both the system and surroundings are defined relative to the *system*.The input and output terms for the system and surroundings together must add to zero. We say the extensive property is conserved if there is no generation or consumption of the extensive property. Therefore, there is no net change in the amount of the extensive property contained within the system and surroundings combined. *The amount of a conserved property in the system and surroundings together is constant.* Equivalently, the amount of a conserved property in an isolated system (no input or output) is constant.

Table 2-4: Conservation Equations for the System and Surroundings

System	$(E.P._{sys})_{end}$	$-$	$(E.P._{sys})_{beg}$	$=$	$E.P._{in}$	$-$	$E.P._{out}$
Surroundings	$(E.P._{sur})_{end}$	$-$	$(E.P._{sur})_{beg}$	$=$	$E.P._{out}$	$-$	$E.P._{in}$
System and Surroundings	$(E.P._{sys+sur})_{end}$	$-$	$(E.P._{sys+sur})_{beg}$	$=$	0	$-$	0

For such a property, the amount contained within the system changes only as the result of input or output. Equivalently, *the only way that the amount of a conserved property in a system can change is by a one-for-one exchange with the surroundings.* So, by saying that mass is conserved, we mean that whatever mass is gained by the system during the time period must have been given up by the surroundings during the same time period. Likewise, because total energy is a conserved property, we know that whatever energy was gained by the system during the time period was given up by the surroundings during the same time period.

For extensive properties which are not conserved (such as mechanical energy, thermal energy, entropy), the total change in the amount of that property in the defined system is not necessarily exactly cancelled by the change in that property for the surroundings. Consequently, if the two accounting equations for the system and surroundings are added together, then the total change is not necessarily zero. A prime example is the second law of thermodynamics for which the entropy of the system and surroundings combined is always increasing. In this case, we really do not have a complete second law statement unless we look at both the system plus surroundings. This difference between conserved and non-conserved properties can be very important and is one that will be reemphasized throughout this text.

One further distinction must be made concerning accounting statements versus conservation laws. We are able to write an accounting statement about anything we can count. There is not anything special about them although they can be very useful. Conservation laws, however, are much more restrictive. Only a limited number of extensive properties have been observed (by very special and precise experiments) to obey conservation laws. It is this uniqueness which makes them (mass, charge, linear and angular momentum, and energy) important. The conservation laws, along with the second law constraint, enable us to understand the behavior of processes in our world and to create processes or machines to do our bidding.

Science and engineering, although governed by laws, are quite unlike our legal society. We live by our scientific laws whether we want to or not. Furthermore, we cannot change them! What we can do, however, is learn to make maximum use of them for a better world.

Example 2-3: Conservation of Mass. Consider the continuous flash distillation of a binary (two-component) mixture. The mixture of two components to be separated is fed continuously as a single-phase liquid at elevated temperature and pressure into a flash column. Once inside, the liquid is allowed to expand through a valve or nozzle into the column to a lower pressure. At this lower pressure, equilibrium (a property of the mixture) dictates the existence of liquid and vapor phases of different composition. The liquid phase will be richer in the less volatile component, i.e. will have a higher mole fraction of the less volatile component whereas the gas phase will be richer in the more volatile component. The vapor phase and the liquid phase can then be removed from the column as vapor and liquid products, each of which contains both of the feed components but in different proportions. We would like to know how these entering and leaving flow rates relate to each other.

For this problem it is convenient to define the distillation column contents as the system. Input of mass to the system then occurs through the feed stream and output of mass from the system occurs as the vapor and liquid product streams. The process is depicted schematically in Figure 2-5.

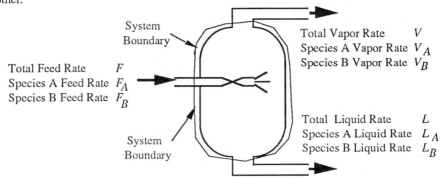

Figure 2-5: Flash Distillation

The verbal statement of the accounting of mass is given below:

$$\left\{\begin{array}{c}\text{Mass entering}\\\text{system during}\\\text{time period}\end{array}\right\} - \left\{\begin{array}{c}\text{Mass leaving}\\\text{system during}\\\text{time period}\end{array}\right\} = \left\{\begin{array}{c}\text{Accumulation of mass}\\\text{within system during}\\\text{time period}\end{array}\right\} \qquad (2-7)$$

Where the accumulation of mass is defined by the following equation:

$$\left\{\begin{array}{c}\text{Accumulation}\\\text{of mass within}\\\text{system during}\\\text{time period}\end{array}\right\} = \left\{\begin{array}{c}\text{Amount of mass}\\\text{contained in system}\\\text{at end of}\\\text{time period}\end{array}\right\} - \left\{\begin{array}{c}\text{Amount of mass}\\\text{contained in system}\\\text{at beginning of}\\\text{time period}\end{array}\right\} \qquad (2-8)$$

If the process is continuous and it is assumed that there is no accumulation of material within the column and that the flow rates are steady, then the conservation of mass is simple. Also, we will assume that no consumption or generation of mass is occurring in the system (no chemical reactions). Actually, we can consider accounting statements for each of the two species (which in this case, with no chemical reactions, become conservation statements even though they are not laws), and we can consider a conservation of total mass.

With the above assumptions, the verbal accounting statement for the mass is:

$$\left\{\begin{array}{c}\text{Mass entering}\\\text{system during}\\\text{time period}\end{array}\right\} - \left\{\begin{array}{c}\text{Mass leaving}\\\text{system during}\\\text{time period}\end{array}\right\} = 0$$

A steady-state mass accounting, for this system, says that for each component (species) the amount that enters the system during any given time period in the feed stream must equal the amount of that component that leaves the system during that same time period in the two product streams combined. Also, the total mass entering the system must equal the total mass which leaves in these two streams combined. These accounting statements are shown in Table 2-5.

Table 2-5: Flash Distillation (Mass)

Description	Rate of Mass Accumulation $\left(\dfrac{d(m_i)_{sys}}{dt}\right)$	$=$	Rate of Mass input $(\dot{m}_i)_{in}$	$-$	Rate of Mass output $(\dot{m}_i)_{out}$		
Species	(kg / s)		Stream F (kg / s)		Stream V (kg / s)		Stream B (kg / s)
Species A	0	$=$	F_A	$-$	V_A	$-$	L_A
Species B	0	$=$	F_B	$-$	V_B	$-$	L_B
Total	0	$=$	F	$-$	V	$-$	L

The terms F, V, and L represent the total mass flow rates in the feed, vapor, and liquid entering and leaving the distillation column. The subscripts A and B denote the species. For example, the liquid mass flow rate of species A from the distillation column is denoted as L_A.

From Table 2-5, the two accounting equations and one conservation law yield:

$$\text{Species A:} \qquad F_A - (V_A + L_A) = 0$$

$$\text{Species B:} \qquad F_B - (V_B + L_B) = 0$$

$$\text{Total Mass:} \qquad F - (V + L) = 0$$

It is important to realize that the total mass equation is the sum of the two species mass equations; therefore, only two of the three equations are independent. Total mass conservation can be expressed as:

$$\text{Total Mass} \qquad F_A + F_B - (V_A + V_B + L_A + L_B) = 0$$

In this example, we have not concerned ourselves with the compositions which represent equilibrium. These compositions must be known independently, from the thermodynamic equilibrium data for this mixture of components at the prescribed temperature and pressure. Data of these sort are the subject of the properties of matter and mixtures and are not the focus of our attention at this time.

Example 2-4: Conservation of Linear Momentum. Consider the motion of two balls of known mass which undergo a collision in the absence of a gravitational field (Figure 2-6). Before the collision they each have a certain velocity (i.e. momentum) and after the collision the velocities are changed.

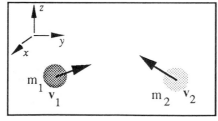

Before Collision

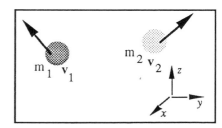

After Collision

Figure 2-6: Collision of Two Balls

If the system is chosen as the two balls together, then the total momentum of the system is simply the sum of the momenta of the two balls. If no other forces are acting on these balls, then once they collide, each ball's momentum changes, but the momentum of the two taken together remains the same. That is, in accordance with the conservation of linear momentum (LM) the change in the momentum of the system in a given time period is equal to the sum of momentum inputs and external forces acting on the system (forces acting at the system boundary exchange momentum between the system and surroundings) each integrated over the time period.

$$\left\{ \begin{array}{c} \text{LM entering} \\ \text{system with mass} \\ \text{during time period} \end{array} \right\} - \left\{ \begin{array}{c} \text{LM leaving} \\ \text{system with mass} \\ \text{during time period} \end{array} \right\} + \left\{ \begin{array}{c} \text{Resultant forces} \\ \text{on system integrated} \\ \text{over time period} \end{array} \right\} = \left\{ \begin{array}{c} \text{Accumulation of LM} \\ \text{within system during} \\ \text{time period} \end{array} \right\} \quad (2-9)$$

Where the accumulation of linear momentum is defined as follows:

$$\left\{ \begin{array}{c} \text{Accumulation of LM} \\ \text{within system} \\ \text{during time period} \end{array} \right\} \equiv \left\{ \begin{array}{c} \text{Amount of LM} \\ \text{contained in system} \\ \text{at end of} \\ \text{time period} \end{array} \right\} - \left\{ \begin{array}{c} \text{Amount of LM} \\ \text{contained in system} \\ \text{at beginning of} \\ \text{time period} \end{array} \right\} \quad (2-10)$$

In this situation, there is no momentum entering or leaving the system with mass because no mass crosses the system's boundaries. Also, there are no forces acting on the system (the two balls taken together) by the surroundings because there is no gravitational field. The only interactions are between two parts of the system, i.e., between the two balls, and these interactions do not contribute to a change in the momentum of the system as a whole. These observations are given in Table 2-6.

Table 2-6: Collision of Two Balls (Linear Momentum)

LM of system at end	−	LM of system at beginning	=	LM entering with mass during time period	−	LM leaving with mass during time period	+	Sum of External Forces integrated over time period
$(\mathbf{p})_{end}$	−	$(\mathbf{p})_{beg}$	=	$\sum (\mathbf{p})_{in}$	−	$\sum (\mathbf{p})_{out}$	+	$\int \left(\sum \mathbf{f}_{ext} \right) dt$
$(lb_m \ ft \ / \ s)$		$(lb_m \ ft \ / \ s)$		$(lb_m \ ft \ / \ s)$		$(lb_m \ ft \ / \ s)$		$(lb_m \ ft \ / \ s)$
$(\mathbf{p})_{end}$	−	$(\mathbf{p})_{beg}$	=	0	−	0	+	0

This reduces the verbal statement to:

$$0 = \left\{ \begin{array}{c} \text{Amount of LM} \\ \text{contained in system} \\ \text{at end of} \\ \text{time period} \end{array} \right\} - \left\{ \begin{array}{c} \text{Amount of LM} \\ \text{contained in system} \\ \text{at beginning of} \\ \text{time period} \end{array} \right\}$$

This can also be expressed as a mathematical expression in terms of vectors as:

$$\mathbf{0} = \left(\sum_{i=1}^{2} (m\mathbf{v})_i \right)_{end} - \left(\sum_{i=1}^{2} (m\mathbf{v})_i \right)_{beg}$$

or in tabular form as given in Table 2-7.

Table 2-7: Linear Momentum of Two Balls

Description	System LM at the end $(\mathbf{p})_{end}$ (lb$_m$ ft / s)	= = =	System LM at the beginning $(\mathbf{p})_{beg}$ (lb$_m$ ft / s)
x-direction	$(m_1 v_{1x} + m_2 v_{2x})_{end}$	=	$(m_1 v_{1x} + m_2 v_{2x})_{beg}$
y-direction	$(m_1 v_{1y} + m_2 v_{2y})_{end}$	=	$(m_1 v_{1y} + m_2 v_{2y})_{beg}$
z-direction	$(m_1 v_{1z} + m_2 v_{2z})_{end}$	=	$(m_1 v_{1z} + m_2 v_{2z})_{beg}$
Total	$(m_1 \mathbf{v}_1 + m_2 \mathbf{v}_2)_{end}$	=	$(m_1 \mathbf{v}_1 + m_2 \mathbf{v}_2)_{beg}$

Linear momentum (LM) is a vector quantity and is described by both magnitude and direction. Therefore, three equations are required to account for linear momentum in the three spatial coordinates. m_1 and m_2 represent the masses of ball 1 and ball 2 respectively and $\mathbf{v}_1$ and $\mathbf{v}_2$ are the velocity vectors of ball 1 and ball 2. The subscripts x, y, and z, denote the three spatial coordinates, i.e., the linear momentum of ball 2 in the z direction is $m_2 v_{2z}$.

In the table, the directional components of the momentum of each ball sum to give the components of the system total momentum. The value of each component total before the collision must equal the total after the collision. This provides three equations which must be satisfied in this collision:

$$\left((m_1 v_{1x} + m_2 v_{2x})\right)_{beg} = \left((m_1 v_{1x} + m_2 v_{2x})\right)_{end}$$

$$\left((m_1 v_{1y} + m_2 v_{2y})\right)_{beg} = \left((m_1 v_{1y} + m_2 v_{2y})\right)_{end}$$

$$\left((m_1 v_{1z} + m_2 v_{2z})\right)_{beg} = \left((m_1 v_{1z} + m_2 v_{2z})\right)_{end}$$

Hence, if we know each momentum except one (i.e. the velocity of both balls before collision and the velocity of one of the balls after collision) then we can calculate the momentum (the three components of momentum) of the other ball after the collision from the three independent equations. Note that this conservation statement requires nothing whatsoever of the type of collision, that is, whether it is fully elastic or partially or even totally inelastic. Momentum is still conserved; it is neither generated nor consumed.

Example 2-5: Conservation of Linear Momentum. As a final example, consider the situation of rigid body statics depicted in Figure 2-7. The body is subjected to a gravitational field. In this situation, the appropriate form of the momentum conservation law to consider is a rate equation which says that the rate of change of momentum of the system is equal to the sum of the external forces acting upon the system.

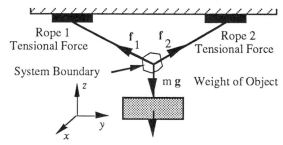

Figure 2-7: Rigid Body Statics

$$\left\{ \begin{array}{c} \text{Resultant external} \\ \text{forces on system} \end{array} \right\} = \left\{ \begin{array}{c} \text{Time rate of change} \\ \text{of linear momentum} \\ \text{of the system} \end{array} \right\}$$

If the system is chosen to be a single point, that point where the two supporting ropes meet and connect to the single rope supporting the weight, then (for this static situation where the rate of change of momentum of this system is zero) the forces in the three ropes must balance each other.

The mathematical expression of this idea is:

$$\sum \mathbf{f}_{\text{ext}} = \mathbf{f}_1 + \mathbf{f}_2 + m\mathbf{g} = 0$$

This statement is also presented in Table 2-8 where the individual directional components of the forces are shown to balance each other.

Table 2-8: Rigid Body Statics (Linear Momentum)

Description	Time Rate of Change of System LM $\left(\dfrac{d\mathbf{p}_{\text{sys}}}{dt}\right)$ (N)	=	Sum of Resultant External Forces $\sum \mathbf{f}_{\text{ext}}$					
			Force 1 (N)		Force 2 (N)		Gravity (N)
x-direction	0	=	f_{1x}	+	f_{2x}	+	mg_x
y-direction	0	=	f_{1y}	+	f_{2y}	+	mg_y
z-direction	0	=	f_{1z}	+	f_{2z}	+	mg_z
Total	**0**	=	$\mathbf{f}_1$	+	$\mathbf{f}_2$	+	$m\mathbf{g}$

The notation to describe this problem is similar to the previous example. Because linear momentum and forces are vector quantities, three equations are necessary. The terms $\mathbf{f}_1$, $\mathbf{f}_2$ and $m\mathbf{g}$ represent the tensional forces in the ropes and the weight of the body. The subscript 1 and 2 distinguish between the two supporting ropes. Again, the x, y, and z denotes the spatial coordinates, e.g., the tensional force in rope 2 in the y-direction is f_{2y}.

Again there are three equations which must hold for this vector equation. Also, as for the previous momentum problem with the two balls in collision, which also had three equations, we could use these to solve for three unknown components of the various vector forces. For example, given any two forces we could solve for the third by solving for each of its three components.

2.3.1 Conservation Versus Balance

In many texts the word *balance* is used to mean *conservation*. For example the terms mass balance, energy balance, momentum balance, etc. are commonly used. The term has also been used, however, for nonconservation situations and for nonconserved properties such as in the term mechanical energy balance, thermal energy balance, and even entropy balance. The distinction between conservation and accounting, and between those properties which are necessarily conserved and those which are not, is so important that we have chosen to use the term *conservation law for conserved properties* and the term *accounting statement for non-conserved* properties and we have avoided using the term balance totally.

2.3.2 Alternate Definition of Conservation

The concept of conservation is not universally stated in the same way. It is frequently stated that a certain property is conserved if the amount of that property contained in the system is constant. We prefer to refer to this situation as steady state and then use conservation in the broader sense of constancy with respect to the system plus surroundings or, equivalently, that there can be no generation or consumption of a conserved property. Linear momentum and angular momentum as far as we know are conserved absolutely. Other quantities such as the amount of ethanol may be conserved under certain situations but, in general, the amount of ethanol in the universe is not constant and so it is not a conserved property. Total mass and total energy are truly conserved, at least in a practical sense in normal engineering applications. In a broader sense, however, we must consider that only mass and energy taken together are conserved as energy.

2.4 CONSERVATION LAWS AND THE STRUCTURE OF ENGINEERING

2.4.1 Objectives

Engineering is a *quantitative profession*. Through calculations, we endeavor to understand, design, or improve processes or machines. These calculations must have a rational basis, however, in order to be able to achieve these goals and this process of mathematical description and calculation we call engineering analysis. In this section we present an overview of the structure of engineering and analysis.

The objectives of this engineering analysis process, are first, to obtain a quantitative (mathematical) description of the physical process which we are trying to understand, second, to solve this mathematical problem, and third, to use the resulting description to make predictive calculations of process behavior for the purpose of improving design or optimizing process operation. Such a quantitative description we refer to as a *model*. The extent to which model calculations agree with actual experimental observations provides information which may be used to improve either the model or the solution procedures, or both so that the predictive calculations will be better.

2.4.2 The Picture

With these objectives in mind, Figure 2-8 presents a picture of the foundations and structure of engineering and analysis. This structure consists of a modeling foundation, a process or device (machine) to be modeled, and solution methodologies.

As discussed previously, the conservation laws (mass, electrical charge, linear momentum, angular momentum, energy) plus the second law of thermodynamics are the heart and soul of the structure and provide the modeling foundation for analysis. Additionally, accounting statements are used to gain further insight into processes. Playing an integral role in the foundation are mathematics, concept definitions (such as mass, energy, work, etc.), properties of materials, and constitutive relations (to describe the transfer of mass, thermal energy, or momentum as the result of diffusion, conduction, and stresses, respectively). Apart from mathematics, these all require properties for completing the model descriptions using the conservation laws. If we are going to talk about conservation of mass, for example, then mass properties such as density are important.

The choice of a process or device to be modeled is guided by the needs and desires of society and our own ingenuity in meeting these needs. We can draw on previous experience with similar situations but, in addition, must always have an open mind to search for new processes and designs.

The quantitative model of this process, embodied in the equations which represent the physical laws and properties, must be solved using analytical (closed-form) or numerical methods and engineering methodologies. Additionally, we must have data representing numbers for the properties and behavior of materials and data having to do with parameters of the process. Combining all of this information and all of these skills, we devise a solution to the problem at hand. Along the way, simplifying assumptions concerning material behavior, process geometry, or process operation may be appropriate or even necessary to obtain solutions in a reasonable length of time. Depending on the nature of the assumptions and the extent to which they are valid the solution obtained in this manner may be very good or it may give only a "ballpark" estimate of the actual solution. A significant part of engineering methodology is to be able to take a problem that is too difficult to solve

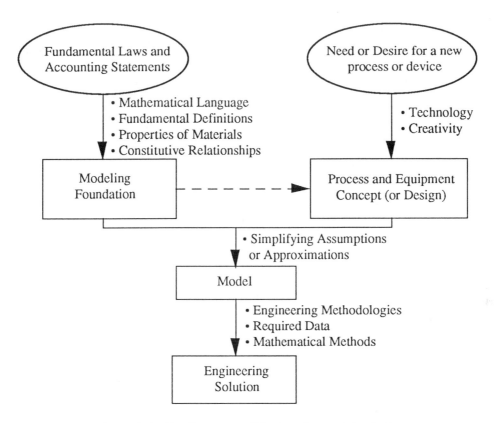

Figure 2-8: The Structure of Engineering and Analysis

in the given amount of time (or even at all) and change it to one that can be solved and that still contains the important elements that govern the situation.

From this picture outlined in Figure 2-8, it is seen that, just as the physical laws motivate a need for knowledge of property behavior, so too the quantitative model of our process motivates a need for a knowledge and command of mathematical, numerical, and solution techniques, the ability to make appropriate assumptions and simplifications, and even a physical feel for the problem.

Obtaining a reasonable solution to our model of the process also requires, at some point, a comparison of our solution to observed experimental data, be this actual rigorous experimental data or simply an ad hoc physical feeling for what takes place. By comparing our answer to an understanding of what is physically realistic, we obtain, finally, an understanding of whether our quantitative model is reasonable. If we find that our solution is unreasonable, then we must go back and sort out where the difficulties lie, whether at some point there is an error in describing the behavior of materials, whether there are erroneous assumptions and simplifications, whether the solution is wrong because of faulty analytical or numerical procedures, or even whether our belief in what is realistic is wrong.

Keeping this picture of the modeling and analysis process in mind throughout your academic career will be a big help in understanding your various engineering courses. Some courses will dwell more on the engineering foundation while others will dwell more on solution procedures, and still others may focus on designing physical processes to meet certain needs in the context of your knowledge of technology and your creativity.

2.4.3 Further Comments

A few additional points concerning conservation laws and accounting statements need to be made.

Summary of Equations. Table 2-9 summarizes the property, type of statement (either accounting or conservation) the number of scalar equations, and the number of independent equations which are at our disposal for analysis.

Table 2-9: Summary of Properties and Types of Statements

Name of Property	Accounting or Conservation	Number of Scalar Equations	Number of Independent Equations
Total Mass	Conservation	1	1
Elemental Mass	Conservation	n	$n - 1$
Species Mass	Accounting	m	$m - 1$
Net Charge	Conservation	1	1
Linear Momentum	Conservation	3	3
Angular Momentum	Conservation	3	3
Total Energy	Conservation	1	1
Mechanical Energy	Accounting	1	1
Thermal Energy	Accounting	1	0
Electrical Energy	Accounting	1	1
Entropy*	Accounting	1	1

* The entropy expression is an inequality for the universe.

The first property is mass which provides one *accounting* equation for each non-elemental species in a multicomponent situation plus one *conservation* equation for total mass and one for each element. The second property is electric charge which provides one scalar *conservation* equation. The third property is linear momentum, a vector quantity, providing three scalar *conservation* equations. The fourth property is angular momentum, also a vector, which consequently provides three scalar *conservation* equations. The fifth property is energy which provides one scalar *conservation* equation. Furthermore, it is useful to classify energy into mechanical and thermal forms and to consider the respective *accounting* equations. Interconversions can occur between the different forms of energy so that none, by itself, is conserved even though total energy is conserved. Furthermore, these subset forms of energy are not independent of each other, due to total energy conservation. If one is concerned with conversions between mass and energy, i.e. nuclear reactions, then actually the mass and energy equations must be considered together as a conservation statement of energy recognizing that mass itself is energy. In this context, relativistic effects become important as well. The final property is entropy, a non-conserved property. Considering the entropy of the universe, this property can only be generated and not consumed; the entropy of the universe (system plus surroundings) can only increase or remain constant. This inequality makes entropy a special property and this accounting for the universe is known as the Second Law of Thermodynamics.

Macro- versus Micro-scale. It has also been mentioned that accounting and conservation statements can be made for systems of different sizes. If the system being considered is finite in size, then we can refer to this as a macro-scale or macroscopic equation. If the system is infinitesimal in size, then we can refer to it as a micro-scale or differential equation.

The macroscopic situation is sometimes referred to as a *lumped parameter system*; the differential sized system is referred to as a *distributed parameter system*.

Finite versus Differential Time Period. Likewise, we can consider accounting statements and conservation laws to be made over either a finite time period or over a differential time period. If it is a finite time period, then we are concerned with the total change in the amount of that property contained within the system, and with the total input, output, generation, and consumption of that property which occurred during that time period. If the time period is differential, then the conservation law is normally written as a *rate equation*; each term in the statement is divided by the differential time period.

Conservation laws written for a system of finite size over a finite time period are algebraic equations. Conservation laws written for a finite size system over a differential time period are ordinary differential equations, and conservation laws for a differential size system over a differential time period are partial differential equations. In this course, we will consider processes which give rise to algebraic and ordinary differential equations.

2.5 REVIEW

1) The accounting concept plays an extremely important role in engineering analysis and design. Accounting for things requires defining a system within which the property is accounted for, a time period over which the accounting is made, and a well-defined property which is being counted.

2) An accounting statement is used to count properties which are not conserved, that is, properties for which the total amount in the system plus surroundings may change.

3) Conservation laws are very special forms of accounting statements, that is, they apply to certain special properties which are conserved in the sense that the total amount of these properties contained in the system and surroundings combined is a constant. For conserved properties there is no generation or consumption within the system or within the surroundings. These laws are postulated based on experimental observations.

4) Properties which are conserved and for which we write quantitative conservation law statements are total mass, elemental mass, total or net electrical charge, linear momentum, angular momentum, and total energy.

5) In addition to conservation laws for conserved properties we find it useful to write accounting statements for some properties which are not conserved. Such important statements are for reacting compounds, mechanical energy, thermal energy, electrical energy, and entropy. Individual species are not conserved in that chemical reactions may result in a gain or loss of species for the system plus surroundings. Mechanical and thermal energy are not generally conserved because there may be an interchange of energy from one form to another. Mechanical energy can be converted to internal energy and to some extent internal energy can be converted to mechanical energy.

6) The conservation laws provide the foundation upon which all of our engineering analysis and design is based. Supplementing these laws is information about the properties and behavior of materials, mathematical and numerical solution methods, engineering techniques and methodologies, and our own experiences with physical reality and the physical world.

7) The various conservation and accounting equations can be written for a finite-sized system for a differential time period (rate form) or for a finite time period (integral form). These forms will be discussed and used throughout this course. They may also be written for differential-sized systems to obtain partial differential rate equations.

QUESTIONS

1) Give a verbal statement of the accounting concept.

2) What is the difference between a conservation law and an accounting statement?

3) What kind of properties can you write accounting statements for?

4) Give some examples of properties which can be counted and properties which cannot.

5) What must be defined in order to obtain a quantitative accounting equation?

6) What properties are conserved and therefore provide our fundamental conservation laws?

7) What property is not conserved but yet is used to quantify another fundamental law or observation of science?

8) What are some example subsets of mass which we may count but which are not necessarily conserved?

9) What are some subset forms of energy which we may count but which are not conserved?

10) In an accounting of dollars, service charge and interest payments are transfer or exchange terms; interest payment is money the bank pays you for the use of your money and a service charge is money you pay the bank. What is it then that makes the dollars equation only an accounting statement and not a conservation law?

11) Why are tables useful for accounting and conservation calculations?

12) What happens to electrical energy when current flows through a resistor?

13) Qualitatively discuss the changes in conserved or counted properties for an appropriate system and during an appropriate time period for the following situations:

> A baseball hit by a bat

> A rocket launched into space

> A tank of water with one feed stream and one outlet stream which are at different flow rates

> A tank of pure water with one feed stream and one outlet stream of the same volume flow rate, but the feed stream suddenly changes from pure water to ten percent salt

> On a hot summer day, your house, which has been at a nice comfortable 75 degrees suddenly loses its air conditioning

In each of the above items discuss what happens with respect to the system, and what happens with respect to the surroundings. Also, be sure to state what property (or properties) are involved in each situation.

14) Make a sketch of Figure 2-8 and write on it in the appropriate place the various things which you have learned so far in your high school and freshman classes. Give specific examples and not just general categories.

15) The conservation of linear momentum provides three independent scalar equations. Explain.

16) In Figure 2-8 the process and equipment concept block which represents the problem at hand is shown having a dashed line input from the modeling foundation information. Discuss the extent to which you think this connection is important i.e. to what extent, given a specific apparatus or process, you should be expected to understand and analyze the *behavior* of a given apparatus versus *changes* which you might want to make to the apparatus or process based upon your understanding of a model of the physical laws (i.e. design).

SCALES*

1) Consider a continuous function $f(x) > 0$ for $x > 0$ with continuous first derivative df / dx.

   a) Make a sketch of a possible function for $0 < x < 10$ (i.e., plot $f(x)$ versus x, your choice, but make it non-linear). You need not have an analytical expression for your function, merely a graph.

   b) For your function on your sketch, indicate df / dx for $x = 3$ (i.e., draw a line whose slope has the correct value of df / dx at $x = 3$). How did you determine such a line? What do you think the significance of this procedure is in the context of counting extensive properties? (Hint: think rate equation and replace x with t, for time.)

   c) For your function, what is the approximate value of $\int_2^7 f dx$? How did you determine this? What is the significance of this procedure in the context of counting extensive properties? (Hint: Think finite time period equation and replace x with t for time.)

2) For examples 2-3, 4, and 5, make a sketch indicating the system, state what the extensive property of interest is, and indicate where this property crosses the boundary.

3) For each of the equations below, solve for the unknown variable. You may use a calculator (with built-in software) or an equation-solver computer software package.

   a) $3x^2 + 1.5 \sin(x) = 2$
   b) $x^2 \sin(x) \cos(x) = 2$
   c) $3e^{-2x} = 2$
   d) $4 \ln(x) = 2$

4) If T (for the temperature of a body) is said to be proportional to t (for time), give a mathematical expression which relates T to t.

5) I have a box sitting on the left side of a desk and a bag of tennis balls sitting on the right. If to the box I add two tennis balls, take 1 out, add 5, take 3 out and then take two more out, how many tennis balls are there in the box?

6) I have a box sitting on the left side of a desk and a bag of peanuts sitting on the right. If to the box I add two peanuts, take 1 out, add 5, take 3 out and then add two more, how many peanuts are there in the box?

7) Solve the following sets of linear simultaneous equations. Use both "by-hand" methods and computer software.

   a)
   $$x + 3y = 2$$
   $$3x + 10y = -6$$

   b)
   $$3f_1 + 2f_2 + f_3 = 2$$
   $$-4f_1 + 2f_2 - 7f_3 = 4$$
   $$-6f_1 + 3f_2 + f_3 = -3$$

---

* Skills, Concepts, and Light Exercise Studies. Practicing your SCALES makes for easier PROBLEMS!

8) A first order ordinary differential equation is one which involves only one independent variable (and hence is ordinary, as opposed to a partial differential equation which involves more than one dependent variable) and contains only the first derivative (i.e., there are no higher derivatives) of the dependent variable. Consider the following *first order ordinary differential equation* (this is the only one you will be concerned with in this course):

$$a\frac{dy}{dx} = by + c$$

Obtain an expression for y as a function of x. Hint: This ODE is simple enough that you do not have to have a course in ODEs to solve it. Simply separate the variables (put all terms containing the independent variable on one side of the equation and all terms containing the dependent variable on the other) and then integrate. Without limits of integration, note that you will obtain a constant of integration. This constant can be evaluated if you know a value for the dependent variable for a given value of the independent variable.

9) Solve the following first order ordinary differential equation:

$$m\frac{dv_z}{dt} = mg - kv_z$$

subject to the initial condition that $v_z = 0$ for $t = 0$. Here, v_z is the velocity of a falling ball under the influence of gravity and air resistance which is proportional to its velocity. (Hint: What variable is the independent variable and which is the dependent?

10) Solve the following first order ordinary differential equation:

$$\frac{dq}{dt} = \dot{q}_{in} - kq$$

subject to the initial condition that $q = 0$ for $t = 0$. Here q is the electrical charge on a plate, $\dot{q}_{in}$ is the rate at which charge enters the plate and is constant and kq is the rate at which charge leaves the plate (it leaves at a rate which is proportional to the amount of charge on the plate, see exercise (4).)

11) Solve the following first order ordinary differential equation (obtain an expression for T as a function of time):

$$\rho\hat{C}\frac{dT}{dt} = -h(T - T_s)$$

subject to the initial condition that $T = T_1$ ($> T_s$) for $t = 0$. Here T is the temperture of a steel ball placed in a water quenching bath, T_1 is the ball's initial temperature, T_s is the temperature of the water in the bath (you may assume that it is constant, i.e. that the bath is big enough, relative to the size of the ball, that it does not change temperature even though the ball cools down.

12) The data in the accompanying table show the rate of change of mass contained in a tank at various times (dm_{sys} / dt). Use computer plotting package to make a plot of dm_{sys} / dt versus t. Then, from this graph, estimate the change in the amount of mass in the tank between $t = 1$ hour and $t = 7$ hours. Assume that dm_{sys} / dt is continuous and "smooth."

Data for Exercise 12

Time (hours)	$\dfrac{dm_{sys}}{dt}$ (kg/hr)
0	2
1	3
2	5
3	5
4	6
5	6
6	7
7	7
8	6
9	6
10	5
11	4

ANSWERS TO SELECTED SCALES

3) (a) $x = 0.812$ (b) $x = 4.86$ (c) $x = 0.203$ (d) $x = 1.65$

5) Three, because there were two tennis balls in the box to begin with. No, I didn't tell you this, but you should have asked! This emphasizes the importance of knowing the initial state of the system when counting extensive properties.

6) None. There were not any in the box to begin with, which would have left five. However, the squirrel inside the box ate the remaining peanuts so there were none left (Oh, I forgot to tell you that there was a squirrel inside the box?). This emphasizes the importance of understanding whether there is generation or consumption (within the system) of the extensive property being counted. Note that the peanuts have disappeared but, as has been pointed out by students, the mass is still there.

PROBLEMS

1) Given the following data about Smithville U.S.A, draw a schematic of the town defining the system. What is the extensive property of these data? Write down the accounting equation for the extensive property and defined system. Determine the rate of accumulation of the extensive property.

 20,000 people move in per year.
 40,000 people move out per year.
 15,000 people are born per year.
 13,000 people die per year.

If the population at the end of 1988 was 500,000, what will be the population at the end of 1989?

2) Financial data on Mr. Jones are presented in the table below. (a) If the data represent the activity of Mr. Jones' account throughout the month, in what month was he bankrupt?

Mr. Jones' Financial Data

Descr.	January	February	March	April	May
Deposit	$2,000	$2,000	$2,000	$2,000	$2,000
Mortgage	1,000	1,000	1,000	1,000	1,000
Auto	500	500	500	500	500
Bills	1,000	700	1,100	900	400
Food	200	200	200	200	200
Insurance	100	100	100	100	100

Initial Balance = $2,000.00

(b) If at the end of each month, the money added to the account in the form of interest is given by the following equation:

$$\text{Dollars Added} = Pi$$

where:
P = the dollars in the account at the beginning of the month.
$i = 0.005$

Now, determine which month Mr. Jones will have a negative balance for his account.

3) Consider an aluminum soft drink can as shown in the figure. At the initial time the can is closed but at time t the can is opened allowing some gas and liquid to escape. If we want to know about the liquid and gas, what is the extensive property being counted? In terms of the accounting equations for the system defined as the aluminum can, discuss this physical phenomenon. If the system is defined as some imaginary volume that contains all of the escaped gas and liquid plus the aluminum can and its contents, discuss the verbal accounting equation that results.

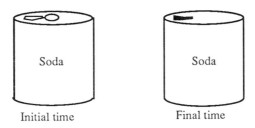

Problem 2-3: Aluminum soft drink can

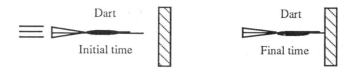

Problem 2-4: Dart and Dart Board

4) At some initial time t_o a dart is thrown at a dart board and at time t the dart embeds in the board as shown in the figure. If we define the system to be the dart, write the verbal linear momentum conservation equation and discuss the results. If the system is defined as the board, discuss the verbal linear momentum conservation statement that results. Now, if we define the system to include both the dart and the board discuss the verbal linear momentum conservation statement. (Note: The linear momentum is not "lost" in the impact with the board. Since linear momentum is conserved the momentum was transferred to the wall, then to the building and finally to the earth. Because the mass of the board, wall, building, and earth is so great compared to the mass of the dart, the change in velocity of these quantities is imperceptible.)

5) If $2,000 is put in a savings account at an interest rate of 9% for 20 years and the dollars added to the account in the form of interest are given by the following equation:

$$\text{Dollars Added} = P\left((1 + i)^n - 1\right)$$

where:
    n = the number of years
    P = the initial amount
    i = the yearly interest rate

(a) Determine the accumulation after 20 years. (b) Calculate the final total amount of money at the end of the time period.

6) Given two balls suspended by two strings as shown in the figure. At the initial time, ball 2 is dropped and is in motion while ball 1 is suspended at rest. At time t after the collision the balls are in motion at some position shown in the figure. First, what extensive property do we want to count? Is this property a conserved property? If the system is defined as ball 1 write the verbal accounting statement for the system and discuss what terms equal zero. Do the same thing if the system is defined as ball 2. Finally, for both balls and the string defined as the system, write the verbal accounting equation and discuss the equation that results.

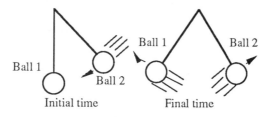

Problem 2-6: Two suspended balls

7) The mass flow rate of crude oil at the entrance of the Alaskan Pipeline was measured to be 76.4 (lb_m / min). There are no leaks, no other entrances, no chemical reactions that take place in the pipeline, and only one outlet. Are you able to tell exactly the mass flow rate at the outlet? Why or why not? If the pipeline is operating at steady-state with no accumulation in the pipeline, what is the mass flow rate at the outlet? Why?

8) There are 20 chickens and 2 foxes present on an acre of land on the morning of June 1, 1990. During the day, half of the chickens leave the acre of land to search for food, but both of the foxes stay. By nightfall, all of the chickens that left returned. At the end of the day, the total number of chickens is 15 and the total number of foxes is 2. Using conservation and accounting (make a table) explain what happened. Remember to explicitly define the counted property, the system and system boundary, and the time period.

9) A cat has a litter of kittens. In the morning, all of the kittens are in the cat box. By afternoon, two-thirds of the total number of kittens are playing in the living room outside the cat box. Two kittens are playing with a ball of string in the dining room outside the cat box. One kitten is still in the cat box sleeping. Using conservation and accounting concepts (make a table), determine how many kittens were in the cat box in the morning. Remember to explicitly define the counted property, the system and system boundary, and the time period. Also, state any assumption that had to be made in order to solve the problem.

10) The local library has 230,000 books on January 1, 1980. During the year, residents donated 50,000 new books, checked out 200,000 books, and returned only 190,000 books. Use conservation and accounting concepts (make a table) to determine how many books are in the library on January 1, 1981. At this average rate of accumulation of books, will the library ever have zero books? Why?

11) A baseball stadium can hold a maximum of 55,000 spectators. Before the gates open, there are only 200 stadium personnel. When the gates open, there are 500 times more spectators than media that enter the stadium. There are 50 baseball players that enter the stadium. During the game, one-fourth of the spectators leave. There are 30,000 spectators at the end of the game. No players, stadium personnel, or media leave. Use conservation and accounting concepts (make a table) and calculate the total number of people in the stadium at the end of the game.

12) A mine in Utah produces ores and extracts metals. The metals are copper and lead. Every day the mine processes 200 tons of ore. Of the entire ore, 2 % by mass is copper and 1 % by mass is lead. If all of the pure metal leaves the mining complex every day, use conservation and accounting to determine the mass of copper, lead, and total metals leaving the complex. Every day, how much of the non-metal ore (tailings) must be stored on the complex site?

13) On a summer day, a house can absorb 1000 (Btu) of energy from the outside environment. In the morning, the house is comfortable and only contains 700 (Btu) of energy. In the evening, the house is comfortable and contains 800 (Btu) of energy. Use conservation and accounting to determine how much energy must be removed from the house during the day.

14) A process is operating at steady-state i.e., there is no accumulation of any property. Mass, charge, energy, momentum, and entropy enter and leave the process. If the following quantities enter the process during a one-day time period,

Mass = 35 (kg)   Charge = 4000 (C)
Energy = 56,000 (kJ)   Momentum = 354 (kg m/s)
Entropy = 0.2 (kJ/K)

use conservation and accounting to determine the quantity of the different properties that left the process during the one-day time period.

15) In a certain region of the universe, scientists are observing a wonderful and unique phenomenon. Although there are many complex processes going on inside the region, the scientists note that over a one year period, 36000 (tons) of mass entered and 20000 (tons) left; 53000 (Mcal) of energy entered and 23000 (Mcal) of energy left; 76000 (ton ft/s) of momentum entered and 54000 (ton ft/s) of momentum left. There are no conversions of mass to energy inside this region of the universe. Use conservation and accounting to determine the accumulation of properties inside this region of the universe during the one year time period.

16) Over a one day period, a person eats and drinks approximately 6 lb_m of food. The energy in the food is about 2000 (kcal). The person does not accumulate any mass or energy at the end of one-day. Use conservation and accounting to determine the amount

of mass that must be excreted by the body in terms of perspiration and other bodily processes. Also, determine the amount of energy that left the body during the one-day period.

17) Billy has a job delivering newspapers every morning before school. Billy is industrious and actually works for two different newspapers, the Herald and the Post. Billy's route covers 5 different streets. The incomplete data for the number of houses and newspapers is given in the following table. Every house gets either the Herald or Post newspaper but not both.

Streets	Total Houses	Herald Houses	Post Houses
Elm	64	34	?
Park	58	27	?
Oak	37	17	?
Main	84	35	?
1st	75	25	?

Complete the table of data. The Herald gives Billy 140 newspapers in the morning, and the Post gives him 190 newspapers in the morning. Use conservation and accounting (make a table) to determine how many Heralds, Posts, and total newspapers Billy accumulates every day?

18) In many western states, the supply of water for the area is of paramount concern. The average water used for some daily activities is listed in the following table.

Showers = 56 (lb_m) Meals = 3 (lb_m)

Clothes Washer = 140 (lb_m)

Every day, a family takes 4 showers, eats 4 meals, and runs the clothes washer 1 time. There are approximately 2 million families in the south-western United States. Use conservation and accounting to determine how much water is used for a 1 year period (365 days) by the families in the south-western United States.

19) In a brick making facility, clay and water are used to make refractory materials. Every day, 2000 lb_m of pure clay and 1000 lb_m of pure water enter the process. The lot of finished bricks contains 2000 lb_m of pure clay and only 200 lb_m of pure water. There is no accumulation of water or clay in the process. Use conservation and accounting (make a table) to determine how much water is removed in the kiln during the brick making process.

20) A satellite in space is rotating around the earth. Momentum is entering the satellite at a rate of 24 $(lb_m ft\,/\,s^2)$ directed through the center of the earth (this is the force due to gravity). No other momentum is entering or leaving the satellite. Use the conservation of momentum to calculate the time rate of change of momentum of the satellite. Discuss and explain this answer in terms of the motion of the satellite. Does it make sense? If the momentum of the satellite is changing, what does conservation and accounting tell you about the momentum of the earth and why?

21) You are a typical college student at your home away from home. As such, you are trying to keep up with your bank account to be sure that you do not bounce any checks. During the past month you wrote checks for: appartment rent ($250), telephone ($30), cable TV ($25), electricity ($100), and cash ($900). Also, as a student worker you worked 50 hours (for the entire month) at $5/hr, had a scholarship of $100 for the month, received money from home ($750), and earned $50 cutting grass. Finally, there were $10 in ATM charges and earned interest of $5. If you started the month with $500, what is your balance at the end of the month? What conclusions might you draw about your immediate financial situation? Is this answer any different for your long-term financial situation?

THE FUNDAMENTAL LAWS

In Part I, a structure of engineering was outlined. This picture is based upon a handful of fundamental conservation laws (total mass, elemental mass, electrical charge, linear momentum, angular momentum, and total energy), the Second Law of Thermodynamics and a number of other accounting relations (species mass such as chemical compounds, positive and negative charge, mechanical energy, thermal energy, electrical energy, money, etc.). The fundamental laws and accounting relations have a common basis: the methodology used for accounting for the amount of an extensive property which is contained within a system. Consequently, rather than learning many diverse and unrelated results spanning several disciplines, you can focus on learning the laws and accounting relations within the context of this single accounting concept. This is an extremely valuable key to learning and understanding engineering and science.

In Part II, we discuss in detail the mathematical statements and physical meaning of these concepts. Examples of their use for a variety of processes or situations are presented.

CHAPTER THREE

THE CONSERVATION OF MASS

3.1 INTRODUCTION

The principle of conservation of mass has already been stated in Chapter 2. In this chapter, the conservation and accounting statements are applied to macroscopic systems and to both total mass and individual species. All of the equations and problems of this chapter deal with situations for which mass-energy conversions are negligibly small.* Consequently, there is no generation or consumption of total mass contained within the system; *total mass is conserved*. Likewise, the chemical elements are conserved. For individual species (compounds, e.g.) there may be generation or consumption due to reaction, and in general, these terms certainly must be included in the species accounting statements.

Conservation of mass is important to a great many areas of engineering. These include the analysis of large chemical plants, bioengineering problems, such as the analysis of vegetation and the design of processes for the use of biomass, and the analysis of our own body processes, for health and research evaluation. Some examples of problems in these and other areas are presented in this chapter.

---

* In the bigger picture, mass *is* energy, "conversion" is an imprecise term, and energy (in all forms, including mass) is conserved. As a practical and historical matter, it is convenient to deal with mass and energy separately in most engineering processes. This is the approach we take in this text.

In accordance with the discussion of the last chapter for a defined system and time period we account for mass through the following statement.

$$\left\{\begin{array}{c}\text{Accumulation of mass}\\ \text{within system during}\\ \text{time period}\end{array}\right\} = \left\{\begin{array}{c}\text{Mass entering}\\ \text{system during}\\ \text{time period}\end{array}\right\} - \left\{\begin{array}{c}\text{Mass leaving}\\ \text{system during}\\ \text{time period}\end{array}\right\}$$

$$+ \left\{\begin{array}{c}\text{Mass generated}\\ \text{in system during}\\ \text{time period}\end{array}\right\} - \left\{\begin{array}{c}\text{Mass consumed}\\ \text{in system during}\\ \text{time period}\end{array}\right\}$$

where the accumulation of mass is defined as follows:

$$\left\{\begin{array}{c}\text{Accumulation of mass}\\ \text{within system}\\ \text{during time period}\end{array}\right\} \equiv \left\{\begin{array}{c}\text{Amount of mass}\\ \text{contained in system}\\ \text{at end of}\\ \text{time period}\end{array}\right\} - \left\{\begin{array}{c}\text{Amount of mass}\\ \text{contained in system}\\ \text{at beginning of}\\ \text{time period}\end{array}\right\}$$

For a defined time period, we can write the verbal statement in symbolic form as:

$$(m_{\text{sys}})_{\text{end}} - (m_{\text{sys}})_{\text{beg}} = m_{\text{in}} - m_{\text{out}} + m_{\text{gen}} - m_{\text{con}} \qquad (3-1)$$

The terms of each equation are further defined:

m_{in}	=	mass which enters the system during the time period
m_{out}	=	mass which leaves the system during the time period
m_{gen}	=	mass generated within the system during the time period
m_{con}	=	mass consumed within the system during the time period
$(m_{\text{sys}})_{\text{end}}$	=	mass contained within the system at the end of the time period
$(m_{\text{sys}})_{\text{beg}}$	=	mass contained within the system at the beginning of the time period

Figure 3-1 shows a system and its surroundings defined by the boundary. Mass entering and leaving the system is simply any which crosses the system boundary. Generation and consumption of mass is of interest only if it occurs inside the system boundary; any generation or consumption outside the system is irrelevant to a statement about this particular system. Usually, there is no magically right system boundary or time period to use.

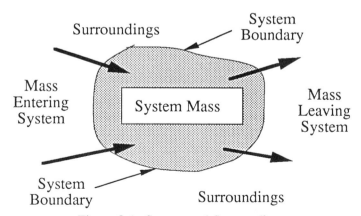

Figure 3-1: System and Surroundings

For mathematical convenience, the input and output terms can be expressed in a single term according to the *definition*:

$$m_{\text{in/out}} \equiv m_{\text{in}} - m_{\text{out}} \qquad (3-2)$$

Mass that enters the system and therefore adds to the system carries a positive sign while mass that leaves the system carries a negative sign.

Similarly, the generation and consumption terms can be represented by a single term:

$$m_{gen/con} \equiv m_{gen} - m_{con} \qquad (3-3)$$

Mass that is generated and therefore adds to the system mass carries a positive sign and mass that is consumed carries a negative sign.

With these definitions, the mass accounting statement is now:

$$(m_{sys})_{end} - (m_{sys})_{beg} = m_{in/out} + m_{gen/con} \qquad (3-4)$$

Note that if total mass is counted, then the generaton/consumption term is zero because (total) mass is conserved.

A simple example illustrating the conservation and accounting of mass is presented below and concerns determining the accumulation of a volatile liquid. Note that accumulation can either be positive or negative depending on whether the amount of mass contained within the system increases or decreases with time.

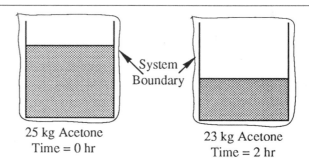

Example 3-1. A container was filled with 25 (kg) of acetone. The container was not sealed, and 2 hours later, 23 (kg) of acetone remained in the container. For the two hour period, determine the accumulation of acetone, and the mass which left the container, shown in Figure 3-2.

25 kg Acetone
Time = 0 hr

23 kg Acetone
Time = 2 hr

Figure 3-2: Acetone in a container

SOLUTION: Choose the system as the container of acetone, and the time period as the two hour observation period. The accumulation of mass in the system is determined using the definition:

$$\text{Accumulation of mass} \equiv (m_{sys})_{end} - (m_{sys})_{beg} = (23) - (25) \text{ (kg)} = -2 \text{ (kg)}$$

To calculate the mass which left the system, we write the accouting of acetone equation applied to the system.

$$(m_{sys})_{end} - (m_{sys})_{beg} = m_{in} - m_{out} + m_{gen} - m_{con}$$

During the time period, acetone is not generated, consumed, or added to the system. These terms of the equation are set equal to zero giving that $-m_{out} = (m_{sys})_{end} - (m_{sys})_{beg} = -2$ (kg) i.e., $m_{out} = 2$ (kg); 2 (kg) of acetone left the system during the time period.

An especially convenient method of organizing the problem data and solution for these conservation and accounting situations is with a table, as was done in Chapter 2. The categories of the accounting or conservation statements are written in columns and each row is for the extensive property being considered (e.g. one for each species and one for total mass). After defining a system and time period, we start by entering the data given in the problem statement in Table 3-1.

The pluses and minuses indicate the relations between the entries in the columns in the context of an accounting statement. In this problem we see immediately that the only term that is unknown is the output and that it must be 2 (kg) in order for the statement to hold.

This is an easy problem (perhaps too easy), but it does illustrate an approach to applying accounting or conservation statements. The value of using tables will become more apparent as more complicated problems are confronted.

Table 3-1: Solution of Example 3-1 (Mass)

$(m_{sys})_{end}$ (kg)	$-$	$(m_{sys})_{beg}$ (kg)	$=$	m_{in} (kg)	$-$	m_{out} (kg)	$+$	m_{gen} (kg)	$-$	m_{con} (kg)
Acetone 23	$-$	25	$=$	0	$-$	?	$+$	0	$-$	0

As stated before, we could choose to call all streams crossing the system boundary "inputs" and, in fact, it is often convenient to do so. In this case, an input stream will have a positive value if it adds mass to the system but a negative value if it removes mass. Likewise, we could define consumption as a negative generation and write only a "generation" column. In problem 3-1, for example, a table using these conventions would be:

Table 3-2: Alternate Solution of Example 3-1 (Mass)

$(m_{sys})_{end}$ (kg)	$-$	$(m_{sys})_{beg}$ (kg)	$=$	$m_{in/out}$ (kg)	$+$	$m_{gen/con}$ (kg)
Acetone 23	$-$	25	$=$	?	$+$	0

and in this sense, the "input" of acetone was -2 (kg).

The example illustrates the accounting of mass principle for a single-component system. Unfortunately (or perhaps fortunately because without mixtures and mixture processes, life would be very drab, even non-existent!), most materials, either fluids or solids, are mixtures of several different components. Common mixtures include air, sea water, gasoline, and fluids such as blood. Multicomponent systems are considered in the next section.

3.2 MULTICOMPONENT SYSTEMS

The principle of *conservation* of mass applies to the *total mass* of a multicomponent mixture and for each chemical *element*. In addition, for each individual component we may write an accounting statement. The term component or species is used interchangeably and identifies a specific type of mass which may undergo reaction. For a system involving n different species, n species accounting statements can be written. The n species statements are not independent of total mass conservation. Because of the law of conservation of total mass, the n species statements must sum to give total conservation. An accounting statement for component i is:

$$(m_{i,sys})_{end} - (m_{i,sys})_{beg} = (m_i)_{in} - (m_i)_{out} + (m_i)_{gen} - (m_i)_{con} \qquad (3-5)$$

The mass of a reacting material in a system changes because of input and output and because of generation and consumption.

Summing the terms of the conservation of mass equation for all n components yields the total mass conservation equation.

$$\sum_{i=1}^{n}(m_{i,\text{sys}})_{\text{end}} - \sum_{i=1}^{n}(m_{i,\text{sys}})_{\text{beg}} = \sum_{i=1}^{n}(m_i)_{\text{in}} - \sum_{i=1}^{n}(m_i)_{\text{out}} + \left(\sum_{i=1}^{n}(m_i)_{\text{gen}} - \sum_{i=1}^{n}(m_i)_{\text{con}}\right) \qquad (3-6)$$

The law of conservation of total mass states that there is no net generation of total mass:

$$\left(\sum_{i=1}^{n}(m_i)_{\text{gen}} - \sum_{i=1}^{n}(m_i)_{\text{con}}\right) = 0 \qquad (3-7)$$

thereby giving:

$$(m_{\text{sys}})_{\text{end}} - (m_{\text{sys}})_{\text{beg}} = m_{\text{in}} - m_{\text{out}} \qquad (3-8)$$

> Conservation of total mass means that the total generation and consumption of mass, considering all species in the system, is zero. Thus, the total mass in the system changes (left hand side) because of an imbalance of input and output (right hand side) only.

Table 3-3: Species Mass Accounting and Total Mass Conservation

Species 1	$(m_{1,\text{sys}})_{\text{end}}$	$-$	$(m_{1,\text{sys}})_{\text{beg}}$	$=$	$(m_1)_{\text{in/out}}$	$+$ $(m_1)_{\text{gen/con}}$
Species 2	$(m_{2,\text{sys}})_{\text{end}}$	$-$	$(m_{2,\text{sys}})_{\text{beg}}$	$=$	$(m_2)_{\text{in/out}}$	$+$ $(m_2)_{\text{gen/con}}$
$\vdots$	$\vdots$		$\vdots$		$\vdots$	$\vdots$
Species i	$(m_{i,\text{sys}})_{\text{end}}$	$-$	$(m_{i,\text{sys}})_{\text{beg}}$	$=$	$(m_i)_{\text{in/out}}$	$+$ $(m_i)_{\text{gen/con}}$
$\vdots$	$\vdots$		$\vdots$		$\vdots$	$\vdots$
Species n	$(m_{n,\text{sys}})_{\text{end}}$	$-$	$(m_{n,\text{sys}})_{\text{beg}}$	$=$	$(m_n)_{\text{in/out}}$	$+$ $(m_n)_{\text{gen/con}}$
Total Mass	$(m_{\text{sys}})_{\text{end}}$	$-$	$(m_{\text{sys}})_{\text{beg}}$	$=$	$m_{\text{in/out}}$	$+$ 0

By combining the input and output terms and generation and consumption terms, the multicomponent species accounting equations and total mass conservation equation are shown in Table 3-3. The following example illustrates the application of the conservation and accounting principles to multicomponent mixtures.

> **A Word on Problem Solving.** Note the primary steps which are followed in the solution: 1) make a sketch showing the process and all given data, 2) define a system, time period, and property(ies) to be counted, 3) make a table showing an accounting of the extensive property, 4) identify the type of math problem (e.g., algebraic equation, ODE with initial conditions, system of equations, etc.), 5) solve the math problem, 6) review (critique) your answer. This procedure shoud be followed with great discipline throughout your work.

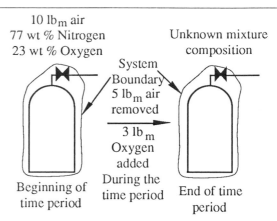

10 lb$_m$ air
77 wt % Nitrogen
23 wt % Oxygen

Unknown mixture composition

System Boundary
5 lb$_m$ air removed

3 lb$_m$ Oxygen added

Beginning of time period

During the time period

End of time period

Figure 3-3: Gas mixture in a cylinder

Example 3-2. Ten pounds mass (lb$_m$) of air is placed in a cylinder. Air consists of a mixture of nitrogen (77 mass %) and oxygen (23 mass %). If 5 (lb$_m$) of air is removed and then 3 (lb$_m$) of oxygen is added, what is the resultant composition of the gas mixture in the cylinder?

SOLUTION: To determine composition we must establish the amount of each component in the gas cylinder. An appropriate system is the "contents of the gas cylinder," and an appropriate time period is the "time required to remove the air and add the pure oxygen" depicted in Figure 3-3. Although we do not know exactly how long this might take, it still is a suitable definition for the time period.

Table 3-4: Mass Accounting for Example 3-2 (Mass)

Species	$(m_{i,\text{sys}})_{\text{end}}$ (lb$_m$)	$-$	$(m_{i,\text{sys}})_{\text{beg}}$ (lb$_m$)	$=$	$(m_i)_{\text{in/out}}$ (lb$_m$)				$+$	$(m_i)_{\text{gen/con}}$ (lb$_m$)
Oxygen	?	$-$	0.23(10)	$=$	3	$-$	0.23(5)		$+$	0
Nitrogen	?	$-$	0.77(10)	$=$	0	$-$	0.77(5)		$+$	0
Total	?	$-$	10	$=$	3	$-$	5		$+$	0

Tabulating the data in the problem statement gives Table 3-4. Note the use of the composition of air to calculate the distribution of the 10 (lb$_m$) and the 5 (lb$_m$) between O_2 and N_2. Also note that it is important in working the problem that air was removed first and then O_2 was added. If the order were reversed, the composition of the mixture removed would be different from that of air.

Each row in this table stands as an equation which can be used in solving the problem. Any two can be treated as independent. For this problem, each equation can be solved separately from the others. Solving the O_2 equation gives $(m_{O_2,\text{sys}})_{\text{end}} = 4.15$ lb$_m$. Solving the N_2 equation gives $(m_{N_2,\text{sys}})_{\text{end}} = 3.85$ lb$_m$. Solving the total mass equation gives $(m_{\text{sys}})_{\text{end}} = 8$ lb$_m$ which matches as it must, the sum of $(m_{O_2,\text{sys}})_{\text{end}}$ and $(m_{N_2,\text{sys}})_{\text{end}}$. Table 3-5 shows the completed problem.

This problem illustrates the power of organizing all of the data together in a table. Note again that each *row* is classified as an accounting equation and that each *column* sums to give an entry for the total mass conservation law. If there is any inconsistency in any row or column, then the solution is wrong.

To calculate the composition of the gas mixture, it is convenient to introduce the term mass fraction. The mass fraction (also called weight fraction) is defined as the mass of one component divided by the total mass. We represent it by the symbol w_i. By definition

Table 3-5: Solution of Example 3-2 (Mass)

Species	$(m_{i,\text{sys}})_{\text{end}}$	$-$	$(m_{i,\text{sys}})_{\text{beg}}$	$=$			$(m_i)_{\text{in/out}}$
	(lb_m)		(lb_m)				(lb_m)
Oxygen	4.15	$-$	2.3	$=$	3	$-$	1.15
Nitrogen	3.85	$-$	7.7	$=$	0	$-$	3.85
Total	8	$-$	10	$=$	3	$-$	5

then, the summation of all mass fractions must equal 1. Mathematically, the mass fraction is represented as:

$$w_i = \left(\frac{m_i}{m}\right) = \left(\frac{m_i}{\sum m_i}\right)$$

(3 − 9)

Note that the sum of the mass fractions is unity:

$$\sum_{i=1}^{n} w_i = \sum_{i=1}^{n}\left(\frac{m_i}{m}\right) = \left(\frac{\sum m_i}{m}\right) = 1$$

(3 − 10)

Alternatively, m_i can be calculated from w_i and the m. Rearranging equation (3-9) gives $m_i = w_i m$.

Another way to represent composition of a mixture is to use the mole fraction (usually denoted by either y_i or x_i). For each species, this is the number of its moles in the mixture M_i given by

$$M_i = \left(\frac{m_i}{(MW)_i}\right)$$

(3 − 11)

divided by the total moles, M. The dimensions of molecular mass (MW_i) are mass per mole, e.g., the molecular mass of O_2 is 32 lb$_m$/lb-mole or 32 kg/kg-mole or 32 g/mole. (There are 0.454 kg/lb$_m$ and also 0.454 kg-mole/lb-mole.) Note that the SI unit of molecular mass is the Dalton with 1 Dalton = 1 g/mole. Generally, $w_i \neq y_i$ because the molecular mass's of the species in a mixture are different. Then,

$$y_i = \left(\frac{M_i}{M}\right) = \left(\frac{M_i}{\sum M_i}\right)$$

(3 − 12)

where the mole fractions sum to unity:

$$\sum_{i=1}^{n} y_i = \sum_{i=1}^{n}\left(\frac{M_i}{M}\right) = \left(\frac{\sum M_i}{M}\right) = 1$$

(3 − 13)

Also, M_i can calculated from y_i and M. A rearrangement of equation (3-12) produces $M_i = y_i M$.

Returning now to the problem at hand, the mass and mole fractions of oxygen and nitrogen at the end of the time period are shown in Table 3-6 ($MW_{O_2} = 32$, $MW_{N_2} = 28$). Note that the mass fraction values are different from the mole fraction values because the molecular masses are different.

Table 3-6: Summary of Mass and Molar Compositions

	mass (lb$_m$)	mass fraction	moles (lb − moles)	mole fraction
O$_2$	4.15	0.52	0.130	0.485
N$_2$	3.85	0.48	0.138	0.515
Total	8	1.00	0.268	1.00

3.3 RATE EQUATION

So far we have considered the conservation and accounting of mass for finite time intervals. The time period was 2 hours for Example 3-1 or the "time required to exchange the gas in the cylinder" whatever that may be in Example 3-2. This type of accounting is known as an integral accounting and is not concerned with instantaneous rates of change of mass in the system or with flow rates. Only time-averaged values are addressed by integral mass accounting equations and conservation laws.

Actually, the integral form is properly thought of as an integration over time of a rate equation. Mass enters the system at rate $\dot{m}_{in}$ and the total amount which enters over a finite time (from t_{beg} to t_{end}) is:

$$\int_{t_{beg}}^{t_{end}} \dot{m}_{in} dt = m_{in}$$

Similar terms can be written for m_{out}, m_{gen}, and m_{con}. Likewise, if the rate of change of the mass in the system is $\left(\dfrac{dm_{sys}}{dt}\right)$ then the total change over the finite time period is:

$$\int_{t_{beg}}^{t_{end}} \left(\frac{dm_{sys}}{dt}\right) dt = \int_{m_{beg}}^{m_{end}} dm_{sys} = (m_{sys})_{end} - (m_{sys})_{beg}$$

so that the integral equation in terms of rates is:

$$\int_{t_{beg}}^{t_{end}} \left(\frac{dm_{sys}}{dt}\right) dt = \int_{t_{beg}}^{t_{end}} \dot{m}_{in} dt - \int_{t_{beg}}^{t_{end}} \dot{m}_{out} dt + \int_{t_{beg}}^{t_{end}} \dot{m}_{gen} dt - \int_{t_{beg}}^{t_{end}} \dot{m}_{con} dt$$

For a differential time period:

$$\left(\frac{dm_{sys}}{dt}\right) dt = \dot{m}_{in} dt - \dot{m}_{out} dt + \dot{m}_{gen} dt - \dot{m}_{con} dt$$

so that the rates are related by the equation (divide by dt)

$$\left(\frac{dm_{sys}}{dt}\right) = \dot{m}_{in} - \dot{m}_{out} + \dot{m}_{gen} - \dot{m}_{con} \qquad (3-14)$$

The concepts of mass flow rates, generation (and consumption) rates and accumulation rates are very important to both pure and multicomponent systems. Several problems illustrate rate equations. It is important also to note the change in notation. The integral mass entering the system is represented by m_{in} while the *rate* of mass in or out of the system is represented by: $\dot{m}_{in}$ and $\dot{m}_{out}$ respectively (note the presence or absence of the over dot). In this case, the dot does not denote change with respect to time but is simply used to distinguish between the integral mass and the mass flow rate. Similarly, $\dot{m}_{gen}$ and $\dot{m}_{con}$ are used to indicate generation and consumption *rates*. The mass of the system is always denoted with the italic m_{sys} whether in the integral equations or the rate equations. The time derivative of m_{sys} denotes the time rate of change of m_{sys}.

We can combine the input and output rates into one term and the generation and consumption rates into one term. The consolidated form of the rate equation is

$$\left(\frac{dm_{sys}}{dt}\right) = \dot{m}_{in/out} + \dot{m}_{gen/con} \qquad (3-15)$$

Note that the term dm_{sys}/dt represents the accumulation rate in that it is the amount of accumulation which occurs over a differential time period divided by the time period:

$$\left(\frac{dm_{sys}}{dt}\right) = \text{accumulation rate} = \lim_{\Delta t \to 0} \frac{m_{sys}|_{t+\Delta t} - m_{sys}|_t}{(t + \Delta t) - t}$$

The operator d( )/dt will always represent an accumulation rate in this text.

Example 3-3. Find the time required for 4 pounds of a species A mass to accumulate in the system. The mass flow rates, and generation and consumption rates are given where a is constant, $a = 1$ (hr^{-1}).

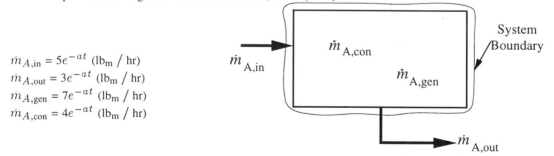

$$\dot{m}_{A,in} = 5e^{-at} \text{ (lb}_m \text{ / hr)}$$
$$\dot{m}_{A,out} = 3e^{-at} \text{ (lb}_m \text{ / hr)}$$
$$\dot{m}_{A,gen} = 7e^{-at} \text{ (lb}_m \text{ / hr)}$$
$$\dot{m}_{A,con} = 4e^{-at} \text{ (lb}_m \text{ / hr)}$$

Figure 3-4: Time dependent conservation of mass

SOLUTION: Figure 3-4 shows a schematic of the system. Because the total accumulation is needed, the integral form of the accounting mass equation is required. However, since the mass flow rates and generation and consumption terms are functions of time, we must integrate the rate form of the equation to produce the integral mass balance. The rate form of a species A accounting equation is given by the table:

	$\left(\dfrac{d(m)_{sys}}{dt}\right)$	$=$	$\dot{m}_{in/out}$	$+$	$\dot{m}_{gen/con}$
Species A:	$\left(\dfrac{d(m_A)_{sys}}{dt}\right)$	$=$	$(5e^{-at} - 3e^{-at})$	$+$	$(7e^{-at} - 4e^{-at})$

Integrating this result over time gives

$$(m_{A,sys})_{end} - (m_{A,sys})_{beg} = \int_0^t (5e^{-at} - 3e^{-at} + 7e^{-at} - 4e^{-at})dt$$

Here, t_{beg} is zero. We could call it t_o in which case we are finding $t - t_o$. Performing the integrations over time gives

$$(m_{A,sys})_{end} - (m_{A,sys})_{beg} = \left(-\frac{5e^{-at}}{a} + \frac{3e^{-at}}{a} - \frac{7e^{-at}}{a} + \frac{4e^{-at}}{a}\right)\Bigg|_0^t$$

We do not know what t (t_{end}) is, but we do know that at the end of the time period:

$$(m_{A,sys})_{end} - (m_{A,sys})_{beg} = 4 \ (lb_m)$$

Evaluating the equation at both limits of integration gives:

$$-\left(\frac{5e^{-at}}{a}\right) - \left(\frac{-5}{a}\right) = 4 \ (lb_m)$$

or:

$$e^{-at} = \left(\frac{1}{5}\right) \qquad \text{(dimensionless)}$$

so that (remember, $a = 1hr^{-1}$)

$$t = 1.61 \ (hr)$$

To summarize, the time required for the accumulation of the mass in the system to reach 4 pounds is 1.61 hours.

3.4 MULTICOMPONENT RATE EXPRESSIONS

The concepts for multicomponent systems introduced with the integral balance are also applicable in differential form. For the component i:

$$\left(\frac{d(m_i)_{sys}}{dt}\right) = (\dot{m}_i)_{in} - (\dot{m}_i)_{out} + (\dot{m}_i)_{gen} - (\dot{m}_i)_{con} \qquad (3-16)$$

Summing over the entire n components in the system yields the total conservation of mass rate equation.

$$\sum_{i=1}^{n}\left(\frac{d(m_i)_{sys}}{dt}\right) = \sum_{i=1}^{n}(\dot{m}_i)_{in} - \sum_{i=1}^{n}(\dot{m}_i)_{out} + \sum_{i=1}^{n}(\dot{m}_i)_{gen} - \sum_{i=1}^{n}(\dot{m}_i)_{con} \qquad (3-17)$$

Table 3-7: Species Mass Accounting and Total Mass Conservation

Species 1	$\left(\dfrac{d(m_1)_{sys}}{dt}\right)$	=	$(\dot{m}_1)_{in/out}$	+	$(\dot{m}_1)_{gen/con}$
Species 2	$\left(\dfrac{d(m_2)_{sys}}{dt}\right)$	=	$(\dot{m}_2)_{in/out}$	+	$(\dot{m}_2)_{gen/con}$
$\vdots$	$\vdots$		$\vdots$		$\vdots$
Species i	$\left(\dfrac{d(m_i)_{sys}}{dt}\right)$	=	$(\dot{m}_i)_{in/out}$	+	$(\dot{m}_i)_{gen/con}$
$\vdots$	$\vdots$		$\vdots$		$\vdots$
Species n	$\left(\dfrac{d(m_n)_{sys}}{dt}\right)$	=	$(\dot{m}_n)_{in/out}$	+	$(\dot{m}_n)_{gen/con}$
Total Mass	$\left(\dfrac{dm_{sys}}{dt}\right)$	=	$\dot{m}_{in/out}$	+	0

Like the integral expression (3-7), the conservation law for the rate equation requires that

$$\left(\sum_{i=1}^{n}(\dot{m}_i)_{gen} - \sum_{i=1}^{n}(\dot{m}_i)_{con}\right) = 0 \qquad (3-18)$$

which gives the conservation of total mass:

$$\left(\frac{dm_{sys}}{dt}\right) = \dot{m}_{in} - \dot{m}_{out} \qquad (3-19)$$

where $\dot{m}_{in} \equiv \sum(\dot{m}_i)_{in}$ and $\dot{m}_{out} \equiv \sum(\dot{m}_i)_{out}$

By combining the input and output terms and generation and consumption terms, the multicomponent species accounting equations and total mass conservation equation are shown in Table 3-7.

Example 3-4. Silica gel (a solid) is used as a drying agent to remove water from processed food. The processed food must be below a certain water content before it can be shipped.

1) Determine the amount of silica gel required per pound mass of wet food if the food is initially 30 wt % H_2O and must be reduced to 20 wt % H_2O. Furthermore, 3.2 (lb$_m$) of silica gel has the capacity to absorb 1 (lb$_m$) of water.

2) Find the time that is required to absorb the water if the silica gel absorbs the water at rate

$$\dot{m}_{H_2O} = be^{-at} \text{ (lb}_m \text{ / min)}$$

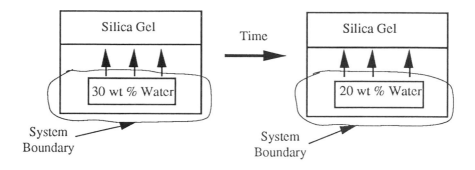

Figure 3-5: Silica gel as a drying agent

where t is time in minutes, $a = 1\text{min}^{-1}$, and $b = 0.13\text{lb}_m / \text{min}$.

SOLUTION: Figure 3-5 depicts the absorbance of water by the silica gel. Initially, the food is 30 wt % water. After a suitable time period, the silica gel has absorbed water, leaving the food to be 20 wt % water. It is convenient to define the food plus the water it contains as the system. Furthermore, because we are interested in the amount of silica gel required *per pound mass of wet food*, we will choose 1 (lb_m) of wet food as the initial amount of food in the system. A logical time period is the amount of time necessary to carry out the required absorption. Eventually, we want to know how long that is but to answer the first question, we do not.

Table 3-8 accounts for the mass in the system using the given data (subscript i is used to indicate the species which is being counted, in this case either "bone-dry food" or water). Note that the 30 wt% composition is used to distribute the 1 (lb_m) of wet food between water and "bone dry" food. Also, note that the silica gel absorbs only water; the food itself (bone-dry food) does not leave the system ((m_{BDF})$_\text{out}$ = 0) and we can enter this in the table. The 20 wt% composition which exists at the end of the process is included as factors applied to the unknown amount of wet food at the end of the process (x).

Table 3-8: Data for Example 3-4 (Mass)

Species	$(m_{i,\text{sys}})_\text{end}$	$-$	$(m_{i,\text{sys}})_\text{beg}$	$=$	$(m_i)_\text{in/out}$		
	(lb_m)		(lb_m)		(lb_m)		(lb_m)
Bone-Dry Food	$0.80x$	$-$	$0.70(1)$	$=$	0	$-$	0
Water	$0.20x$	$-$	$0.30(1)$	$=$	0	$-$	?
Total	x	$-$	1	$=$	0	$-$	?

After entering the data in the table, we can readily see what information we need and whether the equations represented by the table are adequate to solve the problem. Considering the bone dry food, the amount in the system at the end must be the same as at the beginning, 0.70 (lb_m). This lets us solve for x:

$$0.80x = 0.70 \Rightarrow x = 0.875 \ (\text{lb}_m)$$

Table 3-9: Solution of Example 3-4 (Mass)

Species	$(m_{i,sys})_{end}$	$-$	$(m_{i,sys})_{beg}$	$=$	$(m_i)_{in/out}$		
	(lb$_m$)		(lb$_m$)		(lb$_m$)		(lb$_m$)
Bone Dry Food	0.700	$-$	0.70	$=$	0	$-$	0
Water	0.175	$-$	0.30	$=$	0	$-$	0.125
Total	0.875	$-$	1	$=$	0	$-$	0.125

Then, using either the water equation, $-(m_{H_2O})_{out} = (0.20)(0.875) - 0.30$, or the total mass conservation law, $-m_{out} = 0.875 - 1$, we find that m_{out} is 0.125 (lb$_m$). Table 3-9 summarizes the results.

Now we can readily calculate the amount of silica gel required to absorb the 0.125 (lb$_m$) of water:

$$\text{lb}_m \text{ silica gel} = \frac{3.2 \text{ lb}_m \text{ silica gel} \,\big|\, 0.125 \text{ lb}_m \text{ H}_2\text{0}}{1 \text{ lb}_m \text{ H}_2\text{0} \,\big|}$$

$$= 0.400 \text{ (lb}_m \text{ silica gel)}$$

For the system defined as the wet food, to find the time required to absorb this much water start with a rate water accounting equation:

$$\left(\frac{d(m_{H_2O})_{sys}}{dt}\right) = -(\dot{m}_{H_2O})_{out}$$

Note that the accumulation rate of water is negative, indicating that the amount of water in the food decreases with time. Integrating gives

$$(m_{H_2O,sys})_{end} - (m_{H_2O,sys})_{beg} = -\int_0^t (\dot{m}_{H_2O})_{out}dt$$

or

$$(m_{H_2O,sys})_{end} - (m_{H_2O,sys})_{beg} = -0.125 \text{ (lb}_m) = -\int_0^t (\dot{m}_{H_2O})_{out}dt$$

where $(\dot{m}_{H_2O})_{out}$ is the rate of absorption and varies with time:

$$(\dot{m}_{H_2O})_{out} = be^{-at} \text{ (lb}_m / \text{min)}$$

Consequently:

$$(m_{H_2O,sys})_{end} - (m_{H_2O,sys})_{beg} = -0.125 \text{ (lb}_m \text{ H}_2\text{O}) = -\int_0^t be^{-at}dt = \left(\frac{b}{-a}e^{-at}\right)\Big|_0^t = \frac{b}{a}\left(e^{-at} - 1\right)$$

The units for this equation must be pounds mass of water to be homogeneously consistent. (If this is correct, what must be the units of a and b?)

This equation can be rearranged to give an expression for t and then solved after substituting the known values for a and b:

$$t = -\frac{1}{a}\ln\left(1 - \frac{a}{b}\left((m_{H_2O,sys})_{end} - (m_{H_2O,sys})_{beg}\right)\right) = -\frac{1}{1}\ln\left(1 - \frac{1}{0.13}(0.125)\right) = 3.3 \text{ (min)}$$

The time required to absorb the water from the food into the silica gel is about 3.3 minutes.

Note that the equations were manipulated symbolically before any numbers were calculated. This decreases the chance of making a math error and provides an easy check to make sure the units are consistent.

3.5 MULTIPLE INLETS AND OUTLETS

Previously, we have explained the principles of conservation and accounting of mass with explanations that involve only mass crossing the system boundary at one point. What if mass crosses the boundary at multiple points as shown by Figure 3-6?

The terms m_{in}, m_{out}, $\dot{m}_{in}$, and $\dot{m}_{out}$ represent the total contribution of mass or mass flow rate crossing the boundary. For mass entering the system, the contributions of flow streams A and E are summed together. The mass leaving the system is the summation of the contributions of flow streams B, C, and D.

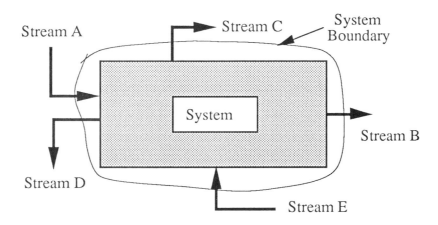

Figure 3-6: Multiple inlets and outlets

3.6 STEADY STATE

Before introducing the next example problem, the concept of steady state must be defined. Steady-state operation means that the state of the system is steady, it does not change. With respect to mass, it means that the time rate of change of mass in the system is zero; i.e the accumulation of mass in the system is zero. The steady-state equation is a special case of the conservation or accounting equations. Mathematically, the steady state condition is:

$$(m_{i,sys})_{end} - (m_{i,sys})_{beg} = 0 \qquad (3-20)$$

or:

$$\left(\frac{d(m_i)_{sys}}{dt}\right) = 0 \qquad (3-21)$$

A continuous flash distillation is the next example. Multiple flow streams leave the system and several species are involved in the separation.

Example 3-5. The feed to a continuous flash distillation consists of 25 (lb$_m$) water, 10 (lb$_m$) ethanol, and 5 (lb$_m$) methanol per hour. The temperature and pressure of the system are 95 °C and 1 atm. The total flow rate from the bottom of the column is 24 (lb$_m$) per hour. The composition of the bottom stream is 83.3 wt% water, 12.5 wt% ethanol, and 4.2 wt% methanol. The system is assumed to be operating at steady state. No reactions are occurring in the system.

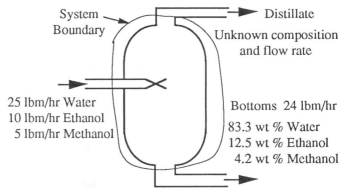

Figure 3-7: Continuous flash distillation

a) Set up the species accounting and total mass conservation equations for the system.

b) Make the proper assumptions, and solve for the unknowns.

c) Determine the composition and mass flow rate of the distillate.

SOLUTION: Figure 3-7 presents the data provided in the problem statement in pictorial or schematic form as an aid to understanding the problem. The system is the distillation column. Table 3-10 shows the flowrate data given in the statement in the context of rate accounting and conservation statements (steady state $\Rightarrow$ all $\left(\dfrac{dm_{sys}}{dt}\right)$ terms are zero, no reactions $\Rightarrow \dot{m}_{gen} = \dot{m}_{con} = 0$).

Table 3-10: Data for Example 3-5 (Mass)

Species	$\left(\dfrac{d(m_i)_{sys}}{dt}\right)$ (lb$_m$ / hr)	=	$(\dot{m}_i)_{in/out}$ Stream F (lb$_m$ / hr)		Stream B (lb$_m$ / hr)		Stream D (lb$_m$ / hr)	+	$(\dot{m}_i)_{gen/con}$ (lb$_m$ / hr)
Water	0	=	25	−	0.833(24)	−	?	+	0
Ethanol	0	=	10	−	0.125(24)	−	?	+	0
Methanol	0	=	5	−	0.042(24)	−	?	+	0
Total	0	=	40	−	24	−	?	+	0

Clearly, each equation can be solved separately for its only unknown entry. The complete results are shown in Table 3-11 (the $\dot{m}_{gen}$, $\dot{m}_{con}$ and $\left(\dfrac{dm_{sys}}{dt}\right)$ terms may be omitted).

Table 3-11: Solution of Example 3-5 (Mass)

| Species | $(\dot{m}_i)_{in}$ | = | $(\dot{m}_i)_{out}$ | | |
	Stream F (lb_m / hr)		Stream B (lb_m / hr)		Stream D (lb_m / hr)
Water	25	=	20.0	+	5.0
Ethanol	10	=	3.0	+	7.0
Methanol	5	=	1.0	+	4.0
Total	40	=	24.0	+	16.0

The composition on a mass basis of the distillate is calculated:

$$\text{Water} \qquad \left(\frac{5}{16}\right) \times 100\% = 31.2\%$$

$$\text{Ethanol} \qquad \left(\frac{7}{16}\right) \times 100\% = 43.8\%$$

$$\text{Methanol} \qquad \left(\frac{4}{16}\right) \times 100\% = 25.0\%$$

The flowrate of distillate is (from Table 3-11) 5.0 (lb_m / hr) water, 7.0 (lb_m / hr) ethanol, and 4.0 (lb_m / hr) of methanol for a total of 16.0 (lb_m / hr).

3.7 MULTIPLE UNIT PROCESSES

Multiple unit processes can be solved using the same type of conservation of mass principles. Each unit separately or in combination with others must satisfy the mass accounting equations for all components. By defining the systems in different ways we can deduce different information on how the system is operating. Also, an overall mass accounting for the entire process can be used to advantage. If the process of each unit is of the same type (e.g., evaporation, distillation, etc.), then each unit is referred to as a stage and, as a whole, they are a multi-stage process. A four-stage evaporator will be used to illustrate how to apply the conservation of mass principles to multi-unit processes. Evaporators are used in the desalination of sea water and concentration of solutions by removing the volatile components.

Example 3-6. A four-stage evaporator is used to desalinate sea water. Determine the percent salt in each solution leaving each evaporator. The amount of water evaporated from each effect is the same. 20,000 lb_m per hour at 3.5 wt % salt enters the first effect. The stream leaving the final effect is 15 wt % salt.

SOLUTION Figure 3-8 is a schematic representing the operation of the evaporators. Only water is removed from the top stream of each evaporator. Assume steady-state operation and that no reactions are occurring in the evaporators. Table 3-12 shows the data in rate form (assuming steady state and no reactions). We are defining the total mass flow rate of the concentrated solution leaving the evaporator D with the symbol D_L. For this accounting, the system is defined to be all four evaporators.

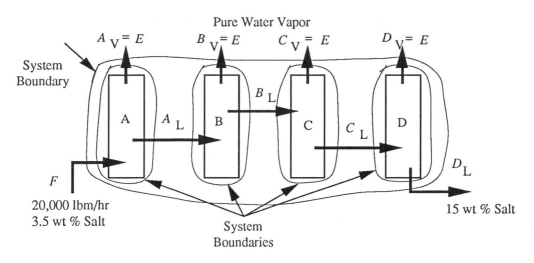

Figure 3-8: Multistage evaporator

Table 3-12: Data for Example 3-6 (Mass, System = All Evaporators)

Species	$(\dot{m}_i)_{in}$ Str. F (lb_m / hr)	=	$(\dot{m}_i)_{out}$ Str. A_V (lb_m / hr)		Str. B_V (lb_m / hr)		Str. C_V (lb_m / hr)		Str. D_V (lb_m / hr)		Str. D_L (lb_m / hr)
Water	0.965(20,000)	=	E	+	E	+	E	+	E	+	$0.85D_L$
Salt	0.035(20,000)	=	0	+	0	+	0	+	0	+	$0.15D_L$
Total	20,000	=	E	+	E	+	E	+	E	+	D_L

As stated, the same amount of water is evaporated from each unit, denoted as E. From the salt equation it is apparent that $D_L = 4,667$ (lb_m / hr) and we can then solve for E using either the water or total mass equation: $E = 3,833$ (lb_m / hr). Similar analyses can be made for each evaporator. For the A evaporator as the system we have Table 3-13 and it is found that: $A_{L,W} = 15,467$ (lb_m / hr), $A_{L,S} = 700$ (lb_m / hr), and $A_L = 16,167$ (lb_m / hr)

Table 3-13: Intermediate Solution of Example 3-6 (Mass, System = Evaporator A)

Species	$(\dot{m}_i)_{in}$ Stream F (lb_m / hr)	=	$(\dot{m}_i)_{out}$ Stream A_V (lb_m / hr)		Stream A_L (lb_m / hr)
Water	0.965(20,000)	=	3833	+	$A_{L,W}$
Salt	0.035(20,000)	=	0	+	$A_{L,S}$
Total	20,000	=	3833	+	A_L

Continuing with the other evaporators, the flowrates and compositions of all remaining streams can be found. A summary of the results is given in Table 3-14. You should complete the problem as an exercise. Before moving on to the next problem you should review this problem by looking at the different systems that were chosen for writing equations, and the information obtained from each. What if you did not include equations for the entire process system? Can you still solve the problem? If so, is the solution procedure easier, harder, or equally involved?

Table 3-14: Solution of Example 3-6 (Mass)

Stage	Total Exiting Brine Flow Rate (lb_m / hr)	Exiting Brine Percent NaCl (lb_m NaCl / lb_m)
A	16,167	4.33
B	12,334	5.68
C	8,501	8.24
D	4,667	15.00

3.8 RECYCLE AND BYPASS

In many processes, such as reactions and separation, some of the material that leaves the system is introduced back into the system as a recycle stream. Recycle streams are used primarily for economic reasons to try and recover and use most of the raw material. Bypass streams are used to remove product that requires no further immediate processing. Figure 3-9 depicts examples of recycle and bypass.

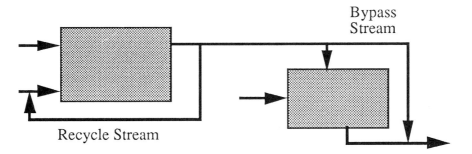

Figure 3-9: Recycle and bypass streams

Example 3-7. A membrane system is used to filter waste products from the blood stream. The membrane can extract 30 (mg/min) of waste, without removing any blood. The waste concentration in the entering blood is 0.17 wt %, and the flow rate of blood is 25 (g/min). A recycle stream is used to help control the concentration of the waste before it enters the membrane. Figure 3-10 depicts the membrane process. (a) If the recycle rate is twice that of the filtered blood rate what is the composition at A? (b) If the composition at A is to be 0.10 % waste, what must be the recycle rate, assuming streams U, W, and F remain the same.

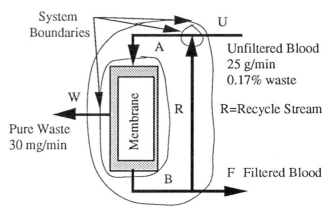

Figure 3-10: Membrane system

SOLUTION: To solve the first problem, we will choose the system to be the membrane and include the recycle stream. The only streams that cross the system boundary are the entering unfiltered blood U, the waste stream W, and the filtered blood F. For the defined system, there is no generation, no consumption, and we assume that the system is operating at steady state.

Table 3-15: Data for Example 3-7a (Mass, System = Membrane plus recycle)

Species	$(\dot{m}_i)_{in}$ Stream U (g / min)	=	$(\dot{m}_i)_{out}$ Stream W (g / min)		Stream F (g / min)
Blood	0.9983(25)	=	0	+	?
Waste	0.0017(25)	=	0.030	+	?
Total	25.0	=	0.030	+	?

Table 3-15 gives the data for this system. We see that the blood flowrate in the filtered blood must be 24.9575 (g/min) and the amount of waste in this stream is 0.0125 (g/min) or 12.5 (mg/min). The total flow rate is 24.9700 g/min. The composition of the filtered blood is found to be 0.05 %. Note the recycle stream for this system boundary did not have to be considered in order to complete the table. The recycle is *inside* the system and does not enter into calculations of streams crossing this system boundary.

To evaluate stream A we can consider the junction of streams U, R, and A as the system. Then Table 3-16 gives the rate equations (for steady state, no reactions) for this system.

Note that this uses the fact that the recycle stream is twice the F flowrate (and obviously the same composition). The A stream values can be found easily from these equations so that Stream A Blood = 74.8725 g / min, Stream A Waste = 0.0675 g / min, Stream A Total = 74.9400 g / min, and the waste in stream A is 0.09 % by mass.

We could also write a set of equations for a system boundary around the membrane but inside the recycle stream as given in Table 3-17. Stream B represents the sum of the recycle stream R and the filtered blood F.

Note 1) that the B values were determined from summing R and F (i.e. a tacit analysis with the junction of B, R, and F as the system) and 2) that these values, all obtained from the previous tables, form a self-consistent table for the membrane system, thus verifying the calculations.

Table 3-16: Intermediate Data for Example 3-7a (Mass, System = Junction U, A, R)

	$(\dot{m}_i)_{in}$				=	$(\dot{m}_i)_{out}$
Species	Stream U (g / min)		Stream R (g / min)			Stream A (g / min)
Blood	24.9575	+	2(24.9575)		=	?
Waste	0.0425	+	2(0.0125)		=	?
Total	25.0	+	2(24.9700)		=	?

Table 3-17: Verification of Example 3-7a (Mass, System = Membrane)

	$(\dot{m}_i)_{in}$	=		$(\dot{m}_i)_{out}$		
Species	Stream A (g / min)		Stream W (g / min)			Stream B (g / min)
Blood	74.8725	=	0	+		74.8725
Waste	0.0675	=	0.03	+		0.0375
Total	74.9400	=	0.03	+		74.9100

Table 3-18: Data for Example 3-7b (Mass, System = Membrane)

	$(\dot{m}_i)_{in}$	=		$(\dot{m}_i)_{out}$		
Species	Stream A (g / min)		Stream W (g / min)	+		Stream B (g / min)
Blood	$0.999A$	=	0	+		$0.9995B$
Waste	$0.001A$	=	0.03	+		$0.0005B$
Total	A	=	0.03	+		B

For the second question of the problem, we begin with Table 3-18, steady-state, no-reaction equations for the membrane system, as defined in the preceding paragraphs. Note that streams U, F, and W, and their composition are still the same.

There are two unknowns and two independent equations. Solving the blood and waste (or Total mass) equations simultaneously gives Stream A Total = 59.9700 g / min and Stream B Total = 59.9400 g / min. The mass flow rates of the blood and waste in streams A and B are Stream A Blood = 59.9100 g / min, Stream A Waste = 0.0600 g / min, Stream B Blood = 59.9100 g / min, and Stream B Waste = 0.0300 g / min.

To establish the recycle rate, we can now look at the junction of B, R, and F given in Table 3-19. Obviously, we can solve for all of the unknown flowrates: Stream R Blood = 34.9525 g / min, Stream R Waste = 0.0175 g / min, and Stream R Total = 34.9700 g / min. One final note: when a completed table is obtained, be sure to check for both column and row consistency. If the sum of the species in each column does not equal the total for that column or if any row equation does not check, then there is an error in the calculations, somewhere. Note, however, that some small discrepancies due to roundoff error may (will) occur, and we should not be concerned about these.

Table 3-19: Intermediate Data for Example 3-7b (Mass, System = Junction B, F, R)

Species	$(\dot{m}_i)_{\text{in}}$	=	$(\dot{m}_i)_{\text{out}}$		
	Stream B (g / min)		Stream R (g / min)		Stream F (g / min)
Blood	59.9100	=	?	+	24.9575
Waste	0.03	=	?	+	0.0125
Total	59.9400	=	?	+	24.9700

3.9 DENSITY IN CONSERVATION OF MASS

In order to fully utilize the power of the rate form of the conservation of mass equation, we must develop useful relationships to represent $(\dot{m}_i)_{\text{in/out}}$, $(\dot{m}_i)_{\text{gen/con}}$, and the time rate of change of mass in the system in terms of a physical property. The density of the material allows us to accomplish this.

The density of a material is defined as the mass of material per unit volume and is denoted by the Greek letter rho, ρ. (The complete Greek alphabet is given in Table A-5).

$$\rho_i = \frac{\text{mass of species } i}{\text{unit volume}} = \frac{m_i}{V} \qquad (3-22)$$

The density can either be mass or mole density. Common examples of densities are listed in Tables C-1 through C-5 in Appendix C.

Since many flow rates are given in volumetric terms, such as the cubic feet per hour of water flow in a river or gallons per minute from a pump, we can determine the mass flow rates with information about the density.

For example, rearranging equation (3-22) and solving for the mass gives

$$m \text{ (lb}_\text{m}) = \rho \left(\frac{\text{lb}_\text{m}}{\text{ft}^3}\right) V \text{ (ft}^3) \qquad (3-22a)$$

If we introduce the idea of a volumetric flow rate as the volume that crosses the system boundary per unit time, we can express the mass entering and leaving the system in terms of the volumetric flow rate.

$$\text{Volumetric flow rate} \equiv \dot{V} = \frac{\text{volume crossing system boundary}}{\text{unit time}} \qquad (3-22b)$$

Now, we can represent the mass flow rate in and out of the system in terms of the volumetric flow rate:

$$\dot{m}_{\text{in}} = (\rho \dot{V})_{\text{in}} = \frac{\text{mass}}{\text{vol}} \left|\frac{\text{vol}}{\text{time}}\right. = \frac{\text{mass}}{\text{time}}$$

$$\dot{m}_{\text{out}} = (\rho \dot{V})_{\text{out}} \qquad (3-22c)$$

For flow in channels when the cross sectional area is known, the volumetric flow rate is usually expressed as the product of the cross sectional area normal to the flow, A, and the average linear velocity of the flow $\langle v \rangle$. The mass flow rates in and out of the system can be expressed as:

$$\dot{m}_{in} = (\rho A \langle v \rangle)_{in}$$
$$\dot{m}_{out} = (\rho A \langle v \rangle)_{out}$$

$$(3-22d)$$

The discussion which addresses calculating the average linear velocity is presented in the fluid dynamics section. Here the important point is that mass flow rates are often expressed as the products of density, cross sectional area and average linear velocity.

The accumulation rate (time rate of increase) of mass in the system is obtained by directly substituting into equation (3-14) the relation

$$\left(\frac{dm_{sys}}{dt} \right) = \left(\frac{d(\rho V)_{sys}}{dt} \right)$$

The resulting conservation of mass equation in terms of density, volumetric flow rates and system volume is:

$$(\rho \dot{V})_{in} - (\rho \dot{V})_{out} = \left(\frac{d(\rho V)_{sys}}{dt} \right)$$

$$(3-23)$$

Of course, generation and consumption rate terms must also be included if we are accounting for reacting species. Example 3-8 describes a lake and reservoir system and uses the concept of volumetric flow rates.

Example 3-8. A lake/reservoir system has a total volumetric capacity of 1,000,000 (ft^3) of water, and the current volume is 300,000 (ft^3) of water. The average temperature in the lake is 15 °C. Two (2) streams feed the lake system and the temperature of each stream is 15 °C. Stream 1 flows at 50,000 (ft^3 / hr) and Stream 2 at 10,000 (ft^3 / hr). An aqueduct diverts water at a constant rate of 5,000 (ft^3 / hr) to supply city needs. The water leaving the aqueduct is the same temperature as the average lake temperature.

a) Determine the mass of water currently in the lake and the mass flow rates of streams 1 and 2 and that of the aqueduct.

b) If the water leaving the dam is 15 °C, find the release rate in (ft^3 / hr) and (lb_m / hr) at the dam to maintain the current volume

c) If the release rate is cut to one-half that of (b), how long will it take to completely fill the reservoir? and

d) If we wish to fill the reservoir in 24 hours, what must the release rate be?

SOLUTION: Figure 3-11 is schematic of the lake and reservoir system.

a) From Table C-1 the density of water at 15°C is found to be 62.3 lb_m / ft^3. In general, the mass of the water in the lake system is the density of the water integrated over the volume.

$$(m_{H_2O})_{sys} = \int_V \rho_{H_2O} dV$$

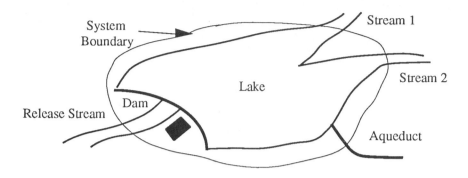

Figure 3-11: Lake or reservoir system

In this case, however, the density is assumed constant and the integration is simply:

$$(m_{H_2O})_{sys} = \rho_{H_2O}V = \frac{62.3 \text{ lb}_m}{\text{ft}^3} \left| \frac{300,000 \text{ ft}^3}{} \right. = 1.87 \times 10^7 \text{ lb}_m \text{ H}_2\text{O}$$

The mass flow rates are calculated from the water density and the volumetric flow rates:

$$\dot{m}_1 = \rho\dot{V}_1 = \frac{62.3 \text{ lb}_m}{\text{ft}^3} \left| \frac{50,000 \text{ ft}^3}{\text{hr}} \right. = 3.12 \times 10^6 \text{ lb}_m / \text{hr}$$

Similarly,

$$\dot{m}_2 = 6.23 \times 10^5 \text{ (lb}_m / \text{hr)}$$
$$\dot{m}_A = 3.12 \times 10^5 \text{ (lb}_m / \text{hr)}$$

b) The next question is to determine the release rate required to maintain constant *volume*. It should be noted that volume is not a system property that can be counted since system volume can be defined totally independent of input and output. Rather, we must ask about the volume of the mass in the system. Consequently, to assess volume of the water in the reservoir, we must look at conservation of mass. Noting the relations between mass, density, and volume of the mass:

$$m = \rho V; mass = \frac{\text{mass}}{\text{volume}} \left| \text{volume} \right.$$

and between density, mass flowrate, and volume flow rate:

$$\dot{m} = \rho\dot{V}; \frac{\text{mass}}{\text{time}} = \frac{\text{mass}}{\text{volume}} \left| \frac{\text{volume}}{\text{time}} \right.$$

the conservation of mass rate equation is

$$\left(\frac{d(\rho V)_{sys}}{dt} \right) = \sum(\rho\dot{V})_{in} - \sum(\rho\dot{V})_{out}$$

Because the density of the water in the lake is the same as that in all of the flow streams and is constant, ρ can be factored, giving

$$\rho\left(\frac{dV_{sys}}{dt}\right) = \rho \sum \dot{V}_{in} - \rho \sum \dot{V}_{out}$$

or

$$\left(\frac{dV_{sys}}{dt}\right) = \sum \dot{V}_{in} - \sum \dot{V}_{out}$$

Then, to maintain constant volume (of water in the reservoir), $dV_{sys}/dt = 0$. Hence,

$$0 = \sum \dot{V}_{in} - \sum \dot{V}_{out}$$

and in terms of mass

$$0 = \rho \sum \dot{V}_{in} - \rho \sum \dot{V}_{out}$$

This last result is simply steady-state conservation of mass. In tabular form, this is

Table 3-20: Data for Example 3-8b (Mass, System = Lake)

Species	$\left(\dfrac{dm_{sys}}{dt}\right)$ (lb_m / hr)	=	$(\dot{m}_i)_{in}$ Stream 1 (lb_m / hr)		$(\dot{m}_i)_{in}$ Stream 2 (lb_m / hr)	−	$(\dot{m}_i)_{out}$ Aqueduct (lb_m / hr)	$(\dot{m}_i)_{out}$ Release (lb_m / hr)
Water	0	=	3.12×10^6	+	6.23×10^5	−	3.12×10^5	?

From this table, the release rate is found to be 3.43×10^6 lbm/hr. In terms of volume flow rate, this is

$$3.43 \times 10^6 \frac{lbm}{hr} \left| \frac{ft^3}{62.3\ lbm} \right. = 5.5 \times 10^4 \frac{ft^3}{hr}$$

c) This part deals with an unsteady-state process. We need to determine the time required to fill the reservoir. From above, as long as density is constant,

$$\left(\frac{dV_{sys}}{dt}\right) = \sum \dot{V}_{in} - \sum \dot{V}_{out} = (\dot{V}_1 + \dot{V}_2 - \dot{V}_A) - \dot{V}_R$$

where $\dot{V}_1 = 50,000$ ft^3 / hr, $\dot{V}_2 = 10,000$ ft^3 / hr, $\dot{V}_A = 10,000$ ft^3 / hr, and $\dot{V}_R = 27,500$ ft^3 / hr (half of the release rate of part (b)). So, with all of the input/output terms known, this ordinary differential equation can be integrated directly (the flow rates are all constant):

$$\int_{V_{sys,beg}}^{V_{sys,end}} dV_{sys} = \int_0^t \left[(\dot{V}_1 + \dot{V}_2 - \dot{V}_A) - \dot{V}_R \right] dt$$

Note that $V_{sys,end} = 10^6$ ft^3, $V_{sys,beg} = 3 \times 10^5$ ft^3. Hence

$$t = \frac{(10^6 - 3 \times 10^5)}{(50,000 + 10,000 - 5,000 - 27,500)}\ hr = 25.5\ hr$$

d) Now the release rate is unknown while the time duration of the release is known and all flowrates are constant. Hence

$$\dot{V}_R = \frac{-(V_{sys,end} - V_{sys,beg}) + (\dot{V}_1 + \dot{V}_2 - \dot{V}_A)t}{t} = \frac{(50,000 + 10,000 - 5,000)(24) - (10^6 - 3 \times 10^5)}{24} \frac{ft^3}{hr} = 25,800 \frac{ft^3}{hr}$$

3.10 CHEMICAL REACTIONS

If chemical reactions occur within a system, the conservation of total mass and the elements still hold. Reactions occur in stoichiometric ratios, the specific ratios in which, atoms or molecules combine to form products. For example, when methane, CH_4 is burned with oxygen, O_2, the products are water, H_2O, and carbon dioxide, CO_2 if complete combustion is achieved. The balanced stoichiometric equation is:

$$CH_4 + 2O_2 \longrightarrow 2H_2O + CO_2$$

Note that the number of atoms (or moles) of each *element* in the reactants is the same as in the products. If we think of the reaction as taking place in a sealed (closed) box so that there is no input or output of any mass, then conservation of total mass and of the elements requires that the amount of total mass and each of the elements in the box after the reaction be the same as before. Conservation of the elements is the basis for obtaining a balanced (stoichiometric) chemical equation.

The following example illustrates how to organize and solve a complex problem involving the conservation of mass with reactions.

Example 3-9. A company is making a product P from a reactant which is governed by the following stoichiometric equation.

$$R \longrightarrow P + W$$

The entire process involves reacting the components to form the products and the waste. This reaction occurs in Unit 1. In Unit 2, only the waste W is removed, and in Unit 3 the product is further purified. Given the data in Table 3-21 and Figure 3-12, determine the flow rates and composition of each stream.

Table 3-21: Data for Problem 3-9

Stream	$\dot{m}$ (lb_m / hr)	w_R (lb_m R / lb_m)	w_P (lb_m P / lb_m)	w_W (lb_m W / lb_m)
A	200	?	0.01	0.01
B	?	?	?	?
C	10	?	?	1.00
D	?	?	?	?
E	150	0.20	0.80	0.00
F	?	0.70	0.30	0.00

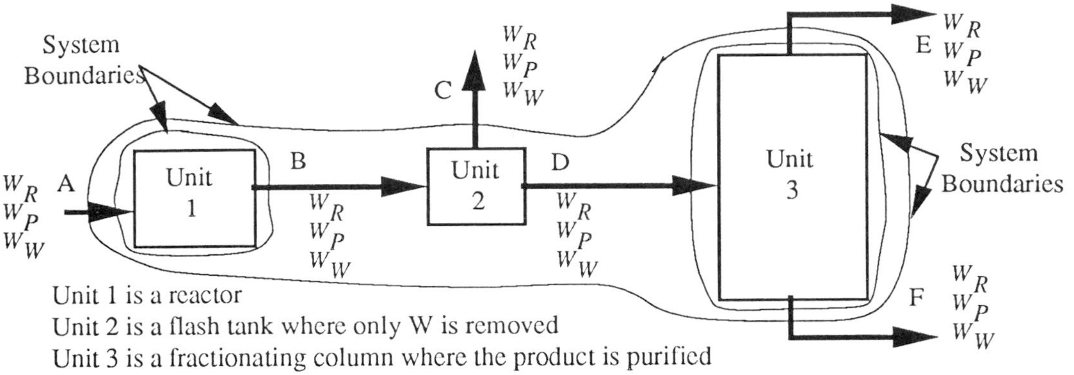
System
Boundaries
W_R
W_P
W_W
A
Unit
1
B
W_R
W_P
W_W
C
Unit
2
D
W_R
W_P
W_W
Unit
3
E
F
System
Boundaries
Unit 1 is a reactor
Unit 2 is a flash tank where only W is removed
Unit 3 is a fractionating column where the product is purified

Figure 3-12: Complex process with reactions

SOLUTION: Figure 3-12 gives a schematic of the process. To solve for all of the unknowns we first extract as much information as possible from the largest system, and then apply total mass conservation and species accounting equations to smaller and smaller systems until all the unknowns have been determined.

First, consider steady-state rate equations around the entire process (Figure 3-12) based on the given data as shown in Table 3-22. Because total mass is conserved, even in the presence of chemcial reactions, $\sum(\dot{m}_i)_{\text{gen}} - \sum(\dot{m}_i)_{\text{con}} = 0$ and we can easily solve for F:

$$F = 40 \ (\text{lb}_\text{m} \ / \ \text{hr})$$

We now have 3 unknowns: $(\dot{m}_P)_{\text{gen}}$, $(\dot{m}_W)_{\text{gen}}$, and $(\dot{m}_R)_{\text{con}}$ in three equations. Solving each equation separately, for these rates of species generation and consumption gives:

$$(\dot{m}_R)_{\text{con}} = 138 \ (\text{lb}_\text{m} \ / \ \text{hr})$$

$$(\dot{m}_P)_{\text{gen}} = 130 \ (\text{lb}_\text{m} \ / \ \text{hr})$$

$$(\dot{m}_W)_{\text{gen}} = 8 \ (\text{lb}_\text{m} \ / \ \text{hr})$$

The completed solution is given in Table 3-23.

Table 3-22: Data for Example 3-9 (Mass, System = Entire Process)

Species	$\left(\frac{d(m_i)_{\text{sys}}}{dt}\right)$ (lb$_\text{m}$ / hr)	=	$(\dot{m}_i)_{\text{in/out}}$ Stream A (lb$_\text{m}$ / hr)		Stream E (lb$_\text{m}$ / hr)		Stream F (lb$_\text{m}$ / hr)		Stream C (lb$_\text{m}$ / hr)	+	$(\dot{m}_i)_{\text{gen/con}}$ (lb$_\text{m}$ / hr)		(lb$_\text{m}$ / hr)
Reactant	0	=	0.98(200)	−	0.2(150)	−	$0.7F$	−	0.0(10)	+	0	−	$(\dot{m}_R)_{\text{con}}$
Product	0	=	0.01(200)	−	0.8(150)	−	$0.3F$	−	0.0(10)	+	$(\dot{m}_P)_{\text{gen}}$	−	0
Waste	0	=	0.01(200)	−	0.0(150)	−	$0.0F$	−	1.0(10)	+	$(\dot{m}_W)_{\text{gen}}$	−	0
Total	0	=	200	−	150	−	F	−	10	+	0		

Table 3-23: Intermediate Solution of Example 3-9 (Mass, System = Entire Process)

Species	$\left(\dfrac{d(m_i)_{\text{sys}}}{dt}\right)$ (lb$_m$ / hr)	=	$(\dot{m}_i)_{\text{in/out}}$					$+$	$(\dot{m}_i)_{\text{gen/con}}$	
			Stream A (lb$_m$ / hr)	Stream E (lb$_m$ / hr)	Stream F (lb$_m$ / hr)	Stream C (lb$_m$ / hr)			(lb$_m$ / hr)	(lb$_m$ / hr)
Reactant	0	=	196	− 30	− 28	− 0	+		0	− 138
Product	0	=	2	− 120	− 12	− 0	+		130	− 0
Waste	0	=	2	− 0	− 0	− 10	+		8	− 0
Total	0	=	200	− 150	− 40	− 10	+		0	

Table 3-24: Intermediate Data for Example 3-9 (Mass, System = Unit 1)

Species	$\left(\dfrac{d(m_i)_{\text{sys}}}{dt}\right)$ (lb$_m$ / hr)	=	$(\dot{m}_i)_{\text{in/out}}$		$+$	$(\dot{m}_i)_{\text{gen}}$	
			Stream A (lb$_m$ / hr)	Stream B (lb$_m$ / hr)		(lb$_m$ / hr)	(lb$_m$ / hr)
Reactant	0	=	196	− $w_{R,B}B$	+	0	− 138
Product	0	=	2	− $w_{P,B}B$	+	130	− 0
Waste	0	=	2	− $w_{W,B}B$	+	8	− 0
Total	0	=	200	− B	+	0	

Table 3-25: Intermediate Data for Example 3-9 (Mass, System = Unit 3)

Species	$(\dot{m}_i)_{\text{in}}$	=	$(\dot{m}_i)_{\text{out}}$		
	Stream D (lb$_m$ / hr)		Stream E (lb$_m$ / hr)		Stream F (lb$_m$ / hr)
Reactant	$w_{R,D}D$	=	30	+	28
Product	$w_{P,D}D$	=	120	+	12
Waste	$w_{W,D}D$	=	0	+	0
Total	D	=	150	+	40

Now we will choose Unit 1 as our system and the equations are presented in Table 3-24. We can solve for B directly from the Total mass equation: $B = 200$ (lb$_m$ / hr). The species equations then can be solved: $w_{R,B} = 0.290$, $w_{P,B} = 0.660$, and $w_{W,B} = 0.050$. Note that $w_{R,B}$, $w_{P,B}$, and $w_{W,B}$ are not independent variables;

$$w_{R,B} + w_{P,B} + w_{W,B} = 1$$

Hence, Table 3-24, which has only three independent equations, also has only three independent unknown variables.

Table 3-26: Solution for Problem 3-9

Stream	$\dot{m}$ (lb$_m$ / hr)	w_R (lb$_m$ R / lb$_m$)	w_P (lb$_m$ P / lb$_m$)	w_W (lb$_m$ W / lb$_m$)
A	200	0.98	0.01	0.01
B	200	0.29	0.66	0.05
C	10	0.00	0.00	1.00
D	190	0.305	0.695	0.00
E	150	0.20	0.80	0.00
F	40	0.70	0.30	0.00

Table 3-25 is for Unit 3 as the system (steady-state, no reactions in this unit). From this set of equations, we see immediately that: $D = 190$ (lb$_m$ / hr). The species equations can be solved to give: $w_{R,D} = 0.305$, $w_{P,D} = 0.695$, and $w_{W,D} = 0.0$. A summary of all the information for each stream is given in Table 3-26.

One final point is noteworthy. The above tables were in terms of mass units; total mass is conserved. If moles were counted, allowance would have to be made for the fact that total moles are not necessarily conserved.

Example 3-10. Butane (C_4H_{10}, MW =58.12) is combusted in a furnace with 20% excess air (based on complete combustion to carbon dioxide (CO_2) and water (H_2O). The furnace is operating at steady-state. The *unbalanced* chemical reaction is

$$C_4H_{10} + O_2 \rightarrow CO_2 + H_2O$$

Calculate the composition (in both mole and mass percentages) of the flue gas if the butane is completely converted to CO_2 and H_2O. You may assume that the entering air is perfectly dry and hence is 79 mole percent nitrogen (N_2) and 21 mole per cent oxygen (O_2).

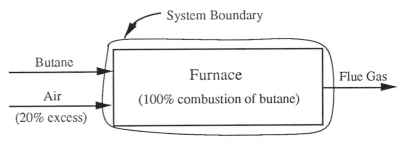

Figure 3-13: Combustion of Butane

SOLUTION: A schematic diagram of the furnace is shown in Figure 3-13. The flue gas composition can be determined by performing an accounting of each species and of total mass. Because a reaction is involved and because reactions occur in molar ratios, an accounting of moles is first performed. Then the moles can be converted to mass by using the compounds' molecular masses.

Begin by balancing the chemical reaction (the elements are conserved; they are neither generated nor consumed by the reaction). Start with C, then, H (they are easy because they appear in only one compound on each side of the reaction equation). Then counting O on the product side of the equation allows you to determine the coefficient of O_2 on the reactants side. This gives the following balanced reaction equation:

$$C_4H_{10} + \frac{13}{2}O_2 \rightarrow 4CO_2 + 5H_2\dot{O}$$

Now you are ready to count each compound. Note that the following compounds are present at some time during the reaction: C_4H_{10}, N_2, O_2, CO_2, and H_2O. Don't forget the nitrogen that enters with the air. Even though it doesn't react, it plays a role in the composition. Proceed by making a table which counts all of these compounds plus total moles. Let the system be the furnace and the time period be "however long it takes to feed, and therefore process, 100 moles of butane." Table 3-27 shows the accounting. Because a reaction is occuring, generation and consumption must be included for the compounds.

Table 3-27: An Accounting of Moles for Example 3-10 (Moles, System = Furnace)

Species	Accum	=	$(moles)_{in}$ Butane + Air		$(moles)_{out}$ Flue Gas	+	$(moles)_{gen}$ Gen		$(moles)_{con}$ Cons
C_4H_{10}	0	=	100	$-$	$FG_{C_4H_{10}}$	+	0	$-$	100
O_2	0	=	$100\left(\dfrac{13}{2}\right)(1.2)$	$-$	FG_{O_2}	+	0	$-$	$100\left(\dfrac{13}{2}\right)$
N_2	0	=	$100\left(\dfrac{13}{2}\right)(1.2)\left(\dfrac{79}{21}\right)$	$-$	FG_{N_2}	+	0	$-$	0
CO_2	0	=	0	$-$	FG_{CO_2}	+	100(4)	$-$	0
H_2O	0	=	0	$-$	FG_{H_2O}	+	100(5)	$-$	0
Total	0	=	3814	$-$	FG	+	100(9)	$-$	750

Several things should be noted about this table. First, steady state means that the accumulation of each species in the reactor is zero. Second, 100 moles of butane entered because we chose this in specifying the time period. Also, 100 moles of butane reacted (was consumed) because conversion is 100%. Next, to calculate the amount of O_2 entering, recognize that it must be supplied in 20% excess (hence the factor 1.2) of the stoichiometric amount (hence 100[13/2]). The amount of O_2 consumed is established by the reaction equation and the conversion. The amount of N_2 that enters is established by the amount of O_2 required and the composition of the air which has 79 moles of N_2 per 21 moles of O_2. Note that the nitrogen is inert and therefore is neither generated nor consumed. Finally, neither CO_2 nor H_2O enters the furnace (according to the assumptions) but that both are generated in accordance with the stoichiometric ratios of the balanced equation and the moles of C_4H_{10} consumed. Note that the total moles equation is determined by adding the species equations.

Each row of this table is an equation which can be solved directly to give the number of moles of each species (or of the total) in the flue gas. The mole percent composition is then easily calculated. The results are in the following table:

Table 3-28: Molar Composition of the Flue Gas

Compound	moles in product	mole %
C_4H_{10}	0	0
O_2	130	3.3
N_2	2,934	74.0
CO_2	400	10.1
H_2O	500	12.6
Total	3,964	100.0

The mass composition is calculated from the mole results. As a check of the calculation, a complete mass accounting table is also made. Moles are converted to mass by multiplying by the molecular mass (mass per mole). The resulting values are shown in Table 3-29. Note that in the mass table the generation and consumption terms sum to zero, as they must (total *mass* is conserved). This serves as a check on the calculations. (Also note that the small discrepancy is due to round-off error.)

Table 3-29: An Accounting of Mass for Example 3-10 (Mass, System = Furnace)

Species	Accum (g)	=	$(mass)_{in}$ Butane + Air (g)		$(mass)_{out}$ Flue Gas (g)	+	$(mass)_{gen}$ Gen (g)		$(mass)_{con}$ Cons (g)
C_4H_{10} (58.12)	0	=	5,812	−	0	+	0	−	5,812
O_2 (32.00)	0	=	24,960	−	4,160	+	0	−	20,800
N_2 (28.01)	0	=	82,189	−	82,189	+	0	−	0
CO_2 (44.01)	0	=	0	−	17,604	+	17,604	−	0
H_2O (18.02)	0	=	0	−	9,010	+	9,010	−	0
Total	0	=	112,961	−	112,961	+	26,614	−	26,612

The flue gas mass percentages are now easily calculated and are shown in the following table. Note that the mass and mole percentages are different due to the different species molecular masses.

Table 3-30: Mass Composition of the Flue Gas

Compound	mass in product	mass %
C_4H_{10}	0	0
O_2	4,160	3.7
N_2	82,189	72.8
CO_2	17,604	15.6
H_2O	9,010	8.0
Total	112,961	100.0

Example 3-11. Rework the previous problem if only 89% of the butane is reacted, everything else remaining the same.

SOLUTION: The procedure is much the same as for the previous problem but must allow for the partial reaction. For 100 moles of butane entering the reactor, Table 3-31 gives the accounting of moles and Table 3-32 gives the resulting molar composition of the flue gas. Tables 3-33 and 3-34 give the corresponding mass tables. In the mole table, note that the O_2 consumption and CO_2 and H_2O generation are based on butane reacted and not on the amount entering the furnace.

Table 3-31: An Accounting of Moles for Example 3-11 (Moles, System = Furnace)

Species	Accum	=	$(moles)_{in}$ Butane + Air		$(moles)_{out}$ Flue Gas	+	$(moles)_{gen}$ Gen		$(moles)_{con}$ Cons
C_4H_{10}	0	=	100	−	$FG_{C_4H_{10}}$	+	0	−	100(0.89)
O_2	0	=	$100\left(\dfrac{13}{2}\right)(1.2)$	−	FG_{O_2}	+	0	−	$100\left(\dfrac{13}{2}\right)(0.89)$
N_2	0	=	$100\left(\dfrac{13}{2}\right)(1.2)\left(\dfrac{79}{21}\right)$	−	FG_{N_2}	+	0	−	0
CO_2	0	=	0	−	FG_{CO_2}	+	100(4)(0.89)	−	0
H_2O	0	=	0	−	FG_{H_2O}	+	100(5)(0.89)	−	0
Total	0	=	3814	−	FG	+	801	−	667.5

Table 3-32: Molar Composition of the Flue Gas

Compound	moles in product	mole %
C_4H_{10}	11	0.3
O_2	201.5	5.1
N_2	2,934	74.5
CO_2	356	9.0
H_2O	445	11.3
Total	3,937	100

Table 3-33: An Accounting of Mass for Example 3-11 (Mass, System = Furnace)

Species	Accum (g)	=	$(mass)_{in}$ Butane + Air (g)		$(mass)_{out}$ Flue Gas (g)	+	$(mass)_{gen}$ Gen (g)		$(mass)_{con}$ Cons (g)
C_4H_{10} (58.12)	0	=	5812	−	639	+	0	−	5,173
O_2 (32.00)	0	=	24,960	−	6,448	+	0	−	18,512
N_2 (28.01)	0	=	82,189	−	82,189	+	0	−	0
CO_2 (44.01)	0	=	0	−	15,668	+	15,668	−	0
H_2O (18.02)	0	=	0	−	8,019	+	8,019	−	0
Total	0	=	112,961	−	112,962	+	26,687	−	26,685

Table 3-34: Mass Composition of the Flue Gas

Compound	mass in product	mass %
C_4H_{10}	639	0.6
O_2	6,448	5.7
N_2	82,189	72.8
CO_2	15,664	13.9
H_2O	8,019	7.1
Total	112,962	100

Example 3-12. As a final example, consider the furnace problem again. This time, however, there is 89% reaction of the butane but it is split between two reaction paths. Of the 89% reacted, 66% reacts to produce CO_2 and H_2O and 34% reacts to produce CO (carbon monoxide, a very poisonous gas) and H_2O. Everything else remains the same.

SOLUTION: First, balance the second reaction

$$C_4H_{10} + \frac{9}{2}O_2 \rightarrow 4CO + 5H_2O$$

The mole accounting results, taking into account the incomplete combustion and the two reaction paths, are shown in Table 3-35. Note that air is still provided in 20% excess of that required for complete combustion to CO_2. Also, note that CO (MW=28.01) must be included as an additional species, that the O_2 consumption is split between the two reactions, that CO_2 and CO are each produced by only one of the reactions, that water is produced by both but the stoichiometric coefficient is the same for both reactions, and that butane is consumed by both reactions but its coefficient of unity is the same for both. Calculation of the molar compositon and of the mass table and composition are left as an exercise.

Table 3-35: An Accounting of Moles for Example 3-12 (Moles, System = Furnace)

Species	Accum	=	(moles)$_{in}$ Butane + Air		(moles)$_{out}$ Flue Gas	+	(moles)$_{gen}$ Gen		(moles)$_{con}$ Cons
C_4H_{10}	0	=	100	−	$FG_{C_4H_{10}}$	+	0	−	100(0.89)
O_2	0	=	$100\left(\frac{13}{2}\right)(1.2)$	−	FG_{O_2}	+	0	−	$100(0.89)\left[(0.66)\frac{13}{2} + (0.34)\frac{9}{2}\right]$
N_2	0	=	$100\left(\frac{13}{2}\right)(1.2)\left(\frac{79}{21}\right)$	−	FG_{N_2}	+	0	−	0
CO_2	0	=	0	−	FG_{CO_2}	+	100(4)(0.89)(0.66)	−	0
CO	0	=	0	−	FG_{CO}	+	100(4)(0.89)(0.34)	−	0
H_2O	0	=	0	−	FG_{H_2O}	+	100(5)(0.89)	−	0
Total	0	=	3814	−	FG	+	801	−	607

3.11 CLOSED AND OPEN SYSTEMS

The terms *closed* and *open* refer to whether mass is crossing the system boundary. A *closed system* is one for which there is no input or output of mass and hence all input and output terms in the total mass or species mass equations are zero. An *open system* is one for which mass is crossing the system boundary.

3.12 REVIEW

1) In the absence of nuclear conversions, processes neither create nor destroy total mass or that of the chemical elements, i.e. total mass and that of the chemical elements for the system plus surroundings are constant.

2) Because of chemical reactions, individual species' masses are not necessarily conserved. However, we may still write accounting equations which allow for generation and consumption due to these reactions. Accounting equations for individual species must be consistent with the conservation of total mass and the chemical elements; all of the chemical reactions added together must not result in any increases or decreases of total mass or of that of the elements.

3) These observations, taken together, provide a basis or structure for analyzing processes occurring within the prescribed system boundaries with respect to total mass and species mass flowrates and compositions. This structure is a system of simultaneous accounting equations for the various species which are involved, plus a statement of the law of conservation of total mass. The sum of the individual species equations is equivalent to the total mass conservation law.

4) A very effective, efficient, and helpful way of organizing the information in a problem and planning a solution to the conservation and accounting mass equations is to formulate a table with each row of the table representing a conservation or accounting equation and each column representing an element of an equation such as input, output, generation, consumption, or accumulation (the mass in the system at the end of the time period minus that in the system at the beginning).

5) For some problems, several different system boundaries may need to be defined in order to obtain a solution to the problem. Different systems will give different information about the various flow streams and process behavior.

6) Sometimes, a problem may be solved in more than one way, that is by selecting different sets of systems. Frequently, even though different sets of systems can give the complete solution, the ease of solution can be affected greatly by the choice of system boundary(ies).

7) Property information about materials and their behavior can play an important role in mass conservation problems. For example, the density of materials may be necessary to convert volume to mass. Concentrations can play a similar role. Also, chemical kinetics, i.e., chemical reaction rates provide information about the generation and consumption of species due to reaction. The details of chemical reaction and material properties are appropriately the subject of other courses. However, it is important to realize in this course the role that knowledge of this sort plays in these problems.

8) The conservation and accounting equations can be applied over a differential time period as rate equations, or over finite time periods as integral equations.

9) The term steady state is used to indicate a special situation, that the state of the system and hence the mass content is not changing with time. Rate equations are often used for steady-state processes.

3.13 NOTATION

For this chapter the following symbols were used in the development of the equations and concepts. It is important to realize throughout your engineering and sciences courses that one symbol such as the letter A can be used to represent many different quantities; for example: A could mean the cross sectional area or A could represent the Helmholtz free energy. Therefore, at the end of each chapter, we will summarize the notation used in that specific chapter. However, be cautious because the symbols might represent different quantities in other chapters. Finally, we have tried to minimize the confusion that notation causes by limiting each quantity to one symbol throughout the book. For example, the symbol for electric current will always be represented by the symbol i. Various letters are used to represent the flow streams throughout the chapter.

Variables and Descriptions		Dimensions
A	cross sectional area	[length^2]
m	total mass	[mass]
m_i	mass of species i	[mass of i]
$\dot{m}$	total mass flow rate	[mass / time]
$\dot{m}_i$	mass flow rate of species i	[mass of i / time]
M	total moles	[moles]
M_i	moles of species i	[moles of i]
$(MW)_i$	Molecular mass (weight) of species i	[mass of i / mole of i]
t	time	[time]
$\dot{V}$	volumetric flow rate	[length^3 / time]
V	total volume	[length^3]
$\langle v \rangle$	average linear velocity	[length / time]
w_i	mass fraction of species i	[mass of i / total mass]
x_i or y_i	mole fraction of species i	[mole of i / total mole]
ρ	density	[mass / length^3]
ρ_i	density of species i	[mass of i / length^3]

Subscripts	
beg	evaluated at the beginning of the time period
con	consumption or usage
end	evaluated at the end of the time period
gen	generation or production
in	input or entering
out	output or leaving
sys	system or within the system boundary

3.14 SUGGESTED READING

Felder, R. M., and R. W. Rousseau, *Elementary Principles of Chemical Processes*, John Wiley & Sons, New York, 1978
Himmelblau, D. M., *Basic Principles and Calculations in Chemical Engineering*, 3rd edition, Prentice-Hall, Englewood Cliffs, New Jersey, 1974

Hougen, O. A., K. M. Watson, and R. A. Ragatz, *Chemical Process Principles, Part I Material and Energy Balances*, 2nd edition, John Wiley & Sons, New York, 1954

Luyben, W. L., and L. A. Wenzel, *Chemical Process Analysis, Mass and Energy Balances*, Prentice Hall, Englewood Cliffs, New Jersey, 1988

QUESTIONS

1) Give a verbal statement of the conservation of mass equation.

2) What two things must be defined in order to have a meaningful conservation of mass statement (in addition to the fact that mass is the property being counted)?

3) What kind of equation (algebraic or integral, or ordinary differential equation) is obtained for the conservation of mass if the system which is defined is of finite size and the time period is also finite in size?

4) What kind of equation is obtained for a finite-size system and a differential time period when the equation is divided by the time period?

5) What is the difference (if any) between an equation written for the conservation of total mass (in the absence of nuclear conversions or relativistic processes) and an equation written for an individual species mass?

6) A tank contains a liquid and has one inlet pipe for introducing more liquid into the tank and one outlet pipe for removing liquid. Discuss the changes which occur in the amount of liquid in the tank as a result of fluctuations which occur in the inlet and outlet flow streams.

7) For the preceding question, consider changes in the volume of the material in the tank rather than mass.

8) Consider an automobile travelling at constant speed on a level highway. With the car taken to be the system, discuss the conservation of total mass and an accounting of pertinent species mass equations. What are the input and output streams and how do the total mass and species mass of the system change?

9) Discuss similar considerations with a person taken as the system.

10) Discuss total mass conservation and species accounting for a living tree (or other plant) as the system.

11) Consider the oxygen-CO_2 cycle on earth. Humans and animals take in oxygen and convert it to carbon dioxide. Plants take in carbon dioxide to produce growth and convert it to oxygen. Discuss the balance between these two processes and the resulting steady-state concentrations of oxygen and carbon dioxide in the atmosphere. As part of your discussion, consider two extreme cases: 1) all plant life ceases, resulting in no further conversion of CO_2 to oxygen and 2) all animal life ceases and oxygen is no longer converted to carbon dioxide. In each of these cases, what is the ultimate steady-state concentration level of carbon dioxide and oxygen in the atmosphere (assuming that these are the only two processes which are occurring)? (The evaluation of easy-to-solve limiting cases is an extremely useful analysis technique.)

12) In light of question 11, how will a major change in the amount of plant life on the planet (for example, significant removal of the rain forests) affect the steady-state CO_2 and oxygen levels, again assuming that no other processes occur. Present your discussion in the context of conservation and accounting of mass principles.

13) At the beginning of each decade in the United States, a census of the population is obtained as required by the Constitution. Discuss how this census (accounting of people) should be done in the context of obtaining appropriate information for an accounting statement. For the 1990 census, all data that were reported were to be as of April 1. What role do you think this date played in the context of counting?

14) If you wanted to propose a quantitative model (predictive equation) which would calculate the population of the United States as a function of time, what factors would have to be included in this model? Your answer should be placed in the context of the requirements for an accounting statement.

SCALES

1) Make sketches of Figures 3-2, 3-3, 3-5, 3-6, 3-7, 3-8, 3-10, 3-11, and 3-12. On each figure indicate one system boundary and identify input and output to and from this system (i.e., where is the input/output on the figure and what enters and leaves at this point?).

2) What are the dimensions and units of molecular "weight" (more properly called molecular mass) in SI and engineering systems?

3) What is the molecular mass of oxygen? Nitrogen? Water? Methane?

4) A brine (salt water) solution contains 0.010 kg of salt and 0.20 kg of water. What is its mass composition: (a) as mass fraction of salt and water and (b) as mass per cent salt and water?

5) A flue gas consists of 25 moles of CO_2, 50 moles of H_2O, 3 moles of O_2, and 10 moles of N_2. What is its molar composition: (a) as mole fraction of each compound and (b) as mole per cent of each compound.

6) For question (4), what is the molar composition of the brine? (i.e., what are the mole fractions and molar percentages of each compound?)

7) For question (5), what is the mass composition of the flue gas? (i.e., what are the mass fractions and mass percentages of each compound?)

8) Air is 79 mole % nitrogen and 21 mole % oxygen. What is its compositon by mass?

9) A mixture of gases exiting a process has the following molar composition:

Compound	Mole %
CO_2	20.8%
CO	4.2%
H_2O	50.0%
N_2	20.8%
O_2	4.2%

What is its mass compositon (percentage of each compound)?

10) There are 18 g of water in 1 gram mole. How many pounds (mass) are there in 1 lb-mole (read "pound mole") of water? How many kg of water in a kg-mole?

11) The molecular mass of sulfur dioxide (SO_2) is 64. How many g-moles of SO_2 are there in 1 g? How many lb-moles are there in 1 lbm? How many g-moles in 1 lbm?

12) From memory, write a mathematical expression for the conservation of total mass, using the symbolic notation of the text. As you write it, think about the *physical meaning* of each term. By doing so, it will be more to you than a mathematical equation, and this is very important.

13) A system of algebraic equations having 5 unknowns must contain exactly _____ independent equations in order to solve for these unknowns.

14) List three ways (including specific computer software products) by which *you* can solve a system of simultaneous linear algebraic equations. Be specific in your answer.

15) Solve the following set of simultaneous equations for the unknown quantities using each of the three methods which you listed in (14):

$$40 - 0.8D - 0.1B = 0$$
$$60 - 0.2D - 0.9B = 0$$

16) When performing an accounting of total mass for a system, what is the value of the generation/consumption term in the accounting equation? Why?

17) When performing an accounting of a compound which is undergoing a chemical reaction, what can you say about the generation/consumption term in the accounting equation?

18) What is the physical significance or meaning of the term dm_{sys} / dt in the context of an accouting equation?

19) For the ordinary differential equation

$$\frac{dh}{dt} = \dot{m}_{in} - kh$$

where $\dot{m}_{in}$ is a constant, what variable is independent and which is dependent?

20) Example 3-5 in section 3.6 discusses a continuous distillation process. What does distillation mean? Give an example process which uses distillation.

21) What are the meanings of the terms "steady-state," "closed," and "open" with respect to the operation of a process?

22) Volume is an extensive property but it is not necessarily conserved. If this is true, then how might the total volume of two miscible (they form a single-phase solution upon mixing) liquids after they are mixed compare to the total volume of the two before they are mixed? Are they the same or not? Explain.

23) Make a schematic sketch of a process which has a bypass stream. Make a schematic sketch of a process which has a recycle stream.

24) What is the meaning of excess air supplied to a process? If 10 moles of air are required for complete stoichiometric combustion of a fuel and air is to be supplied at 20% excess, how much air must be supplied?

25) If 10 moles of *oxygen* are required for complete stoichiometric combustion of a fuel and *air* is to be supplied at 20% excess, how many moles of oxygen must be supplied? How many moles of air must be supplied to provide this much oxygen? How many moles of nitrogen were provided along with the oxygen as part of the air?

PROBLEMS

1) Scientists have estimated that the amount of water released from the earth's mantel over geological time is 3.400×10^{24} grams and the current total reserves of water in the mantel today are 2.0×10^{26} grams. If we define the system to be the earth's mantel, determine the amount of water in the earth at the beginning of the geological time period. If we define the system to be the hydrosphere (oceans, seas, lakes, rivers, ice caps, etc.) plus the atmosphere, what is the amount of water in this system today? What was the amount of water in this system at the beginning of the geological period? If the mass of water in the hydrosphere (oceans, seas, lakes etc.) is 1.664×10^{24} grams today, determine the amount of water in the atmosphere today.

2) A hose is used to fill up a 5 gallon bucket and the volumetric flow rate of the water from the hose is 10 L/min. Furthermore the bucket has a leak and loses water at a rate of 100 g/min. If the bucket is initially half full, and the density of water is 62.4 lb_m / ft^3 how long will it take to fill the bucket? Once the bucket is full, the hose is turned off. Determine the amount of time after the hose is turned off for the bucket to empty completely.

3) The mass flow rates of alcohol entering and leaving a tank were measured during a process run and the following data were recorded.

Alcohol mass flow rate data

Time (hr)	$(\dot{m}_i)_{in}$ (lb$_m$ / hr)	$(\dot{m}_i)_{out}$ (lb$_m$ / hr)
0	5	3
1	6	4
2	7	5
3	8	6
4	9	6
5	10	6
6	11	6

The amount of alcohol in the tank before the process began was 10 (lb$_m$). If there are no reactions occuring in the tank, and the mass flow rates are continuous, calculate the accumulation in the tank between 2 (hr) and 5 (hr) that the process has been running. The problem is much easier if you first graph the data.

4) It is required to have 2000 kg of a solution containing 10% ethyl alcohol (ethanol) by weight in water. Tanks containing 7% ethanol by weight in water and 30% ethanol by weight in water are provided to make the final solution. Choose the tank which contains the final solution as the system and write the accounting and conservation statements for this system in tabular form. What is the extensive property and the time period? What can be said about the entering, leaving, generation, and consumption terms with respect to this defined system? How many independent equations are required to solve this problem? Finally, determine the weight of each solution that is required to form the final solution.

5) A two stage evaporator is used to concentrate a salt solution of potassium chloride. The solution entering the first evaporator is 6.7 weight % potassium chloride.

a) If the process is required to make 10,000 lb$_m$ per day of the concentrated solution of 23.2 weight % and the evaporation rates (the rate of pure water leaving the evaporator) are the same for both evaporators, what is the required mass flow rate to the first evaporator. Also, find the concentration and the total mass flow rate entering the second evaporator.

b) A modification to the evaporator process is proposed which involves a recycle stream. The process is still required to produce 10,000 lb$_m$ per day of the concentrated solution; however, some of the concentrated solution is recycled back into the second evaporator. It was discovered that the second evaporator operates most efficiently when the concentration entering the second evaporator is 20 weight % potassium chloride. Determine the required recycle rate to achieve this concentration entering the second evaporator.

6) It is desired to produce 35 (kg/hr) of a mixture which contains equal parts (by mass) of benzene and cyclohexane. You have available two different streams; stream A is 80 % benzene and 20 % cyclohexane and stream B is 50 % cyclohexane and 50 % toluene (all compositions are mass percent). What flowrates are required for streams A and B in order to achieve the desired ratio of benzene to cyclohexane in stream C (the desired product leaving the mixer) and what is the product composition? Do not assume that the amount of toluene in the product is zero.

7) A reverse osmosis cell is used in a process to desalinate sea water, as shown in the figure. The feed F to the process is 100 (kg/hr) of 3.0% NaCl, the product desalinated water P is 0.04% NaCl, and the waste brine stream B is 5.15% NaCl. A recycle stream R increases the concentration of the stream entering the reverse osmosis cell to 3.8%. What are flow rates and compositions of all streams if the system is operating at steady state? (All compositions are mass percent.)

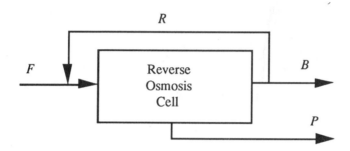

Problem 3-7: Reverse osmosis cell

8) A distillation column is used to separate three components of a single feed stream to produce three product streams, none of which is pure, however. The feed stream is fed at 100 (lb$_m$ / hr) and has a composition of 50% A, 20% B, and 30% C. The more volatile product stream which leaves the top of the distillation column, is 90% A and contains no C. The middle product stream has a composition of 40% A, 50% B, and 10% C. The bottom, least volatile stream, leaves from the bottom of the column and is 90% C. Additionally, 20% of component B which is in the feed stream leaves in the top stream. What are the flowrates and complete compositions of each product stream? Clearly state any assumptions that are required to solve this problem. (All compositions are mass percent.)

9) The three-unit separator shown in the figure is operating at steady state. The feed rate is 50 (kg/min). The feed consists of three components: pentane, cyclohexane, and toluene. There are equal amounts of each component in the feed. In separator A, two-thirds of the toluene fed is removed, and no pentane is removed. In separator B, one-fourth of the toluene fed is removed, one-half of the cyclohexane fed is removed, and one-fourth of the pentane fed is removed. In separator C, one-fourth of the cyclohexane fed is removed out the bottom, and two-thirds of the pentane is removed out the bottom. Only pentane leaves out the top of separator C. There are no chemical reactions occuring in the separators. Calculate the total flow rates and composition in each flow stream. Redraw the figure and clearly label each flow stream.

10) Air is flowing at steady state through a duct at 10 (kg/hr). The molecular mass of nitrogen is 28 (kg/kg-mol), and the molecular mass of oxygen is 32 (kg/kg-mol). If air is 23 mass percent O_2 and 77 mass percent N_2, calculate the molar flow rates of oxygen and nitrogen flowing out of the duct. Furthermore, calculate the composition on a mole basis.

11) A three-component mixture of pentane (MW=72.15), cyclohexane (MW=84.16) and toluene (MW=92.14) is being produced in a chemical plant from feed streams, A and B. Stream A has a mass flow rate of 10 (lb$_m$ / hr) of pentane and 5 (lb$_m$ / hr) of cyclohexane and no toluene. Stream B has a mass flow rate of 3 (lb$_m$ / hr) which is pure toluene. There is only one product stream. If the process is operating at steady state, and there are no chemical reactions, determine the total mass flow rate and

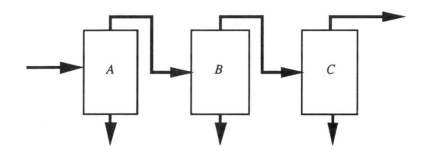

Problem 3-9: Three-unit separator

composition on a mass basis for each stream. Also, calculate the total molar flow rate and composition on a mole basis for each stream.

12) A quantity of pure sodium chloride is added to 5 (lb_m) of brine that is 5.2% salt on a mass basis. How many pound moles of sodium chloride must be added to produce a 10.5% brine by mass. Assume that no water evaporates during the process.

13) It is desired to make a 5.3% solution of HCl acid on a mass basis. To start with, we have 250 grams of a dilute solution of 1.2% HCl solution. Using conservation and accounting concepts, calculate how much concentrated acid at 15.2% must be added to the dilute acid to make the final desired solution. Also, how many grams of the final solution is made?

14) When the gates open, 42,000 spectators, 200 media personnel, and 50 players enter a baseball stadium stadium. Before the gates open, there were 5 gloves, 280 pencils, and 200 baseballs in the stadium. Half the spectators, all the players and no media personnel bring gloves to the game. All the media personnel, two-thirds of the spectators and one-tenth of the players bring pencils. Ten spectators bring baseballs, and each of the players brings 30 baseballs. During the game, there are 30 foul balls hit and collected by spectators. Also, a number of the players gave away an unknown number of balls to spectators. Furthermore, there are 5 homeruns hit out the stadium. No media personnel or baseball players leave the stadium. During the game, one-fourth of the spectators who entered leave leave, taking some pencils, gloves, and baseballs with them. It turns out that the number of gloves that they took equals the number which remain at the end of the game divided by four; the number of pencils that they took equals the number which remain at the end divided by four; and the number of baseballs that they took equals the number of baseballs which remain at the end, divided by ten. Use conservation and accounting to determine the number of people, gloves, pencils, and baseballs in the stadium at the end of the game. Did all the spectators who left with baseballs acquire them from foul balls? Why or why not?

15) A sugar syrup (20 wt % water) is being concentrated in a two-stage evaporator. The syrup enters the first evaporator in which 75% of the entering water is removed and leaves as a pure water stream. The product rich syrup stream is further concentrated in the second stage where 60 % of the entering water is removed. If 100 kg/hr of the syrup are fed to the first evaporator, calculate the flowrates and compositions of each of the other flow streams.

16) The Sun Shine Orange Juice and Concentrate Factory can process one ton of fresh oranges per hour at steady state. The compositions of the fresh oranges, orange juice and concentrate are given in the table below. All percentages are based on mass.

Sun Shine Orange Products Composition Information Sheet

	Solids	Liquids
Fresh Oranges	30%	70%
Orange Juice	12%	88%
Concentrate	42%	58%

The factory has the option to either sell fresh oranges, make orange juice, or make concentrate depending on the price of each item.

    a) If the factory is making orange juice from crushed fresh oranges, calculate the mass flow rate of water required to meet specifications on the orange juice. Draw a schematic of the process and label all the flow streams and compositions.

    b) If the factory is making concentrate from the crushed oranges, the process is a little more complex. Part of the crushed oranges are fed to an evaporator. In the evaporator, the volatile components (primarily water and flavor molecules) are removed. The composition of the product stream from the evaporator is 60% solids. The crushed oranges that were not fed to the evaporator are bypassed around and recombined with the evaporator product stream. This combined stream must have the same composition as the desired concentrate product. Draw a schematic of the process. Calculate all the flow rates and compositions in each stream and label the schematic.

17) You are preparing a mixture of methanol, ethanol, and water. To do so, you add 1) an unspecified amount of a methanol/water mixture (no ethanol) which contains 10 grams of methanol and 2) 120 grams of a mixture of ethanol/water (no methanol) which contains 80 grams of ethanol to a container. When you are done, there is a total of 200 g of methanol/ethanol/water mixture. What was the amount of each component (i.e. of methanol, ethanol, and water) in each of the two make-up solutions, what is the amount of each component in the final mixture, and what is the total amount of each of the make-up solutions.

18) Two feed streams are being mixed in a process vessel to provide a single solution which is then taken to downstream processing. No chemical reactions occur during the mixing process. Stream 1 consists of 15 (mass) % component A and 85 % component B. Stream 2 consists of 67% component A and 33% component B. If you produce 100 kg of mixture by using 30 kg of stream 1, what is the composition of the blended product?

19) Three stock solutions are mixed to provide two solutions which are then taken to downstream processing. Up to three components may be in the stock solutions. No chemical reactions occur during the mixing process. Thirty grams of Solution 1 containing 10 grams of component A and no component C, 20 grams of solution 2 containing 5 grams of component A and 12 grams of component B, and 40 grams of solution 3 containing no A are used to produce 40 grams of product 1 (which contains 20 g of B and 10 g of C) and an unspecified amount of product 2 which contains 30 g of component B. Find all of the unspecified amounts (of each component and of the total) for each of the stock and product solutions.

20) Grapes containing 85% water are sun dried in which 80% of the water in the grapes is removed. If the rate at which water is removed from the grapes on a cloudless day is a function of the exposure time given by the equation: Water removal rate $= k(1 - e^{-bt}) + c$ where: $k = 21.2(kg / hr)$, $b = 4.2(1 / hr)$, and $c = 3.4(kg / hr)$ then how long will it take to dry 100 lb_m of grapes. Choose the system as the grapes and write the rate form of the appropriate accounting and conservation laws. What can be said about the entering, leaving, generation, and consumption terms? Is the system operating at steady state? If 1000 tons of grapes are sun dried, how much additional water must be removed from the partially dried grapes to dry them completely? After drying the 1000 tons of grapes completely, what can you say about the amount of water in the atmosphere or surroundings.

21) The breathing rate of an adult is approximately 12 breaths per minute and with each breath approximately 0.1818 moles of air are inhaled. The *molar* compositions of the inhaled and exhaled gases are given in the table below:

Data for Problem 3-21

Species	Inspired Gas (%)	Expired Gas (%)
O_2	20.6	15.1
CO_2	0.0	3.7
N_2	77.4	75.0
H_2O	2.0	6.2

    a) Determine the yearly production of CO_2 per person due to respiration.

    b) Now, if the average energy usage (electricity, transportation, etc. not including food) per person is equivalent to completely oxidizing 6 lb_m of benzene per week as given by the following unbalanced chemical equation:

$$C_6H_6 + O_2 \longrightarrow CO_2 + H_2O$$

    then determine the yearly production of CO_2 per person due to energy usage.

    c) The total annual fixation rate (conversion of a gas to a solid through chemical reaction) of carbon on land and sea by plants is 2.8×10^{10} tons per year. With the hypothetical energy usage per person of part (b) plus CO_2 production by respiration from part (a), how many people can be accomodated by this plant fixation rate.

22) Propane is being oxidized in a furnace operating at steady state. The mass flow rate of propane ($MW = 44.11$) is 5 (kg/min) and the mass flow rate of oxygen ($MW = 32$) is 25 (kg/min). In the reactor, the propane is completely oxidized to carbon dioxide and water as given by the following unbalanced chemical reaction.

$$C_3H_8 + O_2 \longrightarrow CO_2 + H_2O$$

There is only one product stream.

    a) Balance the chemical reaction (elements are conserved).

    b) Define the system to be the reactor, and calculate the total molar flow rates of propane and oxygen entering the reactor. Make an accounting of moles in a table.

    c) Calculate the molar production rate of carbon dioxide and water in the reactor from the consumption of propane and oxygen. Is all of the oxygen consumed? Why or why not?

    d) Calculate the composition of the exit stream on a molar basis and a mass basis.

    e) Make a table accounting for each species mass and total mass conservation and compare total mass generation and consumption (summed together) with total moles generation and consumption. What do you conclude?

23) Ethane (C_2H_6, $MW = 30.07$) is burned in a furnace with 50% excess air. The excess air is based on total and complete combustion as given by the following unbalanced chemical reation.

$$C_2H_6 + O_2 \longrightarrow CO_2 + H_2O$$

The furnace is operating at steady state.

    a) Balance the chemical reaction.

    b) On a one mole basis of ethane (the time period is however long it takes to process 1 mole of ethane), calculate the composition of the product stream, both mole and mass percents. Assume complete and total combustion. Note, even though the nitrogen does not react, it must be included in the composition of the product stream. (Make molar and mass tables.)

    c) If only 90% of the ethane entering the reactor is consumed (90% conversion of ethane), calculate the composition of the product stream on a molar basis and mass basis. (Make tables!)

24) A compound AB dissociates in a steady-state flow process to produce A_2 and B according to the following balanced chemical reaction

$$2AB \longrightarrow A_2 + 2B$$

If 70% (on a molar basis) of the entering AB dissociates (70% conversion), then

a) What is the molar compositon of the product stream? (For each component in the product stream, what is its per cent of the total number of moles?)

b) What is the mass composition of the product stream? (For each component in the product stream, what is its per cent of the total mass?)

The molecular masses (molecular "weights") of A and B are 25 and 45, respectively.

25) Two compounds, A and B, are reacted in a continuous flow process (operating at steady-state) to produce A_2B according to the following balanced chemical reaction

$$2A + B \longrightarrow A_2B$$

The reaction is conducted with 20% excess moles of B fed to the reactor. In the reaction, 85% (on a molar basis) of A which enters the reactor is converted to A_2B. The molecular masses (molecular "weights") of A and B are 32 and 52, respectively.

a) What is the molar composition of the product stream? (For each component in the product stream, what is its per cent of the total number of moles?)

b) What is the mass compositon of the product stream? (For each component in the product stream, what is its per cent of the total mass?)

c) The same process is now conducted with a recycle stream. For this process, assume that the same overall conversion is achieved as above. Also, assume that the feed to the overall process and the product from the overall process occur in the same molar flowrate ratios as above. If the recycle molar flowrate (R) is 0.1P where P is the product molar flowrate downstream of the recycle stream, then what is the actual conversion of A in the reactor? (The overall conversion is for a system with the recycle stream inside, i.e., it is based upon only the fresh or makeup feed to the reactor. The actual reactor conversion is based upon all the feed to the reactor, fresh feed plus recycle.) Compare this reactor conversion to the overall process conversion. What is the effect of the recycle stream on the conversion?

26) Two compounds, A and B, are reacted in a continuous flow process (operating at steady-state) to produce AB_2 according to the following balanced chemical reaction

$$A + 2B \longrightarrow AB_2$$

The reaction is conducted with 10% excess moles of B fed to the reactor. In the reaction, 90% (on a molar basis) of A which enters the reactor is converted to AB_2. The molecular masses (molecular "weights") of A and B are 25 and 48, respectively.

a) What is the molar composition of the product stream? (For each component in the product stream, what is its per cent of the total number of moles?)

b) What is the mass compositon of the product stream? (For each component in the product stream, what is its per cent of the total mass?)

27) Butane (C_4H_{10}, $MW = 58.12$) is combusted in a furnace with 20% excess air (based on complete combustion to CO_2). The furnace is operating at steady state. In the process only 85% of the butane is reacted (85% conversion). Of the 85% reacted, 66% follows this chemical reaction:

$$C_4H_{10} + O_2 \longrightarrow CO_2 + H_2O$$

and the remaining 34% follows this reaction:

$$C_4H_{10} + O_2 \longrightarrow CO + H_2O$$

Note that these chemical reactions are not complete in that you need to balance them (elements are conserved). If there is only one product stream, calculate the composition of the exit stream on a molar basis and a mass basis.

FOUR

THE CONSERVATION OF CHARGE

4.1 INTRODUCTION

In this chapter we introduce the concept of electric charge and present a statement of the conservation of charge. Conservation of charge finds its chief engineering applications in the study of electric circuits, electrochemical reactions, and nuclear reactions. When written for a finite time period, we obtain the integral conservation of charge equation. The rate equation for conservation of charge also will be discussed, and this finds application in the form of Kirchoff's Current Law for electric circuits.

4.2 ACCOUNTING OF POSITIVE OR NEGATIVE CHARGE

Net charge is conserved; the total amount of charge in an isolated system remains constant. Note however, that it is possible to create within an isolated system a *pair* of electric charges, one of which is negative and one of which is positive. For example, an electron (e^-) *and* positron (e^+, a particle not found in ordinary matter but which is identical to the electron except that its charge is positive) can be created. However, this does not contradict the conservation of charge statement because the total charge of the created pair is zero. Thus the net charge in the system is unchanged. No positive or negative charge has ever been observed to be created by itself. There are always equal amounts of positive and negative charge created. Likewise, if an electron and positron are brought close together the two particles may disappear, converting their rest mass to energy in the form of gamma rays. Charge is conserved since the net charge is zero both before and after

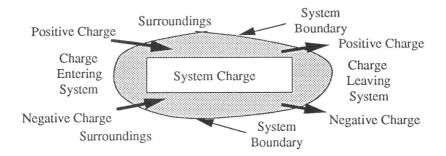

Figure 4-1: System and surroundings for an accounting of positive or negative charge

the annihilation. *Note however that in this case mass is not conserved*; the rest masses of the particles are turned completely into energy according to the relation $E = mc^2$.

A schematic representation of a system is given in Figure 4-1. Charges entering and leaving the system cross the system boundary that we have defined. Generation and consumption of positive and negative charge *can* occur inside the system boundary (as described above in special circumstances), but any generation or consumption outside the system is irrelevant to the statement about this particular system. We show separate inputs and outputs of positive and negative charge in Figure 4-1, but we could have just as well shown one arrow for input and one for output by considering *net* charge (which we will define as the amount of positive charge minus the amount of negative charge).

We account for positive or negative (+/−) charge by the statement

$$\left\{\begin{array}{c} \text{Accumulation of (+/−)} \\ \text{charge within system} \\ \text{during time period} \end{array}\right\} = \left\{\begin{array}{c} \text{(+/−) charge} \\ \text{entering system during} \\ \text{time period} \end{array}\right\} - \left\{\begin{array}{c} \text{(+/−) charge} \\ \text{leaving system during} \\ \text{time period} \end{array}\right\}$$

$$\left\{\begin{array}{c} \text{(+/−) charge generated} \\ \text{in system during} \\ \text{time period} \end{array}\right\} - \left\{\begin{array}{c} \text{(+/−) charge consumed} \\ \text{in system during} \\ \text{time period} \end{array}\right\}$$

where the accumulation of positive charge is defined as

$$\left\{\begin{array}{c} \text{Accumulation of (+/−)} \\ \text{within system during} \\ \text{time period} \end{array}\right\} \equiv \left\{\begin{array}{c} \text{Amount of (+/−) charge} \\ \text{contained in system} \\ \text{at end of time period} \end{array}\right\} - \left\{\begin{array}{c} \text{Amount of (+/−) charge} \\ \text{contained in system at} \\ \text{beginning of time period} \end{array}\right\}$$

For a defined time period we can write these positive or negative charge accounting statements symbolically as:

$$(q_{+,\text{sys}})_{\text{end}} - (q_{+,\text{sys}})_{\text{beg}} = q_{+,\text{in}} - q_{+,\text{out}} + q_{+,\text{gen}} - q_{+,\text{con}} \qquad (4-1)$$

or:

$$(q_{-,\text{sys}})_{\text{end}} - (q_{-,\text{sys}})_{\text{beg}} = q_{-,\text{in}} - q_{-,\text{out}} + q_{-,\text{gen}} - q_{-,\text{con}} \qquad (4-2)$$

where the notation, for positive charge, is defined as follows:

$(q_{+,sys})_{end}$	=	pos. charge in the system at the end of the time period
$(q_{+,sys})_{beg}$	=	pos. charge in the system at the beginning of the time period
$q_{+,in}$	=	positive charge entering the system during the time period
$q_{+,out}$	=	positive charge leaving the system during the time period
$q_{+,gen}$	=	positive charge generated in the system during the time period
$q_{+,con}$	=	positive charge consumed in the system during the time period

For negative charge, the same notation holds except the subscript is a minus sign (–).

It is convenient to define *net* charge as the amount of positive charge minus the amount of negative charge. Then, if we subtract equation (4-2) from equation (4-1), we can write an accounting statement for net charge as follows.

$$(q_{sys})_{end} - (q_{sys})_{beg} = q_{in} - q_{out} + q_{gen} - q_{con} \qquad (4-3)$$

Mathematically, the net charge in, out, generated and consumed are defined by the following expressions. It will usually be easier to work with equation (4-3) than equations (4-1) through (4-2), but we must note that all terms in equations (4-1) and (4-2) are positive (or zero) numbers, while some terms in equation (4-3) could be negative numbers (e.g. if there is more negative charge entering the system than positive charge entering the system then q_{in} will be a negative number).

$$(q_{sys})_{end} = (q_{+,sys})_{end} - (q_{-,sys})_{end} \qquad\qquad (q_{sys})_{beg} = (q_{+,sys})_{beg} - (q_{-,sys})_{beg}$$

$$q_{in} = q_{+,in} - q_{-,in} \qquad\qquad q_{out} = q_{+,out} - q_{-,out}$$

$$q_{gen} = q_{+,gen} - q_{-,gen} \qquad\qquad q_{con} = q_{+,con} - q_{-,con}$$

4.3 CONSERVATION OF NET CHARGE

The observation that net charge cannot be generated or consumed (the conservation of net charge) places constraints on the accounting of net charge, equation (4-3). In equations (4-1) and (4-2) since no *net* charge can be created or consumed within the system we have that

$$q_{+,gen} = q_{-,gen}$$

and

$$q_{+,con} = q_{-,con}$$

which states that only *pairs* of oppositely charged particles can be created or consumed.

Because $q_{+,gen} = q_{-,gen}$ and $q_{+,con} = q_{-,con}$, equation (4-3) becomes the law of conservation of charge:

$$\boxed{(q_{sys})_{end} - (q_{sys})_{beg} = q_{in} - q_{out}} \qquad (4-4)$$

The input and output terms can be combined, and the generation and consumption terms can be combined. Table 4-1 gives a summary of the positive and negative charge accounting equations and net charge conservation equations. We now apply the conservation of charge to several situations. These examples cover a wide range of situations including nuclear fission, nuclear fusion, electrochemical cells, capacitors, and junction diode elements.

Table 4-1: Summary of Positive/Negative Charge Accounting
and Net Charge Conservation

Positive	$(q_{+,sys})_{end}$	$-$	$(q_{+,sys})_{beg}$	$=$	$q_{+,in/out}$	$+$	$q_{+,gen/con}$
Negative	$(q_{-,sys})_{end}$	$-$	$(q_{-,sys})_{beg}$	$=$	$q_{-,in/out}$	$+$	$q_{-,gen/con}$
Net Charge	$(q_{sys})_{end}$	$-$	$(q_{sys})_{beg}$	$=$	$q_{in/out}$	$+$	0

Example 4-1. Interesting examples of charge conservation occur in radioactive decay, nuclear fission, and nuclear fusion. Uranium-238 (atomic number 92) is radioactive and decays according to:

$$^{238}_{92}U \longrightarrow {}^{234}_{90}Th + {}^{4}_{2}He$$

Thorium-234 has an atomic number 90 and helium-4 has atomic number 2.

For this situation the time period is chosen to span the time for the reaction to take place. We have two reasonable choices for our system boundary. Suppose we first choose the system to be a volume in space such that at the beginning of the time period $^{238}_{92}U$ enters the system and at the end of the time period, $^{234}_{90}Th$ and $^{4}_{2}He$ leave the system. This process is shown in Figure 4-2a. Then the amount of charge *into* the system would be the 92 positive and 92 negative charges of the uranium atom. Within the system, no charge pairs would be created or consumed, and the amount of charge *out* of the system would consist of the 90 positive and 90 negative charges of the thorium atom plus the two positive and two negative charges of the helium atom. We could also use net charge to describe the situation (equation (4-4)). The information is presented in Table 4-2. The positive, negative and net charge on the system at the end is zero.

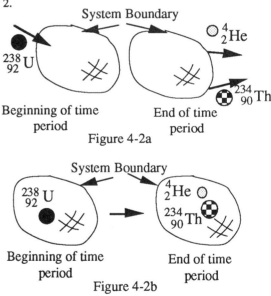

Figure 4-2: System boundary as a volume in space and enclosing the mass

On the other hand, as a second system, we could consider a volume of space enclosing the $^{238}_{92}U$ atom as the system (Figure 4-2b). The time period is however long it takes for the atom to decay. For this system no charge enters or leaves the system. Again a tabular presentation of the accounting and conservation laws yields Table 4-3. Notice how the non-zero terms are different because of the different system that we defined.

Table 4-2: Data for Example 4-1 (Charge)

Description	$(q_{i,sys})_{end}$ (C)	$-$	$(q_{i,sys})_{beg}$ (C)	$=$	$q_{i,in/out}$ Uranium (C)		Thorium (C)		Helium (C)	$+$	$q_{i,gen/con}$ (C)
Positive	0	$-$	0	$=$	92	$-$	90	$-$	2	$+$	0
Negative	0	$-$	0	$=$	92	$-$	90	$-$	2	$+$	0
Net Charge	0	$-$	0	$=$	0	$-$	0	$-$	0	$+$	0

Table 4-3: Alternate Solution for Example 4-1 (Charge)

Description	$(q_{i,sys})_{end}$ Thorium (C)		Helium (C)	$-$	$(q_{i,sys})_{beg}$ Uranium (C)	$=$	$q_{i,in/out}$ (C)	$+$	$q_{i,gen/con}$ (C)
Positive	90	$+$	2	$-$	92	$=$	0	$+$	0
Negative	90	$+$	2	$-$	92	$=$	0	$+$	0
Net Charge	0	$+$	0	$-$	0	$=$	0	$+$	0

Example 4-2. If a uranium-235 atom absorbs a slow neutron, a fission reaction can take place as described: $^{235}_{92}U + n \longrightarrow {}^{138}_{56}Ba + {}^{95}_{36}Kr + 3n + energy$ where n represents a neutron that carries no charge. Consider a volume which encloses the $^{235}_{92}U$ atom as the system as shown in Figure 4-3. Accounting for positive and negative charge along with net charge yields Table 4-4.

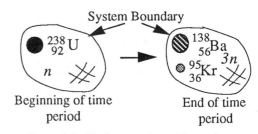

Figure 4-3: Fission reaction with uranium

Table 4-4: Data for Example 4-2 (Charge)

Description	$(q_{i,sys})_{end}$ Barium (C)		Krypton (C)	$-$	$(q_{i,sys})_{beg}$ Uranium (C)	$=$	$q_{i,in/out}$ (C)	$+$	$q_{i,gen/con}$ (C)
Positive	56	$+$	36	$-$	92	$=$	0	$+$	0
Negative	56	$+$	36	$-$	92	$=$	0	$+$	0
Net Charge	0	$+$	0	$-$	0	$=$	0	$+$	0

Table 4-5: Data for Example 4-3 (Charge)

Description	$(q_{i,\text{sys}})_{\text{end}}$ Helium (C)	$-$	$(q_{i,\text{sys}})_{\text{beg}}$ Hydrogen (C)	$-$	Hydrogen (C)	$=$	$q_{i,\text{in/out}}$ (C)	$+$	$q_{i,\text{gen/con}}$ (C)
Positive	2	$-$	1	$-$	1	$=$	0	$+$	0
Negative	2	$-$	1	$-$	1	$=$	0	$+$	0
Net Charge	0	$-$	0	$-$	0	$=$	0	$+$	0

Example 4-3. A fusion reaction that involves the deuterium isotope of hydrogen ($_1^2\text{H}$) is the following $_1^2\text{H} + _1^2\text{H} \longrightarrow _2^3\text{He} + n +$ energy where n represents a neutron that carries no charge. Considering the the volume of space surrounding the two hydrogen atoms depicted in Figure 4-4, we can generate Table 4-5 by accounting for positive and negative charge and using the definition of net charge.

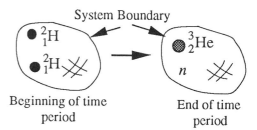

Figure 4-4: Fusion reaction with hydrogen

Example 4-4. Batteries are devices capable of causing charge to flow through a wire. They are usually formed by immersing two dissimilar metals in an ionic solution. For example, the typical automobile battery is a series of six lead-sulfuric acid cells, each similar in operation to that described below.

Figure 4-5 shows a schematic of what is called a Daniell cell. On the left side of the porous barrier is a dilute solution of $ZnSO_4$ which exists almost exclusively as ions Zn^{2+} and SO_4^{2-}. A zinc rod is inserted into the solution. On the right side of the barrier there is a saturated solution of $CuSO_4$ which exists almost exclusively as the ions Cu^{2+} and SO_4^{2-}. A copper rod is inserted into this solution. The porous barrier allows SO_4^{2-} to pass through it but blocks the passage of the zinc or copper ions. Finally, a wire connects the zinc and copper rods.

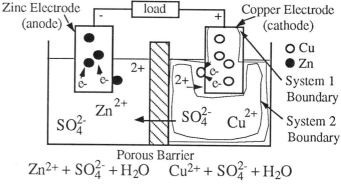

Figure 4-5: Daniell Cell

The operation of the cell is as follows: Zinc atoms leave the zinc electrode and enter the left side solution as Zn^{2+} ions, leaving behind two electrons. These electrons travel through the wire and end up at the copper electrode. At the copper electrode a Cu^{2+} ion in solution absorbs these electrons at the electrode surface and the copper atom thus formed is plated onto the electrode. A SO_4^{2-} ion crosses the porous barrier from the saturated solution on the right to the dilute solution on the left. The process continues as other zinc atoms leave the zinc electrode and enter the solution as ions.

Let us define the system to be the copper electrode and saturated cell on the right and the time period the time between consecutive zinc atoms entering the solution on the left (one cycle of the cell's operation). For this situation the negative charge entering the system through the wire from the load during the time period is:

$$q_{-,\text{in}} = 2$$

Also,

$$q_{-,\text{out}} = 2$$

because a SO_4^{2-} ion leaves the system through the membrane during the time period. The amount of negative charge that is generated in the system during the time period is zero. However, the amount of negative charge that is consumed is 2 when one copper ion is deposited at the anode surface.

The beginning and ending net charge is zero, although there are distinct positive and negative charges which exist. Let the number of negative charges (due to SO_4^{2-}) in the system at the beginning of the time period be n and denoted by n^- and the number of positive charges (due to Cu^{2+}) be n and denoted by n^+. Table 4-6 summarizes the positive and negative charge accounting equations and the net charge conservation law.

Table 4-6: Data for Example 4-4a (Charge, System = Copper Electrode and Cell)

Description	$(q_{i,\text{sys}})_{\text{end}}$	$-$	$(q_{i,\text{sys}})_{\text{beg}}$	$=$	$q_{i,\text{in/out}}$ Wire	$-$	$q_{i,\text{in/out}}$ Membrane	$+$	$q_{i,\text{gen/con}}$
	(C)		(C)		(C)		(C)		(C)
Positive	$n^+ - 2^+$	$-$	n^+	$=$	0	$-$	0	$+$	-2^+
Negative	$n^- - 2^-$	$-$	n^-	$=$	2^-	$-$	2^-	$+$	-2^-
Net Charge	0	$-$	0	$=$	(-2)	$-$	(-2)	$+$	0

If we had considered just the copper solution as our system, one copper ion would be seen as leaving the system during the time period, while one SO_4^{2-} ion also leaves the system; thus the accounting equation and conservation law are given in Table 4-7. Again notice that the terms in the columns change according to the system that we define.

Table 4-7: Data for Example 4-4b (Charge, System = Copper Solution)

Description	$(q_{i,\text{sys}})_{\text{end}}$	$-$	$(q_{i,\text{sys}})_{\text{beg}}$	$=$	$q_{i,\text{in/out}}$ Copper	$-$	$q_{i,\text{in/out}}$ Membrane	$+$	$q_{i,\text{gen/con}}$
	(C)		(C)		(C)		(C)		(C)
Positive	$(n^+ - 2^+)$	$-$	n^+	$=$	-2^+	$-$	0	$+$	0
Negative	$(n^- - 2^-)$	$-$	n^-	$=$	-0	$-$	-2^-	$+$	0
Net Charge	0	$-$	0	$=$	$-(+2)$	$-$	(-2)	$+$	0

Example 4-5. A pair of parallel plates separated by a distance d forms a capacitor (Figure 4-6). If equal and opposite charges (q_+ and q_-) are placed on the plates, an electric field is set up in the region between the plates as depicted in the diagram.

Electric field

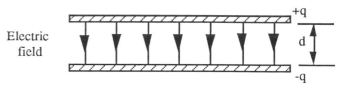

Figure 4-6: Capacitor

The strength of the electric field depends on the amount of (positive and negative) charge on the plates. Note that the charge refers to the magnitude of the equal and opposite charge that is on each of the plates. It is not the *net* charge of the capacitor taken as a whole, since this would be zero ($q_{+,sys} - q_{i,sys} = 0$). Energy is stored in the electric field set up by the charged plates, and so capacitors can be used as energy storage devices. A capacitor is said to be *charging* if more positive charge is being added to the positive plate and more negative charge is being added to the negative plate. This increases the size of the electric field. Conversely if positive charge is flowing from the positive plate (and thus also negative charge is flowing from the negative plate), the electric field is weakened and the capacitor is said to be *discharging*.

Suppose at a time $t = 0$ a capacitor has a charge of +4 (C) on one plate and −4 (C) on the other plate. In the time period between $t = 0$ and $t = 3$ seconds, an additional charge of +2.4 (C) is placed on the positive plate and −2.4 (C) is placed on the negative plate. We wish to account for charge for the 3 second time period of interest. The capacitor is shown in Figure 4-7.

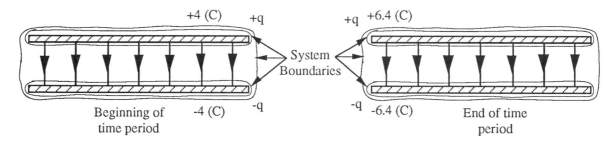

Figure 4-7: Capacitor with defined systems

SOLUTION: Suppose we take as the system the positive plate only. The accounting equations for positive and negative charge and the conservation of net charge gives Table 4-8. There is an accumulation of positive charge of 2.4 (C) on the positive plate and the final net charge on the positive plate is found to be 6.4 (C). Next, we can define the system to be the negative plate only. This system gives Table 4-9 when counting positive and negative charge and applying the conservation of net charge. There is an accumulation of negative charge of −2.4 (C) on the negative plate and the final net charge on the negative plate is found to be −6.4 (C). Finally, considering the entire capacitor as the system, the accounting of positive and negative charge yields Table 4-10. Notice that even though there is an accumulation of negative and positive charge, there is no accumulation of net charge when the entire capacitor is taken as the system.

Table 4-8: Data for Example 4-5a (Charge, System = Positive plate)

	$(q_{i,sys})_{end}$	−	$(q_{i,sys})_{beg}$	=	$q_{i,in/out}$	+	$q_{i,gen/con}$
Description	(C)		(C)		(C)		(C)
Positive	?	−	4^+	=	2.4^+	+	0
Negative	?	−	0	=	0	+	0
Net Charge	?	−	4	=	2.4	+	0

Table 4-9: Data for Example 4-5b (Charge, System = Negative plate)

Description	$(q_{i,sys})_{end}$ (C)	$-$	$(q_{i,sys})_{beg}$ (C)	$=$	$q_{i,in/out}$ (C)	$+$	$q_{i,gen/con}$ (C)
Positive	?	$-$	0	$=$	0	$+$	0
Negative	?	$-$	4^-	$=$	2.4^-	$+$	0
Net Charge	?	$-$	(-4)	$=$	(-2.4)	$+$	0

Table 4-10: Data for Example 4-5c (Charge, System = Entire Capacitor)

Description	$(q_{i,sys})_{end}$ (C)	$-$	$(q_{i,sys})_{beg}$ (C)	$=$	$q_{i,in/out}$ (C)	$+$	$q_{i,gen/con}$ (C)
Positive	6.4^+	$-$	4^+	$=$	2.4^+	$+$	0
Negative	6.4^-	$-$	4^-	$=$	2.4^-	$+$	0
Net Charge	0	$-$	0	$=$	0	$+$	0

Example 4-6. A sample of pure silicon will not conduct electricity especially well. However if small amounts of "impurity atoms" are added to the crystal lattice of the silicon, the conduction properties can be greatly altered. Silicon is atomic number 14 in the periodic table and it has four valence electrons per atom. A sample of silicon is electrically neutral since each atom has 14 protons and 14 electrons. The silicon sample is depicted in Figure 4-8(a). Note that each silicon atom is covalently bonded to four other silicon atoms as shown in Figure 4-8(a).

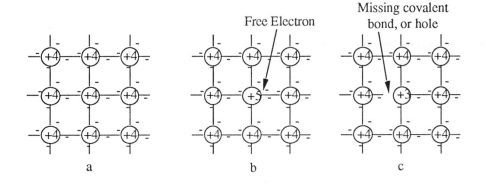

Figure 4-8: Silicon crystal with Type V and Type III Impurities

Arsenic is element 33 in the periodic chart. Each atom has 33 protons and 33 electrons, and 5 of these are in the valence band. A specimen of arsenic would likewise be electrically neutral. If a very small concentration of arsenic atoms (in comparison with the number of silicon atoms) is added to a sample of pure silicon, the arsenic atoms do not disrupt the overall structure of the crystal lattice, but merely enter the lattice where there had been a silicon atom as is shown in Figure 4-8(b). Note that the material formed is still electrically neutral since there are equal numbers of positive and negative charges.

Each silicon atom in the lattice is supposed to bond with four neighbors, but the arsenic atom has five valence electrons. Four of these electrons bond with the silicon valence electrons, while the fifth valence electron is loosely bound and can become a charge carrier. The conductivity of the new material is greater than before since there are more charge carriers. This type of material is called n-type since there are more negative charge carriers than would exist in the pure material.

On the other hand, suppose the impurity atom is gallium, which has atomic number 31 and three valence electrons per atom. Each atom has 31 protons and 31 electrons; a specimen of gallium would likewise be electrically neutral. When a small concentration of gallium atoms is added to a sample of pure silicon, each gallium atom enters the lattice where there had been a silicon atom. Now instead of an extra electron not participating in a bond, there is an unfilled bond needing an electron to complete it. In effect, a *hole* is created waiting to be filled by some electron. See Figure 4-8(c). This type of material is called p-type since the positive holes can act as charge carriers.

Suppose p-type and n-type materials are brought into contact forming a p-n junction. The n-type material has a high concentration of free electrons while the *p-type* has a high concentration of "holes". Some of the free electrons from the n-type will diffuse across the boundary into the p-type where they will fill holes. This diffusion will make the n-side of the boundary become slightly positively charged, while the p-side becomes negatively charged. An electric field is set up in the junction region that limits the amount of diffusion since electrons that would wish to diffuse across the boundary now see an electric field against them (Figure 4-9). Such a junction device is called a diode.

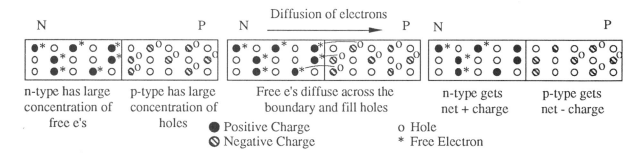

Figure 4-9: The p-n junction

Let us describe the situation at the junction in conservation of charge terms. If we consider the entire diode as the system: (a) no net charge enters or leaves, (b) the beginning net charge is zero since each part was electrically neutral, and (c) the final net charge is also zero since whatever negative charge the n-side has lost $(-q^-)$, the p-side has gained $(+q^-)$. The accounting equations for entire p-n diode give Table 4-11. In the table n_p^+ and n_n^+ are the initial number of positive charges in the p and n sides, respectively. Similarly, n_p^- and n_n^- are the corresponding negative charges. Note, because both p and n materials initially are electrically neutral, that $n_p^+ = n_p^-$ and $n_n^+ = n_n^-$; i.e., $n_p^+ - n_p^- = 0$ and $n_n^+ - n_n^- = 0$

Table 4-11: Data for Example 4-6a (Charge, System = Entire diode)

Description	$(q_{i,sys})_{end}$	$-$	$(q_{i,sys})_{beg}$	$=$	$q_{i,in/out}$	$+$	$q_{i,gen/con}$
	(C)		(C)		(C)		(C)
Positive	$(n_p^+ + n_n^+)$	$-$	$(n_p^+ + n_n^+)$	$=$	0	$+$	0
Negative	$(n_p^- + n_n^-)$	$-$	$(n_p^- + n_n^-)$	$=$	0	$+$	0
Net Charge	0	$-$	0	$=$	0	$+$	0

If we were to consider the p-type material alone as the system then, for q^- negative charges transferred to the p-type material, the accounting equations for charge give Table 4-12. Instead, if we were to consider the n-type material as the system, the accounting equations give Table 4-13. Notice that for each defined system, the solution for the unknown charge, either negative, positive, or net, is easily accomplished using the tabular format.

Table 4-12: Data for Example 4-6b (Charge, System = p-type material)

Description	$(q_{i,sys})_{end}$	$-$	$(q_{i,sys})_{beg}$	$=$	$q_{i,in/out}$	$+$	$q_{i,gen/con}$
	(C)		(C)		(C)		(C)
Positive	n_p^+	$-$	n_p^+	$=$	0	$+$	0
Negative	$(n_p^- + q^-)$	$-$	n_p^-	$=$	q^-	$+$	0
Net Charge	$(-q)$	$-$	0	$=$	$(-q)$	$+$	0

Table 4-13: Data for Example 4-6c (Charge, System = n-type material)

Description	$(q_{i,sys})_{end}$	$-$	$(q_{i,sys})_{beg}$	$=$	$q_{i,in/out}$	$+$	$q_{i,gen/con}$
	(C)		(C)		(C)		(C)
Positive	n_n^+	$-$	n_n^+	$=$	0	$+$	0
Negative	$n_n^- - q^-$	$-$	n_n^-	$=$	$-q^-$	$+$	0
Net Charge	$-(-q)$	$-$	0	$=$	$-(-q)$	$+$	0

4.4 RATE EQUATION

The examples presented so far for conservation and accounting of charge all involved finite time intervals, and the equations developed were integral charge equations. To consider instantaneous changes in the system, we use the rate form for the conservation of net charge and the accounting of positive and negative charge.

The positive charge that enters a system can be thought of as an integral over the time period of the positive-charge flow rate into the system.

$$q_{+,in} = \int_{t_{beg}}^{t_{end}} \dot{q}_{+,in} dt$$

Similarly, the terms for the positive-charge output, generation, and consumption can be expressed as an integration of positive charge output, generation, and consumption rates over a finite time period.

Paralleling the development in Chapter 3, the rate equation for the accounting of positive charge for a given system is:

$$\left(\frac{dq_{+,sys}}{dt}\right) = \dot{q}_{+,in} - \dot{q}_{+,out} + \dot{q}_{+,gen} - \dot{q}_{+,con} \qquad (4-5)$$

where $\dot{q}_{+,in}$ represents a positive-charge flow rate into the system and $\dot{q}_{+,out}$ represents a positive-charge flow rate out of the system. The rate of positive-charge generation and consumption are $\dot{q}_{+,gen}$ and $\dot{q}_{+,con}$ respectively.

The rate equation accounting of negative charge is:

$$\left(\frac{dq_{-,sys}}{dt}\right) = \dot{q}_{-,in} - \dot{q}_{-,out} + \dot{q}_{-,gen} - \dot{q}_{-,con} \qquad (4-6)$$

where $\dot{q}_{-,in}$ denotes the negative-charge flow rate into the system.

Finally, the rate equation for net charge is obtained by subtracting (4-6) from (4-5) giving:

$$\left(\frac{dq_{sys}}{dt}\right) = \dot{q}_{in} - \dot{q}_{out} + \dot{q}_{gen} - \dot{q}_{con} \qquad (4-7)$$

Because, net charge is conserved, the rates of generation and rates of consumption are equal yielding:

$$\dot{q}_{gen} - \dot{q}_{con} = 0$$

and the rate equation for net charge reduces to:

$$\left(\frac{dq_{sys}}{dt}\right) = \dot{q}_{in} - \dot{q}_{out} \qquad (4-8)$$

4.4.1 Net Charge Flow Rate–Current

The instantaneous rate at which net charge passes some point is called the electric *current*. We define the direction of the current to be the direction that *positive* charge would flow to cause the observed rate of change of charge. Thus in metals, since the free electrons are the charge carriers that move in the presence of a field, the direction of the current is *opposite* to that of the physical movement of the electrons through the conductor.

Mathematically in the conservation of net charge, the current is represented by the symbol $\dot{q}$. However, because the symbol i is used widely to denote the net charge flow rate or current, we will use this notation from now on. The initial development of the rate equation for net charge used the dot notation similar to the development in Chapter 3 to provide consistency in explaining the concepts. The unit of electric current (C/s) is given the name *ampere* (A). A one ampere current exists if one coulomb of net charge is moving past a given point each second.

The conservation of net charge for multiple inlets and outlets is, in terms of current,

$$\left(\frac{dq_{sys}}{dt}\right) = \sum i_{in} - \sum i_{out} \tag{4-9}$$

where the notation i stands for the instantaneous rate of *net* charge entering or leaving the system. This result says that if the current entering the system is not the same as the current leaving the system, then net charge will build up or accumulate inside the system.

We can conveniently combine the current entering and the current leaving into one summation.

$$\boxed{\left(\frac{dq_{sys}}{dt}\right) = \sum i_{in/out}}$$

In this case, the current *in* would add net charge to the system and carry a positive sign while the current *out* would subtract net charge from the system and carry a negative sign.

By combining the input and output terms together and the generation and consumption terms together, we have provided a summary of the rate accounting equations and net charge conservation in Table 4-14.

Table 4-14: Summary of Positive/Negative Charge Accounting
and Net Charge Conservation

Positive	$\left(\dfrac{dq_{+,sys}}{dt}\right)$	=	$\dot{q}_{+,in/out}$	+	$\dot{q}_{+,gen/con}$
Negative	$\left(\dfrac{dq_{-,sys}}{dt}\right)$	=	$\dot{q}_{-,in/out}$	+	$\dot{q}_{-,gen/con}$
Net Charge	$\left(\dfrac{dq_{sys}}{dt}\right)$	=	$i_{in/out}$	+	0

Example 4-7. The current entering a system is given by $i(t) = ae^{-bt}$ (A) where $a = 3.2$ (A), and $b = 40$ (1 / s) (with t in seconds) and puts net charge on the positive plate of a capacitor that initially has no net charge as shown in Figure 4-10. Find the amount of charge on the plate after 0.03 seconds have passed.

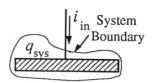

Figure 4-10: A charging plate

SOLUTION: Choose the system to be the positive plate of the capacitor (Figure 4-10) and the time period as 0.03 seconds. The conservation of net charge is given in Table 4-15.

Table 4-15: Data for Example 4-7 (Charge, System = Capacitor plate)

Description	$\left(\dfrac{dq_{sys}}{dt}\right)$ (C/s)	$=$	$\sum i_{in/out}$ (C/s)
Net Charge	$\left(\dfrac{dq_{sys}}{dt}\right)$	$=$	ae^{-bt}

Now we can substitute the mathematical expression for the entering current into the equation and integrate.

$$\left(\frac{dq_{sys}}{dt}\right) = 3.2e^{-40t}$$

Separating variables and integrating

$$\int_{q_{sys}(t)|_0}^{q_{sys}(t)|_{0.03}} dq_{sys} = \int_{t=0}^{t=0.03} 3.2e^{-40t}\,dt$$

or

$$q_{sys}(t)|_{0.03} - q_{sys}(t)|_0 = \frac{3.2}{-40}(e^{-1.2} - 1)$$

Initially, there is no charge on the system ($q_{sys}(t)|_0 = 0$) and:

$$q_{sys}(t)|_{0.03} = \left(\frac{3.2}{-40}\right)(e^{-1.2} - 1) = 0.056 \ (C)$$

The charge on the plate at the end of the time period is 0.056 C.

Example 4-8. The positive plate of a capacitor initially ($t = 0$) has a charge of 10 (C). The plate discharges at a rate proportional to the net charge on the positive plate given by $i_{out} = kq_{sys}$ (A) where $k = 0.5$ (1 / s). For the system defined as the postive plate given in Figure 4-11.

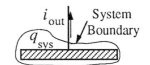

Figure 4-11: A discharging plate

a) Determine the net charge on the plate as a function of time, $q_{sys}(t)$.

b) Determine the net charge on the plate at 2, 4, and 6, seconds.

c) Determine the time it takes for half of the net charge initially on the plate to leave.

Table 4-16: Data for Example 4-8 (Charge, System = Positive plate)

Description	$\left(\dfrac{dq_{sys}}{dt}\right)$ (C/s)	$=$	$\sum i_{in/out}$ (C/s)
Net Charge	$\left(\dfrac{dq_{sys}}{dt}\right)$	$=$	$-kq_{sys}$

SOLUTION: Choose the system to be the positive plate of the capacitor. The conservation of net charge is given in Table 4-16. From the table, the conservation of net charge for this system gives:

$$\left(\frac{dq_{sys}}{dt}\right) = -kq_{sys}$$

This equation can be separated by multiplying both sides of the equation by dt and dividing both sides of the equation by q_{sys} giving:

$$\frac{dq_{sys}}{q_{sys}} = -kdt$$

Now the equation can be integrated with the upper limit of integration being the unknown and the lower limit of integration being the initial condition.

$$\int_{q_o}^{q}\left(\frac{dq_{sys}}{q_{sys}}\right) = \int_{t_o}^{t}(-k)dt$$

Integration and evalution gives:

$$\ln\left(\frac{q}{q_o}\right) = -k(t - t_o)$$

Because the initial time is zero ($t_o = 0$), exponentiating both sides of the equation yields:

$$q = q_o e^{-kt}$$

The initial charge on the plate is 10 (C) and the constant k is 0.5 (1/s).

$$q_{sys}(t) = 10e^{-0.5t} \text{ (C)}$$

This equation gives the net charge on the plate as a function of time.

The solution of part (b) is simply evaluating this function at 2, 4, and 6 seconds.

$$q_{sys}(t)|_2 = 3.679 \text{ (C)} \qquad q_{sys}(t)|_4 = 1.353 \text{ (C)} \qquad q_{sys}(t)|_6 = 0.498 \text{ (C)}$$

The time it takes for half of the initial net charge to leave the system is when the value $q_{sys}(t) = 5$. This can also be expressed as:

$$e^{-kt} = \frac{q_{sys}(t)}{q_o} = 0.5$$

Rearranging this equation gives:

$$t = -\left(\frac{\ln(0.5)}{k}\right)$$

Evaluating this expression yields:

$$t = -\left(\frac{\ln(0.5)}{-0.5}\right)$$

$$= 1.39 \ (s)$$

The plate lost half of its initial charge after 1.39 seconds.

4.5 STEADY STATE – KIRCHHOFF'S CURRENT LAW

The rate equation (4-9) can be further simplified for the special case of electrical networks made of conductors. Since electrons in a conductor are free to move in response to an applied field, in the steady state they will do so and charge cannot *accumulate* at any point in the conductor. Similarly, at a junction formed by two or more conducting paths, no net charge will accumulate at the junction (Figure 4-12).

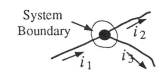

Figure 4-12: System for Kirchhoff's Current Law

Suppose we consider any part of an electrical circuit made of conductors as the system and apply equation (4-9) to this system. The term $\left(\frac{dq_{sys}}{dt}\right)$ must be zero since net charge will not accumulate. Therefore, for any part of an electrical conductor

$$0 = \sum i_{in/out}$$

or "current in equals current out" for any portion of the circuit.

In particular we can state that:

> The sum of the currents flowing toward any point in a circuit is equal to the sum of the currents flowing away from that point.

This statement of conservation of charge for electric circuits is called Kirchhoff's Current Law (KCL), and it may be applied at each connection point in a circuit. These connection points may be thought of as individual systems.

Example 4-9. Write the steady-state conservation of net charge equation for the currents defined in Figure 4-13. We see that the currents labelled i_1, i_2, and i_4 are towards the junction while i_3 and i_5 are away from the junction. The information is presented in Table 4-17.

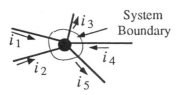

Figure 4-13: Example system for the conservation of net charge

Table 4-17: Data for Example 4-9 (Charge, System = Junction)

$\left(\dfrac{dq_{sys}}{dt}\right)$	=	$\sum i_{in/out}$					
Description		Wire 1 (A)	Wire 2 (A)	Wire 4 (A)	Wire 3 (A)		Wire 5 (A)
Net Charge	0 =	i_1 +	i_2 +	i_4 −	i_3 −		i_5

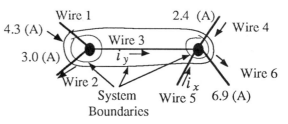

Figure 4-14: Multiple systems for charge conservation equations

Example 4-10. Find the value of the current i_x for the circuit given in Figure 4-14.

Table 4-18: Data for Example 4-10 (Charge, System-Left junction)

	i_{in}	=	i_{out}	
Description	Wire 1 (A)		Wire 2 (A)	Wire 3 (A)
Net Charge	4.3	=	3.0 +	i_y

SOLUTION: At the left junction point (node), we see that a 4.3 (A) current is going towards the node while a 3.0 (A) current and i_y are going away. Table 4-18 gives the conservation of net charge for this system. From this we can see that i_y must be 1.3 (A). At the right node, the conservation of net charge is given in Table 4-19. It is easy to see that i_x must have a value of 3.2 (A). If we look at

Table 4-19: Solution of Example 4-10 (Charge, System-Right junction)

Description	i_{in}			=	i_{out}
	Wire 3 (A)	Wire 4 (A)	Wire 5 (A)		Wire 6 (A)
Net Charge	1.3 +	2.4 +	i_x =		6.9

Table 4-20: Alternate Solution of Example 4-10 (Charge, System-Both junctions)

Description	i_{in}			=	i_{out}	
	Wire 1 (A)	Wire 4 (A)	Wire 5 (A)		Wire 2 (A)	Wire 6 (A)
Net Charge	4.3 +	2.4 +	i_x =		3.0 +	6.9

the boundary drawn around both nodes, (Table 4-20) the solution for i_x is 3.2 (A), the same result that was obtained by considering the two nodes separately as systems.

Example 4-11. A circuit consisting of a 2 ampere *current source* and several circuit elements (resistors) is shown in Figure 4-15. A current source has the property that the current through it takes on the specified value, independent of the rest of the circuit. In this case 2 amperes enter the bottom of the source and two amperes leave the top. Names and directions for the currents in each branch of the circuit are given. Write the relations among these currents implied by Kirchhoff's Current Law applied at each node of the circuit. The circuit is shown in Figure 4-15.

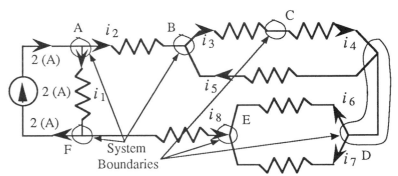

Figure 4-15: Multiple systems and multiple currents in a circuit

Table 4-21: Data for Example 4-11 (Charge)

System Node	$\left(\dfrac{dq_{sys}}{dt}\right) =$ (A)		$i_{in/out}$								
		Lead (A)	Wire 1 (A)	Wire 2 (A)	Wire 3 (A)	Wire 4 (A)	Wire 5 (A)	Wire 6 (A)	Wire 7 (A)	Wire 8 (A)
A	0 =	2	$+ (-i_1)$	$+ (-i_2)$						
B	0 =			i_2	$+ (-i_3)$		$+ \quad i_5$			
C	0 =				i_3	$+ (-i_4)$				
D	0 =					i_4	$+ (-i_5)$	$+ (-i_6)$	$+ (-i_7)$	
E	0 =							i_6	$+ \quad i_7$	$+ \quad i_8$
F	0 =	(-2)	$+ \quad i_1$							$+ (-i_8)$

SOLUTION: Net charge equations for the nodes are summarized in Table 4-21. The columns represent the various wires in the network and the rows represent the different systems (nodes). Note that for this problem there are *eight* currents and we were able to write *six* equations relating them (in fact it turns out that only five of these are independent equations). Thus, unless some of the current values are known, we do not yet have enough information to uniquely determine each of the currents. Notice that even though there are elements in the circuit, the conservation of charge in the form of Kirchhoff's Current Law still is valid. The extra information needed for and provided by the circuit elements will come from an application of the conservation of energy and the electrical energy accounting statement. This is covered in Chapter 7.

4.6 REVIEW

1) The are two types of charge, and we label these as positive and negative.

2) Matter in its normal state contains equal amounts of positive and negative charge (each atom has an equal number of protons and electrons) and is electrically neutral. Such an object becomes positively charged if some electrons are removed from it, and negatively charged if extra electrons are added to it.

3) Net charge is conserved. The total amount of net charge in an isolated system remains constant. No net charge is created or destroyed.

4) Movement of charge in a conductor is called current. The amount of electric current is given by the net charge flow rate, which is in the units Amperes; 1 A = 1 Coulomb/second.

5) Kirchhoff's Current Law is a statement of the rate form of the conservation equation at steady state for the special case of electric circuits made up of conductors. Since net charge cannot accumulate at any point on a conductor, the net-charge flow rate into any junction or system equals the net-charge flow rate out. That is, at any connection point, the sum of the currents flowing in must be equal to the sum of the currents flowing out.

4.7 NOTATION

The following symbols were used in the development of the equations and concepts for this chapter.

	Variables and Descriptions	Dimensions
i	net charge flow rate or current	[charge / time]
q	net charge	[charge]
q_+	positive charge	[charge]
q_-	negative charge	[charge]
$\dot{q}$	net-charge flow rate	[charge / time]
$\dot{q}_+$	positive-charge flow rate	[charge / time]
$\dot{q}_-$	negative-charge flow rate	[charge / time]
t	time	[time]

Subscripts

beg	evaluated at the beginning of the time period
con	consumption or usage
end	evaluated at the end of the time period
gen	generation or production
in	input or entering
out	output or leaving
sys	system or within the system boundary

4.8 SUGGESTED READING

Fitzgerald, A., D. Higginbotham, and A. Grabel, *Basic Electrical Engineering*, 5th edition, McGraw-Hill, New York, 1981

Halliday, D. and R. Resnick, *Physics for Students of Science and Engineering, Part II*, 2nd edition, John Wiley & Sons, New York, 1962

Nilsson, J., *Electrical Circuits*, 3rd edition, Addison-Wesley, Reading, Massachussetts, 1989

Paul, C., S. Nasar, and L. Unnewehr, *Introduction to Electrical Engineering*, McGraw-Hill, New York, 1986

Roadstrum, W. H., and D. H. Wolaver, *Electrical Engineering for All Engineers*, Harper and Row, New York, 1987

QUESTIONS

1) Give a verbal statement of the conservation of net charge equation.

2) What two things must be defined in order to have a meaningful conservation of net charge statement (in addition to the fact that charge is the property being counted)?

3) What kind of equation (algebraic or ordinary differential equation) is obtained for the conservation of charge if the system which is defined is of finite size and the time period is also finite in size?

4) What kind of equation is obtained for a finite-size system and a differential time period when the equation is divided by the time period?

5) What is the difference (if any) between an equation written for the conservation of net charge and an equation written for positive and negative charges?

6) A battery contains both positive and negative charges. By connecting the leads of the battery, the charges will flow from the negative terminal of the battery to the positive terminal of the battery. Discuss the changes which occur in the amount of net charge, positive charge, and negative charge contained in the battery. (Remember net charge is conserved)

SCALES

1) Solve the following sets of equations for the unknowns:

a) $2i_1 + i_2 - 3i_3 = 4$
$i_1 - 3i_2 + i_3 = -2$
$-i_1 + 2i_2 - i_3 = 0$

b) $2i_1 + i_2 - 3i_3 + i_4 - i_5 = 2$
$i_1 - 3i_2 + i_3 + i_5 = -1$
$-i_1 + 2i_2 - i_3 = 0$
$i_3 + i_4 - i_5 = -2$
$i_1 - i_2 + 2i_3 - 5i_4 = 0$

2) Give at least one method (which may include a specific computer software product) that *you* know how to use to find the root(s) of an equation.

3) Find the root(s) of the following equations:

a) $t^3 + t^2 - 3t - 5 = 0$

b) $t\dfrac{e^t}{(t+1)^2} - t^2 + 5 = 0$

4) Give at least one method (which may include a specific computer software product) that *you* know how to use to solve a system of non-linear algebraic equations.

5) Solve the following system of non-linear equations for $t > 0$:

a) $t^3 + t^2 - 3t - 5 = 0$

$t\dfrac{e^t}{(t+1)^2} - t^2 + 5 = 0$

6) Name at least one computer software product which *you* can use to graph a linear or non-linear function (and to plot multiple functions on the same graph).

7) Plot the following functions over the range $0 < t < 5$ using a computer software program:

a) $t^3 + t^2 - 3t - 5 = 0$

b) $t\dfrac{e^t}{(t+1)^2} - t^2 + 5 = 0$

8) If the charge (coulombs, C) on a plate as a function of time (seconds) is given by

$$q(t) = t^3 + t^2 - 3t - 5 = 0$$

then what is the accumulation rate of charge at $t = 1.5$ s?

9) If the charge (coulombs, C) on a plate as a function of time (seconds) is given by

$$q(t) = t^3 + t^2 - 3t - 5 = 0$$

then what is the accumulation of charge from $t = 0.5$s to $t = 3$ s ?

10) The charge on a plate as a function of time is measured and found to be:

Charge (C)	Time (seconds)
2	0
3	1
5	2
5	3
6	4
6	5
7	6
7	7
6	8
6	9
5	10
4	11

a) Plot the data for charge versus time, using a computer software product.

b) Choose an appropriate function to represent the data and use the computer plotting package to estimate parameters for this function.

c) Estimate the accumulation rate of charge at $t = 4$ s.

d) What is the accumulation of charge during the time period from 3 seconds to 9 seconds?

11) What are the dimensions of current? What are SI units for current?

12) A differential equation for the charge on a plate is found to be $\dfrac{dq_{sys}}{dt} = \dot{q}_{in} - kq_{sys}$ where $\dot{q}_{in} = 5$ C/s and $k = 2$ s^{-1}. If, at $t = 0$ $q_{sys} = 2$ C, then how much is the charge on the plate at $t = 4$ s?

13) True or False: For a given system, if 3 C of net charge enters the system, then 3 C also must have left the surroundings.

14) Both current and charge accumulation rate have the same dimensions but yet, physically and conceptually, they are different quantities. Explain.

PROBLEMS

1) Solve the following circuits.

    a) Given the figure below, solve for the unknown electrical currents x and y. What assumption is required in order to solve this problem.

    b) For the figure below determine the unknown currents through a, b, c, and d.

    c) Given the figure below, solve for the unknown currents.

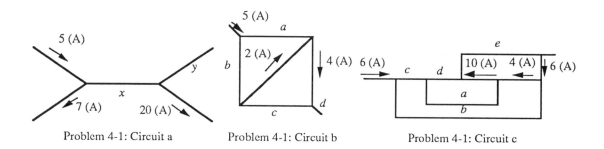

Problem 4-1: Circuit a   Problem 4-1: Circuit b   Problem 4-1: Circuit c

2) If the currents entering and leaving a system are given by the following equations:

$$i_{in}(t) = 5\sin(3t) \text{ (A)}$$

$$i_{out}(t) = 4\cos(2t) \text{ (A)}$$

what is the accumulation of net charge after 1, 2, and 3 seconds? The constants 2 and 3 in the expressions for the current have the units of (1/s). If the initial net charge on the system is zero, graph the net charge on the plate as a function of time.

3) Solve the following circuits:

    a) For the given figure, solve for the unknown currents.

    b) For the given figure, determine the currents a, b, c, and d. Draw a diagram and clearly show which direction the current is flowing. Can you solve for the current in b writing only one equation? If so, explain.

    c) For the given figure, determine the currents in wires a, b, and c.

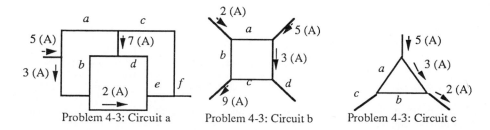

Problem 4-3: Circuit a   Problem 4-3: Circuit b   Problem 4-3: Circuit c

4) Solve the following problems from the given data.

a) For the given data, approximate the time rate of change of charge at $t = 9$ (s) of the system by graphical techniques.

Charge Data

t (s)	0	1	2	3	4	5	6	7	8	9	10
q (C)	0	3.5	7	8	8.5	9	9.5	9.75	9.8	9.9	9.95

b) Given the following data for current entering a system:

Current Data

t (s)	0	1	2	3	4	5
$i(t)$ (A)	0	2	2.5	2.7	2.8	3

graph the data and determine the accumulation of net charge in the system from $t = 0$ to $t = 4$ (s). The initial charge or the system is 1 (C).

5) The accompanying figure shows a system in which current enters through one wire A and leaves through two wires B and C The currents through A and B are known functions of time:

$$i_A(t) = 6\sin(2t) \text{ (A)}$$

$$i_B(t) = 6\left(1 - \exp(-0.5t)\right) \text{ (A)}$$

The constants 2 and 0.5 carry the units (1/s). If the system does not accumulate net charge, determine the unknown current C and display the current in a plot or graph as a function of time.

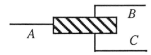

Problem 4-5: Circuit

6) For the circuit shown in the figure, if there is no accumulation of net charge anywhere, determine the unknown currents (d, e, f and g). The following currents are functions of time:

$$a = \exp(2t) \text{ (mA)}$$

$$b = \sin(1t) \text{ (mA)}$$

$$c = \cos(1t) \text{ (mA)}$$

The constants 2 and 1 in the equation carry the units (1/s). Redraw the figure and write the current (function of time) at each point (a through g) on your figure.

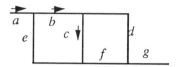

Problem 4-6: Circuit

7) The current to the positive plate of a capacitor is given by the data in the following table. Assume the function is continuous.

Problem 4-7 Data

Time (s)	0	1	3	5	7	9
Current (A)	7.38	6.61	5.07	3.53	2	0.46

a) Plot the data.
b) At what rate is the plate charging at 4 (s)?
c) At 2 (s) the charge on the capacitor is 5 (C). What is the charge at 7 (s)?
d) Sketch the charge on the plate as a function of time.

8) The current to a plate is given by the following equation:

$$i(t) = k\left(1 - \ln(t + 0.01)\right)$$

where $k = 25$ (A) and t is in seconds. If the charge on a plate cannot exceed 25 (C), determine the time when the current source must be shut down. The initial charge on the plate ($t = 0$) is 2 (C). If allowed to continue charging (instead of being shut off at 25 C), at what time would the charge on the plate start to decrease, and what would be its corresponding maximum value?

9) A capacitor plate has an initial charge of 10 (C). The plate is also being supplied with a constant current of 5 (A). At any time, the capacitor "leaks" charge at a rate that is proportional to the charge on the plate at that time, and the proportionality constant is $k = 2$ (1 / s).

a) Determine the charge on the plate as a function of time.
b) What is the charge on the plate as the time becomes very large ($t \to \infty$)
c) Sketch the graph of the charge on the plate.

Hint $$\int \frac{dx}{a + bx} = \frac{1}{b}\ln(a + bx) + C$$

10) For the following nuclear reactions, account for negative charge, positive charge and show that net charge is conserved. The ν represents a neutrino which carries no charge. The e^+ represents a positron that carries a positive charge. The η is a neutron which carries no charge.

a) $^{11}_{6}C \longrightarrow ^{11}_{5}B^- + e^+ + \nu$

b) $^{16}_{8}O + ^{2}_{1}H \longrightarrow ^{14}_{7}N + ^{4}_{2}He + energy$

c) $^{9}_{4}Be + ^{4}_{2}He \longrightarrow ^{12}_{6}C + \eta$

d) $^{63}_{29}Cn + ^{1}_{1}H \longrightarrow ^{63}_{30}Zn + \eta$

e) $^{40}_{18}Ar + ^{107}_{47}Ag \longrightarrow ^{147}_{65}Tb + \eta \longrightarrow ^{16}_{8}O + ^{131}_{57}La$

11) The sun converts mass to energy by the following solar cycle. For all of the reactions, account for negative charge, positive charge and net charge. The ν represents a neutrino which carries no charge. The e^+ represents a positron that carries a positive charge. The γ represents energy which is neutral.

a) $^{12}_{6}C + ^{1}_{1}H \longrightarrow ^{13}_{7}N + \gamma$

b) $^{13}_{7}N \longrightarrow ^{13}_{6}C^{-} + e^{+} + \nu + \gamma$

c) $^{13}_{6}C + ^{1}_{1}H \longrightarrow ^{14}_{7}N + \gamma$

d) $^{14}_{7}C + ^{1}_{1}H \longrightarrow ^{15}_{8}O + \gamma$

e) $^{15}_{8}O \longrightarrow ^{15}_{7}N^{-} + e^{+} + \nu + \gamma$

f) $^{15}_{7}N + ^{1}_{1}H \longrightarrow ^{12}_{6}C + ^{4}_{2}He$

12) A typical stadium light draws about 3 (A) of current when turned on. There are 5 banks of lights surrounding a baseball stadium. Banks A and B have 50 lights each. Banks C and D have 100 lights each, and bank E has 150 lights. Given the following table of data:

Time schedule for stadium lights

Bank	6:00 p.m.	7:00 p.m.	8:00 p.m.	9:00 p.m.	10:00 p.m.	11:00 p.m.
A	Off	On	On	On	Off	Off
B	Off	Off	On	On	On	Off
C	Off	Off	On	On	On	Off
D	Off	On	On	On	On	Off
E	On	Off	Off	On	On	On

Determine the following:

a) Calculate the total amperes the stadium lights draw as a function of time and graph the function. During what time period are the stadium lights drawing the maximum current.

b) Calculate the total net charge that passed through the banks of lights during the period between 7:00 p.m. and 11:00 p.m.

c) How does the graph that you constructed in part (a) relate to the answer to part (b)?

CHAPTER FIVE

THE CONSERVATION OF LINEAR MOMENTUM

5.1 INTRODUCTION

In this chapter we consider the conservation of linear momentum. The linear momentum of a system is the product of the velocity and mass of the system. As such, linear momentum is an extensive property, proportional to the mass of the system and it is a 3-dimensional vector quantity requiring three (3) scalar components for complete definition.

In Chapter 2 the conservation of linear momentum was introduced and two examples were given. In this chapter, the conservation of linear momentum is discussed in more detail. Emphasis is on the linear momentum of rigid bodies. Furthermore, the equations will be developed using vector notation, to allow learning in a way which can be applied to 3-D problems as easily as to 2-D (or nearly as easily).

The principle of conservation of linear momentum is used to solve a multitude of engineering problems. The trajectories of rockets, the load carrying capabilities of support beams, the forces caused by flowing fluids in pipes, and the force of impact during vehicle collision are several situations that are governed by the conservation of linear momentum. We must understand this law and its impact in order to understand the world around us.

A verbal statement of the conservation of linear momentum (LM) is made below for a given system and time period. Because momentum is a conserved property, there is no generation or consumption and we have:

$$\left\{ \begin{array}{c} \text{Accumulation of LM} \\ \text{within system during} \\ \text{time period} \end{array} \right\} = \left\{ \begin{array}{c} \text{LM entering} \\ \text{system during} \\ \text{time period} \end{array} \right\} - \left\{ \begin{array}{c} \text{LM leaving} \\ \text{system during} \\ \text{time period} \end{array} \right\}$$

where the accumulation of linear momentum is defined as follows:

$$\left\{ \begin{array}{c} \text{Accumulation of LM} \\ \text{within system} \\ \text{during time period} \end{array} \right\} \equiv \left\{ \begin{array}{c} \text{Amount of LM} \\ \text{contained in system} \\ \text{at end of} \\ \text{time period} \end{array} \right\} - \left\{ \begin{array}{c} \text{Amount of LM} \\ \text{contained in system} \\ \text{at beginning of} \\ \text{time period} \end{array} \right\}$$

Before proceeding further, it is essential to understand the ways in which momentum may enter (or leave) the system.

5.1.1 Momentum Possessed by Mass

First, of course, momentum may enter or leave the system with mass. Mass enters with a certain velocity and when it crosses the system boundary, that much momentum is added to the system. The rate at which momentum enters the system due to mass traveling at velocity **v** and entering the system at rate $\dot{m}_{\text{in}}$ is:

$$(\mathbf{v}\dot{m})_{\text{in}}$$

Likewise, the rate of momentum leaving is:

$$(\mathbf{v}\dot{m})_{\text{out}}$$

Note that the correct way to calculate the rate at which momentum enters (or leaves) is the rate at which mass enters (or leaves) multiplied by the momentum (per unit mass) possessed by that mass:

$$\frac{\text{momentum}}{\text{mass}} \bigg| \frac{\text{mass}}{\text{time}} = \left(\frac{\text{momentum}}{\text{time}} \right)$$

Where there are multiple entering (m, say) and leaving (n, say) streams having different velocities then we must write:

$$\sum_{i=1}^{m} \left((\mathbf{v}\dot{m})_i \right)_{\text{in}}$$

and

$$\sum_{i=1}^{n} \left((\mathbf{v}\dot{m})_i \right)_{\text{out}}$$

Note that within this context of momentum exchange with mass it is very useful to think of the velocity, **v**, as *momentum per mass*.

5.1.2 Momentum in Transit: Forces

Second, momentum enters as the result of forces which act upon mass in the system by its surroundings; an external force **f** acting on a system provides momentum to the system at rate **f**. These forces may act as surface or contact forces *at the boundary* or they may act as body forces as the result of phenomena such as gravity or electrical fields *across the boundary*. In any case, these forces are mutual between the system and surroundings, in accordance with Newton's laws of motion; forces acting on the system provide an oppositely directed but equal in magnitude force on the surroundings.

A schematic representation of a system and surroundings to apply the conservation of linear momentum principle is given in Figure 5-1. The linear momentum entering and leaving the system must cross the system boundary that we have defined. Furthermore, only the external forces which act on the mass in the system are of interest. Forces acting between mass elements, both of which are inside the system boundary, do not contribute to momentum changes of the system as a whole.

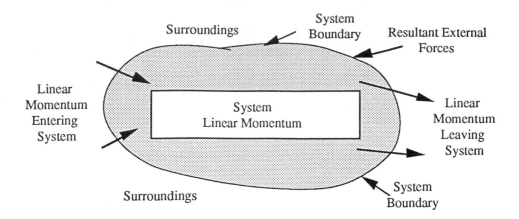

Figure 5-1: System and surroundings

As examples, the external forces can be the pressure exerted over the surface area of the system, the supporting columns or cables to suspend a system, drag or frictional forces, and gravitational and electrostatic body forces. Other examples of sources of external forces are springs attached to a surface. An ideal spring exerts a force which varies linearly with the amount of its extension. The proportionality constant is known as the spring constant.

5.2 CONSERVATION OF LINEAR MOMENTUM EQUATIONS

5.2.1 Rate Equation

We are now prepared to combine the verbal statement of linear momentum with these quantitative rate expressions. Accordingly, a rate equation for linear momentum conservation is:

$$\left(\frac{d\mathbf{p}_{sys}}{dt}\right) = \sum(\dot{m}\mathbf{v})_{in} - \sum(\dot{m}\mathbf{v})_{out} + \sum \mathbf{f}_{ext} \qquad (5-1)$$

$\left(\dfrac{d\mathbf{p}_{sys}}{dt}\right)$ = the time rate of change of the system linear momentum.

$(\dot{m}\mathbf{v})_{in}$ = the rate at which linear momentum enters the system with mass.

$(\dot{m}\mathbf{v})_{out}$ = the rate at which linear momentum leaves the system with mass.

$\sum \mathbf{f}_{ext}$ = the sum of the external forces acting on the system.

The momentum of the system may change because of mass that enters and leaves and because of external forces which act on the system. Note again, as stated above, that as mass enters the system, the amount of momentum that comes into the system is this mass flow rate times the momentum per unit mass (velocity) of the mass.

Table 5-1: Components of the Linear Momentum Equation

	$\left(\dfrac{d\mathbf{p}_{sys}}{dt}\right)$	=	$\sum(\dot{m}\mathbf{v})_{in/out}$	+	$\sum\mathbf{f}_{ext}$
Description	[LM/time]		[LM/time]		[LM/time]
x-direction	$\left(\dfrac{d(p_x)_{sys}}{dt}\right)$	=	$\sum(\dot{m}v_x)_{in/out}$	+	$\sum(f_x)_{ext}$
y-direction	$\left(\dfrac{d(p_y)_{sys}}{dt}\right)$	=	$\sum(\dot{m}v_y)_{in/out}$	+	$\sum(f_y)_{ext}$
z-direction	$\left(\dfrac{d(p_z)_{sys}}{dt}\right)$	=	$\sum(\dot{m}v_z)_{in/out}$	+	$\sum(f_z)_{ext}$

Because this is a vector equation, the three component equations give scalar equations which are given in Table 5-1 for a rectangular cartesian coordinate system. Furthermore, the input and output terms are combined as in Chapters 3 and 4. In solving problems, it is the component equations which must be used.

5.2.2 Momentum of the System

So far we have considered the momentum of mass entering and leaving the system and the transfer of momentum to and from the system through forces acting between the system and surroundings. However, we have not yet said anything about the momentum of the system and how this is determined.

If the system is a single particle, then the momentum is easily calculated. By definition, a particle is assumed to occupy a very small volume and therefore only one velocity exists to describe the motion of that particle. In this case, the momentum is the mass of that particle times its velocity and the velocity of a particle is an unambiguous quantity.

For a larger system however, the calculation is complicated by the fact that different pieces of the system may be moving differently. Elements of the body may be moving relative to each other or the body itself may be rotating in addition to translating. To calculate the momentum of such a body, we divide it up into individual elements on a small enough scale that each element can be described in terms of a single velocity and then we sum over all such elements within the system. Consequently we may write (for n particles in the system):

$$\mathbf{p}_{sys} = \sum_{i=1}^{n}\left((m\mathbf{v})_i\right)_{sys} \tag{5-2}$$

where subscript i represents the ith element of the system, and this may now be used in the above equations to represent the momentum of the system.

The Rate of Change of Linear Momentum of the System. To use the conservation of linear momentum rate equation, we must calculate the time rate of change of the momentum of the system. Since the linear momentum is the product of the mass of the system and the linear velocity of the system, the time derivative is calculated using the product rule as shown:

$$\left(\frac{d\mathbf{p}_{sys}}{dt}\right) = \left(\frac{d(m\mathbf{v})_{sys}}{dt}\right)$$

$$= \mathbf{v}_{sys}\left(\frac{dm_{sys}}{dt}\right) + m_{sys}\left(\frac{d\mathbf{v}_{sys}}{dt}\right)$$

If the mass of the system is not constant (varies with time), the time rate of change of the mass of the system is determined from the conservation of mass. If the mass in the system is constant, the time derivative of the mass in the system is zero.

Example 5-1. A hopper is placed above a conveyor belt. The mass from the hopper drops onto the conveyor belt at a rate of $\dot{m}_{in}$. The mass of the conveyor belt alone is m_b. The mass of the conveyor belt is constant. What additional (due to the falling mass) force is required to keep the conveyer belt moving at a constant velocity $\mathbf{v}$?

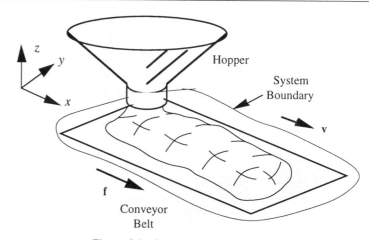

Figure 5-2: Conveyor belt with hopper

SOLUTION: Figure 5-2 gives a schematic of the hopper and conveyor belt. We will choose our system to be the belt and the material being conveyed by it. As material falls onto the hopper, mass is added to the system. Before contacting the belt, its x-direction velocity is zero; afterwards its velocity is $\mathbf{v}$. First for the defined system, let us look at the conservation of total mass for this problem as given in Table 5-2. From the table, the time rate of change of the mass of the system is equal to the mass flow rate into the system.

Table 5-2: Data for Example 5-1 (Mass, System = Conveyor belt)

Description	$\left(\dfrac{dm_{sys}}{dt}\right)$ [mass/time]	=	$\sum \dot{m}_{in/out}$ [mass/time]
Total Mass	$\left(\dfrac{dm_{sys}}{dt}\right)$	=	$\dot{m}_{in}$

The conservation of linear momentum, tells about the momentum entering with mass and the forces that are acting on the system. The momentum equation for this system is given in Table 5-3. Note that the table includes the unit vectors **i**, **j**, and **k** to emphasize the different coordinate directions.

Table 5-3: Data for Example 5-1 (Linear Momentum, System = Conveyor belt)

Description	$\left(\dfrac{d\mathbf{p}_{sys}}{dt}\right)$ (N)	=	$\sum(\dot{m}\mathbf{v})_{in/out}$ (N)	+	$\sum\mathbf{f}_{ext}$ $\mathbf{f}$ (N)
x-direction	$\left(\dfrac{d(p_x)_{sys}}{dt}\right)\mathbf{i}$	=	$0\mathbf{i}$	+	$f_x\mathbf{i}$
y-direction	$0\mathbf{j}$	=	$0\mathbf{j}$	+	$0\mathbf{j}$
z-direction	$0\mathbf{k}$	=	$(\dot{m}v_z)_{in}\mathbf{k}$	+	$f_z\mathbf{k}$

The mass that enters the system from the hopper is falling in the z-direction but has no x-direction velocity. The x-direction momentum of the system changes because the mass of the system is changing. The velocity $(v_x)_{sys}$ is constant; however, a force is required because the momentum is changing.

$$v_x\left(\frac{dm_{sys}}{dt}\right) = f_x$$

From the conservation of mass, the time derivative of the mass of the system is the mass flow rate in, and the force in the x direction is:

$$f_x = v_x\dot{m}_{in}$$

The force required to keep the belt moving is f_x.

A force in the z-direction also is required to bring the falling material to zero z-direction velocity. However, this does not involve a change in the system momentum.

$$(\dot{m}v_z)_{in} + f_z = 0$$

Furthermore, f_z is not a force required to keep the belt moving.

Velocity of the Center of Mass. For closed systems, especially for rigid bodies, it is convenient to define a velocity which characterizes the system as a whole. This velocity we call the system average velocity or velocity of the center of mass. We define this velocity in such a way that the total mass of the system times this average velocity is equal to the total momentum of the system, a logical and convenient definition. Accordingly, we have

$$m_{sys}\mathbf{v}_G = \sum_{i=1}^{n}\left((m\mathbf{v})_i\right)_{sys} \tag{5-3}$$

or, by rearranging

$$\mathbf{v}_G \equiv \frac{\displaystyle\sum_{i=1}^{n}\left((m\mathbf{v})_i\right)_{sys}}{m_{sys}} \tag{5-4}$$

f the size of the mass elements m_i becomes differentially small, the summation process becomes an integration and the elocity of the center of mass is defined as:

$$\mathbf{v}_G \equiv \frac{\int_{\text{sys}} \mathbf{v}\,dm}{m_{\text{sys}}}$$

(5 – 5)

his may also be called the mass-averaged velocity of the system or, of course, the momentum per mass of the system.

Now because the velocity of an element of mass is defined to be the time rate of change of the position vector escribing the location of that element of mass, we can write

$$\mathbf{v}_G = \left(\frac{d\mathbf{r}_G}{dt}\right)$$

(5 – 6)

nd hence we can define the center of mass position in terms of the position vectors of all of the elements of the system:

$$\mathbf{r}_G = \frac{\sum_{i=1}^{n} \left((m\mathbf{r})_i\right)_{\text{sys}}}{m_{\text{sys}}}$$

(5 – 7)

As the size of the m_i particles becomes differentially small, this summation again becomes an integration yielding:

$$\mathbf{r}_G \equiv \frac{\int_{\text{sys}} \mathbf{r}\,dm}{m_{\text{sys}}}$$

(5 – 8)

his definition of the center of mass applies only for a *closed system* for which there is no exchange of mass between the ystem and the surroundings. For such a system then we can write the conservation of linear momentum as

$$\sum \mathbf{f}_{\text{ext}} = \left(\frac{d\mathbf{p}_{\text{sys}}}{dt}\right) = \left(\frac{d(m_{\text{sys}}\mathbf{v}_G)}{dt}\right) = m_{\text{sys}}\left(\frac{d\mathbf{v}_G}{dt}\right)$$

(5 – 9)

Note again that there is nothing magic about this velocity of the center of mass. It is defined in such a way that when we o the straight-forward calculation of the mass of the system times this center-of-mass velocity we obtain what we know as to be the proper momentum of the system. This is a definition, not a proof. The following example illustrates the rate quation.

5.2.3 Integral (Finite Time Period) Equation

The rate equation (5-1) with the input and output terms combined can be integrated to obtain an equation for a finite time period.

$$\int_{t_{beg}}^{t_{end}} \left(\frac{d\mathbf{p}_{sys}}{dt} \right) dt = \int_{t_{beg}}^{t_{end}} (\dot{m}\mathbf{v})_{in/out} dt + \int_{t_{beg}}^{t_{end}} \sum \mathbf{f}_{ext} dt \qquad (5-10)$$

which can then be written in symbolic form with the input and output terms combined as:

$$(\mathbf{p}_{sys})_{end} - (\mathbf{p}_{sys})_{beg} = \mathbf{p}_{in/out} + \int \sum \mathbf{f}_{ext} dt \qquad (5-11)$$

The terms in the integral equation are:

$(\mathbf{p}_{sys})_{end}$	=	the L.M. of the system at the end of the time period
$(\mathbf{p}_{sys})_{beg}$	=	the L.M. of the system at the beginning of the time period
$\mathbf{p}_{in}$	=	L.M. entering the system with mass during the time period.
$\mathbf{p}_{out}$	=	L.M. leaving the system with mass during the time period.
$\int \sum \mathbf{f}_{ext} dt$	=	L.M. entering due to external forces acting during the time period. This is also called the impulse.

Example 5-2. We will consider the collision of two (2) balls in the absence of a gravitational field. Figure 5-3 depicts the collision. The mass of ball 1 is 2.5 (lb$_m$) , and the mass of ball 2 is 5 (lb$_m$) The velocity vectors are given in Table 5-4 for both balls before the collision.

a) If the components of the velocity vector of Ball 2 after the collision are $(v_{2x})_{end} = 1$ ft/s, $(v_{2y})_{end} = -2$ ft/s, and $(v_{2z})_{end} = 4$ ft/s, what is the velocity vector of Ball 1 after the collision?

b) If, instead, the velocity components of Ball 1 after the collision were $(v_{1x})_{end} = -1$ ft/s, $(v_{1y})_{end} = 12$ ft/s, and $(v_{1z})_{end} = 2$ ft/s, what would be the magnitude of the external forces which must have acted on the two balls if the time period was 0.1 s for the collision? Assume that the forces are constant during the time period and the other data are the same as given in part (a).

Table 5-4: Velocity Data Before Collision

	$(v_i)_{beg}$		
Ball #	$(v_x)_{beg}$ (ft / s)	$(v_y)_{beg}$ (ft / s)	$(v_z)_{beg}$ (ft / s)
Ball 1	4	7	5
Ball 2	-2	1	3

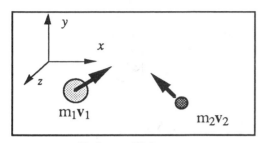
Before collision

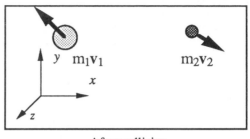
After collision

Figure 5-3: Collision of two balls

SOLUTION: (a) The system is the space containing the two balls before and after the collision. Because the balls do not cross the system boundary, the system is closed. Equation (5-11) is used to describe the system:

$$(\mathbf{p}_{sys})_{end} - (\mathbf{p}_{sys})_{beg} = \mathbf{p}_{in/out} + \int \sum \mathbf{f}_{ext} dt$$

Because the system is closed, the first term of the equation which represents the momentum entering and leaving due to mass entering and leaving the system is zero. Furthermore, no external forces are acting on the system since we are in the absence of a gravitational field.

As with mass accounting and conservation equations, we will write the x-,y-,and z-component equations in tabular form in such a way as to emphasize the role of each term in the conservation equation. The linear momentum $\mathbf{p}$ of the system is the sum of the mass-velocity product for each mass in the system. There are three component equations shown in Table 5-5. This gives us three equations in terms of the three unknown velocity components of ball 1 after the collision. The solution is $(v_{1x})_{end} = -2$ (ft / s), $(v_{1y})_{end} = 13$ (ft / s), and $(v_{1z})_{end} = 3$ (ft / s). The velocity vector at the end of the collision for Ball 1 is:

$$(\mathbf{v}_1)_{end} = -2\mathbf{i} + 13\mathbf{j} + 3\mathbf{k} \text{ (ft / s)}$$

Table 5-5: Data for Example 5-2a (Linear Momentum, System = Two balls)

	$(\mathbf{p}_{sys})_{end}$		$-$	$(\mathbf{p}_{sys})_{beg}$		$=$	$\sum \mathbf{p}_{in/out}$	$+$	$\int \left(\sum \mathbf{f}_{ext}\right) dt$
Descr.	$\left((m\mathbf{v})_1\right)_{end}$	$\left((m\mathbf{v})_2\right)_{end}$		$\left((m\mathbf{v})_1\right)_{beg}$	$\left((m\mathbf{v})_2\right)_{beg}$				
	$\dfrac{lb_m \ ft}{s}$	$\dfrac{lb_m \ ft}{s}$		$\dfrac{lb_m \ ft}{s}$	$\dfrac{lb_m \ ft}{s}$		$\dfrac{lb_m \ ft}{s}$		$\dfrac{lb_m \ ft}{s}$
x-dir	$2.5(v_{1x})_{end}\mathbf{i}$ +	$5(1)\mathbf{i}$	$-$	$2.5(4)\mathbf{i}$ $-$	$5(-2)\mathbf{i}$	$=$	$0\mathbf{i}$	$+$	$0\mathbf{i}$
y-dir	$2.5(v_{1y})_{end}\mathbf{j}$ +	$5(-2)\mathbf{j}$	$-$	$2.5(7)\mathbf{j}$ $-$	$5(1)\mathbf{j}$	$=$	$0\mathbf{j}$	$+$	$0\mathbf{j}$
z-dir	$2.5(v_{1z})_{end}\mathbf{k}$ +	$5(4)\mathbf{k}$	$-$	$2.5(5)\mathbf{k}$ $-$	$5(3)\mathbf{k}$	$=$	$0\mathbf{k}$	$+$	$0\mathbf{k}$

Table 5-6: Data for Example 5-2b (Linear Momentum, System = Two balls)

Descr.	$(p_{sys})_{end}$			$-$	$(p_{sys})_{beg}$			$=$	$\sum p_{in/out}$	$+$	$f_{ext}\Delta t$
	$\left((mv)_1\right)_{end}$		$\left((mv)_2\right)_{end}$		$\left((mv)_1\right)_{beg}$		$\left((mv)_2\right)_{beg}$				
	$\dfrac{lb_m\ ft}{s}$		$\dfrac{lb_m\ ft}{s}$		$\dfrac{lb_m\ ft}{s}$		$\dfrac{lb_m\ ft}{s}$		$\dfrac{lb_m\ ft}{s}$		$\dfrac{lb_m\ ft}{s}$
x-dir	$2.5(-1)i$	$+$	$5(1)i$	$-$	$2.5(4)i$	$-$	$5(-2)i$	$=$	$0i$	$+$	$f_x\Delta ti$
y-dir	$2.5(12)j$	$+$	$5(-2)j$	$-$	$2.5(7)j$	$-$	$5(1)j$	$=$	$0j$	$+$	$f_y\Delta tj$
z-dir	$2.5(2)k$	$+$	$5(4)k$	$-$	$2.5(5)k$	$-$	$5(3)k$	$=$	$0k$	$+$	$f_z\Delta tk$

For part (b), For the closed system, Equation (5-11) is used again

$$(p_{sys})_{end} - (p_{sys})_{beg} = \int_t f_{ext}dt$$

For the constant forces, the integration over the time period yields:

$$(p_{sys})_{end} - (p_{sys})_{beg} = f_{ext}\Delta t$$

The three components of the linear momentum conservation equation are given in Table 5-6 and are easily solved to give: $f_x = 25.0\ (lb_m\ ft\ /\ s^2)$, $f_y = -25.0\ (lb_m\ ft\ /\ s^2)$, and $f_z = -25.0\ (lb_m\ ft\ /\ s^2)$. The force vector would be represented as:

$$f_{ext} = 25.0i - 25.0j - 25.0k\ (lb_m\ ft\ /\ s^2)$$

5.3 LINEAR MOMENTUM AND NEWTON'S LAWS OF MOTION

The concept of conservation of linear momentum (Equation 5-1) applied to a rigid body, together with the realization that forces *exchange* momentum between the surroundings and the body, is actually a statement of Newton's three laws of motion. Newton's *first law of motion* says that a body which is at rest will stay at rest unless acted upon by an external force. In other words, a body does not spontaneously generate momentum; momentum is conserved. Newton's *second law of motion* says that for a body (system) of constant mass the sum of the external forces is equal to the mass times the acceleration, clearly a result of the conservation law stated above when $\dot{m}_{in} = \dot{m}_{out} = 0$ so that, by conservation of mass, $\left(\dfrac{dm_{sys}}{dt}\right) = 0$. Newton's *third law of motion* says that for every force there is an opposite and equal reaction force and this also implies conservation of momentum. That is, a force acting on a system by the surroundings provides an oppositely directed but equal-in-magnitude force exerted on the surroundings. Therefore, momentum transferred to the system must have been provided or given up by the surroundings, implying absolute conservation of momentum due to the action of forces.

Example 5-3. A mass of 50 (kg) is being lifted with a cable. a) If the mass is to be accelerated at 1 (m / s^2) upwards, what force is required? b) If the mass is to rise at a constant velocity of 5 (m/s), what is the required force?

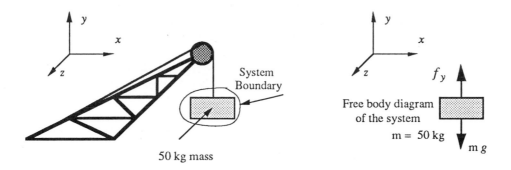

Figure 5-4: Rigid body dynamics

Table 5-7: Data for Example 5-3a (Linear Momentum, System = Mass under crane)

Description	$\left(\dfrac{d\mathbf{p}_{sys}}{dt}\right)$	=	$\sum(\dot{m}\mathbf{v})_{in/out}$	+	$\sum \mathbf{f}_{ext}$		
					$m\mathbf{g}$		$\mathbf{f}_1$
	(N)		(N)		(N)		(N)
x-direction	$50(0)\mathbf{i}$	=	$0\mathbf{i}$	+	$50(0)\mathbf{i}$	+	$f_{1x}\mathbf{i}$
y-direction	$50(1)\mathbf{j}$	=	$0\mathbf{j}$	+	$50(-9.8)\mathbf{j}$	+	$f_{1y}\mathbf{j}$
z-direction	$50(0)\mathbf{k}$	=	$0\mathbf{k}$	+	$50(0)\mathbf{k}$	+	$f_{1z}\mathbf{k}$

SOLUTION: The problem is illustrated in Figure 5-4. The system is the 50 (kg) mass. We will draw a free body diagram and label the forces. The forces on the body include the weight of the mass and force lifting the mass. The weight of the body is the mass of the body times the force per mass due to gravity, 9.8 (m / s^2).

The linear momentum conservation equations in the x-,y-, and z-directions are shown in Table 5-7.

The force exerted by the cable is:

$$f_y = 540 \text{ (kg m / s}^2) = 540 \text{ (N)}$$

The forces in the x-direction and z-direction are zero. The force required to lift the mass and accelerate it upwards at 1 (m / s^2) is 540 Newtons.

For the second part of the question, with the mass rising at a constant velocity, a similar analysis of the conservation of linear momentum is presented in Table 5-8.

The y-direction equation gives, the force in the y-direction as:

$$f_y = 490 \text{ (N)}$$

Again, the forces in the x and z direction are zero. The force required to lift the mass at a constant velocity is equal to the weight of the mass.

Table 5-8: Data for Example 5-3b (Linear Momentum, System = Mass under crane)

Description	$\left(\dfrac{d\mathbf{p}_{sys}}{dt}\right)$	=	$\sum(\dot{m}\mathbf{v})_{in/out}$	+	$\sum \mathbf{f}_{ext}$		
	(N)		(N)		mg (N)		$\mathbf{f}_1$ (N)
x-direction	50(0)$\mathbf{i}$	=	0$\mathbf{i}$	+	50(0)$\mathbf{i}$	+	$f_{1x}\mathbf{i}$
y-direction	50(0)$\mathbf{j}$	=	0$\mathbf{j}$	+	50(-9.8)$\mathbf{j}$	+	$f_{1y}\mathbf{j}$
z-direction	50(0)$\mathbf{k}$	=	0$\mathbf{k}$	+	50(0)$\mathbf{k}$	+	$f_{1z}\mathbf{k}$

5.4 STEADY STATE

The *steady-state* condition for a system means that the system is not changing with time and hence the accumulation within the system is zero for any time period. Thus, for a finite time period, steady state means that for momentum:

$$(\mathbf{p}_{sys})_{end} - (\mathbf{p}_{sys})_{beg} = 0 \qquad (5-12)$$

and at any instant in time:

$$\left(\frac{d\mathbf{p}_{sys}}{dt}\right) = 0, \qquad \text{steady-state} \qquad (5-13)$$

Then, through the conservation of linear momentum, the steady-state assumption requires:

$$0 = \sum(\dot{m}\mathbf{v})_{in/out} + \sum \mathbf{f}_{ext} \qquad (5-14)$$

Note also that steady state implies that $\left(\dfrac{dm_{sys}}{dt}\right) = 0$ in the mass conservation law.

5.5 OTHER SPECIAL CASES

Other special cases are important. For example, rigid body statics addresses closed systems (no mass entering or leaving the system) which are at steady state. Consequently, the resulting momentum conservation equation is

$$\sum \mathbf{f}_{ext} = 0 \qquad (5-15)$$

Example 5-4. A fifty (50) pound mass (lb$_m$) is suspended from two cables as shown in Figure 5-5. Cable 1 makes an angle of 25° and cable 2 makes an angle of 60° with the horizontal. Solve for the forces in the cables.

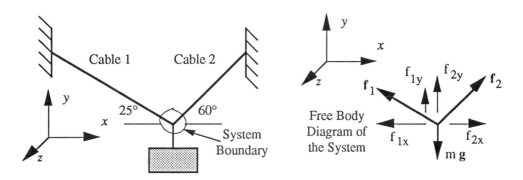

Figure 5-5: Rigid body statics

SOLUTION: Choose the system as the point of contact of the three cables. No linear momentum is entering or leaving the system in the form of mass entering and leaving the system and the system is at steady state. The components of the conservation of linear momentum equation are given in Table 5-9.

Table 5-9: Data for Example 5-4 (Linear Momentum, System = Cable junction)

Description	$\left(\dfrac{d\mathbf{p}_{sys}}{dt}\right)$ (lb$_m$ ft / s^2)	=	$m\mathbf{g}$ (lb$_m$ ft / s^2)		$\mathbf{f}_1$ (lb$_m$ ft / s^2)		$\mathbf{f}_2$ (lb$_m$ ft / s^2)
x-direction	$0\mathbf{i}$	=	$50(0)\mathbf{i}$	+	$f_{1x}\mathbf{i}$	+	$f_{2x}\mathbf{i}$
y-direction	$0\mathbf{j}$	=	$50(-32.174)\mathbf{j}$	+	$f_{1y}\mathbf{j}$	+	$f_{2y}\mathbf{j}$
z-direction	$0\mathbf{k}$	=	$50(0)\mathbf{k}$	+	$f_{1z}\mathbf{k}$	+	$f_{2z}\mathbf{k}$

From the table, you can see that there are two useful independent equations, the x-and y-direction equations. However, you should also notice that there are four (4) unknowns, f_{1x}, f_{1y}, f_{2x}, and f_{2y}. We cannot solve these equations in this form because the number of unknowns does not equal the number of equations. If we can express the components of the force vectors as an unknown magnitude and a known direction, then we can proceed with a solution.

The force $\mathbf{f}_1$ can be expressed as a product of its magnitude and the unit directional vector.

$$\mathbf{f}_1 = |\mathbf{f}_1|\mathbf{u}_1$$

The unit directional vector is calculated using the direction cosines as given in Appendix B. Furthermore, the unit directional vector has a magnitude of one (1) by definition. Now, the force vector $\mathbf{f}_1$ is given as:

$$\mathbf{f}_1 = f_{1x}\mathbf{i} + f_{1y}\mathbf{j} + f_{1z}\mathbf{k}$$

$$= |\mathbf{f}_1|\left(\cos(155°)\mathbf{i} + \cos(65°)\mathbf{j} + \cos(90°)\mathbf{k}\right)$$

The force vector $\mathbf{f}_2$ is expressed as:

$$\mathbf{f}_2 = f_{2x}\mathbf{i} + f_{2y}\mathbf{j} + f_{2z}\mathbf{k}$$

$$= |\mathbf{f}_2|\left(\cos(60°)\mathbf{i} + \cos(-30°)\mathbf{j} + \cos(90°)\mathbf{k}\right)$$

With the forces expressed in this manner, the conservation of linear momentum equations only contain two (2) independent unknowns, the magnitude of the two forces. Table 5-10 shows the components of the conservation of linear momentum equation.

Table 5-10: Intermediate Data for Example 5-4 (Linear Momentum, System = Cable junction)

Description	$\left(\dfrac{d\mathbf{p}_{sys}}{dt}\right)$ $(lb_m \ ft \ / \ s^2)$	$=$	$m\mathbf{g}$ $(lb_m \ ft \ / \ s^2)$		$\mathbf{f}_1$ $(lb_m \ ft \ / \ s^2)$		$\mathbf{f}_2$ $(lb_m \ ft \ / \ s^2)$
x-direction	$0\mathbf{i}$	$=$	$0\mathbf{i}$	$+$	$|\mathbf{f}_1|\cos(155°)\mathbf{i}$	$+$	$|\mathbf{f}_2|\cos(60°)\mathbf{i}$
y-direction	$0\mathbf{j}$	$=$	$50(-32.174)\mathbf{j}$	$+$	$|\mathbf{f}_1|\cos(65°)\mathbf{j}$	$+$	$|\mathbf{f}_2|\cos(-30°)\mathbf{j}$
z-direction	$0\mathbf{k}$	$=$	$0\mathbf{k}$	$+$	$0\mathbf{k}$	$+$	$0\mathbf{k}$

Solving for the two independent unknowns gives the magnitudes of the forces as:

$$|\mathbf{f}_1| = 807.6 \ (lb_m \ ft \ / \ s^2) \qquad |\mathbf{f}_2| = 1,463 \ (lb_m \ ft \ / \ s^2)$$

Because of the conversion factor that between $(lb_m \ ft \ / \ s^2)$ and (lb_f) i.e.,

$$1 \ (lb_f) = 32.174 \ (lb_m \ ft \ / \ s^2)$$

the magnitudes of the force vectors can expressed in units of pounds force as:

$$|\mathbf{f}_1| = 25.1 \ (lb_f) \qquad |\mathbf{f}_2| = 45.5 \ (lb_f)$$

The magnitude of the force vectors are 25.1 (lb_f) and 45.5 (lb_f) respectively.

For fluid statics, we also have no entering or leaving mass and will have a static system and so again our sum of the external forces will be zero. The external forces in this case will be pressure forces acting throughout the fluid.

The dimensions of pressure are force per area. Engineering units are psi ($lb_f \ / \ in^2$), psia (psi, absolute, i.e. relative to a true zero pressure), or psig (psi, relative to, i.e., in excess of, the ambient pressure: psig = psia − ambient pressure). SI units of pressure are N $/ \ m^2$ or pascals (Pa).

Example 5-5. We want to derive the equation that will determine the pressure in a vertical column of static fluid as a function of the fluid density and position.

SOLUTION: To evaluate pressure changes in a static fluid, consider a rectangular element of fluid having dimensions Δx, Δy, Δz, oriented with the faces as shown in Figure 5-6 with respect to the gravitational field.

Opposing faces are defined to be of equal area, an assumption which places no limitation on the end result. We will assume that we can represent a force on each face of this element by an average pressure appropriate for that face. We then consider how pressure changes with position in the fluid. The forces on this element of fluid are the pressure-area product for each face and the weight of the element of fluid. Because the fluid is static, there is no mass entering or leaving the system and the mass of the system is constant so that momentum conservation establishes that the sum of the forces is zero. The force of gravity acts in the negative z-direction; $g_z = -g$.

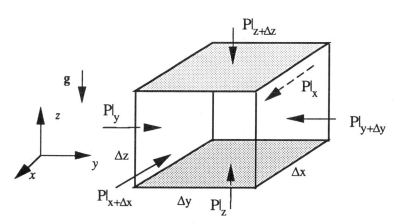

Figure 5-6: Pressure on a fluid element

The linear momentum conservation equations in the x-,y-, and z-directions are given in Table 5-11. Note that the pressure is compressive on each face of the cube, acting towards the surface and perpendicular to it.

Table 5-11: Data for Example 5-5 (Linear Momentum, System = Fluid element)

	$\left(\dfrac{d\mathbf{p}_{sys}}{dt}\right)$	$=$		$\sum \mathbf{f}_{ext}$
Description			$m\mathbf{g}$	(Pressure)(Area)
	[force]		[force]	[force]
x-direction	$0\mathbf{i}$	$=$	$m(0)\mathbf{i}$ +	$(PA)_x\mathbf{i} - (PA)_{x+\Delta x}\mathbf{i}$
y-direction	$0\mathbf{j}$	$=$	$m(0)\mathbf{j}$ +	$(PA)_y\mathbf{j} - (PA)_{y+\Delta y}\mathbf{j}$
z-direction	$0\mathbf{k}$	$=$	$m(g_z)\mathbf{k}$ +	$(PA)_z\mathbf{k} - (PA)_{z+\Delta z}\mathbf{k}$

Hence, in the z-direction the pressure changes according to:

$$0 = (PA)|_z - (PA)|_{z+\Delta z} + mg_z$$

The areas can be calculated from cube dimensions so that

$$0 = \Delta x \Delta y P|_z - \Delta x \Delta y P|_{z+\Delta z} + mg_z$$

Furthermore, for a constant density system, the mass of the system (from Chapter 3) can be represented as the volume of the system times the density of the system.

$$m = \rho V = \rho \Delta x \Delta y \Delta z$$

For the z-component, after the substitutions, we have

$$0 = \Delta x \Delta y P|_z - \Delta x \Delta y P|_{z+\Delta z} + \rho g_z \Delta x \Delta y \Delta z$$

Dividing both sides of the equation by $\Delta x \Delta y$ and rearranging gives:

$$P|_z - P|_{z+\Delta z} = -\rho g_z \Delta z$$

Similarly, for the x and y directions, because $g_x = 0$ and $g_y = 0$,

$$P|_x = P|_{x+\Delta x}$$
$$P|_y = P|_{y+\Delta y}$$

Changes in pressure, then, in the x- and y-direction are zero, whereas in the z-direction, which is aligned with gravity, there must be a change in pressure to accommodate the weight of the fluid. That is, pressure changes in a static fluid because of the weight of the fluid; moving around in a static fluid normal to the gravity field (i.e. without changing depth) does not cause a change in pressure.

   This conclusion holds true for arbitrarily shaped vessels or bodies of fluid as well. As long as we can find a continuous, static path through the fluid, we can calculate the difference in pressure from one point to another as being simply the weight of the fluid (per unit area) which corresponds to the given change in elevation i.e. for a constant density fluid, $\rho g \Delta z$. Engineers frequently talk about the "head" of fluid or equivalent height of fluid that would create a certain pressure or pressure change.

   Now, we can divide the equation by Δz yielding:

$$\left(\frac{P|_z - P|_{z+\Delta z}}{\Delta z} \right) = -\rho g_z$$

As the limit of Δz goes to zero:

$$\lim_{\Delta z \to 0} \left(\frac{P|_z - P|_{z+\Delta z}}{\Delta z} \right) = -\rho g_z$$

and from the definition of a derivative, and by recognizing that $g_z = -g$, we have

$$\left(\frac{dP}{dz} \right) = \rho g_z = -\rho g \qquad\qquad (5-16)$$

The change in pressure can be determined for a tank of fluid as a function of the height. Notice that the cross-sectional area of the fluid column does not appear in the equation. Also, the pressure increases with decreasing z, i.e., with increasing depth; the farther we decrease in elevation into a fluid the greater the pressure.

   Example 5-6. Using the equation derived in the previous example, determine the pressure in the tank (Figure 5-7) if (a) the density is constant and (b) if the density is a function of position z. For part (a), the fluid will be water and the density is:

$$\rho_{H_2O} = 62.4 \ (\text{lb}_\text{m} / \text{ft}^3)$$

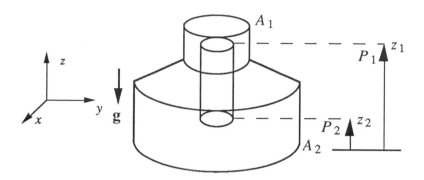

Figure 5-7: Pressure in a vertical column

In part (b), the fluid density changes with position described by the following equation

$$\rho = \frac{5}{z} \ (\text{lb}_\text{m} \ / \ \text{ft}^3)$$

where the variable z has units of feet and the constant 5 has units of ($\text{lb}_\text{m} \ / \ \text{ft}^2$). From the reference point, $z_1 = 50$ (ft) and $z_2 = 30$ (ft). The cross sectional area $A_1 = 4$ (ft^2) and $A_2 = 8$ (ft^2). Pressure at point 1, (P_1) is equal to 1 (atm). Determine the pressure at point 2 (P_2) for the two different fluids.

SOLUTION: To solve this problem we must rearrange and integrate equation (5-16). Because the equation was derived using only the conservation of linear momentum with no assumptions about how the density varies, the equation is completely valid. Rearranging equation (5-16) yields

$$dP = -\rho g dz$$

Integrating the equation from P_1 to P_2 and z_1 to z_2 produces:

$$\int_{P_1}^{P_2} dP = -\int_{z_1}^{z_2} \rho g dz$$

Evaluating the left side of the integral and assuming that g remains constant gives:

$$P_2 - P_1 = -g \int_{z_1}^{z_2} \rho dz$$

a) For the first case, where the density is constant, the right hand side of the integral is evaluated to give

$$P_2 - P_1 = -\rho g(z_2 - z_1)$$

and rearranging yields P_2:

$$P_2 = P_1 - \rho g(z_2 - z_1)$$

In this uniform-density case, the pressure increases linearly with increasing depth. Now we can substitute the numerical values into the equation:

$$P_2 = 1 \ (\text{atm}) - 62.4 \left(\frac{\text{lb}_\text{m}}{\text{ft}^3}\right) \times 32.174 \left(\frac{\text{ft}}{\text{s}^2}\right) \times (30 \ \text{ft} - 50 \ \text{ft})$$

Evaluation yields:

$$P_2 = 1 \text{ (atm)} + 40,153 \left(\frac{\text{lb}_m \text{ ft}^2}{\text{ft}^3 \text{ s}^2} \right)$$

Now we must convert to the proper units:

$$P_2 = 1 \text{ (atm)} + \frac{40,153 \text{ lb}_m \text{ ft}^2}{\text{ft}^3 \text{ s}^2} \left| \frac{1 \text{ lb}_f \text{ s}^2}{32.174 \text{ lb}_m \text{ ft}} \right| \frac{1 \text{ ft}^2}{144 \text{ in}^2} \left| \frac{1 \text{ atm}}{14.7 \text{ psi}} \right.$$

$$= 1.6 \text{ (atm)}$$

b) For the second case, where the density of the fluid is a function of the position, the function must be substituted into the equation before integration:

$$P_2 - P_1 = -g \int_{z_1}^{z_2} \rho \, dz$$

$$= -g \int_{z_1}^{z_2} \left(\frac{5}{z} \right) dz$$

$$= -5g \ln(z) \Big|_{z_1}^{z_2}$$

$$= -5g \ln \left(\frac{z_2}{z_1} \right)$$

Rearranging:

$$P_2 = P_1 - 5g \ln \left(\frac{z_2}{z_1} \right)$$

Substituting the numerical values:

$$P_2 = 1 \text{ (atm)} - 5 \left(\frac{\text{lb}_m}{\text{ft}^2} \right) \times 32.174 \left(\frac{\text{ft}}{\text{sec}^2} \right) \times \ln \left(\frac{30 \text{ ft}}{50 \text{ ft}} \right)$$

$$= 1 \text{ (atm)} + \frac{82.18 \text{ lb}_m \text{ ft}}{\text{ft}^2 \text{ s}^2} \left| \frac{1 \text{ lb}_f \text{ s}^2}{32.174 \text{ lb}_m \text{ ft}^2} \right| \frac{1 \text{ ft}^2}{144 \text{ in}^2} \left| \frac{1 \text{ atm}}{14.7 \text{ psi}} \right.$$

$$= 1.001 \text{ (atm)}$$

A comparison of the two answers is given below.

$$\text{Case a)} \qquad P_2 = 1.6 \text{ (atm)}$$

$$\text{Case b)} \qquad P_2 = 1.001 \text{ (atm)}$$

In fluid dynamics one application of momentum conservation is to calculate the forces required to support fluid flowing through a pipe. The momentum entering and leaving a pipe must be balanced by the forces exerted by the support members on the pipe.

Example 5-7. A stream of liquid is flowing through a section of stationary tubing or hose as shown in Figure 5-8. If the linear momentum of the streams entering and leaving the section of the hose are given by $(\dot{m}v)_{in} = 25\mathbf{i} + 0\mathbf{j} + 0\mathbf{k}$ (lb$_m$ ft / s^2) and $(\dot{m}v)_{out} = 15\mathbf{i} + 20\mathbf{j} + 0\mathbf{k}$ (lb$_m$ ft / s^2) respectively, calculate the force **f** associated with this change in momentum.

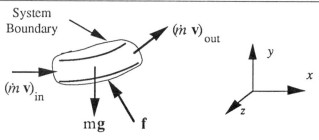

Figure 5-8: Fluid dynamics

SOLUTION: Choose the system to be the section of stationary tubing. The conservation of linear momentum for this system is:

$$\left(\frac{d\mathbf{p}_{sys}}{dt}\right) = \sum(\dot{m}\mathbf{v})_{in/out} + \sum \mathbf{f}_{ext}$$

Furthermore, this system is at steady state with only one inlet and one outlet so that $d\mathbf{p}_{sys} / dt = 0$. The vector equation is presented in Table 5-12 to organize the three coordinate directions. In the table, the force due to gravity, the weight of the mass in the system, also is included. Although the mass of the system is not stated in the problem, we should recognize that it must still be included in the analysis.

Table 5-12: Data for Example 5-7 (Linear Momentum, System = Section of tubing)

Description	$\left(\dfrac{d\mathbf{p}_{sys}}{dt}\right)$ (lb$_m$ ft / s^2)	=	$\sum(\dot{m}\mathbf{v})_{in/out}$			+	$\sum \mathbf{f}_{ext}$		
			$(\dot{m}v)_{in}$ (lb$_m$ ft / s^2)		$(\dot{m}v)_{out}$ (lb$_m$ ft / s^2)		**f** (lb$_m$ ft / s^2)		$m_{sys}\mathbf{g}$ (lb$_m$ ft / s^2)
x-direction	$m(0)\mathbf{i}$	=	$25\mathbf{i}$	−	$15\mathbf{i}$	+	$f_x\mathbf{i}$	+	$0\mathbf{i}$
y-direction	$m(0)\mathbf{j}$	=	$0\mathbf{j}$	−	$20\mathbf{j}$	+	$f_y\mathbf{j}$	−	$m_{sys}g\mathbf{j}$
z-direction	$m(0)\mathbf{k}$	=	$0\mathbf{k}$	−	$0\mathbf{k}$	+	$f_z\mathbf{k}$	+	$0\mathbf{k}$

The components of the force vector are solved directly from the table:

$$f_x = -10 \ (\text{lb}_m \text{ ft } / \text{ s}^2)$$
$$f_y = 20 + m_{sys}g \ (\text{lb}_m \text{ ft } / \text{ s}^2)$$
$$f_z = 0 \ (\text{lb}_m \text{ ft } / \text{ s}^2)$$

In vector notation this is

$$\mathbf{f} = -10\mathbf{i} + (20 + m_{sys}g)\mathbf{j} + 0\mathbf{k} \ (\text{lb}_m \text{ ft } / \text{ s}^2)$$

In (lb$_f$) units, the force on the system required to change the momentum of the mass entering the system and to support the system weight is:

$$\mathbf{f} = -0.31\mathbf{i} + \left(0.62 + \frac{m_{sys}g}{g_c}\right)\mathbf{j} + 0\mathbf{k} \ (\text{lb}_f)$$

It should be noted that this force is the total of all forces applied to the system except gravity. Thus it includes 1) pressure acting on the flow cross section at the entrance and exit of the system, 2) forces acting through the pipe wall at the system boundary, and 3) any external support provided to the pipe.

Note again that conservation of momentum is a vector equation. As such it consists of three independent component equations; conservation of linear momentum provides three independent scalar equations.

5.6 REFERENCE FRAME

One more point needs to be made about the conservation of momentum. In order to define a velocity, and hence momentum, we must first define a coordinate system. Position and velocity are then measured with respect to the origin of this system.

It turns out that as far as satisfying physical laws is concerned, the exact reference frame which is chosen is immaterial, so long as it is an *inertial frame*, i.e., it is nonacelerating (constant velocity and non-rotating).

Consider momentum conservation equation (5-1). The velocities (and acceleration) are measured with respect to a specific coordinate system. If we wish to express this law in terms of a coordinate system which is moving at (constant) velocity v_{ref} with respect to the first coordinate system, then each velocity v, for example is replaced by $(v - v_{ref})$. Then equation (5-1) would be expresssed as:

$$\frac{d}{dt}\left(m_{sys}(v_{sys} - v_{ref})\right) = \sum(v_{in/out} - v_{ref})\dot{m}_{in/out} + \sum f_{ext} \qquad (5-17)$$

These velocity vectors are exactly the same vectors that would have been defined had the second reference frame been used in the first place.

Now compare this result to equation (5-1). Expanding equation (5-17) and collecting terms appropriately gives:

$$\left(\frac{d(mv)_{sys}}{dt}\right) - \left\{\frac{d}{dt}(m_{sys}v_{ref}) + v_{ref}(\sum \dot{m}_{in/out})\right\} = \sum(\dot{m}v)_{in/out} + \sum f_{ext}$$

Now v_{ref} is constant so

$$\left(\frac{d(mv)_{sys}}{dt}\right) - v_{ref}\left\{\left(\frac{dm_{sys}}{dt}\right) - \sum \dot{m}_{in/out}\right\} = \sum(\dot{m}v)_{in/out} + \sum f_{ext}$$

and the term in braces is zero through mass conservation. Consequently, equation (5-17), which is expressed in terms of the second reference frame is exactly the same as equation (5-1) which is expressed in terms of the first reference frame so long as:

(1) v_{ref} the velocity of the second frame with respect to the first is constant

(2) mass is conserved, and

(3) the external forces which act on a body are independent of the (inertial) reference frame, i.e., they are *frame indifferent* In fact, this last statement is the crux of Newton's Second Law of Motion, i.e., of the conservation of linear momentum. The physical observation that leads to the notion of momentum conservation is that forces are, in fact, inertial-frame indifferent

5.7 REVIEW

1) The total momentum of a system plus its surroundings is conserved, that is, it is constant. Equivalently, there is no generation or consumption of momentum within a given system.

2) Momentum may enter or leave a system by virtue of mass entering or leaving a system; as mass enters or leaves, the momentum of that mass enters or leaves. Momentum may also be transferred between the system and surroundings as a result of mutual forces which act between the system and surroundings. Because the forces and motions are mutual (opposite and equal), the momentum lost by the surroundings is gained by the system and vice versa in accordance with true conservation of momentum.

3) The linear momentum conservation equation includes momentum transferred between the system and its surroundings by virtue of mass crossing the system boundary and by virtue of forces which act upon the system. The equation can be written for a differential time period as a rate equation, or it can be integrated over time to give a finite time period integral equation. The rate of momentum entering or leaving the system with mass is calculated as momentum per unit mass times the rate at which mass enters or leaves the system.

4) The equation of conservation of linear momentum may be thought of as an equation of motion in that it is used to describe the motion (or lack thereof) of a system. The steady-state equation is the equation used for analyzing nonaccelerating systems and the unsteady state equation is used for accelerating systems.

5) For rigid bodies, the conservation of linear momentum concept contains the familiar three laws of Motion of Newton. For fluid systems, the momentum of the system may change as a result of change in mass of the system as well as a result of a change in its velocity.

5.8 NOTATION

The following notation was used for this chapter. The boldface lowercase letters denote vector quantities.

Scalar Variables and Descriptions		Dimensions
A	cross sectional area	[length^2]
f	magnitude of the vector $\mathbf{f}$	[mass length / time^2]
g	magnitude of the acceleration of gravity	[length / time^2]
m	total mass	[mass]
$\dot{m}$	total mass flow rate	[mass / time]
P	pressure	[mass / length time^2] or [force / length^2]
t	time	[time]
v_x, v_y, v_z	components of velocity vector	[length / time]
x, y, z	spatial rectangular cartesian variables	[length]
ρ	density	[mass / length^3]

Vector Variables and Descriptions		Dimensions
a	acceleration	[length / time^2]
f	force	[mass length / time^2] or [force]
i, j, k	cartesian unit directional vectors	
p	linear momentum	[mass length / time] or [momentum]
r	position	[length]
u	general unit directional vector	
v	velocity	[length / time] or [momentum / mass]

Subscripts

beg	evaluated at the beginning of the time period
end	evaluated at the end of the time period
ext	external to the system or acting at the system boundary
G	corresponding to the center of mass of the system
in	input or entering
out	output or leaving
ref	corresponding to the defined coordinate system
sys	system or within the system boundary

5.9 SUGGESTED READING

Beer, F. P., and E. R. Johnston, Jr., *Mechanics for Engineers, Statics and Dynamics*, 4th edition, McGraw-Hill, New York, 1987

Halliday, D. and R. Resnick, *Physics for Students of Science and Engineering, Part I*, John Wiley & Sons, 1960

Hibbeler, R. C., *Engineering Mechanics, Statics and Dynamics*, 3rd edition, MacMillan, New York, 1983

Sears, F. W., M. W. Zemansky, and H. D. Young, *University Physics*, Addison-Wesley, Reading, Massachussetts, 1982

QUESTIONS

1) In what ways can momentum be exchanged between a system and surroundings?

2) Why are forces an exchange of momentum (rate) rather than a generation of momentum?

3) Give a verbal statement of the conservation of linear momentum.

4) Write the equation of conservation of linear momentum.

5) There is not necessarily any unique or single value of the momentum of a finite-size system. Explain.

6) In light of the previous question, how do we define the momentum of a finite-size body?

7) How is the velocity of the body defined so as to be consistent with the previous question?

8) What is the linear momentum per unit mass of a particle?

9) What is the physical meaning of force in the context of the conservation of linear momentum law? (Hint: What are the dimensions of force expressed in terms of momentum?)

10) Discuss the motion of a baseball as it is hit with a baseball bat in the context of the conservation of linear momentum.

11) Discuss the motion of a rocket launched from Cape Canaveral in the context of the conservation of linear momentum.

12) Discuss the motion of the moon in the context of the conservation of linear momentum.

13) List and describe or explain the kinds of forces which may be exerted on a rigid body from the surroundings.

14) List 10 situations or problems which may be analyzed using the conservation of linear momentum.

SCALES

1) For the three vectors $\mathbf{a} = 2\mathbf{i} + 3\mathbf{j} - \mathbf{k}$, $\mathbf{b} = -3\mathbf{i} + 2\mathbf{k}$, $\mathbf{c} = \mathbf{i} - \mathbf{j} - \mathbf{k}$:

    a) determine the three unique dot products between the three vectors,
    b) determine the six unique cross products between the three vectors,
    c) determine the two unique triple scalar products,
    d) determine the magnitudes of $\mathbf{a}$ and $\mathbf{b}$, and
    e) determine the angle between vectors $\mathbf{a}$ and $\mathbf{b}$.

2) For the points A and B having (x, y, z) coordinates (2, 1, 3) and (-1, 3, -4), respectively, determine:

    a) the position vectors from the origin to A, and to B
    b) the vector representing the position of B relative to A
    c) a unit vector pointing from A to B
    d) a vector having magnitude 2 and pointing from A to B.

3) Write the vector $\mathbf{f} = 2\mathbf{i} + 3\mathbf{j} - \mathbf{k}$ in terms of its magnitude and an appropriate unit vector.

4) Resolve the vector $\mathbf{f} = 2\mathbf{i} + 3\mathbf{j} - \mathbf{k}$ into two vectors, the first which is in the direction of the vector $\mathbf{i} + \mathbf{j}$ and the second vector which is perpendicular to the first. (Hint: Convert $\mathbf{i} + \mathbf{j}$ to a unit vector.)

5) Suppose you have a vector which acts along a line between the two points A (which is at $(x, y, z) = (1, 2, 4)$) and B (which is at (0, -3, -2). Express this vector in terms of its magnitude (say $\mathbf{F}$) and an appropriate unit vector.

6) Given the x-coordinate of a particle as a function of time. In general, what is the relation between $x(t)$, $v_x(t)$ (the x-component of velocity), and $a_x(t)$ (the x-component of acceleration)? If $x(t) = 2t - 3t^2 + 5\ln(t)$, then what are expressions for v_x and a_x.

7) What are the units of force (a) in SI units and (b) in engineering units?

8) What are the dimensions of pressure? What are the units of pressure in (a) SI, (b) engineering units?

9) What is the difference between gauge and absolute pressure?

10) What is the value of $\mathbf{g}$ in (a) SI units and (b) engineering units (give two cases: mass in units of lbm and mass in units of slugs)?

11) What is the value of g_c in engineering units for mass in units of lbm?

12) Sketch a free-body diagram for:

   a) a roller coaster going over a hill
   b) a roller coaster going around a sharp turn
   c) an Indy 500 car on a turn of the track
   d) a bucket of water being swung in a vertical circle, at the top of the circle
   e) a bucket of water being swung in a vertical circle, at the bottom of the circle
   f) a skydiver in free fall
   g) a bungee jumper after the cord is stretched

In each case, indicate on your diagram momentum input and output and the momentum of the body (system). Remember that at each point of contact of the body with its surroundings, a force may be present and should, in general, be included on the free-body diagram.

13) A baseball hit by a bat, under the assumption of no air resistance, may have a trajectory given by:

$$x(t) = 132t$$
$$y(t) = 2 + 95.2t - 16.1t^2$$

where x is the horizontal position of the ball from home plate and y is the vertical distance above the ground, both in feet. Plot this trajectory, i.e., make a plot of y versus x. Use a computer spreadsheet/plotting package.

14) A baseball hit by a bat, under the assumption that air resistance is proportional to the velocity, may have a trajectory given by:

$$x(t) = 426.5(1 - e^{-0.31t})$$
$$y(t) = 641.8(1 - e^{-0.31t}) - 103.8t + 2$$

where x is the horizontal position of the ball from home plate and y is the vertical distance above the ground, both in feet. Plot this trajectory, i.e., make a plot of y versus x.

PROBLEMS

1) For the following velocity vectors,

$$\mathbf{a} = 5.2\mathbf{i} + 3.7\mathbf{j} + 6.1\mathbf{k} \text{ (ft / s)}$$
$$\mathbf{b} = 3.7\mathbf{i} - 6.1\mathbf{j} - 5.2\mathbf{k} \text{ (ft / s)}$$
$$\mathbf{c} = 2.5\mathbf{i} + 7.3\mathbf{j} + 1.6\mathbf{k} \text{ (m / s)}$$
$$\mathbf{d} = 1.6\mathbf{i} - 2.5\mathbf{j} - 7.3\mathbf{k} \text{ (m / s)}$$

calculate the following vectors:

(a) $\mathbf{u}_a = \mathbf{a} / |\mathbf{a}|$ (e) $\mathbf{e} = \mathbf{a} + \mathbf{b}$ (i) $\mathbf{i} = \mathbf{a} - \mathbf{b}$

(b) $\mathbf{u}_b = \mathbf{b} / |\mathbf{b}|$ (f) $\mathbf{f} = \mathbf{c} + \mathbf{d}$ (j) $\mathbf{j} = \mathbf{c} - \mathbf{d}$

(c) $\mathbf{u}_c = \mathbf{c} / |\mathbf{c}|$ (g) $\mathbf{g} = \mathbf{a} + \mathbf{c}$ (ft / s)   (k) $\mathbf{u}_e = \mathbf{e} / |\mathbf{e}|$

(d) $\mathbf{u}_d = \mathbf{d} / |\mathbf{d}|$ (h) $\mathbf{h} = \mathbf{b} + \mathbf{d}$ (m / s)   (l) $\mathbf{u}_h = \mathbf{h} / |\mathbf{h}|$

2) For the following force vectors,

$$f_1 = 5i + 0j - 1k \ (N)$$

$$f_2 = 3i - 2j + 5k \ (N)$$

$$f_3 = 1i + 4j + 0k \ (lb_f)$$

$$f_4 = 7i - 0j + 0k \ (lb_f)$$

calculate the following vectors:

(a) $u_1 = f_1 \ / \ |f_1|$ (e) $f_5 = f_1 + f_2$ (i) $f_9 = \sum_{i=1}^{4} f_i \ (N)$

(b) $u_2 = f_2 \ / \ |f_2|$ (f) $f_6 = f_3 + f_4$ (j) $f_{10} = f_1 - f_2$

(c) $u_3 = f_3 \ / \ |f_3|$ (g) $f_7 = f_1 + f_3 \ (N)$ (k) $u_5 = f_5 \ / \ |f_5|$

(d) $u_4 = f_4 \ / \ |f_4|$ (h) $f_8 = f_2 + f_4 \ (lb_f)$ (l) $u_9 = f_9 \ / \ |f_9|$

3) A system has three masses, having the velocities given below:

$$m_1 = 10 \ (kg) \qquad v_1 = 5i + 2j + 3k \ (m \ / \ s)$$

$$m_2 = 5 \ (kg) \qquad v_2 = -2i + 3j - 4k \ (m \ / \ s)$$

$$m_3 = 15 \ (kg) \qquad v_3 = -1i - 2j + 2k \ (m \ / \ s)$$

a) Calculate the momentum of the system and the velocity of the center of mass for this system. b) If the third mass m_3 leaves the system and the velocities of m_1 and m_2 remain unchanged, calculate the momentum of the system and the velocity of the center of mass.

4) From figure Problem 5-4, determine the force in the member AB, BC and BD in the static structure. A 10 (kg) block is suspended below point B. Assume that the beams are massless and the force in each member is collinear with it.

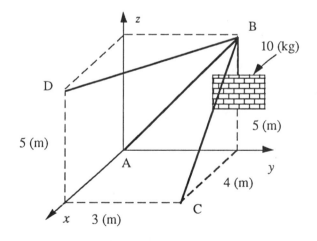

Problem 5-4: Three-dimensional static structure

5) A car with mass of 2000 (kg) is traveling at 88 (km/hr) in the x-direction. If a truck with mass of 6000 (kg) and going 44 (km/hr) in the y-direction impacts the car, calculate the velocity vectors of the car and truck after the collision. Assume that no other forces act on the car in the x- and y-directions and that the collision is completely inelastic, i.e., that the car and truck are stuck together after the collision.

6) An airplane is flying at 30,000 ft above sea level at a constant velocity of 600 (mph) in the horizontal, x-direction. If the plane has a mass of 4,000 (lb$_m$) and the magnitude of gravity at 30,000 ft is 32.095 (ft $/$ s^2), use the conservation of linear momentum to determine the lift on the wings. Furthermore, if the drag on the plane is proportional to the square of the velocity of the plane and acts in the opposite direction as given in the following equation:

$$\mathbf{f}_{drag} = -kv^2\mathbf{i}$$

where $k = 0.022$ (lb$_f$s^2 $/$ ft^2), calculate the thrust (lb$_f$) of the engines to keep the plane flying at constant velocity.

7) A particle with mass of 2 kg has an initial momentum ($t = 0$) given by $\mathbf{p}|_{t=0} = 3\mathbf{i} + 5\mathbf{j} + 6\mathbf{k}$ kg m $/$ s. A constant force vector $\mathbf{f} = 2\mathbf{i} + 4\mathbf{j} + 3\mathbf{k}$ (N) acts on the particle. Determine the initial velocity vector of the particle. Use conservation and accounting to determine the velocity of the particle after 9 seconds.

8) A stationary particle with mass of 3 kg is acted on by three (3) different forces $\mathbf{f}_1$, $\mathbf{f}_2$, and $\mathbf{f}_3$. Two of the forces are known $\mathbf{f}_1 = 5\mathbf{i} + 2\mathbf{j} + 1\mathbf{k}$ N, and $\mathbf{f}_2 = -3\mathbf{i} - 4\mathbf{j} + 5\mathbf{k}$ N. Gravity is not acting on the particle. Use conservation and accounting to determine the unknown force vector acting on the particle.

9) A static body with mass of 14 (lb$_m$) is acted on by four (4) different forces $\mathbf{f}_1$, $\mathbf{f}_2$, $\mathbf{f}_3$, and $\mathbf{f}_4$. Three of the forces are known $\mathbf{f}_1 = 10\mathbf{i} - 13\mathbf{j} - 9\mathbf{k}$ lb$_f$, $\mathbf{f}_2 = -15\mathbf{i} + 5\mathbf{j} - 3\mathbf{k}$ lb$_f$, and $\mathbf{f}_3 = -4\mathbf{i} + 7\mathbf{j} + 7\mathbf{k}$ lb$_f$. Gravity is not acting on the body. Use conservation and accounting to determine $\mathbf{f}_4$.

10) A ball with mass 10 (lb$_m$) is subjected to a constant force of unknown magnitude and direction for 5 (s). The initial velocity of the ball is given by the following vector:
$$\mathbf{v}_{beg} = 5\mathbf{i} + 6\mathbf{j} + 8\mathbf{k} \ (\text{ft} / \text{s})$$

The final velocity is:
$$\mathbf{v}_{end} = -2\mathbf{i} + 3\mathbf{j} - 10\mathbf{k} \ (\text{ft} / \text{s})$$

If there are no other forces acting on the ball, use the conservation of linear momentum to determine the unknown force vector.

11) A pitcher hurls a baseball at 90 (mph) in the horizontal direction at the batter. After being struck by the bat, the ball's speed is 110 (mph) in the opposite direction at a 35° incline from the horizontal. The ball has a mass of 5-1/8 ounces. If the collision only lasts 1/1000 of a second, calculate the average force on the ball during the time period. What is the average force exerted by the ball on the bat during the time period? If the batter and the ball are defined as the system, other than the force of gravity, are there any other forces acting on the system? (Remember, force is a vector quantity and linear momentum must be conserved)

12) As shown in the figure, hockey puck B rests on a frictionless, smooth ice surface and is struck by puck A. The mass of a puck is 7 (oz). The velocity vector of puck A before the collision is:

$$\mathbf{v}_A = 45\mathbf{i} + 0\mathbf{j} + 0\mathbf{k} \ (\text{mph})$$

If the collision takes 1/100 of a second and the force on puck B from puck A during the impact is given by the following function of time:

$$\mathbf{f}_{AB} = 45\sin\left(\frac{100\pi}{2}t\right)\mathbf{i} - 45\sin^2\left(\frac{100\pi}{2}t\right)\mathbf{j} + 0\mathbf{k} \ (\text{lb}_f)$$

Determine the final velocity vector of puck A and the final velocity vector of puck B. What is the average force exerted on puck A from puck B during the impact?

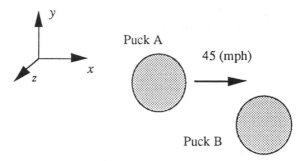

Problem 5-12: Two hockey pucks

13) Consider figure Problem 5-13 in which the pulleys are frictionless and massless, and the surfaces are frictionless. For the system defined as mass 1, discuss the conservation of linear momentum, remembering to include all 3 vector components. For the system defined as mass 2, discuss the conservation of linear momentum, including all 3 components. If the system is defined as both the masses together, what is the resultant conservation of linear momentum equation, including all 3 components?

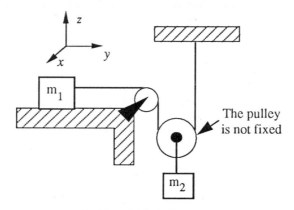

Problem 5-13: Pulley system

14) A particle has an initial momentum ($t = 0$) given by the following vector.

$$\mathbf{p}(t_o) = 200\mathbf{i} + 300\mathbf{j} + 400\mathbf{k} \ (\text{lb}_\text{m} \ \text{ft} / \text{s})$$

A variable force acts on the particle over a time period given by the data in the table. Determine the momentum vector of the particle at 4 (s).

Data for Problem 5-7

t (s)	f_x (lb$_\text{f}$)	f_y (lb$_\text{f}$)	f_z (lb$_\text{f}$)
0	0	−2	10
1	1	−2	8
2	2	−2	6
3	3	−2	4
4	4	−2	2
5	5	−2	0

15) A small ball of steel with mass 2 (kg) is dropped from a helicopter hovering at 2000 meters above sea level. After releasing the ball, it starts to fall to the earth. If the force due to air resistance is proportional to the velocity of the ball and acts in the opposite direction, i.e.,

$$\mathbf{f} = -k\mathbf{v}$$

where $k = 0.3$ (kg/s), then

    a) determine the velocity of the ball as a function of time and sketch the function,

    b) determine the maximum or terminal velocity, and

    c) what is the air resistance on the ball when this terminal velocity is reached?

16) Liquid ethanol at room temperature is flowing through a $90°$ horizontal elbow at a mass flow rate of 2 (lb_m /s). The inside diameter of the entrance of the elbow is 4 (in) and the inside diameter of the exit of the elbow is 2 (in). The inside volume of the elbow is 32.2 (in³). The pressure at the entrance is 100 (psi) and the pressure at the exit is 99.98 (psi), nearly the same as the entrance.

    a) If the flow is steady state, use the conservation of mass to calculate the average linear velocity of the entering and leaving ethanol. (Remember velocity is a vector quantity.)

    b) Calculate the mass of enthanol in the inside volume of the $90°$ elbow.

    c) Define the system to be the ethanol inside the $90°$ elbow and use the conservation of linear momentum to calculate the force that acts on this fluid to change its direction. (Remember to include the momentum that enters and leaves with mass and the pressure forces on the entrance and exit.) This force must be provided by the pipe and its support.

17) The block at point A in figure Problem 5-17 is 10 (kg) and the total length of the rope is 8 (m). If the system is at steady state (equilibrium), and the length of the rope between point B and the block at A is 2 (m), what is the mass of the block at point C? Assume that the pulleys are massless and frictionless and neglect the mass of the rope. Hence, the tension in the rope is everywhere the same.

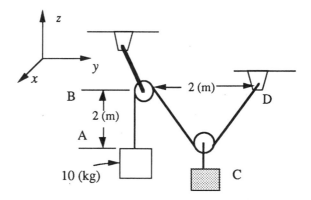

Problem 5-17: Pulley system

18) A particle with mass of 5 lb_m has initial momentum ($t = 0$) $\mathbf{p}|_{t=0} = 50\mathbf{i} + 75\mathbf{j} - 83\mathbf{k}$ lb_m ft / s. Two (2) constant force vectors $\mathbf{f}_1$ and $\mathbf{f}_2$ are acting on the particle $\mathbf{f}_1 = 2\mathbf{i} - 5\mathbf{j} - 3\mathbf{k}$ lb_f, and $\mathbf{f}_2 = -2\mathbf{i} + 3\mathbf{j} + 3\mathbf{k}$ (lb_f). Determine the initial velocity vector of the particle. Use conservation and accounting to determine the velocity of the particle after 12 seconds.

19) For the following static situations shown in figures a) through e), use the conservation of linear momentum to calculate the unkown force. All members shown are rigid and all connections are pin connections.

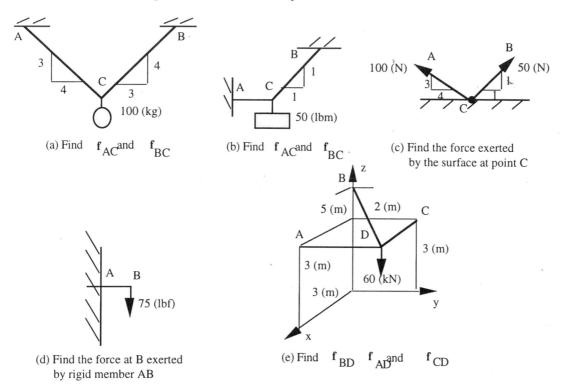

(a) Find f_{AC} and f_{BC}

(b) Find f_{AC} and f_{BC}

(c) Find the force exerted by the surface at point C

(d) Find the force at B exerted by rigid member AB

(e) Find f_{BD} f_{AD} and f_{CD}

20) A 10 (kg) bucket which has the capacity to hold 10 liters is suspended by a rope. The bucket is initially empty.

a) Determine the force that the rope exerts on the bucket to suspend the bucket.

b) Water is added to the bucket at a mass flow rate of 1 (kg/s) and the water enters the bucket at a linear velocity of 5 (m/s). Determine the mass in the bucket as a function of time and the time required to fill the bucket.

c) Now, if the rope can only support a force of 150 (N) determine at what time the rope breaks and the amount of water in the bucket at the time the rope breaks.

CHAPTER
SIX

THE CONSERVATION OF ANGULAR MOMENTUM

6.1 INTRODUCTION

In this chapter we consider the conservation of angular momentum. The angular momentum of a particle is the cross product of the particle's position vector and its momentum. As such, angular momentum is an extensive property, proportional to the mass of the system and it is a 3-dimensional vector quantity requiring three (3) scalar components for complete definition.

The angular momentum of rigid bodies is the primary concern of this chapter. The equations and mathematical representations will be developed using vector notation as a powerful means of expressing the concepts and solving problems. In addition to the vector operations of the preceeding chapter, the cross product is necessary for representing angular momentum problems. An approach to the analysis of *general motion* is followed, where mathematically feasible (through section 6.3), for the sake of providing a complete picture of angular momentum and the impact of this important law. However, for rigid body motion (Sections 6.4 and 6.5), the discussion is narrowed to planar motion for the sake of providing a mathematically palatable picture. This picture, while less comprehensive than is needed for the most general applications, is certainly sufficient for first-time engineering applications and still follows a line of reasoning that is readily extended to general motion.

Example applications of the conservation of angular momentum to engineering calculations are the determination of torque on rotating bodies, and static and dynamic analysis of structures and machines. The mathematics are more complex for angular momentum equations than linear momentum and many of the concepts such as moment of inertia and torque are not as familiar as mass and force.

A verbal statement of the conservation of angular momentum (AM) is given below. Because angular momentum is a conserved property, there is no generation or consumption and hence

$$\left\{\begin{array}{c} \text{Accumulation of AM} \\ \text{within system during} \\ \text{time period} \end{array}\right\} = \left\{\begin{array}{c} \text{AM entering} \\ \text{system during} \\ \text{time period} \end{array}\right\} - \left\{\begin{array}{c} \text{AM leaving} \\ \text{system during} \\ \text{time period} \end{array}\right\}$$

where the accumulation of angular momentum is defined as follows:

$$\left\{\begin{array}{c} \text{Accumulation of AM} \\ \text{within system} \\ \text{during time period} \end{array}\right\} \equiv \left\{\begin{array}{c} \text{Amount of AM} \\ \text{contained in system} \\ \text{at end of} \\ \text{time period} \end{array}\right\} - \left\{\begin{array}{c} \text{Amount of AM} \\ \text{contained in system} \\ \text{at beginning of} \\ \text{time period} \end{array}\right\}$$

Before proceeding further, it is essential to understand the ways in which angular momentum may enter the system.

6.1.1 Angular Momentum Possessed by Mass

First, of course, angular momentum may enter with mass. Mass enters with a certain velocity (and hence angular momentum) and when it crosses the system boundary, that much angular momentum is added to the system. The rate at which angular momentum enters the system at position $\mathbf{r}_{\text{in}}$ due to mass traveling at velocity $\mathbf{v}_{\text{in}}$ and entering the system at rate $\dot{m}_{\text{in}}$ is:

$$(\mathbf{r} \times \mathbf{v}\dot{m})_{\text{in}}$$

Likewise, the rate of momentum leaving is:

$$(\mathbf{r} \times \mathbf{v}\dot{m})_{\text{out}}$$

Note that the correct way to calculate the rate at which angular momentum enters (or leaves) the system with mass is the angular momentum (per unit mass) possessed by that mass multiplied by the rate at which mass enters (or leaves) the system

$$\frac{\text{angular momentum}}{\text{mass}} \left| \frac{\text{mass}}{\text{time}} \right. = \left(\frac{\text{angular momentum}}{\text{time}}\right)$$

Where there are multiple entering (p, say) and leaving (q, say) streams having different velocities, simply add them together

$$\sum_{i=1}^{p} \left((\mathbf{r} \times \mathbf{v}\dot{m})_i\right)_{\text{in}}$$

and

$$\sum_{i=1}^{q} \left((\mathbf{r} \times \mathbf{v}\dot{m})_i\right)_{\text{out}}$$

6.1.2 Angular Momentum in Transit: Torque (Moment)

Second, angular momentum enters as the result of forces which act upon the system mass by its surroundings; an external force $\mathbf{f}$ acting on a system exerts a torque (twisting force or *moment*) about a point due to the moment arm (distance from the point to the line of action of the force), thereby providing angular momentum to the system at rate $\mathbf{r} \times \mathbf{f}$, where $\mathbf{r}$ is the position vector from the point of interest to the point of application of the force. Note that the torque is a vector which is normal (perpendicular) to the plane of $\mathbf{r}$ and $\mathbf{f}$ and that this direction of torque defines an axis about which rotation due to the torque will tend to occur. These forces may act as surface or contact forces at the boundary or they may act as body forces as the result of phenomena such as gravitational or electrical fields. The torques resulting from these forces are mutual between the system and surroundings, in accordance with Newton's laws of motion; forces acting on the system provide an oppositely directed but equal-in-magnitude force on the surroundings and the position vector is the same at the point of contact between the system and surroundings.

A schematic representation of a system and surroundings for applying the conservation of angular momentum principle is given in Figure 6-1. The angular momentum entering and leaving the system must cross the system boundary that we have defined. Furthermore, only the external forces which act on the mass in the system are of interest. *Forces acting between mass elements which are inside the system boundary do not contribute to angular momentum changes of the system.*

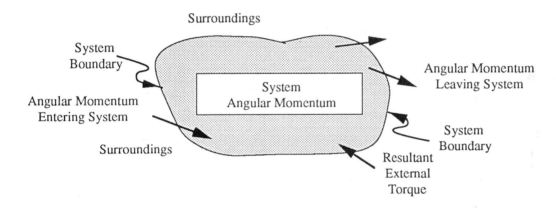

Figure 6-1: System and surroundings

Example external forces are the pressure exerted over the surface area of the system, the supporting columns or cables to suspend a system, the drag or frictional forces on a moving system, and gravitational and electrostatic body forces.

6.2 CONSERVATION OF ANGULAR MOMENTUM EQUATIONS

6.2.1 Angular Momentum Rate Equation

We are now prepared to combine the verbal statement of angular momentum with these quantitative rate expressions. Angular momentum of the system changes because of mass entering or leaving and because of torques caused by external forces. Accordingly, a rate equation for angular momentum conservation is:

$$\left(\frac{d\mathbf{l}_{\text{sys}}}{dt} \right) = \sum (\mathbf{r} \times \dot{m}\mathbf{v})_{\text{in}} - \sum (\mathbf{r} \times \dot{m}\mathbf{v})_{\text{out}} + \sum (\mathbf{r} \times \mathbf{f})_{\text{ext}}$$

$$(6-1)$$

where:

$$\left(\frac{d\mathbf{l}_{sys}}{dt}\right) \quad = \quad \text{the time rate of change of AM of the system}$$

$(\mathbf{r} \times \dot{m}\mathbf{v})_{in} \quad = \quad$ the rate of AM entering the system with mass

$(\mathbf{r} \times \dot{m}\mathbf{v})_{out} \quad = \quad$ the rate of AM leaving the system with mass

$(\mathbf{r} \times \mathbf{f})_{ext} \quad = \quad$ the rate of AM entering the system due to external forces

Note again, as stated above, that as mass enters the system, the amount of angular momentum that it brings into the system is this mass flow rate times the angular momentum per unit mass $(\mathbf{r} \times \mathbf{v})$ of the mass. Futhermore, because the conservation of angular momentum contains position vectors relative to some defined origin, the position information and the terms in the equation are dependent on the defined origin. However, the observation that angular momentum is conserved is completely independent of where the origin is defined, so long as an inertial frame is used.

Table 6-1: Components of the Angular Momentum Equation

Description	$\left(\dfrac{d\mathbf{l}_{sys}}{dt}\right)$ [force length]	$=$	$\sum(\mathbf{r} \times \dot{m}\mathbf{v})_{in/out}$ [force length]	$+$	$\sum(\mathbf{r} \times \mathbf{f})_{ext}$ [force length]
x-direction	$\left(\dfrac{d(l_x)_{sys}}{dt}\right)$	$=$	$\sum\left((\mathbf{r} \times \dot{m}\mathbf{v})_x\right)_{in/out}$	$+$	$\sum\left((\mathbf{r} \times \mathbf{f})_x\right)_{ext}$
y-direction	$\left(\dfrac{d(l_y)_{sys}}{dt}\right)$	$=$	$\sum\left((\mathbf{r} \times \dot{m}\mathbf{v})_y\right)_{in/out}$	$+$	$\sum\left((\mathbf{r} \times \mathbf{f})_y\right)_{ext}$
z-direction	$\left(\dfrac{d(l_z)_{sys}}{dt}\right)$	$=$	$\sum\left((\mathbf{r} \times \dot{m}\mathbf{v})_z\right)_{in/out}$	$+$	$\sum\left((\mathbf{r} \times \mathbf{f})_z\right)_{ext}$

The individual angular momentum component equations in terms of a rectangular cartesian coordinate system are given in Table 6-1. Here the input and output terms have been combined for conciseness. The vector cross products of the position vector and the rates of linear momentum entering and leaving the system are evaluated first, and then the component of that resulting vector is used in the component equation.

Example 6-1. For a grindstone in the form of a solid cylinder which is initially at rest, what constant torque must be exerted for 10 seconds if the final angular momentum of the system is 424 (lb_m ft^2 / sec^2) in the z-direction?

SOLUTION: Figure 6-2 shows the grindstone initially at rest and then after 10 (sec) rotating about the z-axis. The system that we define is the grindstone, and the conservation of angular momentum is given in Table 6-2.

No mass enters or leaves the system during the time period. Also, the only torque is in the z-direction. From Table 6-2:

$$\left(\frac{d(l_z)_{sys}}{dt}\right) = (\mathbf{r} \times \mathbf{f})_z$$

If we assume the torque is constant during the time period, the equation can be integrated as:

$$(l_z)_{end} - (l_z)_{beg} = (\mathbf{r} \times \mathbf{f})_z \Delta t$$

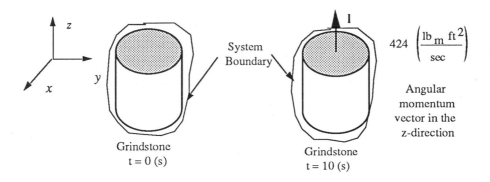

Figure 6-2: Rotating grindstone

Table 6-2: Data for Example 6-1 (Angular Momentum, System = Grindstone)

Description	$\left(\dfrac{d\mathbf{l}_{\text{sys}}}{dt}\right)$	=	$\sum (\mathbf{r} \times \dot{m}\mathbf{v})_{\text{in/out}}$	+	$\sum (\mathbf{r} \times \mathbf{f})_{\text{ext}}$
	(lb$_f$ ft)		(lb$_f$ ft)		Torque (lb$_f$ ft)
x-direction	$0\mathbf{i}$	=	$0\mathbf{i}$	+	$0\mathbf{i}$
y-direction	$0\mathbf{j}$	=	$0\mathbf{j}$	+	$0\mathbf{j}$
z-direction	$\left(\dfrac{d(l_z)_{\text{sys}}}{dt}\right)\mathbf{k}$	=	$0\mathbf{k}$	+	$(\mathbf{r} \times \mathbf{f})_z\mathbf{k}$

Because the grindstone is not initially rotating, the initial angular momentum is zero so the torque can be calculated as:

$$(\mathbf{r} \times \mathbf{f})_z = \left(\frac{1}{\Delta t}\right)(l_z)_{\text{end}} = \left(\frac{1}{10 \text{ s}}\right) 424 \text{ (lb}_m \text{ ft}^2 / \text{ s)}$$

or:

$$(\mathbf{r} \times \mathbf{f})_z = \frac{42.40 \text{ lb}_m \text{ ft}^2}{\text{s}^2} \left| \frac{1 \text{ lb}_f \text{ s}^2}{32.174 \text{ lb}_m \text{ ft}} \right. = 1.32 \text{ (lb}_f \text{ ft)}$$

The torque in the z-direction is 1.32 (lb$_f$ ft).

Incidentally, to calculate the torque, we did not have to know anything about the external force ($\mathbf{f}_{\text{ext}}$) or its point of application ($\mathbf{r}$).

6.2.2 Angular Momentum Integral (Finite Time Period) Equation

This rate equation for the conservation of angular momentum can be integrated to obtain an equation for a finite time period.

$$\int_{t_{\text{beg}}}^{t_{\text{end}}} \left(\frac{d\mathbf{l}_{\text{sys}}}{dt}\right) dt = \int_{t_{\text{beg}}}^{t_{\text{end}}} \sum (\mathbf{r} \times \dot{m}\mathbf{v})_{\text{in}} dt - \int_{t_{\text{beg}}}^{t_{\text{end}}} \sum (\mathbf{r} \times \dot{m}\mathbf{v})_{\text{out}} dt + \int_{t_{\text{beg}}}^{t_{\text{end}}} \sum (\mathbf{r} \times \mathbf{f})_{\text{ext}} dt \qquad (6-2)$$

which can then be written in symbolic form as

$$(I_{sys})_{end} - (I_{sys})_{beg} = \sum I_{in} - \sum I_{out} + \int_t \sum (\mathbf{r} \times \mathbf{f})_{ext} dt$$

(6 – 3)

where:

$(I_{sys})_{end}$	=	the AM of the system at the end
$(I_{sys})_{beg}$	=	the AM momentum of the system at the beginning
I_{in}	=	AM entering the system with mass
I_{out}	=	AM leaving the system with mass
$\int_t \sum (\mathbf{r} \times \mathbf{f})_{ext} dt$	=	the AM entering the system with external forces

6.3 STEADY STATE

The steady state assumption for a defined system means that the accumulation of angular momentum in the system is zero for any time period, finite or differential. For a finite time period, steady state means

$$(I_{sys})_{end} - (I_{sys})_{beg} = 0, \qquad \text{steady state}$$

(6 – 4)

and for a differential time period steady state means

$$\left(\frac{dI_{sys}}{dt}\right) = 0, \qquad \text{steady state}$$

(6 – 5)

Consequently, the steady state assumption implies that (from equation (6-1)):

$$0 = \sum (\mathbf{r} \times \dot{m}\mathbf{v})_{in/out} + \sum (\mathbf{r} \times \mathbf{f})_{ext}$$

(6 – 6)

Furthermore, for rigid body statics (no mass enters or leaves the system and the body is at steady state), the conservation of angular momentum gives that:

$$0 = \sum (\mathbf{r} \times \mathbf{f})_{ext}$$

(6 – 7)

or

$$0 = \sum (_B\mathbf{r} \times \mathbf{f})_{ext}$$

(6 – 8)

Note that if the body is static, the inertial frame origin can be anywhere, which may or may not be the center of mass. Consequently, this second result is introduced to emphasize that the torques can be calculated with respect to *any* convenient origin (point B).

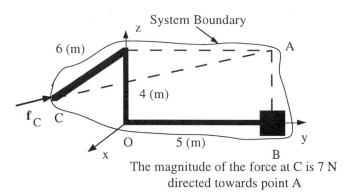

The magnitude of the force at C is 7 N
directed towards point A

Figure 6-3: Steady state conservation
of angular momentum

Example 6-2. Determine the torque exerted about point O by a combined force and moment at point B in order for the body to be static. Force $\mathbf{f}_C$, acting at point C, is directed at point A. The magnitude of $\mathbf{f}_C$ is 7 (N) as shown in Figure 6-3.

SOLUTION: The system is the body depicted in the figure. The conservation of angular momentum about point O is given in Table 6-3.

Table 6-3: Data for Example 6-2 (Angular Momentum, System = Static Structure)

	0	**=**	$\sum (\mathbf{r} \times \mathbf{f})_{\text{ext}}$		
Description			Torque C		Torque B
	(N m)		(N m)		(N m)
x-direction	$0\mathbf{i}$	$=$	$(\mathbf{r} \times \mathbf{f})_{C,x}\mathbf{i}$	$+$	$T_{B,x}\mathbf{i}$
y-direction	$0\mathbf{j}$	$=$	$(\mathbf{r} \times \mathbf{f})_{C,y}\mathbf{j}$	$+$	$T_{B,y}\mathbf{j}$
z-direction	$0\mathbf{k}$	$=$	$(\mathbf{r} \times \mathbf{f})_{C,z}\mathbf{k}$	$+$	$T_{B,z}\mathbf{k}$

The conservation of angular momentum gives three independent scalar equations in which the torque exerted by the force at C must be included as known. First, we write the position vector from point O to point C as:

$$\mathbf{r}_C = 6\mathbf{i} + 0\mathbf{j} + 4\mathbf{k} \text{ (m)}$$

Since we were only given the magnitude of the force and the point of application of the force we must calculate the unit direction vector or line of action of the force (from point A to C) as:

$$_C\mathbf{u}_A = \frac{_C\mathbf{r}_A}{|_C\mathbf{r}_A|} = \frac{-6\mathbf{i} + 5\mathbf{j} + 0\mathbf{k}}{\sqrt{61}}$$

Now the cross product of the position vector $\mathbf{r}_C$ and the force at point C is:

$$(\mathbf{r}_C \times \mathbf{f}_C) = (\mathbf{r}_C \times |\mathbf{f}_C|_C\mathbf{u}_A) = |\mathbf{f}_C|(\mathbf{r}_C \times {}_C\mathbf{u}_A)$$

$$= \left(\frac{7}{\sqrt{61}}\right) \begin{vmatrix} \mathbf{i} & \mathbf{j} & \mathbf{k} \\ 6 & 0 & 4 \\ -6 & 5 & 0 \end{vmatrix} = \left(\frac{7}{\sqrt{61}}\right)(-20\mathbf{i} - 24\mathbf{j} + 30\mathbf{k}) \text{ (N m)}$$

Table 6-4: Solution of Example 6-2 (Angular Momentum, System = Static Structure)

Description	**0** (N m)	=	$\sum (\mathbf{r} \times \mathbf{f})_{ext}$		
			Torque C (N m)		Torque B (N m)
x-direction	$0\mathbf{i}$	=	$(-17.9)\mathbf{i}$	+	$T_{B,x}\mathbf{i}$
y-direction	$0\mathbf{j}$	=	$(-21.5)\mathbf{j}$	+	$T_{B,y}\mathbf{j}$
z-direction	$0\mathbf{k}$	=	$(26.9)\mathbf{k}$	+	$T_{B,z}\mathbf{k}$

This is substituted back into Table 6-3 to give Table 6-4.

Consequently, the torque exerted about point O by the combined force and moment at B must be:

$$\mathbf{T}_B = 17.9\mathbf{i} + 21.5\mathbf{j} - 26.9\mathbf{k} \text{ (N m)}$$

6.4 ANGULAR MOMENTUM OF THE SYSTEM

So far we have considered the angular momentum of mass entering and leaving the system and the transfer of angular momentum to and from the system as the result of forces acting between the system and surroundings. However, we have not yet said anything about the angular momentum of the system and how this is determined. In what follows, especially beginning with section 6.4.2, the discussion will narrow to consider only planar motion (rotation about an axis of constant direction) and, for that, only one component of angular momentum, that which is in the direction of the axis of rotation.

If the system is a single particle, then the angular momentum is easily calculated. By definition, a particle is assumed to occupy such a small volume that only one velocity exists to describe the motion of that particle. In this case, the angular momentum is the cross product of that particle's position (a vector) with its momentum, an unambiguous calculation.

For a larger system however, the calculation is complicated, as it was for linear momentum, by the fact that different pieces of the system may be moving differently. Different elements of the body may be moving relative to each other, if the body is not rigid, or the body itself may be rotating in addition to translating. To calculate the angular momentum of such a body, we divide it into individual elements on a small enough scale that each element can be described in terms of a single momentum and then sum over all such elements within the system. Consequently we may write:

$$\mathbf{l}_{sys} = \sum_{i=1}^{n} \Big((\mathbf{r} \times m\mathbf{v})_i \Big)_{sys} \tag{6-9}$$

where subscript i represents the ith element of the system. This may now be used in the above equations to represent the momentum of the system.

Because mass is a scalar, it can be factored out of the cross product. The cross product of the two vectors is calculated using the vectors' components and the base vectors of the coordinate system as described by the following matrix operation:

or a rectangular cartesian coordinate system (In a rectangular cartesian coordinate system, the x, y, and z components of
e position vector are $r_x = x$, $r_y = y$, and $r_z = z$):

$$\mathbf{r} \times \mathbf{v} = \begin{vmatrix} \mathbf{i} & \mathbf{j} & \mathbf{k} \\ x & y & z \\ v_x & v_y & v_z \end{vmatrix} = (yv_z - zv_y)\mathbf{i} + (zv_x - xv_z)\mathbf{j} + (xv_y - yv_x)\mathbf{k}$$

Appendix B contains more information about vectors and vector operations and is suggested additional reading.

Example 6-3. For the given rigid body
onsisting of three masses connected by massless
upports, as shown in Figure 6-4, calculate the
angular momentum. The masses and their vectors
e given below.

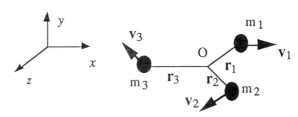

Figure 6-4: Angular momentum
of several particles

SOLUTION: The system is defined to be the three masses connected to each other by the structure. The angular momentum
f the body will be the sum of the angular momenta of each individual particle. The position vectors (with respect to point O), velocity
ectors and masses of each particle are:

$$\mathbf{r}_1 = 3\mathbf{i} + 9\mathbf{j} + 0\mathbf{k} \text{ (cm)} \qquad \mathbf{v}_1 = 2\mathbf{i} + 0\mathbf{j} + 0\mathbf{k} \text{ (cm / s)} \qquad m_1 = 5 \text{ (g)}$$
$$\mathbf{r}_2 = 4\mathbf{i} - 3\mathbf{j} + 0\mathbf{k} \text{ (cm)} \qquad \mathbf{v}_2 = -7\mathbf{i} - 7\mathbf{j} + 0\mathbf{k} \text{ (cm / s)} \qquad m_2 = 3 \text{ (g)}$$
$$\mathbf{r}_3 = -6\mathbf{i} + 0\mathbf{j} + 0\mathbf{k} \text{ (cm)} \qquad \mathbf{v}_3 = -3\mathbf{i} + 2\mathbf{j} + 0\mathbf{k} \text{ (cm / s)} \qquad m_3 = 2 \text{ (g)}$$

Now we will use the definition of angular momentum to calculate the angular momentum of each particle:

$$\mathbf{l}_1 = \mathbf{r}_1 \times m_1\mathbf{v}_1 = 5 \begin{vmatrix} \mathbf{i} & \mathbf{j} & \mathbf{k} \\ 3 & 9 & 0 \\ 2 & 0 & 0 \end{vmatrix} = 5\,(0\mathbf{i} + 0\mathbf{j} - 18\mathbf{k}) = -90\mathbf{k} \text{ (g cm}^2 \text{/ s)}$$

$$\mathbf{l}_2 = \mathbf{r}_2 \times m_2\mathbf{v}_2 = 3 \begin{vmatrix} \mathbf{i} & \mathbf{j} & \mathbf{k} \\ 4 & -3 & 0 \\ -7 & -7 & 0 \end{vmatrix} = 3\,(0\mathbf{i} + 0\mathbf{j} - 49\mathbf{k}) = -147\mathbf{k} \text{ (g cm}^2 \text{/ s)}$$

$$\mathbf{l}_3 = \mathbf{r}_3 \times m_3\mathbf{v}_3 = 2 \begin{vmatrix} \mathbf{i} & \mathbf{j} & \mathbf{k} \\ -6 & 0 & 0 \\ -3 & 2 & 0 \end{vmatrix} = 2\,(0\mathbf{i} + 0\mathbf{j} - 12\mathbf{k})r = -24\mathbf{k} \text{ (g cm}^2 \text{/ s)}$$

he angular momentum of the system (masses m_1, m_2, and m_3) is:

$$\mathbf{l}_{sys} = \mathbf{l}_1 + \mathbf{l}_2 + \mathbf{l}_3 = -90\mathbf{k} - 147\mathbf{k} - 24\mathbf{k} \text{ (g cm}^2 \text{/ s)} = -261\mathbf{k} \text{ (g cm}^2 \text{/ s)}$$

he angular momentum of the system has a magnitude of 261 (g cm^2 / s) in the negative z-direction.

6.4.1 Components of the Velocity of a Particle

At any instant of time we can resolve the velocity of a particle into a component in the direction of its position vector (v_R, the radial component) in the given coordinate system and a component that is normal to this position vector (v_T):

$$\mathbf{v} = \mathbf{v}_T + \mathbf{v}_R \tag{6 - 10}$$

The normal component is called the *transverse component* of the velocity and can be written in terms of a vector which is termed the angular velocity ω and defined according to:

$$\mathbf{v}_T = \omega \times \mathbf{r} \tag{6 - 11}$$

Figure 6-5 depicts this process of resolving the particle velocity into its transverse and radial components and of writing the transverse velocity in terms of an angular velocity ω. The angular velocity is itself a vector such that the transverse velocity vector is normal to both the position vector and this angular velocity. Note that ω and $\mathbf{r}$ are not necessarily mutually perpendicular

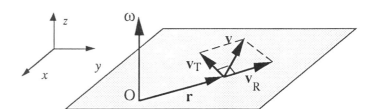

Figure 6-5: Radial and transverse components of the velocity vector

6.4.2 Angular Momentum and the Moment of Inertia

Angular Momentum of a Particle. The angular momentum of a particle can be described in terms of its angular velocity ω. The angular momentum is given by

$$\mathbf{l} = \mathbf{r} \times m\mathbf{v}$$

which, by expressing the velocity in terms of its transverse and radial components is

$$\mathbf{l} = \mathbf{r} \times m(\mathbf{v}_T + \mathbf{v}_R) \tag{6 - 12}$$

or, because the cross product is distributive,

$$\mathbf{l} = (\mathbf{r} \times m\mathbf{v}_T) + (\mathbf{r} \times m\mathbf{v}_R) \tag{6 - 13}$$

Now because v_R is collinear with $\mathbf{r}$ (by definition), $\mathbf{r} \times \mathbf{v}_R \equiv 0$ and hence

$$\mathbf{l} = (\mathbf{r} \times m\mathbf{v}_T) \tag{6 - 14}$$

Then by writing the transverse velocity in terms of the angular velocity ($\mathbf{v}_T = \omega \times \mathbf{r}$), the *angular momentum of the particle* becomes

$$\boxed{\mathbf{l} = \mathbf{r} \times m(\omega \times \mathbf{r})} \tag{6 - 15}$$

Now we could stop here because we have expressed the angular momentum in terms of the mass, the position of the mass, and the angular velocity of the system. However, there is a more convenient form for this expression and it is obtained by using a vector identity for the cross product of three vectors, namely that

$$\mathbf{a} \times (\mathbf{b} \times \mathbf{c}) = (\mathbf{a} \cdot \mathbf{c})\mathbf{b} - (\mathbf{a} \cdot \mathbf{b})\mathbf{c}$$

This vector identity is always true, much like an identity in trigonometry. For more such identities, see Appendix B. Using this identity gives

$$\mathbf{l} = \mathbf{r} \times m(\boldsymbol{\omega} \times \mathbf{r}) = m(\mathbf{r} \cdot \mathbf{r})\boldsymbol{\omega} - m(\mathbf{r} \cdot \boldsymbol{\omega})\mathbf{r} \qquad (6-16)$$

Remember that mass is a scalar quantity and can be factored out of the cross product operation.

The discussion to this point has addressed general, 3-dimensional motion; it has not been restricted to planar motion. Now, for the sake of mathematical convenience, we address this simplification.

By planar motion, we mean that $\boldsymbol{\omega}$ is always normal (perpendicular) to the same plane, i.e., that the motion consists of rotation about an axis of constant direction, defined by unit vector $\mathbf{u}_a$ (say), see Figure 6-6. For such a motion of a particle, $\boldsymbol{\omega} = \omega \mathbf{u}_a$ and hence

$$\mathbf{l} = m(\mathbf{r} \cdot \mathbf{r})\omega \mathbf{u}_a - m(\mathbf{r} \cdot \omega \mathbf{u}_a)\mathbf{r} \qquad (6-17)$$

Now, look at $\mathbf{l} \cdot \mathbf{u}_a = l_a$ which is the component of $\mathbf{l}$ which is in the direction of $\mathbf{u}_a$:

$$\begin{aligned} l_a &= [m(\mathbf{r} \cdot \mathbf{r})\omega \mathbf{u}_a - m(\mathbf{r} \cdot \omega \mathbf{u}_a)\mathbf{r}] \cdot \mathbf{u}_a \\ &= m\omega(r^2 - r_a^2) \qquad\qquad\qquad (6-18) \\ &= m\omega r_n^2 \end{aligned}$$

Note that $r_n^2 = r^2 - r_a^2$ by the Pythagorean theorem (see Figure 6-6). Consequently, we see that for a particle undergoing such a motion, the component of angular momentum in the direction of the axis of rotation is proportional to the square of the distance of the particle from this axis. Because of this, the value of l_a does not depend on where, along the axis, the origin of the coordinate system is located. Note also that, in general, this is not the only non-zero component of $\mathbf{l}$ (unless $\mathbf{u}_a$ coincides with a coordinate axis). It is, nevertheless, the component which we will focus on for the rest of this chapter.

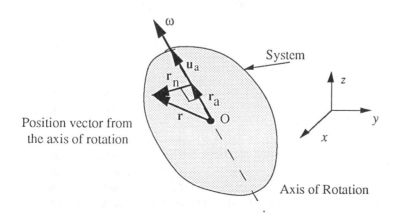

Position vector from
the axis of rotation

Figure 6-6: Rotation about a constant-direction axis

Example 6-4. There is a massless string of length 20 (cm) with a mass of 100 (g) attached at the end. The initial angular velocity of the mass in the z-direction (the axis about which rotation occurs is in the z direction) about point O is 30 (rad/s). What is the magnitude of the initial transverse velocity? If the mass of the rope is neglected, what is the final angular velocity if the rope is suddenly shortened to 10 (cm) and still revolves about the point O. Assume that there are no external torques acting on the system.

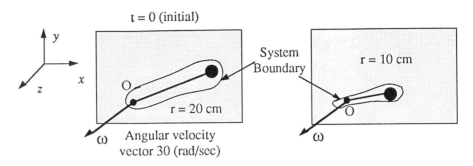

Figure 6-7: Angular momentum of a rotating body

SOLUTION: The system is the string and the attached mass which is revolving about point O as shown in Figure 6-7. For convenience we will use a cylindrical coordinate system as defined in Appendix B. The transverse velocity vector is calculated as

$$\mathbf{v}_T = \boldsymbol{\omega} \times \mathbf{r}$$

where:

$$\boldsymbol{\omega} = 0\mathbf{e}_r + 0\mathbf{e}_\theta + 30\mathbf{e}_z \text{ (rad / s)}$$

$$\mathbf{r} = 20\mathbf{e}_r + 0\mathbf{e}_\theta + 0\mathbf{e}_z \text{ (cm)}$$

The transverse component of the velocity then is:

$$\mathbf{v}_T = \begin{vmatrix} \mathbf{e}_r & \mathbf{e}_\theta & \mathbf{e}_z \\ 0 & 0 & 30 \\ 20 & 0 & 0 \end{vmatrix}$$

$$= 0\mathbf{e}_r + 600\mathbf{e}_\theta + 0\mathbf{e}_z \text{ (cm / s)}$$

The magnitude of the transverse velocity is 600 (cm/s) and it is in the θ direction in a cylindrical coordinate system. (Be careful to distinguish between the direction of $\mathbf{v}_T$, the *transverse* velocity, and the direction of $\boldsymbol{\omega}$, the *angular* velocity.)

The conservation of angular momentum is given in Table 6-5. This says that there is no change in the angular momentum of the system as the rope is shortened. Hence, over a finite time period,

$$(I_{sys})_{end} = (I_{sys})_{beg}$$

and of particular interest

$$(l_{z,sys})_{end} = (l_{z,sys})_{beg}$$

where $l_z = m\omega r_n^2$ (from equation 6-18) so that

$$(m\omega r_n^2)_{end} = (m\omega r_n^2)_{beg}$$

The mass, of course, is the same at the beginning and end. Also, $r_{n,\text{beg}} = 20$ cm and $r_{n,\text{end}} = 10$ cm. Hence,

$$\omega_{\text{end}} = \left(\frac{r_{n,\text{beg}}}{r_{n,\text{end}}}\right)^2 \omega_{\text{beg}} = 120 \text{ rad/s}$$

The magnitude of the angular velocity after the string was shortened is 120 (rad/s).

Table 6-5: Data for Example 6-4 (Angular Momentum, System = Ball on String)

	$\left(\dfrac{d\mathbf{l}_{\text{sys}}}{dt}\right)$	$=$	$\sum(\mathbf{r} \times \dot{m}\mathbf{v})_{\text{in/out}}$	$+$	$\sum(\mathbf{r} \times \mathbf{f})_{\text{ext}}$
Description	(g cm^2 / s^2)		(g cm^2 / s^2)		Torque (g cm^2 / s^2)
r-direction	$0\mathbf{e}_r$	$=$	$0\mathbf{e}_r$	$+$	$0\mathbf{e}_r$
θ-direction	$0\mathbf{e}_\theta$	$=$	$0\mathbf{e}_\theta$	$+$	$0\mathbf{e}_\theta$
z-direction	$\left(\dfrac{d(l_z)_{\text{sys}}}{dt}\right)\mathbf{e}_z$	$=$	$0\mathbf{e}_z$	$+$	$0\mathbf{e}_z$

Angular Momentum for a Rotating Rigid Body. For a collection of particles, such as a rigid body of finite volume, we are interested in calculating the angular momentum of the body as the sum of the angular momenta of the individual elements of that body. Equivalently this can be expressed as an integral of the individual momenta per unit mass over the mass of the body

$$\mathbf{l} = \int (\mathbf{r} \times \mathbf{v})dm \qquad (6-19)$$

where the scalar mass has been factored out of the cross product and reduced to a differential size dm.

   This is a general result but we now want to consider the special case of the body rotating (with angular velocity $\boldsymbol{\omega}$) about a single stationary (or at least non-accelerating) point within the body (or even a point outside the body so long as it is fixed with respect to each point in the body) with that point used as the origin of the (inertial) coordinate system. By fixed, is meant that the distance from that point to each point of the body is constant as the body rotates.

   For this special case, each mass element's velocity can be broken into a transverse and radial component, $\mathbf{v} = \mathbf{v}_T + \mathbf{v}_R$, where $\mathbf{v}_T = \boldsymbol{\omega} \times \mathbf{r}$ and $\boldsymbol{\omega}$ is the same for each element of the rigid body. Hence (see equations 6-12 through 6-15) $\mathbf{r} \times \mathbf{v} = \mathbf{r} \times \boldsymbol{\omega} \times \mathbf{r}$ for each element dm so that $\mathbf{l} = \int(\mathbf{r} \times \boldsymbol{\omega} \times \mathbf{r})dm$. As was done for a single particle, we now consider planar motion of the rigid body. Integrating the particle result over the entire body gives the component of angular momentum about the axis of rotation.

$$l_a = \int \omega r_n^2 dm \qquad (6-20)$$

In this way, the rotational-axis component of the angular momentum of the rigid body is expressed in terms of the distance of each of the mass elements from the axis of rotation and the angular velocity of each element. Remember that for a rigid body all elements of mass have the same angular velocity so that ω can be moved outside the integral which then depends only on the position of each differential element of mass in the body. Hence, the a-component of the angular momentum of the body is

$$l_a = \omega \int r_n^2 dm \qquad (6-21)$$

or:

$$l_a = I_a \omega \qquad (6-22)$$

where I_a, the *moment of inertia* for rotation of the body about the a axis for the finite size rigid body is defined by:

$$I_a = \int r_n^2 dm \qquad (6-23)$$

and has the dimensions (mass)(length)^2.

Example 6-5. Calculate the moment of inertia of the 3-body system (A, B, C) for the geometry shown in Figure 6-8 about each point, A, B, and C. If the angular velocity for the rigid-body rotation about stationary point A is 10 (rad/s) in the positive z-direction, what is the angular momentum of the system with respect to point A (point A is used as the origin of the coordinate system and $v_A = 0$)? What is the angular momentum for rotation about point B (with $v_B = 0$) with $\omega = 10k$ rad/s?

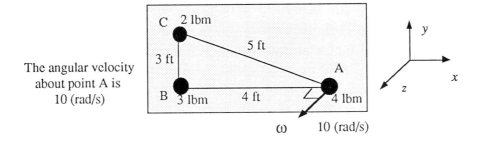

Figure 6-8: Angular momentum of a rotating rigid body

SOLUTION: The moment of inertia about point A is $_A I_z$ and is calculated as:

$$_A I_z = \sum (_A r_n^2 m)_i$$

where the distance of each mass from the axis of rotation (with point A as the origin) are:

$$_A r_{n,A} = 0 \text{ (ft)} \qquad m_A = 4 \text{ (lb}_m)$$
$$_A r_{n,B} = 4 \text{ (ft)} \qquad m_B = 3 \text{ (lb}_m)$$
$$_A r_{n,C} = 5 \text{ (ft)} \qquad m_C = 2 \text{ (lb}_m)$$

Hence,

$$_A I_z = (4 \text{ ft})^2 (3 \text{ lb}_m) + (5 \text{ ft})^2 (2 \text{ lb}_m) = 98 \text{ lb}_m \text{ft}^2$$

and therefore

$$_A l_z = \omega(_A I_z) = 980 \text{ lb}_m \text{ft}^2 / \text{s}$$

Similarly, if point B is stationary ($v_B = 0$), then

$$_B I_z = (4 \text{ ft})^2 (4 \text{ lb}_m) + (3 \text{ ft})^2 (2 \text{ lb}_m) = 82 \text{ lb}_m \text{ft}^2$$

and therefore

$$_B l_z = \omega(_B I_z) = 820 \text{ lb}_m \text{ft}^2 / \text{s}$$

Note that the two angular momenta are different because $_A I_z \neq _B I_z$.

6.4.3 Angular Momentum of a Translating and Rotating Rigid Body

Consider now a rigid body which is undergoing both translation and rotation. Figure 6-9 shows a body undergoing pure translation (a), pure rotation (b), and both translation and rotation simultaneously (c). The total motion, then, is the sum of the separate translation and rotation motions and both must be considered in calculating angular momentum. It is helpful to rewrite the expression for the angular momentum of a rigid body in terms of the position vectors for the various elements of the body with respect to the center of mass. For this result we recognize

$$\mathbf{r} = \mathbf{r}_G + _G\mathbf{r} \qquad\qquad (6-24)$$

The position vector $\mathbf{r}$ could be denoted $_o\mathbf{r}$ but we omit the leading subscript when the vector (or moment of inertia) is with respect to the origin of the coordinate system. The vector $_G\mathbf{r}$ is the position vector of a particle *with respect to the center of mass*. The magnitude of this vector, because the body is a rigid body, is a constant for each element of the body regardless of what the body is doing in terms of translation and rotation.

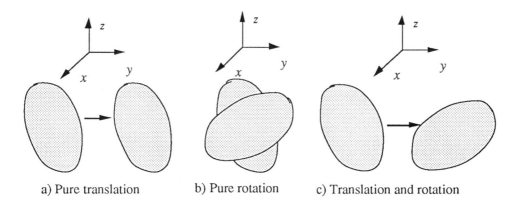

a) Pure translation   b) Pure rotation   c) Translation and rotation

Figure 6-9: Translation and rotation of a rigid body

Each element is fixed in distance with respect to every other element of the mass including the center of mass. The center of mass, however, certainly may be moving as the body translates and otherwise moves about in space. Each of these vectors as well as the angular velocity and other quantities mentioned previously are expressed in terms of the same inertial coordinate system as shown in Figure 6-10.

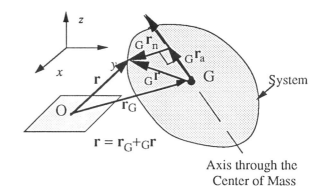

Figure 6-10: Position vectors in terms of the center of mass

Using this substitution for the position vector in equation (6-19) we can obtain an expression for the angular momentum of a rigid body and integrate that expression over the entire body to obtain a simplified result. (For now, we will again follow a general 3-D approach, as this is actually easier. At the appropriate point in the development, we will again restrict our attention to rotation about a constant-direction axis.) Accordingly, the integral becomes:

$$\mathbf{l} = \int (\mathbf{r} \times \mathbf{v})dm = \int \left((\mathbf{r}_G + {}_G\mathbf{r}) \times \mathbf{v} \right) dm$$
$$= \int (\mathbf{r}_G \times \mathbf{v})dm + \int ({}_G\mathbf{r} \times \mathbf{v})dm \tag{6-25}$$

Now, $\mathbf{r}_G$ at any instant of time, does not depend upon which piece of mass within the body is of interest; i.e. $\mathbf{r}_G$ is constant with respect to an integral over mass. Hence

$$\int (\mathbf{r}_G \times \mathbf{v})dm = \mathbf{r}_G \times \int \mathbf{v}dm$$
$$= m\mathbf{r}_G \times \frac{\int \mathbf{v}dm}{m} \tag{6-26}$$
$$= m\mathbf{r}_G \times \mathbf{v}_G$$

This latter result arises because of the definition of the velocity of the center of mass equation (5-5).

Now consider $\int ({}_G\mathbf{r} \times \mathbf{v})dm$. Because $\mathbf{r} = \mathbf{r}_G + {}_G\mathbf{r}$, by differentiation $\mathbf{v} = \mathbf{v}_G + {}_G\mathbf{v}$ so that

$$\int ({}_G\mathbf{r} \times \mathbf{v})dm = \int {}_G\mathbf{r} \times (\mathbf{v}_G + {}_G\mathbf{v})dm \tag{6-27}$$

Now $\mathbf{v}_G$, the velocity of the center of mass, is a constant with respect to the integral over the body so that

$$\int ({}_G\mathbf{r} \times \mathbf{v})dm = \int ({}_G\mathbf{r} \times \mathbf{v}_G)dm + \int ({}_G\mathbf{r} \times {}_G\mathbf{v})dm \tag{6-28}$$

Then, by the definition of the center of mass, equation (5-8), $\int {}_G\mathbf{r}\,dm = \mathbf{0}$, giving

$$\int ({}_G\mathbf{r} \times \mathbf{v})dm = \int ({}_G\mathbf{r} \times {}_G\mathbf{v})dm \qquad (6-29)$$

Then, writing each particle's velocity relative to the center of mass as ${}_G\mathbf{v} = {}_G\mathbf{v}_T + {}_G\mathbf{v}_R$ and ${}_G\mathbf{v}_T = \boldsymbol{\omega} \times {}_G\mathbf{r}$ gives

$$\int ({}_G\mathbf{r} \times \mathbf{v})dm = \int {}_G\mathbf{r} \times (\boldsymbol{\omega} \times {}_G\mathbf{r})dm \qquad (6-30)$$

We now return to the case of motion about a constant-direction axis having direction $\mathbf{u}_a$. In this case, we will need only the a-component of ${}_G\mathbf{r} \times (\boldsymbol{\omega} \times {}_G\mathbf{r})$ and from previous discussions (see equations 6-15 to 6-18), we recognize that this is

$$\int [{}_G\mathbf{r} \times (\boldsymbol{\omega} \times {}_G\mathbf{r})]_a\, dm = {}_G I_a \omega \qquad (6-31)$$

where

$$\boxed{{}_G I_a \equiv \int [{}_G\mathbf{r} \times (\boldsymbol{\omega} \times {}_G\mathbf{r})]_a\, dm = \int {}_G r_n^2\, dm} \qquad (6-32)$$

and is the moment of inertia of the body with respect to the center of mass (G) for rotation about an axis with direction defined by unit vector $\mathbf{u}_a$, as indicated by the pre-subscript G and the subscript a.

Combining equations (6-25), (6-26), (6-30), and (6-31) gives

$$\boxed{l_a = m(\mathbf{r}_G \times \mathbf{v}_G)a + {}_G I_a \omega} \qquad (6-33)$$

which is a very useful result. The a component of angular momentum of a rigid body undergoing combined translation and rotation about a constant direction axis can be calculated from the angular momentum of the center of mass (in terms of its velocity, $\mathbf{v}_G$ and position $\mathbf{r}_G$), plus the angular momentum due to rotation of the body about the center of mass (due to its angular velocity $\boldsymbol{\omega}$ and the center-of-mass moment of inertia, ${}_G I_a$).

This important result is true even though the motion of the body may be more naturally viewed as translation of a point A which is fixed in the body at velocity $\mathbf{v}_A$, plus rotation about A at angular velocity $\boldsymbol{\omega}$, as we now show, again for general 3-dimensional motion. In this case we can write $\mathbf{v}$ in terms of translation of point A ($\mathbf{v}_A$) and rotation about A at angular velocity $\boldsymbol{\omega}$, and then write $\mathbf{v}_A$ and $\mathbf{r}_A$ in terms of $\mathbf{v}_G$, ${}_G\mathbf{v}_A$ and $\mathbf{r}_G$ and ${}_G\mathbf{r}_A$ (Note ${}_G\mathbf{v}_A = \boldsymbol{\omega} \times {}_G\mathbf{r}_A$):

$$\mathbf{l} = \int \mathbf{r} \times \mathbf{v} dm$$

$$= \int \mathbf{r} \times \left(\mathbf{v}_A + (\boldsymbol{\omega} \times {}_A\mathbf{r}) \right) dm$$

$$= \int \mathbf{r} \times \left(\mathbf{v}_G + {}_G\mathbf{v}_A + \left[\boldsymbol{\omega} \times \left(\mathbf{r} - (\mathbf{r}_G + {}_G\mathbf{r}_A) \right) \right] \right) dm$$

$$= \int \mathbf{r} \times \left(\mathbf{v}_G + {}_G\mathbf{v}_A + \left[\boldsymbol{\omega} \times ({}_G\mathbf{r} - {}_G\mathbf{r}_A) \right] \right) dm \qquad (6-34)$$

$$= \int (\mathbf{r}_G + {}_G\mathbf{r}) \times \left(\mathbf{v}_G + (\boldsymbol{\omega} \times {}_G\mathbf{r}) \right) dm$$

$$= (\mathbf{r}_G \times \mathbf{v}_G)m + \int {}_G\mathbf{r} dm \times \mathbf{v}_G + \mathbf{r}_G \times \boldsymbol{\omega} \times \int {}_G\mathbf{r} dm + \int ({}_G\mathbf{r} \times \boldsymbol{\omega} \times {}_G\mathbf{r}) dm$$

$$= m\mathbf{r}_G \times \mathbf{v}_G + \int ({}_G\mathbf{r} \times \boldsymbol{\omega} \times {}_G\mathbf{r}) dm$$

where again $\int {}_G\mathbf{r} dm = 0$ and hence, equation (6-33) is recovered.

Note that ${}_G I_a$ is totally independent of the origin of the coordinate system. It depends only upon the mass distribution of the body with respect to the relevant axis through the center of mass. This moment of inertia can be calculated independent of any specific problem or motion so long as the body in question and its orientation with respect to the rotational axis are defined.

Tabulations of such integral calculations have been made for common symmetric bodies and are found in Appendix C Table C-7. Note that values are given for rotation about each of the coordinate axes. Also various texts on statics and dynamics give these values and they exist for the purpose of calculating the angular momentum (and rotational kinetic energy, as we will see in Chapter 7) for use in calculations of the conservation of angular momentum.

Example 6-6. Calculate the moment of inertia about the y axis through the center of mass for a solid sphere of radius R and constant density ρ as shown in Figure 6-11. Note that due to symmetry of the sphere, ${}_G I_x$, ${}_G I_y$, and ${}_G I_z$ are all equal.

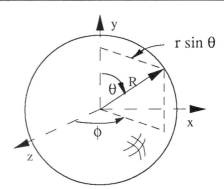

Figure 6-11: Constant density sphere with radius R

SOLUTION: The moment of inertia for a rigid body about an axis through the center of mass is given by:

$$_G I_a = \int {}_G r_n^2 \, dm$$

where $_Gr_n$ is the distance of differential mass element dm from the axis of rotation which passes through the center of mass. The differential mass element is expressed as the density times the differential volume element. For spherical coordinates, the differential volume element is $dV = (r \sin \theta d\phi)(rd\theta)dr$ and so

$$dm = \rho dV = \rho(r \sin \theta d\phi)(rd\theta)dr = \rho r^2 \sin \theta dr d\theta d\phi$$

which we can verify by integrating over the volume of the sphere to give the well-known result

$$m = \int_0^\pi \sin \theta d\theta \int_0^{2\pi} d\phi \int_0^R \rho r^2 dr = \rho \left(\frac{4}{3}\pi R^3 \right)$$

In analogous fashion, with $_Gr_n = r \sin \theta$, we can obtain $_GI_y$:

$$_GI_y = \int_0^{2\pi} \int_0^\pi \int_0^R (r \sin \theta)^2 \rho r^2 \sin \theta dr d\theta d\phi = \frac{2\pi R^5}{5} \left(\frac{\cos^3 \theta}{3} - \cos \theta \right) \Big|_0^\pi = \frac{8\pi R^5}{15}\rho$$

which, in terms of the total mass of the sphere, gives

$$_GI_y = \frac{2}{5}mR^2$$

Example 6-7. The mass of the earth is 5.9763×10^{24} (kg) and has a radius of 6,371 (km). The mass of the moon is 0.0123 times the mass of the earth, and the radius of the moon is 0.27 times the radius of the earth. The orbital radius of the moon from the center of mass of the earth to the center of mass of the moon is 3.8×10^8 (m). The period of the moon's revolution about the earth is 27.3 (days), as is the period for its rotation about its center of mass. If the center of the earth is an inertial reference frame, calculate the angular momentum of the moon with respect to the center of the earth. Note that the axis of revolution of the moon about the Earth is parallel to the axis of rotation of the Moon about its center of mass.

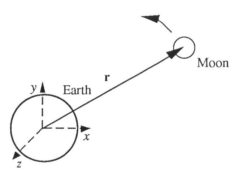

Figure 6-12: Moon's revolution about the earth

SOLUTION: Choose the z axis so that it is the axis of revolution of the Moon about the Earth. Then the z-component of the angular momentum vector of the moon with respect to the origin of the coordinate system located at the center of the mass of the earth is:

$$l_z = m(\mathbf{r}_G \times \mathbf{v}_G)_z + {}_GI_z\omega_{rot}$$

where ω_{rot} is the angular velocity of the Moon about its axis of rotation (as opposed to its angular velocity for revolution about the Earth). For the rotation of the moon on its axis, the z-component of the angular momentum vector is the only nonzero component as shown in Figure 6-12. From the given information, the mass and radius of the moon are calculated:

$$m_{moon} = 0.0123(5.9763 \times 10^{24}) = 7.3508 \times 10^{22} \text{ (kg)}$$

$$r_{moon} = 0.27(6371) = 1720 \text{ (km)}$$

The moment of inertia of the moon about the center of mass of the moon is:

$$_GI_z = \frac{2}{5}mR^2 = \frac{2}{5}(7.3508 \times 10^{22})(1720)^2 \text{ (kg km}^2)\left(\frac{1000 \text{ m}}{1 \text{ km}}\right)^2 = 8.7 \times 10^{34} \text{ (kg m}^2)$$

The angular velocity of the moon is calculated from the rotation period:

$$\text{Angular velocity for rotation} = \frac{2\pi \text{ radians}}{\text{rotation}}\left|\frac{1 \text{ rotation}}{27.3 \text{ days}}\right. = \frac{2\pi \text{ radians}}{27.3 \text{ days}}$$

Converting the values:

$$\omega_{rot} = \frac{2\pi \text{ radians}}{27.3 \text{ days}}\left|\frac{1 \text{ day}}{24 \text{ hrs}}\right|\frac{1 \text{ hrs}}{3600 \text{ s}} = 2.66 \times 10^{-6} \text{ (rad / s)}$$

so that

$$_GI_z\omega_{rot} = (8.7 \times 10^{34} \text{ kg m}^2)(2.66 \times 10^{-6} \text{ rad / s}) = 2.31 \times 10^{29} \text{ (kg m}^2 \text{ / s)}$$

Because the moon is revolving about the earth at angular velocity ω_{rev} the velocity of the center of mass of the moon is:

$$\mathbf{v}_G = \boldsymbol{\omega}_{rev} \times \mathbf{r}_G$$

Substituting the numbers (remember, for the moon $\boldsymbol{\omega}_{rev} = \boldsymbol{\omega}_{rot}$):

$$\mathbf{v}_G = 2.66 \times 10^{-6}\mathbf{e}_z \times 3.8 \times 10^8\mathbf{e}_r \text{ (m / s)} = 1011\mathbf{e}_\theta \text{ (m / s)}$$

Then the contribution to $\mathbf{l}$ by $\mathbf{v}_G$ is

$$\left|m\mathbf{r}_G \times \mathbf{v}_G\right| = (7.3508 \times 10^{22} \text{ kg})(3.8 \times 10^8 \text{ m})(1011 \text{ m / s}) = 2.82 \times 10^{34} \text{ (kg m}^2 \text{ / s)}$$

Evaluating the angular momentum of the moon with respect to the earth gives:

$$l_z = 2.82 \times 10^{34} + 2.31 \times 10^{29} \text{ (kg m}^2 \text{ / s)} = 2.82 \times 10^{34} \text{ (kg m}^2 \text{ / s)}$$

Note that the contribution to the angular momentum from the rotation about the center of mass of the moon is insignificant compared with the contribution from the revolution of the center of mass of the moon about the center of mass of the earth.

6.4.4 Rate of Change of Rigid Body Angular Momentum

Using the above results for calculating the momentum of a system, we can obtain an expression for the rate of change of angular momentum of a translating and rotating rigid body for use in a rate expression of the conservation of angular momentum. Again, we restrict attention to planar motion so that $\mathbf{r}_G$ and $\mathbf{v}_G$ are always in the same plane, which is perpendicular to the axis of rotation of the body. Differentiating both sides of equation (6-33) with respect to time,

$$\left(\frac{dl_a}{dt}\right) = \frac{d}{dt}\left(m(\mathbf{r}_G \times \mathbf{v}_G)a + _GI_a\omega\right) \tag{6-35}$$

Now, for the sake of convenience, we recognize that for this planar motion, the *only* component of $\mathbf{r}_G \times \mathbf{v}_G$ is the a component and hence,

$$\frac{d(\mathbf{r}_G \times \mathbf{v}_G)a}{dt} = \left[\frac{d(\mathbf{r}_G \times \mathbf{v}_G)}{dt}\right]_a$$

Then, from the product rule (which works for vectors and tensors provided their order is preserved),

$$\left(\frac{dl_a}{dt}\right) = m\left[\mathbf{v}_G \times \mathbf{v}_G + \mathbf{r}_G \times \left(\frac{d\mathbf{v}_G}{dt}\right)\right]_a + \frac{d}{dt}\left({}_G I_a \omega\right)$$

Now the cross product of a vector with itself is zero, identically. Hence, $\mathbf{v}_G \times \mathbf{v}_G \equiv 0$ and

$$\left(\frac{dl_a}{dt}\right) = m\left[\mathbf{r}_G \times \left(\frac{d\mathbf{v}_G}{dt}\right)\right]_a + \frac{d}{dt}\left({}_G I_a \omega\right) \tag{6 - 36}$$

6.5 RIGID BODY ANGULAR MOMENTUM CONSERVATION

We are now in a position to formulate equations for the a-component of the conservation of angular momentum of a rigid body in both rate and finite-time-period forms. Equation (6-36) combined with equation (6-1) for a rigid body (mass entering and leaving are zero) gives

$$\left(\frac{dl_a}{dt}\right) = m\left[\mathbf{r}_G \times \left(\frac{d\mathbf{v}_G}{dt}\right)\right]_a + \frac{d}{dt}\left({}_G I_a \omega\right) = \sum(\mathbf{r} \times \mathbf{f})_{\text{ext},a} \tag{6 - 37}$$

In this form, the a-components of the torques, position vectors, and velocity vectors are written with respect to the origin of the inertial coordinate system. This equation, while correct, is not always the most convenient form.

Torques with Respect to the Center of Mass. Now writing $\mathbf{r} = \mathbf{r}_G + {}_G\mathbf{r}$ gives

$$m\left[\mathbf{r}_G \times \left(\frac{d\mathbf{v}_G}{dt}\right)\right]_a + \frac{d}{dt}\left({}_G I_a \omega\right) = \sum(\mathbf{r}_G \times \mathbf{f})_{\text{ext},a} + \sum({}_G\mathbf{r} \times \mathbf{f})_{\text{ext},a} \tag{6 - 38}$$

where $\sum(\mathbf{r}_G \times \mathbf{f})_{\text{ext},a} = \left[\mathbf{r}_G \times \sum \mathbf{f}_{\text{ext}}\right]_a = m\left[\mathbf{r}_G \times \left(\frac{d\mathbf{v}_G}{dt}\right)\right]_a$ as a consequence of the conservation of linear momentum for a rigid body, equation (5-9). Note also that ${}_G I_a$ is independent of time and so $d({}_G I_a \omega)\,/\,dt = {}_G I_a \alpha$ where α is the angular acceleration ($\alpha \equiv d\omega\,/\,dt$). Using these results gives

$$\boxed{{}_G I_a \alpha = \sum({}_G\mathbf{r} \times \mathbf{f})_{\text{ext},a}} \tag{6 - 39}$$

After all this work, this is a wonderfully simple result. The conservation of angular momentum statement for planar motion of a rigid body does not depend upon the point in space chosen as the origin of the inertial coordinate system. It depends simply upon the forces acting on the body, their points of application *measured with respect to the center of mass*, the moment of inertia about the axis of rotation, but also *measured with respect to the center of mass*, and the angular acceleration of the body about a point which is fixed relative to all points in the body.

Note, however, that the term $_GI_a\omega$ is no longer correctly identified as the complete angular momentum of the body (even apart from the fact that it represents only the a-component). This terminology has meaning only with respect to an inertial frame and if the center of mass is accelerating, then the actual angular momentum must be calculated according to equation (6-33) from which it is clear that $_GI_a\omega$ is only that part of angular momentum which is due to rotation of the body with respect to the inertial frame.

We can also write the conservation of angular momentum for a rigid body for a finite time period. In this case we have

$$_GI_a(\omega_{end} - \omega_{beg}) = \int \sum (_Gr \times f)_{ext,a} dt \qquad (6-40)$$

Of course this result is an integration of the rate equation over time from the beginning time to the ending time. If the situation at hand is steady state, then the change in the angular momentum of the system, i.e., of the rigid body, is zero over any period of time. Then we have simply that the total torque integrated over time applied by the external forces is zero or equivalently the sum of the instantaneous external torques about the center of mass must be zero.

One further point should be noted. Equation (6-39) looks as though the conservation of angular momentum holds for a non-inertial frame. Nothing in reaching this result required that the center of mass be non-accelerating and the only position vectors which have survived ($_Gr$) are measured with respect to the center of mass. However, the forces must be those which would be observed (measured) in an inertial frame and the result is exactly equal to that which would be calculated using position vectors which are relative to the inertial frame origin, i.e., $\frac{d}{dt}\int(r \times v)dm = \sum(r \times f)_{ext}$. Hence, conservation of angular momentum is, in fact, with respect to an inertial frame.

Torques with Respect to an Arbitrary Point (B). Sometimes it is convenient to calculate external torques with respect to a point other than the center of mass. For example, if one of the external forces is unknown (friction, e.g.) then its unknown torque can be eliminated by calculating torques about its point of contact. This point, in general, may be non-inertial (i.e., accelerating) and is not necessarily fixed in the body. In this sense it is an arbitrary point. Suppose point B is such a point as shown in Figure 6-13. In equation (6-39) $_Gr$ can be expressed in terms of r_B, the position vector of point B:

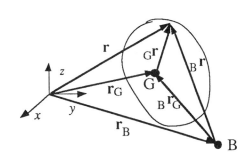

Figure 6-13: Coordinates with respect to point B

$$_Gr = r - r_G$$
$$= r - (r_G - r_B) - r_B \qquad (6-41)$$
$$= (r - r_B) - (r_G - r_B)$$

Then defining $_Br \equiv r - r_B$ (i.e., denoting posistion vectors measured from point B as $_Br$) gives

$$_Gr = _Br - _Br_G \qquad (6-42)$$

This substitution in equation (6-39) gives

$$\frac{d}{dt}\left({}_GI_a\omega\right) = \sum\left(\left({}_B\mathbf{r} - {}_B\mathbf{r}_G\right) \times \mathbf{f}\right)_{\text{ext},a} \tag{6-43}$$

or

$$\frac{d}{dt}\left({}_GI_a\omega\right) = \sum({}_B\mathbf{r} \times \mathbf{f})_{\text{ext},a} - \sum({}_B\mathbf{r}_G \times \mathbf{f})_{\text{ext},a} \tag{6-44}$$

Now, through conservation of *linear* momentum for a rigid body ($\sum \mathbf{f}_{\text{ext}} = d(m\mathbf{v}_G) / dt$),

$$\sum({}_B\mathbf{r}_G \times \mathbf{f})_{\text{ext},a} = \left[{}_B\mathbf{r}_G \times \sum \mathbf{f}_{\text{ext}}\right]_a = \left[{}_B\mathbf{r}_G \times \left(\frac{dm\mathbf{v}_G}{dt}\right)\right]_a \tag{6-45}$$

Substituting this result in equation (6-44) gives

$$\frac{d}{dt}\left({}_GI_a\omega\right) + m\left[({}_B\mathbf{r}_G \times \mathbf{a}_G)\right]_a = \sum({}_B\mathbf{r} \times \mathbf{f})_{\text{ext},a}$$

$$\frac{d}{dt}\left({}_GI_a\omega\right) + m\left[{}_B\mathbf{r}_G \times ({}_B\mathbf{a}_G + \mathbf{a}_B)\right]_a = \sum({}_B\mathbf{r} \times \mathbf{f})_{\text{ext},a} \tag{6-46}$$

where the expanded form arises because $\mathbf{r}_B + {}_B\mathbf{r}_G = \mathbf{r}_G$ and hence $\mathbf{a}_G = \mathbf{a}_B + {}_B\mathbf{a}_G$ by differentiation. Note also that this last form is equivalent to

$$\boxed{\frac{d}{dt}\left({}_GI_a\omega + m\left({}_B\mathbf{r}_G \times {}_B\mathbf{v}_G\right)_a\right) + m\left({}_B\mathbf{r}_G \times \frac{d\mathbf{v}_B}{dt}\right)_a = \sum({}_B\mathbf{r} \times \mathbf{f})_{\text{ext},a}} \tag{6-47}$$

because ${}_B\mathbf{v}_G \times {}_B\mathbf{v}_G \equiv \mathbf{0}$. In these equations all position and velocity vectors are measured with respect to point B, except for ${}_GI_a$ (which is with respect to the center of mass) and $\mathbf{v}_B$ (which is with respect to the origin of the inertial frame). In equation (6-46) $\mathbf{a}_G$ and $\mathbf{a}_B$ also are measured with respect to the inertial frame.

Torques with Respect to Point B Fixed in the Body. Note that if point B is fixed in the body (or at least fixed relative to every point in the body), then ${}_B\mathbf{v}_G = \omega \times {}_B\mathbf{r}_G$ and a result called the parallel axis theorem is obtained.

Parallel Axis Theorem. Consider an axis through the center of mass and a parallel axis through point B, a point which is fixed in the body. This is shown in Figure 6-14. In this case

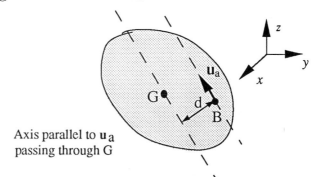

Axis parallel to $\mathbf{u}_a$
passing through G

Figure 6-14: Parallel Axis Theorem

$$\frac{d}{dt}\left(m(_B\mathbf{r}_G \times \boldsymbol{\omega} \times {_B}\mathbf{r}_G)a + {_G}I_a\omega\right) + m\left(_B\mathbf{r}_G \times \frac{d\mathbf{v}_B}{dt}\right)_a = \sum(_B\mathbf{r} \times \mathbf{f})_{ext,a}$$

$$\frac{d}{dt}\left((md^2 + {_G}I_a)\omega\right) + m\left(_B\mathbf{r}_G \times \frac{d\mathbf{v}_B}{dt}\right)_a = \sum(_B\mathbf{r} \times \mathbf{f})_{ext,a}$$

(6 – 48)

where $(_B\mathbf{r}_G \times \boldsymbol{\omega} \times {_B}\mathbf{r}_G)a = md^2$ by following an argument analogous to that for a particle, which led to equation (6-18).

Note that $_GI_a + md^2 = {_B}I_a$, where $_BI_a$ is the moment of inertia for the body relative to point B. This can be seen by starting with the definition of $_BI_a$:

$$_BI_a \equiv \int {_B}r_n^2\,dm$$

Then from equation (6-42) $_B r_n = d + {_G}r_n$ and so

$$_BI_n = \int (d + {_G}r_n)^2\,dm$$

and expanding terms and re-collecting gives

$$_BI_n = \int d^2\,dm + \int 2(_Gr_n)(d)\,dm + \int (_Gr_n^2)\,dm$$

Now in this form, the middle term is zero, identically, because d, the distance between an axis through the center of mass and a parallel axis through B is constant over the integral and $\int {_G}r_n\,dm = 0$ because of the definition of the center of mass. Consequently, by definition of $_BI_a$ and $_GI_a$,

$$_BI_a = md^2 + {_G}I_a$$

as was to be proven. Hence, equation (6-48) can also be written

$$\frac{d}{dt}(_BI_a\omega) + m(_B\mathbf{r}_G \times \mathbf{a}_B)a = \sum(_B\mathbf{r} \times \mathbf{f})_{ext,a}$$

provided point B is fixed in the body. This should be contrasted with equation (6-39) in which the acceleration of the center of mass does not appear even though it may be non-zero.

Furthermore, *if point B is non-accelerating*, then

$$\frac{d}{dt}\left(_BI_a\omega\right) = {_B}I_a\alpha = \sum(_B\mathbf{r} \times \mathbf{f})_{ext,a}$$

(6 – 49)

a result which also follows directly from equation (6-22) for the angular momentum of a body rotating about a fixed (or at least inertial) point together with equation (6-1) applied to a rigid body.

Table 6-6: Equivalent forms of the Conservation of Angular Momentum
for a Translating and Rotating Rigid Body: Planar Motion

Equation†	Comments
$$\frac{d}{dt}\left(\int_{\text{sys}} \mathbf{r} \times \mathbf{v}\,dm\right)_a = \sum(\mathbf{r} \times \mathbf{f})_{\text{ext},a}$$	$\mathbf{r}$, $\mathbf{v}$ are measured with respect to an inertial frame
$$_G I_a \alpha + m(\mathbf{r}_G \times \mathbf{a}_G)a = \sum T_a$$	$\mathbf{r}_G$, $\mathbf{a}_G$, and T_a are measured with respect to an inertial frame; $_G I_a$ is with respect to the center of mass, which is not necessarily an inertial point
$$_G I_a \alpha = \sum {}_G T_a$$	$_G T_a$ and $_G I_a$ are measured with respect to the center of mass which is not necessarily an inertial point
$$_G I_a \alpha + m[_B\mathbf{r}_G \times \mathbf{a}_G]_a = \sum {}_B T_a$$ $$_G I_a \alpha + m[_B\mathbf{r}_G \times ({}_{Ba}\mathbf{a}_G + \mathbf{a}_B)]_a = \sum {}_B T_a$$	$_B\mathbf{r}$, $_B\mathbf{r}_G$, $_B\mathbf{v}_G$ are measured with respect to an arbitrary point B which is not necessarily an inertial point and is not necessarily fixed in the body. $\mathbf{v}_B$ is the velocity of point B with respect to the inertial frame
$$_B I_a \alpha + m[_B\mathbf{r}_G \times \mathbf{a}_B]_a = \sum {}_B T_a$$ $$(_G I_a + md^2)\alpha + m[_B\mathbf{r}_G \times \mathbf{a}_B]_a = \sum {}_B T_a$$	Point B is not necessarily inertial but is fixed in the body.* d is the distance between an axis through the center of mass and a parallel axis through point B.
$$_B I_a \alpha = \sum {}_B T_a$$	Point B is inertial and is fixed in the body.*

†Subscript a indicates the component of the vector which is in the direction of the axis of rotation.
*Point B may also be outside the body so long as it is fixed relative to each point in the body.

This is the *parallel axis theorem. When assessing the conservation of angular momentum for a rigid body for rotation about a constant-direction axis by using torques about a point other than the center of mass but which is still fixed in the body and inertial, the moment of inertia must be shifted from the center of mass to that point according to*

$$_B I_a = {_G I_a} + md^2$$

(6 – 50)

Obviously, there are a number of different forms of the angular momentum conservation law which can be used to analyze any given problem. The forms which have been discussed in the chapter for planar motion are summarized in Table 6-6. Note that in this table the a component of the torque ($\mathbf{r} \times \mathbf{f}$) is represented as T_a. Which of these forms is most useful depends upon the problem at hand. Experience and ingenuity will be your guide.

Example 6-8. Consider a solid right cylinder with length 2 (m), radius 0.25 (m) and mass 30 (kg) as shown in Figure 6-15. There is an unknown frictional force between the surface of the cylinder and the horizontal surface. There is a 20 N horizontal force exerted on a frictionless axle and applied through the center of mass of the cylinder. Also, there is no slip at the contact of the horizontal plane and the cylinder. a) Calculate the moment of inertia of the cylinder with respect to its center of mass and for rotation about the z axis. b) Consider rigid body angular momentum conservation equations for i) torques with respect to the center of mass, ii) torques with respect to a stationary point B in the flat surface, iii) torques with respect to a point B fixed in the cylinder on its surface but not at the point of contact with the surface, and iv) a point B that is fixed in the cylinder and in contact with the surface.

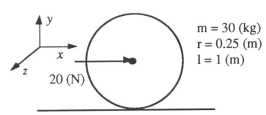

$$m = 30 \text{ (kg)}$$
$$r = 0.25 \text{ (m)}$$
$$l = 1 \text{ (m)}$$

20 (N)

Figure 6-15: Cylindrical roller

SOLUTION: a) The moment or inertia of the cylinder about its center of mass for rotation about the z axis is (Table C-7):

$$_G I_z = \frac{1}{2} m R^2 = \frac{1}{2}(30)(0.25)^2 \text{ (kg m}^2) = 0.94 \text{ (kg m}^2)$$

For case (i) we consider torques with respect to the center of mass as shown in Figure 6-16. The z-component of the conservation of angular momentum is:

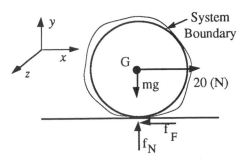

System Boundary

G

mg   20 (N)

f_F

f_N

Figure 6-16: Cylindrical roller, torques about the center of mass

$$_G I_z \alpha = \sum (_G \mathbf{r} \times \mathbf{f})_{\text{ext}, z}$$

The forces that act on the body are gravity, the normal force at the surface, the frictional force at the surface, and the force at the axle. However, the only force that contributes to the torque about the center of mass is the unknown frictional force because the position vector to the force in all other cases is either zero or collinear with the force. This reduces the conservation of angular momentum equation to:

$$_G I_z \alpha = (_G \mathbf{r} \times \mathbf{f})_{\text{frict}, z} = -R f_F$$

where f_F is the magnitude of the frictional force. (Note, therefore, that α, the angular acceleration will be negative. This is because a clockwise rotation of the cylinder will be in the negative angular direction.) However, we cannot proceed further with this equation alone because the frictional force is unknown. By using conservation of linear momentum and a kinematic relation between v_G and ω, α can be determined.

For the case (ii) we consider torques about a point B that is stationary as shown in Figure 6-17. The z-component of the conservation of angular momentum is:

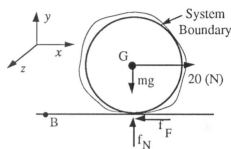

Figure 6-17: Cylindrical roller, torques about stationary point B

$$\frac{d}{dt} \left(_G I_z \omega + m(_B \mathbf{r}_G \times {_B}\mathbf{v}_G)_z \right) + m \left(_B \mathbf{r}_G \times \left(\frac{d\mathbf{v}_B}{dt} \right) \right)_z = \sum (_B \mathbf{r} \times \mathbf{f})_{\text{ext}, z}$$

The force of gravity and the normal force torques cancel because they are equal and opposite forces, are collinear, and have the same moment arm. Furthermore, the unknown frictional force does not contribute any torque because the line of action is collinear with the position vector. The only force that contributes to the rotation is the constant force on the axle. Also, $d\mathbf{v}_B / dt$ is zero since B is stationary. The velocity of the center of mass with respect to point B is simply the velocity of the center of mass since B is stationary. Thus, the z-component of the conservation of angular momentum equation is:

$$\frac{d}{dt} \left(_G I_z \omega + m(_B \mathbf{r}_G \times \mathbf{v}_G)_z \right) = (_B \mathbf{r} \times \mathbf{f})_{\text{axle}, z}$$

This equation can be solved by relating the velocity of the center of mass to the angular velocity through a kinematic relationship that will be discussed in Chapter 10, $v_G = -R\omega$.

For the case (iii) we consider torques about point B that is in the body as shown in Figure 6-18. The conservation of angular momentum is:

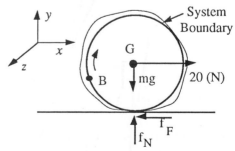

Figure 6-18: Cylindrical roller, torques about point B on the body

$$_GI_z\alpha + m(_B\mathbf{r}_G \times \mathbf{a}_B)_z = \sum(_B\mathbf{r} \times \mathbf{f})_{ext,z}$$

In this case, the only forces that do not contribute to the torques are the gravitational and normal forces because the magnitudes and line of action are opposite. However, depending on where B is during the rotation, the unknown frictional force and force at the axle both contribute torque. Furthermore, point B does not have constant velocity so $\mathbf{a}_B$ is not zero. Because the equation contains the unknown frictional force and the unknown acceleration $\mathbf{a}_B$, this approach is not very productive.

It should be noted, however, that at the instant that point B is in contact with the surface, its acceleration is perpendicular to the surface so that $\mathbf{r}_B \times \mathbf{a}_B = \mathbf{0}$ and equation 6-49 is recovered. It turns out that this result is then the same as for case (ii).

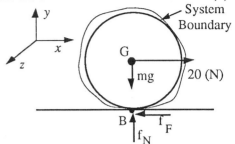

For case (iv) we consider torques about point B that is in the flat surface and that is moving with the contact point as shown in Figure 6-19. The z-component of the conservation of angular momentum in this case is:

Figure 6-19: Cylindrical roller, torques about point B at the contact

$$\frac{d}{dt}\left(_GI_z\omega + m(_B\mathbf{r}_G \times _B\mathbf{v}_G)_z\right) + m\left(_B\mathbf{r}_G \times \left(\frac{d\mathbf{v}_B}{dt}\right)\right)_z = \sum(_B\mathbf{r} \times \mathbf{f})_{ext,z}$$

The only force that contributes to the rotation is the force at the axle. The friction force contributes no torque because the position vector is zero, and the gravity and normal force vectors, besides being equal and opposite in line of action and of equal magnitude, have zero moment arm. Furthermore, the velocity of point B is equal to the velocity at point G so $_B\mathbf{v}_G = \mathbf{0}$ and $\mathbf{v}_B = \mathbf{v}_G$. This reduces the conservation of angular momentum equation to:

$$\frac{d}{dt}\left(_GI_z\omega\right) + m\left(_B\mathbf{r}_G \times \left(\frac{d\mathbf{v}_G}{dt}\right)\right)_z = (_B\mathbf{r} \times \mathbf{f})_{axle,z}$$

Furthmore, the position vector from point B to the center of mass is constant so:

$$\frac{d}{dt}\left(_GI_z\omega\right) + \frac{d}{dt}\left(m(_B\mathbf{r}_G \times \mathbf{v}_G)\right)_z = (_B\mathbf{r} \times \mathbf{f})_{axle,z}$$

or:

$$\frac{d}{dt}\left(_GI_z\omega + m(_B\mathbf{r}_G \times \mathbf{v}_G)\right)_z = (_B\mathbf{r} \times \mathbf{f})_{axle,z}$$

This statement is identical to the conservation of angular momentum for case (ii) and a solution for α is readily obtained. Incidentally, the result for α is

$$\alpha = \frac{-Rf}{_GI_z + mR^2}$$

where the minus sign occurs because $+\theta$ is in the counterclockwise direction.

The point of this example is that different approaches to the same problem are possible and, furthermore, that some approaches may be more direct than others. Do not be affraid to try different methods.

Example 6-9. Rework example 6-7 by using the parallel axis theorem.

SOLUTION: Because the moon's rotation rate is the same as its revolution rate, the center of mass of the earth is a point (B) fixed with respect to each point of the moon. Consequently, if it is a reasonable assumption that the earth's center of mass is an inertial point, then (m = the mass of the moon and r_G = distance from the earth to the moon)

$$l_z = {}_BI_z\omega_z = ({}_GI_z + mr_G^2)\omega_z$$
$$= \left[8.7 \times 10^{34}\text{kg m}^2 + (7.35 \times 10^{22} \text{ kg})(3.8 \times 10^8 \text{ m})^2\right] (2.66 \times 10^{-6} \text{ rad/s})$$
$$= 2.31 \times 10^{29}(\text{kg m}^2 / \text{s}) + 2.82 \times 10^{34}(\text{kg m}^2 / \text{s})$$
$$= 2.82 \times 10^{34}(\text{kg m}^2 / \text{s})$$

as was obtained in example 6-6.

Bodies with Planes of Symmetry. The preceeding results are independent of any symmetry (or lack thereof) of the body and, as a result, torques about axes other than the a axis are not necessarily zero; the other two equations of angular momentum are not necessarily trivial. Due to assymetry, torques may be required to maintain the direction of the axis of rotation constant. However, if the body is symmetric about a plane perpendicular to the axis of rotation (say the z-axis) through the center of mass, then the x- and y-components of torque (with respect to the center of mass) must be zero, i.e.,

$$0 = \sum({}_G\mathbf{r} \times \mathbf{f})_{\text{ext},x}$$
$$0 = \sum({}_G\mathbf{r} \times \mathbf{f})_{\text{ext},y} \tag{6 − 51}$$
$${}_GI_z\alpha = \sum({}_G\mathbf{r} \times \mathbf{f})_{\text{ext},z}$$

As a second symmetry case, if a body is symmetric about two of the planes which are defined by the coordinate axes and which contain the center of mass of the body, then rotation about an axis parallel to *any* of these coordinate axes produces one non-trivial equation of rotational motion and two trivial ones (having no motion and no torques with respect to those axes), just as above. In this case, the coordinate axes are also the *principal* axes of the body.

Example 6-10. A ball with radius 1.5 in and mass of 8 ounces is rotating about an axis through the center of mass of the ball at 1,800 revolutions per minute. Determine the angular momentum of the ball. If a constant force is applied in the θ direction on the ball and stops the rotation in 1/1000 of a second, what is the position and magnitude of the minimum constant force to stop the ball's rotation. The angular velocity vector is given as:

$$\mathcal{W} = 0\mathbf{e}_r + 0\mathbf{e}_\theta + 188.5\mathbf{e}_z \text{ (rad / s)}$$

SOLUTION: Figure 6-20 shows a schematic of the spinning ball. Choose the ball as the system with a cylindrical coordinate system situated at the center of mass of the ball and the z-axis aligned with the axis of rotation. The time period is 1/1000 of a second. The conservation of angular momentum says:

$$(l_{\text{sys}})_{\text{end}} - (l_{\text{sys}})_{\text{beg}} = \sum l_{\text{in/out}} + \int_t \sum(\mathbf{r} \times \mathbf{f})_{\text{ext}}dt$$

The system is closed so there is no mass entering or leaving the system, the body is rigid, and the force is constant over the time period giving

$$(l_{\text{sys}})_{\text{end}} - (l_{\text{sys}})_{\text{beg}} = \sum(\mathbf{r} \times \mathbf{f})_{\text{ext}}\Delta t$$

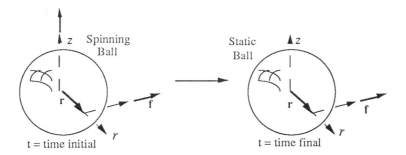

Figure 6-20: Spinning ball problem

The force vector and position vectors are given as:

$$\mathbf{r} = r\mathbf{e}_r + 0\mathbf{e}_\theta + z\mathbf{e}_z$$

$$\mathbf{f} = 0\mathbf{e}_r + f_\theta\mathbf{e}_\theta + 0\mathbf{e}_z$$

and the cross product of the position vector and the force vector is evaluated:

$$(\mathbf{r} \times \mathbf{f}) = \begin{vmatrix} \mathbf{e}_r & \mathbf{e}_\theta & \mathbf{e}_z \\ r & 0 & z \\ 0 & f_\theta & 0 \end{vmatrix} = (-f_\theta z)\mathbf{e}_r + 0\mathbf{e}_\theta + (rf_\theta)\mathbf{e}_z$$

Now if we assume that the angular momentum of the ball is only in the z-direction then the only component of the angular momentum vector that is non-zero is l_z. Table 6-7 displays the three coordinate equations for the consevation of angular momentum.

Table 6-7: Data for Example 6-10 (Angular Momentum, System = Spinning Ball)

Description	$(l_{sys})_{end}$ (lb$_m$ ft^2 / s)	$-$	$(l_{sys})_{beg}$ (lb$_m$ ft^2 / s)	$=$	$\sum(\mathbf{r} \times \mathbf{f})_{ext}\Delta t$ Angular Impulse (lb$_m$ ft^2 / s)
r-direction	$0\mathbf{e}_r$	$-$	$0\mathbf{e}_r$	$=$	$(-z)(f_\theta)\Delta t\mathbf{e}_r$
θ-direction	$0\mathbf{e}_\theta$	$-$	$0\mathbf{e}_\theta$	$=$	$0\mathbf{e}_\theta$
z-direction	$0\mathbf{e}_z$	$-$	$(l_z)_{sys}\mathbf{e}_z$	$=$	$r(f_\theta)\Delta t\mathbf{e}_z$

If we choose the value of z (distance from the equatorial plane to the point of application of the force) to be exactly zero, then the top row of the table is correct. However, if z is not zero than the assumption that the angular momentum was only in the z direction is wrong or we did not consider all the forces that were acting on the body. For now, we will say that z is zero and consider the bottom row of the table.

The z-component of the angular momentum can be expressed as the dot product of the unit vector in the z-direction with the angular momentum vector. Furthermore, the coordinate system was defined to be at the center of mass of the ball so:

$$_GI_z\omega = l_z$$

We can determine the moment of inertia of the ball about the z-axis from Table C-7 in Appendix C. For a sphere:

$$_GI_z = \left(\frac{2}{5}\right)mR^2$$

Substituting the values gives:

$$_G I_z = \left(\frac{2}{5}\right)(0.5)(1.5)^2 \text{ (lb}_m \text{ in}^2) \left(\frac{1 \text{ ft}}{12 \text{ in}}\right)^2 = 3.12 \times 10^{-3} \text{ (lb}_m \text{ ft}^2)$$

The initial angular momentum in the z-direction is:

$$(l_z)_{beg} = (_G I_z \omega_z)_{beg} = (3.12 \times 10^{-3})(188.5) = 5.88 \times 10^{-1} \text{ (lb}_m \text{ ft}^2 / s)$$

The final angular momentum in the z-direction is zero because the ball has stopped rotating. Looking at the z-component of the conservation of angular momentum, we obtain:

$$0 - (l_z)_{beg} = f_\theta r \Delta t$$

To find the minimum force in the θ direction, the value of r must be a maximum or the value of r is the radius of the ball and the force is acting at the skin of the ball as the ball slips.

$$r = 1.5 \text{ (in)}$$

Now we can solve for the magnitude of the force:

$$f_\theta = \left(\frac{-l_z}{r \Delta t}\right) = \left(\frac{-5.88 \times 10^{-1} \text{ (lb}_m \text{ ft}^2 / s)}{(1.5 \text{ in})(1 / 1000 \text{ s})}\right) \frac{12 \text{ in}}{1 \text{ ft}} \left| \frac{1 \text{ lb}_f}{32.174 \text{ lb}_m \text{ ft} / s^2} \right. = -146.4 \text{ (lb}_f)$$

A constant force of 146.4 (lb$_f$) directed in the negative θ direction and acting at the skin of the ball will stop the rotation of the ball in 1/1000 of a second.

5.6 THE EQUATION OF ROTATIONAL MOTION

The equation representing the conservation of angular momentum is used ultimately to describe the forces which act upon a body, their placement and the resulting motion of the body. For example, the rate equation is described in terms of the rotational inertia for that body and the angular acceleration. The angular acceleration involves the angular position of the body. Consequently this equation could be used to solve for the components of the angular acceleration in terms of the forces that are involved (i.e., the torques) and then the resulting rotational motion of the body could be analyzed by solving these equations. Obtaining such solutions and analyses is part of the broader subject of rigid body dynamics which we will study in more detail later in this text. The point here is that the equation of angular momentum provides a result along with the conservation of linear momentum which can be used to analyze the motion of a body.

5.7 LINEAR MOMENTUM VERSUS ANGULAR MOMENTUM

For the motion of an idealized particle, we can see that linear momentum and angular momentum are really the same statement. Because a particle is idealized to be of finite mass and zero volume it possesses no rotation and its angular momentum arises simply from its own velocity with respect to its position. More formally we may write the equation of linear momentum as:

$$\left(\frac{d\mathbf{p}_{sys}}{dt}\right) = m_{sys}\left(\frac{d\mathbf{v}_{sys}}{dt}\right) = \sum \mathbf{f}_{ext} \tag{5-9}$$

Here we are considering a single particle so we have no mass entering or leaving. The equation of angular momentum for this single particle is given by:

$$\frac{d}{dt}\left(\mathbf{r} \times m\mathbf{v}\right)_{sys} = \mathbf{r} \times m_{sys}\left(\frac{d\mathbf{v}_{sys}}{dt}\right) = \sum(\mathbf{r} \times \mathbf{f}_{ext}) = \mathbf{r} \times \sum \mathbf{f}_{ext} \qquad (6-52)$$

Note that because we are talking about a single particle with zero volume (a hypothetical situation), the position vector for the particle and for application of whatever external forces there may be are all the same and hence $\sum(\mathbf{r} \times \mathbf{f}_{ext}) = \mathbf{r} \times \sum \mathbf{f}_{ext}$. Also, note that $\mathbf{v}_{sys} \times \mathbf{v}_{sys} \equiv 0$. Consequently, for a single particle the conservation of angular momentum can be obtained directly from the statement of conservation of linear momentum, that is, conservation of angular momentum contains no additional information over that contained in a statement of conservation of linear momentum for a single particle. Nevertheless, angular momentum can provide a useful alternate perspective, as in example 6-4.

However, when we deal with a collection of particles, i.e., bodies of mass which are finite in size, then we must treat the individual pieces of mass which comprise the body as separate particles having different position vectors. In that case, then, in order to adequately describe the application of the forces and the relative motion of the elements of the system, we must use the law of conservation of angular momentum which now contains additional information above and beyond conservation of linear momentum. That is, linear momentum describes the momentum of the system and the way it changes in accordance with the action of external forces. Angular momentum, however, involves not just the velocity of the individual elements of the system and the external forces which are applied but also information about the position of each of these elements and about the position of application of the external forces. This is some important additional information which means that the statement of conservation of angular momentum is a result which is completely independent of the conservation of linear momentum.

6.8 REVIEW

1) The total angular momentum of a system *plus its surroundings* is conserved, that is, it is constant. Accordingly, there is no generation or consumption of angular momentum within a given system.

2) Angular momentum may enter or leave a system by virtue of mass entering or leaving a system in that as the mass enters or leaves, the angular momentum of that mass enters or leaves the system. Angular momentum may also be transferred between the system and surroundings as a result of mutual forces which exert torques between the system and surroundings. Because the forces and motions are mutual (opposite and equal), the angular momentum lost by the surroundings is gained by the system and vice versa in accordance with true conservation of momentum.

3) The angular momentum conservation equation includes angular momentum transferred between the system and its surroundings by virtue of mass crossing the system boundary and by virtue of forces which exert torques upon the system. The equation can be written for a differential time period as a rate equation or it can be integrated over time to give a finite time period integral equation. The rate of angular momentum entering or leaving the system with mass is calculated as angular momentum per unit mass times the rate at which mass enters or leaves the system.

4) The equation of conservation of angular momentum may be thought of as an equation of rotational motion in that it is used to describe the rotational motion (or lack thereof) of a system. The steady state equation is the equation used for analyzing nonaccelerating systems and the unsteady state equation is used for accelerating systems.

5) For planar motion (rotation about a constant-direction axis) the component of the conservation of angular momentum equation for a rigid body which is in the direction of the rotation axis (component a) reduces to

$$_{G}I_{a}\alpha = \sum (_{G}\mathbf{r} \times \mathbf{f})_{\text{ext},a} = \sum {_{G}T_{a}} \qquad (6-39)$$

6) For fluid systems, the angular momentum of the system may change as a result of change in mass of the system as well as a result of a change in its velocity and position.

6.9 NOTATION

The following notation was used for this chapter. The lower case boldface letters represent vector quantities and the upper case boldface letters represent tensor or matrix quantities. Angular vectors are represented by lower case greek letters with and arrow over the letter.

Scalar Variables and Descriptions		Dimensions
f	magnitude of the vector $\mathbf{f}$	[mass length / time^2]
g	magnitude of the acceleration of gravity	[length / time^2]
I	moment of inertia	[mass length^2]
m	total mass	[mass]
$\dot{m}$	total mass flow rate	[mass / time]
t	time	[time]
T	torque	[mass length^2 / time^2]
v_x, v_y, v_z	components of the velocity	[length / time]
x, y, z	spatial rectangular cartesian variables	[length]
ρ	density	[mass / length^3]

Vector Variables and Descriptions		Dimensions
$\mathbf{a}$	acceleration	[length / time^2]
$\mathbf{f}$	force	[mass length / time^2] or [force]
$\mathbf{i}, \mathbf{j}, \mathbf{k}$	cartesian unit directional vectors	
$\mathbf{e}_r, \mathbf{e}_\theta, \mathbf{e}_z$	cylindrical unit directional vectors	
$\mathbf{l}$	angular momentum	[mass length^2 / time] or [energy]
$\mathbf{p}$	linear momentum	[mass length / time] or [momentum]
$\mathbf{r}$	position	[length]
$\mathbf{u}$	general unit directional vector	
$\mathbf{v}$	velocity	[length / time] or [momentum / mass]
ω	angular velocity	[1 / time]
α	angular acceleration	[1 / time^2]

Pre-Subscripts

A, B, G position (or velocity or acceleration) is relative to
                 a point A, B, or G

Subscripts

a	a-component of a vector
A, B	corresponding to a point A or B
beg	evaluated at the beginning of the time period
end	evaluated at the end of the time period
ext	external to the system or acting at the system boundary
G	corresponding to the center of mass of the system
in	input or entering
out	output or leaving
sys	system or within the system boundary
R	radial component
T	transverse component

6.10 SUGGESTED READING

Beer, F. P., and E. R. Johnston, Jr., *Mechanics for Engineers, Statics and Dynamics*, 4th edition, McGraw-Hill, New York, 1987

Halliday, D. and R. Resnick, *Physics for Students of Science and Engineering, Part I*, John Wiley & Sons, 1960

Hibbeler, R. C., *Engineering Mechanics, Statics and Dynamics*, 3rd edition, MacMillan, New York, 1983

Sears, F. W., M. W. Zemansky, and H. D. Young, *University Physics*, Addison-Wesley, Reading, Massachussetts, 1982

QUESTIONS

1) Give a verbal statement of the conservation of angular momentum.

2) What is (how do you calculate) the angular momentum per unit mass for a particle?

3) How do you calculate the exchange of angular momentum between the system and its surroundings as a result of a force acting at the system boundary?

4) The conservation of angular momentum for a body of zero size is no different from the conservation of linear momentum. Explain.

5) The angular momentum of a finite-sized system may be broken into subsets of mass each having a different angular momentum. Explain why the momenta of these subsets may be different and explain how then to calculate the total angular momentum of the system.

6) Write a vector equation for the conservation of angular momentum for a system of finite-size.

7) What role does the moment of inertia play in the conservation of angular momentum?

8) Discuss the changes in the rotational speed and angular momentum of a figure skater spinning on the toes of the ice skates as the arms and legs are pulled in towards the axis of rotation or moved out away from the axis of rotation in the context of conservation of angular momentum. Assume that the contact between the ice and the ice skates is frictionless.

9) Discuss the use of a gyroscope on a ship or rocket to provide a measure of changes in direction to be used by the propulsion systems to maintain or change the direction of motion.

10) Consider the action of a spaceship in circular orbit around the earth. Treat it as a point mass with no thrusters firing. Does its angular momentum change as it orbits the earth? In what ways can the thrusters which are fired change the angular momentum?

11) Discuss the torque applied to a nut when it is being tightened onto a bolt using a wrench. In what ways can you change the amount of torque?

12) If a cat is held upside down with its back to the floor and dropped while motionless, it will somehow right itself and land on its feet. Now according to the conservation of angular momentum, because there are no external torques applied to the cat its angular momentum connot change. Yet, somehow, it manages to turn its body, thereby landing on its fet. How do you think it manages to do this? (Hint: Have faith in the conservation of angular momentum and its implications on the changes of the cat's angular momentum. Then what does this imply? Take heart in the fact that in order to fully understand this researchers photographed a falling cat with high speed motion photography).

SCALES

1) For the position vector $\mathbf{r} = 2\mathbf{i} + 3\mathbf{j} - 4\mathbf{k}$ and the force vector $\mathbf{f} = 2\mathbf{i} + 3\mathbf{j} - \mathbf{k}$ which acts at the point $(2, 3, -4)$, calculate $\mathbf{r} \times \mathbf{f}$ which is the torque exerted by $\mathbf{f}$ with respect to the origin of the coordinate system.

2) For $\mathbf{r} \times \mathbf{f}$ in exercise (1), what is $|\mathbf{r} \times \mathbf{f}|$?

3) For $\mathbf{r} \times \mathbf{f}$ in exercise (1), what is the angle θ between $\mathbf{r}$ and $\mathbf{f}$? (use a sketch to help you get the angle in the right quadrant.)

4) For $\mathbf{r} \times \mathbf{f}$ in exercise (1), compare the answer to exercise (2) to $|\mathbf{r}||\mathbf{f}| \sin \theta$.

5) For exercise (1) sketch (you should have already done this) $\mathbf{r}$ and $\mathbf{f}$ in a RCCS coordinate system. Also, sketch a vector which depicts $|\mathbf{r}| \sin \theta$, the magnitude of the moment arm of $\mathbf{f}$ with respect to the origin.

6) For $\mathbf{r} \times \mathbf{f}$ in exercise (1), what is the z-component of the torque? What is the torque exerted by $\mathbf{f}$ about the z axis?

7) For $\mathbf{r} \times \mathbf{f}$ in exercise (1), what is the x-component of the torque? What is the torque exerted by $\mathbf{f}$ about the x axis?

8) For $\mathbf{r} \times \mathbf{f}$ in exercise (1), what is the component of the torque in the direction of the vector $\mathbf{i} + \mathbf{j}$? (Don't forget to normalize $\mathbf{i} + \mathbf{j}$.) What is the torque exerted by $\mathbf{f}$ about an axis that passes through the origin and points in the direction of $\mathbf{i} + \mathbf{j}$?

9) How would your answers to exercises (1) through (8) change if you wanted to know about the torque with respect to the point $(1, 2, -2)$?

10) What is the expression for the moment of inertia of a thin disk with respect to an axis of symmetry which is perpendicular to the disk? What is the value of the moment of inertia if the disk radius is 2 cm, its thickness is 0.1 cm, and the density of the material is 1.5 g/cm^3? (Look the expression up in a table of moments of inertia.)

11) For the disk of exercise (10), what is an expression for its moment of inertia about an axis which is perpendicular to the disk and is tangent to its circumference? (Hint: this axis is parallel to that of exercise (10).) What is the value of the moment of inertia?

12) What is the expression for the moment of inertia of a right circular cone about an axis which contains its apex and is perpendicular to its axis of symmetry?

13) By integration and using the definition of moment of inertia (equation (6-23) of Chapter 6), show that the expression which you found in exercise (10) is correct. (Let the disk thickness be L and obtain a result for a cylinder of length L. Then, let $(L/R) \to 0$.) Do the integration also for the expression which you found for exercise (12).

14) For a given change in angular momentum (starting from zero), which object has the bigger ω for rotation about an axis through its center: a cylinder of radius R or a sphere of radius R? (The cylinder rotates about a longitudinal axis.)

15) Which is bigger, I_x, I_y, or I_z for a cylinder of radius R and length L. The x, y, and z axes are as defined in Table C-7.

16) Show by appropriate integration that the volume of a cylinder is $\pi R^2 L$, that the volume of a sphere is $4\pi R^3 / 3$, and that the volume of a cone of radius R at the base and height H is $\pi R^2 H / 3$.

17) What are the dimensions of angular momentum? Torque? Moment of inertia?

18) What are appropriate SI units of angular momentum, torque, and moment of inertia?

19) What are appropriate engineering units of angular momentum, torque, and moment of inertia for mass in slugs?

20) What are appropriate engineering units of angular momentum, torque, and moment of inertia for mass in pounds-mass?

PROBLEMS

1) For the given position vectors and force vectors, calculate the torque with respect to the origin.

  a) $\mathbf{r} = 1\mathbf{i} + 1\mathbf{j} + 1\mathbf{k}$ (ft), $\mathbf{f} = 1\mathbf{i} + 2\mathbf{j} + 3\mathbf{k}$ (lb$_f$)
  b) $\mathbf{r} = 1\mathbf{i} + 2\mathbf{j} + 0\mathbf{k}$ (ft), $\mathbf{f} = 2\mathbf{i} + 4\mathbf{j} + 0\mathbf{k}$ (lb$_f$)
  c) $\mathbf{r} = 0\mathbf{i} - 3\mathbf{j} - 2\mathbf{k}$ (m), $\mathbf{f} = 5\mathbf{i} - 1\mathbf{j} + 1\mathbf{k}$ (N)
  d) $\mathbf{r} = 2\mathbf{i} - 2\mathbf{j} + 1\mathbf{k}$ (m), $\mathbf{f} = 7\mathbf{i} + 1\mathbf{j} - 2\mathbf{k}$ (N)
  e) $\mathbf{r} = x\mathbf{i} + y\mathbf{j} + z\mathbf{k}$ (in), $\mathbf{f} = x^2\mathbf{i} + y^2\mathbf{j} + z^2\mathbf{k}$ (lb$_f$)
  f) $\mathbf{r} = y\mathbf{i} - e^y\mathbf{j} + y^2\mathbf{k}$ (in), $\mathbf{f} = y^2\mathbf{i} + e^y\mathbf{j} + y\mathbf{k}$ (lb$_f$)

2) For the forces in item a) through f) in problem 1, calculate the torque with respect to point $(1,-1,1)$.

3) Given the following vectors,

$$a = 1\mathbf{i} + 2\mathbf{j} + 0\mathbf{k} \text{ (m)}$$

$$b = 2\mathbf{i} - 0\mathbf{j} + 1\mathbf{k} \text{ (m)}$$

$$c = 0\mathbf{i} + 1\mathbf{j} - 2\mathbf{k} \text{ (m)}$$

$$d = -2\mathbf{i} + 0\mathbf{j} - 1\mathbf{k} \text{ (m)}$$

calculate the following vectors or scalars:

(a) $\mathbf{u}_a$	(e) $\mathbf{e} = \mathbf{a} \times \mathbf{b}$	(i) $\mathbf{a} \cdot \mathbf{b}$
(b) $\mathbf{u}_b$	(f) $\mathbf{f} = \mathbf{b} \times \mathbf{a}$	(j) $\mathbf{b} \cdot \mathbf{a}$
(c) $\mathbf{u}_c$	(g) $\mathbf{g} = \mathbf{c} \times \mathbf{a}$	(k) $\mathbf{c} \cdot \mathbf{a}$
(d) $\mathbf{u}_d$	(h) $\mathbf{h} = \mathbf{d} \times \mathbf{b}$	(l) $\mathbf{d} \cdot \mathbf{b}$

4) Given the following coordinates in meters

O (0,0,0)   Q (2,4,1)

  P (−1,2,3)   R (1,−1,−1)

calculate the following vectors. ($\mathbf{u}_e$ is a unit vector in the direction of $\mathbf{e}$)

(a) $_O\mathbf{r}_P$ (e) $\mathbf{e} = {_O\mathbf{r}_P} \times {_O\mathbf{r}_Q}$ (i) $\mathbf{u}_e$

(b) $_O\mathbf{r}_Q$ (f) $\mathbf{f} = {_O\mathbf{r}_P} \times {_P\mathbf{r}_Q}$ (j) $\mathbf{u}_f$

(c) $_P\mathbf{r}_Q$ (g) $\mathbf{g} = {_P\mathbf{r}_Q} \times {_P\mathbf{r}_R}$ (k) $\mathbf{u}_g$

(d) $_P\mathbf{r}_R$ (h) $\mathbf{h} = {_Q\mathbf{r}_P} \times {_R\mathbf{r}_P}$ (l) $\mathbf{u}_h$

5) For the given torque vectors, which are with respect to the origin, and the known force vectors given in items a) through c), determine a possible position vector (also with respect to the origin). (Multiple answers exist, representing the set of position vectors which provide the same moment arm with respect to $\mathbf{f}$.)

a) $\mathbf{r} \times \mathbf{f} = 12\mathbf{i} - 3\mathbf{j} + 11\mathbf{k}$ (N m), $\mathbf{f} = 3\mathbf{i} + 1\mathbf{j} - 3\mathbf{k}$ (N)

b) $\mathbf{r} \times \mathbf{f} = 15\mathbf{i} + 20\mathbf{j} + 7\mathbf{k}$ (lb$_f$ ft), $\mathbf{f} = 3\mathbf{i} - 4\mathbf{j} + 5\mathbf{k}$ (lb$_f$)

c) $\mathbf{r} \times \mathbf{f} = 2\mathbf{i} - 2\mathbf{j} + 1\mathbf{k}$ (N m), $\mathbf{f} = 1\mathbf{i} + 3\mathbf{j} + 4\mathbf{k}$ (N)

Also in each case, find the unique position vector which is normal to $\mathbf{f}$.

6) A system has three masses and each mass has a velocity and position vector (with respect to the origin) at time $t = 0$, as given below.

$m_1 = 10$ (kg) $\mathbf{v}_1 = 5\mathbf{i} + 2\mathbf{j} + 3\mathbf{k}$ (m / s) $\mathbf{r}_1 = 1\mathbf{i} + 1\mathbf{j} + 1\mathbf{k}$ (m)

$m_2 = 5$ (kg) $\mathbf{v}_2 = -2\mathbf{i} + 3\mathbf{j} - 4\mathbf{k}$ (m / s) $\mathbf{r}_2 = 2\mathbf{i} - 2\mathbf{j} + 1\mathbf{k}$ (m)

$m_3 = 15$ (kg) $\mathbf{v}_3 = -1\mathbf{i} - 2\mathbf{j} + 2\mathbf{k}$ (m / s) $\mathbf{r}_3 = 1\mathbf{i} - 1\mathbf{j} - 1\mathbf{k}$ (m)

The total linear momentum of the system, the velocity of the center of mass was determined in Chapter 5 Problem 3.

a) At $t = 0$ calculate the position of the center of mass with respect to the origin.

b) Calculate the angular momentum of the system with respect to the origin at $t = 0$.

c) Calculate the angular momentum of the system with respect to the center of mass.

7) Calculate the three moments of inertia (one for each axis of symmetry) about the center of mass for a right cylinder with mass m, length L, and radius r.

8) If the radius of the grindstone in example 6-1 is 1 (ft) and its moment of inertia about the axis of rotation is 60 lb$_m$ft^2, calculate the mass of the grindstone. If the force acting on the lateral surface of the grindstone is both normal to the position vector and angular momentum vector, calculate the average constant force that changed the momentum of the grindstone.

9) Given the static rigid body shown in Figure Problem 6-9, calculate the forces at point A and point B. The surface at A is smooth, and the force is normal to the surface. Do you need to calculate the moment of inertia for this problem? Why or why not?

10) The angular velocity with respect to the center of a flywheel decreases from 1000 (rev/min) to 400 (rev/min) as given by the vectors:

$$\boldsymbol{\omega}_{beg} = 1000\mathbf{i} + 0\mathbf{j} + 0\mathbf{k} \text{ (rpm)}$$

$$\boldsymbol{\omega}_{end} = 400\mathbf{i} + 0\mathbf{j} + 0\mathbf{k} \text{ (rpm)}$$

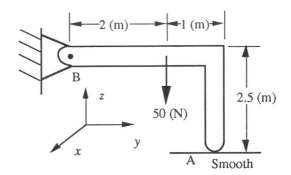

Problem 6-9: Static rigid body

The flywheel, as shown in Figure Problem 6-10, can be approximated by a thin plate with a diameter of 2 (ft) and mass of 20 (lb$_m$). Use the tables in Appendix C to calculate the moment of inertia about the x-axis. For the rotation about the x-axis, calculate the final and initial angular momentum. If the flywheel is defined as the system, must there be an external torque present. If so, why? If the torque exerted on the fly wheel is given as a function of time by the following vector:

$$(\mathbf{r} \times \mathbf{f})_{ext} = -b\,(\ln(at + 1))\,\mathbf{i} + 0\mathbf{j} + 0\mathbf{k}$$

where b = 50 (lb$_f$ ft), and a = 25 (1 / s). Determine the time required to decrease the flywheel's angular velocity from 1000 (rev/min) to 400 (rev/min).

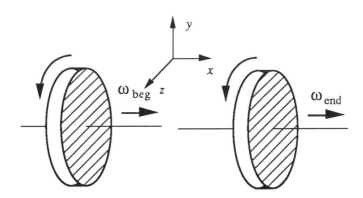

Problem 6-10: Rotating flywheel

11) Given a ball of string unraveling while rolling down a hill. If the system is chosen only as the string wound in the ball, discuss, in terms of the conservation of angular and linear momentum, what is happening to the system. Remember to include the force of gravity. Now consider a snowball that is started from the top of a steep hill. The snowball accumulates snow as it rolls down the hill. Discuss the conservation of linear and angular momentum equations that result if we define the snowball as the system.

12) A pitcher throws a curveball at 70 (mph) and the time that it takes for the ball to travel the 60'6" distance from the pitcher to the catcher is about 0.6 (s) if we neglect air resistance. The mass of the ball is 5 1/8 ounces.

a) If the ball is defined as the system, what is the linear momentum of the system while the ball is in motion. If the pitcher's hand exerts a force directed through the center of mass of the ball during the pitching motion, and that force is in contact with the ball for 0.5 (s), calculate the average force on the the ball. Does this force contribute to the rotation of the ball? Why or why not?

b) If the circumference of the ball is 9 (in), use the tables in Appendix C to calculate the moment of inertia about the center of the ball. If the angular velocity with respect to the center of the ball is 178 (rad/s), (the ball rotates 17 times from leaving the pitcher's hand to hitting the catcher's glove) calculate the angular momentum of the baseball with respect to the center of mass. During the release of the ball, the pitchers hand exerts a force on the skin of the ball causing it to rotate. If this force is present for 1/10 of a second right before the ball is released, determine the average torque exerted on the ball.

c) Based on your understanding of the conservation of linear and angular momentum, if both the pitcher and the ball are defined as the system and neglecting the force of gravity, is there a resultant external force on this system? Is there a resultant torque on this system?

13) For the given 2 dimensional static massless structure and cases a) through c), use the conservation of linear and angular momentum to calculate the reaction forces at points A and B. The surface at point B is smooth so the force is normal to the surface. Show that taking torques about point O in all cases, also satisfies the conservation of angular momentum.

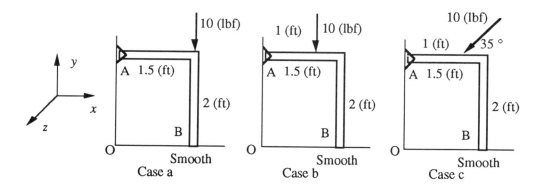

Problem 6-13: Static massless structure with force at different locations

14) A slender rod with length 3 ft and mass 15 lb$_m$ is rotating about its center of mass like a propeller with an initial angular velocity $(t = 0)$ of 2 k (rad / s). A constant torque acts on the rotating rod,

$$(r \times f)_{ext} = 0i + 0j - 2k \text{ (ft lb}_f)$$

Use the conservation of angular momentum to determine the angular velocity of the rod at 8 seconds.

15) A cylinder with length 1 m, radius 0.3 m and mass 20 kg is rotating about its center of mass like a propeller with an initial angular velocity $(t = 0)$ of -5k rad/s. A constant torque acts on the rotating cylinder

$$(r \times f)_{ext} = 0i + 0j + 2k \text{ (N m)}$$

Use conservation of angular momentum to determine at what time the angular velocity of the cylinder is zero.

16) Given a right cylinder with mass m, radius r, and length L, that is rolling down an inclined plane. The motion is perpendicular to the axis of the cylinder. The angle of inclination of the plane is β. The cylinder does not slip and gravity does act on the cylinder causing it to roll. Discuss the conservation of angular momentum equations for torques about the following points. State what forces would not contribute to rotation for each case. State what velocity, and acceleration terms are zero and why.

   a) The center of mass of the cylinder.

   b) Stationary point B on the surface of the incline plane.

   c) Stationary point B in the incline plane.

   d) Point B located on the surface of the moving cylinder.

   e) Point B located at the point of contact between the cylinder and incline plane.

17) In the previous problem, what complications arise if the incline plane is not of constant slope (such as a parabola, or exponential curve). In the conservation of angular momentum equations for torques about the different points, what is the only point that removes the unknown friction force from the conservation of angular momentum equations and why?

THE CONSERVATION OF ENERGY

.1 INTRODUCTION

In concept, the conservation of total energy for a process is no different from the conservation principles which we ave discussed previously. Similar to the conservation of mass and momentum, we define a system and a time period and onsider energy entering and leaving the system and the accumulation of energy within the system.

Hence,

$$\left\{ \begin{array}{c} \text{Accumulation of energy} \\ \text{within system during} \\ \text{time period} \end{array} \right\} = \left\{ \begin{array}{c} \text{Energy entering} \\ \text{system during} \\ \text{time period} \end{array} \right\} - \left\{ \begin{array}{c} \text{Energy leaving} \\ \text{system during} \\ \text{time period} \end{array} \right\}$$

here the accumulation of energy is defined as:

$$\left\{ \begin{array}{c} \text{Accumulation of energy} \\ \text{within system} \\ \text{during time period} \end{array} \right\} = \left\{ \begin{array}{c} \text{Amount of energy} \\ \text{contained in system} \\ \text{at end of} \\ \text{time period} \end{array} \right\} - \left\{ \begin{array}{c} \text{Amount of energy} \\ \text{contained in system} \\ \text{at beginning of} \\ \text{time period} \end{array} \right\}$$

ecause energy is conserved, generation and consumption of energy are zero. Also, as was true for the conservation of omentum, we must consider energy which is possessed by mass and therefore which crosses the system boundaries with at mass. Finally, we must consider other ways in which energy can cross the system boundaries independent of mass. So, y analogy to mass, charge, and momentum conservation, you already are familiar with the conservation of energy.

However, in other ways the conservation of energy is more complicated than either mass or momentum. The primary complication is that energy may exist in a variety of forms and this makes it somewhat more difficult to recognize and to account for. Mass, although it may exist as different species or elements or compounds, is still readily identified as mass. Likewise, momentum even though it may be transferred between the system and surroundings by forces, is still manifested solely in the form of motion of mass. Energy, on the other hand, can exist as kinetic energy, potential energy, and internal energy. Furthermore, internal energy can assume a variety of forms, leading to confusion. Apart from these classifications of energy, we can be confronted with mechanical energy (having to do with the motion of bodies), thermal energy (having to do with temperature), electrical energy, and nuclear energy.

The problems that involve the conservation of energy are numerous and encompassing. The generation of electric power from reservoirs, fossil fuels, and nuclear fuels involve the conservation of energy. The heat required to vaporize a liquid, and the work of an air conditioner (required to make heat flow from cold to hot!) are other examples of conservation of energy principles applied to everyday situations. Understanding the energy associated with bending beams and the behavior of electrical circuits also require conservation of energy principles.

Figure 7-1 gives a schematic of a system. Energy enters and leaves the system as the result of mass, heat, and work crossing the system boundary. The energy that the mass possesses is in the form of kinetic energy, potential energy, and internal energy of the mass.

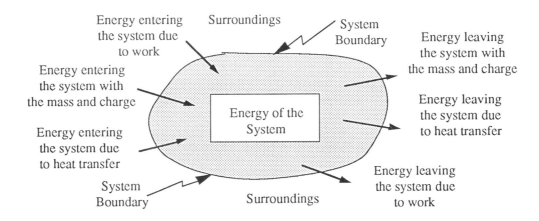

Figure 7-1: System and surroundings

Before we can proceed further with a detailed statement of the conservation of energy we must discuss these various forms of energy and the ways in which we will describe them quantitatively. This discussion will be presented in two sections: one on energy which is possessed by mass and the other on energy which exists *only* in transition (heat and work). Following these sections will be quantitative statements of the conservation of energy in both rate and integral forms and a more common form in terms of flow stream enthalpies (a thermodynamic property related to internal energy). Following these statements will be discussion and examples for closed systems, i.e. those which have no mass crossing the system boundaries, and for open systems where mass does cross the system boundaries. Finally, we will introduce the mechanical energy, thermal energy, and electrical energy accounting statements which count subset forms of energy. Total energy may be divided into mechanical, thermal and electrical energy forms, none of which is conserved, but which, nevertheless, provide useful information for evaluating processes. Example problems using these equations also are given. Finally, a review summarizes the main points of this chapter concerning energy conservation and its use in engineering problem solving.

7.2 ENERGY POSSESSED BY MASS

By energy possessed by mass we mean that because of its state of existence (temperature, pressure, composition, phase state) and motion and position in space, mass possesses energy. If we change any of these conditions, then we change its energy. While these relations are easily stated and can seem trivial, we must realize that discovering these forms of energy in the first place was not a trivial task.

It is essential that we understand the forms of energy possessed by mass for two reasons. First, we must be able to count the amount of energy contained within the system for application of the conservation of total energy; to the extent mass has energy, a system that contains that mass will have energy. Second, we must be able to count the amount of energy entering (or leaving) the system; to the extent mass has energy, that amount of energy can be carried into (or out of) the system by the mass. In the sections which follow we consider three forms of energy possessed by mass: kinetic, potential, and internal energy.

7.2.1 Kinetic Energy

Kinetic energy (KE) is the energy that a body possesses as a result of its motion. If a body of mass m is moving at velocity v, then it has kinetic energy $mv^2/2$. The velocity, of course, must be described with respect to a specific reference frame and consequently the amount of kinetic energy is relative. For example, the velocity of a particle (and hence kinetic energy) with respect to the surface of the earth will be different from the velocity (and hence kinetic energy) of the particle with respect to a train moving at 100 mph.

The *kinetic energy of a rigid body* can be separated into two forms as a matter of convenience: translational kinetic energy of the center of the mass and rotational kinetic energy with respect to the center of mass. This will be shown below. The total kinetic energy will be the sum of these two. If the system involves multiple bodies, then the total kinetic energy for each body must be summed over the number of bodies to obtain the total system kinetic energy.

The kinetic energy of the entire rigid body is the integral of the kinetic energies of the elements of the rigid body over the entire rigid body ($v^2 = \mathbf{v} \cdot \mathbf{v}$):

$$KE = \int \left(\frac{\mathbf{v} \cdot \mathbf{v}}{2} \right) dm \qquad (7-1)$$

As in Chapter 6, we can express the position vector of each element in terms of the position of the center of mass and the additional displacement from this center (Figure 6-10):

$$\mathbf{r} = \mathbf{r}_G + {}_G\mathbf{r}$$

then:

$$\mathbf{v} = \mathbf{v}_G + {}_G\mathbf{v}$$

and

$$\mathbf{v} \cdot \mathbf{v} = \mathbf{v}_G \cdot \mathbf{v}_G + {}_G\mathbf{v} \cdot {}_G\mathbf{v} + 2\mathbf{v}_G \cdot {}_G\mathbf{v}$$

so that:

$$
\begin{aligned}
KE &= \int \left(\frac{\mathbf{v} \cdot \mathbf{v}}{2} \right) dm \\
&= \int \left(\frac{\mathbf{v}_G \cdot \mathbf{v}_G}{2} \right) dm + \int \left(\frac{{}_G\mathbf{v} \cdot {}_G\mathbf{v}}{2} \right) dm + \int (\mathbf{v}_G \cdot {}_G\mathbf{v}) dm
\end{aligned}
\qquad (7-2)
$$

Because the velocity of the center of mass is independent of position within the mass, this result is equivalent to:

$$KE = \left(\frac{mv_G^2}{2}\right) + \int\left(\frac{_G\mathbf{v}\cdot{}_G\mathbf{v}}{2}\right)dm + \mathbf{v}_G\cdot\left(\int {}_G\mathbf{v}dm\right) \qquad (7-3)$$

The velocity relative to the center of mass is defined as the time derivative of the position vector relative to the center of mass or:

$$KE = \left(\frac{mv_G^2}{2}\right) + \int\left(\frac{_G\mathbf{v}\cdot{}_G\mathbf{v}}{2}\right)dm + \mathbf{v}_G\cdot\left(\int\left(\frac{d_G\mathbf{r}}{dt}\right)dm\right)$$

Because the mass of the system and its distribution *relative to the center of mass* does not change with time for a rigid body, the time derivative can be moved across the integral sign to give

$$KE = \left(\frac{mv_G^2}{2}\right) + \int\left(\frac{_G\mathbf{v}\cdot{}_G\mathbf{v}}{2}\right)dm + \mathbf{v}_G\cdot\frac{d}{dt}\left(\int {}_G\mathbf{r}dm\right)$$

This last integral, $\int {}_G\mathbf{r}dm$ is identically zero by the definition of the center of mass (see Chapter 5). The kinetic energy possessed by mass then is:

$$KE = \left(\frac{mv_G^2}{2}\right) + \int\left(\frac{_G\mathbf{v}\cdot{}_G\mathbf{v}}{2}\right)dm \qquad (7-4)$$

It now remains to calculate $\int {}_G\mathbf{v}\cdot{}_G\mathbf{v}dm$. Now $_G\mathbf{v}$ is the time rate of change of the position (displacement) vector with respect to the center of mass. Because the body is rigid, this displacement for each element cannot change in magnitude, only direction. Consequently, $_G\mathbf{v}$ consists only of a transverse component and not radial (see Chapter 6). Hence

$$_G\mathbf{v} = \boldsymbol{\omega}\times{}_G\mathbf{r} \qquad (6-6)$$

and

$$_G\mathbf{v}\cdot{}_G\mathbf{v} = (\boldsymbol{\omega}\times{}_G\mathbf{r})\cdot(\boldsymbol{\omega}\times{}_G\mathbf{r}) \qquad (7-5)$$

Now we have the vector identity (Appendix B):

$$(\mathbf{a}\times\mathbf{b})\cdot(\mathbf{c}\times\mathbf{d}) = (\mathbf{a}\cdot\mathbf{c})(\mathbf{b}\cdot\mathbf{d}) - (\mathbf{a}\cdot\mathbf{d})(\mathbf{b}\cdot\mathbf{c})$$

so that:

$$\begin{aligned}_G\mathbf{v}\cdot{}_G\mathbf{v} &= (\boldsymbol{\omega}\times{}_G\mathbf{r})\cdot(\boldsymbol{\omega}\times{}_G\mathbf{r}) \\ &= (\boldsymbol{\omega}\cdot\boldsymbol{\omega})({}_G\mathbf{r}\cdot{}_G\mathbf{r}) - (\boldsymbol{\omega}\cdot{}_G\mathbf{r})({}_G\mathbf{r}\cdot\boldsymbol{\omega})\end{aligned} \qquad (7-6)$$

Now $\boldsymbol{\omega}\cdot\boldsymbol{\omega} = \omega^2$ and $_G\mathbf{r}\cdot{}_G\mathbf{r} = {}_Gr^2$. Furthermore, $\boldsymbol{\omega}\cdot{}_G\mathbf{r} = \omega({}_Gr_a)$ [see figure 6-10 and note that $\boldsymbol{\omega}$ is a vector of magnitude ω and direction $\mathbf{u}_a$ where $\mathbf{u}_a$ is a unit vector aligned with the axis of rotation. Hence, $\boldsymbol{\omega}\cdot{}_G\mathbf{r} = \omega\mathbf{u}_a\cdot{}_G\mathbf{r} = \omega({}_Gr_a)$]. This gives that

$$_G\mathbf{v}\cdot{}_G\mathbf{v} = \omega^2{}_Gr^2 - \omega^2{}_Gr_a^2 \qquad (7-7)$$

which from Figure 6-10 and the Pythagorean theorem gives

$$_G\mathbf{v}\cdot{}_G\mathbf{v} = \omega^2{}_Gr_n^2 \qquad (7-8)$$

This result in equation (7-4) then gives the kinetic energy of the rigid body as

$$KE = \frac{1}{2}mv_G^2 + \frac{\omega^2}{2} \int {}_Grn^2 dm \tag{7-9}$$

The integral in this equation is defined to be the moment of inertia of the body about the axis of rotation through the center of mass:

$${}_GI_a \equiv \int {}_Grn^2 dm \tag{7-10}$$

Note the use of two subscripts on the moment of inertia to indicate that the moment of inertia is with respect to a *specific axis of rotation* which passes *through the center of mass*. For a given element of mass in the body (dm), ${}_Gr_n$ is the (perpendicular or normal [subscript n]) distance of this element to the axis of rotation.

To summarize, the kinetic energy is the sum of that due to translation of the center of mass at velocity $\mathbf{v}_G$ and that due to rotation at angular velocity $\boldsymbol{\omega}$ about the body's center of mass, as was to be shown. This rotational kinetic energy is a scalar and is usually expressed using the scalar rotation rate ($\omega^2 = \boldsymbol{\omega} \cdot \boldsymbol{\omega}$) and the moment of inertia of the body with respect of an axis passing through the mass center and collinear with $\boldsymbol{\omega}$:

$$\text{Rigid Body:}\quad KE = \frac{1}{2}mv_G^2 + \frac{1}{2}{}_GI_a\omega^2 \tag{7-11}$$

where ${}_GI_a$ is the moment of inertia of the body about uni-directional axis a which passes through the center of mass. Note that the dimensions of ${}_GI_a$ are (mass)(length^2). Table C-7 gives moments of inertia for some common bodies.

These separate considerations of rotational and translational kinetic energy arise from the fact that for finite sized bodies individual elements or pieces of that body can be doing different things even for the case of rigid bodies. For example, in the case of a spinning wheel undergoing no translational motion (for example, for a flywheel) particles at different radial positions will have different magnitudes of velocity. At the center of the wheel the velocity will be zero whereas at the perimeter it could be very large. Consequently the kinetic energy of a particle at the center may be drastically different from that of a particle at the rim of the wheel. The body can be treated as a whole, however, by treating translational and rotational motion separately and, in the case of the stationary flywheel, the translational kinetic energy is zero while the rotational kinetic energy may be quite large.

As shown above, the definitions of translational and rotational kinetic energies are done in such a way that they correspond exactly to calculations of energy done by considering each element separately and integrating over the entire body. To avoid having to do this in each instance, the rotational moment of inertia and the translational center of mass are defined. In this way the translational kinetic energy can be defined with respect to the motion of the center of mass and the rotational kinetic energy may be calculated in terms of the rotational moment of inertia. The fundamental concept, that for a single particle the kinetic energy is equal to $mv^2/2$, is the basis for rigid body definitions of ${}_Gv$, ω, and ${}_GI_a$.

Next, we will define the *specific* kinetic energy, i.e., the kinetic energy per unit mass according to

$$KE = m\widehat{KE} \tag{7-12}$$

and therefore, the specific kinetic kinetic energy of a rigid body is

$$\text{Rigid Body: } \widehat{KE} = \frac{1}{2}v_G^2 + \frac{1}{2}\left(\frac{_GI_a}{m}\right)\omega^2 \tag{7 – 13}$$

Example 7-1. Calculate the kinetic energy of a non-rotating mass of 5 lb$_m$ moving with a linear velocity of 3 ft/s.

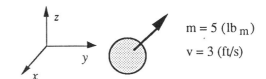

$$m = 5 \ (\text{lb}_m)$$
$$v = 3 \ (\text{ft/s})$$

Figure 7-2: Kinetic energy of a mass

SOLUTION: Figure 7-2 describes the kinetic energy of the mass. Since there is no rotational motion of the mass, $KE = mv_G^2/2$. Straightforward substitution yields: $KE = (5)(3)^2 / 2 = 22.5 \ (\text{lb}_m \ \text{ft}^2 / \text{s}^2)$ Converting to pounds force gives:

$$KE = \frac{22.5 \ \text{lb}_m \ \text{ft}^2}{\text{s}^2} \left| \frac{1 \ \text{lb}_f}{32.174 \ \text{lb}_m \ \text{ft} / \text{s}^2} \right. = 0.70 \ \text{ft} - \text{lb}_f$$

The kinetic energy of the mass is 0.7 (ft − lb$_f$). Note that while the units for energy and torque are the same, torque is a vector while energy is a scalar and they are, conceptually, very different properties.

7.2.2 Potential Energy

Potential energy (PE) is the energy that a body possesses as the result of its position in a potential force field. Work (energy) is associated with moving the body through the field and, *because of the nature of the field*, this work can be described in terms of a potential energy possessed by the body. Not all force fields can be treated in this way. The ones that can are special and are called *conservative force fields*.

With a change in potential energy of a body viewed within the context of work done to overcome a conservative force (f_C), the change in potential energy in moving from reference point *o* to the point of interest is defined to be

$$PE - PE_o = -\int_o f_C(s) \cdot d\mathbf{s} \tag{7 – 14}$$

The minus sign means that the force performing the work is acting against (opposite to) the conservative force f_C and increases the body's potential energy when it acts in the same direction as the displacement (e.g., when you lift a ball from the floor to the table, you provide a force which is opposite to gravity and in the direction of the ball's motion; you increase the ball's potential energy). Conversely, if the applied force and displacement act in the opposite direction, a decrease in potential energy results (When you lower the ball from the table to the floor, you provide a force which opposes gravity but the displacment is opposite this force and you lower the ball's potential energy).

The primary and most familiar example of a conservative force field and hence of potential energy is a gravitational force. In this case, a force is required to do work against the gravitational force as the body moves through the force field and the gravitational potential energy change then is:

$$PE - PE_o = -\int_o mg \cdot d\mathbf{s} \tag{7-15}$$

where the gravitational force per mass (a vector) in cartesian coordinates (with the z-axis aligned with gravity and the positive direction "up") is

$$\mathbf{g} = 0\mathbf{i} + 0\mathbf{j} - g\mathbf{k}$$

where $g = |\mathbf{g}| > 0$; the only non-zero component is in the z-direction. The differential displacement vector is

$$d\mathbf{s} = dx\mathbf{i} + dy\mathbf{j} + dz\mathbf{k}$$

Then, the dot product between $\mathbf{g}$ and $d\mathbf{s}$ is

$$PE - PE_o = -m\int_o (-g)dz \tag{7-16}$$

and if g is assumed to be constant over the displacement then

$$\boxed{PE - PE_o = mg(z - z_o)} \tag{7-17}$$

which is the familiar result for calculating changes in potential energy.

As with kinetic energy, it is useful to define a specific potential energy, the potential energy per unit mass such that

$$PE - PE_o = m(\widehat{PE} - \widehat{PE}_o) \tag{7-18}$$

Hence,

$$\widehat{PE} - \widehat{PE}_o = -\frac{1}{m}\int_o \mathbf{f}_C(\mathbf{s}) \cdot d\mathbf{s} \tag{7-19}$$

which for gravitational potential energy gives:

$$\widehat{PE} - \widehat{PE}_o = g(z - z_o) \tag{7-20}$$

Example 7-2. Determine the change in potential energy due to moving 10 (kg) from a height of 1 meter to a height of 20 meters.

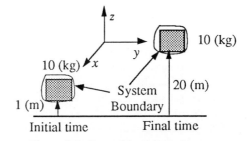

Figure 7-3: Potential energy of a mass

SOLUTION: Figure 7-3 describes the change in potential energy. The change in potential energy for the mass as the system is calculated as follows:

$$PE_{end} - PE_{beg} = \int_{beg}^{end} mgdz$$

For a constant mass and force per mass due to gravity:

$$PE_{end} - PE_{beg} = mg(z_{end} - z_{beg})$$

Using the previous equation and substituting the numerical values yields:

$$PE_{end} - PE_{beg} = (10)(9.8)(20 - 1) \text{ (kg m}^2 / \text{s}^2)$$

Evaluation gives:

$$PE_{end} - PE_{beg} = 1,862 \text{ (kg m}^2 / \text{s}^2) = 1,862 \text{ (J)}$$

The potential energy of the mass increased by 1,862 (J) by raising the mass from 1 (m) to 20 (m).

There are other conservative forces besides gravity and these also *could* be treated as giving rise to potential energy changes. For example, the potential energy associated with the displacement of a spring can be determined if we know the (conservative) force exerted by the spring. If the force is linear and is aligned in the x-direction the force is given by:

$$\mathbf{f}_C(s) = -kx\mathbf{i} + 0\mathbf{j} + 0\mathbf{k} \tag{7-21}$$

where k is defined as the spring constant and x is the amount of extension or compression from the equilibrium position ($x = 0$ is chosen as the point where $\mathbf{f}_C = 0$). Note that the spring force *resists* either extension or compression and hence the minus sign in equation (7-21).

The direction of the force exerted by the spring is collinear with the spring's compression or elongation direction. Equation (7-14), then gives for the potential energy

$$PE - PE_o = \int_o^x kxdx \tag{7-22}$$

and integration yields:

$$PE - PE_o = \frac{1}{2}k(x^2 - x_o^2) \tag{7-23}$$

7.2.3 Internal Energy

The two forms of energy possessed by mass (kinetic energy and potential energy) discussed above were the result of the motion of a body *as a whole* as opposed to the motions of individual molecules. Even where we consider smaller elements of a body of mass and add them up to obtain translational and rotational kinetic energy we are not talking about individual pieces of mass on the molecular scale. Instead we are still talking about pieces of mass which possess the properties of the matter as a whole.

On a smaller scale, however, we can think of mass in terms of the individual motions of the molecular and atomic particles. Each *individual* molecule can possess translational, rotational, and vibrational kinetic energy. Furthermore, a *collection* of molecules can possess different levels of potential energy which depend upon intermolecular separations, i.e., the distances between molecules. Force fields exist between molecules and consequently, the process of moving them closer together against these force fields requires work or, equivalently, changes in potential energy (because of the nature of the force fields the work can be considered as a type of potential energy).

All of these forms of molecular energy can be changed by changes in state of the mass (changes in temperature, composition, phase, etc.) and hence we must be able to include such changes in our energy accounting. To do so rigorously and in complete detail atom-by-atom and molecule-by-molecule is an impossible task. Consequently, we take a different tack. All of these forms of molecular motion (and atomic motion and subatomic motion as well) we lump together in a form of energy called *internal energy* and denote by the symbol U. Then, through appropriate measurements of temperature, pressure, volume, and energy transfers, it turns out that we can calculate changes in U.

Consider now how the internal energy of a body may be changed. Because internal energy depends upon molecular motion and intermolecular distances it becomes clear that changes in internal energy come about because of anything that changes these motions or distances. For example, as we add energy (say heat) to a material at constant volume the energy which is transferred to it goes into increased motion of the molecules and the internal energy increases. We observe this as an increase in the temperature of the material. As we compress the material at constant temperature (and thereby without increasing molecular motion), to the extent force fields exist between molecules we change the potential energy of the system and therefore its internal energy (if forces do not exist between molecules, as is the case for an ideal gas, then moving them closer together at constant temperature does not increase its internal energy. Finally, it should be noted that different materials generally have different values of internal energy even though they are at the same temperature and pressure.

Two more points concerning internal energy are noteworthy. First, as with both kinetic and potential energy, internal energy is calculated with respect to a reference point. We (rather arbitrarily) select a reference point and define the amount of internal energy at that point. Then we calculate the value of internal energy for conditions different from that reference state as the value at the reference state plus the amount of change as we move away from that reference state. Second, just as with kinetic and potential energies, and because internal energy is an extensive property, it is convenient to define a specific internal energy (internal energy per unit mass) such that the specific internal energy times the amount of mass in the system gives the total internal energy for the system:

$$U = m\hat{U}$$

$$(7 - 24)$$

We defer further discussion of how to calculate U (and other thermodynamic properties) to Chapter 15.

7.3 ENERGY POSSESSED BY CHARGE

Similar to the gravitational energy possessed by mass, charged particles can possess potential energy because of their relative positions in an electric field. In the presence of an electric field, a positive charge (q_+) is subjected to a force given by:

$$\mathbf{f}_C = q_+\mathbf{E}$$

$$(7 - 25)$$

where:

$$\mathbf{E} = \text{electric field vector} = \text{force per charge}$$

Let A and B be any two points in an electric field.

Any charge (q) at either of these points is subjected to the force as shown in Figure 7-4. Suppose a test charge of size (q_o) is at point B and we attempt to move it to point A. The external agent causing the test charge to move each infinitesimal distance ds in the field will do an amount of work given by $\mathbf{f}_C \cdot ds$, and the change in electrical potential energy (EPE) with respect to point B is given by the line integral

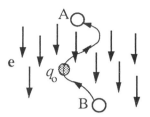

Figure 7-4: Electric field

$$_B EPE_A = -\int_B^A \mathbf{f}_C \cdot d\mathbf{s}$$

$$= -q_o \int_B^A \mathbf{E} \cdot d\mathbf{s}$$

$$(7-26)$$

The negative sign arises because the force that the external agent supplies is opposite the force on the test charge. Note that any displacement normal to the electric field does not contribute to the change in potential energy of the charged particle.

Furthermore we can define a potential energy per unit *charge* (V) so that the total electrical potential energy possessed by a charge is that potential energy per net charge times the amount of net charge:

$$\boxed{EPE - EPE_o = q(V - V_o)}$$

$$(7-27)$$

This potential energy per charge is given the name voltage. The unit of potential difference (voltage) is the volt (of dimensions energy/charge), which is defined as

$$1 \text{ (V)} = 1 \left(\frac{\text{Joule}}{\text{Coulomb}} \right)$$

$$(7-28)$$

In terms of voltage, then,

$$\left(\frac{_B EPE_A}{q_o} \right) = -\int_B^A \mathbf{E} \cdot d\mathbf{s}$$

$$= V_A - V_B$$

$$(7-29)$$

where $_B EPE_A$ is the amount of work required (or the change in potential energy) to move a positive charge of size $+q_o$ from point B to point A. Because the force is a conservative force, it can be demonstrated that the amount of work or the change in potential energy is independent of the path taken between B and A.

The change in electrical potential energy may be positive, negative, or zero depending on the characteristics of the electric field and the positions of the points A and B. If a positive change in potential energy is accomplished in moving the test charge from B to A, we say the electric potential at point A is higher than at point B, or the voltage $_B V_A$ is positive. If a negative change in potential energy is accomplished in moving the test charge from B to A, then $_B V_A$ will be negative. We say that $_B V_A = 0$ if there is no net change in potential energy. The net change in potential energy in moving a test charge from point B by any path and ending at point B will be zero, showing that the force due to an electric field is conservative.

Finally, the rate at which electrical potential energy enters or leaves a system is defined as the current (charge flow rate) entering of leaving the system times the specific potential energy (voltage, i.e., electrical energy per charge) for that current:

$$\text{electrical energy flow rate} = (\text{charge flow rate}) \times (\text{electrical energy per charge})$$

$$= (\text{power}) \qquad\qquad (7-30)$$

$$= iV$$

As suggested by the term power in equation (7-30), this electrical energy flow rate should be viewed as a work rate (see section 7.4.2).

Example 7-3. A battery is a two-terminal device that maintains a constant potential difference between its terminals. Connected to other devices in a circuit, it can provide electrical power. For example, a 12 volt automobile battery is constructed so that $_BV_A$ = 12 (V) where we have labelled the (+) terminal as point A and the (−) terminal as point B. It would take +12 (J) of work to move +1 (C) of charge from the negative to the positive terminal. In an electric circuit, charge moves from the battery to the circuit, thereby providing electrical energy (power).The rate of energy (power) supplied by a battery which is delivering current $i(t)$ at voltage V is $i(t)V$.

7.4 ENERGY IN TRANSITION

The previous section dealt with forms of energy possessed by mass or charge and which therefore may be conveyed into a system or out of a system by virtue of mass or charge exchange. In this section we are concerned with other forms of energy which may be transferred to or from the system independent of the transfer of mass or charge. These forms of energy are heat and work and they are forms of energy which exist only in transition. When we talk about heat or work, we are talking about energy that is moving from one place to another and, in the context of our discussions (which are concerned with analyzing systems and their behavior) it is energy which is transferred to or from a *system*. In the absence of such transfers we do not have heat or work. Also "heat" and "work" have no meaning in the absence of a system definition.

7.4.1 Heat

Heat refers to the transfer of thermal energy by virtue of a temperature difference, i.e., by virtue of a temperature driving force. For example, a cup of hot coffee sitting in a cold room will get colder because of the heat which is transferred from the cup to its surroundings. The first part of this statement refers to heat as the energy which is being transferred. Before the energy is transferred we do not say that the cup contains heat. Heat exists only in transition. The second part of this statement is that in order for heat to be transferred there must be a temperature difference which acts as a driving force. Without this difference there is no tendency for energy to move from one body to another as heat.

We will use the symbol Q to denote heat transfer, $\dot{Q}$ to denote the *rate* of heat transfer. We will also adopt the convention that:

> Heat is defined to be positive when energy is added to the system and negative when energy leaves the system.

Note that our definition of Q is tied directly to the definition of a system and surroundings; without a system boundary defined, the meaning of Q is undefined. If we define a different system, then the value of Q may be different. If to a defined system Q = 0, then the system (and process) is said to be *adiabatic*.

How the rate of heat transfer is related to the temperature difference between the system and the surroundings or to the temperature gradient at the boundary is the subject of a heat flux constitutive equation which will be covered more extensively in heat transfer courses. With the proper relationships one can evaluate the heat transfer between the system and surroundings for use in the energy conservation and accounting equations.

7.4.2 Work

Work is defined as energy transferred that is not heat; if it is not transferred due to a temperature difference, then it is work. The reason for this distinction lies in the Second Law of Thermodynamics and entropy generation, to be discussed in Chapter 8.

Additionally, work may be viewed as the result of a force acting *on the system* so as to move it (or part of it) *through a distance*. Force, of course, is a vector and distance is properly viewed as a displacement vector whereas energy is a scalar quantity. Work is calculated from these force and displacement vectors as the dot product of the two:

$$W = \int \mathbf{f}_{ext} \cdot d\mathbf{s}$$

(7 – 31)

Note that if the external force acts *with* the displacement (i.e., $\mathbf{f}_{ext} \cdot d\mathbf{s} > 0$), then $W > 0$ and energy enters the system.

No matter what phenomenon is responsible for the force, to the extent that force acts over a distance, work is performed. The force may be gravitational (in which case the work may, *alternatively* be counted as a change in potential energy); it may be the force (tension) in a cable being used to pull an object; it may be a force which acts on an electrical charge and which is created by an electric field (in which case the work may be counted as a change in electric potential energy); the force may be the interfacial tension of a fluid; or it may be any one of a number of other forces. The point is that if there is force and if that force acts through a displacement, then work is performed.

Note that it is not sufficient for a force to exist for work to be performed; a displacement in the direction of the force must also accompany that force. If a force exists in a cable suspending a body, for example, and that body is at rest because the force in the cable exactly balances the weight of that body, then no work is done. Likewise if a body is moved around in space in a direction normal (perpendicular) to the gravitational force, then the dot product of the gravitational force with the displacement is zero and no work is done by gravity.

Again as with heat, work is energy in transition and its definition is dependent upon the definition (implied or explicit) of a system boundary. If we change the definition of our system, then the amount of work that is performed may also change. We will choose the symbol W to denote work and the dot notation $\dot{W}$ to denote the rate of work (power). Furthermore we will adopt the convention that:

> Work is defined to be positive when energy is added to the system and negative when energy leaves the system.

This latter statement is contrary to classical thermodynamic conventions which say that work is positive when work is done by the system, i.e., when energy is transferred from the system to the surroundings. However, the convention adopted here, that work is positive when work is added to the system, is consistent with that adopted for heat; both are positive when they result in an increase in the energy of the system. Further discussions of work will elaborate upon the different forms which work may take as a result of the different causes of forces and these will be explained in various example problems.

Note again that *work rate* has a special name. The rate at which work is done we call *power*.

7.5 THE CONSERVATION OF ENERGY EQUATIONS

The conservation of energy equations in the rate form, integral form and the enthalpy form are given in this section. Note in these equations that, because total energy is conserved, there are no generation or consumption terms.

7.5.1 Rate Form of the Conservation Equation

With recognition of these forms of energy we are now ready to formulate a statement of the conservation of energy. We will assume that in general there are multiple input and output streams and the total amount of energy entering or leaving will then be calculated as the sum over these individual streams. As with previous laws, the mass input and output mass flow rate terms are combined into a single term with output having negative values.

The statement of conservation of energy then becomes:

$$\left(\frac{dE_{sys}}{dt}\right) = \sum\left(\dot{m}(\widehat{U} + \widehat{PE} + \widehat{KE})\right)_{in/out} + \sum\dot{Q} + \sum\dot{W}_{tot} \qquad (7-32)$$

In words, the energy of the system changes because of input and output due to mass exchange, and because of energy transfers as heat and work. The total energy of the system is represented as the sum of of all of the forms of energy that the system can possess. The system cannot possess heat or work since those quantities exist only as energy transfer across a system boundary. The energy of the system is given as:

$$E_{sys} = E_U + E_P + E_K \qquad (7-33)$$

where:

E_U	=	the internal energy of the system
E_P	=	the potential energy of the system
E_K	=	the kinetic energy of the system
$\sum\left(\dot{m}(\widehat{U} + \widehat{PE} + \widehat{KE})\right)_{in/out}$	=	the rate of energy entering/leaving with mass
$\sum\dot{Q}$	=	the net rate of heat entering
$\sum\dot{W}_{tot}$	=	the net rate of work entering (due to all forces except gravity, which is included as gravitational potential energy)

In accordance with the verbal statement given in the introduction, the left side of this equation represents energy entering and leaving the system, both with mass that enters and leaves, and as heat and work, while the right side represents changes in the total energy of the system (accumulation). Again, heat and work are not included in the system energy because these are not forms of energy which are possessed by the system; instead they exist only as energy transfer across the system boundaries. Note that this is a rate equation and represents the *rate* at which energy enters and leaves the system and the *rate* of accumulation of energy within the system at any instant in time.

7.5.2 Integral Form of the Conservation Equation

We can obtain a finite-time-period form of equation (7-32) by integrating each term over the time period

$$\int_{t_{beg}}^{t_{end}} \left(\frac{dE_{sys}}{dt} \right) dt = \int_{t_{beg}}^{t_{end}} \left(\sum \left(\dot{m}(\hat{U} + \widehat{PE} + \widehat{KE}) \right)_{in/out} \right) dt$$
$$+ \int_{t_{beg}}^{t_{end}} \left(\sum \dot{Q} \right) dt + \int_{t_{beg}}^{t_{end}} \left(\sum \dot{W}_{tot} \right) dt$$

(7 – 34)

Note that if the energies of the flow streams are constant (not changing with time), then the energy that enters and leaves is simply the specific energy quantities (energy per mass) times the total amount of mass that enters or leaves in that flow stream during that time interval. More generally, however, we write

$$(E_{sys})_{end} - (E_{sys})_{beg} = \sum \left(U + PE + KE \right)_{in/out} + \sum Q + \sum W_{tot}$$

(7 – 35)

Note the relationships between Q and $\dot{Q}$ and between W and $\dot{W}$:

$$Q = \int_{t_{beg}}^{t_{end}} \dot{Q} \, dt$$

$$W = \int_{t_{beg}}^{t_{end}} \dot{W} \, dt$$

7.5.3 Enthalpy Form of the Conservation Equation

It is common practice to write this total energy equation using the *enthalpy* of the flow streams in place of the internal energy (enthalpy is a thermodynamic property related to the internal energy as defined below). In this form of the equation the flow stream terms include the work which is required to introduce the entering mass into the system and that associated with removing the leaving mass from the system while omitting this work from the work term within the equation. This form of the equation is convenient for fluid streams.

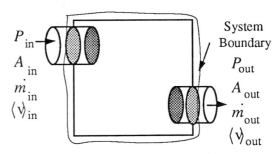

Figure 7-5: Flow work

To obtain this form of the energy equation from the rate form above (equation (7-32)) we must consider the work which is associated with moving mass into and out of the system across the system boundary. The total work (rate) can be split into two terms which account separately for the flow work and the non-flow work:

$$\sum \dot{W}_{\text{tot}} = \sum \dot{W}_{\text{flow}} + \sum \dot{W}_{\text{non-flow}}$$ (7 − 36)

The flow work is expressed in terms of the pressure, the cross-sectional area of the flow stream, and the magnitude of the average velocity normal to this area (Figure 7-5):

$$\sum \dot{W}_{\text{flow}} = \mathbf{f} \cdot d\mathbf{s} / dt = \mathbf{f} \cdot \mathbf{v}$$
$$= \sum (P A \langle v \rangle)$$ (7 − 37)
$$= \sum (P \dot{V})$$

Note that the force is calculated as pressure (i.e., force/area) times area and that distance/time is the velocity. Also, (area)(velocity) = $\dot{V}$ = volumetric flow rate. The reader should verify, by noting the dimensions of pressure and volumetric flow rate that the product has the dimensions of energy/time and therefore is dimensionally consistent with the rest of the energy equation. For entering flow and exiting flow the subscripts "in" and "out" are added

$$\sum \dot{W}_{\text{flow}} = \sum_{i=1}^{m} \left(\dot{W}_{\text{flow}} \right)_{\text{in}} - \sum_{i=1}^{n} \left(\dot{W}_{\text{flow}} \right)_{\text{out}}$$ (7 − 38)

and the injection and ejection work are:

Injection $$\sum_{i=1}^{m} \left(\dot{W}_{\text{flow}} \right)_{\text{in}} = \sum_{i=1}^{m} (P\dot{V})_{\text{in}}$$

(7 − 39)

Ejection $$\sum_{i=1}^{n} \left(\dot{W}_{\text{flow}} \right)_{\text{out}} = \sum_{i=1}^{n} (P\dot{V})_{\text{out}}$$

Through use of the material property of density, we can express the volumetric flow rate as the mass flow rate divided by the density as done in Chapter 3:

$$\sum_{i=1}^{m} \left(\dot{W}_{\text{flow}} \right)_{\text{in}} = \sum_{i=1}^{m} \left(P\frac{\dot{m}}{\rho} \right)_{\text{in}}$$

(7 − 40)

$$\sum_{i=1}^{n} \left(\dot{W}_{\text{flow}} \right)_{\text{out}} = \sum_{i=1}^{n} \left(P\frac{\dot{m}}{\rho} \right)_{\text{out}}$$

or

$$\sum_{i=1}^{m} \left(\dot{W}_{\text{flow}} \right)_{\text{in}} = \sum_{i=1}^{m} (\dot{m} P \widehat{V})_{\text{in}}$$

(7 − 41)

$$\sum_{i=1}^{n} \left(\dot{W}_{\text{flow}} \right)_{\text{out}} = \sum_{i=1}^{n} (\dot{m} P \widehat{V})_{\text{out}}$$

where the specific volume is defined as:

$$\hat{V} = \left(\frac{1}{\rho}\right)$$

$$= \text{the specific volume}$$

(7 – 42)

Moving the flow work from the work term to the entering and leaving mass streams then gives:

$$\left(\frac{dE_{sys}}{dt}\right) = \sum\left(\dot{m}(\hat{U} + \widehat{PE} + \widehat{KE} + P\hat{V})\right)_{in/out} + \sum\dot{Q} + \sum\dot{W}_{non-flow}$$

(7 – 43)

Now, the flow work terms can be combined with the internal energy terms and when this is done a new thermodynamic property is obtained called the enthalpy (per unit mass), defined to be

$$\widehat{H} = \hat{U} + P\hat{V}$$

$$= \hat{U} + \frac{P}{\rho}$$

(7 – 44)

The enthalpy form of the total energy conservation rate equation, then, is

$$\left(\frac{dE_{sys}}{dt}\right) = \sum\left(\dot{m}(\widehat{H} + \widehat{PE} + \widehat{KE})\right)_{in/out} + \sum\dot{Q} + \sum\dot{W}_{non-flow}$$

(7 – 45)

Note that in these latter equations the work term is not the same work as in the internal energy forms of the equation because the flow work has been separated out. The work term in this case is all work except the flow work. This enthalpy rate equation is exactly the same equation as the first rate equation, only in a different form. Examples of non-flow work include work associated with the expansion of the system volume against an external force or pressure, the work of pumps, electric motors, or compressors, and work associated with electrical current, microwaves, surface tension forces, etc.

The finite-time-period form of the conservation of energy in terms of enthalpy is:

$$(E_{sys})_{end} - (E_{sys})_{beg} = \sum\left(H + PE + KE\right)_{in/out} + \sum Q + \sum W_{non-flow}$$

(7 – 46)

7.5.4 Closed Systems

A closed system is defined to be one which has no mass crossing the system boundary, Because heat and work can still cross the boundary these remain as forms of energy entering and leaving the system. Again, heat is positive when it enters the system or flows from the surroundings to the system. Work is positive when the surroundings performs work on the system (energy is added to the system). For a closed system, the rate conservation of energy equation becomes:

$$\left(\frac{dE_{sys}}{dt}\right) = \sum\dot{Q} + \sum\dot{W}_{non-flow}$$

(7 – 47)

Note that since no mass enters or leaves, the only work is non-flow work; we cannot have flow work if no mass flows across the system boundary. The work does include, of course, any work associated with the system boundaries expanding (or contracting) against the surroundings pressure or force. This equation can be integrated over a finite time period to obtain the corresponding integral equation.

7.5.5 Isolated systems

An isolated system is defined to be one for which there is no exchange of anything between the system and the surroundings. The conservation of total energy for an islolated system is:

$$\left(\frac{dE_{sys}}{dt}\right) = 0$$

$$(7-48)$$

The total energy in a completely isolated system is constant; it is not a function of time. Because we consider the whole universe to be an isolated system, its energy is constant (in the absence of nuclear reactions and relativistic effects).

Table 7-1: Total Energy Conservation Equations[a]

Equation	Comment
$\left(\dfrac{dE_{sys}}{dt}\right) = \sum\left(\dot{m}(\hat{U} + \widehat{PE} + \widehat{KE})\right)_{in/out} + \sum\dot{Q} + \sum\dot{W}_{tot}$	U Form
$\left(\dfrac{dE_{sys}}{dt}\right) = \sum\left(\dot{m}(\widehat{H} + \widehat{PE} + \widehat{KE})\right)_{in/out} + \sum\dot{Q} + \sum\dot{W}_{non-flow}$	H Form
$\left(\dfrac{dE_{sys}}{dt}\right) = \sum\dot{Q} + \sum\dot{W}_{non-flow}$	Closed System
$\left(\dfrac{dE_{sys}}{dt}\right) = 0$	Isolated System

[a] $E_{sys} = E_U + E_P + E_K$
E_U = system internal energy
E_P = system potential energy
E_K = system kinetic energy

Table 7-1 summarizes these different forms of the total energy conservation equation. Use the equations only after you have clearly defined a system boundary, and time period. Also, keep in mind the physical significance and importance of each term in the equation; frequently, many of them will be zero.

Example 7-4. Consider the gas in a piston-cylinder chamber. Calculate the change in internal energy of the gas if 100 (J) of heat is transferred to the gas and it does 50 (J) of work by expanding against the piston and moving it to the right.

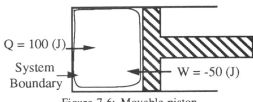

Figure 7-6: Movable piston

SOLUTION: Figure 7-6 gives a schematic of the gas system. The system is a closed system with no mass, charge or other energy crossing the system boundary. Furthermore, of the total energy of the system, only the internal energy of the gas changes. The potential and kinetic energies are constant. The time period is however long it takes to transfer the specified amounts of energy. The change in energy of the gas is an increase of 50 (J). The gas does not change potential energy and does not change velocity so the total change in energy of the system appears only as a change in internal energy. Table 7-2 is used to account for the energy terms during the process:

Table 7-2: Data for Example 7-4 (Total Energy, System = Gas in chamber)

$(E_{sys})_{end}$	$-$	$(E_{sys})_{beg}$	$=$	$\sum Q$	$+$	$\sum W_{non-flow}$
(J)		(J)		(J)		(J)
$(U_{sys})_{end}$	$-$	$(U_{sys})_{beg}$	$=$	100	$+$	(-50)

Example 7-5. A cylinder with a movable piston contains an ideal gas at 25 °C. The cylinder is immersed in a hot bath and 7 (kcal) of heat is transferred to the gas without moving the piston. The temperature of the gas increases to 100 °C. The piston is released and does 150 (J) of work in moving the piston. The temperature of the gas is maintained at 100 °C by the transfer of additional heat. The internal energy of the gas is assumed to be only a function of the temperature of the gas only. For the gas system, write the conservation of energy for each of the two steps and solve for the unknown energy term.

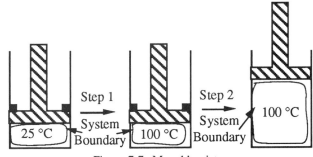

Figure 7-7: Movable piston

SOLUTION: Figure 7-7 shows the two-step process in which the gas is initially heated and the clamp is then released. The system is the gas contained in the cylinder. No mass crosses the system boundary. The conservation of energy for the first step is stated in Table 7-3. The time period to accomplish this transfer of heat is "however long it takes".

The system does not change potential or kinetic energy. The change in internal energy is 7 (kcal):

$$(U_{sys})_{end} - (U_{sys})_{beg} = \frac{7 \text{ kcal} \mid 4184 \text{ J}}{\mid 1 \text{ kcal}} = 29,300 \text{ J}$$

Table 7-3: Data for Example 7-5a (Total Energy, System = Gas in chamber)

$(E_{sys})_{end}$	$-$	$(E_{sys})_{beg}$	$=$	$\sum Q$	$+$	$\sum W_{non-flow}$
(kcal)		(kcal)		(kcal)		(kcal)
$(U_{sys})_{end}$	$-$	$(U_{sys})_{beg}$	$=$	7	$+$	0

Table 7-4: Data for Example 7-5b (Total Energy, System = Gas in Chamber)

$(E_{sys})_{end}$	$-$	$(E_{sys})_{beg}$	$=$	$\sum Q$	$+$	$\sum W_{non-flow}$
(J)		(J)		(J)		(J)
$(U_{sys})_{end}$	$-$	$(U_{sys})_{beg}$	$=$	?	$+$	(-150)

For Step 2, the system is still closed with no mass or charge crossing the system boundary and the process takes however long it needs to to perform the work. Table 7-4 gives the conservation of energy for this process and this system.

Because the temperature of the gas does not change and its internal energy depends only on temperature, its internal energy does not change; (a result which, although not generally true, does hold for ideal gases, as will be shown in Chapter 15):

$$(U_{sys})_{end} - (U_{sys})_{beg} = 0$$

Solving for the heat transfer term:

$$Q = 150 \text{ (J)}$$

To maintain the system temperature at 100 °C, 150 (J) of heat must have been transferred to the system after or during the expansion.

7.6 STEADY STATE

For *steady-state processes*, the system properties do not change with time and the resulting rate equation is:

$$0 = \sum \left(\dot{m}(\widehat{H} + \widehat{PE} + \widehat{KE}) \right)_{in/out} + \sum \dot{Q} + \sum \dot{W}_{\substack{non-flow \\ non-exp}} \qquad (7-49)$$

The important thing to note is that for steady-state processes, there is no expansion work because the system volume is constant (Everything about the system is steady state.). The steady-state conditions can apply to both open and closed systems.

Example 7-6. A hydroelectric plant operating at steady state is fed by a stream of falling water from a height of 100 (ft). How many tons of water per hour are needed to keep a 60 watt light bulb burning.

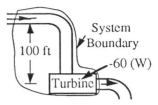

Figure 7-8: Hydroelectric plant

SOLUTION: Figure 7-8 shows the turbine system with the inlet and exit streams. The system is the turbine plus the 100 ft of inlet pipe above the turbine. Table 7-6 gives the conservation of energy for the open system in which no charge is entering of leaving. Along with the conservation of energy, a conservation of total mass for the system is presented in Table 7-5.

Table 7-5: Data for Example 7-6 (Mass, System = Turbine and inlet stream)

Description	$\left(\dfrac{dm_{sys}}{dt}\right)$	=	$\sum \dot{m}_{in/out}$		
	(ton/hr)		(ton/hr)		(ton/hr)
Total	0	=	$\dot{m}_{in}$	$-$	$\dot{m}_{out}$

From the conservation of total mass we know (steady-state):

$$\dot{m}_{in} = \dot{m}_{out}$$

The change in velocity of the water is assumed negligible so the difference in the kinetic energy entering and leaving is zero. The difference in the entering and leaving enthalpies is approximately zero since the water stays at the same temperature. No heat transfer occurs.

Table 7-6: Data for Example 7-6 (Total Energy, System = Turbine and inlet stream)

$\left(\dfrac{dE_{sys}}{dt}\right)$	=	$\sum\left(\dot{m}(\widehat{H} + \widehat{PE} + \widehat{KE})\right)_{in/out}$		+	$\sum \dot{Q}$	+	$\sum \dot{W}_{non-flow}$
		$(\dot{m}\widehat{PE})_{in}$	$(\dot{m}\widehat{PE})_{out}$				$\sum \dot{W}_{non-flow}^{non-exp}$
(W)		(W)	(W)		(W)		(W)
0	=	$(\dot{m}gz)_{in}$ $-$	$(\dot{m}gz)_{out}$	+	0	+	(-60)

From Tables 7-5 and 7-6, the conservation of energy and mass says:

$$\dot{m}g(z_{in} - z_{out}) = 60 \text{ (W)}$$

The only unknown is the mass flow rate of water. Solving for $\dot{m}$ gives:

$$\dot{m} = \frac{60 \text{ W}}{\left|\begin{array}{c|c} 1 & s^2 \\ \hline 100 \text{ ft} & 32.174 \text{ ft} \end{array}\right|} = 0.0186 \text{ (W s}^2 / \text{ft}^2)$$

Using the proper conversion factors:

$$\dot{m} = \frac{0.0186 \text{ W s}^2}{\text{ft}^2} \left| \frac{0.7376 \text{ lb}_f \text{ ft}}{1 \text{ W s}} \right| \frac{32.174 \text{ lb}_m \text{ ft}}{1 \text{ lb}_f \text{ s}^2} \left| \frac{3600 \text{ s}}{1 \text{ hr}} \right| \frac{1 \text{ ton}}{2000 \text{ lb}_m} = 0.79 \text{ (ton} / \text{hr)}$$

To keep the light bulb going, 0.79 tons of water needs to fall per hour. This is about 1 cubic yard of water per hour.

Example 7-7. For the system operating at steady state described by Figure 7-9, solve for the change in specific enthalpy of the gas. The following table describes the entering and exiting streams

$\dot{m}_{in}$ = 230 kg / hr
$\dot{m}_{out}$ = 230 kg / hr
v_{in} = 53 ft / s
v_{out} = 352 ft / s
$\widehat{H}_{in}$ =? kJ / kg
$\widehat{H}_{out}$ =? kJ / kg
$\dot{Q}$ = 1000 kcal / hr leaving the turbine
$\dot{W}$ = 300 kW leaving the turbine

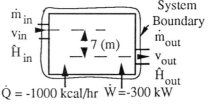

Figure 7-9: Gas Turbine

SOLUTION: The system is the gas turbine and the conservation of energy for the open system is presented in Table 7-8. Along with the steady-state conservation of energy, the steady state conservation of total mass is given in Table 7-7.

Table 7-7: Data for Example 7-7 (Mass, System = Turbine)

Description	$\left(\dfrac{dm_{sys}}{dt}\right)$ (kg/hr)	=	$\sum \dot{m}_{in/out}$ (kg/hr)		(kg/hr)
Total	0	=	$\dot{m}_{in}$	−	$\dot{m}_{out}$

Table 7-8: Data for Example 7-7 (Total Energy, System = Turbine)

$\left(\dfrac{dE_{sys}}{dt}\right)$	$=$	$\sum\left(\dot{m}(\widehat{H}+\widehat{PE}+\widehat{KE})\right)_{in/out}$	$+$	$\sum\dot{Q}$	$+$	$\sum\dot{W}_{non-flow}$
$\left[\dfrac{\text{Energy}}{\text{time}}\right]$		$\left[\dfrac{\text{Energy}}{\text{time}}\right]$		$\left[\dfrac{\text{Energy}}{\text{time}}\right]$		$\left[\dfrac{\text{Energy}}{\text{time}}\right]$
0	$=$	$\left(\dot{m}(\widehat{H}+gz+\dfrac{v^2}{2})\right)_{in/out}$	$+$	$\dot{Q}$	$+$	$\dot{W}_{non-flow\ non-exp}$

From the conservation of total energy and mass together, the following result is obtained:

$$0 = \dot{m}\left(\widehat{H}_{in} - \widehat{H}_{out} + gz_{in} - gz_{out} + \frac{v_{in}^2}{2} - \frac{v_{out}^2}{2}\right) + \dot{Q} + \dot{W}_{non-flow\ non-exp}$$

Dividing both sides of the equation by the mass flow rate and rearranging gives:

$$\widehat{H}_{in} - \widehat{H}_{out} = -\frac{\dot{W}_{non-flow\ non-exp}}{\dot{m}} - \frac{\dot{Q}}{\dot{m}} - g(z_{in} - z_{out}) - \left(\frac{v_{in}^2}{2} - \frac{v_{out}^2}{2}\right)$$

Because each of the terms is in different units, it will be easier to convert units term-by-term. The only unknown is the change in specific enthalpy of the gas; all the terms on the right side of the equation are known. The rate of work done by the system per mass flow rate is

$$-\frac{\dot{W}_{non-flow\ non-exp}}{\dot{m}} = -\frac{-300}{230} \quad \frac{-300\text{ kW}}{230\text{ kg}}\left|\frac{\text{hr}}{}\right|\frac{1\text{ kJ}/\text{s}}{1\text{ kW}}\left|\frac{3600\text{ s}}{1\text{ hr}}\right| = 4,695.6 \ (\text{kJ}/\text{kg})$$

The rate of heat transfer per mass flow rate is:

$$-\frac{\dot{Q}}{\dot{m}} = -\frac{-10^3\text{ kcal}}{\text{hr}}\left|\frac{\text{hr}}{230\text{ kg}}\right|\frac{1\text{ kJ}}{0.2391\text{ kcal}} = 18 \ (\text{kJ}/\text{kg})$$

The change in specific potential energy is:

$$-g(z_{in} - z_{out}) = -\frac{9.8\text{ m}}{s^2}\left|\frac{7\text{ m}}{}\right|\frac{1\text{ N}}{1\text{ kg m}/\text{s}^2}\left|\frac{1\text{ kJ}}{1000\text{ N m}}\right| = -0.069 \ (\text{kJ}/\text{kg})$$

And the change in specific kinetic energy:

$$-\left(\frac{v_{in}^2}{2} - \frac{v_{out}^2}{2}\right) = -\left(\frac{53^2}{2} - \frac{352^2}{2}\right)\frac{\text{ft}^2}{s^2}\left|\frac{1\text{ m}^2}{3.281^2\text{ ft}^2}\right|\frac{1\text{ N}}{1\text{ kg m}/\text{s}^2}\left|\frac{1\text{ kJ}}{1000\text{ N m}}\right| = 5.63 \ (\text{kJ}/\text{kg})$$

Substituting these numeric values in to the specific enthalpy equation gives:

$$\widehat{H}_{in} - \widehat{H}_{out} = 4695.6 + 18 - 0.069 + 5.63 \ (\text{kJ}/\text{kg}) = 4,720 \ (\text{kJ}/\text{kg})$$

The change in specific enthalpy of the gas is $-4,720$ (kJ/kg). Notice the relative magnitudes of the energy terms. The total change in specific enthalpy is almost equal to the work of the system per unit mass while the change in potential energy was negligible in this problem.

7.7 ENERGY ACCOUNTING EQUATIONS

Because total energy is conserved it becomes meaningful and useful to sometimes consider the accounting of certain subsets of energy. Mechanical energy, thermal energy, and electrical energy are three forms of the total energy picture which exist but which are not totally separate entities in that conversions between them can occur. Frictional processes, for example, convert mechanical energy to thermal energy (rub your hands together, exerting a force over a distance, and your hands will warm up implying an increase in their internal energy). Before proceeding further, however, we need to clarify the concepts of mechanical, thermal, and electrical energy.

Mechanical energy is kinetic energy, potential energy, and work. These forms of energy involve the motion of bodies as a whole, the displacement of mass, and forces involving these motions and displacements. Thermal energy includes internal energy and heat transfer. Electrical energy refers to the energy associated with the flow of electrical current and with electric and magnetic fields. Clear examples of conversions from one form of energy to another are resistance heating in a resistor caused by electrical current (as the resistor heats up its internal energy or thermal energy increases; electrical energy has been converted to thermal energy) and the expansion of a gas to move a piston and exert a force, thereby performing work (the conversion of thermal energy in the form of internal energy of the gas to mechanical energy in the form of work).

The development of the individual accounting equations in a rigorous way is well beyond the scope of this course. To do so for mechanical energy, for example, requires integrating the linear momentum equations which are differential in both time and position over the finite volume of the system and expressing fluid stresses in a tensor form throughout the integration.

7.7.1 Mechanical Energy Accounting Equation

Let us first write the verbal statement of the accounting of mechanical energy for a given system and time period as:

$$\left\{ \begin{array}{c} \text{Accumulation of mechanical} \\ \text{energy within system} \\ \text{during time period} \end{array} \right\} = \left\{ \begin{array}{c} \text{Mechanical energy} \\ \text{entering system} \\ \text{during time period} \end{array} \right\} - \left\{ \begin{array}{c} \text{Mechanical energy} \\ \text{leaving system} \\ \text{during time period} \end{array} \right\}$$

$$+ \left\{ \begin{array}{c} \text{Mechanical energy} \\ \text{generated in system} \\ \text{during time period} \end{array} \right\} - \left\{ \begin{array}{c} \text{Mechanical energy} \\ \text{consumed in system} \\ \text{during time period} \end{array} \right\}$$

As was discussed before, the mechanical energy entering and leaving the system is in the form of kinetic, potential energy and work. Mechanical energy generation could be thought of in terms of an electric motor in which mechanical energy is generated through the conversion of electrical energy. Mechanical energy consumption represents the conversion of mechanical energy to thermal or electrical energy such as frictional losses in fluid flow or conversions to electric energy by turbines.

The macroscopic statement of the accounting of mechanical energy is obtained by integrating the linear momentum equation on the continuum level over the volume of the defined system and along the specified path (Bird, 1957). We will not present the derivation as it is well beyond the scope of this text. Furthermore, the mechanical energy accounting equation cannot be obtained by integrating the macroscopic linear momentum equation as given in Chapter 5 over the system volume. Because we have to integrate the continuum equation over the system volume, the macroscopic mechanical energy accounting equation is independent of the macroscopic linear momentum conservation law.

Because of the complexity of the accumulation term in the mechanical energy accounting equation we will limit further discussion to steady-state situations with only one entering mass flow stream and one leaving mass flow stream. If a situation arises involving an unsteady-state problem, it is advised to use the conservation of total energy. Furthermore, we will now consider situations that involve only mechanical and thermal energy; we will not include electrical or other forms of energy that may convert to mechanical energy.

The steady-state mechanical energy accounting rate equation for *one entering* and *one leaving* stream on the macroscopic level is:

$$0 = \dot{m}(\widehat{PE}_{in} - \widehat{PE}_{out}) + \dot{m}(\widehat{KE}_{in} - \widehat{KE}_{out}) + \dot{m}\left(\int_{P_{out}}^{P_{in}} \widehat{V} dP\right)$$
$$+ \sum \dot{W}_{shaft} - \sum \dot{F} \qquad (7-50)$$

or in alternate form obtained by noting that $d(P/\rho) = Pd\widehat{V} + \widehat{V}dP$ because $\rho = 1/\widehat{V}$

$$0 = \dot{m}\left(\widehat{PE} + \widehat{KE} + \frac{P}{\rho}\right)_{in/out} - \dot{m}\int_{out}^{in} Pd\widehat{V} + \sum \dot{W}_{shaft} - \sum \dot{F} \qquad (7-50 \text{ alt})$$

This alternate form is especially convenient for understanding the meanings of the terms in the mechanical energy equation. Here $\dot{W}_{shaft}$ is the non-flow and non-expansion work, and in a mechanical energy setting this is brought about by the operation of a shaft (e.g., compressor, pump, turbine, etc.) and hence the term *shaft work*. Note also that $\widehat{PE}_{in/out}$, $\widehat{KE}_{in/out}$, and $(P/\rho)_{in/out}$ are the differences between inlet and outlet flow stream values of potential energy, kinetic energy, and flow work (all per unit mass). The integral $-\int Pd\widehat{V}$ is the reversible conversion between internal energy and mechanical energy of the fluid due to its expansion or compression as it flows through the pipe while the term $\dot{F}$ is the irreversible conversion from mechanical energy to internal energy due to friction which occurs while the fluid flows inside the pipe. Hence, Equation (7-50) or its alternate, Equation (7-50 alt) is an accounting of mechanical energy for the fluid as it flows through the pipe at steady state; *mechanical energy enters and leaves the section of pipe with the fluid in the form potential, kinetic, and pressure energy (flow work) and enters or leaves as work. Additionally, there is generation or consumption of mechanical energy due either to a reversible compression and expansion of the fluid as it flows through the pipe or due to irreversible friction losses.*

Either of the integral terms $\int \widehat{V}dP$ or $\int Pd\widehat{V}$ must be integrated along the flow path of the fluid through the pipe. This requires knowing the relation between P and $\widehat{V}$ for the flow, something that will not necessarily be known. Three examples for which the integral can be calculated are isothermal flow of an ideal gas (for which $P\widehat{V}$ is a constant due to the ideal gas law), isentropic (constant entropy) flow of an ideal gas (for which $P\widehat{V}^{\gamma}$ is constant, where γ is approximately 1.4), and flow of an incompressible fluid (liquids at constant temperature, e.g.). For an incompressible fluid, e.g., the integral in Equation (7-50 alt) is identically zero because there is not change in fluid volume as it flows through the pipe. All other terms remain. The use of this equation will be discussed more in Chapter 12.

This equation is known as the extended, or engineering Bernoulli equation, and can be used for designing fluid flow systems and sizing pump requirements. If the mechanical energy losses are assumed small and the fluid is incompressible, then:

$$0 = \dot{m}(\widehat{PE}_{in} - \widehat{PE}_{out}) + \dot{m}(\widehat{KE}_{in} - \widehat{KE}_{out}) + \dot{m}\left(\int_{P_{out}}^{P_{in}} \widehat{V}dP\right) + \sum \dot{W}_{shaft} \qquad (7-51)$$

This result says that the changes in the sum of the kinetic energy, potential energy, and pressure energy of a fluid as it flows at steady state through a pipe are due to the energy added through a pump, compressor, or other device. If a pump is not present, then it says that the sum of these forms of energy does not change, i.e., that the total mechanical energy in the form of kinetic energy, potential energy, and pressure energy does not change:

$$0 = \left(\widehat{PE} + \widehat{KE} + \frac{P}{\rho}\right)_{in} - \left(\widehat{PE} + \widehat{KE} + \frac{P}{\rho}\right)_{out} \tag{7 - 52}$$

This result, which is for no friction losses, no work, and for an incompressible fluid, is known as the Bernoulli equation.

Example 7-8. Water flows at steady state through the system shown in Figure 7-10 at a rate of 5 gallons per minute. Estimate the pressure required at the inlet *if friction is negligible*. Determine the pressure in pounds force per square inch. The following data are given:

$$d_{in} = 0.5 \text{ cm}$$
$$d_{out} = 1.0 \text{ cm}$$
$$\dot{V}_{in} = 5 \text{ gal} / \text{min}$$
$$\dot{V}_{out} = ? \text{ gal} / \text{min}$$
$$P_{in} = ? \text{ kJ} / \text{kg}$$
$$P_{out} = 1 \text{ kJ} / \text{kg}$$

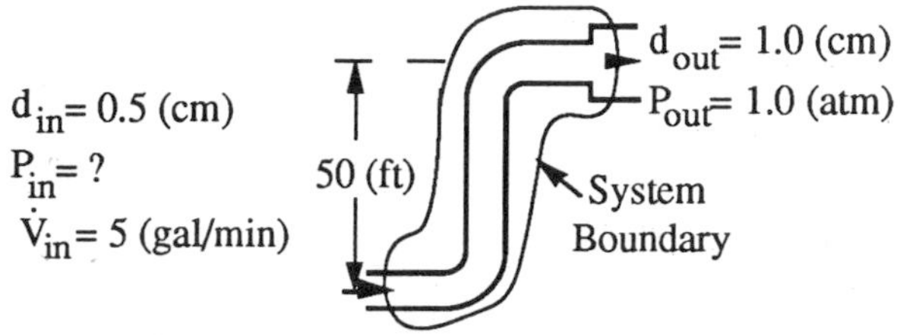

Figure 7-10: Steady state flow in a pipe

SOLUTION: The process is operating at steady state and there is no work done on the system and no irreversible conversion of mechanical to thermal energy. The accounting of mechanical energy is given in Table 7-9. From the table, the following equation can be written (at steady state $\dot{m}_{in} = \dot{m}_{out} = \dot{m}$):

$$0 = \dot{m}g(z_{in} - z_{out}) + \frac{\dot{m}}{2}(v_{in}^2 - v_{out}^2) + \dot{m}\left(\int_{P_{out}}^{P_{in}} \widehat{V} dP\right)$$

Now dividing both sides of the equation by the mass flow rate, recognizing that $\widehat{V} = 1 / \rho$ where ρ is the water density and that ρ is constant gives

$$0 = g(z_{in} - z_{out}) + \frac{1}{2}(v_{in}^2 - v_{out}^2) + \frac{1}{\rho}(P_{in} - P_{out})$$

Solving for P_{in} gives:

$$P_{in} = P_{out} - \rho g(z_{in} - z_{out}) - \frac{\rho}{2}(v_{in}^2 - v_{out}^2)$$

The change in potential energy can be calculated from the known information:

$$-\rho g(z_{in} - z_{out}) = -\frac{62.4 \text{ lb}_m}{\text{ft}^3}\left|\frac{32.174 \text{ ft}}{s^2}\right|\frac{(-50) \text{ ft}}{}\left|\frac{1 \text{ lb}_f}{32.174 \text{ lb}_m \text{ ft} / s^2}\right|\frac{1 \text{ ft}^2}{12^2 \text{ in}^2}\left|\frac{1 \text{ psi}}{1 \text{ lb}_f / \text{in}^2}\right| = 21.67 \text{ (psi)}$$

To solve for the change in kinetic energy we need to know the average linear velocity at the inlet and the outlet. To determine the average velocity of the outlet we will need to apply the conservation of mass to the system given by Table 7-10.

Table 7-9: Data for Example 7-8 (Mechanical Energy, System = Section of pipe)

0	$=$	$\dot{m}(\widehat{PE}_{in} - \widehat{PE}_{out})$	$+$	$\dot{m}(\widehat{KE}_{in} - \widehat{KE}_{out})$	$+$	$\dot{m}\left(\int_{P_{out}}^{P_{in}} \widehat{V} dP\right)$	$+$	$\sum W_{shaft}$	$-$	$\sum \dot{F}$
$\left[\dfrac{Energy}{time}\right]$		$\left[\dfrac{Energy}{time}\right]$		$\left[\dfrac{Energy}{time}\right]$		$\left[\dfrac{Energy}{time}\right]$		$\left[\dfrac{Energy}{time}\right]$		$\left[\dfrac{Energy}{time}\right]$
0	$=$	$\dot{m}g(z_{in} - z_{out})$	$+$	$0.5\dot{m}(v_{in}^2 - v_{out}^2)$	$+$	$\dot{m}\left(\int_{P_{out}}^{P_{in}} \widehat{V} dP\right)$	$+$	0	$-$	0

Table 7-10: Data for Example 7-8 (Mass, System = Section of pipe)

Description	$\left(\dfrac{dm_{sys}}{dt}\right)$	$=$	$\sum \dot{m}_{in}$		
	(lb_m / min)		(lb_m / min)		(lb_m / min)
Total	0	$=$	$(\rho A v)_{in}$	$-$	$(\rho A v)_{out}$

From Table 7-10, for constant density:

$$v_{out} = v_{in}\left(\frac{A_{in}}{A_{out}}\right)$$

The average linear velocity in the system is the volumetric flow rate divided by the cross sectional area of the tube:

$$v_{in} = \left(\frac{\dot{V}}{A}\right)_{in} = \left(\frac{4\dot{V}}{\pi d^2}\right)_{in}$$

Substituting the values gives:

$$v_{in} = \frac{4}{\pi}\left.\frac{5\ gal}{min}\right|\frac{1}{(0.5)^2\ cm^2}\left|\frac{1\ min}{60\ s}\right|\frac{1\ ft^3}{7.48\ gal}\left|\frac{(30.48)^2\ cm^2}{1\ ft^2}\right.$$

Evalution and conversion of units yields:

$$v_{in} = 52.71\ (ft\ /\ s)$$

Now the exit velocity can be determined;

$$v_{out} = v_{in}\left(\frac{A_{in}}{A_{out}}\right) = v_{in}\left(\frac{d_{in}}{d_{out}}\right)^2 = 13.18\ (ft\ /\ s)$$

The change in kinetic energy now can be calculated:

$$-\frac{\rho}{2}(v_{in}^2 - v_{out}^2) = -\frac{62.4\ lb_m}{2\ ft^2}\left((52.71)^2 - (13.18)^2\right)\left.\frac{ft^2}{s^2}\right|\frac{1\ lb_f}{32.174\ lb_m\ ft\ /\ s^2}\left|\frac{1\ ft^2}{12^2\ in^2}\right|\frac{1\ psi}{1\ lb_f\ /\ in^2}$$

$$= -17.54\ (psi)$$

Now, by combining the above results, the pressure at the inlet is found to be:

$$P_{in} = (14.7) + (21.67) - (17.54)\ (psi) = 18.8\ (psi)$$

If we neglect all losses of mechanical energy due to frictional dissipation, the pressure at the inlet must be 18.8 (lb_f / in^2) in order t deliver 5 (gal / min) of water.

7.7.2 Thermal Energy Accounting Equation

The general verbal statement of the accounting of thermal energy is

$$\left\{ \begin{array}{c} \text{Accumulation of thermal} \\ \text{energy within system} \\ \text{during time period} \end{array} \right\} = \left\{ \begin{array}{c} \text{Thermal energy} \\ \text{entering system} \\ \text{during time period} \end{array} \right\} - \left\{ \begin{array}{c} \text{Thermal energy} \\ \text{leaving system} \\ \text{during time period} \end{array} \right\}$$

$$+ \left\{ \begin{array}{c} \text{Thermal energy} \\ \text{generated in system} \\ \text{during time period} \end{array} \right\} - \left\{ \begin{array}{c} \text{Thermal energy} \\ \text{consumed in system} \\ \text{during time period} \end{array} \right\}$$

The thermal energy entering and leaving the system can be in the form of the internal energy or enthalpy of the mass entering and leaving the system as well as in the form of the heat transfer that occurs because of temperature differences between the system and the surroundings. The generation and consumption of thermal energy arise due to conversions of mechanical, electrical, or other forms of energy to thermal energy; because total energy is conserved, any thermal energy that is generated must come from other forms of energy. Because of the complexity of the unsteady-state thermal energy accounting equation, we will limit our discussion of thermal energy accounting to a system that is operating at steady state and having only one entering and one leaving stream and with no conversion to or from electrical or other forms of energy except mechanical.

The steady-state thermal energy accounting equation for one entering stream and one leaving stream is calculated by taking the conservation of total energy and subtracting the mechanical energy. Because of this, the thermal energy accounting equation is not independent of the previously discussed total energy and mechanical energy equations. The conservation of total energy in enthalpy form for a steady-state process having one entering and one leaving stream is:

$$0 = \dot{m}(\widehat{H} + \widehat{PE} + \widehat{KE})_{\text{in}} - \dot{m}(\widehat{H} + \widehat{PE} + \widehat{KE})_{\text{out}} + \sum \dot{Q} + \sum \dot{W}_{\substack{\text{non-flow} \\ \text{non-exp}}} \tag{7-53}$$

The accounting of mechanical energy for the same system and with the same restrictions is:

$$0 = \dot{m}(\widehat{PE}_{\text{in}} - \widehat{PE}_{\text{out}}) + \dot{m}(\widehat{KE}_{\text{in}} - \widehat{KE}_{\text{out}}) + \dot{m}\left(\int_{P_{\text{out}}}^{P_{\text{in}}} \widehat{V} dP \right) + \sum \dot{W}_{\substack{\text{non-flow} \\ \text{non-exp}}} - \sum \dot{F} \tag{7-50}$$

If we subtract the macroscopic mechanical energy accounting equation from the total energy conservation law, the macroscopic thermal energy accounting equation results:

$$\boxed{0 = \dot{m}(\widehat{H}_{\text{in}} - \widehat{H}_{\text{out}}) - \dot{m}\left(\int_{P_{\text{out}}}^{P_{\text{in}}} \widehat{V} dP \right) + \sum \dot{Q} + \sum \dot{F}} \tag{7-54}$$

An alternate, internal energy form of this result is

$$0 = \dot{m}(\widehat{U}_{\text{in}} - \widehat{U}_{\text{out}}) - \dot{m}\left(\int_{\widehat{V}_{\text{in}}}^{\widehat{V}_{\text{out}}} P d\widehat{V} \right) + \sum \dot{Q} + \sum \dot{F}$$

If the irreversible conversions from mechanical to thermal energy ($\sum \dot{F}$) are small as compared to the other thermal energy terms, the equation reduces to:

$$0 = \dot{m}(\widehat{H}_{in} - \widehat{H}_{out}) - \dot{m}\left(\int_{P_{out}}^{P_{in}} \widehat{V}dP\right) + \sum \dot{Q} \tag{7-55}$$

Example 7-9. In a refrigerator, fluid flows through the cooling coils at a mass flow rate of 1 (lb$_m$ / min) as shown in Figure 7-11. The inlet and exit pressure are the same at 43.2 (psia). Furthermore, the entering and exiting temperatures are 30 °F. Physically, the liquid in the cooling coils is being vaporized to a gas at constant temperature. If the system is the inside volume of the coil, from the given enthalpy and specific volume data for the fluid, calculate the rate of heat transferred to the system. Assume there are no conversions of mechanical energy to thermal energy.

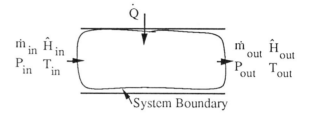

$\widehat{H}_{in}$ = 15.06 Btu / lb$_m$
$\widehat{H}_{out}$ = 80.42 Btu / lb$_m$
$\widehat{V}_{in}$ = 0.01137 ft^3 / lb$_m$
$\widehat{V}_{out}$ = 0.9188 (ft^3 / lb$_m$)

Figure 7-11: Refrigerator coil

SOLUTION: The system is the inside volume of the cooling coils with one inlet and one outlet. Furthermore, the system is operating at steady state. The thermal energy accounting equation is given in Table 7-11. $\int_{out}^{in} \widehat{V}dP = 0$ because the inlet and outlet pressures are the same.

Table 7-11: Data for Example 7-9 (Thermal Energy, System = Refrigerator coil)

$\dot{m}(\widehat{H}_{in} - \widehat{H}_{out})$		$\dot{m}\left(\int_{P_{out}}^{P_{in}} \widehat{V}dP\right)$		$\sum \dot{Q}$		$\sum \dot{F}$		
(Btu/min)		(Btu/min)		(Btu/min)		(Btu/min)		(Btu/min)
(1)(15.06 − 80.42)	−	0	+	$\dot{Q}$	+	0	=	0

From the table we see:

$$m(\widehat{H}_{in} - \widehat{H}_{out}) + \dot{Q} = 0$$

The only unknown in this equation is the rate of heat transfer. Rearranging and solving for the rate of heat transfer gives:

$$\dot{Q} = \dot{m}(\widehat{H}_{out} - \widehat{H}_{in})$$

$$= \frac{1 \text{ lb}_m}{\text{min}} \left| \frac{(80.42 - 15.06) \text{ Btu}}{\text{lb}_m} \right| = 65.36 \text{ (Btu / min)}$$

The rate of heat transfer to the refrigerator coils is 65.35 (Btu/min). Note that even though the inlet and exit temperatures were the same, energy, in the form of heat, was transferred to the system. The same result is obtained if total energy conservation is used.

7.7.3 Electrical Energy Accounting Equation

The general verbal statement of the accounting of electrical energy is:

$$\left\{ \begin{array}{c} \text{Accumulation of electrical} \\ \text{energy within system} \\ \text{during time period} \end{array} \right\} = \left\{ \begin{array}{c} \text{Electrical energy} \\ \text{entering system} \\ \text{during time period} \end{array} \right\} - \left\{ \begin{array}{c} \text{Electrical energy} \\ \text{leaving system} \\ \text{during time period} \end{array} \right\}$$

$$+ \left\{ \begin{array}{c} \text{Electrical energy} \\ \text{generated in system} \\ \text{during time period} \end{array} \right\} - \left\{ \begin{array}{c} \text{Electrical energy} \\ \text{consumed in system} \\ \text{during time period} \end{array} \right\}$$

Electrical energy enters and leaves the system in the form of the electrical energy of entering and leaving charges (current) and in the form of electromagnetic energy (radio/television, e.g.). Electrical energy can be generated by a conversion from mechanical energy (generator, alternator, turbine), thermal energy (thermocouple), internal energy (battery or fuel cell) and it can be consumed by conversion to thermal energy (resistors), mechanical energy (motors), and internal energy (electrolysis or battery charging). A system can accumulate electrical energy in a capacitor as an electric field or an inductor as a magnetic field.

Now voltage is electrical energy per charge so that current (net charge flow rate) at voltage V provides an electrical energy flow rate of iV. The verbal statement then can be written in symbolic form as:

$$\left(\frac{d(E_E)_{sys}}{dt} \right) = \sum (iV)_{in/out} + \sum \dot{W}_{ele} - \sum \dot{F} \qquad (7-56)$$

The term $\sum \dot{W}_{ele}$ is used to represent the rate of generation of electrical energy through power supplies, conversions from chemical energy by electro-chemical cells, or from mechanical turbines and generators. Likewise, the term $\sum \dot{F}$ denotes the consumption rate of electrical energy with the sign convention denoting positive $\dot{F}_i$ for consumption. Energy stored in a capacitor $(E_C)_{sys}$ or inductor $(E_L)_{sys}$ is represented by $(E_E)_{sys}$ where:

$$E_E = E_L + E_C \qquad (7-57)$$

If there is no current crossing the system boundary, then the accounting of electrical energy reduces to:

$$\left(\frac{d(E_E)_{sys}}{dt}\right) = \sum \dot{W}_{ele} - \sum \dot{F} \tag{7 – 58}$$

That is, the total rate of electrical energy generated with (supplied by) the system (by batteries, generators, etc.) minus the total rate of electrical energy consumed in (used by) the system (by resistors, motors, etc.), must equal the accumulation of electrical energy within the system (stored as an electric field in capacitors or stored in a magnetic field in inductors).

If there is no current crossing the system boundary and there is no acculumation of electrical energy in the system, the sum of the electrical energy generated in the system minus the sum of the electrical energy consumed in the system must equal zero:

$$0 = \sum \dot{W}_{ele} - \sum \dot{F} \tag{7 – 59}$$

The term "system" is a generic term, and when accounting for electrical energy we usually use the specific term "circuit" or "electrical circuit" or "electrical device" or circuit element. To keep with the philosophy and focus of this text we will continue to use the term "system" throughout this section.

The amount of power used or supplied by any element of a circuit is given by:

$$\text{Power supplied or used} = (iV)_{in} - (iV)_{out}$$

where V is the voltage across the element and i is the current through it, in accordance with equation 7-56 written for a single element. If a positive current enters the higher potential side of an element that has a potential difference across it, then electrical energy is consumed by that element (more electrical energy iV enters than leaves because $i_{in} = i_{out}$ and $V_{in} > V_{out}$). This is depicted below for some arbitrary circuit element. On the other hand if the positive current enters the lower potential end, that element is a generator of electrical energy as shown in Figure 7-12 (less electrical energy enters iV than leaves because $i_{in} = i_{out}$ and $V_{in} < V_{out}$).

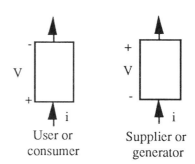

Figure 7-12: Suppliers and users of power

Batteries are usually generators of power (they force charge to move through a potential difference and so generate a power iV). Batteries convert chemical (internal) energy to electrical energy. Resistors are always consumers of electrical energy, converting it to thermal (internal) energy. Capacitors and inductors are users of power when they are strengthening their fields (storing more energy), and are suppliers of power when their fields are weakening (returning the stored electric and magnetic energy).

Kirchhoff's Voltage Law. As a direct consequence of net charge conservation and electrical energy accounting in a loop, Kirchhoff's Voltage Law (KVL) is obtained:

$$\sum_{\text{Loop}} V = 0 \tag{7 – 60}$$

f we start at any point in an electrical circuit and follow any path through the circuit to the same starting point, then all ·lectrical energy which leaves the circuit along the way (as dissipation, work, current, etc.) must be replenished (as work, ·urrent, etc.) or accumulated in order to return to the same energy flowrate (power) as at the start. It may be applied for ·ach complete loop that can be found in a circuit or system.

From the point of view of work associated with moving charge in an electric field, taking +1 (C) of charge from one ·oint in a circuit, and traversing any loop so as to arrive at the same place we started, the net work must be zero. When ·arrying the charge from a lower to higher potential energy, positive work is done; the charge increases in potential energy, ·ut when going from a higher to lower potential energy, the charge decreases in potential energy and the work is negative.

If the net change in potential energy per unit charge must be zero, the positive change in potential energy in going ·rom lower to higher potentials (voltage rises) must be equal to the negative change in potential energy in going from higher ·o lower potentials (voltage drops). Since the voltage is the electrical potential energy per unit charge, we can state that for ·ny complete loop in an electrical circuit or system:

> The sum of the voltage drops encountered in traversing any complete loop is equal to the sum of the voltage rises encountered; the algebraic sum of the potential differences across each element encountered in traversing any complete loop must be zero.

Example 7-10. A circuit consisting of several ·rbitrary circuit elements and two batteries is shown in ·igure 7-13. Each element has a potential difference across ·as labelled in the diagram. Obtain relations between the ·oltages for the different paths in the circuit.

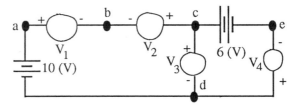

Figure 7-13: Circuit diagram

SOLUTION: There are three loops in the diagram (abcda, cedc, abceda). The Kirchhoff's Voltage Law equations for this circuit ·re given in Table 7-12 where we have chosen to traverse each loop in a clockwise direction and have assigned (-) to voltage drops and +) to voltage rises. Note that KVL does not necessarily allow calculation of the value of each voltage in a circuit. In this example there ·re four unknown voltages to be determined V_1, V_2, V_3 and V_4 but only two (2) independent equations. The last equation in the table is ·ot independent because it is the sum of the two inner loops.

Table 7-12: Data for Example 7-10 (Electric Energy-Voltage)

			$\sum V$						=	0
Description	10 (V)		V_1 (V)		V_2 (V)		V_3 (V)		V_4 (V)		6 (V)	(V)

Description	10 (V)		V_1 (V)		V_2 (V)		V_3 (V)		V_4 (V)		6 (V)		(V)
Loop (abcde)	10	+	$(-V_1)$	+	V_2	+	$(-V_3)$					=	0
Loop (cedc)							V_3	+	V_4	+	(-6)	=	0
Loop (abceda)	10	+	$(-V_1)$	+	V_2			+	V_4	+	(-6)	=	0

Resistance. Resistance represents an irreversible conversion or consumption of electrical energy to thermal energy. The defining relationship for a resistor is:

$$V_{in} - V_{out} = iR$$
$$V_R = iR$$

$$(7-61)$$

or the voltage across a resistor is proportional to the current i through the resistor. The proportionality constant is known as the resistance and has units of voltage per ampere or *ohms*. While this ratio could vary depending on the size of the applied voltage, for metallic conductors a plot of V_R vs. i gives a straight line. The slope of this line is the resistance of the conductor. The relation is known as Ohm's Law. It must be pointed out that Ohm's Law (the linear relation between voltage and current) does not hold for all conductors. Some important devices have non-linear relations between the voltage across and the current through the device. The symbol for a resistor in a circuit diagram is denoted by the series of alternating diagonal lines.

To calculate the rate of electrical energy conversion to thermal energy, we will choose the resistor as the system as shown in Figure 7-14 and apply the conservation of net charge and the accounting of electrical energy. There is only one inlet and one outlet, and there is no generation and no accumulation of electrical energy in this system. Applying the electrical energy accounting rate equation we obtain:

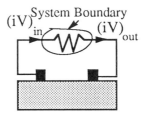

Figure 7-14: Resistor

$$(iV)_{in} - (iV)_{out} - \dot{F}_R = 0$$

$$(7-62)$$

From the conservation of net charge for the resistor (a resistor does not accumulate net charge):

$$i_{in} - i_{out} = 0$$

or

$$i_{in} = i_{out} = i$$

which, substituting into equation 7-61 gives

$$i(V_{in} - V_{out}) = i^2 R$$

Comparing this result to equation 7-62 we see that:

$$\dot{F}_R = i^2 R$$

$$(7-63)$$

A resistor dissipates electrical energy at a rate equal to the product of the square of the current through the resistor and the resistance. Ohm's Law allows the energy dissipated in a resistor to be written in any of the forms:

$$\dot{F}_R = i^2 R$$
$$= iV_R$$
$$= \frac{V_R^2}{R}$$

An analogy using mechanical energy accounting is helpful in remembering the relations between current, voltage, and resistance for materials that obey Ohm's Law. Consider the steady-state flow of an incompressible fluid (constant $\widehat{V}$ through a pipe with no change in $\widehat{PE}$, $\widehat{KE}$, no other energy terms, no electrical energy input or output, and no pump or turbine work. The mechanical energy equation is (see equation 7-50).

$$\dot{m}(P_{in}\widehat{V}_{in} - P_{out}\widehat{V}_{in}) - \sum \dot{F} = 0$$

Current is analogous to the mass flow rate ($\dot{m}$) of fluid through the pipe and voltage across the resistor (potential difference) is analogous to the flow work ($P\widehat{V}$) between the inlet and outlet of the pipe. Resistance heating (electrical energy losses) is analogous to the mechanical energy losses in pipe flow. Furthermore, the effect of area and length on resistance are, at least qualitatively, analogous to their effects on friction loss in pipe flow; smaller area and greater length (everything else being equal) give greater losses. Ohm's law provides a simple relationship between the rate of electrical energy entering the system and the losses; however, such a simple relation for calculating the dissipation of mechanical energy in pipe flow is not available.

Example 7-11. For the following circuit as shown below in Figure 7-15, use the conservation of net charge (Kirchhoff's Current Law) and the accounting of electrical energy (Kirchhoff's Voltage Law) to determine all of the currents, voltage differences and rates of electrical energy consumption or generation of each circuit element. Also show that the rate of energy supplied by the battery is exactly equal to the rate of energy dissipated in the resistors.

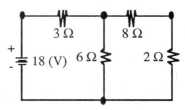

Figure 7-15: Resistive circuit

SOLUTION: The currents through each wire and circuit element are given a name and direction. The direction is completely arbitrary at this point. Furthermore, for each circuit element the voltage across each of the elements is named. Also, the polarity of the voltage is chosen so that the higher voltage side receives the entering current. The named currents and voltages are given in Figure 7-16.

For the first case we will define 3 systems A, B, and as shown in Figure 7-16. The systems are at steady state and there is no accumulation of net charge at the junction points. For the three systems, the conservation of net charge given in Table 7-13. From the table, the equation for system C can be used to eliminate i_4 from the system B equation. The new sets of equations for system A and B are given in Table 7-14.

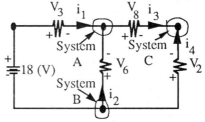

Figure 7-16: Labelled circuit diagram

The first thing that you should notice is that there are three unknowns in these equations, i_1, i_2, and i_3, and there are two (2) equations. However, the equations are *not* independent. In fact you can obtain the system B equation by multiplying the system A equation by -1. So the conservation of net charge for this circuit produced 1 independent equation and 3 unknowns. Before we can solve the problem we must have two more independent equations.

Table 7-13: Data for Example 7-11 (Charge)

System (Node)	$\left(\dfrac{dq_{sys}}{dt}\right)$ (A)	=	i_1 (A)		i_2 (A)		i_3 (A)		i_4 (A)
System A	0	=	i_1	+	i_2	+	$(-i_3)$		
System B	0	=	$(-i_1)$	+	$(-i_2)$			+	$(-i_4)$
System C	0	=					i_3	+	i_4

Table 7-14: Data for Example 7-11 (Charge)

System (Node)	$\left(\dfrac{dq_{sys}}{dt}\right)$ (A)	=	i_1 (A)		i_2 (A)		i_3 (A)
System A	0	=	i_1	+	i_2	+	$(-i_3)$
System B	0	=	$(-i_1)$	+	$(-i_2)$	+	i_3

Now we will define paths or loops around the circuit as shown below in Figure 7-17. Using the accounting of electrical energy in the form of Kirchhoff's Voltage Law for the 2 loops in the circuit we are able to generate Table 7-15. Now we have 4 more unknowns V_3, V_6, V_8 and V_2. However, we have 2 independent equations. Finally, we have to relate the unknown currents in the conservation of net charge to the unknown voltages in the accounting of electrical energy to reach a solution. The relationship that we will use is the defining relationship for resistors or Ohm's law. Table 7-16 gives the voltages across each of the resistive elements in terms of the unknown currents through the resistors.

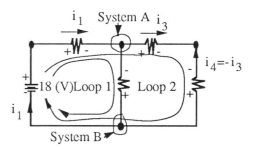

Figure 7-17: Circuit diagram with systems and loops

Combining the steady state conservation of net charge with the electrical energy accounting equation through Kirchhoff's Voltage Law we now have 3 independent equations in 3 unknowns as given in Table 7-17.

The simultaneous solution of the three linearly independent equations gives:

$$i_1 = 8\,/\,3 \qquad i_2 = -5\,/\,3 \qquad i_3 = 1 \text{ (A)}$$

The current i_4 is given as:

$$i_4 = -i_3$$
$$= -1 \text{ (A)}$$

Table 7-15: Data for Example 7-11 (Electric Energy-Voltage)

		$\sum V$				=	0
	18	V_3	V_6	V_8	V_2		
Description	(V)	(V)	(V)	(V)	(V)		(V)
Loop 1	18 +	$(-V_3)$ +	V_6			=	0
Loop 2	18 +	$(-V_3)$		$(-V_8)$ +	V_2	=	0

Table 7-16: Data for Example 7-11 (Electric Energy-Voltage)

		$\sum V$				=	0
	18	V_3	V_6	V_8	V_2		
Description	(V)	(V)	(V)	(V)	(V)		(V)
Loop 1	18 +	$(-3i_1)$ +	$(+6i_2)$			=	0
Loop 2	18 +	$(-3i_1)$		$(-8i_3)$ +	$-2i_3$	=	0

Table 7-17: Equations for Example 7-11

Description	3-Independent Equations				
System A	i_1 +	i_2 +	$(-i_3)$	=	0
Loop 1	$3i_1$ −	$6i_2$		=	18
Loop 2	$3i_1$		+ $10i_3$	=	18

The voltages across each of the resistors are calculated using Ohm's law:

$$V_3 = 8 \qquad V_6 = -10 \qquad V_8 = 8 \qquad V_2 = -2 \text{ (V)}$$

The circuit with all voltages and currents labelled is redrawn as Figure 7-18. This was only one way to solve this problem using (KCL), (KVL) and Ohm's Law for our defined systems and loops. There are other systems and loops which can be defined and the use of Ohm's relationship could have been substituted into the (KVL) equations yielding three independent equations in terms of the unknown voltages.

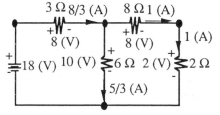

Figure 7-18: Solution of circuit

Table 7-18: Data for Example 7-11 (Electrical Energy)

$\left(\dfrac{dE_{sys}}{dt}\right)$	=	$\sum (iV)_{in/out}$	+	$\sum \dot{W}_{ele}$	-	$\sum \dot{F}$			
		$(iV)_{in/out}$				$\dot{F}_3$	$\dot{F}_6$	$\dot{F}_8$	$\dot{F}_2$
(W)		(W)		(W)		(W)	(W)	(W)	(W)
0	=	$(iV)_{in}$	+	0	-	$3i_1^2$	$6i_2^2$	$8i_3^2$	$2i_4^2$
						Substituting in the known values:			
0	=	$18(8/3)$	+	0	-	$3(8/3)^2$	$6(-5/3)^2$	$8(1)^2$	$2(-1)^2$
						Evaluation of the sum yields:			
0	=	48	+	0	-		48		

Now, using the explicit accounting of electrical energy for the circuit shown in Figure 7-19 show that the rate of energy supplied by the battery is equal to the rate of energy dissipated in the resistors. The accounting of electrical energy for this system is given in Table 7-18. The voltage leaving the system is defined as zero. The table shows that the power supplied by the battery $(iV)_{in} - (iV)_{out}$ is completely converted to thermal energy in the resistors $\sum \dot{F}$.

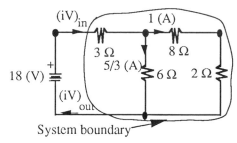

Figure 7-19: System for the accounting of electrical energy

Magnetic Fields and Inductance. Electric current was defined as the amount of net charge per unit time that moves past some point in a conductor. Any wire that has a current flowing in it sets up a magnetic field (**B** field) around the wire. The strength of the field at any point in space depends on the amount of current, the distance from the wire, and the geometric configuration of the wire. For a straight wire it can be shown that $|\mathbf{B}|$ is directly proportional to the size of the current and inversely proportional to distance from the wire. A closely wound coil of wire produces a directed magnetic field inside the coil whose strength is proportional to the size of the current in the coil. Such a coil of wire is called an inductor. The geometrical configuration of the coil (closeness of the turns, radius, etc.) is described by a number called the inductance (L) of the inductor. The unit of inductance is the Henry where 1 (H) = 1 (Volt s^2 / Coulomb) .

The defining relationship for the voltage across an inductor as related to the current through the inductor is given by:

$$V_{in} - V_{out} = L\left(\frac{di}{dt}\right)$$

$$V_L = L\left(\frac{di}{dt}\right)$$

$$(7-64)$$

where L is the inductance of the coil and $(di\,/\,dt)$ is the time rate of change of the current through the coil. The voltage across an inductor is proportional to the the time rate of change of the current through the inductor and the proportionality constant is the inductance L.

Now let us look at the energy storage in an inductor by applying the accounting of electrical energy and choosing the inductor as the system. This is shown in Figure 7-20.In this system there is only one inlet and one outlet. There is no generation of electrical energy and no consumption; however, there is an accumulation of electrical energy in the inductor in the form of a magnetic field. The accounting of electrical energy for this system is:

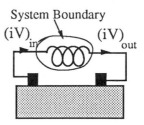

System Boundary

Figure 7-20: Inductor

$$\left(\frac{d(E_L)_{\text{sys}}}{dt}\right) = (iV)_{\text{in}} - (iV)_{\text{out}} \qquad (7-65)$$

Furthermore, the conservation of net charge for this system gives (an inductor does not accumulate net charge):

$$i_{\text{in}} - i_{\text{out}} = 0$$

or

$$i_{\text{in}} = i_{\text{out}} = i$$

which, combined with equation 7-65 gives

$$\left(\frac{d(E_L)_{\text{sys}}}{dt}\right) = i(V_{\text{in}} - V_{\text{out}})$$

Then, comparing this last result with equation 7-64,

$$\left(\frac{dE_L}{dt}\right) = Li\left(\frac{di}{dt}\right) = \left(\frac{d(Li^2\,/\,2)}{dt}\right) \qquad (7-66)$$

An inductor stores (accumulates) electrical energy in a magnetic field without accumulating net charge. The storage rate depends on the inductance, the current, and the time rate of change of the current through the inductor.

Example 7-12. Consider the case where there is an energy source, a resistor R and an inductor L as shown in Figure 7-21. Assume that before the switch is moved as shown in the top figure, the system reached steady state. Use the conservation of net charge and the accounting of electrical energy through Kirchhoff's Voltage Law to derive the current through the inductor and resistor, and the voltage across the inductor and resistor as a function of time after the switch has been moved. Use the accounting of electrical energy to explain where the energy is going in the circuit.

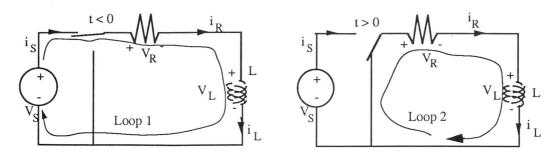

Figure 7-21: Resistive and Inductive circuit

SOLUTION: First we will look at the circuit before the switch is moved as given by the figure on the bottom. With the circuit in this configuration, and at steady state, the current through the energy soure is equal to the current through the resistor and inductor.

$$\text{for} \qquad t < 0, \qquad i_S = i_R = i_L = i_o$$

Furthermore, at steady state, this current is constant. Using Kirchhoff's Voltage Law around Loop 1 to account for the electrical energy in the circuit, Table 7-19 is generated:

Table 7-19: Data for Example 7-12 (Electrical Energy-Voltage)

| Description | $\sum V$ | | | = | 0 |
	V_S (V)	V_R (V)	V_L (V)		(V)
Loop 1	V_S	$- \quad i_o R$	$- \quad L\left(\dfrac{di_o}{dt}\right)$	=	0

Because the current is constant for $t < 0$ and the time derivative of a constant is zero the equation in Table 7-19 reduces to:

$$V_S - i_o R = 0$$

or the initial current through the circuit elements before switch is thrown is:

$$i_o = (V_S \,/\, R)$$

At $t = 0$ the switch goes down and the circuit changes to the one on the right. For $t > 0$ the conservation of net charge still says that:

$$i(t)_R = i(t)_L = i$$

Table 7-20: Data for Example 7-12 (Electrical Energy-Voltage)

Description	$\sum V$			$=$	0
	V_R (V)		V_L (V)		(V)
Loop 2	iR	$+$	$L\left(\dfrac{di}{dt}\right)$	$=$	0

since net charge cannot accumulate in the resistor or the inductor. Even though the currents are functions of time, this relationship is still true. For Loop 2 the accounting of electrical energy through Kirchhoff's Voltage Law is given in Table 7-20.

For constant inductance L and resistance R this equation can be rearranged as:

$$\frac{di}{dt} = -\left(\frac{R}{L}\right)i$$

and separating the variables gives:

$$\frac{di}{i} = -\left(\frac{R}{L}\right)dt$$

Now, if two expressions are equal, then their definite integrals are equal and integrating from the lower limit at $t = 0$ to the upper limit of t gives:

$$\int_{i_o}^{i} \frac{di}{i} = -\int_{0}^{t} \left(\frac{R}{L}\right)dt$$

Evaluating the integrals between the upper and lower limits gives:

$$\ln\left(\frac{i}{i_o}\right) = -\left(\frac{R}{L}\right)(t - 0)$$

or

$$i(t) = i_o \cdot e^{-\frac{R}{L}t}$$

Note that the current decreases from its initial value (i_o) to zero at a rate determined by (R / L) with increasing time. Since the current through the resistor is equal to the current through the inductor:

$$i_L(t) = i_R(t) = i_o \cdot e^{-\frac{R}{L}t}$$

The voltage across the resistor as a function of time is calculated from Ohm's Law:

$$V_R(t) = Ri_R(t) = Ri_o \cdot e^{-\frac{R}{L}t}$$

and the voltage across the inductor is calculated from the defining relationship:

$$V_L(t) = L\left(\frac{di_L}{dt}\right) = L\left(\frac{-R}{L}\right)i_o \cdot e^{-\frac{R}{L}t} = -Ri_o \cdot e^{-\frac{R}{L}t}$$

which equals $-V_R(t)$ as it must.

Table 7-21: Data for Example 7-12 (Electrical Energy)

$\left(\dfrac{dE_{sys}}{dt}\right)$	$=$	$\sum(iV)_{in/out}$	$+$	$\sum \dot{W}_{ele}$	$-$	$\sum \dot{F}$
$\left(\dfrac{d(E_L)_{sys}}{dt}\right)$						$\dot{F}_R$
(W)		(W)		(W)		(W)
System 1 $\quad \left(\dfrac{d(Li_{sys}^2/2)}{dt}\right)$	$=$	0	$+$	0	$-$	Ri^2

An accounting for energy in the circuit is given in Table 7-21. There is no inlet or outlet and no electrical generation devices in this system.

Integrating this equation from the initial time to a final time shows that initially the inductor stored energy in the amount $(Li_o^2/2)$, but for $t > 0$ this energy is being dissipated as heat in the resistor, and the amount of energy stored decreases with time towards zero. As time goes towards infinity all of the energy originally stored in the magnetic field of the inductor will have been dissipated as heat in the resistor.

Capacitor. A capacitor is formed by a pair of parallel plates separated by some distance (call it d). If equal and opposite charge (q) is placed on the plates, a uniform electric field **e** is set up between the plates. It would take work to move a positive test charge from the negatively charged plate to the positive plate since this would take a constant positive force acting through the distance d. Thus the potential difference across the capacitor V_C is positive.

The *voltage across* a capacitor C is related to the magnitude of positive charge on the two plates (q_+) by the following defining relationship.

$$V_{in} - V_{out} = \left(\frac{q_+}{C}\right)$$

$$V_C = \left(\frac{q_+}{C}\right)$$

(7 − 67)

The units of capacitance (C) are Farads 1 (F) = 1 (Coulomb / Volt).

We now look at the energy storage in a capacitor by an accounting of electrical energy with the capacitor as shown in Figure 7-22 as the system. For this system, there is only one inlet and one outlet, there are no electrical energy generation elements or electrical energy consumption elements. There is an electrical energy storage element, the capacitor. The accounting of electrical energy gives:

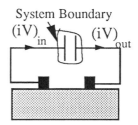

Figure 7-22: Capacitor

$$\left(\frac{d(E_C)_{sys}}{dt}\right) = (iV)_{in} - (iV)_{out}$$

(7 − 68)

Now because the net charge the two plates at all times is always zero for the system, there is no accumulation of net charge on the two plates *together*, i.e.,

$$0 = \dot{i}_{in} - \dot{i}_{out}$$

or

$$\dot{i}_{in} = \dot{i}_{out} = \dot{i}$$

and equation 7-68 gives

$$\left(\frac{d(E_C)_{sys}}{dt}\right) = i(V_{in} - V_{out}) \tag{7 – 69}$$

Comparing this result to equation 7-69, we see that

$$\left(\frac{d(E_C)_{sys}}{dt}\right) = \left(\frac{iq_+}{C}\right) \tag{7 – 70}$$

Consider, now, the positively charged plate as a system. The conservation of net charge ($i_{out} = 0$ and $i_{in} = i$) gives

$$i = \left(\frac{dq_+}{dt}\right) \tag{7 – 71}$$

Consequently

$$\left(\frac{d(E_C)_{sys}}{dt}\right) = \left(\frac{q_+}{C}\right)\left(\frac{dq_+}{dt}\right) \tag{7 – 72}$$

or, equivalently, in terms of voltage (using equation 7-67)

$$\left(\frac{d(E_C)_{sys}}{dt}\right) = CV_C\left(\frac{dV_C}{dt}\right) = \left(\frac{d(CV_C^2 / 2)}{dt}\right) \tag{7 – 73}$$

Electrical energy is stored in the electric field that exists between the plates of a capacitor. The storage rate of this energy depends on the capacitance C, the *voltage across* the plates, and the time rate of change of the voltage across the plates.

Example 7-13. Consider the case where there is an energy source, a resistor R and a capacitor C as shown in Figure 7-22. Assume that before the switch is moved as shown in the top figure, the system reached steady state. Use the conservation of net charge and the accounting of electrical energy through Kirchhoff's Voltage Law to derive the voltage across the capacitor and resistor, and the current through the capacitor and resistor as a function of time after the switch has been moved. Use the accounting of electrical energy to explain where the energy is going in the circuit.

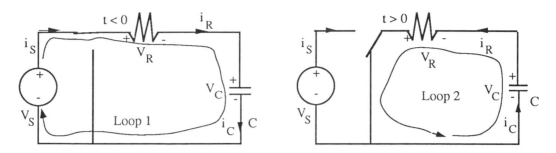

Figure 7-23: Resistive and Capacitive circuit

SOLUTION: First we will look at the circuit before the switch is moved as given by the figure on the bottom. With the circuit in this configuration, and at steady state, the current through the energy source is equal to the current through the resistor and capacitor.

$$\text{for} \qquad t < 0, \qquad i_S = i_R = i_C = i_o$$

Furthermore, at steady state, this current is constant and it is exactly zero. Using Kirchhoff's Voltage Law around Loop 1 to account for the electrical energy in the circuit, Table 7-22 is generated.

Table 7-22: Data for Example 7-13 (Electrical Energy-Voltage)

| Description | $\sum V$ | | | | | = | 0 |
	V_S (V)		V_R (V)		V_C (V)		(V)
Loop 1	$-V_S$	+	$i_o R$	+	V_C	=	0

Because the current is zero for $t < 0$ the equation in Table 7-22 reduces to:

$$V_S - V_C = 0$$

or the initial voltage across the capacitor before the switch is thrown is:

$$V_C = V_S$$

At $t = 0$ the switch goes down and the circuit changes to the one on the right. Notice that the current is traveling in the opposite direction and that the energy source is not in this system or loop. For $t > 0$ the conservation of net charge still says that:

$$i(t)_R = i(t)_C = i$$

since net charge cannot accumulate in the resistor of the capacitor. Even though the currents are functions of time, this relationship is still true. From the defining relationship for a capacitor equation (7-67) and the current relation, equation (7-71), the current through the resistor and capacitor is:

$$i = C \left(\frac{dV_C}{dt} \right)$$

Table 7-23: Data for Example 7-12 (Electrical Energy-Voltage)

Description	V_R (V)		V_C (V)		(V)
		$\sum V$		$=$	0
Loop 2	iR	$+$	V_C	$=$	0
Loop 2	$RC\left(\dfrac{dV_C}{dt}\right)$	$+$	V_C	$=$	0

For Loop 2, the accounting of electrical energy through Kirchhoff's Voltage Law is given in Table 7-23. For constant capacitance C and resistance R this equation can be rearranged as:

$$\frac{dV_C}{dt} = -\left(\frac{1}{RC}\right)V_C$$

and separating the variables gives:

$$\frac{dV_C}{V_C} = -\left(\frac{1}{RC}\right)dt$$

Now, if two expressions are equal, then their definite integrals are equal and integrating from the lower limit at $t = 0$ to the upper limit of t gives:

$$\int_{V_S}^{V_C} \frac{dV_C}{V_C} = -\int_0^t \left(\frac{1}{RC}\right)dt$$

Evaluating the integrals between the upper and lower limits gives:

$$\ln\left(\frac{V_C}{V_S}\right) = -\left(\frac{1}{RC}\right)(t - 0)$$

or

$$V_C(t) = V_S \cdot e^{-\frac{t}{RC}}$$

Note that the voltage across the capacitor decreases from its initial value (V_S) to zero at a rate determined by $(1 / RC)$ with increasing time.

Because the voltage across the resistor is equal and opposite to the voltage across the capacitor from Kirchhoff's Voltage Law, V_R as a function of time is calculated from Ohm's Law:

$$V_R(t) = -V_C(t) = -V_S \cdot e^{-\frac{t}{RC}}$$

The current through the resistor is calculated from the defining relationship using Ohm's law:

$$i_R(t) = \left(\frac{V_R(t)}{R}\right) = -\left(\frac{V_S}{R}\right) \cdot e^{-\frac{t}{RC}}$$

and the current through the capacitor is calculated from the defining relationship:

$$i_C(t) = C\left(\frac{dV_C}{dt}\right) = C\left(\frac{-1}{RC}\right)V_S \cdot e^{-\frac{t}{RC}} = -\left(\frac{V_S}{R}\right) \cdot e^{-\frac{t}{RC}}$$

Table 7-24: Data for Example 7-13 (Electrical Energy)

	$\left(\dfrac{dE_{sys}}{dt}\right)$	$=$	$\sum(iV)_{in/out}$	$+$	$\sum\dot{W}_{ele}$	$-$	$\sum\dot{F}$
	$\left(\dfrac{d(E_C)_{sys}}{dt}\right)$						$\dot{F}_R$
Description	(W)		(W)		(W)		(W)
System 1	$\left(\dfrac{d(CV_{sys}^2/2)}{dt}\right)$	$=$	0	$+$	0	$-$	(V^2/R)

which equals $i_R(t)$, as it must.

An accounting of energy in the circuit is given in Table 7-24. There is no inlet or outlet and no electrical generation devices in this system.

Integrating the equation in Table 7-24 from the initial time to some final time gives the accounting of electrical energy in the circuit. This would show that initially the capacitor stored energy in the amount $(CV_S^2/2)$, but for $t > 0$ this energy is being dissipated as heat in the resistor, and the amount of energy stored decreases with time towards zero. As time goes towards infinity all of the energy originally stored in the electric field of the capacitor will have been dissipated as heat in the resistor.

An accounting of charge in the circuit would show that the initial positive charge on the top plate of the capacitor will have left that plate and circulated around to the bottom plate canceling negative charge there. Note that the current $i_C(t)$ (net or positive charge going onto the top plate) is *negative*; the plate is discharging. Over any interval of time, the integral of the current describes the net change of charge on the plates during that interval.

7.8 REVIEW

1) In the absence of nuclear conversions, the total energy of a system plus its surroundings is conserved. There is no generation or consumption of total energy; there are only interconversions between forms of energy.

2) Energy may be possessed by mass in the form of kinetic energy, potential energy, and internal energy. It exists in transit between a system and its surroundings in the form of heat and work. Heat and work are forms of energy that are not possessed by mass. The rate at which work is done is called power.

3) The energy conservation equation includes energy associated with mass entering and leaving the system in the form of kinetic energy, potential energy, and internal energy and energy transferred to or from the system in the form of heat and work. The accumulation of energy within the system is in the form of kinetic energy, potential energy, and internal energy possessed by the mass contained within the system. Useful equations for a macroscopic system may be in a rate form for a differential time period or an integral form for a finite time period.

4) For fluid systems it is convenient to separate the work associated with mass entering or leaving the system from other work. When this is done it is combined with the internal energy of the mass of the flow streams to give a quantity called enthalpy. Enthalpy is a well-defined thermodynamic property just as is internal energy. When the equation is written in this form, the work term includes all work except this flow work.

5) It is sometimes convenient to classify energy as mechanical, thermal, or electrical. Mechanical energy includes kinetic energy, potential energy, and work. Thermal energy includes internal energy and heat transfer. Electrical

energy includes energy associated with moving charged species in an electric field. None of these forms of energy is conserved in that interconversions between the different forms may occur. However, it still can be useful to write accounting equations for these subsets of energy.

6) The steady-state mechanical energy accounting equation for fluids is called the Bernoulli equation when it neglects losses of mechanical energy due to dissipative friction. If these frictional losses are included, then it is called the extended Bernoulli equation. These equations are especially useful for analyzing and sizing fluid flow in conduits and piping systems.

7) The thermal energy accounting equation generally is not used for macroscopic-scale systems. It is not independent from the total energy equation and the extended Bernoulli equation. It can provide, however, an alternative conceptual understanding of the interconnections and interrelations of the different forms of energy.

8) The electrical energy accounting equation is used for the analysis of electrical circuits. In these systems, the conversion of electrical energy to and from other forms of energy play an important role.

7.9 NOTATION

The following notation is used in this chapter. The boldface lowercase letters denote vector quantities and the boldface uppercase letters denote tensor quantities.

Scalar Variables and Descriptions		Dimensions
A	cross sectional area	[length^2]
C	capacitance	[charge / volt]
d	diameter	[length]
E	energy	[energy]
EPE	electrical potential energy	[energy]
$\dot{E}$	energy flow rate	[energy / time]
f	magnitude of the vector $\mathbf{f}$	[mass length / time^2]
$\dot{F}$	rate of mech/elec energy consumption	[energy / time]
g	magnitude of the acceleration of gravity	[length / time^2]
H	enthalpy	[energy]
i	net charge flow rate	[charge / time]
k	spring constant	[force / length]
KE	kinetic energy	[energy]
L	inductance	[volts time^2 / charge]
m	total mass	[mass]
$\dot{m}$	total mass flow rate	[mass / time]
P	pressure	[force / length^2]
PE	potential energy	[energy]
q	net charge	[charge]
Q	heat	[energy]
$\dot{Q}$	heat flow rate	[energy / time]
R	resistance	[volts time / charge]

t	time	[time]
T	temperature	[temperature]
U	interal energy	[energy]
v_x, v_y, v_z	components of velocity	[length / time]
V	volume	[length^3]
$\dot{V}$	volumetric flow rate	[length^3 / time]
V	voltage	[energy / charge]
W	work	[energy]
$\dot{W}$	rate of work or power	[energy / time]
x, y, z	spatial rectangular cartesian variables	[length]
ρ	density	[mass / length^3]

Vector Variables and Descriptions		Dimensions
B	magnetic field vector	[force time / charge length]
E	electric field vector	[force / charge]
f	force	[mass length / time^2] or [force]
g	acceleration of gravity	[length / time^2]
i, j, k	cartesian unit directional vectors	
r	position	[length]
s	displacement	[length]
v	velocity	[length / time] or [momentum / mass]
$\boldsymbol{\omega}$	angular velocity	[1 / time]

Tensor Variables and Descriptions		Dimensions
I	the identity tensor	
R	the moment of inertia tensor	[mass length^2]

Subscripts

beg	evaluated at the beginning of the time period
C	corresponding to a conservative force
ele	work associated with electrical energy
end	evaluated at the end of the time period
exp	expansion work
ext	external to the system or acting at the system boundary
flow	flow work
G	corresponding to the center of mass of the system
in	input or entering
non-flow	non-flow work

non-exp	not associated with the system boundary expanding or contracting
out	output or leaving
shaft	work associated with rotating equipment
sys	system or within the system boundary
tot	total work

Superscripts

| $\frown$ | property per mass |

7.10 SUGGESTED READING

Bird, R. B., *Chem. Eng. Sci.*, 6, 123 (1957).

Felder, R. M., and R. W. Rousseau, *Elementary Principles of Chemical Processes*, John Wiley & Sons, New York, 1978

Halliday, D. and R. Resnick, *Physics for Students of Science and Engineering, Part I*, John Wiley & Sons, 1960

Himmelblau, D. M., *Basic Principles and Calculations in Chemical Engineering*, 3rd edition, Prentice-Hall, Englewood Cliffs, New Jersey, 1974

Hougen, O. A., K. M. Watson, and R. A. Ragatz, *Chemical Process Principles, Part I Material and Energy Balances*, 2nd edition, John Wiley & Sons, New York, 1954

Luyben, W. L., and L. A. Wenzel, *Chemical Process Analysis, Mass and Energy Balances*, Prentice Hall, Englewood Cliffs, New Jersey, 1988

QUESTIONS

1) Write out a verbal statement of the conservation of total energy. Be complete in your statement, citing the elements which must be defined in order to have a complete and meaningful term and also state the nature of each term, that is, the specific means or mechanism by which energy enters or leaves the system.

2) Write out a symbolic statement of the conservation of total energy. Define the physical meaning of each of the terms in this equation in the context of the conservation law.

3) In what ways can energy enter and leave a system as energy associated with mass. Specify and describe each form of energy. Also give examples of different ways in which work can be performed on the system.

4) The gravitational force may either be considered as resulting in changes in potential energy as a body moves around within a gravitational field, or as performing work on the body. Explain.

5) Explain the internal energy of matter and what it represents or accounts for in terms of molecular motion and interaction. What intensive property of matter would you be very familiar with and hear about every day that is a direct measure of internal energy?

6) In considering the conservation of total energy applied to a system we do not have to allow for or consider the conversion of energy from one form to another (for example conversion between mechanical and thermal energy). Explain why this is true.

7) In making an accounting of mechanical or thermal energy for a system, the conversion from mechanical energy to thermal energy or vice versa which occur within the system must be included in the equation. Why is this so for this result whereas it was not for the total energy statement.

8) Give some examples in which energy is converted from one form to another. For example, when mechanical energy is converted to electrical energy or vice versa, electrical energy is converted to thermal, electromagnetic energy converted to electrical, etc.

Give as many specific processes or examples as you can covering as many different types of conversion as you can. What must be true of all of these conversion processes with respect to total energy?

9) In considering the motion of a zero-volume rigid body, a statement of mechanical energy contains no more information than the conservation of linear momentum. Explain.

10) The mechanical energy equation for a rigid body of zero volume contains less information than the conservation of linear momentum equation. Explain.

11) For a finite-sized rigid body the conservation of linear momentum may be conveniently expressed in terms of the velocity of the center of mass of that body. Furthermore, if the dot product of this momentum equation with the velocity of the center of mass is obtained, then an energy-like equation is obtained in that the terms all carry the dimensions of energy. In fact this result is sometimes referred to as the work-energy theorem. However, a rigorous interpretation of the terms which are obtained in terms of true work and energy is not accurate. In the case of a cylinder rolling down an inclined plane, for example, a term is obtained which is the frictional force exerted at the point of contact with the cylinder and the plane dotted with the velocity of the center of mass. This term might be interpreted to be power exerted by the frictional force in the context of this equation. Explain why this is not a correct physical interpretation (in spite of difficulties with physically interpreting this result, it can still be very useful for understanding the motion of rigid bodies, but one must remember that it is useful because it contains information from the conservation of linear momentum).

12) A tennis ball is dropped from rest to the floor. Discuss the changes and conversions between different forms of energy as the ball leaves your hand, strikes the floor, and rebounds to a level which is lower than the initial drop height. At the peak of its bounce, then, its potential energy is less than it was when first released. What has happened to this potential energy and what has happened to the total energy of the ball? What is the motion of the earth in response to this motion of the ball as it falls toward the earth and bounces, turns around, and moves away from the earth?

13) In what ways can the amount of energy of an isolated system change?

14) In what ways can the amount of energy in a closed but not otherwise isolated system change?

15) In what ways can the amount of energy in an open system change?

16) How does an accounting of mechanical energy differ from the total energy conservation law? What term exists in this equation that has no place in a conservation equation?

17) How does an accounting of thermal energy differ from the conservation of total energy law? What term does it include which has no place in a conservation law?

18) What quantity or term is common to both the mechanical energy accounting statement and the thermal energy accounting statement?

19) What other kinds of accounting statements can we have for other forms of energy?

20) What name is given to the equation which counts electrical energy in an electrical circuit?

21) In this chapter we have discussed equations for counting total energy, mechanical energy, thermal energy, electrical energy, and other forms of energy. Which of these equations may be considered to be independent equations, thereby providing unique information about the behavior of a system.

22) In what conservation law does Bernoulli's mechanical energy equation have its roots? What rigid-body equation is it analogous to (with respect to both the energy which is counted and its origin)?

23) Consider a bullet shot from a rifle as a system. What forces act upon that system and therefore are responsible for its motion within the barrel of the gun? What forces are responsible for its motion after it leaves the barrel of the gun? What physical law would you use to analyze and understand the motion of the bullet which occurs as a result of these forces? Why does a bullet fired from a rifle have a higher muzzle velocity than one fired from a pistol?

24) In the previous question, if you consider the gasses which are expanding as a result of the explosion in the barrel of the gun so as to propel the bullet, what factors affect this expansion process? What equation or equations would you use to understand this expansion process and to describe and quantify the energy of this gas as it is related to the motion and position of the bullet in the rifle?

25) In the previous question consider the conversion between various forms of energy starting with the unreacted materials in the gunpowder of the bullet and ending with the reacted and fully expanded product gasses and the bullet exiting the muzzle of the rifle.

SCALES

1) Problem 19, Chapter 7.

2) What are the dimensions of energy? What are SI energy units? Engineering energy units?

3) A uniform (everwhere the same) pressure acts against area A. What is the magnitude of the force exerted by this pressure?

4) The velocity of a particle is given by $\mathbf{v} = 2\mathbf{i} + 3\mathbf{j} - \mathbf{k}$. Calculate its kinetic energy, per unit mass ($\widehat{KE} = v^2 / 2 = \mathbf{v} \cdot \mathbf{v} / 2$).

5) A baseball is hit by a bat at home plate ($x = 0$, $y = 2$). Its subsequent distance from home plate (x, feet) and elevation above the ground (y, feet), if there is no wind resistance, are given (as functions of time) by the relations:

$$x(t) = 132t$$
$$y(t) = 2 + 95.2t - 16.1t^2$$

Hence, its potential energy per unit mass, as a function of time, is given by $\widehat{PE} = gy(t) = 2g + 95.2gt - 16.1gt^2$ where $g = 1$ lb$_f$/lb$_m$ and its kinetic energy per unit mass is $\widehat{KE} = \mathbf{v} \cdot \mathbf{v} / 2$, where $\mathbf{v} = d\mathbf{r} / dt$ and $\mathbf{r} = x(t)\mathbf{i} + y(t)\mathbf{j}$. Plot, using spreadsheet / plotting software, $\widehat{PE}$, $\widehat{KE}$, and ($\widehat{PE} + \widehat{KE}$). Is energy conserved?

6) A baseball is hit by a bat at home plate ($x = 0$, $y = 2$). Its subsequent distance from home plate (x, feet) and elevation above the ground (y, feet), if there is wind resistance which is proportional to its velocity, are given (as functions of time) by the relations:

$$x(t) = 426.5(1 - e^{-0.31t})$$
$$y(t) = 641.8(1 - e^{-0.31t}) - 103.8t + 2$$

Hence, its potential energy per unit mass, as a function of time, is given by $\widehat{PE} = gy(t) = 2g + 95.2gt - 16.1gt^2$ where $g = 1$ lb$_f$/lb$_m$ and its kinetic energy per unit mass is $\widehat{KE} = \mathbf{v} \cdot \mathbf{v} / 2$, where $\mathbf{v} = d\mathbf{r} / dt$ and $\mathbf{r} = x(t)\mathbf{i} + y(t)\mathbf{j}$. Plot, using spreadsheet / plotting software, $\widehat{PE}$, $\widehat{KE}$, and ($\widehat{PE} + \widehat{KE}$). Is energy conserved?

7) Consider a cylinder rolling down an inclined plane without slip. At the point of contact of the cylinder with the plane a normal force (perpendicular to the plane) counteracts the cylinder's weight and a friction force (acting parallel to the plane) keeps it from slipping. Sketch a free-body diagram for this rolling cylinder. How much work is done by the friction force?

8) Consider a piston and cylinder. The piston may move within the cylinder and there is a pressure P_i inside the piston-cylinder chamber.

(a) Make a sketch of the piston-cylinder arrangement.

b) If a constant external pressure P_o acts against the area A of the piston while the piston moves through distance L, how much is the work associated with this movement of the piston, in terms of P_i (or P_o), A, and L? Assume that the piston is massless and that the there is no friction between the piston and the walls of the cylinder.

9) If $\widehat{U}$ is the internal energy per unit mass for mass entering a system at mass flowrate $\dot{m}$, then what is the rate at which internal energy enters the system due to this mass flow?

10) Convert 19 Btu to Joules. Convert 19 Btu to ft-lb$_f$.

11) Convert 100 ft^2/s^2 to Btu/lb$_m$

12 What is the internal energy per kg of steam at 300°C and 600 kPa pressure? (Hint: See Table C-23.) What is its enthalpy per mass? Its specific volume (volume per mass)? Its entropy per mass?

13 What is the change in internal energy for 2 kg of steam as it changes state from saturated steam at 200 kPa to superheated steam at 600°C and 800 kPa.

PROBLEMS

1) A fastball is thrown at a batter at 90 (mph) in the horizontal direction. The mass of the ball is 5 1/8 ounces. After the batter has hit the ball, the speed of the ball is 110 (mph) at an angle $35°$ above the horizontal. The time of the impact was 1/1000 second. If the coordinate system is defined to be at home plate with the **i** direction towards and the pitcher and the **k** direction towards the sky, the average force calculated was:

$$\mathbf{f} = 2500\mathbf{i} + 0\mathbf{j} + 100\mathbf{k} \text{ (lb}_f)$$

If the displacement of the ball in contact with the bat during the impact is given by the following displacement vector,

$$\mathbf{s} = 1\mathbf{i} + 0\mathbf{j} + 1\mathbf{k} \text{ (in)}$$

calculate the work that the bat did on the ball. During the course of the impact, is there a change in the potential energy? Calculate the initial and final kinetic energy of the ball. Did the internal energy of the ball change and if so how much? Some time after the collision, what happens to the internal energy of the ball.

2) High pressure steam at 100 (psia) and 1000 °F enters a turbine at a mass flow rate of 100 (kg/min). The inside diameter of the pipe entering the turbine is 12 (in). The steam leaves the turbine at 20 (psia) and 250 °F. The inside diameter of the pipe leaving the turbine is 24 (in). The inlet stream to the turbine is 10 (ft) higher than the exit steam. The entering and leaving steam condition are given in the table below:

Entering	Leaving
$T = 1000$ °F	$T = 250$ °F
$P = 100$ (psia)	$P = 20$ (psia)
$\widehat{V} = 8.659$ (ft^3 / lb$_m$)	$\widehat{V} = 20.81$ (ft^3 / lb$_m$)
$\widehat{H} = 1529.2$ (Btu / lb$_m$)	$\widehat{H} = 1168.0$ (Btu / lb$_m$)

a) Calculate the kinetic and potential energy (per unit mass) entering and leaving the steam turbine.

b) If the turbine is operating at steady state and there is no heat transfer, calculate the rate (in Btu/hr) of energy (power) provided by the turbine.

c) From the definition of enthalpy, calculate the specific internal energy of the entering and leaving steam.

3) A 2 gram bullet is fired from a rifle horizontally at a velocity of 335 (m/s) into a block of wood with mass 1000 (g) that is constrained to be stationary. The bullet embeds to a distance of 10 (cm) in the wood and the time it takes to embed is 1/1000 of a second.

a) Consider the block and the bullet as the system, using the conservation of linear momentum, determine the average force that acted on this sytem.

b) Define the system as the bullet and determine the average force from the wood acting on the bullet.

c) Define the system to be the block and bullet together and write the energy conservation statement for the period from the moment of impact to the moment the bullet stops. Is there any work being done on this system and why or why not? Assume that in this brief period a negligible amount of heat is lost to the surroundings. Calculate the change in internal energy for this sytem.

d) If the bullet is the system, calculate the work that the wood did on the bullet during the impact. What is the change in the internal energy of the bullet?

e) If just the block of wood is the system calculate the change in internal energy of the block of wood.

f) For the system defined as the bullet and the block, write a conservation of energy statement for the period from the moment of impact to a time when the system temperature has reached its final (preimpact) value. For simplicity, assume that the bullet and block are at the same temperature prior to impact. What is the ultimate fate of the kinetic energy initially possessed by the bullet?

4) Heat is transferred to a 1 (kg) metal plate. The rate of heat transfer is given as a function of time as $\dot{Q} = b(e^{at} - 1)$ (J / min) where $b = 500$ (J / min) and $a = 5$ (1 / min). The plate has an initial internal energy of 100 (kJ). The plate will start to melt when the value of the internal energy of the plate reaches 10,000 (kJ).

a) Plot the heat transfer rate as a function of time.

b) What can you say about the potential and kinetic energy of the plate during this process?

c) Determine the time when the plate just starts to melt. On the figure that you plotted, graphically show what you calculated to determine the time.

5) Write and simplify the energy conservation equation for the following open systems.

a) Water passes through the sluice gate of a dam and falls on a turbine rotor, which turns a shaft connected to a generator. The change in fluid velocity from the intake of the turbine to its outlet is negligible, and the water undergoes insignificant pressure and temperature changes between the system inlet and outlet.

b) Crude oil is pumped through a cross-country pipeline. The pipe inlet is 200 (m) higher than the outlet, the pipe diameter is constant, and the pump is located near the midpoint of the pipeline. Heat generated by friction in the line is lost through the wall.

c) An electric fan is operating and increases the linear velocity of the air that passes through the fan. Assume that there is negligible heat transfer and the fan is operating at steady state.

6) Consider an automobile traveling at constant speed on a level highway. With the car defined as the system, discuss the conservation of mass, linear momentum, angular momentum and total energy. A frictional force acts on the car. Discuss how mass enters and leaves the system, and how linear and angular momentum are exchanged with the earth. Also discuss an accounting of mechanical, thermal and electrical energy. In what forms does the energy enter and leave this system and how does the energy in this system change?

7) Liquid water is pumped from a lake to an elevated tank for storage. The pipe is a 2 inch schedule 40 with an inside diameter (I.D.) equal to 2.067 (in). The required volumetric flow rate is 100 (gal/min). A pump that delivers 10 (hp) is available for you to use. The frictional losses throughout the pipe are proportional to the length of the pipe according to: $\dot{F} = kL$ where $k = 0.04$ (hp / ft). The pressure at the entrance of the pipe is 16 (psia) and the pressure at the outlet of the pipe is 14.7 (psia).

a) Define the system to be the pipe and its contents. Is this system operating at steady state? Calculate the mass flow rate entering and leaving this system. Calculate the rate at which kinetic energy enters and leaves this system. Why are or why aren't these terms equal? Write the mechanical energy accounting equation and reduce the equation based on your previous calculations. Calculate $\int \hat{V} dP$.

b) If all of the frictional losses are neglected, determine the minimum length of the pipe and the height of the elevated tank.

c) If the frictional losses are not neglected, determine the minimum length of the pipe and the height of the tank above the entrance of the pipe. Why is there a difference in the length, and discuss the ultimate fate of the mechanical energy that is consumed or lost in this process.

8) A cylinder with a movable frictionless piston contains 5 (L) of an ideal gas at 30 °C and 10 (atm). The piston is slowly displaced, compressing the gas to 15 (atm).

a) Considering the system to be the gas in the cylinder and neglecting the changes in potential energy, write and simplify the conservation of energy equation. Is this process steady state? What does the conservation of mass say about this system?

b) Suppose the process is carried out at constant temperature (isothermally), and the compression work done on the gas is 20 (L atm). If the internal energy of the gas is only a function of temperature ($\hat{U}(T)$), how much heat (in joules) is transferred or exchanged with the surroundings?

c) Suppose instead that the process is completely insulated from the surroundings and no heat transfer occurs (adiabatic), and that the specific internal energy $\hat{U}(T)$ increases as T increases. Is the final system temperature greater than, equal to, or less than 30 °C? (Briefly state your reasoning.)

9) A rigid slender rod of 2 kg with length 0.5 m is rotating about its center of mass like a propeller. The initial angular velocity ($t = 0$) is 0.4 rad/s in a constant direction. There is no translation, no change in potential energy or internal energy, and no heat transfer. After some time period, the angular velocity is 0.9 rad/s.

a) Calculate the initial and final kinetic energy of the slender rod.

b) For the system defined as the slender rod, calculate the work required to change the kinetic energy.

10) For the given electrical circuit, determine the current through each of the resistors. Also, calculate the consumption or loss of electrical energy through each of the resistors. Calculate the total power supplied by the voltage source or battery. What can be said about the power supplied and the power consumed for this circuit? Discuss possible ways that the power supplied might have

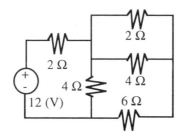

Problem 7-10: Electrical circuit

been generated through conversions of other forms of energy. From a total energy standpoint, in what way does energy leave the electrical circuit?

11) A refrigerent, R134a, at $T = 95°C$ and $P = 3.6$ (MPa) is passed through a turbine. The turbine is operating adiabatically (there is no heat transfer from the turbine to the surroundings) and at steady state. The inlet and outlet inside pipe diameters are 1 cm. The R134a leaving the turbine is $T = -70$ °C and $P = 0.008$ (MPa). Given the table of data and with a mass flow rate of 5 (kg / hr), calculate the rate of work that the fluid is doing on the turbine.

Data for Problem 7-11

Inlet Conditions	Outlet Conditions
$T = 95°C$	$T = -70°C$
$P = 3.6$ MPa	$P = 0.008$ MPa
$\rho = 267$ kg / m^3	$\rho = 0.573$ kg / m^3
$\hat{H} = 420.6$ kJ / kg	$\hat{H} = 318.6$ kJ / kg

12) A truck with mass 3000 kg is traveling 30 km/hr on a horizontal plane in the x-direction. A car with mass 1400 kg is traveling with a velocity of:

$$\mathbf{v} = 20\mathbf{i} + 30\mathbf{j} + 0\mathbf{k} \text{ (km / hr)}$$

The car and truck collide, and there are no other forces acting on the car or the truck.

a) If the collision is totally elastic (there is no loss of kinetic energy by conversion to internal energy) and if the truck's direction of travel remains the same, calculate the final velocity vectors of the car and truck.

b) If the collision is totally inelastic and the car "sticks" to the truck, calculate the final velocity vector.

c) For cases a) and b), calculate the initial and final kinetic energy of each vehicle and of the two together. What can you say about both cases in terms of energy? (Consider both mechanical and total energy of the car and truck together.)

13) A projectile with mass of 80 lb$_m$ is launched with an initial velocity $\mathbf{v} = 0\mathbf{i} + 70\mathbf{j} + 0\mathbf{k}$ mph.

a) If gravity acts in the $-\mathbf{j}$ direction and there are no other forces acting on the projectile, use the conservation of energy to calculate the maximum height.

b) If the projectile is launched at an angle such that the initial velocity vector is $\mathbf{v} = 30\mathbf{i} + 50\mathbf{j} + 0\mathbf{k}$ mph and no other forces act on the projectile except gravity, then calculate the maximum height. Also, at the maximum height, calculate the kinetic energy of the projectile.

c) Suppose now that the effect of the air resistance (drag force) causes 20% of the initial kinetic energy to the projectile to convert to internal energy of the projectile (this conversion results in an increase in temperature of the projectile). Calculate the maximum height of the projectile for both (a) and (b) velocities. For the latter case, assume that the kinetic energy at maximum height is 80% of that in part (b).

14) A roller coaster at Funland Amusement Park has an initial drop of 120 ft. The mass of the empty cars is 2000 (lb_m) and the mass of the loaded cars is 3000 (lb_m). The degree of inclination of the first hill is 27°. The motor that hauls the cars up is not 100% efficient but requires 20% more work to move the cars to the top of the first hill.

a) Calculate the change in potential energy of the empty cars from the bottom to the apex of the highest hill.

b) If the motor is 100% efficient, how much work is required to move the empty cars up to the apex of the highest hill.

c) If the motor really requires 20% more energy, calculate the actual work.

15) Given the circuits shown in the figure, calculate the rate of electrical energy conversion to internal energy of the resistors.

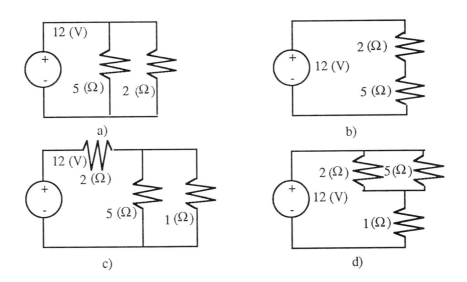

Problem 7-15: Resitive circuits

16) For the given circuits, write the steady-state KVL equations in terms of current for each path.

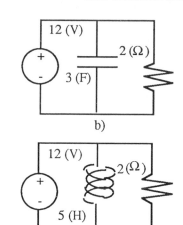
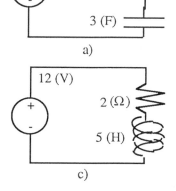

Problem 7-16: Capacitive and inductive circuits

17) Water from a reservoir passes through a turbine in a dam, and discharges from a 70 (cm) I.D. pipe at a point 50 (m) below the reservoir surface. The turbine delivers 0.8 Mega Watts of power. Calculate the required mass flow rate of water in (kg/min) if frictional losses are neglected. If friction were included, would a higher flow rate be required? If the diameter of the pipe is constant calculate the rate at which kinetic energy enters and leaves the system? If the system is defined to include the water of the lake to the lake level, and the kinetic energy of the mass entering this system is so small because the velocity of the mass entering the system is close to zero, calculate the rate of kinetic energy leaving this system. Now for this system, discuss why the kinetic energy term can be the neglected in the accounting of mechanical energy in this problem.

18) A particle of 5 lb_m has an initial velocity ($t = 0$) of $2i + 3j - 3k$ ft / s. After some time period, the velocity is $-3i + 2j - 3k$ ft / s. No mass enters the particle, there is no change in the particle's potential energy or internal energy, and there is no heat transfer.

   a) Calculate the initial and final kinetic energy of the particle.

   b) Calculate the work that was required to change the kinetic energy.

19) Given the following displacement vectors and force vectors

$$s_1 = 3i - 1j + 0k \text{ ft}$$
$$s_2 = 0i + 2j + 3k \text{ ft}$$
$$s_3 = -2i + 1j - 4k \text{ m}$$

$$f_1 = 2i + 0j + 1k \text{ lb}_f$$
$$f_2 = 0i + 2j + 0k \text{ lb}_f$$
$$f_3 = 3i - 2j + 1k \text{ N}$$

calculate the work

(a) $w_1 = f_1 \cdot s_1$ lb$_f$ ft

(b) $w_2 = f_2 \cdot s_2$ lb$_f$ ft

(c) $w_3 = f_3 \cdot s_3$ N m

(d) $w_4 = f_1 \cdot s_2$ lb$_f$ ft

(e) $w_5 = f_1 \cdot s_3$ lb$_f$ ft

(f) $w_6 = f_3 \cdot s_1$ N m

20) A 10 kg projectile is fired at a 45° angle above the horizontal. The initial speed ($t = 0$) is 60 m/s. Using linear momentum conervation in the presence of gravity, the velocity of the projectile as a function of time is $60\cos(45^\circ)\mathbf{i} + (60\sin(45^\circ) - 9.8t)\mathbf{j} + 0\mathbf{k}$ m / s. The gravity is in the j-direction. The projectile has no rotational kinetic energy. There is no heat transfer, and there is no work done except that due to gravity which is considered potential energy.

   a) Calculate the kinetic energy of the projectile as function of time. Graph the function.

   b) The time period that we are interested in and needs to be determined is from the initial firing time to the time when the projectile has the minimum kinetic energy. Calculate the time period and the mimimum kinetic energy.

   c) Based on the conservation of energy for the system defined as the projectile, determine the change in potential energy over the time period and the maximum height of the projectile.

21) A house in the morning is approximately 25 °C and contains 50,000 (J) of energy. During the day, 30,000 (J) of energy is added to the house in the form of heat transfer. The temperature of the house at the end of the day is 30 °C. No mass enters or leaves the house and no work is done. There is no change in the kinetic or potential energy of the house. Calculate the change in internal enegy of the house.

22) The average person consumes about 6 lb_m of food in one day. The food contains about 2000 kcal of energy. The person does not accumulate mass or energy during the day. Calculate the amount of mass that has left the body by the end of the day. Determine the amount of energy that has left the body and discuss the forms that the energy takes.

23) A complex process is operating at steady state. There are two entering mass streams and 1 exiting mass stream. No work is being done on the process. Neglect the entering and leaving kinetic energies. Given the following table of data, calculate the leaving mass flow rate and the heat transfer rate required to operate the process.

Stream 1	Stream 2	Stream 3
$\dot{m}_1 = 5$ kg / s	$\dot{m}_2 = 2$ kg / s	?
$\widehat{H}_1 = 2.3$ kJ / kg	$\widehat{H}_2 = 1.5$ kJ / kg	$\widehat{H}_3 = 3.7$ kJ / kg
$h_1 = 10$ m	$h_2 = 15$ m	$h_3 = 3$ m

24) The specific internal energy ($\widehat{U}$) of steam at 200°C is listed below for three pressures:

P (bar)	$\widehat{U}$ (J/g)
1	2658
2	2654
4	2647

If steam in a closed cylinder is changed isothermally (at constant temperature) from 2 bar to 4 bar at 200°C, what can you say, if anything, about how this was accomplished (i.e., about the heat and work)? For this same process, if the work was $\widehat{W} = -50$ J/g, then how much heat (per unit mass) was transferred?

CHAPTER
EIGHT

THE SECOND LAW OF THERMODYNAMICS

8.1 INTRODUCTION

So far, we have considered conservation laws (mass, energy, momentum, and charge) and accounting statements which are associated with them (species mass, positive or negative charge, mechanical energy, thermal energy, and electrical energy). Because these accounting statements count subsets of mass, charge, or energy they are not totally independent equations, but still have special utility for analyzing problems.

At this point, one more property, entropy, must be addressed quantitatively. This is done by writing an accounting statement for entropy. However, it should be noted at the outset that *entropy is not a conserved property* and hence, the entropy statement is *not* a conservation law but rather an accounting statement. However, it should also be noted that the *entropy statement is special* in that even though entropy is not conserved, there is a very special constraint on what can happen to it. This also should be well noted at the outset: *entropy can be generated but it can never, in any way or by any means, be consumed.* It is this constraint that makes an accounting of entropy another special law, the 2nd Law of Thermodynamics.

As with the statements of conservation of mass, charge, momentum, and energy, this entropy statement is based upon extensive physical observations; it is not a statement which can be proven mathematically. Also, as with mass, charge, momentum, and energy, these are observations with which you already have considerable experience and qualitative understanding. What remains now is to quantify the concept, to learn how to apply it, and to deduce further implications.

The general observation is that some events do not happen even though to do so would not violate the conservation laws. As a first example, heat will not flow from a cold temperature to higher temperature spontaneously (in the absence of external influences). A hot cup of coffee sitting on a table in a cooler room will not get hotter at the expense of its surroundings. Instead, it will become cooler and the room will become warmer. This happens in spite of the fact that if the reverse occurred (heat flow from cold to hot), none of the conservation laws need be violated.

A second observation is that a single-phase mixture of components will not spontaneously unmix and in fact, if a single phase mixture is not already uniform, then as an isolated system, concentrations will adjust to become uniform (i.e. so that each species concentration will not be a function of position). This is the basis of diffusion; to the extent a mole fraction gradient exists in a mixture, diffusion (migration of species due to a mole fraction difference) will occur in the direction which will eventually produce a uniform composition even though to occur in the opposite direction (unmixing) would not necessarily violate any of the conservation laws.

A third example is that mechanical energy, through viscous dissipation or friction, is converted to internal energy, a conversion which is manifested by a temperature rise. The reverse process of internal energy being converted to mechanical energy through friction does not occur even though to do so would not be a violation of conservation of mass, energy, or momentum. This physical observation is equivalent to saying that friction exerts a drag on a body (a force acting in the direction opposite to its motion) instead of a thrust even though to do so would not violate the conservation of momentum (if counteracted by a force on the surroundings).

The specific directionality of these events is very important and must hold true in all processes. If we are considering a process for which the opposite directions are implied, then we know *a priori* that the process is not viable; it will not happen.

Note, however, that the above statements do not preclude some events happening in the presence of external influences. By performing work we can take heat from a cold reservoir and exhaust it to a hot one (a heat pump, refrigerator, or air conditioner), or we can separate species in solution by membranes having special properties (kidney dialysis machines). *Spontaneous* processes, however, occur in only one direction.

It was a major triumph of the science of thermodynamics to discover a property which quantifies the directionality in which events occur. Equivalently, it quantifies the extent to which irreversibility occurs during a process. This property is entropy and denoted with the symbol S. Through calculations of entropy change during processes (actually calculations which account for the transfer of entropy and its generation during processes) we are able to understand and quantify the above statements concerning the direction and irreversibility of processes. A precise definition of entropy will be delayed; however, we will proceed with other statements, as appropriate, in the meantime to establish a physical understanding and feel about entropy.

First, entropy is an extensive property which can be counted. As such, we can write verbal statements which account for it within a system and within the system's surroundings:

$$\left\{ \begin{array}{c} \text{Accumulation of entropy} \\ \text{in system during} \\ \text{the time period} \end{array} \right\} = \left\{ \begin{array}{c} \text{Entropy entering} \\ \text{system during} \\ \text{the time period} \end{array} \right\} - \left\{ \begin{array}{c} \text{Entropy leaving} \\ \text{system during} \\ \text{the time period} \end{array} \right\} + \left\{ \begin{array}{c} \text{Entropy generated} \\ \text{in system during} \\ \text{the time period} \end{array} \right\}$$

We can also write an accounting of entropy for the surroundings:

$$\left\{ \begin{array}{c} \text{Accumulation of entropy} \\ \text{in surroundings during} \\ \text{the time period} \end{array} \right\} = \left\{ \begin{array}{c} \text{Entropy entering} \\ \text{surroundings during} \\ \text{the time period} \end{array} \right\} - \left\{ \begin{array}{c} \text{Entropy leaving} \\ \text{surroundings during} \\ \text{the time period} \end{array} \right\} + \left\{ \begin{array}{c} \text{Entropy generated in} \\ \text{surroundings during} \\ \text{the time period} \end{array} \right\}$$

These accounting statements assume exactly the same form as the statements which have been written previously except that we have omitted a term allowing for the consumption of entropy in that, as stated above, *entropy can only be generated and not consumed.*

To restate this very important observation: the only processes which occur spontaneously in an *isolated* system are those which result in either an *increase* in entropy or no change in entropy of the system, *i.e., entropy can only be generated; it cannot be consumed.* Equivalently, the entropy of the universe is always increasing! The limiting case of zero entropy increase defines a *reversible process* and consequently the increase in entropy of the system plus its surroundings (i.e. of the universe) during a process is a direct measure of the irreversibilities which have occurred during the process. This statement,

$$\left\{ \begin{array}{c} \text{Accumulation of entropy} \\ \text{in system during} \\ \text{the time period} \end{array} \right\} + \left\{ \begin{array}{c} \text{Accumulation of entropy} \\ \text{in surroundings during} \\ \text{the time period} \end{array} \right\} \geq 0 \qquad (8-1)$$

is written symbolically as:

$$\left(\frac{dS_{\text{sys}}}{dt} \right) + \left(\frac{dS_{\text{sur}}}{dt} \right) \geq 0$$

$$\boxed{\left(\frac{dS_{\text{uni}}}{dt} \right) \geq 0} \qquad (8-2)$$

and is known as the *second law of thermodynamics*. It quantifies the directional constraints which have been observed experimentally and mentioned above.

Because entropy is not conserved, accounting for entropy in the system is not enough; we must consider the entropy in the surroundings together with that in the system to have a complete statement. This is different from those properties which are conserved. For conserved properties we will normally write statements only for the system (i.e., we write mass, charge, energy, or momentum equations for the system and do not write a separate equation for the surroundings). For entropy, however, we must write equations for both the surroundings and the system in order to check that the two together give a total entropy change of the universe which is greater than or equal to zero.

The next section describes the different ways that entropy can enter and leave a system, and also discusses the entropy generation due to irreversibilities.

8.1.1 Entropy possessed by mass

Now entropy, in addition to being an extensive property, is a state propert. As such, the entropy which is possessed by a mass is defined totally in terms of the state of existence of that mass (i.e. its temperature, pressure and composition, phase state). Therefore, the entropy contained within the system (or surroundings) is that associated with the mass contained within the system at its conditions of temperature, pressure, and composition. If the state of the system is defined, then so also is its entropy. Furthermore, entropy enters or leaves the system as mass enters and leaves the system. As such we can define a specific entropy (like specific internal energy or specific kinetic energy) such that the specific entropy times the mass represents the entropy of the mass:

$$S = m\widehat{S} \qquad (8-3)$$

8.1.2 Entropy in transit

In addition to entropy associated with mass, entropy may also enter and leave the system as the result of heat transfer; the transfer of heat transfers entropy across the system boundaries (The exact definition of entropy with respect to heat transfer will be given later. For the time being it will suffice to recognize simply that when heat is transferred, entropy is transferred.):

$$\dot{S}_Q = \text{rate of entropy transfer due to heat transfer} \qquad (8-4)$$

8.1.3 Entropy generation

Furthermore, within a system or its surroundings, or as the result of heat transfer between the system and surroundings, *irreversibilities* may occur. *Within* the system (or surroundings) entropy is generated as a result of dissipation of mechanical energy (i.e. the conversion of mechanical energy to internal energy through viscous heating), the mixing of a multicomponent miscible solution to produce a uniform composition, the redistribution of thermal energy to result in a uniform temperature distribution, and other irreversible processes:

$$\dot{S}_{gen(non-Q)} = \quad \begin{array}{l} \text{rate of entropy generation due to irreversibilities} \\ \text{within the system (or surroundings)} \end{array} \qquad (8-5)$$

Entropy is also generated by the transfer of heat between the system and surroundings. This generation will automatically be accounted for by the entropy transfer terms due to heat transfer. To the extent entropy leaving (entering) the surroundings does not equal the entropy entering (leaving) the system, entropy is generated.

8.2 ENTROPY ACCOUNTING EQUATIONS

Now that we know the different ways that entropy may enter and leave a system, we can formulate a mathematical expression of the verbal statement given at the beginning of the chapter.

Figure 8-1 depicts a system and surroundings and the many ways that entropy enters and leaves the system.

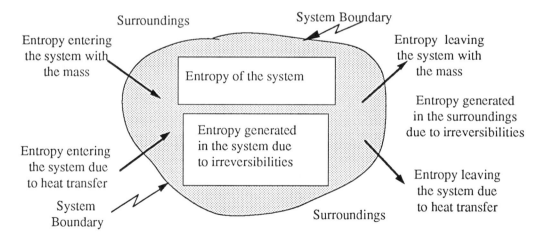

Figure 8-1: System and surroundings

Entropy enters and leaves the system as mass enters and leaves the system. Also, entropy enters and leaves the system by heat transfer across the system boundary. Furthermore, entropy can be generated in the system due to irreversibilities. The sum of all the rates of entering, leaving and generating entropy must be equal to the time rate of change of the entropy of the system. Combining the input and output terms due to mass, and the input and output terms due to heat transfer gives:

$$\left(\frac{dS_{sys}}{dt}\right) = \sum (\dot{m}\widehat{S})_{in/out,sys} + \sum (\dot{S}_Q)_{in/out,sys} + \sum \dot{S}_{gen(non-Q),sys} \qquad (8-6)$$

where:

$\left(\dfrac{dS_{sys}}{dt}\right)$ =   the time rate of change of entropy in the system

$(\dot{m}\widehat{S})_{in/out,sys}$ =   the entropy entering/leaving the system with mass

$(\dot{S}_Q)_{in/out,sys}$ =   the entropy entering/leaving the system due to heat transfer

$\dot{S}_{gen(non-Q),sys}$ =   the rate of entropy generation in the system

Similarly, we may write an accounting of entropy equation for the surroundings:

$$\left(\frac{dS_{sur}}{dt}\right) = -\sum (\dot{m}\widehat{S})_{in/out,sys} + \sum (\dot{S}_Q)_{in/out,sur} + \sum \dot{S}_{gen(non-Q),sur} \qquad (8-7)$$

where:

$\left(\dfrac{dS_{sur}}{dt}\right)$ =   the time rate of change of the entropy of the surroundings due to the process

$(\dot{m}\widehat{S})_{in/out,sys}$ =   the entropy entering/leaving the system with mass

$(\dot{S}_Q)_{in/out,sur}$ =   the entropy entering/leaving the surroundings due to heat transfer

$\dot{S}_{gen(non-Q),sur}$ =   the rate of entropy generation within the surroundings due to the process

When the two equations are added together, the flow streams, of course, cancel out and we are left with the statement that the change in the total entropy of the system plus surroundings is the sum of the heat transfer terms for both the system and the surroundings plus the entropy generation terms for both the system and the surroundings due to irreversibilities. By the second law, this sum of terms must be greater than or equal to zero.

$$\frac{d}{dt}\left(S_{sys} + S_{sur}\right) = \sum (\dot{S}_Q)_{in/out,sys} + \sum \dot{S}_{gen(non-Q),sys}$$
$$+ \sum (\dot{S}_Q)_{in/out,sur} + \sum \dot{S}_{gen(non-Q),sur} \qquad (8-8)$$

or

$$\left(\frac{dS_{uni}}{dt}\right) = \sum (\dot{S}_Q)_{in/out,sys} + \sum \dot{S}_{gen(non-Q),sys}$$
$$+ \sum (\dot{S}_Q)_{in/out,sur} + \sum \dot{S}_{gen(non-Q),sur} \qquad (8-9)$$

In later chapters on thermodynamics, we will discuss in some detail the calculation of entropy input to the system as a result of heat transfer. For now, we will simply state that

$$\sum (\dot{S}_Q)_{in/out,sys} = \sum \left(\frac{\dot{Q}}{T_{sys}} \right)_{in/out} \tag{8-10}$$

where:

$$\dot{Q} \quad = \quad \text{the rate of heat transfer across the system boundary}$$
$$T_{sys} \quad = \quad \text{the absolute thermodynamic temperature of the system}$$
$$\text{at which the heat transfer occurs}$$

For the heat transfer to the surroundings we state:

$$\sum (\dot{S}_Q)_{in/out,sur} = \sum \left(\frac{-\dot{Q}}{T_{sur}} \right)_{in/out} \tag{8-11}$$

where:

$$\dot{Q} \quad = \quad \text{the rate of heat transfer across the system boundary}$$
$$\text{(having the sign which is appropriate for the } system\text{)}$$
$$T_{sur} \quad = \quad \text{the absolute thermodynamic temperature of the surroundings}$$
$$\text{at which the heat transfer occurs}$$

Thus, using these substitutions:

$$\left(\frac{dS_{uni}}{dt} \right) = \left[\sum \left(\frac{\dot{Q}}{T_{sys}} \right) - \sum \left(\frac{\dot{Q}}{T_{sur}} \right) \right] + \dot{S}_{gen(non-Q),sys} + \dot{S}_{gen(non-Q),sur} \geq 0 \tag{8-12}$$

The heat transfer terms combine to give entropy generation *due to the heat transfer* in the system and surroundings *combined*:

$$\dot{S}_{gen(Q)} = \left[\sum \left(\frac{\dot{Q}}{T_{sys}} \right) - \sum \left(\frac{\dot{Q}}{T_{sur}} \right) \right] \geq 0$$

Thus, we can see that if the temperature of the surroundings is not equal to the temperature of the system, then these two terms do not cancel, that is to say, the extent to which heat transfer occurs across a finite temperature difference determines the amount of the irreversibility (entropy generation) due to the heat transfer. In the limit of zero temperature difference, the heat transfer is reversible and the entropy of the universe does not increase due to the transfer of heat. Furthermore, we can see that in the absence of other kinds of irreversibilities, the heat transfer irreversibility itself must be positive since in order for the entropy of the universe to increase due to the transfer of heat, the heat transfer must occur in the direction of low from hot to cold. That is to say, if $\dot{Q}_{\text{sys}} \geq 0$, then $\dot{Q}_{\text{sur}} \leq 0$, $(\dot{Q}_{\text{sur}} = -\dot{Q}_{\text{sys}})$. Substituting into the previous equation gives

$$\left(\frac{\dot{Q}_{\text{sys}}}{T_{\text{sys}}}\right) + \left(\frac{\dot{Q}_{\text{sur}}}{T_{\text{sur}}}\right) = \left(\frac{\dot{Q}_{\text{sys}}}{T_{\text{sys}}}\right) - \left(\frac{\dot{Q}_{\text{sys}}}{T_{\text{sur}}}\right)$$

$$= \left(\frac{\dot{Q}_{\text{sys}}(T_{\text{sur}} - T_{\text{sys}})}{T_{\text{sur}}T_{\text{sys}}}\right)$$

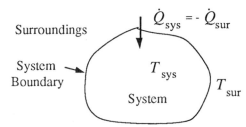

and so $T_{\text{sur}} \geq T_{\text{sys}}$ for the second law to be satisfied, i.e. $\dot{Q}$ goes from hot to cold. Figure 8-2 depicts the direction of heat flow from the surroundings to the system as a result of the temperature difference.

Figure 8-2: Direction of heat flow for a system and surroundings if $T_{\text{sur}} > T_{\text{sys}}$

The other entropy generation terms ($\dot{S}_{\text{gen(non-Q)}}$) represent other kinds of irreversibility besides that due to heat transfer: the irreversible conversion of mechanical energy to thermal (internal) energy (frictional dissipative heating), the diffusional mixing of components, the equilibration of temperature *inside the system*, etc. Each of these generation terms must, *by itself*, be positive (or zero).

One further point should be noted. Work does not directly affect entropy calculations. Only through the connection between heat and work seen in the conservation of total energy or through entropy-generating irreversibilities associated with work does work affect entropy.

This, then, is the essence of the 2nd Law of Thermodynamics:

> Entropy can only be generated; it can never, in any way or by any means, ever be consumed.

A summary of the entropy accounting equations for the system and the surroundings is given in Table 8-1. Note, however, that the separate statements for the system and surroundings are only stepping stones to a statement for the universe. Because there is entropy generation that results from heat transfer and because the amount of this generation is not known until the transfer terms for both the system and surroundings are summed together, *the single accounting statement for the universe must stand as the final result for evaluating the second law.*

Table 8-1: Entropy Accounting Equations[a]

System	$\left(\dfrac{dS_{sys}}{dt}\right)$	=	$\sum(\dot{m}\widehat{S})_{in/out,sys}$	+	$\sum(\dot{S}_Q)_{in/out,sys}$	+	$\sum\dot{S}_{gen(non-Q),sys}$
Surroundings	$\left(\dfrac{dS_{sur}}{dt}\right)$	=	$-\sum(\dot{m}\widehat{S})_{in/out,sys}$	+	$\sum(\dot{S}_Q)_{in/out,sur}$	+	$\sum\dot{S}_{gen(non-Q),sur}$
Universe	$\left(\dfrac{dS_{uni}}{dt}\right)$			=	$\sum\left(\dfrac{\dot{Q}}{T_{sys}}\right)-\sum\left(\dfrac{\dot{Q}}{T_{sur}}\right)$	+	$\sum\dot{S}_{gen(non-Q),uni}$

[a] $\dot{Q}$ is that for the system, i.e., $\dot{Q} > 0$ for the energy added to the system.

Example 8-1. Using only saturated steam at 212 ($^\circ$F), a process to make heat continuously available at a temperature of 270 ($^\circ$F) (higher than the incoming steam) is proposed. For every pound mass of steam taken into the process, 250.4 (Btu) of energy as heat is liberated to an external temperature of 32 $^\circ$F. Show whether the process is possible. The following table gives the states and thermodynamic properties of the entering and leaving steam.

T_{in} = 212 $^\circ$F
P_{in} = 1 atm
$\widehat{H}_{in}$ = 1,150. 4 Btu / lb$_m$
$\widehat{S}_{in}$ = 1.7566 Btu / lb$_m$ R
T_{out} = 32 $^\circ$F
P_{out} = 1 atm
$\widehat{H}_{out}$ = 0. 0 Btu / lb$_m$
$\widehat{S}_{out}$ = 0. 0 Btu / lb$_m$ R
$\widehat{Q}_1$ at T_{sur} = 32 $^\circ$F = 250. 4 Btu / lbm
   leaving the defined system
$\widehat{Q}_2$ at T_{sur} = 270 $^\circ$F =?

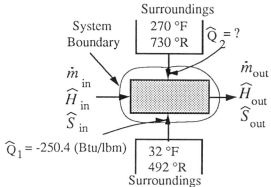

Figure 8-3: Complex exchange of heat

SOLUTION: Figure 8-3 gives a schematic of the process in question. For any process to be possible, it must satisfy the conservation of mass, energy and the second law of thermodynamics. We will choose the system to be the process and write conservation of mass, energy and an accounting of entropy. The conservation of mass is given in Table 8-2.

Table 8-2: Data for Example 8-1 (Mass, System = The process)

	$\left(\dfrac{dm_{sys}}{dt}\right)$		$\sum\dot{m}_{in/out}$	
Description	[mass/time]	=	[mass/time]	[mass/time]
Total	0	=	$\dot{m}_{in}$	$\dot{m}_{out}$

Table 8-3: Data for Example 8-1 (Total Energy, System = The process)

$\left(\dfrac{dE_{sys}}{dt}\right)$	=	$\sum\left(\dot{m}(\widehat{H} + \widehat{PE} + \widehat{KE})\right)_{in/out}$		+	$\sum\dot{Q}$	+	$\sum\dot{W}_{non-flow}$
$\left[\dfrac{energy}{time}\right]$		$\left[\dfrac{energy}{time}\right]$	$\left[\dfrac{energy}{time}\right]$		$\left[\dfrac{energy}{time}\right]$ $\left[\dfrac{energy}{time}\right]$		$\left[\dfrac{energy}{time}\right]$
0	=	$(\dot{m}\widehat{H})_{in}$	$-\quad(\dot{m}\widehat{H})_{out}$	+	$\dot{Q}_1$ $\quad + \quad \dot{Q}_2$	+	0

For a system operating at steady state, the total mass flow rate in equals the total mass flow rate out.

$$\dot{m}_{in} = \dot{m}_{out} = \dot{m}$$

The conservation of energy for the system is given in Table 8-3. The changes in the kinetic and potential energy entering and leaving are assumed to be negligible. Also, no non-flow work is accomplished in the process. The resulting equation is:

$$0 = \dot{m}\widehat{H}_{in} - \dot{m}\widehat{H}_{out} + \dot{Q}_1 + \dot{Q}_2$$

Because the mass flow rate of the entering stream is exactly equal to the mass flow rate leaving, according to the conservation of mass, we can divide each side of the equation by the mass flow rate:

$$0 = \widehat{H}_{in} - \widehat{H}_{out} + \left(\frac{\dot{Q}_1}{\dot{m}}\right) + \left(\frac{\dot{Q}_2}{\dot{m}}\right)$$

Rearranging the equation:

$$\left(\frac{\dot{Q}_2}{\dot{m}}\right) = \widehat{H}_{out} - \widehat{H}_{in} - \left(\frac{\dot{Q}_1}{\dot{m}}\right)$$

Substituting values gives:

$$\left(\frac{\dot{Q}_2}{\dot{m}}\right) = \left[(0.00) - (1,150.4) - (-250.4)\right] \text{ Btu / lb}_m = -900.0 \text{ Btu / lb}_m$$

We have satisfied the conservation of total energy for this system, now we must see if the the second law of thermodynamics is satisfied. We must show that:

$$\left(\frac{dS_{uni}}{dt}\right) = \frac{d}{dt}\left(S_{sys} + S_{sur}\right) \geq 0$$

Because we have assumed that the system undergoes steady-state operation:

$$\left(\frac{dS_{sys}}{dt}\right) = 0$$

Table 8-4: Data for Example 8-1 (Entropy of Surroundings)

$\left(\dfrac{dS_{sur}}{dt}\right)$	=	$-\sum(\dot m\widehat S)_{in/out,sys}$		+	$\sum(\dot S_Q)_{in/out,sur}$		+	$\sum \dot S_{gen(non-Q),sur}$
$\left[\dfrac{entropy}{time}\right]$		$\left[\dfrac{entropy}{time}\right]$	$\left[\dfrac{entropy}{time}\right]$		$\left[\dfrac{entropy}{time}\right]$	$\left[\dfrac{entropy}{time}\right]$		$\left[\dfrac{entropy}{time}\right]$
$\left(\dfrac{dS_{sur}}{dt}\right)$	=	$-(\dot m\widehat S)_{in}$	$+\quad(\dot m\widehat S)_{out}$	$-$	$\left(\dfrac{\dot Q}{T_{sur}}\right)_1$	$-\left(\dfrac{\dot Q}{T_{sur}}\right)_2$	$+$	0

The accounting of entropy in the surroundings determines the total entropy change for the universe. This accounting is presented in Table 8-4. We have assumed that there is no generation of entropy because of irreversiblities (other than heat transfer), i.e., $\dot S_{gen(non-Q),sur} = 0$. From the table, the time rate of change of entropy in the surroundings is calculated:

$$\left(\frac{dS_{sur}}{dt}\right) = -(\dot m\widehat S)_{in} + (\dot m\widehat S)_{out} - \left(\frac{\dot Q}{T_{sur}}\right)_1 - \left(\frac{\dot Q}{T_{sur}}\right)_2$$

Because the mass flow rate entering the system is equal to the mass flow rate leaving the system, we can divide both sides of the equation by the mass flow rate.

$$\left(\frac{1}{\dot m}\right)\left(\frac{dS_{sur}}{dt}\right) = -\widehat S_{in} + \widehat S_{out} - \left(\frac{\dot Q}{\dot m T_{sur}}\right)_1 - \left(\frac{\dot Q}{\dot m T_{sur}}\right)_2$$

Substituting values into the equation gives:

$$\left(\frac{1}{\dot m}\right)\left(\frac{dS_{sur}}{dt}\right) \text{ (Btu / lb}_m\text{ R)} = -1.7566 + 0.0000 + \left(\frac{250.4}{492}\right) + \left(\frac{900}{730}\right)$$

The temperature of the surroundings must be written as an *absolute* temperature either in degrees Rankine or degrees Kelvin. The time rate of change of entropy per unit mass in the surroundings is:

$$\left(\frac{1}{\dot m}\right)\left(\frac{dS_{sur}}{dt}\right) = -0.014 \text{ (Btu / lb}_m\text{ R)}$$

And the total time rate of change of entropy is:

$$\left(\frac{dS_{uni}}{dt}\right) = \left(\frac{dS_{sys}}{dt}\right) + \left(\frac{dS_{sur}}{dt}\right) = 0.0 - 0.014\dot m \text{ (Btu / R s)} = -0.014\dot m \text{ (Btu / R s)}$$

The mass flow rate is a positive quantity, and because the time rate of change of the total entropy is less than zero, *the process is not possible.*

Example 8-2. Consider an isolated system where a cell is divided into V_1 and V_2 by a thin membrane as shown in Figure 8-4. Volume 1 contains a single-component gas at P_1. In volume 2 there is a vacuum. Discuss what will happen and how the entropy will change if the membrane is punctured.

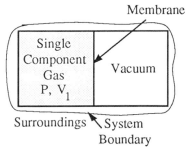

Figure 8-4: Free expansion of a gas

SOLUTION: Choose the system as both of the divided compartments together. No heat is transferred, no work is done and no mass crosses the system boundary. For the system, the accounting of entropy says that the system entropy increases only due to non-Q irreversibilities:

$$\left(\frac{dS_{\text{sys}}}{dt}\right) = \sum \dot{S}_{\text{gen(non}-Q),\text{sys}}$$

When the membrane is punctured, the gas in V_1 will begin to fill V_2. The pressure in V_1 will also decrease. In the end, the gas will completely fill both volumes at some equilibrium pressure P_{eq}. Because this process occurred spontaneously and the state of the gas has definitely changed, the total entropy of the universe must have increased. The system is completely isolated so we know that the entropy of the *surroundings* did not change as a result of this process:

$$\left(\frac{dS_{\text{sur}}}{dt}\right) = 0$$

Therefore, *the entropy of the system had to increase* after the membrane was punctured and the gas expanded to fill the entire volume.

Example 8-3. Consider an isolated system where a cell is divided into V_1 and V_2 by a thin membrane. Volume 1 contains a single component gas at pressure P, and volume 2 contains a different single component gas at pressure P as depicted in Figure 8-5. Discuss what will happen and how the entropy will change if the membrane is punctured.

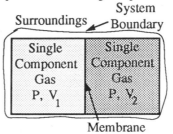

Figure 8-5: Mixing of two gases

SOLUTION: Again, choose the system as the entire cell. No heat is transferred, no work is done, and no mass crosses the system boundary. Also, the pressure in both compartments is the same. The accounting of entropy for this system is:

$$\left(\frac{dS_{\text{sys}}}{dt}\right) = \sum \dot{S}_{\text{gen(non}-Q),\text{sys}}$$

The entropy of the system, it being an isolated system, changes only due to entropy generation, a positive quantity; the system entropy can only increase. After the membrane is punctured, the components of the gases will mix by a process known as diffusion. After a sufficient time has passed, the gases will be completely "mixed" and the concentration of the gas mixture will be the same throughout the cell. Because this process is in an isolated system, this process must increase the entropy of the system. The mixing of two gases or miscible liquids is another example of how entropy increases.

In the above discussion, entropy was described in terms of two separate concepts. On the one hand, it is a property which is used to quantify irreversibilities which occur. Consequently, the entropy of the universe is always increasing. On the other hand, entropy represents an intrinsic property of matter, that is, it is a state function and for any material, given the state of existence (e.g., temperature, pressure, composition, and phase state) of that material, the entropy is then defined. No matter what happens to that material or what has happened to it in the past, every time that it exists in that state, its entropy is fixed at precisely the same value.

Though these two different concepts of entropy are not inconsistent, they may pose some considerable confusion. We must always be clear in our use of entropy as to whether we are calculating changes in the property of the material (which we can do without knowing anything at all about how the process occurs so long as we know the beginning and the end points with respect to properties which establish the state) or calculations of the entropy change of the universe for a process. These latter changes, in order to be totally quantified, are, in fact, dependent upon the path which is followed. However, for a check of a second law we do not necessarily need to know how much the entropy of the universe has increased or the exact causes of the increase, but only that it has increased. If our calculations show that the entropy of the universe has decreased during any process, then there is something wrong either with our calculations or with the process. In fact, rarely will we be able to calculate exactly the irreversibilities which occur in a process.

8.3 OTHER CONCEPTS RELATED TO ENTROPY AND THE SECOND LAW

Because of the second law and the fact that it dictates a direction that all isolated processes will follow, we can make statements about the quality of energy in its different forms, the difference between two types of processes (reversible and irreversible), and use the state of maximum entropy for an isolated system as a definition of equilibrium. These concepts are discussed further below.

8.3.1 Quality of Energy

Another concept with which you have some experience is that all forms of energy are not equivalent in terms of quality. For example, we can use mechanical energy in the form of either potential energy or kinetic energy to perform equivalent amounts (equivalent energy) of work. Also, we can even convert work entirely to thermal or internal energy. On the other hand, we are not able to convert thermal energy (internal energy) entirely into work any more than we are able to transfer heat from a lower temperature to a higher temperature spontaneously. Evidently internal energy or thermal energy is a lower quality of energy and is not as accessible for performing work as is mechanical energy.

This concept imposes an inherent and insurmountable limitation on heat engines (cyclic processes which use heat, transferred from a high temperature, to perform work – e.g. electric power plants, which use hot steam to produce work, or internal combustion engines which use the heat of combustion to perform work). Not all of the heat transferred can be converted to work. In fact, typical values for mechanically reversible processes (the best we can do) are only about 50 % conversion! The remainder of the heat must be rejected to a reservoir at a lower temperature. This is a consequence of the second law of thermodynamics.

8.3.2 Reversible versus Irreversible

Another important concept in thermodynamics is that of reversible and irreversible processes. A process is reversible if it can be reversed in the sense that when it is undone everything exists as it did before the process took place. *This includes both the system and the surroundings.* From our previous discussions of entropy and processes that cannot take place without increasing the entropy of the universe, we can say that a process is reversible if it does not result in an increase in the entropy of the universe. That is, if the sum of the change in entropy of the system plus surroundings is zero, then a reversible process has occurred. The processes which we discussed above which are recognized as going in only one direction give an indication of irreversibilities, that is, the heat transfer from a hot body to a cold body is an irreversible process if it takes place across a finite temperature difference. Likewise, the mixing of two gases spontaneously is an irreversible process in that the unmixing process will not occur spontaneously. In order to bring about the separation of the gases and return them to their original state, we must perform some process which is incapable of reducing or eliminating the entropy increase which occurred during the spontaneous mixing. Another example is the dissipation of mechanical energy by its conversion to thermal energy. Thermal energy still exists as energy and we have lost no energy, but the form of the energy has changed. However, we cannot recover all of that thermal energy as mechanical energy. Consequently, the state of the universe has irreversibly changed.

A general guideline in assessing whether a process is reversible or irreversible is related to the driving forces which make the process go. In general we would say that if the driving forces are infinitesimal, that is, if the forces between the system and the surroundings are at all times equally balanced, then we have a reversible process. Driving forces include physical forces, temperature differences (as driving forces for heat transfer), concentration differences (as driving forces for molecular mass transfer), and velocity gradients (as driving forces for momentum transfer). If there are unbalanced forces in a process, then things tend to happen much more quickly and more irreversibly. Irreversible processes, in summary, are those which result in a nonzero entropy generation term. In other words, they are those situations which result in entropy not being a conserved property.

8.3.3 Equilibrium

In engineering and science, the term equilibrium is used to denote a situation in which forces for change are in balance. In mechanical situations we say a body is in mechanical equilibrium if the sum of the forces acting on it is zero. In this case, the body is static or unchanging. In chemical or thermodynamic situations, equilibrium refers to a situation in which the system is isolated and undergoes no changes in macroscopic properties such as temperature, pressure, or composition. A single phase is at equilibrium if its temperature, pressure, and composition are uniform (everywhere the same). If the temperature is not uniform then heat will flow until the temperature is uniform, i.e., until there are no gradients of temperature within the phase. Likewise, if mole fractions are not uniform, then species will migrate until mole fractions have become uniform.

In multicomponent multiphase systems, the situation is complicated by the fact that different phases which are in equilibrium, while having the *same temperature and pressure*, will be of *different compositions*. For example, in vapor-liquid equilibrium, we know that the liquid phase is of different composition from the vapor phase in equilibrium with it. (This is the basis for distillation processes as in the refining of crude oil to gasoline.) Each phase by itself, however, when at equilibrium, will be of uniform composition, temperature, and pressure. If equilibrium does not exist between the two phases, then transfer of species between phases will occur until equilibrium is attained. The departure from equilibrium establishes a driving force for mass transfer and in the presence of this driving force, species will transfer until equilibrium is established.

The concept of chemical and mechanical equilibrium is quantified by entropy. In the discussion of the previous section, it was established that in an isolated system, that is with no heat, mass, or work transfer across system boundaries, that entropy will either increase or stay the same; the only processes which occur spontaneously (in an isolated system) are those which result in an increase in entropy. A single phase which is isolated (in the absence of heat or work or mass crossing the system boundaries) and is not at uniform temperature will, with time, equilibrate to a uniform temperature. This kind of process in an isolated system must result in an increase in entropy so that evidently an equilibrium state is the state which has maximum entropy. Likewise, for an isolated system consisting of two phases, entropy will be maximized when those two phases have reached equilibrium with each other. This is also true for additional phases and for a large number of species.

> Equilibrium in an isolated system corresponds to a state of maximum entropy.

This concept of maximum entropy for an isolated system at equilibrium is used extensively in thermodynamics. It appears in other forms (For a system at constant T and P, Gibbs free energy $(G = H - TS)$ is a minimum and at constant T and V the Helmholtz free energy $(A = U - TS)$ is a minimum. While these functions may be used later in this course (and in the properties course) and in a variety of contexts to describe or quantify equilibrium, remember that entropy is always the fundamental variable for understanding equilibrium.

8.4 REVIEW

1) Certain processes have been observed to occur spontaneously in only one direction.

2) Entropy was invented by Clausius to quantify this directional aspect of mass, momentum, charge, and heat transfer.

3) Processes which occur spontaneously within an isolated system result in an increase of the entropy of the system. For all processes, the total entropy of the universe either remains constant (reversible process) or increases. The entropy of the universe never decreases.

4) Irreversibilities occur when there are finite driving forces.

5) Irreversibilities cause an increase in entropy.

6) Entropy is an extensive, state property of matter. Specifying temperature, pressure, and composition of a (single phase) body is equivalent to specifying its entropy (per unit mass).

7) Different forms of energy are not equivalent in quality. Work can be totally converted to heat but heat cannot be totally converted to work.

8) Power cycles are used to convert heat to work (heat engine) and to transfer heat from a low temperature to a high temperature using work (heat pump or refrigerator) under the constraints of 1 through 7.

8.5 NOTATION

The following notation was used throughout this chapter.

Scalar Variables and Descriptions		Dimensions
E	energy	[energy]
H	enthalpy	[energy]
KE	kinetic energy	[energy]
m	total mass	[mass]
$\dot{m}$	total mass flow rate	[mass / time]
P	pressure	[force / length^2]
PE	potential energy	[energy]
Q	heat	[energy]
$\dot{Q}$	heat flow rate	[energy / time]
t	time	[time]
S	entropy	[energy / temperature]
$\dot{S}$	entropy flow or generation rate	[energy / temperature time]
T	temperature	[temperature]
V	volume	[length^3]

Subscripts

in	input or entering
irr	associated with irreversibilities
gen	generation or production
gen(non-Q)	generated by other means except heat transfer
gen(Q)	generated by heat transfer
out	output or leaving
sur	surroundings or external to the system
sys	system or within the system boundary
uni	universe or system plus surroundings

Superscripts

⌒   property per mass

3.6 SUGGESTED READING

Abbott, M. M., and H. C. Van Ness, *Theory and Problems of Thermodynamics*, 2nd edition, McGraw-Hill, New York, 1989

Cengel, Y. A., and M. A. Boles, *Thermodynamics, An Engineering Approach*, McGraw-Hill New York 1989

Halliday, D. and R. Resnick, *Physics for Students of Science and Engineering, Part I*, John Wiley & Sons, 1960

Hougen, O. A., K. M. Watson, and R. A. Ragatz, *Chemical Process Principles, Part II Thermodynamics*, 2nd edition, John Wiley & Sons, New York, 1959

Sandler, S. I., *Chemical and Engineering Thermodynamics*, John Wiley & Sons New York 1989

Smith, J. M., and H. C. Van Ness, *Introduction to Chemical Engineering Thermodynamics*, 4th edition, McGraw-Hill, New York, 1987

Van Wylen, G. J., and R. E. Sonntag, *Fundamentals of Classical Thermodynamics*, 3rd edition, John Wiley & Sons, New York, 1986

QUESTIONS

1) What observations concerning physical processes have motivated defining entropy?

2) In what ways can the entropy of a system change?

3) Does work associated with a system cause a change in entropy? a) directly b) indirectly - Explain.

4) Entropy is not a conserved property. Explain.

5) What must happen to the entropy of an isolated system or, if you prefer, what can never happen to the entropy of an isolated system?

6) The entropy of an open system can decrease. True or false. Explain.

7) The entropy of a closed system can decrease. True or false. Explain.

8) Give a verbal statement of an accounting of entropy for an open system.

9) Give a symbolic equation accounting for the entropy of a system.

10) Give a verbal statement accounting for the entropy of the system plus its surroundings.

11) Give a symbolic equation accounting for the entropy of the system plus its surroundings.

12) What processes can occur within an isolated system that lead to an increase of entropy and therefore are irreversible processes? Answer by listing specific examples.

13) What are the characteristics of a reversible process with respect to the speed at which the processes occur and the driving forces which exist?

14) You shake up a mixture of oil and vinegar to use on your salad. After shaking, it is in an emulsified state consisting of many smaller drops of oil and vinegar. After pouring some on your salad, you set the bottle down and notice that after a few minutes the oil and vinegar have completely separated with the small drops having coalesced into two single continuous phases with the oil on top. From the time that you set the bottle down to the time that it was completely separated, how has the entropy of the oil and vinegar changed (you may assume that there is no heat transfer between the bottle and the surroundings)? Explain.

15) You are in a closed room and notice that because the cap has been left off a felt tip marker, the smell of the pen writing fluid permeates through the air. What has happened to the entropy of the room from the time that the cap was left off until the time that you detect the writer fluid? You may assume that there is no heat transfer between the room and its surroundings and again the door is closed. Explain your answer.

16) You have been stirring a bucket of water vigorously and now the water is whirling around and around. You suddenly stop stirring and gradually the water ceases to flow until finally after a long enough period of time, the water is motionless in the bucket. What has happened to the entropy of the water? Explain. There is no heat transfer between the bucket and the surroundings and no water is added or subtracted from the bucket.

17) In the preceding three questions, what has happened to the total energy of the system under consideration? Explain.

18) How do you calculate the exchange of entropy between the system and its surroundings as a result of heat transfer?

19) How do you calculate the exchange of entropy between the system and its surroundings as a result of mass exchange?

20) What is the change in entropy of a system undergoing a steady-state process?

21) An accounting of the entropy of a system is not an adequate statement of the second law of thermodynamics for the process associated with that system. Explain why not and what is required for a complete statement of the second law.

SCALES

1) Entropy has dimensions of energy/temperature. What is 1 J/K in units of BTU/R?

2) Tables C-22 (English units) and C-23 (SI units) give thermodynamic properties $(\widehat{U}, \widehat{H}, \widehat{S}, P, T, \widehat{V})$ of steam as saturated liquid, saturated vapor, and superheated vapor (vapor that is either too hot for the given pressure or at too low a pressure for the given temperature to be saturated). Find the value of entropy for superheated steam at a pressure of 200 psi and a temperature of 600°F, what is $\widehat{S}$?

3) What is the value of $\widehat{S}$ for superheated steam at 200 kPa and 500°C?

4) What is the value of $\widehat{S}$ for saturated steam at 200 kPa? (Note that a temperature need not be specified. What is the saturation temperature at 200 kPa?)

5) What is the value of $\widehat{S}$ for saturated liquid water at 200 kPa?

6) Steam enters a boiler at 200 kPa and 200°C and leaves the boiler at 200 kPa and 600°C. How much is $\widehat{S}_{out} - \widehat{S}_{in}$ for the boiler defined as the system?

7) In exercise 6, if $\dot{m}_{in} = \dot{m}_{out} = 0.7$ kg/s, then what is $\dot{S}_{in/out}$ due to mass exchange?

8) For a given system, $\widehat{Q} = 5$ kJ/kg.

    a) Is this energy entering or leaving the system? How do you know?
    b) As a result of this heat transfer, is entropy added to the system or removed from it?
    c) What, if anything, can you say about the temperature of the system compared to that of the surroundings where this heat transfer occurs?

9) If a process operates at steady-state, what, if anything, can you say about its entropy change over time?

10) For a system $\dot{Q} = k_1(T_{sur} - T_{sys})$ where $k_1 = 0.02$ kJ/(min K), $T_{sys} = 300$ K+(5 K/min)t, and T_{sur} is constant at 400 K. Calculate

    a) Q for t from 5 to 10 min,
    b) $\dot{S}_Q(t)$ at $t = 0, 2,$ and 5 min,
    c) $S_{in,Q}$ for t from 5 to 10 min.

11) Given the data in the accompanying table, calculate:

    a) Q for t from 5 to 10 min,
    b) $\dot{S}_Q(t)$ at $t = 0, 2,$ and 5 min,
    c) $S_{in,Q}$ for t from 5 to 10 min.

Data for exercise 11.

t	$\dot{Q}$	T_{sys}	T_{sur}
0	5	100	170
1	6	102	172
2	7	104	174
3	7.5	106	176
4	7	110	172
5	6	112	171
6	4	114	170
7	3	115	170
8	2.5	116	174
9	3	118	178
10	4	120	180

PROBLEMS

1) A steady state process produces 1000 (kW) of power. The process involves steam entering the process through one flow stream at a mass flow rate of 100 (kg/min) and leaving through one flow stream. The entering and leaving conditions of the steam are given in the table.

Entering	Leaving
$T = 1000\ °F$	$T = 250\ °F$
$P = 100$ (psia)	$P = 20$ (psia)
$\widehat{H} = 1529.2$ (Btu / lb$_m$)	$\widehat{H} = 1168.0$ (Btu / lb$_m$)
$\widehat{S} = 1.9205$ (Btu / lb$_m$ R)	$\widehat{S} = 1.7475$ (Btu / lb$_m$ R)

a) Define the system to be the internal volume of the process. Calculate the rate that heat is being transferred. With respect to the system, is the heat transfer negative or positive. What is the time rate of change of the total energy of the system and why?

b) Again, define the system to be the internal volume of the process. What is the time rate of change of the entropy of the system and why?

c) Define the system to be the surroundings. Calculate the rate at which entropy enters the surroundings due to the mass entering the surroundings. Calculate the rate at which entropy leaves the surroundings due to mass leaving the surroundings. If the surroundings is always at a constant temperature of 100 °F, calculate the rate at which entropy enters or leaves the surroundings due to heat transfer. Finally, calculate the time rate of change of the entropy of the surroundings.

d) Based on these calculations, can this process happen and why?

2) An inventor claims to have developed a steady-state engine that converts heat to work. There is no mass that flows in or out of the engine. It takes in heat at the rate of 25,000 (J/s) from part of the surroundings which is at a temperature of 400 (K), rejects heat at a rate of 5,000 (J/s) to a part of the surroundings which is at temperature of 200 (K), and delivers 20 (kW) of mechanical power. Would you advise investing money to put this engine on the market? Explain your reasons for investing or not investing your money.

3) Given the temperature of the system (T_{sys}) and the temperature of the surroundings (T_{sur}) and the heat transfer rate with respect to the system, calculate the net rate of entropy entering the system and surrounding if this is the only process. Also, calculate the entropy generation in the universe due to this process.

a) $T_{sys} = 100\ °F$, $T_{sur} = 100\ °F$, $\dot{Q}_{in,sys} = 50$ (Btu / hr)
b) $T_{sys} = 279\ K$, $T_{sur} = 230\ K$, $\dot{Q}_{in,sys} = -350$ (J / s)
c) $T_{sys} = 400\ R$, $T_{sur} = 600\ R$, $\dot{Q}_{in,sys} = 200$ (Btu / s)
d) $T_{sys} = 25\ °C$, $T_{sur} = 0\ °C$, $\dot{Q}_{in,sys} = -200$ (J / s)
e) $T_{sys} = 40\ °F$, $T_{sur} = 0\ °F$, $\dot{Q}_{in,sys} = -30$ (Btu / min)
f) $T_{sys} = 20\ °C$, $T_{sur} = 20\ °C$, $\dot{Q}_{in,sys} = 50$ (J / hr)

4) A baseball is struck by a bat. The bat performs work on the ball and the velocity of the ball changes and eventually the ball travels out of the ball park. All of the work that the bat did on the ball was not required for ball to travel out of the ball park and the ball heats up because of the increase in internal energy. The ball eventually returns to the preimpact temperature and preimpact shape.

Use the conservation of energy and the Second Law of Thermodynamics to determine what happened to the entropy of the ball, the entropy of the surroundings, and the entropy of the universe. Why?

5) A process is operating at steady state with the following entering and leaving fluid properties. The mass flow rate of the fluid is 20 (lb$_m$ / hr).

	Entering		Leaving
	$P = 30$ (psia)		$P = 5$ (psia)
	$\widehat{H} = 670$ (Btu / lb$_m$)		$\widehat{H} = 520$ (Btu / lb$_m$)
	$\widehat{S} = 1.56$ (Btu / lb$_m$ R)		$\widehat{S} = 1.45$ (Btu / lb$_m$ R)

If the temperature of the surroundings is 75 °F, and the entering and leaving fluid properties are fixed, calculate the maximum theoretical work rate that this process could produce and why. (Neglect changes in kinetic and potential energy)

6) A car is going 60 (mph) and has a mass of 3000 (lb$_m$). The driver applies the brakes and stops the car. Define the car as the system. From the conservation of linear momentum, how much momentum was transferred to the earth. This transfer of momentum results in practically zero transfer of the initial kinetic energy of the car to the earth. Where did the kinetic energy of the car go? The brakes are initially at 75 °F and the outside air is 75 °F. Right after the car stops the brakes are 500 °F. If the car sits there long enough, the temperature of the brakes will eventually reach 75 °F. If all of the energy in the brakes is tranferred to the surroundings, calculate the entropy increase of the surroundings due to stopping the car. What is the ultimate fate of the the kinetic energy of the car?

7) A complex process is operating at steady state. There is one entering mass flow stream and two exiting streams. Within the system, there is no entropy generation from irreversibilities. The data are given in the table below.

Stream 1	Stream 2	Stream 3
$\dot{m}_1 = 5$ kg / s	$\dot{m}_2 = 2$ kg / s	?
$\widehat{S}_1 = 1.23$ kJ / kg K	$\widehat{S}_2 = 0.5$ kJ / kg K	$\widehat{S}_3 = 2.10$ kJ / kg K

Calculate the rate of entropy entering the system due to heat transfer. (Remember to use the conservation of mass.)

8) An isolated process has initial entropy value of 20.3 kJ/K. Over some time period, an irreversible process occurs. The final entropy value is 25.7 kJ/K. Calculate the entropy generation in the system from the irreversibility.

9) A system contains 20 kg at a constant temperature of 30°C. The surroundings temperature is 35 °C. At the beginning of the time period, the entropy per unit mass is 1.235 kJ/kg K. At the end of the time period, the entropy per unit mass is 1.724 kJ/kg K. No mass crosses the system boundary and the process is reversible.

a) Calculate the change in entropy for the system.

b) Use the entropy accounting equation to determine the entropy increase due to heat transfer.

c) Calculate the heat transfer.

d) Use the entropy accounting equation to determine the entropy change in the surroundings.

e) Calculate the change in entropy for the universe.

10) A process contains 15 lb_m at a constant temperature of 90°F. The surrounding temperature is 75°F. At the beginning of the time period, the entropy per unit mass is 2.193 Btu / lb_m R. At the end of the time period, the entropy per unit mass is 1.527 Btu / lb_m R. No mass crosses the system boundary and the process is mechanically reversible (reversible except for heat transfer).

a) Calculate the change in entropy for the system.

b) Use the entropy accounting equation to determine the entropy entering with heat transfer.

c) Calculate the heat transfer.

d) Use the entropy accounting equation to determine the entropy change in the surroundings.

e) Calculate the change in entropy for the universe.

11) A process is operating at steady state. There is only one inlet and one outlet. The following data are given.

$$\dot{m}_{in} = 5 \text{ kg / hr} \qquad\qquad \dot{m}_{out} = ?$$
$$\widehat{S}_{in} = 1.374 \text{ kJ / kg K} \qquad\qquad \widehat{S}_{out} = 1.574 \text{ kJ / kg K}$$

The process is reversible. Calculate the entropy entering/leaving the system due to heat transfer.

12) An unsteady state process is proposed. The initial conditions are

$$m_{beg} = 2 \text{ lb}_m$$
$$\widehat{U}_{beg} = 1024 \text{ Btu / lb}_m$$
$$\widehat{S}_{beg} = 1.214 \text{ Btu / lb}_m \text{ R}$$

Over a one minute time period mass is added to the system. The mass has the following properties.

$$m_{in} = 3 \text{ lb}_m$$
$$\widehat{H}_{in} = 1221 \text{ Btu / lb}_m$$
$$\widehat{S}_{in} = 1.314 \text{ Btu / lb}_m \text{ R}$$

The process is adiabatic and reversible. There is no non-flow work, no change in kinetic or potential energy for the process. Determine the final internal energy and entropy.

13) A reversible adiabatic process is operating at steady state. The first two streams enter the process and the third leaves the process. The flow streams contain the same material; however, the material is at different conditions.

Stream 1	Stream 2	Stream 3
$\dot{m}_1 = 150$ kg / hr	$\dot{m}_2 = 75$ kg / hr	$\dot{m}_3 = ?$
$\widehat{H}_1 = 1230$ kJ / kg	$\widehat{H}_2 = 650$ kJ / kg	$\widehat{H}_3 = 860$ kJ / kg
$\widehat{S}_1 = 1.271$ kJ / kg K	$\widehat{S}_2 = 1.053$ kJ / kg K	$\widehat{S}_3 = ?$

Neglect the kinetic and potential energy terms. The entropy generation from irreversibilities is zero. Calculate the rate of mass leaving the process, the rate of work done on the process and the specific entropy of the mass leaving the process.

Summary of Conservation and Accounting Relationships

Extensive Property		Statement
Mass	Total	$\left(\dfrac{dm_{sys}}{dt}\right) = \sum \dot{m}_{in/out}$
	Species i	$\left(\dfrac{d(m_i)_{sys}}{dt}\right) = \sum (\dot{m}_i)_{in/out} + \sum (\dot{m}_i)_{gen/con}$
Charge	Net	$\left(\dfrac{dq_{sys}}{dt}\right) = \sum i_{in/out}$
	Positive	$\left(\dfrac{d(q_+)_{sys}}{dt}\right) = \sum (\dot{q}_+)_{in/out} + \sum (\dot{q}_+)_{gen/con}$
	Negative	$\left(\dfrac{d(q_-)_{sys}}{dt}\right) = \sum (\dot{q}_-)_{in/out} + \sum (\dot{q}_-)_{gen/con}$
Linear Momentum	General	$\left(\dfrac{d\mathbf{p}_{sys}}{dt}\right) = \sum (\dot{m}\mathbf{v})_{in/out} + \sum \mathbf{f}_{ext}$
	Particle	$m_{sys}\mathbf{a} = \sum \mathbf{f}_{ext}$
	Rigid Body	$m_{sys}\mathbf{a}_G = \sum \mathbf{f}_{ext}$
Angular Momentum	General	$\left(\dfrac{d\mathbf{l}_{sys}}{dt}\right) = \sum (\mathbf{r} \times \dot{m}\mathbf{v})_{in/out} + \sum (\mathbf{r} \times \mathbf{f})_{ext}$
	Rigid Body–Planar Motion	$_G I_a \alpha + m(\mathbf{r}_G \times \mathbf{a}_G)_a = \sum T_a$
	Rigid Body about G–Planar Motion	$_G I_a \alpha = \sum {}_G T_a$
	Rigid Body about B–Planar Motion	$_G I_a \alpha + m[_B\mathbf{r}_G \times ({}_B\mathbf{a}_G + \mathbf{a}_B)]_a = \sum {}_B T_a$
	Planar Rigid Body about B Inertial and fixed in the body–Planar Motion	$_B I_a \alpha = \sum {}_B T_a$

Total Energy — Internal Energy Form

$$\left(\frac{dE_{sys}}{dt}\right) = \sum\left(\dot{m}(\hat{U}+\widehat{PE}+\widehat{KE})\right)_{in/out} + \sum\dot{W}_{tot} + \sum\dot{Q}$$

Enthalpy Form

$$\left(\frac{dE_{sys}}{dt}\right) = \sum\left(\dot{m}(\hat{H}+\widehat{KE}+\widehat{PE})\right)_{in/out} + \sum\dot{W}_{non-flow} + \sum\dot{Q}$$

Closed System

$$\left(\frac{dE_{sys}}{dt}\right) = \sum\dot{W}_{non-flow} + \sum\dot{Q}$$

Isolated System

$$\left(\frac{dE_{sys}}{dt}\right) = 0$$

Mechanical Energy — Steady State (1 Inlet/Outlet)

$$0 = \dot{m}(\widehat{PE}+\widehat{KE})_{in} - \dot{m}(\widehat{PE}+\widehat{KE})_{out} + \dot{m}\left(\int_{P_{out}}^{P_{in}}\hat{V}dP\right) + \sum\dot{W}_{shaft} - \sum\dot{F}$$

Bernoulli

$$0 = \dot{m}(\widehat{PE}+\widehat{KE})_{in} - \dot{m}(\widehat{PE}+\widehat{KE})_{out} + \dot{m}\left(\int_{P_{out}}^{P_{in}}\hat{V}dP\right)$$

Thermal Energy — Steady State (1 Inlet/Outlet)

$$0 = \dot{m}(\hat{H}_{in}-\hat{H}_{out}) - \dot{m}\left(\int_{P_{out}}^{P_{in}}\hat{V}dP\right) + \sum\dot{Q} + \sum\dot{F}$$

Electrical Energy

$$\left(\frac{d(E_E)_{sys}}{dt}\right) = \sum(iV)_{in/out} + \sum\dot{W}_{ele} - \sum\dot{F}$$

Entropy — System

$$\left(\frac{dS_{sys}}{dt}\right) = \sum(\dot{m}\hat{S})_{in/out,sys} + \sum(\dot{S}_Q)_{in/out,sys} + \sum\dot{S}_{gen(non-Q),sys}$$

Surroundings

$$\left(\frac{dS_{sur}}{dt}\right) = -\sum(\dot{m}\hat{S})_{in/out,sys} + \sum(\dot{S}_Q)_{in/out,sur} + \sum\dot{S}_{gen(non-Q),sur}$$

Universe

$$\left(\frac{dS_{uni}}{dt}\right) = \sum\left(\frac{\dot{Q}}{T_{sys}}\right) - \sum\left(\frac{\dot{Q}}{T_{sur}}\right) + \sum\dot{S}_{gen(non-Q),uni} \geq 0$$

RIGID BODY MECHANICS

In Part II you were introduced to the foundation laws which govern our physical world. These laws consist of a handful of conservation laws together with the Second Law of Thermodynamics which imposes an additional constraint on processes. In addition to these laws we also find it useful to account for certain subsets of the conserved properties. For example, it is useful to account for individual species or compounds, even though they are not, in general, conserved, whereas total mass is conserved. Additionally, particular forms of energy such as mechanical energy, thermal energy, or electrical energy, while not conserved, still can provide useful insight into processes through accounting equations. When using these fundamental laws and accounting equations it was essential to define a specific time period and system in addition to the particular extensive property being considered; all terms of the accounting or conservation statements had to hold for this same time period and system.

With these fundamental laws and accounting statements in hand, together with some basic knowledge of vector operations, integral and differential calculus, and linear algebra, you are now equipped to handle a great many problems in science and engineering. As you learn more about the properties and behavior of materials you will be able to address even more problems. Whatever problems you may confront, you now know that certain things must be conserved and that furthermore, by accounting for subsets of these conserved properties you can glean additional useful information. It is the purpose of Part III to provide you with practice and experience in applying these principles to a traditional area of engineering, rigid body mechanics.

9.1 INTRODUCTION

In Chapter 9 the topic of rigid body statics is addressed. This material includes a summary of the governing equations for these situations and applications to particle (or particle-like) situations, finite-sized rigid body situations, and to the analysis of structures and machines. In some of these situations, especially for machines, it will be useful to consider, instead of (or in addition to) linear and angular momentum, a mechanical energy accounting equation.

9.2 PARTICLE VERSUS RIGID BODY

In this text we will be concerned with particle masses and with rigid bodies. These terms need to be clearly understood in terms of the ramifications that they have on forces and the effects caused by forces which act on these bodies. Figure 9-1 shows a particle and a rigid body with forces acting on each.

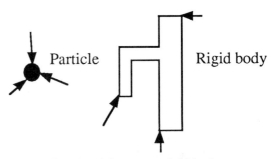

Figure 9-1: Particle versus rigid body

A *particle* is defined as an idealized mass having finite mass but zero volume. In that way all forces acting upon the particle (gravitational and contact forces) are acting at exactly the same point. This simplifies the consideration of their action and, for a single particle system, results in an angular momentum equation which is identical to the linear momentum equation.

A *rigid body* is a mass of finite size. In this case different elements of the body may be doing different things. Furthermore, forces may act at different points on the body. There may be different points of contact of the forces on the body and furthermore, gravity forces are distributed throughout the body. Both linear and angular momentum must be considered. However, a rigid body is simplified considerably by the fact that it is rigid, i.e. there is no relative displacement of any two elements of the body; the body undergoes no deformation. This implies that all elements of the rigid body are undergoing the same angular velocity and, in the absence of rotation, all elements of the body are translating in exactly the same way, i.e. they have the same velocity. Another implication of the term rigid body is that the mass of the body is constant; there is no mass entering or leaving the body.

9.3 APPLICABILITY OF THE BASIC EQUATIONS

Although we have now at our disposal for process or system analysis five conservation laws, one accounting law, and a number of accounting equations, they do not all provide independent or non-trivial information for static situations. Some are important and others are not.

The conservation of linear and angular momentum, in their roles as the equations of translational and rotational motion (or lack thereof) respectively, play key roles in the analysis of the mechanics of rigid bodies. Conservation of linear momentum says that the total amount (vector sum) of force acting on a body is vital information for understanding its motion (or lack thereof). Conservation of angular momentum says that in addition to the (vector) values of the forces, their points of application also are important.

Conservation of total mass and charge, however, are trivial. For example, because our concern in this chapter is with rigid bodies having no entering and leaving mass, conservation of mass provides us with no useful information. The result is simply that the total mass of the body is a constant. Likewise, we are not interested in charged systems or processes involving exchange of charge and therefore conservation of charge in and of itself provides us with no useful information. (If the body is charged, however, and subjected to an electric field, it will experience forces due to the charge. These forces must be included, of course, in calculations of linear and angular momentum.)

Finally, energy considerations, although not providing independent information, may be useful. There is no mass entering or leaving a system and, therefore, there is no enthalpy, kinetic energy, or potential energy entering or leaving the system. Also, for a steady-state (static) process, there are no changes in kinetic energy, potential energy or internal energy within the system. Finally, we are considering situations where there is no heat transfer between the system and surroundings. Consequently, the only total energy conservation terms that might remain in rigid body mechanics are those involving the exchange of energy between the system and surroundings as work. However, such a statement, involving only mechanical energy, contains no additional information not already incorporated in momentum conservation; in fact it contains less. Therefore, we also do not need to consider energy conservation (however, although energy provides less information, it turns out that this can still be a useful way of analyzing certain situations). Finally, for a static situation the second law of thermodynamics provides no useful information. Nothing is going on, nothing is happening; therefore there is no entropy transfer or entropy generation to be considered.

So, for rigid body statics we are left with the conservation of linear and angular momentum as the two laws which can provide us with useful information; the other laws also are obeyed, of course, but they are satisfied trivially and give us no additional information about the process. For static situations these two laws, in rate form, reduce to the result that (remember, the left side of the equation represents the rate of change of system momentum (accumulation rate) while the right side represents input and output rates of linear momentum)

$$0 = \sum \mathbf{f}_{ext} \tag{5 – 15}$$

and that

$$0 = \sum (\mathbf{r} \times \mathbf{f})_{ext} \tag{6 – 53}$$

or

$$0 = \sum (_B\mathbf{r} \times \mathbf{f})_{ext} \tag{6 – 54}$$

where a torque is expressed by $\mathbf{r} \times \mathbf{f}$. The remainder of this chapter deals with applying these laws to a number of particle and rigid body situations. Additionally, as mentioned above, we will have some occasions to use a mechanical energy equation in lieu of the conservation of linear momentum equation.

This chapter is divided into problems considering the statics of particles, the statics of rigid bodies, and finally the statics of structures such as frames, trusses, and machines. Along the way, considerations of applied forces other than those acting at a point will have to be considered. These include distributed loads such as the weight of a finite body distributed over the entire volume of the body and the distribution of frictional forces acting at a finite surface of the body. Throughout this chapter, and for all problems, the tabular formulation and presentation of the concepts and calculations will be important as a way of organizing the conservation information. Additionally, procedures and concepts using vector operations will be employed. *In this way problems in both three dimensions and two dimensions can be handled with virtually identical conceptual ease and only marginally different computational difficulty.*

9.4 PARTICLE STATICS

As discussed previously, for a single particle there is no difference between conservation of linear momentum and conservation of angular momentum. As a result the consideration of single particle systems is done using only the conservation of linear momentum. For a static particle this result is simply that

$$0 = \sum \mathbf{f}_{ext} \tag{5 – 15}$$

The vector addition of all of the external forces acting on the single particle must be zero. This law provides three independent scalar equations (one for each coordinate direction or component of the vector equation). Consequently, with three independent scalar equations, we can solve for three unknowns which can be components of any of the vector forces, angular directions of any of the forces, magnitudes of any of the forces or combinations of these. Example applications of the conservation of linear momentum to particle or point statics follow.

Example 9-1. Determine the required length of cord AC in Figure 9-2. The fan is suspended in the position shown. The unstretched or equilibrium length of the spring AB is $l'_{AB} = 0.3$ (m), and the spring has a stiffness of $k_{AB} = 200$ (N / m).

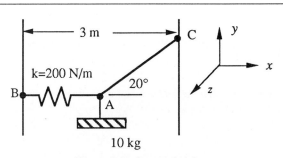

Figure 9-2: Suspended fan

SOLUTION: Choose the system to be the point A where the cord AC, spring AB and fan are attached. A *free body diagram* of the system is shown. The free body diagram shows all the external forces acting on the system. The conservation of linear momentum for this system is shown in Table 9-1 where f_{AB}, f_{AC}, and mg are the *magnitudes* of the corresponding forces. The force in AC is written in terms of an unknown magnitude f_{AC} and known unit vector $\mathbf{u}_{AC}$ such that:

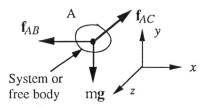

Figure 9-2a: Suspended fan
free body diagram

$$\mathbf{f}_{AC} = f_{AC}\mathbf{u}_{AC}$$
$$= f_{AC}\cos(20°)\mathbf{i} + f_{AC}\sin(20°)\mathbf{j} + 0\mathbf{k} \text{ (N)}$$

Table 9-1: Example 9-1 (Linear Momentum, System = Point A)

Description	**0**	**=**	$\sum \mathbf{f}_{\text{ext}}$		
			$\mathbf{f}_{AB}$	$\mathbf{f}_{AC}$	$m\mathbf{g}$
	(N)		(N)	(N)	(N)
x-direction	$0\,\mathbf{i}$	$=$	$-f_{AB}\mathbf{i}$ +	$f_{AC}\cos(20°)\mathbf{i}$ +	$0\mathbf{i}$
y-direction	$0\,\mathbf{j}$	$=$	$0\mathbf{j}$ +	$f_{AC}\sin(20°)\mathbf{j}$ −	$mg\mathbf{j}$
z-direction	$0\,\mathbf{k}$	$=$	$0\mathbf{k}$ +	$0\mathbf{k}$ +	$0\mathbf{k}$

From the table, we can solve for f_{AC} from the y-direction equation.

$$f_{AC} = \left(\frac{mg}{\sin(20°)}\right) = 286.5 \text{ (N)}$$

Back substitution in the x-direction equation yields:

$$f_{AB} = f_{AC}\cos(20°) = 269.2 \text{ (N)}$$

The force vectors are given:

$$\mathbf{f}_{AC} = 269.2\mathbf{i} + 98\mathbf{j} + 0\mathbf{k} \text{ (N)}$$
$$\mathbf{f}_{AB} = -269.2\mathbf{i} + 0\mathbf{j} + 0\mathbf{k} \text{ (N)}$$

Now the force exerted by the spring on point A is proportional to the extension of the spring in the x-direction. If l'_{AB} is the unstreched length of the spring, then

$$-f_{AB}\mathbf{i} = -k_{AB}(l_{AB} - l'_{AB})\mathbf{i}$$

The minus sign says that the force resists the extension when $k_{AB} > 0$. Rearranging:

$$l_{AB} - l'_{AB} = \left(\frac{f_{AB}}{k_{AB}}\right) = \left(\frac{269.2}{200}\right) = 1.346 \text{ (m)}$$

The total length of the stretched spring is:

$$l_{AB} = l'_{AB} + (1.346) = (0.3) + (1.346) = 1.646 \text{ (m)}$$

The total horizontal distance from C to B is 3 meters or:

$$3 = l_{AC}\cos(20°) + l_{AB} \text{ (m)} = l_{AC}\cos(20°) + 1.646 \text{ (m)}$$

Rearranging this equation gives:

$$l_{AC} = \left(\frac{1}{\cos(20°)}\right)(3 - 1.646) = 1.44 \text{ (m)}$$

The length of the cord AC is 1.44 (m).

Example 9-2. The 80 (lb_m) cylinder shown in Figure 9-3 is supported by two cables and a spring having a stiffness of $k = 300$ (lb_f / ft). Determine the force in the cables and the extension of the spring. Cord AC lies in the y-z plane and cord AD lies in the x-z plane.

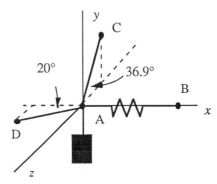

Figure 9-3: Supported cylinder

SOLUTION: Choose the point A as the system since the cable forces are concurrent at this point. Figure 9-3a shows the free body diagram of the isolated system with all the external forces shown. The conservation of linear momentum is given in Table 9-2; f_B, f_C, and f_D are the *magnitudes* of the corresponding vectors. The vectors $\mathbf{f}_C$ and $\mathbf{f}_D$ are expressed in terms of unknown magnitudes and known unit directional vectors such as:

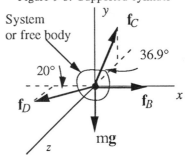

Figure 9-3a: Free body diagram of supported cylinder

$$\mathbf{f}_C = f_C\mathbf{u}_B$$
$$= 0\mathbf{i} + f_C\sin(36.9°)\mathbf{j} - f_C\cos(36.9°)\mathbf{k} \text{ (}lb_f\text{)}$$
$$\mathbf{f}_D = f_D\mathbf{u}_D$$
$$= -f_D\cos(20°)\mathbf{i} + 0\mathbf{j} + f_D\sin(20°)\mathbf{k} \text{ (}lb_f\text{)}$$

Table 9-2: Example 9-2 (Linear Momentum, System = Point A)

	0	=					$\sum \mathbf{f}_{\text{ext}}$				
Description			$\mathbf{f}_D$		$\mathbf{f}_C$		$\mathbf{f}_B$		$m\mathbf{g}$		
	(lb$_f$)		(lb$_f$)		(lb$_f$)		(lb$_f$)		(lb$_f$)		
x-direction	0 i	=	$-f_D \cos(20°)$i	+	0i	+	f_Bi	+	0i		
y-direction	0 j	=	0j	+	$0.60 f_C$j	+	0j	−	mgj		
z-direction	0 k	=	$f_D \sin(20°)$k	−	$0.80 f_C$k	+	0k	+	0k		

The solution of the y-direction equation gives: $f_C = 133.3$ lb$_f$. Back substitution in the z-direction equation yields: $f_D = 311.9$ lb$_f$. And finally the x-direction gives: $f_B = 293.1$ lb$_f$. The vectors are written out in complete form.

$$\mathbf{f}_B = 293.1\mathbf{i} + 0\mathbf{j} + 0\mathbf{k} \text{ (lb}_f)$$

$$\mathbf{f}_C = 0\mathbf{i} + 80\mathbf{j} + -106.6\mathbf{k} \text{ (lb}_f)$$

$$\mathbf{f}_D = -293.1\mathbf{i} + 0\mathbf{j} + 106.6\mathbf{k} \text{ (lb}_f)$$

The spring force is proportional to the extension of the spring in the x-direction $(l_{AB} - l'_{AB})$

$$f_B\mathbf{i} = k(l_{AB} - l'_{AB})\mathbf{i}$$

Rearranging and substitution yields:

$$l_{AB} - l'_{AB} = \left(\frac{f_B}{k}\right)$$

$$= \left(\frac{293.1}{300}\right) = 0.977 \text{ ft}$$

The stretch of the spring in the x-direction is 0.977 feet.

9.5 RIGID BODY STATICS

The situation of rigid body statics differs from that of particle statics in that the body is now of finite size. As a consequence, not only is the force itself important as it acts on a body, but also its point of action is important. Two forces acting at different points of the body, even though they are equal forces may impart a different torque or tendency to rotate upon the rigid body. This torque is accounted for through the conservation of angular momentum as discussed in Chapter 6. Thus, the conservation of linear momentum and the conservation of angular momentum taken together and the fact that the body is rigid (there is no relative motion of any parts of the body) are used to analyze the forces which act on a static rigid body.

Because a rigid body is finite in size, the forces which act upon it may act as *concentrated forces* at single points on the body or they may act as *distributed forces* throughout the body or across an entire surface. For example, a cable may be attached to a particular point on a rigid body and therefore exert a force at that point. A support member for the body exerts a force at a single point, where the support member is connected to the rigid body. However, the gravitational force acting upon the body is distributed throughout the entire body according to the distribution of the mass. Each force acting upon ach element of the body produces a different effect due to the different torque exerted upon the body. Frictional forces lso are distributed over a finite portion of the body namely the surface at which the frictional force occurs. Again, as with gravity, each point of contact on the surface exerts a different effect upon the body as a whole. We will consider further the reatment of these distributed forces before we continue with specific analyses of rigid bodies.

9.5.1 Distributed Forces - Gravity

Unless the body is quite large, we may con-ider the gravitational force exerted on each element f the body to be acting in the same direction and vith the same force per unit mass (gravitational ac-eleration). Consequently, the fact that the mass is istributed has no effect upon the conservation of *linear* momentum equation; we can simply substitute he force exerted by the entire body for the sum of all he individual forces. For the conservation of *angular* nomentum, however, we must consider a little more arefully the effect on the body of the forces on the arious elements inasmuch as they act at different ositions and therefore exert different torques, with espect to a common point, on the body as a whole. Figure 9-4 shows a body subjected to the distributed orce of gravity.

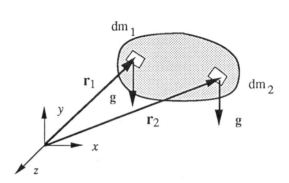

Figure 9-4: Distributed force of gravity

Considering each element of the body dm we calculate the combined torque (with respect to the origin of the oordinate system) exerted on the body as (subscript **g** indicates the torque due to **g**)

$$(\mathbf{r} \times \mathbf{f})_g = \int_{sys} (\mathbf{r} \times \mathbf{g}) dm \tag{9-1}$$

Because of the assumption of constant **g** and because of the distributive law for vector cross products, we may remove **g** rom the integral to obtain:

$$(\mathbf{r} \times \mathbf{f})_g = \left(\int_{sys} \mathbf{r} dm \right) \times \mathbf{g} \tag{9-2}$$

Note that for a number of individual finite-sized particles this integral would be a summation:

$$(\mathbf{r} \times \mathbf{f})_g = \sum_{sys} (\mathbf{r} \times \mathbf{g}m)_i$$
$$= (\mathbf{r} \times \mathbf{g}m)_1 + (\mathbf{r} \times \mathbf{g}m)_2 + \cdots (\mathbf{r} \times \mathbf{g}m)_n \tag{9-3}$$

or which we can clearly factor out the gravitational force per mass and then sum the position vector-mass products for all he individual particles:

$$(\mathbf{r} \times \mathbf{f})_g = \left(\sum_{sys} (\mathbf{r}m)_i \right) \times \mathbf{g} \tag{9-4}$$

Moving the gravitational force per mass outside the integral for the case of a continuum is analogous to moving it outside the summation for a body consisting of a number of individual particles. The integral may now be divided by the total mass of the body with the result unchanged, provided we multiply by the mass of the body elsewhere:

$$(\mathbf{r} \times \mathbf{f})_g = \frac{\int_{sys} \mathbf{r}\,dm}{m_{sys}} \times g m_{sys} \qquad (9-5)$$

Referring to the definition of the center of mass (Chapter 5)

$$\mathbf{r}_G = \frac{\int_{sys} \mathbf{r}\,dm}{m_{sys}} \qquad (5-8)$$

this result now can be written as the cross product of the position vector of the center of mass of the body with the weight of the body

$$\boxed{(\mathbf{r} \times \mathbf{f})_g = (\mathbf{r}_G \times g m_{sys})} \qquad (9-6)$$

This result simplifies considerably calculation of the effect of the distributed gravity force. The distributed weight is equivalent to a concentrated force equal to the body's weight which is acting at the center of mass of the body. This is shown in Figure 9-5. The two representations (distributed weight versus weight concentrated at the center of mass) are equivalent because they give the same result in both linear and angular conservation laws. Again, this result is contingent upon the body being small enough that the gravitational force per unit mass exerted upon each element of the body is everywhere the same in both direction and magnitude. If this is not the case, then the direct integral representation must be used.

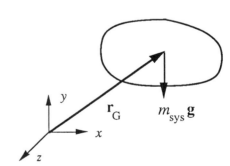

Figure 9-5: Equivalent force of gravity

9.5.2 Distributed Forces - Friction

Friction is a force which acts on a body as a result of contact of that body with another surface. In most cases, two surfaces in contact do not slide freely, that is there is a force which resists the relative sliding motion of the two surfaces. If weighted block is pulled across a flat surface, a resistance is felt which depends upon the weight of the block. Experimentally it has been found that a block at rest will not be moved until the pull parallel to the surface exceeds a certain value. As the pull increases up to that level the frictional force adjusts upward as well, always balancing the pull, so that the block remains stationary. At some point the force reaches the maximum capability of the friction to resist the movement of the block. Further increases in the pull will cause the block to move and in fact to accelerate. A further experimental observation is that once a body is in motion, the force required to keep it in motion in the presence of a frictional resistance is smaller han that required to initiate the motion starting from a static position. These experimental observations are summarized by an equation for the static force which states that the force required to *initiate* motion is

$$|\mathbf{f}_{s,\max}| = \mu_s |\mathbf{f}_N|$$

$$(9-7)$$

and the force required to *maintain* motion is given by

$$|\mathbf{f}_k| = \mu_k |\mathbf{f}_N|$$

$$(9-8)$$

Two points should be noted about these observations. First, in both of these equations the frictional resistance (force) is directly proportional to the load of the body on the surface, i.e. the normal force exerted against the surface. This is a rather interesting result; that no matter how big or how large the area, the same frictional force is observed for the same normal force (e.g. weight) acting on the surface, provided, of course, that the nature of the surfaces remains the same. Second, even though both the static and kinetic frictional resistances are proportional to the loading, they have different proportionality factors or different "coefficients of friction." Experimentally it has been found that the proportionality factor for the case of a moving object (the coefficient of kinetic friction, μ_k) is about 25% less than the coefficient of static friction (μ_s); it takes less force to keep a block in motion against a frictional resistance than it does to begin its motion. The actual values of the coefficients of friction depend upon the nature of the two surfaces (i.e. the materials and the roughness of the surface) which are in contact. Typical values of coefficients of static friction are given in Table C-8.

One other implication of the above discussion should be emphasized and understood. In a *static* situation the amount of the frictional resistance depends upon the other forces involved. If there is no force tending to move a block parallel to a surface contact, then the frictional resistance in that direction is zero. On the other hand if the force tending to move the block parallel to the surface is sufficient to almost start the motion of the block (*impending motion*), then the frictional resistance is equal to μ_s times the normal force. This is the maximum frictional force and hence the *max* subscript in equation 9-7. Values anywhere between these extremes are possible depending upon the magnitude of the force parallel to the surface. Within the limits of these properties of friction, frictional forces are treated no differently from other forces acting upon a rigid body.

Example 9-3. The box shown in Figure 9-6 has a mass of 30 (kg). Determine the force P so that the crate is on the verge of moving up the inclined plane (impending motion). The coefficient of static friction between the box and the plane is $\mu_s = 0.2$.

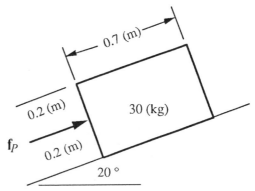

Figure 9-6: Impending motion for a box

SOLUTION Choose the system as the box and coordinate system aligned with the axis of the box as shown in the free body diagram Figure 9-6b. The conservation of linear momentum for this static system is given in Table 9-3. The force due to gravity is expressed in terms of the components with respect to the defined coordinate system. From the table, the magnitude of the nomal force is calculated from the y-direction equation, $f_N = 276.3$ (N). Back substitution into the x-direction equation gives that $f_P = 155.8$ (N). In vector form, then, the forces are found to be:

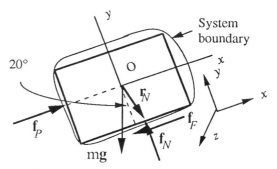

Figure 9-6a: Free body diagram of the box

$$f_F = -55.3i + 0j + 0k$$

$$f_P = 155.8i + 0j + 0k$$

$$f_N = 0i + 276.3j + 0k$$

and the crate requires a force of 155.8 (N) to start it in motion.

Table 9-3: Data for Example 9-3 (Rigid Body Statics-Linear Momentum)

Description	0	=	$\sum f_{ext}$						
			f_P		f_F		f_N		mg
	(N)		(N)		(N)		(N)		(N)
x-direction	0 i	=	$f_P i$	−	$\mu_s f_N i$	+	0i	−	$mg \sin(20°)i$
y-direction	0 j	=	0j	+	0j	+	$f_N j$	−	$mg \cos(20°)j$
z-direction	0 k	=	0k	−	0k	+	0k	+	0k

To calculate the position where the equivalent normal force is applied, we look at the conservation of angular momentum about point O of the defined coordinate system. Because the forces acting through the origin do not contribute to the angular momentum (the position vectors to these forces are of zero length), only the frictional force and the normal force contribute to the angular momentum equation. The position vectors from the origin to the points of application of the forces are:

$$r_F = 0i - 0.2j + 0k \text{ (m)}$$

$$r_N = xi - 0.2j + 0k \text{ (m)}$$

The cross products of the position vectors with the forces then are:

$$(r \times f)_F = \begin{vmatrix} i & j & k \\ 0 & (-0.2) & 0 \\ (-55.3) & 0 & 0 \end{vmatrix} = 0i + 0j - 11.0k \text{ (N m)}$$

$$(r \times f)_N = \begin{vmatrix} i & j & k \\ x & (-0.2) & 0 \\ 0 & 276.3 & 0 \end{vmatrix} = 0i + 0j + 276.3xk \text{ (N m)}$$

Table 9-4: Example 9-3 (Angular Momentum, System = box)

Description	0 (N m)	=	$\sum(\mathbf{r} \times \mathbf{f})_{\text{ext}}$		
			$(\mathbf{r} \times \mathbf{f})_F$ (N m)		$(\mathbf{r} \times \mathbf{f})_N$ (N m)
x-direction	0 i	=	0i	+	0i
y-direction	0 j	=	0j	+	0j
z-direction	0k	=	$(-11.0)\mathbf{k}$	+	$276.3x\mathbf{k}$

The conservation of angular momentum is given in Table 9-4. From the z-direction equation in the table, solving for x yields

$$x = \left(\frac{11.0}{276.3}\right) = 0.04 \text{ m}$$

The position vector from the origin to the point of application of the equivalent normal force is $\mathbf{r}_N = 0.04\mathbf{i} - 0.2\mathbf{j} + 0\mathbf{k}$ m.

9.5.3 Distributed Forces - Other Forces

We have considered the distribution of the weight of an object throughout the body and found that it is equivalent to considering that the entire weight of the object acts at the center of mass of the object. We can perform a similar treatment for other kinds of distributed loads. As the name implies, a distributed load is one for which the entire force is applied over a finite area or finite length instead of at a single point (concentrated load). For example, a supporting beam could have weight evenly distributed along its length. The question then would be how does this even distribution of load compare with that same total load applied at a single point; i.e. at what point should an equal weight acting as a single force, be applied in order to produce the same effect upon the body?

In either of these cases we are interested in obtaining a single applied force which is equivalent to a given force distribution. The first question to be understood is in what sense do we mean equivalent. Obviously because conservation of linear and angular momentum govern the motion (or lack thereof) of a rigid body, *we must say that two systems of forces are equivalent if they result in the same equations for the conservation of linear momentum and angular momentum.* This is exactly the same criterion that we used in considering the distributed force of the weight of the body. In that case we found that the weight distributed throughout the entire body was equivalent to that same weight applied at the center of mass of the body.

Now in general, a distributed load may be a function of position, that is the force that is exerted at a certain point of an area or at a certain position on a beam may depend upon that position. If we denote $\mathbf{d}_A(\mathbf{r})$ to be the force per area acting upon a given surface, then the total force acting on that surface due to $\mathbf{d}_A$ will be given by the integral:

$$\mathbf{f}_{\text{eq}} = \int_A \mathbf{d}_A(\mathbf{r})dA$$

$$(9-9)$$

A distributed force per unit area is depicted in Figure 9-7. The conservation of linear momentum involves the external forces acting upon a body; this total force acting at the surface of the body will have the same effect in the conservation of linear momentum if we use this integrated force to represent the total force on that surface.

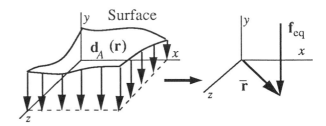

Figure 9-7: Distributed force per unit area

Similarly, the conservation of angular momentum requires calculation of the torque exerted by each force acting on the body. The force per unit area acting on a surface exerts a total (equivalent) torque which is calculated by forming the integral

$$\text{Total Torque} = (\mathbf{r} \times \mathbf{f})_{eq} = \int_A \mathbf{r} \times \mathbf{d}_A(\mathbf{r})dA \tag{9 – 10}$$

If we are going to define a single force acting at a single point on the surface as equivalent to the distributed force acting on the surface, then obviously the total distributed force must equal the equivalent single force (in the vector sense) in order for the conservation of linear momentum to be equivalent. Then the total torque exerted by this single force acting at some position $\bar{\mathbf{r}}$ must be the same total torque as the distributed force acting over the entire area. Hence,

$$\bar{\mathbf{r}} \times \mathbf{f}_{eq} = \int_A \mathbf{r} \times \mathbf{d}_A(\mathbf{r})dA \tag{9 – 11}$$

This last result in combination with the integrated force over the entire area (equation 9-9) is properly thought of as a defining relation for the point (position vector) at which the integrated force must act in order to be an equivalent force. If we know $\mathbf{d}_A(\mathbf{r})$ as a function of position on the surface, then both the integrated force and the integrated torque can be calculated; these are known quantities. This last equation is a vector equation and therefore consists of up to three scalar non-trivial equations. The position vector $\bar{\mathbf{r}}$ can be determined, at least to within a line of action.

A similar result can be obtained for a distributed load over one dimension, for example a load distributed over a beam. In this case we can write that the force per unit length of the beam is given by $\mathbf{d}_L(\mathbf{r})$. That is, the force per unit length is a function of position along the beam. Figure 9-8 shows a distributed force per unit length. Then the total force acting on the beam due to this distributed load is simply the integral over the length of the beam of this force per unit length:

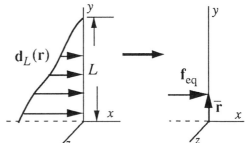

Figure 9-8: Distributed force per unit length

$$\mathbf{f}_{eq} = \int_L \mathbf{d}_L(\mathbf{r})dL \tag{9 – 12}$$

Furthermore, the total (equivalent) torque created by this distributed load is given by

$$\text{Total Torque} = (\mathbf{r} \times \mathbf{f})_{eq} = \int_L \mathbf{r} \times \mathbf{d}_L(\mathbf{r}) dL \tag{9-13}$$

and again this total torque can be represented in terms of an average position along the length of the beam and the total force exerted by this distributed load:

$$\bar{\mathbf{r}} \times \mathbf{f}_{eq} = \int_L \mathbf{r} \times \mathbf{d}_L(\mathbf{r}) dL \tag{9-14}$$

This result, along with equation 9-12, is the defining relation for the position at which the total force exerted by this distributed load must act in order to be equivalent, again in the context of both linear and angular momentum, to the distributed loading.

For the one-dimensional case we can write the loading in terms of a one-dimensional unit vector and also the position vector in terms of the unit vector along the length of the beam. Accordingly we may write (for the case where $\mathbf{d}_L(\mathbf{r})$ is everywhere normal to the longitudinal direction of the beam)

$$f_{eq}\mathbf{j} = \int_L d_L(L)\mathbf{j} dL \tag{9-15}$$

and

$$\bar{r}_L \mathbf{i} \times f_{eq}\mathbf{j} = \int_L \left(L\mathbf{i} \times d_L(L)\mathbf{j} \right) dL \tag{9-16}$$

which gives

$$\bar{r}_L f_{eq}\mathbf{k} = \left(\int_L L d_L(L) dL \right) \mathbf{k} \tag{9-17}$$

In this one-dimensional situation we see that the position of the point of action of the equivalent force is:

$$\bar{r}_L = \left(\frac{\int L d_L(L) dL}{\int d_L(L) dL} \right) \tag{9-18}$$

because the only component of this vector equation is in the z-direction

This result has a nice geometrical interpretation in that if we plot the force loading $d_L(L)$, versus the length or the position along the beam, then $\bar{r}_L$ is the coordinate in the L-direction of the *centroid of the area* which lies under this force versus length figure; if we draw the distributed force figure then we can easily visualize the equivalent force which would be acting upon that beam.

Similarly, for a uni-directional loading over an area we can write the total force exerted on that area

$$f_{eq}\mathbf{j} = \int_A d_A(\mathbf{r})\mathbf{j}\,dA$$

(9 – 19)

and the total moment may be written as

$$\bar{\mathbf{r}} \times f_{eq}\mathbf{j} = \int_A \left(\mathbf{r} \times d_A(\mathbf{r})\mathbf{j} \right) dA$$

(9 – 20)

Because the force is acting in the **j** direction, the position vector can only have components in the **i** and **k** direction. (The cross product of $\mathbf{j} \times \mathbf{j} = 0$.) From these results we have that:

$$\left((\bar{r}_x\mathbf{i} + \bar{r}_z\mathbf{k}) \times f_{eq}\mathbf{j} \right) = \int_A \left((x\mathbf{i} + z\mathbf{k}) \times d_A(\mathbf{r})\mathbf{j} \right) dA$$

(9 – 21)

Evaluating the cross products yields:

$$(\bar{r}_x f_{eq})\mathbf{k} - (\bar{r}_z f_{eq})\mathbf{i} = \int_A \left(x d_A(\mathbf{r}) \right) dA\,\mathbf{k} - \int_A \left(z d_A(\mathbf{r}) \right) dA\,\mathbf{i}$$

(9 – 22)

The components of the position vector to the point of application of the equivalent concentrated force is given by:

$$\bar{r}_x = \left(\frac{\int_A \left(x d_A(\mathbf{r}) \right) dA}{\int_A d_A(\mathbf{r})dA} \right)$$

$$\bar{r}_z = \left(\frac{\int_A \left(z d_A(\mathbf{r}) \right) dA}{\int_A d_A(\mathbf{r})dA} \right)$$

(9 – 23)

which for a flat surface, is the projection of the position vector to the geometric center of the volume under the force per area loading surface onto the surface.

Example 9-4. Replace the system of forces on the beam shown in Figure 9-9 by an equivalent single resultant force. Specify the distance along the beam from point A at which the force acts.

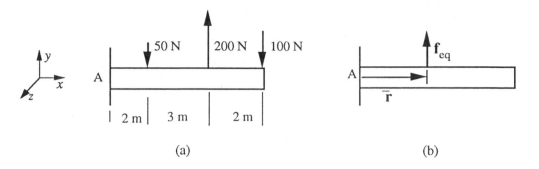

Figure 9-9: Loaded beam (a) and equivalent force (b)

SOLUTION: The resultant force is calculated as the net sum of the forces as shown in Table 9-5. The equivalent force is $f_{eq} = 0i + 50j + 0k$ (N). All the forces were in the **j** direction.

Table 9-5: Example 9-4 Equivalent force

	$\sum f_i$					=	f_{eq}
	f_{50} (N)		f_{200} (N)		f_{100} (N)		f_{eq} (N)
x-direction	$0i$	+	$0i$	+	$0i$	=	$(f_{eq})_x i$
y-direction	$(-50)j$	+	$200j$	−	$100j$	=	$(f_{eq})_y j$
z-direction	$0k$	+	$0k$	+	$0k$	=	$(f_{eq})_z k$

The moments, the cross products of all the position vectors (measured from point A) with their respective force vectors, will be summed. Then the moment of the equivalent force f_{eq} must be equal to the total moment of all these individual forces. Each of the moments about point A is calculated as:

$$(r \times f)_{50} = \begin{vmatrix} i & j & k \\ 2 & 0 & 0 \\ 0 & (-50) & 0 \end{vmatrix} = 0i + 0j - 100k \text{ (N m)}$$

$$(r \times f)_{200} = \begin{vmatrix} i & j & k \\ 5 & 0 & 0 \\ 0 & 200 & 0 \end{vmatrix} = 0i + 0j + 1000k \text{ (N m)}$$

$$(r \times f)_{100} = \begin{vmatrix} i & j & k \\ 7 & 0 & 0 \\ 0 & (-100) & 0 \end{vmatrix} = 0i + 0j - 700k \text{ (N m)}$$

Note that for these simple cases, the magnitude of the torque is easily calculated as the length of the momentum arm (distance from point A to the line of action of the force) times the magnitude of the force (e.g., (2 m)(500 N) = 100 (N m); (5 m)(200 N) = 1000 (N m):

Table 9-6: Example 9-4 Equivalent torque

	$\sum(\mathbf{r} \times \mathbf{f})_i$					=	$\bar{\mathbf{r}} \times \mathbf{f}_{eq}$
	$(\mathbf{r} \times \mathbf{f})_{50}$ (N m)		$(\mathbf{r} \times \mathbf{f})_{200}$ (N m)		$(\mathbf{r} \times \mathbf{f})_{100}$ (N m)		$\bar{\mathbf{r}} \times \mathbf{f}_{eq}$ (N m)
x-direction	$0\mathbf{i}$	$+$	$0\mathbf{i}$	$+$	$0\mathbf{i}$	$=$	$(\bar{\mathbf{r}} \times \mathbf{f}_{eq})_x\mathbf{i}$
y-direction	$0\mathbf{j}$	$+$	$0\mathbf{j}$	$+$	$0\mathbf{j}$	$=$	$(\bar{\mathbf{r}} \times \mathbf{f}_{eq})_y\mathbf{j}$
z-direction	$(-100)\mathbf{k}$	$+$	$1000\mathbf{k}$	$-$	$700\,\mathbf{k}$	$=$	$(\bar{\mathbf{r}} \times \mathbf{f}_{eq})_z\mathbf{k}$

(7 m)(100 N) = 700 (N m)). The sign must then be selected according to whether the torque is clockwise (−) or counter-clockwise (+). The cross products are now summed to yield an equivalent torque about point A as shown in Table 9-6.

The z-direction equation in the table reduces to:

$$200\mathbf{k} = \bar{x}\mathbf{i} \times 50\mathbf{j} = 50\bar{x}\mathbf{k} = 200\mathbf{k} \text{ (N m)}$$

Rearranging this equation, we can solve for the position vector to the point of application of the equivalent force

$$\bar{x}\mathbf{k} = \left(\frac{200}{50}\right)\mathbf{k}$$

$$\bar{x} = 4 \text{ (m)}$$

(Again, note the momentum arm–force-torque relation.) In vector form, the position vector to the point of application of the equivalent force is $\bar{\mathbf{r}} = 4\mathbf{i} + 0\mathbf{j} + 0\mathbf{k}$. The equivalent force acts 4 meters to the right of point A along the x-axis as shown in Figure 9-9.

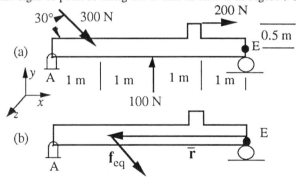

Figure 9-10: Coplanar loaded beam (a) and equivalent force (b)

Example 9-5. The beam AE in Figure 9-10 is subjected to several coplanar forces. Determine the magnitude, direction and position of the equivalent single force.

SOLUTION: The origin is located at point E for this system (It could be any place!). The total equivalent force is calculated a the sum of the forces on the beam. The components of the 300 (N) force are given in Table 9-7.

Evaluation yields the components :

$$(f_{eq})_x = 459.8 \text{ (N)}$$

$$(f_{eq})_y = -50.0 \text{ (N)}$$

$$(f_{eq})_z = 0.0 \text{ (N)}$$

Table 9-7: Example 9-5 Equivalent force

	$\sum \mathbf{f}_i$					=	$\mathbf{f}_{eq}$
	$\mathbf{f}_{300}$ (N)		$\mathbf{f}_{100}$ (N)		$\mathbf{f}_{200}$ (N)		$\mathbf{f}_{eq}$ (N)
x-direction	$300\cos(30°)\mathbf{i}$	+	$0\mathbf{i}$	+	$200\mathbf{i}$	=	$(f_{eq})_x\mathbf{i}$
y-direction	$(-300)\sin(30°)\mathbf{j}$	+	$100\mathbf{j}$	+	$0\mathbf{j}$	=	$(f_{eq})_y\mathbf{j}$
z-direction	$0\mathbf{k}$	+	$0\mathbf{k}$	+	$0\mathbf{k}$	=	$(f_{eq})_z\mathbf{k}$

And in vector form as:

$$\mathbf{f}_{eq} = 459.8\mathbf{i} - 50\mathbf{j} + 0\mathbf{k}$$

The moments will be summed about point E (the origin) to determine the point of application (as defined by position vector $\mathbf{r}$) of the equivalent force $\mathbf{f}_{eq}$. First, all of the individual cross products are evaluated as:

$$(\mathbf{r} \times \mathbf{f})_{300} = \begin{vmatrix} \mathbf{i} & \mathbf{j} & \mathbf{k} \\ (-3) & 0 & 0 \\ 259.8 & (-150) & 0 \end{vmatrix} = 0\mathbf{i} + 0\mathbf{j} + 450\mathbf{k} \text{ (N m)}$$

$$(\mathbf{r} \times \mathbf{f})_{100} = \begin{vmatrix} \mathbf{i} & \mathbf{j} & \mathbf{k} \\ (-2) & 0 & 0 \\ 0 & 100 & 0 \end{vmatrix} = 0\mathbf{i} + 0\mathbf{j} - 200\mathbf{k} \text{ (N m)}$$

$$(\mathbf{r} \times \mathbf{f})_{200} = \begin{vmatrix} \mathbf{i} & \mathbf{j} & \mathbf{k} \\ (-1) & 0.5 & 0 \\ 200 & 0 & 0 \end{vmatrix} = 0\mathbf{i} + 0\mathbf{j} - 100\mathbf{k} \text{ (N m)}$$

We will look for a position vector of an equivalent force which acts on the beam, i.e., with y and z components equal to zero. The equivalent torque is evaluated as:

$$\bar{\mathbf{r}} \times \mathbf{f}_{eq} = \begin{vmatrix} \mathbf{i} & \mathbf{j} & \mathbf{k} \\ \bar{x} & 0 & 0 \\ 459.8 & (-50) & 0 \end{vmatrix} = 0\mathbf{i} + 0\mathbf{j} - \bar{x}50\mathbf{k} \text{ (N m)}$$

Summing the torques in Table 9-8 and equating gives the equivalent torque.

Table 9-8: Example 9-5 Equivalent torque

	$\sum(\mathbf{r} \times \mathbf{f})_i$					=	$\bar{\mathbf{r}} \times \mathbf{f}_{eq}$
	$(\mathbf{r} \times \mathbf{f})_{300}$ (N m)		$(\mathbf{r} \times \mathbf{f})_{100}$ (N m)		$(\mathbf{r} \times \mathbf{f})_{200}$ (N m)		$\bar{\mathbf{r}} \times \mathbf{f}_{eq}$ (N m)
x-direction	$0\mathbf{i}$	+	$0\mathbf{i}$	+	$0\mathbf{i}$	=	$0\mathbf{i}$
y-direction	$0\mathbf{j}$	+	$0\mathbf{j}$	+	$0\mathbf{j}$	=	$0\mathbf{j}$
z-direction	$450\mathbf{k}$	−	$200\mathbf{k}$	−	$100\mathbf{k}$	=	$(-\bar{x})50\mathbf{k}$

From the table of equations we can see that:

$$(-\bar{x})50 = 150 \text{ (N m)}$$

Rearranging and solving for the position of the equivalent force gives:

$$\bar{x} = \left(\frac{150}{-50}\right) = -3 \text{ (m)}$$

Actually, a force which acts anywhere along the line of action of this equivalent force will also be equivalent. The position vector to the point of application of the equivalent force shown in Figure 9-10 is $\bar{r} = -3\mathbf{i} + 0\mathbf{j} + 0\mathbf{k}$ (m).

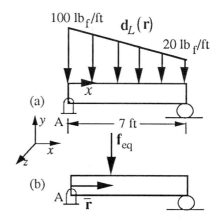

Example 9-6. Determine the magnitude and location of the equivalent force for the distributed load on the beam as shown in Figure 9-11.

Figure 9-11: Distributed load on a beam (a) and equivalent force (b)

SOLUTION: The equivalent force $\mathbf{f}_{eq}$ is equal to the integral of the distributed force over the length of the beam:

$$\mathbf{f}_{eq} = \int_L \mathbf{d}_L(\mathbf{r})dL$$

which can be viewed as the area under the distributed force versus distance curve. The direction of the distributed force and the total force is in the $\mathbf{j}$ direction yielding

$$f_{eq}\mathbf{j} = \int_L d_L(x)\mathbf{j}dL$$

The function of the distributed force for this problem is linear and has the general form

$$d_L(x) = ax + b$$

The values of the constants a and b are obtained by evaluating the function at the end point of the beam $x = 0$ and $x = 7$ (ft) (the origin is chosen to be at point A):

$$d_L(0) = a0 + b = -100 \text{ (lb}_f \text{ / ft)}$$

or

$$b = -100 \text{ (lb}_f \text{ / ft)}$$

and at the other limit

$$d_L(7) = a7 - 100 = -20 \text{ (lb}_f \text{ / ft)}$$

so that

$$a = \left(\frac{80}{7}\right) \text{ (lb}_f \text{ / ft}^2)$$

The component of the distributed force in the **j** direction is given by the equation

$$d_L(x) = \left(\frac{80x}{7}\right) - 100 \text{ (lb}_f \text{ / ft)}$$

The equivalent force is found by integrating the distributed force over the length of the beam:

$$f_{eq}\mathbf{j} = \int_0^7 \left(\frac{80x}{7} - 100\right) dx\mathbf{j}r = \left(\frac{80x^2}{14} - 100x\right)\Bigg|_0^7 \mathbf{j} \text{ (lb}_f) = \left((-420) - (0)\right)\mathbf{j} \text{ (lb}_f)$$

$$= -420\mathbf{j} \text{ (lb}_f)$$

If we calculated the area by using formula for a trapezoid we would get the exact same answer:

$$f_{eq}\mathbf{j} = \left(\frac{1}{2}\right)\left((-100) + (-20)\right)(7)\mathbf{j}$$

$$= -420\mathbf{j} \text{ (lb}_f)$$

Integrating the moments about point A gives the position vector of the equivalent force:

$$\bar{\mathbf{r}} \times \mathbf{f}_{eq} = \int_L \left(\mathbf{r} \times d_L(\mathbf{r})\right) dL$$

For a position vector in the **i** direction substitution yields:

$$(\bar{x}\mathbf{i} \times -420\mathbf{j}) = \int_0^7 \left(x\mathbf{i}\right) \times \left(\frac{80x}{7} - 100\right)\mathbf{j}dx$$

Evaluating the cross products gives

$$-420\bar{x}\mathbf{k} = \int_0^7 \left(\frac{80x^2}{7} - 100x\right)dx\mathbf{k}$$

or

$$-420\bar{x} = \int_0^7 \left(\frac{80x^2}{7} - 100x\right)dx = \left(\frac{80x^3}{21} - 50x^2\right)\Bigg|_0^7 \text{ (ft lb}_f) = (-1143 - 0) \text{ (ft lb}_f)$$

$$= -1143 \text{ (ft lb}_f)$$

Solving for the component of the position vector yields:

$$\bar{x} = \left(\frac{-1143}{-420}\right) = 2.72 \text{ (ft)}$$

In vector form and shown in Figure 9-11, the position vector to the equivalent force is $\bar{r} = 2.72\mathbf{i} + 0\mathbf{j} + 0\mathbf{k}$ (ft).

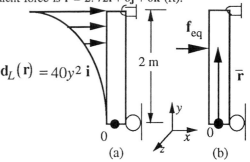

Example 9-7. Determine the magnitude and position of the equivalent force due to the distributed load acting on the beam as shown in Figure 9-12. The distributed force is given by $d_L(y)\mathbf{i} = 40y^2\mathbf{i}$ (N / m).

Figure 9-12: Distributed force over a beam (a) and equivalent force (b)

SOLUTION: The system is the beam and the origin of the coordinate system is chosen to be at point O. The equivalent force acting on the beam is the integral of the distributed force or loading over the length of the beam:

$$\mathbf{f}_{eq} = \int_L \mathbf{d}_L(\mathbf{r})dL$$

The equivalent force and the distributed force act only in the $\mathbf{i}$ direction yielding:

$$f_{eq}\mathbf{i} = \int_L d_y(y)\mathbf{i}dy$$

Substituting the distributed force into the integral gives:

$$f_{eq}\mathbf{i} = \int_0^2 40y^2\mathbf{i}dy = \left(\frac{40y^3}{3}\right)\Bigg|_0^2 \mathbf{i} \text{ (N)} = 106.7\mathbf{i} \text{ (N)}$$

The position vector to the point of application of the equivalent force is calculated from:

$$\bar{r} \times \mathbf{f}_{eq} = \int_L \mathbf{r} \times \mathbf{d}_L(\mathbf{r})dL$$

The position vector is in the $\mathbf{j}$ direction only yielding:

$$(\bar{y}\mathbf{j} \times 106.7\mathbf{i}) = \int_0^2 (y\mathbf{j}) \times (40y^2\mathbf{i})dy$$

Evaluating the cross products gives:

$$-106.7\bar{y}\mathbf{k} = -\int_0^2 40y^3 dy\mathbf{k}$$

or

$$106.7\bar{y} = \int_0^2 40y^3 dy = \left(\frac{40y^4}{4}\right)\Bigg|_0^2 \text{ (N m)} = 160 \text{ (N m)}$$

Solving for the component of the position vector:

$$\bar{y} = \left(\frac{160}{106.7} \right) = 1.5 \ (m)$$

In vector form and shown in Figure 9-12, the position vector to the equivalent force is $\bar{r} = 0i + 1.5j + 0k$ (m).

9.6 ANALYSIS OF STRUCTURES AND MACHINES

You now have all of the tools which you need to analyze and understand the forces which exist in static structures such as trusses and frames and those in simple machines consisting of interconnected rigid bodies which may undergo some limited motion. Each of these devices consists of members connected together to perform a certain function. Consequently, they can be analyzed in a similar way. We can look at the point where several members join together and treat that point as we would a single particle. We can also look at larger sections of the entire device and treat them as rigid bodies, in which case both conservation of linear and angular momentum are important, just as they are for finite-sized rigid bodies. Finally, for the case of machines where the connected members may undergo movement we find it useful to consider the conservation of energy either alone or in addition to linear and angular momentum. Each of these types of devices and their analysis are considered in more detail below.

In structures and machines the connections and supports at some of the members are made in such a way that the forces are constrained to act only in certain directions. For example, at one end of a bridge the support may be by means of a roller so as to accommodate changes in the length of the bridge due to expansion or contraction with temperature changes. In such a case, the only force that would be experienced at this support would be a normal force. Forces which would develop as the result of expansion are neutralized by the free-rolling end. Table C-9 and Table C-10 in Appendix C show a number of typical supports and connections which appear in structures and machines and the resulting forces which can be transmitted through these supports or connections. Where such supports exist, one must be careful to count only the appropriate number of force components (i.e., the non-zero components) in the linear and angular momentum equations.

9.6.1 Trusses

Trusses are structures which are designed to provide support for a load. Examples are trusses for roofs and bridges. The connecting members of a truss are relatively much longer than they are wide. Furthermore, *by definition the members of a truss are connected only at their ends*, and there may be more than two members connected at a point.

In a planar truss all members lie in the same plane. For additional support and stability, multiple trusses of the same design may be used in parallel planes with the load transmitted to each truss through load transmitting members. The load may be transmitted from above as in the case of a roof truss or it may be transmitted from below in the case of a bridge deck. A space truss is a 3-dimensional truss with forces (or components of forces) which act in all three coordinate directions.

Because of the design of a truss, two major assumptions are appropriate. 1) Members are connected by smooth pins so that the members can transmit no torque at their ends. 2) The forces on members are applied only at two points, e.g., only at their ends (they are "two-force" members). Also, it is normally assumed that the weight of the members is small compared to the load.

As a result of these assumptions and as a consequence of the conservation of linear and angular momentum for a *static* body, it is seen that:

> The vector sum of all the forces acting at one point of a truss member must be collinear with the vector sum of all the forces acting at the other point.

Examples follow and illustrate the analysis procedures for trusses using linear and angular momentum conservation principles. One procedure uses an analysis, the *method of joints*, which selects systems defined to be the points of contact of the various members, treating each point of contact as a single point or particle, and analyzing using the conservation of linear momentum. Alternatively, the analysis may involve choosing a finite-sized section of the truss, the *method of sections*, which may include several joints and cut through individual members. In this case, because it is a finite-sized body, we must consider both linear and angular momentum.

Example 9-8. Determine the force in each member of the truss as shown in Figure 9-13. Also indicate whether the member is in tension or compression from the direction of the force in the member. Note that there is no support at D and that the 400 N force is an applied load.

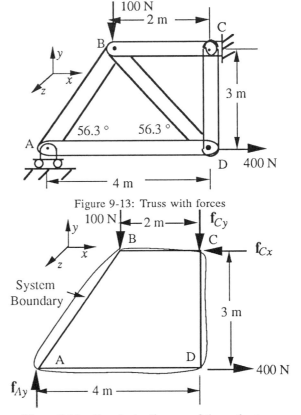

Figure 9-13: Truss with forces

Figure 9-13a: Free body diagram of the entire truss

SOLUTION: First we need to determine the support reactions on the truss at points C (pin support) and A (roller support). The system (free body) we will choose is the entire truss as shown in Figure 9-13a. The conservation of linear momentum for this free body diagram is given in Table 9-9. From the table, there are three (3) unknowns and two (2) independent equations and the only unknown that we can determine is $f_{C,x}$ from the x-direction equation: $f_{C,x} = 400$ (N). We cannot solve for all the reaction forces by looking only at linear momentum.

Table 9-9: Example 9-8 (Linear Momentum, System = Entire Truss)

	0	$=$					$\sum f_{ext}$			
Description			f_A		f_B		f_C		f_D	
	(N)		(N)		(N)		(N)		(N)	
x-direction	$0\,\mathbf{i}$	$=$	$0\mathbf{i}$	$+$	$0\mathbf{i}$	$-$	$f_{C,x}\mathbf{i}$	$+$	$400\mathbf{i}$	
y-direction	$0\,\mathbf{j}$	$=$	$f_{A,y}\mathbf{j}$	$-$	$100\mathbf{j}$	$-$	$f_{C,y}\mathbf{j}$	$+$	$0\mathbf{j}$	
z-direction	$0\,\mathbf{k}$	$=$	$0\mathbf{k}$	$-$	$0\mathbf{k}$	$+$	$0\mathbf{k}$	$+$	$0\mathbf{k}$	

The conservation of angular momentum about point C for the system will be determined next. The position vectors from point C to the points of application of the external forces are:

$$r_{CA} = -4i - 3j + 0k \text{ (m)}$$

$$r_{CB} = -2i + 0j + 0k \text{ (m)}$$

$$r_{CD} = 0i - 3j + 0k \text{ (m)}$$

The torques exerted by the forces acting at these points is calculated as the cross products of these position vectors with their forces:

$$(r \times f)_A = \begin{vmatrix} i & j & k \\ (-4) & (-3) & 0 \\ 0 & f_{A,y} & 0 \end{vmatrix} = 0i + 0j - 4f_{A,y}k \text{ (N m)}$$

$$(r \times f)_B = \begin{vmatrix} i & j & k \\ (-2) & 0 & 0 \\ 0 & (-100) & 0 \end{vmatrix} = 0i + 0j + 200k \text{ (N m)}$$

$$(r \times f)_D = \begin{vmatrix} i & j & k \\ 0 & (-3) & 0 \\ 400 & 0 & 0 \end{vmatrix} = 0i + 0j + 1200k \text{ (N m)}$$

Table 9-10: Example 9-8 (Angular Momentum, System = Entire Truss)

Description	0	=	$\sum (r \times f)_{ext}$				
	(N m)		$(r \times f)_A$ (N m)		$(r \times f)_B$ (N m)		$(r \times f)_D$ (N m)
x-direction	0 i	=	0i	+	0i	+	0i
y-direction	0 j	=	0j	+	0j	+	0j
z-direction	0 k	=	$(-4)f_{A,y}k$	+	200k	+	1200k

The conservation of angular momentum about point C is given in Table 9-10 with the sum of these torques being the zero vector. From the table we can solve for the only unknown $f_{A,y}$ = 350 (N). Substituting this value back into the linear momentum table for the y-direction yields a solution for the force at C in the y-direction $f_{C,y}$ = 250 (N). So, the reaction vectors at points A and C are: $f_A = 0i + 350j + 0k$ and $f_C = -400i - 250j + 0k$.

Now we need to solve for the forces in all the members of the truss. We will choose joint A as the first system as shown in Figure 9-13b. For this system, the conservation of linear momentum is given in Table 9-11. From the table there are two (2) unknowns and two independent equations. The simultaneous solution yields the values for the unknowns: f_{BA} = 421 (N) and f_{AD} = 233 (N). The force vectors then are $f_{BA} = -233i - 350j + 0k$ N and $f_{AD} = 233i + 0j + 0k$ N.

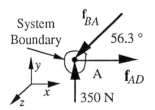

Figure 9-13b: Free body diagram of joint A

Table 9-11: Example 9-8 (Linear Momentum, System = Joint A)

	0	**=**					$\sum \mathbf{f}_{ext}$			
Description	(N)		$\mathbf{f}_A$ (N)			$\mathbf{f}_{BA}$ (N)			$\mathbf{f}_{AD}$ (N)	
x-direction	0 i	=	0i	$-$		$f_{BA}\cos(56.3°)$i	$+$		f_{AD}i	
y-direction	0 j	=	350j	$-$		$f_{BA}\sin(56.3°)$j	$+$		0j	
z-direction	0 k	=	0k	$+$		0k	$+$		0k	

Table 9-12: Example 9-8 (Linear Momentum, System = Joint D)

	0	**=**					$\sum \mathbf{f}_{ext}$			
Description	(N)		$\mathbf{f}_D$ (N)		$\mathbf{f}_{DA}$ (N)		$\mathbf{f}_{BD}$ (N)			$\mathbf{f}_{CD}$ (N)
x-direction	0 i	=	400i	$-$	233i	$+$	$f_{BD}\cos(56.3°)$i	$+$		0i
y-direction	0 j	=	0j	$+$	0j	$-$	$f_{BD}\sin(56.3°)$j	$-$		f_{CD}j
z-direction	0 k	=	0k	$-$	0k	$+$	0k	$+$		0k

Now we choose joint D as the system as shown in Figure 9-13c. The conservation of linear momentum for this free body is given in Table 9-12. From the table there are two (2) unknowns and two (2) independent equations. The simultaneous solution yields the magnitudes of the unknown forces: $f_{BD} = -300$ (N) and $f_{CD} = 250$ (N). Since the magnitude of vector $\mathbf{f}_{DB}$ was negative, the assumed direction of the force was incorrect and the force really acts in the opposite direction. The force vectors are: $\mathbf{f}_{BD} = -166\mathbf{i} + 250\mathbf{j} + 0\mathbf{k}$ N and $\mathbf{f}_{CD} = 0\mathbf{i} - 250\mathbf{j} + 0\mathbf{k}$ N.

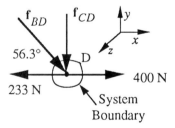

Figure 9-13c: Free body diagram of joint D

Finally, we choose the system to be the joint C as given in Figure 9-13d. The conservation of linear momentum for this system is shown in Table 9-13. From the table there is only one (1) unknown. The unknown magnitude is solved from the x-direction equation, $f_{BC} = 400$ N from which: $\mathbf{f}_{BC} = 400\mathbf{i} + 0\mathbf{j} + 0\mathbf{k}$ N

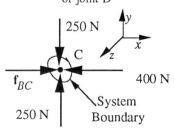

Figure 9-13d: Free body diagram of joint C

The forces in each of the members of the truss are shown in Figure 9-13e.

Table 9-13: Example 9-8 (Linear Momentum, System = Joint C)

Description	$\mathbf{0}$ (N)	=	$\mathbf{f}_C$ (N)		$\mathbf{f}_{DC}$ (N)		$\mathbf{f}_{BC}$ (N)
x-direction	$0\,\mathbf{i}$	=	$(-400)\mathbf{i}$	+	$0\mathbf{i}$	+	$f_{BC}\mathbf{i}$
y-direction	$0\,\mathbf{j}$	=	$(-250)\mathbf{j}$	+	$250\mathbf{j}$	+	$0\mathbf{j}$
z-direction	$0\,\mathbf{k}$	=	$0\mathbf{k}$	+	$0\mathbf{k}$	+	$0\mathbf{k}$

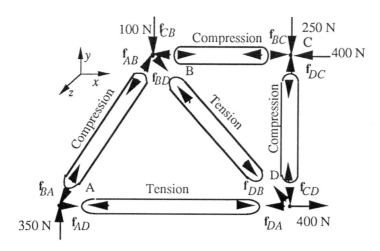

Figure 9-13e: Diagram of the solution

Example 9-9. For the massless space truss shown in Figure 9-14 that is supporting the 100 (N) load at point C, calculate the reaction forces on the truss at A, B, and D. The truss is hinged collinear with the z axis at points A and B, and supported by a horizontal cable to the wall at D. Also, determine the forces in each of the members of the truss and state whether the members are in tension or compression.

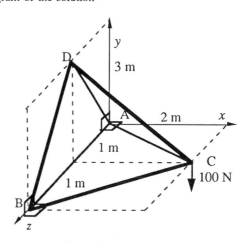

Figure 9-14: Space truss

SOLUTION: To determine the reaction forces at points A, B, and C we will pick the system to be the entire truss. This system or *free body diagram* is shown in Figure 9-14a. Because of the hinge support at A and B, there are two (2) independent forces acting at A and B and they are $f_{A,x}$, $f_{A,y}$, $f_{B,x}$, and $f_{B,y}$. The cable at point D can only support a force in the x-direction; $f_{D,x}$ is the only non-zero component at point D. Furthermore, because the truss is massless, there is no effect of gravity on the system.

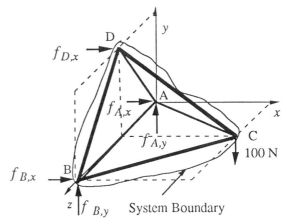

Figure 9-14a: Space truss system (free body)

The conservation of linear momentum for the cartesian coordinate system defined at point A is given in Table 9-14. From the table we can see that there are two (2) independent equations and five (5) unknowns. Using only linear momentum, we cannot solve this problem.

Table 9-14: Example 9-9 (Linear Momentum, System = Entire Truss)

Description	0	=	$\sum \mathbf{f}_{ext}$							
			$\mathbf{f}_A$		$\mathbf{f}_B$		$\mathbf{f}_C$		$\mathbf{f}_D$	
	(N)		(N)		(N)		(N)		(N)	
x-direction	$0\,\mathbf{i}$	=	$f_{A,x}\mathbf{i}$	+	$f_{B,x}\mathbf{i}$	+	$0\mathbf{i}$	+	$f_{D,x}\mathbf{i}$	
y-direction	$0\,\mathbf{j}$	=	$f_{A,y}\mathbf{j}$	+	$f_{B,y}\mathbf{j}$	−	$100\mathbf{j}$	+	$0\mathbf{j}$	
z-direction	$0\,\mathbf{k}$	=	$0\mathbf{k}$	+	$0\mathbf{k}$	+	$0\mathbf{k}$	+	$0\mathbf{k}$	

Now, we will use the conservation of angular momentum about point A for this same system or *free body*. Since angular momentum requires the cross product of the position vector and the force vector we will first determine all of the position vectors:

$$\mathbf{r}_{AA} = 0\mathbf{i} + 0\mathbf{j} + 0\mathbf{k} \text{ (m)}$$

$$\mathbf{r}_{AB} = 0\mathbf{i} + 0\mathbf{j} + 2\mathbf{k} \text{ (m)}$$

$$\mathbf{r}_{AC} = 2\mathbf{i} + 0\mathbf{j} + 1\mathbf{k} \text{ (m)}$$

$$\mathbf{r}_{AD} = 0\mathbf{i} + 3\mathbf{j} + 1\mathbf{k} \text{ (m)}$$

The cross products of the position vectors with respect to point A and the application of the force vectors are then calculated:

$$(\mathbf{r} \times \mathbf{f})_A = \begin{vmatrix} \mathbf{i} & \mathbf{j} & \mathbf{k} \\ 0 & 0 & 0 \\ f_{A,x} & f_{A,y} & 0 \end{vmatrix} = 0\mathbf{i} + 0\mathbf{j} + 0\mathbf{k} \text{ (N m)}$$

$$(\mathbf{r} \times \mathbf{f})_B = \begin{vmatrix} \mathbf{i} & \mathbf{j} & \mathbf{k} \\ 0 & 0 & 2 \\ f_{B,x} & f_{B,y} & 0 \end{vmatrix} = (-2f_{B,y})\mathbf{i} + 2f_{B,x}\mathbf{j} + 0\mathbf{k} \text{ (N m)}$$

$$(\mathbf{r} \times \mathbf{f})_C = \begin{vmatrix} \mathbf{i} & \mathbf{j} & \mathbf{k} \\ 2 & 0 & 1 \\ 0 & (-100) & 0 \end{vmatrix} = 100\mathbf{i} + 0\mathbf{j} + (-200)\mathbf{k} \text{ (N m)}$$

$$(\mathbf{r} \times \mathbf{f})_D = \begin{vmatrix} \mathbf{i} & \mathbf{j} & \mathbf{k} \\ 0 & 3 & 1 \\ f_{D,x} & 0 & 0 \end{vmatrix} = 0\mathbf{i} + f_{D,x}\mathbf{j} - 3f_{D,x}\mathbf{k} \text{ (N m)}$$

Notice that the force acting at point A does not contribute to the angular momentum about point A. Now that the cross products have been evaluated, conservation of angular momentum is given in Table 9-15.

Table 9-15: Example 9-9 (Angular Momentum, System = Entire Truss)

Description	0 (N m)	=	$\sum(\mathbf{r} \times \mathbf{f})_{\text{ext},A}$						
			$(\mathbf{r} \times \mathbf{f})_A$ (N m)		$(\mathbf{r} \times \mathbf{f})_B$ (N m)		$(\mathbf{r} \times \mathbf{f})_C$ (N m)		$(\mathbf{r} \times \mathbf{f})_D$ (N m)
x-direction	$0\,\mathbf{i}$	=	$0\mathbf{i}$	−	$2f_{B,y}\mathbf{i}$	+	$100\mathbf{i}$	+	$0\mathbf{i}$
y-direction	$0\,\mathbf{j}$	=	$0\mathbf{j}$	+	$2f_{B,x}\mathbf{j}$	+	$0\mathbf{j}$	+	$f_{D,x}\mathbf{j}$
z-direction	$0\,\mathbf{k}$	=	$0\mathbf{k}$	+	$0\mathbf{k}$	−	$200\mathbf{k}$	−	$3f_{D,x}\mathbf{k}$

From Table 9-15 the solution of the unknown quantities is: $f_{B,x} = 33.3$ (N), $f_{B,y} = 50$ (N), and $f_{D,x} = -66.7$ (N). Taking these results and substituting the known quantities into the linear momentum equation now gives 2 independent equations and 2 unknowns $f_{A,x}$, $f_{B,y}$. The solution of these unknowns is: $f_{A,x} = 33.3$ (N), and $f_{A,y} = 50$ (N). In vector form, the reaction forces at A, B, and D are:

$$\mathbf{f}_A = 33.3\mathbf{i} + 50\mathbf{j} + 0\mathbf{k} \text{ (N)}$$

$$\mathbf{f}_B = 33.3\mathbf{i} + 50\mathbf{j} + 0\mathbf{k} \text{ (N)}$$

$$\mathbf{f}_D = -66.7\mathbf{i} + 0\mathbf{j} + 0\mathbf{k} \text{ (N)}$$

To determine the forces in each of the members, we will define systems about different joints or intersection points and apply the conservation of linear momentum to these points. The conservation of angular momentum will not be used because our systems or *free bodies* will be of zero volume in principle. The first system will be the junction at point C as shown in Figure 9-14b.

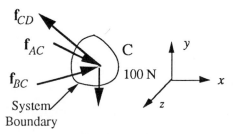

Figure 9-14b: Junction C system (free body)

The direction of the forces is assumed in the figure. Because the structure is a truss, the forces act in the direction of the members of the truss, so from the geometry of the structure we can calculate the unit directional vectors along which the forces act. By definition, the unit directional vectors have magnitude of one (1) and give only direction information:

$$\mathbf{u}_{AC} = \left(\frac{\mathbf{r}_{AC}}{|\mathbf{r}_{AC}|}\right) = \left(\frac{2\mathbf{i} + 0\mathbf{j} + 1\mathbf{k}}{\sqrt{5}}\right)$$

$$\mathbf{u}_{BC} = \left(\frac{\mathbf{r}_{BC}}{|\mathbf{r}_{BC}|}\right) = \left(\frac{2\mathbf{i} + 0\mathbf{j} - 1\mathbf{k}}{\sqrt{5}}\right)$$

$$\mathbf{u}_{CD} = \left(\frac{\mathbf{r}_{CD}}{|\mathbf{r}_{CD}|}\right) = \left(\frac{-2\mathbf{i} + 3\mathbf{j} + 0\mathbf{k}}{\sqrt{13}}\right)$$

The force in each of the members is expresed as the magnitude of the force times the unit directional vector:

$$\mathbf{f}_{AC} = f_{AC}\mathbf{u}_{AC}$$
$$\mathbf{f}_{BC} = f_{BC}\mathbf{u}_{BC}$$
$$\mathbf{f}_{CD} = f_{CD}\mathbf{u}_{CD}$$

The magnitudes of the vectors are unknown. The conservation of linear momentum for the system or *free body* defined at point C is given in Table 9-16.

Table 9-16: Example 9-9 (Linear Momentum, System = Joint C)

	$\mathbf{0}$	=	$\sum \mathbf{f}_{ext}$						
Description	(N)		$\mathbf{f}_{AC}$ (N)		$\mathbf{f}_{BC}$ (N)		$\mathbf{f}_C$ (N)		$\mathbf{f}_{CD}$ (N)
x-direction	0 i	=	$f_{AC}\left(\frac{2}{\sqrt{5}}\right)\mathbf{i}$	+	$f_{BC}\left(\frac{2}{\sqrt{5}}\right)\mathbf{i}$	+	0i	−	$f_{CD}\left(\frac{2}{\sqrt{13}}\right)\mathbf{i}$
y-direction	0 j	=	0j	+	0j	−	100j	+	$f_{CD}\left(\frac{3}{\sqrt{13}}\right)\mathbf{j}$
z-direction	0 k	=	$f_{AC}\left(\frac{1}{\sqrt{5}}\right)\mathbf{k}$	−	$f_{BC}\left(\frac{1}{\sqrt{5}}\right)\mathbf{k}$	+	0k	+	0k

From Table 9-16, there are three (3) unknowns, the magnitudes of the forces in the members, and three (3) independent equations. Simultaneous solution of the equations gives the solution of the unknown magnitudes as: $f_{AC} = 37.3$ N, $f_{BC} = 37.3$ N, and $f_{CD} = 120.2$ N. The magnitudes are positive because the assumed direction of the forces was correct. If the forces really operated in the opposite direction, the magnitude would carry a negative sign. The force vectors are:

$$\mathbf{f}_{AC} = 33.4\mathbf{i} + 0\mathbf{j} + 16.7\mathbf{k} \text{ N}$$
$$\mathbf{f}_{BC} = 33.4\mathbf{i} + 0\mathbf{j} - 16.7\mathbf{k} \text{ N}$$
$$\mathbf{f}_{CD} = -66.7\mathbf{i} + 100\mathbf{j} + 0\mathbf{k} \text{ N}$$

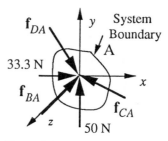

Figure 9-14c: Junction A system
(free body)

The next system is the junction point at A as shown is figure 9-14c with the assumed direction of the forces in the members. The unit directional vectors for the members BA and DA are

$$u_{BA} = \left(\frac{r_{BA}}{|r_{BA}|} \right) = \left(\frac{0i + 0j - 2k}{\sqrt{4}} \right)$$

$$u_{DA} = \left(\frac{r_{DA}}{|r_{DA}|} \right) = \left(\frac{0i - 3j - 1k}{\sqrt{10}} \right)$$

and the force vectors in the members are

$$f_{BA} = f_{BA}u_{BA}$$

$$f_{DA} = f_{DA}u_{DA}$$

Now the forces can be used in the conservation of linear momentum given in Table 9-17.

Table 9-17: Example 9-9 (Linear Momentum, System = Joint A)

Description	0 (N)	=	$\sum f_{ext}$						
			f_A (N)		f_{CA} (N)		f_{BA} (N)		f_{DA} (N)
x-direction	0 i	=	33. 3i	−	33.3 i	+	0i	+	0i
y-direction	0 j	=	100j	+	0j	+	0j	−	$f_{DA} \left(\frac{3}{\sqrt{10}} \right) j$
z-direction	0 k	=	0k	−	16. 7k	−	$f_{BA}k$	+	$f_{DA} \left(\frac{1}{\sqrt{5}} \right) k$

The solution of the table gives the magnitudes of the unknown forces as: $f_{BA} = 16.6$ (N) and $f_{DA} = 105.4$ (N) and the forces, written as vectors are

$$f_{BA} = 0i + 0j - 16.6k \text{ (N)}$$

$$f_{DA} = 0i - 100j - 33.3k \text{ (N)}$$

Because the structure is symmetrical, the force in member BD is equal to the force in AD, and the structure is solved. (If you do not believe this symmetry argument, we suggest that you choose the system as point D and use the conservation of linear momentum and solve for the unknown force.) The solved structure is shown in Figure 9-14d and the members are labeled according to whether they are in tension or compression. Also, the magnitudes of the forces in the members are given in the table

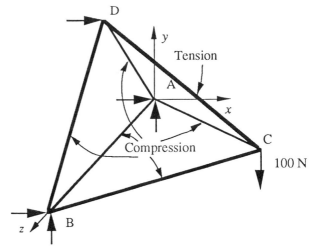

Figure 9-14d: Solved space truss

$f_{AB} = 16.6$ (N)		Compression
$f_{AC} = 37.3$ (N)		Compression
$f_{BC} = 37.3$ (N)		Compression
$f_{CD} = 120.2$ (N)		Tension
$f_{AD} = 105.4$ (N)		Compression
$f_{BD} = 105.4$ (N)		Compression

9.6.2 Frames

Like trusses, frames are static structures which are designed to carry loads. However, they have some significant differences. Because of the design of a frame, two major assumptions are appropriate. 1) *Like* the members of a truss, the members of a frame are connected by smooth pins so that the members can transmit no torque at their ends. 2) *Unlike* the forces on the members of a truss, the forces on the members of a frame *may* be applied at *more* than two points. Also, *like* the members of a truss, it is normally assumed that the weight of the members of a frame is small compared to the load. Note that a member subjected to a distributed load is a multi-force member, and hence must be a frame member.

As a result of these assumptions and as a consequence of the conservation of linear and angular momentum for a *static* body, it is seen that:

> For a frame member, which has forces applied at more than two different points (multi-force member), the vector sum of the forces acting at a point are *not* necessarily collinear with the member. This must be considered in designating forces on a free-body diagram.

Actually, a more general definition is that a frame has members which are subjected to both axial and bending stresses as opposed to a truss for which (two-force) members sustain only axial stresses. By this definition, the members could be connected with fixed supports, which transmit both load and torque, and the members could be non-straight.

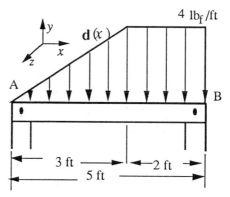

Figure 9-15: Distributed load on a beam

Example 9-10. A distributed load is acting on the massless beam between the points A and B as shown in Figure 9-15. Calculate the reaction forces at pin supported A and B. Clearly, the beam is a member which has forces acting at more than two points. The distributed load is given as a function of the scalar position x and acts in the negative **j** direction by:

$$\mathbf{d}(x) = -\left(\frac{4}{3}\right)x\mathbf{j} \; (\text{lb}_f / \text{ft}) \qquad 0 \le x \le 3$$

$$\mathbf{d}(x) = -4\mathbf{j} \; (\text{lb}_f / \text{ft}) \qquad 3 < x \le 5$$

SOLUTION: Because we have a distributed load in this problem, we will first simplify the distributed load and torque to a single equivalent force and torque. The equivalent force is calculated as the integral of the distributed force over the length.

$$\mathbf{f}_{eq} = \int_0^5 \mathbf{d}(x)dx$$

In the **j** direction this equation reduces to a scalar:

$$(f_{eq})_y = \int_0^5 d(x)dx$$

Now because the function changes at $x = 3$ (ft) the integral is divided as:

$$(f_{eq})_y = \int_0^3 -\left(\frac{4}{3}\right)x\,dx + \int_3^5 -4\,dx$$

Integrating the expressions and evaluating gives:

$$(f_{eq})_y = -\left(\frac{4}{6}\right)(3^2 - 0) - 4(5 - 3) \; (\text{lb}_f)$$

$$= -14 \; (\text{lb}_f)$$

This process of integrating the distributed force over the length of the beam is the same process as calculating the area under the distributed load in the figure. Calculate this area by calculating the area of the triangle and rectangle to convince yourself that this is true. The equivalent force has a magnitude of 14 (lb$_f$) and acts in the negative **j** direction.

$$\mathbf{f}_{eq} = 0\mathbf{i} - 14\mathbf{j} + 0\mathbf{k} \; (\text{lb}_f)$$

To determine where the force is acting, the same torque that the distibuted load exerts on the system must be equal to the equivalent torque on the system or:

$$(\bar{r} \times f)_{eq} = \int \left(r \times d(x) \right) dx$$

The origin of the coordinate system is chosen as point A and the only non-zero component of the position vector is in the x-direction or

$$r = x\mathbf{i} + 0\mathbf{j} + 0\mathbf{k} \text{ (ft)}$$

The equivalent position vector is:

$$\bar{r} = \bar{x}\mathbf{i} + 0\mathbf{j} + 0\mathbf{k} \text{ (ft)}$$

The cross product of the position vector in the $\mathbf{i}$ direction and the force and distributed force vector in the $\mathbf{j}$ direction yields a vector in the $\mathbf{k}$ direction as:

$$\bar{x}(f_{eq})_y \mathbf{k} = \int_0^5 \left(x d(x) \right) dx \mathbf{k}$$

Since the distributed force changes functionality over the limits of integration, the integral is divided into two integrals as:

$$\bar{x}(f_{eq})_y = \int_0^3 -\left(\frac{4}{3} \right) x^2 dx + \int_3^5 -4x\, dx$$

The only unknown in this equation is the value of $\bar{x}$. All of the other values are known. Evaluating the integral:

$$\bar{x}(f_{eq})_y = -\left(\frac{4}{9} \right)(3^3 - 0) - \left(\frac{4}{2} \right)(5^2 - 3^2) \text{ (lb}_f \text{ ft)} = -44 \text{ (lb}_f \text{ ft)}$$

And the value of $\bar{x}$ is calculated as:

$$\bar{x} = \left(\frac{-44}{(f_{eq})_y} \right) = \left(\frac{-44}{-14} \right) = 3.14 \text{ (ft)}$$

The equivalent force system is shown in Figure 9-15a and this is the system that we have defined to solve for the unknown forces at points A and B. For this defined system, the conservation of linear momentum is given in Table 9-18. Because the structure is massless, the effect of gravity is not included in conservation of linear momentum. From Table 9-18, there are two (2) unknowns, f_A and f_B and only one (1) independent equation. The problem cannot be solved by only using linear momentum.

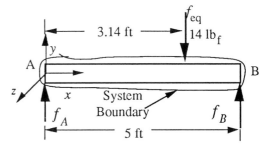

Figure 9-15a: Simplified load and system boundary

Table 9-18: Example 9-10 (Linear Momentum, System = Beam)

Description	0	=	$\sum \mathbf{f}_{ext}$					
	(lb$_f$)		$\mathbf{f}_A$ (lb$_f$)		$\mathbf{f}_B$ (lb$_f$)		$\mathbf{f}_{eq}$ (lb$_f$)	
x-direction	0 i	=	0i	+	0i	+	0i	
y-direction	0 j	=	f_Aj	+	f_Bj	−	14j	
z-direction	0 k	=	0k	+	0k	+	0k	

Now we will use the conservation of angular momentum about point A. The position vectors from point A to the point of application of each of the forces is:

$$\mathbf{r}_A = 0\mathbf{i} + 0\mathbf{j} + 0\mathbf{k} \text{ (ft)}$$

$$\mathbf{r}_B = 5\mathbf{i} + 0\mathbf{j} + 0\mathbf{k} \text{ (ft)}$$

$$\mathbf{r}_{eq} = 3.14\mathbf{i} + 0\mathbf{j} + 0\mathbf{k} \text{ (ft)}$$

The cross products of the position vectors from point A to the point of application of the force and the force vectors is:

$$(\mathbf{r} \times \mathbf{f})_A = \begin{vmatrix} \mathbf{i} & \mathbf{j} & \mathbf{k} \\ 0 & 0 & 0 \\ 0 & f_A & 0 \end{vmatrix} = 0\mathbf{i} + 0\mathbf{j} + 0\mathbf{k} \text{ (ft lb}_f)$$

$$(\mathbf{r} \times \mathbf{f})_B = \begin{vmatrix} \mathbf{i} & \mathbf{j} & \mathbf{k} \\ 5 & 0 & 0 \\ 0 & f_B & 0 \end{vmatrix} = 0\mathbf{i} + 0\mathbf{j} + 5f_B\mathbf{k} \text{ (ft lb}_f)$$

$$(\mathbf{r} \times \mathbf{f})_{eq} = \begin{vmatrix} \mathbf{i} & \mathbf{j} & \mathbf{k} \\ 3.14 & 0 & 0 \\ 0 & (-14) & 0 \end{vmatrix} = 0\mathbf{i} + 0\mathbf{j} - 44\mathbf{k} \text{ (ft lb}_f)$$

The conservation of angular momentum for this system is given in Table 9-19. The magnitude of the force at point B is calculated from Table 9-19 $f_B = 8.8$ lb$_f$. Substituting this result back into the conservation of linear momentum table gives the magnitude of the force at point A: $f_A = 5.2$ (lb$_f$). The force vectors at points A and B are: $\mathbf{f}_A = 0\mathbf{i} + 5.2\mathbf{j} + 0\mathbf{k}$ (lb$_f$), and $\mathbf{f}_B = 0\mathbf{i} + 8.8\mathbf{j} + 0\mathbf{k}$ (lb$_f$)

Table 9-19: Example 9-10 (Angular Momentum, System = Beam)

Description	0	=	$\sum (\mathbf{r} \times \mathbf{f})_{ext,A}$					
	(ft lb$_f$)		$(\mathbf{r} \times \mathbf{f})_A$ (ft lb$_f$)		$(\mathbf{r} \times \mathbf{f})_B$ (ft lb$_f$)		$(\mathbf{r} \times \mathbf{f})_{eq}$ (ft lb$_f$)	
x-direction	0 i	=	0i	+	0i	+	0i	
y-direction	0 j	=	0j	+	0j	+	0j	
z-direction	0 k	=	0k	+	$5f_B$k	−	44k	

Example 9-11. A 10 (N) load directed in the negative **j** direction is acting on the three (3) member massless frame as shown in Figure 9-16 (not to scale). The members include AC, CE, and BD. The points A and C are pin supported and the point E is roller supported. Determine the external reaction forces at points A and E and the internal reaction forces in the members at points C, B and D.

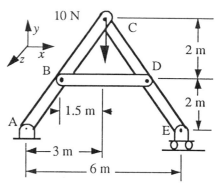

Figure 9-16: Load bearing frame

SOLUTION: We will pick as the system or *free body* the entire structure and apply the conservation of linear momentum. The system is shown in figure 9-16a and the conservatin of linear momentum for the system is given in Table 9-20. There are three (3) unknowns and only two independent equations so the problem cannot be solved using only linear momentum. However, by inspection, we know that: $f_{A,x} = 0$ N.

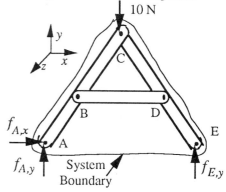

Figure 9-16a: System of the whole frame

Table 9-20: Example 9-11 (Linear Momentum, System = Entire Frame)

Description	0 (N)	=	$\sum \mathbf{f}_{ext}$ $\mathbf{f}_A$ (N)		$\mathbf{f}_C$ (N)		$\mathbf{f}_E$ (N)
x-direction	0 **i**	=	$f_{A,x}\mathbf{i}$	+	0**i**	+	0**i**
y-direction	0 **j**	=	$f_{A,y}\mathbf{j}$	−	10**j**	+	$f_{E,y}\mathbf{j}$
z-direction	0 **k**	=	0**k**	+	0**k**	+	0**k**

The conservation of angular momentum about point A is used for this system. First we will calculate the position vectors with respect to point A to the point of application of each of the external forces.

$$\mathbf{r}_A = 0\mathbf{i} + 0\mathbf{j} + 0\mathbf{k} \text{ (m)}$$

$$\mathbf{r}_C = 3\mathbf{i} + 4\mathbf{j} + 0\mathbf{k} \text{ (m)}$$

$$\mathbf{r}_E = 6\mathbf{i} + 0\mathbf{j} + 0\mathbf{k} \text{ (m)}$$

The cross products of the position vectors and the point of application of the forces are calculated:

$$(r \times f)_A = \begin{vmatrix} i & j & k \\ 0 & 0 & 0 \\ 0 & f_A & 0 \end{vmatrix} = 0i + 0j + 0k \text{ (ft lb}_f)$$

$$(r \times f)_C = \begin{vmatrix} i & j & k \\ 3 & 4 & 0 \\ 0 & (-10) & 0 \end{vmatrix} = 0i + 0j - 30k \text{ (ft lb}_f)$$

$$(r \times f)_E = \begin{vmatrix} i & j & k \\ 6 & 0 & 0 \\ 0 & f_{B,y} & 0 \end{vmatrix} = 0i + 0j + 6f_{E,y}k \text{ (ft lb}_f)$$

Table 9-21: Example 9-11 (Angular Momentum, System = Entire Frame)

Description	0	=	$\sum(r \times f)_{ext,A}$				
			$(r \times f)_A$		$(r \times f)_C$		$(r \times f)_E$
	(N m)		(N m)		(N m)		(N m)
x-direction	0 i	=	0i	+	0i	+	0i
y-direction	0 j	=	0j	+	0j	+	0j
z-direction	0 k	=	0k	−	30k	+	$6f_{E,y}k$

The conservation of angular momentum for this system is given in Table 9-21. The magnitude of the force at point E is calculated from Table 9-21 $f_{E,y} = 5$ (N). Substituting this result back into the conservation of linear momentum table gives the magnitude of the force at point A: $f_{A,y} = 5$ (N). The force vectors at points A and B are: $\mathbf{f}_A = 0i + 5j + 0k$ (N) $\mathbf{f}_E = 0i + 5j + 0k$ (N).

Next, we will choose as the system the member EC as shown in the *free body diagram* in Figure 9-16b. The force at point C consists of the loading in the negative direction and the force due to member AC. The force from member AC does not necessarily act in the direction of member (member AC is not a two-force member) and consequently there are two independent variables f_{ACx} and f_{ACy} required:

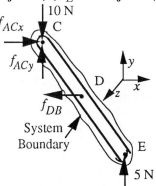

Figure 9-16b: Member EC
free body diagram

$$\mathbf{f}_{AC} = f_{ACx}i + f_{ACy}j + 0k \text{ (N)}$$

The force at point D is only in the x-direction because member BD is a two-force member:

$$\mathbf{f}_{BD} = -f_{BD}i + 0j + 0k \text{ (N)}$$

Table 9-22: Example 9-11 (Linear Momentum, System = Member CE)

Description	0 (N)	$=$	$\sum \mathbf{f}_{\text{ext}}$						
			$\mathbf{f}_{AC}$ (N)		$\mathbf{f}_{DB}$ (N)		$\mathbf{f}_E$ (N)		$\mathbf{f}_C$ (N)
x-direction	$0\,\mathbf{i}$	$=$	$f_{ACx}\mathbf{i}$	$-$	$f_{BD}\mathbf{i}$	$+$	$0\mathbf{i}$	$+$	$0\mathbf{i}$
y-direction	$0\,\mathbf{j}$	$=$	$f_{ACy}\mathbf{j}$	$+$	$0\mathbf{j}$	$+$	$5\mathbf{j}$	$-$	$10\mathbf{j}$
z-direction	$0\,\mathbf{k}$	$=$	$0\mathbf{k}$	$+$	$0\mathbf{k}$	$+$	$0\mathbf{k}$	$+$	$0\mathbf{k}$

Using these force representations, the conservation of linear momentum for the system CE is shown in Table 9-22. There are three independent (3) unknowns and two (2) independent equations. Only f_{ACy} can be solved: $f_{ACy} = 5$ (N).

To determine the other unknown force components, we will use the conservation of angular momentum about point C. The position vectors with respect to point C are given:

$$\mathbf{r}_{CD} = 1.5\mathbf{i} - 2\mathbf{j} + 0\mathbf{k} \text{ (m)}$$

$$\mathbf{r}_{CE} = 3\mathbf{i} - 4\mathbf{j} + 0\mathbf{k} \text{ (m)}$$

The cross products of the position vectors and the points of application of the forces is calculated. The forces at C do not contribute to the angular momentum and the unknown is the force f_{DB}.

$$(\mathbf{r} \times \mathbf{f})_{CD} = \begin{vmatrix} \mathbf{i} & \mathbf{j} & \mathbf{k} \\ 1.5 & -2 & 0 \\ (-f_{DB}) & 0 & 0 \end{vmatrix} = 0\mathbf{i} + 0\mathbf{j} - 2f_{DB}\mathbf{k} \text{ (N m)}$$

$$(\mathbf{r} \times \mathbf{f})_{CE} = \begin{vmatrix} \mathbf{i} & \mathbf{j} & \mathbf{k} \\ 3 & -4 & 0 \\ 0 & 5 & 0 \end{vmatrix} = 0\mathbf{i} + 0\mathbf{j} + 15\mathbf{k} \text{ (N m)}$$

The conservation of angular momentum is given in Table 9-23. The table shows only one independent equation and one unknown. The solution is $f_{DBx} = 7.5$ (N). Substituting this result back into the linear momentum equation gives the force vector $\mathbf{f}_{AC} = 7.5\mathbf{i} + 5\mathbf{j} + 0\mathbf{k}$ (N). The solved member is shown in Figure 9-16c.

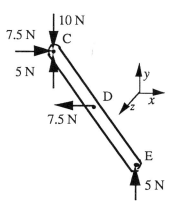

Figure 9-16c: Member EC solution

Table 9-23: Example 9-11 (Angular Momentum, System = Member CE)

	0	=	$\sum (\mathbf{r} \times \mathbf{f})_{\text{ext},C}$		
Description			$(\mathbf{r} \times \mathbf{f})_{CD}$		$(\mathbf{r} \times \mathbf{f})_{CE}$
	(N m)		(N m)		(N m)
x-direction	0 $\mathbf{i}$	=	0$\mathbf{i}$	+	0$\mathbf{i}$
y-direction	0 $\mathbf{j}$	=	0$\mathbf{j}$	+	0$\mathbf{j}$
z-direction	0 $\mathbf{k}$	=	$(-2f_{DBx})\mathbf{k}$	+	15$\mathbf{k}$

Next we will choose the member BD as the system as shown in Figure 9-16d. The force acting at point D must be equal and opposite to the force that is acting on member CE because momentum is conserved so $\mathbf{f}_D = 7.5\mathbf{i} + 0\mathbf{j} + 0\mathbf{k}$ (N). Now by inspection for the system to be at steady state, the force at point B must be: $\mathbf{f}_B = -7.5\mathbf{i}+0\mathbf{j}+0\mathbf{k}$ (N). We could have done this in a tabular format but the solution is trivial.

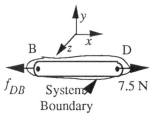

Figure 9-16d: Member BD
free body diagram

The last system that we will choose is the member AC as given in Figure 9-16e. The only unknown forces are f_{ECx} and f_{ECy}. The external loading of 10 (N) is included. The conservation of linear momentum is shown in Table 9-24. This table has two independent equations and two unknowns. The unknown force vector is $\mathbf{f}_{EC} = -7.5\mathbf{i} + 5\mathbf{j} + 0\mathbf{k}$ (N).

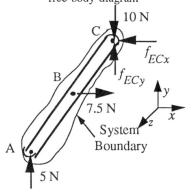

Figure 9-16e: Member AC
free body diagram

Table 9-24: Example 9-11 (Linear Momentum, System = Member AC)

	0	=	$\sum \mathbf{f}_{\text{ext}}$						
Description			$\mathbf{f}_A$		$\mathbf{f}_{BD}$		$\mathbf{f}_C$		$\mathbf{f}_{EC}$
	(N)		(N)		(N)		(N)		(N)
x-direction	0 $\mathbf{i}$	=	0$\mathbf{i}$	+	7.5$\mathbf{i}$	+	0$\mathbf{i}$	−	$f_{ECx}\mathbf{i}$
y-direction	0 $\mathbf{j}$	=	5$\mathbf{j}$	+	0$\mathbf{j}$	−	10$\mathbf{j}$	+	$f_{ECy}\mathbf{j}$
z-direction	0 $\mathbf{k}$	=	0$\mathbf{k}$	+	0$\mathbf{k}$	+	0$\mathbf{k}$	+	0$\mathbf{k}$

Figure 9-16f gives the entire solved frame. The loading in the vertical direction at point C on both members AC and EC is a net -5 (N) (the 10 (N) load minus the vertical component of the force exerted by the other member). As a check, you should calculate torques about point A in member AC to show that the body is truly static.

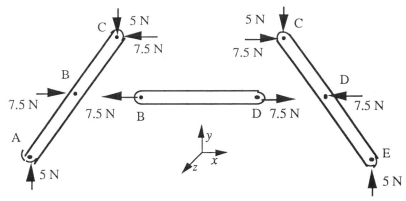

Figure 9-16f: Solution of the frame

9.6.3 Machines

Machines are devices whose purpose is not to support a load but rather to gain some mechanical advantage or leverage so that with the application of a fairly small force, a much larger force can be used to perform some function. Again, analysis may be by any of the previously mentioned methods: conservation of linear and angular momentum (methods of joints or sections) or conservation of energy.

From an energy viewpoint, in the absence of thermal effects (no heat transfer) and conversions of mechanical energy to thermal or other forms of energy, the conservation of total energy for a static rigid body (a closed system without any changes in kinetic or potential energy) is

$$0 = \sum_i \dot{W}_i \qquad (9-24)$$

(Friction can be present, of course, resulting in losses of mechanical energy. Such losses decrease the useful work extracted from the machine.)

Work is a force vector dotted with a displacement vector, the rate of work is the force vector dotted with a velocity vector at the point of application of the force vector:

$$\dot{W}_i = (\mathbf{f} \cdot \mathbf{v})_i \qquad (9-25)$$

The system rate equation can be integrated over time giving:

$$\boxed{0 = \sum_i W_i} \qquad (9-26)$$

and for a constant force over a small displacement the work is calculated as:

$$W_i = (\mathbf{f} \cdot \mathbf{s})_i \qquad (9-27)$$

Now for a *truly* static body, there is no movement at its boundaries and, therefore, no work. However, if there is a small amount of movement as the machine is used to apply a concentrated force on an object (e.g. you use a pair of pliars to apply a large force on a bolt by applying a smaller force with your hand), then this movement certainly is work. The energy equation then says that the work at the pliars grip, in our example, is balanced by the work done at the bolt. Therefore, if the displacement ratios can be determined based on geometric arguments, then the force ratios can be determined through energy conservation. This is called the method of *virtual work*.

Example 9-12. A 300 (lb$_m$) rock is setting at the end of a lever as shown in Figure 9-17. The rock, at point A, is 3 (ft) from the centerline of the base of fulcrum and an unknown force is applied perpendicular to the massless lever at point B. Point B is 6 (ft) in the horizontal direction from the centerline of the fulcrum. The fulcrum is 1 (ft) high. Use the conservation of energy to determine the magnitude of the force applied at point B to just lift the rock at point A.

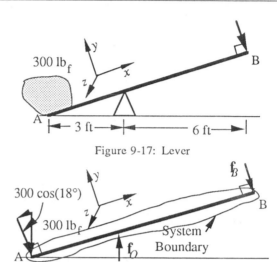

Figure 9-17: Lever

Figure 9-17a: Lever system
(free body)

SOLUTION: The system is the lever and a *free body* diagram is given in Figure 9-17a. The conservation of energy for this static system says: $\sum W = 0$. Since work is expressed as the dot product of a force and a displacement we will dot the external forces at A, B, O with the displacement vector such that:

$$\sum W = \mathbf{f}_B \cdot \mathbf{s}_B + \mathbf{f}_O \cdot \mathbf{s}_O + \mathbf{f}_A \cdot \mathbf{s}_A$$

The displacement vector at the fulcrum tip is zero:

$$\mathbf{s}_O = \mathbf{0}$$

so the the sum of the work is:

$$\sum W = \mathbf{f}_B \cdot \mathbf{s}_B + \mathbf{f}_A \cdot \mathbf{s}_A$$

now the individual dot products are evaluated as:

$$\mathbf{f}_A \cdot \mathbf{s}_A = -300\cos(18°)s_A \ \text{(lb}_f \ \text{ft)}$$
$$\mathbf{f}_B \cdot \mathbf{s}_B = f_B s_B \text{(lb}_f \ \text{ft)}$$

The negative sign is required because the displacement is in the opposite direction of the force. We know that the displacement at point A and the displacement at point B are related, because the lever is rigid. From similar triangles the displacement at B is related to the displacement A by:

$$s_A = \left(\frac{1}{2}\right)s_B$$

Table 9-25: Example 9-12 (Energy, System = Lever)

0	=	$\sum W$		
		$\mathbf{f}_A \cdot \mathbf{s}_A$ (lb$_f$ ft)		$\mathbf{f}_B \cdot \mathbf{s}_B$ (lb$_f$ ft)
(lb$_f$ ft)				
0	=	$(-285)s_B$	+	$2f_B s_B$

The conservation of energy for this static system is given in Table 9-25. Dividing by the displacement gives the magnitude of the force at point B as: $f_B = 142.5$ (lb$_f$). The direction is in the direction of the displacement which was perpendicular to the lever.

Example 9-13. Determine the tension in the cables and also the force $\mathbf{f}_P$ required to support the 400 (N) force using the frictionless and massless pulley system when the system is static. If the 400 (N) load is moving vertically up at a constant speed of 5 (m/s), determine the constant velocity or speed of the rope. Figure 9-18 shows the pulley system.

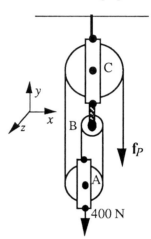

Figure 9-18: Pulley system

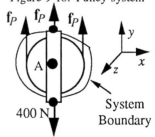

Figure 9-18a: Free body diagram of pulley A

SOLUTION. First choose pulley A as the system as shown in the Figure 9-18a. Since all the forces are in the y direction, the conservation of linear momentum for this system is given in Table 9-26 as only a y-direction equation. Therefore, the force in the y direction in each rope is: $f_P = 133.3$ (N).

Table 9-26: Example 9-13 (Linear Momentum, System = Pulley A)

	0	**=**				$\sum \mathbf{f}_{ext}$				
			$\mathbf{f}_P$		$\mathbf{f}_P$		$\mathbf{f}_P$		$\mathbf{f}_{400}$	
Description	(N)		(N)		(N)		(N)		(N)	
y-direction	0 **j**	=	$f_P\mathbf{j}$	+	$f_P\mathbf{j}$	+	$f_P\mathbf{j}$	−	400 **j**	

Table 9-27: Example 9-13 (Linear Momentum, System = Pulley B)

	0	**=**			$\sum \mathbf{f}_{ext}$			
			$\mathbf{f}_P$		$\mathbf{f}_P$		$\mathbf{f}_T$	
Description	(N)		(N)		(N)		(N)	
y-direction	0 **j**	=	$(-133.3)\mathbf{j}$	−	$133.3\mathbf{j}$	+	$f_T\mathbf{j}$	

Now we can choose pulley B as the system as shown in Figure 9-18b. Applying the conservation of linear momentum to this system gives Table 9-27 from which $f_T = 266.7$ (N).

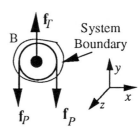

Figure 9-18b: Free body diagram of pulley B

Finally we can choose pulley C as the system as shown in the figure 9-18c. The conservation of linear momentum for this system is shown in Table 9-28 and f_R is found to be $f_R = 533.3$ (N). The tension in the cable to support the force is 133.3 (N).

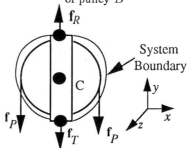

Figure 9-18c: Free body diagram of pulley C

Table 9-28: Example 9-13 (Linear Momentum, System = Pulley C)

	0	**=**	\multicolumn{9}{c}{$\sum \mathbf{f}_{ext}$}
Description	(N)		$\mathbf{f}_P$ (N)		$\mathbf{f}_P$ (N)		$\mathbf{f}_T$ (N)		$\mathbf{f}_R$ (N)
y-direction	$0\,\mathbf{j}$	$=$	$(-133.3)\mathbf{j}$	$-$	$133.3\mathbf{j}$	$-$	$266.7\mathbf{j}$	$+$	$f_R\mathbf{j}$

Now, if the load of 400 (N) is moving in the positive vertical direction at a velocity of 5 (m/s) we need to determine the velocity of the rope v_R. If we define pulley A as the system and use the conservation of energy for this system at steady state the following equation is obtained: $\sum_i \dot{W}_i = 0$. The system is massless so we are not including the effects of gravity. For the pulley system, the rate at which work is done on the system is the dot product of the force and velocity vector. If the velocity vector is in the opposite direction of the force, then the dot product is negative. Pulley A choosen as the system is given in Figure 18d.

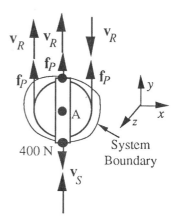

Figure 9-18d: Free body diagram
of pulley A

The conservation of energy for this system is given in Table 9-29. From the table the velocity of the rope is: $v_R = 15$ (m / s). The entire pulley system could also be used as the system. In fact, this is a much easier way to solve the problem. Define the entire pulley system as the system and solve for the velocity v_R and check to see that you obtain the same answer.

Table 9-29: Example 9-13 (Energy, System = Pulley A)

0	**=**	\multicolumn{7}{c}{$\sum \dot{W}$}
(W)		$(\mathbf{f}_P \cdot \mathbf{v})$ (W)		$(\mathbf{f}_P \cdot \mathbf{v})$ (W)		$(\mathbf{f}_P \cdot \mathbf{v})$ (W)		$(\mathbf{f}_{400} \cdot \mathbf{v})$ (W)
0	$=$	$133.3v_R$	$+$	$133.3v_R$	$-$	$133.3v_R$	$-$	2000

Because the equations of statics are steady state, they are algebraic equations or sets or algebraic equations. This means that they can be solved using algebraic techniques. The primary difficulty of solving static systems is in creatively selecting the right system and system boundaries so as to obtain sets of equations which are easy to solve.

9.7 RIGID-BODY STATICS PROBLEM-SOLVING TOOLS AND STRATEGY

The tools which have been presented in this and earlier chapters and which are needed for addressing and understanding rigid-body statics problems are summarized below.

1) **The Laws.** This includes, for rigid-body statics, linear momentum, angular momentum, and, for mechanical advantage machines, the conservation of energy (and/or an accounting of mechanical energy).

2) **Description of Forces.** This includes gravity, static friciton, springs, and distributed loads.

3) **Process Constraints.** This tool refers to the nature of the rigid body. Example considerations are whether the object is a particle or finite-size rigid body, truss, frame, mechanical advantage machine, the type of supports present (e.g., roller, pin, hinge, etc.), 2-dimensional, 3-dimensional, etc.

4) **Calculations of Work.** Although truly static situations do not involve work, mechanical advantage machines do, and this analysis using energy is referred to as the method of virtual work. Therefore, this method requires understanding how work is calculated in terms of the forces associated with it: (work) $= \int \mathbf{f} \cdot d\mathbf{s}$. If a force acts at or across a system boundary, then there must be a displacement associated with that force which has a component collinear with the force in order for there to be work. Remember also that gravity is a force but that work due to the displacements associated with it are normally accounted for through changes in system potential energy. It would be wrong to count gravity as both producing changes in the system potential energy *and* transferring energy as work.

5) **Mathematics.** Mathematics of course, is an important tool in stating and solving problems. It plays two roles. One is as a language for expressing the physical concepts and laws and the other is as a technique for solving the mathematics problem which is thus obtained. For rigid-body statics, the use of vectors to state the laws must be well understood. Also, obtaining solutions to simultaneous algebraic equations and integrating distributed forces are important mathematical skills.

Strategies include the assessment of each of the laws in turn and the selection of different systems for analysis (e.g., the entire body, or the method of joints, or the method of sections). Also, the selection of the point (or points) about which to calculate torques (for angular momentum) can be an important decision with respect to ease of solving the problem, or even as to whether a useful result is obtained. Traditionally, linear and angular momentum conservation for statics is referred to as force and torque equilibrium and conservation of energy (and/or mechanical energy accounting) are referred to as energy methods or virtual work. Applying the various laws to a variety of systems for a given problem provides a wealth of information for understanding the relations between the various forces.

9.8 REVIEW

1) In analyzing the forces which act on static particles, the only governing law or accounting equation which is nontrivial is that of the conservation of linear momentum.

2) Considering the mechanics of static rigid bodies, the only governing laws which are not trivial are the conservation of linear momentum and the conservation of angular momentum.

3) For a system consisting of a single static particle or point in space the conservation of linear momentum provides three independent scalar equations from which can be determined three quantities related to the forces acting upon the

particle. If the forces acting upon the particle are all in the same plane then there are only two nontrivial equations from which two unknowns can be determined. If the forces act collinearly then there is only one nontrivial equation from which one unknown may be determined.

4) For each system which is a rigid body, there are two vector equations and hence six scalar equations which describe the forces on the system. From these six equations there may be determined six unknowns related to the forces acting on the body. If the forces are coplanar then this system of six equations reduces to three nontrivial equations from which can be determined three unknowns. If the forces are all collinear, then there is only one nontrivial equation because the conservation of angular momentum is satisfied identically. Consequently, only one unknown may be determined.

5) Forces that are distributed over a finite volume, area or length may be replaced in the linear and angular momentum equations by an equivalent concentrated force acting at a single point in the volume, on the surface, or along the length of the body. The equivalency of the distributed versus concentrated loads is defined by the requirement that the equations of linear and angular momentum for each case give the same results. For example, the distributed gravitational force (weight) of a body can be replaced by a concentrated force acting at the center of mass of the body.

6) In a system of members connected to form a truss, the forces exerted on the members may be analyzed by selecting, in turn, a variety of systems (free bodies) and applying the conservation of linear and angular momentum equations to each system. The systems selected may be the point of contact of multiple members (the method of joints) or the systems may be finite-sized sections of the truss (the method of sections). By working with a number of systems one may obtain a large set of simultaneous equations which may be solved for the unknown forces acting on the members of the truss.

7) Frames are supporting structures which may have loads at points other than the ends of the members. They may be analyzed similarly to trusses, using either the method of joints or the method of sections. In addition, it may be advantageous to consider an accounting of mechanical energy (the method of virtual work) for analyzing the forces on the structure. This method is not independent of the conservation of linear and angular momentum but may give a simplified analysis or calculation.

8) Several mechanical-advantage machines are similar to frames in terms of their structure. The purpose is to amplify a force exerted on one member into a larger force exerted through another member or set of members. Like frames they are analyzed using conservation of linear and angular momentum for point or particle systems or for finite-sized rigid body systems. An alternative analysis using an accounting of mechanical energy can also be used. Such an equation may provide a simplified analysis provided that the geometrical relations between the various members can be easily interpreted. Otherwise, no advantage is gained over using the momentum equations.

9.9 NOTATION

The following notation has been used throughout this chapter. The bold face lower case letters represent vector quantities.

<u>Scalar Variables and Descriptions</u>		<u>Dimensions</u>
A	cross sectional area	[length^2]

f	magnitude of the vector $\mathbf{f}$	[mass length / time^2]
g	magnitude of the acceleration of gravity	[length / time^2]
k	spring constant	[force / length]
L	length	[length]
m	total mass	[mass]
W	work	[energy]
$\dot{W}$	rate of work or power	[energy / time]
x, y, z	spatial rectangular cartesian variables	[length]
μ	coefficient of friction	[length]

<u>Vector Variables and Descriptions</u>   <u>Dimensions</u>

$\mathbf{d}_A$	distributed force per area	[force / length^2]
$\mathbf{d}_L$	distributed force per length	[force / length]
$\mathbf{f}$	force	[force]
$\mathbf{g}$	acceleration of gravity	[length / time^2]
$\mathbf{i}, \mathbf{j}, \mathbf{k}$	cartesian unit directional vectors	
$\mathbf{r}$	position	[length]
$\mathbf{s}$	displacement	[length]
$\mathbf{v}$	velocity	[length / time] or

<u>Subscripts</u>

ext	external to the system or acting at the system boundary
eq	equivalent or equal to
G	corresponding to the center of mass of the system
k	kinetic
s	static

<u>Superscripts</u>

$-$	position of equivalent force

9.10 SUGGESTED READING

Beer, F. P., and E. R. Johnston, Jr., *Mechanics for Engineers, Statics and Dynamics*, 4th edition, McGraw-Hill, New York, 1987

Halliday, D. and R. Resnick, *Physics for Students of Science and Engineering*, 2nd edition, John Wiley & Sons, 1963

Hibbeler, R. C., *Engineering Mechanics, Statics and Dynamics*, 3rd edition, MacMillan, New York, 1983

Sears, F. W., M. W. Zemansky, and H. D. Young, *University Physics*, Addison-Wesley, Reading, Massachussetts, 1982

QUESTIONS

1) Chapter 9 addresses the analysis of rigid body static systems. Of the fundamental governing laws, which ones are applicable for such systems? Explain why the laws which you have omitted are trivial and offer no useful information for these systems.

2) Of the laws which are applicable and were given in response to question one, how many independent equations does this provide for understanding the system? To what extent does your answer depend upon the dimensionality of the problem?

3) Static systems may be thought of as steady-state. Explain.

4) Write out verbal statements of the laws which govern static systems.

5) Write out symbolic statements of the laws which govern static systems.

6) The number of laws which govern particle statics is reduced in number compared to finite-sized rigid bodies. Explain.

7) What kind of equations are obtained for analyzing static systems (algebraic, ordinary differential or partial differential equations)?

8) What is the name which is traditionally used in statics for the sketch which defines the system and the forces which act upon it?

9) How many independent equations do we have at our disposal for analyzing the forces which act upon a static particle if the situation is a) one dimensional b) two dimensional or c) three dimensional?

10) How many independent equations do we have at our disposal for understanding the forces acting upon a finite-sized static body if those forces are a) one dimensional and planar but not collinear b) one dimensional and collinear c) two dimensional and planar d) three dimensional?

11) Sometimes it is useful to use a (scalar) energy equation to analyze static systems such as machines which are used to obtain a mechanical advantage. Explain why an energy equation is applicable to such a static system and why it can be of advantage over the full momentum equations. Also explain why this energy equation contains no information that the momentum equation does not contain. What name is given to this kind of analysis?

12) A truss and a frame are two kinds of finite-sized static support structures. Explain in what ways these two structures are similar and in what ways they are different. How does this difference affect the way in which the two kinds of structures are analyzed?

13) For any given static system that we define we have at most six independent equations which relate the forces acting on that system. However, for any given structure we can obtain more equations than that by defining a number of systems for each structure for the entire structure. Give an example of a statics problem which requires defining at least two different systems in order to reach a solution.

14) Describe the kind of forces and/or torques which can be sustained by a) pins b) rollers c) fixed support d) etc. Explain why this information is important in interpreting and analyzing a given structural problem.

15) Explain the concept of equivalent forces. For example, in what sense is a single force acting at a single point on a static finite-sized body equivalent to a distributed force or set of forces acting over a number of points of the body?

SCALES

1) Practice your scales for chapters 5 and 6.

2) Given points A at (2, 5, -3) and B at (-3, -1, 4), calculate:

    a) $_A r_B$ (the position of point B relative to point A)
    b) $_B r_A$
    c) $_A u_B$ (the unit vector having direction from point A to B)
    d) $_B u_A$

e) How are the answers to (a) and (b) related?
e) How are the answers to (c) and (d) related?

3) Given position vector $\mathbf{r} = 2\mathbf{i} - 3\mathbf{j} + 0\mathbf{k}$ relative to the origin of the coordinate system and force $\mathbf{F} = 4\mathbf{i}$ which acts at the point $(2, -3, 0)$. Calculate the torque exerted by $\mathbf{F}$ relative to the origin in two ways:
 a) from the appropriate determinant
 b) geometrically, by drawing $\mathbf{r}$ and $\mathbf{F}$ in the x-y plane and using the moment arm together with the magnitude of $\mathbf{F}$

Which method was easiest for this exercise?

4) Repeat exercise (3) for

 a) $\mathbf{r} = 2\mathbf{i} - 3\mathbf{j}$ and $\mathbf{F} = 4\mathbf{i} + \mathbf{j}$
 b) $\mathbf{r} = 2\mathbf{i} - 3\mathbf{j}$ and $\mathbf{F} = 4\mathbf{i} + \mathbf{j} - 5\mathbf{k}$

Is the moment arm method ever easiest? When? Is the determinant calculation ever easiest? When?

5) Solve the following system of equations for the magnitude of the three forces where $m = 80$ lb$_m$:

$$
\begin{array}{ccccccc}
f_B & + & 0 & - & 0.940 f_D & = 0 \\
0 & + & 0.6 f_C & + & 0 & = mg \\
0 & - & 0.8 f_C & + & 0.342 f_D & = 0
\end{array}
$$

You should be able to solve these equations "by-hand" as well as by software on your calculator or computer.

6) Solve the following system of equations for the magnitude of the three forces where $m = 50$ kg:

$$
\frac{2 f_{AB}}{\sqrt{30}} + \frac{f_{AD}}{\sqrt{30}} - \frac{f_{AC}}{\sqrt{27}} = 0
$$

$$
-\frac{f_{AB}}{\sqrt{30}} + \frac{2 f_{AD}}{\sqrt{30}} - \frac{f_{AC}}{\sqrt{27}} = 0
$$

$$
\frac{2 f_{AB}}{\sqrt{30}} + \frac{5 f_{AD}}{\sqrt{30}} + \frac{5 f_{AC}}{\sqrt{27}} = mg
$$

You should be able to do this quickly and efficiently on your calculator and computer.

7) Draw a free-body diagram for the *surroundings* for example problems 1, 2, 3, 8, 9, 10, 11, and 12 in Chapter 9. That is, show a boundary between the object in each problem and its surroundings and indicate the forces which act across the boundary on the surroundings. How do the forces compare in both magnitude and direction to those shown in the text?

8) Calculator review: Be sure that you know how to calculate dot products of two vectors, cross products of two vectors, and solutions to simultaneous equations on your calculator using built-in software, if yours has it.

9) A point is positioned at (r, θ, z) cylindrical coordinates of $(2, 0.8$ radians, $5)$. What is the position vector from the origin to this point in terms of these coordinates and base vectors as defined at this point $(\mathbf{e}_r, \mathbf{e}_\theta, \mathbf{e}_z)$?

10) A point is positioned at (r, θ, ϕ) spherical coordinates $(2, 0.6$ radians, 1.5radians$)$. What is the position vector from the origin to this point in terms of these coordinates and base vectors as defined at this point $(\mathbf{e}_r, \mathbf{e}_\theta, \mathbf{e}_\phi)$?

11) A linearly distributed downward force acts from one end of a 5 m beam to the other. At the left end, its value is 100 N/m and at the right end 100 N/m.

a) Sketch the beam and the distributed force.

b) What is the total equivalent force exerted by this distributed force and what is its equivalent point of application, measured from the left end of the beam. (Do two ways: 1) integration to determine the area and the centroid, 2) geometrically to determine the area and centroid.)

12) Repeat exercise (10) for a linearly distributed downward force which is 0 N/m at the left end of the beam and 100 N/m at the right end.

13) Repeat exercise (10) for a linearly distributed downward force which is 100 N/m at the left end of the beam and 200 N/m at the right end. Note that this is the sum of the two distributed forces of exercises (10) and (11). In addition to the two methods used in (10) and (11), determine the equivalent force as the sum of the two equivalent forces in (10) and (11) and determine the point of application by calculating an appropriate average of the two positions found in exercises (10) and (11).

14) Repeat exercise (10) for a linearly distributed downward force which is 300 N/m at the left end and 50 N/m at the right end.

15) Repeat exercise (10) for a linearly distributed downward force which is 200 N/m at the left end, 300N/m in the middle and 200 N/m at the right end. Note that at the mid-point of the beam there is a discontinuity in the distributed force function. Feel free to use the shortcuts of exercise (12).

16) Repeat exercise (10) for a linearly distributed downward force which is 0 N/m at the left end, 300N/m at a distance 3 m from the left end, and 300 N/m at the right end. Note that at 3 m from the left end of the beam there is a discontinuity in the distributed force function. Feel free to use the shortcuts of exercise (12).

17) A static block of 20 kg rests on a horizontal table and is acted upon by a horizontal force acting to the left. The coefficient of static friction is 0.3.

a) Which direction does the friction force act?

b) Draw a free-body diagram for the block.

c) Draw a free-body diagram for the surroundings

d) Plot the magnitude of the friction force versus the magnitude of the applied horizontal force as the latter varies from zero.

e) What is the magnitude of the friction force when the horizontal force is zero?

f) What is the maximum friction force which may exist for the static situation and what is the magnitude of the corresponding applied force?

18) A truss member connects points at (2,1,0) (point A) and (1,3,2) (point B). The truss member exerts a force on point A towards point B. Make a sketch of the truss member in the coordinate system, another sketch of point A showing the force acting on it, and express the force mathematically in terms of its magnitude f_{AB} times an appropriate unit vector. The unit vector should point in the smae direction as the arrow you showed for the force.

PROBLEMS

1) The massless truss shown in the figure is acted on by a force at point B given by the vector below:

$$f_B = 8.7i - 5j + 0k \ (N)$$

Points A, B, and C are pin supports and the roller at point C is free to roll in the x-direction. Also, the forces in the members act along the axis of the member.

a) Calculate the reactions at points A and C and express the answer as a vector.

b) Determine the force vectors in each member of the truss and state whether the member is in tension or compression.

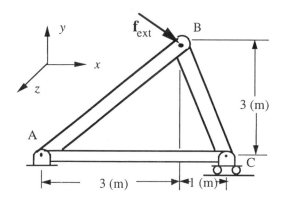

Problem 9-1: Massless truss

c) Redraw the truss and clearly label all of the forces on your figure.

2) For the given 3 dimensional structure shown below, calculate the forces in AC, AB, and AD. The mass of the block being supported is 50 (kg).

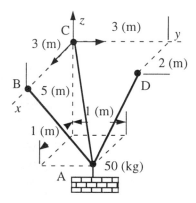

Problem 9-2: 3 Dimensional structure

3) A painter is hired to paint the side of your two story house. His ladder is 40 (ft) long and has a mass of 30 (lb_m) and is made out of aluminum. The siding on your house is also aluminum. The painter places his ladder on your dry cement patio and the aluminum/air/cement contact point gives a static coefficient of friction of 0.9. The painter places the ladder such that it makes a 70 ° angle with the horizontal. If the painter has a mass of 170 (lb_m), how high up the wall do you suggest he paint and why? Remember, if the painter gets hurt on your property he could sue you.

a) Consider the system to be the ladder and account for all of the *external* forces that act on this system. Verbally explain what is physically happening to these forces as the painter moves up the ladder. When will the ladder just begin to move and in what direction will the ladder move.

Problem 9-4: Loaded plank

4) A massless plank AC rests on points A and B and on the plank twenty (20) bricks are placed as shown in the figure below. Each brick is 1 (ft) in length and is 3 (lb_m).

    a) Determine the magnitude and position of the equivalent force acting on AC due to the loading the bricks on the plank AC.

    b) Calculate the reaction force vectors acting at points A and B.

5) You are building a cabin in Colorado and your cabin will have a 10 (ft) ceiling. The width of the cabin is 36 (ft) and the length is 52 (ft). For aesthetic reasons, you wish to have an open ceiling with a pitch of 30 °(the angle of incline of the roof) and the roof truss design is given below.

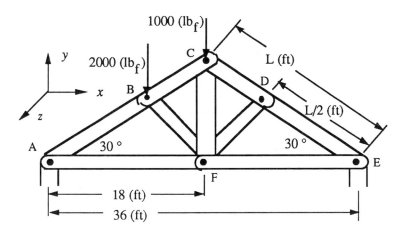

Problem 9-5: Ceiling truss

If each internal truss supports a load at the joints labeled in the figure, determine the reaction forces at the base of the truss and the forces in each of the members. Also, state whether the member is in tension or compression. You may assume that both reaction forces (at A and E) are vertical only.

6) You are transporting a 50 (lb$_m$) barrel with a 2 (ft) radius and a 3 (ft) width from a barn to a shed. You apply the force to move the barrel at 20 $^\circ$ above the horizontal directed towards the center of the barrel. Sometime on your trek you come to a small concrete step that has a height of 0.4 (ft). Determine the force required to "push" the barrel over the step and the force required to "pull" the barrel over the step as shown in the figure below. (This problem can be worked as a constant momentum problem.) Based on your answer, how and why would you overcome this obstacle?

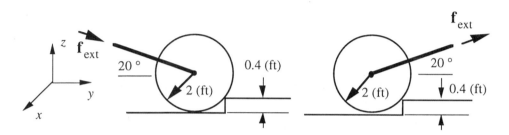

Problem 9-6: Barrel

7) Given the structure and force in Problem 9-1 calculate the reaction forces at A and C if the same force acts at position $2\mathbf{i} + 2\mathbf{j}$ (m) relative to point A. Also, calculate the forces in members AC and BC and whether these members are in tension or compression. (Note: member AB is a three-force member so that its state of stress cannot be characterized simply as tension or compression.)

8) A cube of glass with mass 100 (lb$_m$) is on a 25° inclined glass plane.

   a) From the density of glass calculate the volume of the cube and the length of each side.

   b) Is the block moving or stationary and why?

   c) If the block is moving, what minimum extra force applied to the top of the cube is required to prevent this impending motion

9) Given the structure in the figure with a frictionless contact at point B, calculate the forces at points A and B.

10) Given the truss and the loads at points B, C, and D in the figure, calculate the reaction forces at points A and E and the forces in each member along with the members' tension or compression state.

11) A mass of 10 lb_m is suspended by two ropes as shown in the figure. Calculate the tension in each ropes AC and AB.

12) A 20 kg block is suspended by a rod BC and a wire AC as shown in the figure. The rod is pin supported at B. Calculate the force in the rod and wire.

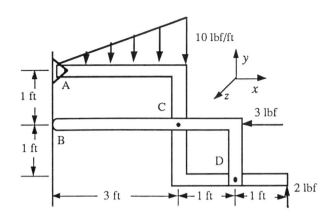

Problem 9-9: Loaded structure

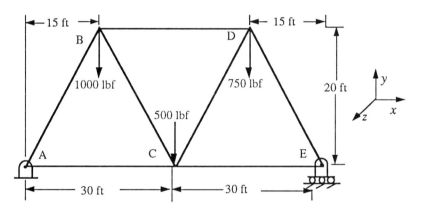

Problem 9-10: Loaded structure

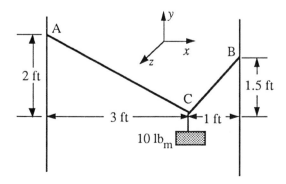

Problem 9-11: Weight suspended by ropes

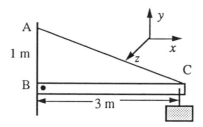

Problem 9-12: Weight suspended by rod and wire

13) Four forces f_1, f_2 f_3, and f_4 are acting on a static particle in space. Gravity is not acting on the mass. The three known forces are

$$f_1 = -5i + 3j - 7k$$
$$f_2 = 3i - 3j + 3k$$
$$f_3 = 2i - 3j + 5k$$

Deterine the unknown force f_4.

14) Calculate the equivalent forces for figures (a), (b), (c), and (d). The length of the bar is always 4 m.

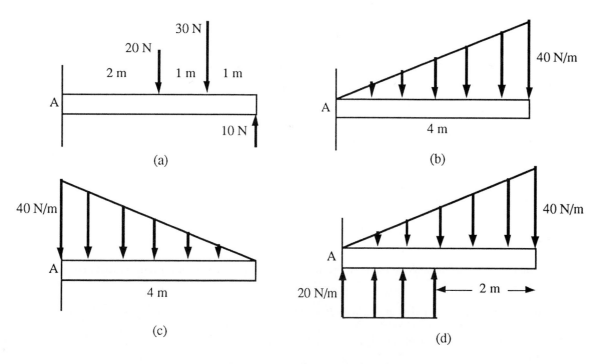

Figure 9-14: Distributed loads

15) In problem 14, calculate the reaction forces and torques (moments) at point A.

16) A 3 kg cube of aluminum is resting on a glass surface. Calculate the angle of inclination of the glass when the cube will start to slide.

17) The structure in the accompanying figure supports a load **P** at point C, which is exerted parallel to member BD. Determine the reaction forces (in terms of **P**) at points A and B (i.e., the forces exerted on the structure by its surroundings) and the force in each member of the structure and whether the member is in tension or compression. When you have finished your calculations, make a summary sketch of the structure with all forces indicated. For each member, it is sufficient on the sketch to indicate the magnitude of the force and whether the member is in tension or compression. What kind of structure is this and why?

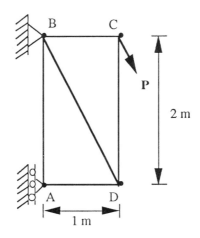

Problem 9-17: Loaded structure

18) The structure in the accompanying figure supports a load **P**, at point D, which is exerted perpendicular to member BD. Determine the reaction forces (in terms of **P**) at points A and B (i.e., the forces exerted on the structure by its surroundings) and the force in each member of the structure and whether the member is in tension or compression. When you have finished your calculations, make a summary sketch of the structure with all forces indicated. For each member, it is sufficient on the sketch to indicate the magnitude of the force and whether the member is in tension or compression. What kind of structure is this and why?

19) The structure in the accompanying figure is used to lift an engine out of a car, shown as load **L** 200 lb$_f$. ADB and BEC are each single members and all connections are with frictionless pins. A block and tackle is used to assist and is shown also. If the mechanic pulls on the chain at the angle shown in the figure, then determine:

   a) the force he must exert to just lift the engine,

   b) the reaction forces at points A and C (i.e., the forces exerted on the structure by its surroundings) when the engine is suspended,

   c) the forces acting on the members at points A, B, C, D, and E.

   d) the force in member DE and whether it is in tension or compression.

What kind of structure is this and why?

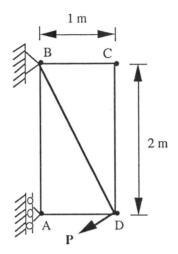

Problem 9-18: Loaded structure

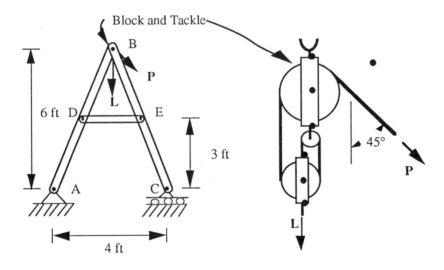

Problem 9-19: Frame

CHAPTER
TEN
RIGID BODY DYNAMICS

10.1 INTRODUCTION

In the first part of this section we discussed the procedures for analyzing static systems. Static means unchanging and refers to those systems for which the motion of the system is unchanging. For these systems the conservation of linear and angular momentum equations were used to obtain information about the relationships which exist between the forces which are acting upon particles or rigid bodies. The equations which were obtained were algebraic equations because of the fact that the system itself was static or unchanging. For the rate equations, which were the forms considered, the rates of change of the linear and angular momentum of the system were zero.

In this chapter we consider dynamic situations. Dynamic refers to change and, appropriately, here we are talking about systems where the linear and angular momentum of the system may change. Accordingly, we are concerned with rate equations for which the sum of the forces and the sum of the torques, instead of being equal to zero, will be equal to the time rate of change of the linear and angular momentum of the system. For a rigid body, which has no mass entering or leaving, and in terms of the center-of-mass velocity the conservation of linear momentum becomes (Chapter 5)

$$\left(\frac{d\mathbf{p}_{sys}}{dt}\right) = m_{sys}\left(\frac{d\mathbf{v}_G}{dt}\right) = \sum \mathbf{f}_{ext}$$

$(5-9)$

and the a-component of the conservation of angular momentum (equation 6-1) for planar motion about the a axis becomes

$$_G I_a \alpha = \sum (_G\mathbf{r} \times \mathbf{f})_{\text{ext},a} = \sum T_a$$

(6 − 39)

or one of the other forms which appears in Table 6-6. Accordingly, the equations which will be obtained will be ordinary differential equations. Details of this analysis and applications to specific situations will be given below. The point here is that the same principles are involved as were applied to the analysis of static systems but now we must use these equations for changing (dynamic) systems.

In this chapter we first consider the subject of kinematics, the mathematics which is used to describe the motion of an object independent of the cause of that motion. This involves the definitions of velocity and acceleration and their relations to each other and to position. Then we consider the dynamics of a single particle. As was the case for particle statics, the linear momentum conservation equation governs the behavior of the particle, providing a relationship between the acceleration of the particle and the forces acting upon it. If the forces are known, then information can be obtained about the acceleration, velocity, and position of a particle as a function of time. Then we consider the dynamics of finite-sized bodies where, in addition to the conservation of linear momentum, we must also consider the conservation of angular momentum. Consequently, in addition to translational velocity and acceleration we consider rotational velocity and acceleration. For both particle and rigid body dynamics we will also consider the conservation of energy which, as for static situations, is not an independent relation but still provides some additional understanding and insight.

10.2 KINEMATICS

As mentioned above, kinematics is the description of motion *per se* without reference to its cause. The conservation of linear momentum provides an equation of motion; the forces which act upon a body affect its motion. That is the physics. However, the manner in which we describe the motion by means of position vectors, velocity vectors, and acceleration vectors for particles, and in addition, in terms of angular velocity and acceleration for rigid bodies, is the subject of kinematics. We will consider first the kinematics of a single particle and then the kinematics of a collection of particles which make up a rigid body. Additionally we will consider the relative motion of two objects and the interdependent motion of two objects connected by a cord of fixed length.

10.2.1 Particle Kinematics

In describing the motion of a particle we are concerned with its position as a function of time, velocity as a function of time, and acceleration as a function of time. All three of these quantities are vectors. They are related to each other through time derivatives according to the equations:

$$\mathbf{v} = \left(\frac{d\mathbf{r}}{dt} \right)$$

(10 − 1)

and

$$\mathbf{a} = \left(\frac{d\mathbf{v}}{dt} \right) = \left(\frac{d^2\mathbf{r}}{dt^2} \right)$$

(10 − 2)

The velocity has the physical interpretation that if we measure the particle's position as it changes with time, then at any instant in time the rate at which the position vector changes with respect to time is the particle's velocity. If we measure the change in the position vector over a short period of time, Δt, and that change is $\Delta \mathbf{r} = \mathbf{r}|_{t+\Delta t} - \mathbf{r}|_t$, then the velocity is given by:

$$\mathbf{v} = \left(\frac{d\mathbf{r}}{dt}\right) = \lim_{\Delta t \to 0}\left(\frac{\mathbf{r}_{t+\Delta t} - \mathbf{r}_t}{\Delta t}\right) \tag{10 - 3}$$

Similarly, if we measure the particle's velocity as a function of time, then its acceleration is the rate at which this velocity changes with respect to time and is given by the equation

$$\mathbf{a} = \lim_{\Delta t \to 0}\left(\frac{\mathbf{v}_{t+\Delta t} - \mathbf{v}_t}{\Delta t}\right) \tag{10 - 4}$$

Figure 10-1 shows the relation between the position, velocity, and acceleration vectors. The fact that position, velocity, and acceleration are related through time derivatives allows us to calculate one from the other if we know the appropriate function of time describing the first one. For example, if we know the position of a particle as a function of time (the position vector) and this is expressed as an explicit function of time, then we can calculate the velocity and acceleration of the particle, also as functions of time, by differentiating the position vector with respect to time to get the velocity, and then differentiating the velocity with respect to time to get the acceleration.

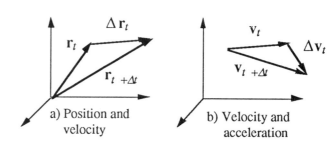

a) Position and velocity   b) Velocity and acceleration

Figure 10-1: Position, velocity and acceleration vectors

Likewise, if the velocity is known as a function of time, then this velocity can be integrated over time to obtain the position as a function of time (to within an integration constant) and the acceleration can be calculated by differentiating this velocity function with respect to time. Finally, if the acceleration is known as a function of time, then integration once gives the velocity function of time which, in turn, can be integrated to obtain the position as a function of time.

We are concerned with such calculations because the equation for the conservation of linear momentum provides just such information for a particle. If the forces acting on that particle are known functions of time and the mass of the particle is known, then we can say that the acceleration is equal to some known function of time and we can integrate to obtain the velocity and subsequently the position of the particle as a function of time. Such applications will appear later in the section on particle dynamics.

Rectangular Cartesian Coordinate Systems. The relations between position, velocity, and acceleration in terms of coordinates are straightforward for a rectangular cartesian coordinate system. The position vector is given by (Appendix):

$$\mathbf{r} = x\mathbf{i} + y\mathbf{j} + z\mathbf{k} \tag{10 - 5}$$

and, recognizing that the base vectors are independent of position, the velocity is given by:

$$\mathbf{v} = \left(\frac{dx}{dt}\right)\mathbf{i} + \left(\frac{dy}{dt}\right)\mathbf{j} + \left(\frac{dz}{dt}\right)\mathbf{k}$$

$$= v_x\mathbf{i} + v_y\mathbf{j} + v_z\mathbf{k}$$

$$(10-6)$$

and the acceleration by

$$\mathbf{a} = \left(\frac{d^2x}{dt^2}\right)\mathbf{i} + \left(\frac{d^2y}{dt^2}\right)\mathbf{j} + \left(\frac{d^2z}{dt^2}\right)\mathbf{k}$$

$$= a_x\mathbf{i} + a_y\mathbf{j} + a_z\mathbf{k}$$

$$(10-7)$$

The dot notation is used to denote differentiation with respect to the variable time in the kinematical relationships and is used to simplify the equations. A single dot is the first derivative with respect to time and a double dot is the second derivative with respect to time:

$$\dot{x} = \left(\frac{dx}{dt}\right) = v_x \qquad \dot{y} = \left(\frac{dy}{dt}\right) = v_y \qquad \dot{z} = \left(\frac{dz}{dt}\right) = v_z$$

and

$$\ddot{x} = \left(\frac{d^2x}{dt^2}\right) = a_x \qquad \ddot{y} = \left(\frac{d^2y}{dt^2}\right) = a_y \qquad \ddot{z} = \left(\frac{d^2z}{dt^2}\right) = a_z$$

the components of the position, velocity and acceleration vectors in a rectangular cartesian coordinate system are given in Table 10-1.

Table 10-1: Components of **r**, **v** and **a**
Rectangular Cartesian Coordinate System

	x-comp.	y-comp.	z-comp.
r	x	y	z
v	$\dot{x}$	$\dot{y}$	$\dot{z}$
a	$\ddot{x}$	$\ddot{y}$	$\ddot{z}$

Curvilinear Coordinate Systems-General. While in rectangular cartesian coordinates the velocity and acceleration are easily calculated from derivatives of the position vector components, the results are not so easily obtained for curvilinear coordinate systems. For these systems the base vectors are functions of position as well as the components of the vector and so, to differentiate the position with respect to time to obtain the velocity and acceleration, we must differentiate both the base vectors and the position vector components. Accordingly, as a general result we would write:

$$\mathbf{v} = \left(\frac{d\mathbf{r}}{dt}\right)$$

$$= \sum_{i=1}^{3}\left(\frac{dr_i}{dt}\right)\mathbf{e}_i + \sum_{i=1}^{3} r_i\left(\frac{d\mathbf{e}_i}{dt}\right)$$

(10 – 8)

and

$$\mathbf{a} = \left(\frac{d^2\mathbf{r}}{dt^2}\right)$$

$$= \sum_{i=1}^{3}\left(\frac{d^2 r_i}{dt^2}\right)\mathbf{e}_i + 2\sum_{i=1}^{3}\left(\frac{dr_i}{dt}\right)\left(\frac{d\mathbf{e}_i}{dt}\right) + \sum_{i=1}^{3} r_i\left(\frac{d^2\mathbf{e}_i}{dt^2}\right)$$

(10 – 9)

In a rectangular cartesian coordinate system the derivatives of the base vectors are zero giving a simplified expression; in a curvilinear system the derivatives are not zero.

Curvilinear Coordinate Systems-Cylindrical. The derivatives of the base vectors can be expressed, for a particular coordinate system, in terms of the derivatives of the coordinates and the base vectors themselves. This was done, for example, in Appendix B for the cylindrical and spherical coordinate systems. This representation is independent of the physics of the problem at hand, i.e. independent of the reasons for the motion of the particle. Accordingly, for a cylindrical system we have that the position vector is written (Appendix B):

$$\mathbf{r} = r\mathbf{e}_r + z\mathbf{e}_z$$

(10 – 10)

which gives the velocity equal to:

$$\mathbf{v} = \left(\frac{d\mathbf{r}}{dt}\right)$$

$$= \left(\frac{dr}{dt}\right)\mathbf{e}_r + r\left(\frac{d\mathbf{e}_r}{dt}\right) + \left(\frac{dz}{dt}\right)\mathbf{e}_z + z\left(\frac{d\mathbf{e}_z}{dt}\right)$$

(10 – 11)

It is shown in Appendix B that:

$$\left(\frac{d\mathbf{e}_z}{dt}\right) = \mathbf{0}$$

and

$$\left(\frac{d\mathbf{e}_r}{dt}\right) = \left(\frac{d\theta}{dt}\right)\mathbf{e}_\theta$$

This reduces the equation to:

$$\mathbf{v} = \left(\frac{dr}{dt}\right)\mathbf{e}_r + r\left(\frac{d\theta}{dt}\right)\mathbf{e}_\theta + \left(\frac{dz}{dt}\right)\mathbf{e}_z$$
$$= \dot{r}\mathbf{e}_r + r\dot{\theta}\mathbf{e}_\theta + \dot{z}\mathbf{e}_z$$
$$= v_r\mathbf{e}_r + v_\theta\mathbf{e}_\theta + v_z\mathbf{e}_z$$

(10 – 12)

Again the dot is used to denote differentiation with respect to the variable time and is used to simplify the equations:

$$\dot{r} = \left(\frac{dr}{dt}\right) \qquad \dot{\theta} = \left(\frac{d\theta}{dt}\right) \qquad \dot{z} = \left(\frac{dz}{dt}\right)$$

The acceleration is:

$$\mathbf{a} = \left(\frac{d\mathbf{v}}{dt}\right) = \left(\frac{dv_r}{dt}\right)\mathbf{e}_r + v_r\left(\frac{d\mathbf{e}_r}{dt}\right) + \left(\frac{dv_\theta}{dt}\right)\mathbf{e}_\theta + v_\theta\left(\frac{d\mathbf{e}_\theta}{dt}\right) + \left(\frac{dv_z}{dt}\right)\mathbf{e}_z + v_z\left(\frac{d\mathbf{e}_z}{dt}\right)$$

(10 – 13)

Appendix B gives the expression for the time derivative of the θ-coordinate vector with respect to time as:

$$\left(\frac{d\mathbf{e}_\theta}{dt}\right) = -\left(\frac{d\theta}{dt}\right)\mathbf{e}_r$$

We can extend the dot notation to include second derivatives with respect to time as:

$$\ddot{r} = \left(\frac{d^2r}{dt^2}\right) \qquad \ddot{\theta} = \left(\frac{d^2\theta}{dt^2}\right) \qquad \ddot{z} = \left(\frac{d^2z}{dt^2}\right)$$

Combining the terms gives:

$$\mathbf{a} = (\ddot{r} - r\dot{\theta}^2)\mathbf{e}_r + (r\ddot{\theta} + 2\dot{r}\dot{\theta})\mathbf{e}_\theta + (\ddot{z})\mathbf{e}_z$$
$$= a_r\mathbf{e}_r + a_\theta\mathbf{e}_\theta + a_z\mathbf{e}_z$$

(10 – 14)

The r, θ, and z components of the position, velocity, and acceleration vectors in a cylindrical coordinate system are summarized in Table 10-2.

Table 10-2: Components of **r**, **v** and **a**
Cylindrical Coordinate System

	r-comp.	θ-comp.	z-comp.
r	r	0	z
v	$\dot{r}$	$r\dot{\theta}$	$\dot{z}$
a	$(\ddot{r} - r\dot{\theta}^2)$	$(r\ddot{\theta} + 2\dot{r}\dot{\theta})$	$\ddot{z}$

Curvilinear Coordinate Systems-Spherical. The velocity and acceleration vectors can be determined for the spherical coordinate system in the same way. The position vector for a spherical system is given by (Appendix B)

$$\mathbf{r} = r\mathbf{e}_r$$

$$(10-15)$$

and the velocity is

$$\mathbf{v} = \left(\frac{d\mathbf{r}}{dt}\right)$$

$$= \left(\frac{dr}{dt}\right)\mathbf{e}_r + r\left(\frac{d\mathbf{e}_r}{dt}\right)$$

$$(10-16)$$

In Appendix B it is shown that:

$$\left(\frac{d\mathbf{e}_r}{dt}\right) = \left(\frac{d\theta}{dt}\right)\mathbf{e}_\theta + \sin\theta\left(\frac{d\phi}{dt}\right)\mathbf{e}_\phi$$

Substituting back

$$\mathbf{v} = \left(\frac{dr}{dt}\right)\mathbf{e}_r + r\left(\frac{d\theta}{dt}\right)\mathbf{e}_\theta + r\sin\theta\left(\frac{d\phi}{dt}\right)\mathbf{e}_\phi$$

$$= \dot{r}\mathbf{e}_r + r\dot{\theta}\mathbf{e}_\theta + r\sin\theta\,\dot{\phi}\mathbf{e}_\phi$$

$$= v_r\mathbf{e}_r + v_\theta\mathbf{e}_\theta + v_\phi\mathbf{e}_\phi$$

$$(10-17)$$

The acceleration is:

$$\mathbf{a} = \left(\frac{d\mathbf{v}}{dt}\right)$$

$$= \left(\frac{dv_r}{dt}\right)\mathbf{e}_r + v_r\left(\frac{d\mathbf{e}_r}{dt}\right) + \left(\frac{dv_\theta}{dt}\right)\mathbf{e}_\theta + v_\theta\left(\frac{d\mathbf{e}_\theta}{dt}\right) + \left(\frac{dv_\phi}{dt}\right)\mathbf{e}_\phi + v_\phi\left(\frac{d\mathbf{e}_\phi}{dt}\right)$$

$$= \ddot{r}\mathbf{e}_r + \dot{r}\left(\frac{d\mathbf{e}_r}{dt}\right) + (\dot{r}\dot{\theta} + r\ddot{\theta})\mathbf{e}_\theta + r\dot{\theta}\left(\frac{d\mathbf{e}_\theta}{dt}\right) +$$

$$(\dot{r}\dot{\phi}\sin\theta + r\ddot{\phi}\sin\theta + r\dot{\theta}\dot{\phi}\cos\theta)\mathbf{e}_\phi + r\dot{\phi}\sin\theta\left(\frac{d\mathbf{e}_\phi}{dt}\right)$$

$$(10-18)$$

The time derivatives of the base vectors are given below (Appendix B):

$$\left(\frac{d\mathbf{e}_r}{dt}\right) = \left(\frac{d\theta}{dt}\right)\mathbf{e}_\theta + \sin\theta\left(\frac{d\phi}{dt}\right)\mathbf{e}_\phi$$

$$\left(\frac{d\mathbf{e}_\theta}{dt}\right) = -\left(\frac{d\theta}{dt}\right)\mathbf{e}_r + \cos\theta\left(\frac{d\phi}{dt}\right)\mathbf{e}_\phi$$

$$\left(\frac{d\mathbf{e}_\phi}{dt}\right) = -\sin\theta\left(\frac{d\phi}{dt}\right)\mathbf{e}_r - \cos\theta\left(\frac{d\phi}{dt}\right)\mathbf{e}_\theta$$

Combining the terms and using the dot notation gives:

$$\mathbf{a} = \left(\ddot{r} - r\dot{\theta}^2 - r(\sin\theta\,\dot{\phi})^2 \right)\mathbf{e}_r +$$
$$\left(r\ddot{\theta} + 2\dot{r}\dot{\theta} - r\dot{\phi}^2 \sin\theta\cos\theta \right)\mathbf{e}_\theta +$$
$$\left(2\dot{r}\dot{\phi}\sin\theta + r\ddot{\phi}\sin\theta + 2r\dot{\theta}\dot{\phi}\cos\theta \right)\mathbf{e}_\phi$$

$$(10-19)$$

Note again, for the velocity and acceleration vectors, contributions occur because of the changes of the base vectors with position as well as directly from changes in the vector components with position (and consequently with time). Remember also that the above expressions are written entirely independently of the cause of the motion, that is, these relations are true kinematic relations in that they are used to describe the motion of a particle without any reference to the physics that brings about that motion. The descriptions are in terms of the coordinates in the coordinate system and their time derivatives. For a specific problem the values of these time derivatives are certainly governed by the physics of the problem but in terms of those derivatives and coordinates, we always express the position vector, velocity, and acceleration in exactly the same way. These results for a spherical system are summarized in Table 10-3. A lot of extra terms arise just because $\mathbf{e}_r$ is a function of position! Instead of full 3-D in spherical coordinates, it may be better to use rectangular cartesian coordinates.

Table 10-3: Components of **r**, **v** and **a**
Spherical Coordinate System

	r-comp.	θ-comp.	ϕ-comp.
r	r	0	0
v	$\dot{r}$	$r\dot{\theta}$	$r\dot{\phi}\sin\theta$
a	$\left(\ddot{r} - r\dot{\theta}^2 - r(\dot{\phi}\sin\theta)^2 \right)$	$\left(r\ddot{\theta} + 2\dot{r}\dot{\theta} - r\dot{\phi}^2\sin\theta\cos\theta \right)$	$\left(2\dot{r}\dot{\phi}\sin\theta + r\ddot{\phi}\sin\theta + 2r\dot{\theta}\dot{\phi}\cos\theta \right)$

Curvilinear Coordinate Systems-Path Orientation. We could also write the kinematical descriptions (velocity and acceleration) in terms of the tangential and normal unit vectors to the particle's path (see Appendix B). For this coordinate system defined in terms of the path, we can write the velocity as:

$$\mathbf{v} = v_T\mathbf{u}_T + v_N\mathbf{u}_N$$

$$(10-20)$$

But, by definition of path coordinates, the normal component of the velocity is zero and $v_T = v$ giving

$$\mathbf{v} = v\mathbf{u}_T$$

$$(10-21)$$

The acceleration is calculated as:

$$\mathbf{a} = \left(\frac{d\mathbf{v}}{dt}\right)$$

$$= \left(\frac{dv}{dt}\right)\mathbf{u}_T + v\left(\frac{d\mathbf{u}_T}{dt}\right)$$

(10 − 22)

From Appendix B the time derivative of the tangential base vector is:

$$\left(\frac{d\mathbf{u}_T}{dt}\right) = \left(\frac{v^2}{\rho}\right)\mathbf{u}_N$$

Substituting into the acceleration equation gives:

$$\mathbf{a} = \left(\frac{dv}{dt}\right)\mathbf{u}_T + \left(\frac{v^2}{\rho}\right)\mathbf{u}_N$$

$$= a_T\mathbf{u}_T + a_N\mathbf{u}_N$$

(10 − 23)

from which we see that the acceleration consists of a normal component and a tangential component. The tangential component of the acceleration is the time rate of change of the speed of the object (magnitude of the velocity) and the normal component is the amount associated with the curvature of the path. For an infinite radius of curvature, the normal component of the acceleration is zero.

The components of the velocity and acceleration for the path oriented coordinate system are given in Table 10-4. Note that for both vectors the component in the binormal direction is zero. This is, of course, a direct consequence of the way the path coordinates are defined. Examples applying these kinematical relationships for position, velocity, and acceleration to rectangular cartesian, cylindrical, spherical, and arc length coordinate systems follow.

Table 10-4: Components of $\mathbf{v}$ and $\mathbf{a}$
Path Coordinate System

	T-comp.	N-comp.	B-comp.
$\mathbf{v}$	v	0	0
$\mathbf{a}$	$\dot{v}$	$\left(\dfrac{v^2}{\rho}\right)$	0

Example 10-1. A small particle has an initial velocity downward (negative $\mathbf{j}$ direction) into a fluid medium of 60 (m/s). If the particle's acceleration is in the positive $\mathbf{j}$ direction (drag force opposes the motion) with value $a_y = -0.4v_y{}^3$ (m/s^2), where v_y is the y-component of the velocity vector in (m/s), determine both the velocity vector $\mathbf{v}$ and position vector $\mathbf{r}$, four seconds after the particle is set into motion.

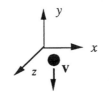

Figure 10-2: 1-Dimensional
motion of a particle

SOLUTION: Figure 10-2 shows a schematic of the motion of the particle. We will choose the coordinate system so the y axis is aligned with the motion of the particle and all the other components are zero. At the initial time, $t = 0$ (s) the position, velocity and acceleration vectors are:

$$\mathbf{r}(t_o) = 0\mathbf{i} + 0\mathbf{j} + 0\mathbf{k} \text{ (m)}$$

$$\mathbf{v}(t_o) = 0\mathbf{i} - 60\mathbf{j} + 0\mathbf{k} \text{ (m / s)}$$

$$\mathbf{a}(t_o) = 0\mathbf{i} - 0.4v_y^3\mathbf{j} + 0\mathbf{k} \text{ (m / s}^2)$$

Notice that the acceleration is a function of the velocity. The motion is in the y-direction only and we can look at the component in that direction independent of the other directions. In the $\mathbf{j}$ direction, and as a kinematic relation,

$$a_y\mathbf{j} = \left(\frac{dv_y}{dt}\right)\mathbf{j}$$

Then, by substituting $a_y = -0.4v_y{}^3$, a differential equation for v_y is obtained

$$\left(\frac{dv_y}{dt}\right) = -0.4v_y^3$$

Separating variables and integrating gives:

$$\int_{v_{yo}}^{v} v_y^{-3}\,dv_y = -\int_{t_o}^{t} 0.4dt \quad \text{or} \quad \left.\frac{-1}{2}\left(\frac{1}{v_y^2}\right)\right|_{v_{yo}}^{v_y} = \left.-0.4t\right|_{t_o}^{t}$$

Substituting the limits of integration:

$$\left(\frac{-1}{2}\right)\left(\frac{1}{v_y^2} - \frac{1}{v_{yo}^2}\right) = -0.4(t - t_o)$$

Rearranging and substituting the initial conditions gives:

$$v_y = \left(\frac{1}{(-60)^2} + 0.8t\right)^{-0.5} \text{ (m / s)}$$

If we evaluate the expression at $t = 4$ (s) the velocity in the y-direction is (we note that it must be true that $v_y < 0$ and pick the negative root):

$$v_y = -0.559 \text{ (m / s)}$$

and the velocity vector is:

$$\mathbf{v}|_{t=4} = 0\mathbf{i} - 0.559\mathbf{j} + 0\mathbf{k} \text{ (m / s)}$$

Since the velocity vector is the time rate of change of the position vector (as a kinematic relation), and the velocity is now known as a function of time, we can look at the time rate of change of the position vector in the $\mathbf{j}$ direction.

$$v_y\mathbf{j} = \left(\frac{dy}{dt}\right)\mathbf{j} = \left(\frac{1}{(-60)^2} + 0.8t\right)^{-0.5}\mathbf{j} \text{ (m / s)}$$

for $0 \le t \le 5$ (s), $v_x = 2t$ (ft/s)   and $a_x = 2$ (ft / s^2)

for $5 \le t \le 10$ (s), $v_x = 10$ (ft/s)   and $a_x = 0$ (ft / s^2)

Separating the variables and integrating yields:

$$\int_{y_o}^{y} dy = \int_{t_o}^{t} \left(\frac{1}{(-60)^2} + 0.8t \right)^{-0.5} dt$$

$$y - 0 = \left(\frac{-2}{0.8} \right) \left(\frac{1}{(-60)^2} + 0.8t \right)^{0.5} \Bigg|_{t_0}^{t}$$

$$y = \left(\frac{-1}{0.4} \right) \left(\left(\frac{1}{(-60)^2} + 0.8t \right)^{0.5} - \frac{1}{(60)^2} \right) \text{ (m)}$$

Evaluating the expression at $t = 4$ (s) gives:

$$y = -4.47 \text{ (m)}$$

and the position vector is: $\mathbf{r}|_{t=4} = 0\mathbf{i} - 4.47\mathbf{j} + 0\mathbf{k}$ (m)

Example 10-2. A particle moves along a straight line such that its position in the x-direction as a function of time is described by Figure 10-3. If the motion is only in the x-direction, graph velocity versus time and acceleration versus time for the time period $0 \le t \le 10$(s).

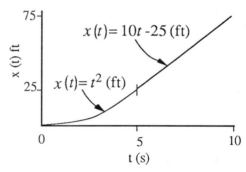

Figure 10-3: Position versus time

SOLUTION: Kinematically, the velocity is the time rate of change of the position vector. Hence, for motion in the **i** direction:

$$v_x \mathbf{i} = \left(\frac{dx}{dt} \right) \mathbf{i}$$

In terms of the graph of postion versus time, the velocity is the slope of the curve.

for $0 \le t \le 5$ (s), $x = t^2$ (ft)   and $v_x = 2t$ (ft/s)

for $5 \le t \le 10$ (s), $x = 10t - 25$ (ft)   and $v_x = 10$ (ft/s)

As another kinematic relation, the acceleration is the time rate of change of the velocity. Hence,

$$a_x \mathbf{i} = \left(\frac{dv_x}{dt} \right) \mathbf{i}$$

and in terms of the graphs, the acceleration is the slope of the velocity versus time curve:

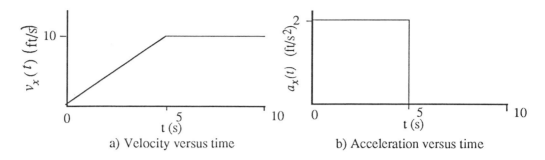

a) Velocity versus time

b) Acceleration versus time

Figure 10-3a: Velocity and Acceleration versus time

These results are shown in Figure 10-3a (a) and (b).

Example 10-3. The motion of a particle A along a spiral path as shown in Figure 10-4 is defined by the position vector given as: $\mathbf{r}(t) = \sin(3t)\mathbf{i} + \cos(3t)\mathbf{j} - 2t\mathbf{k}$ ft where t is given in seconds and arguments of the sine and cosine functions are in radians. Determine the position vector, the velocity vector, and the acceleration vector of particle A at $t = 0.25$ s. Also find the distance from the origin, the magnitude of the velocity and the magnitude of the acceleration at $t = 0.25$ s.

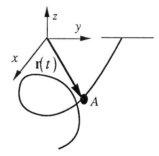

Figure 10-4: Position vector as a function of time

SOLUTION: The position vector at 0.25 seconds is found simply by evaluating the position vector at that time:

$$\mathbf{r}(0.25) = \sin(0.75)\mathbf{i} + \cos(0.75)\mathbf{j} - 0.5\mathbf{k} \text{ (ft)} = 0.682\mathbf{i} + 0.732\mathbf{j} - 0.5\mathbf{k} \text{ (ft)}$$

The magnitude of the position vector is the distance the particle is from the defined origin:

$$r(t) = |\mathbf{r}(t)| = \sqrt{x(t)^2 + y(t)^2 + z(t)^2}$$

The distance from the origin at 0.25 seconds is:

$$r(0.25) = \sqrt{(0.682)^2 + (0.732)^2 + (-0.5)^2} = 1.118 \text{ (ft)}$$

The velocity is defined as:

$$\mathbf{v}(t) = \left(\frac{d\mathbf{r}(t)}{dt}\right) = 3\cos(3t)\mathbf{i} - 3\sin(3t)\mathbf{j} - 2\mathbf{k} \text{ (ft / s)}$$

Evaluating the expression at $t = 0.25$ seconds yields:

$$\mathbf{v}(0.25) = 2.195\mathbf{i} - 2.045\mathbf{j} - 2\mathbf{k} \text{ (ft / s)}$$

The magnitude of the velocity is given by:

$$v(t) = |\mathbf{v}(t)| = \sqrt{v_x(t)^2 + v_y(t)^2 + v_z(t)^2}$$

The magnitude of the velocity at 0.25 seconds is:

$$v(0.25) = \sqrt{(2.195)^2 + (-2.045)^2 + (2)^2} = 3.666 \ \ (\text{ft} / \text{s})$$

The acceleration vector is defined as the time derivative of the velocity vector or:

$$\mathbf{a}(t) = \left(\frac{d\mathbf{v}(t)}{dt} \right) = -9\sin(3t)\mathbf{i} - 9\cos(3t)\mathbf{j} + 0\mathbf{k} \ \ (\text{ft} / \text{s}^2)$$

Evaluating the acceleration vector at 0.25 seconds gives:

$$\mathbf{a} = -6.135\mathbf{i} - 6.585\mathbf{j} + 0\mathbf{k} \ \ (\text{ft} / \text{s}^2)$$

The magnitude of the acceleration vector is given as:

$$a(t) = |\mathbf{a}(t)| = \sqrt{a_x(t)^2 + a_y(t)^2 + a_z(t)^2}$$

The magnitude of the acceleration vector at 0.25 seconds is:

$$a(0.25) = \sqrt{(-6.135)^2 + (-6.585)^2 + (0)^2} = 9.0 \ (\text{ft} / \text{s}^2)$$

Example 10-4. A particle moves around a horizontal circular track that has a radius of 200 (ft) as shown in Figure 10-5. If the particle starts from rest and increases its (tangential) velocity at a constant rate of 8 (ft / s²), determine the time when the magnitude of the acceleration of the particle is 10 (ft/s²). What is the (tangential) velocity at that time?

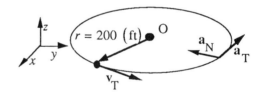

Figure 10-5: Particle traveling
in a circular path

SOLUTION: This problem can be worked in two ways depending on the coordinate system that we choose. The first way the coordinate system will be a path-oriented coordinate system. The acceleration vector in this coordinate system is defined as:

$$\mathbf{a}(t) = a_T(t)\mathbf{u}_T(t) + a_N(t)\mathbf{u}_N(t)$$

or, equivalently,

$$\mathbf{a}(t) = a_T(t)\mathbf{u}_T(t) + \left(\frac{v(t)^2}{r(t)} \right)\mathbf{u}_N(t)$$

Furthermore, we know that the magnitude of the acceleration vector is:

$$a(t) = |\mathbf{a}(t)| = \sqrt{(a_T(t))^2 + \left(\frac{v(t)^2}{r(t)} \right)^2}$$

The only unknown at the time in question in this equation is $v(t)$, and rearrangement gives:

$$v(t) = \left(r(t)^2 \left(a(t)^2 - a_T(t)^2 \right) \right)^{1/4}$$

Substitution gives:

$$v(t) = \left((200)^2 \left((10)^2 - (8)^2 \right) \right)^{1/4} = 34.64 \ (\text{ft} / \text{s})$$

In the tangential direction, the velocity is related to the acceleration as:

$$a_T \mathbf{u}_T = \left(\frac{dv}{dt} \right) \mathbf{u}_T = 8 \mathbf{u}_T \ (\text{ft} / \text{s}^2)$$

Rearranging and integrating gives:

$$\int_{v_o}^{v} dv = \int_{t_o}^{t} 8 dt \quad \text{or} \quad v - v_o = 8(t - t_o)$$

We can assign the beginning time as zero and the particle starts at rest

$$\text{at } t_o = 0 \ (\text{s}), \qquad v_o = 0 \ (\text{ft} / \text{s})$$

Solving for the time gives

$$t = \left(\frac{v}{8} \right) = 4.33 \ (\text{s})$$

The second way to work the problem is to define a cylindrical coordinate system situated at the center of the circle. The position, velocity, and acceleration vectors are

$$\mathbf{r} = r\mathbf{e}_r + z\mathbf{e}_z$$
$$\mathbf{v} = \dot{r}\mathbf{e}_r + r\dot{\theta}\mathbf{e}_\theta + \dot{z}\mathbf{e}_z$$
$$\mathbf{a} = (\ddot{r} - r\dot{\theta}^2)\mathbf{e}_r + (r\ddot{\theta} + 2\dot{r}\dot{\theta})\mathbf{e}_\theta + \ddot{z}\mathbf{e}_z$$

These equations can be simplified by making the following observations. There is no component in the z-direction so the position, velocity and acceleration have no component in the z-direction. The radius of the circle is constant; the time derivatives of r are zero. With these observations the equations reduce to:

$$\mathbf{r} = r\mathbf{e}_r$$
$$\mathbf{v} = r\dot{\theta}\mathbf{e}_\theta$$
$$\mathbf{a} = (-r\dot{\theta}^2)\mathbf{e}_r + (r\ddot{\theta})\mathbf{e}_\theta$$

Now let's look at the acceleration vector more closely,

$$\mathbf{a} = -r\dot{\theta}^2\mathbf{e}_r + r\ddot{\theta}\mathbf{e}_\theta$$

The term $r\ddot{\theta}$ is the acceleration in the θ or transverse direction. Furthermore, the term $r\dot{\theta}$ is the velocity in the θ or transverse direction. If we multiply and divide the $\mathbf{e}_r$ term by the radius r the equation becomes:

$$\mathbf{a} = -\left(\frac{(r\dot{\theta})^2}{r} \right) \mathbf{e}_r + r\ddot{\theta}\mathbf{e}_\theta$$

This equation looks very similar to the equation using the path oriented coordinate system only the base vectors are different and the terms are expressed in different ways; however, this equation when solved for the transverse velocity yields the same results as before.

Furthermore, the relationship between the acceleration and velocity in the θ direction is:

$$a_\theta \mathbf{e}_\theta = \left(\frac{dv_\theta}{dt}\right)\mathbf{e}_\theta = \left(\frac{d(r\dot\theta)}{dt}\right)\mathbf{e}_\theta$$

This is analagous to the expression for the path oriented coordinates, and integrating this expression yields the same answers for the time variable.

10.2.2 Rigid Body Kinematics

In describing the motion of a particle it was sufficient to consider its velocity, position, and acceleration; spin or angular rotation of the particle was not considered because the particle is idealized as having zero volume. For a finite-sized rigid body, however, we must consider its motion in terms of both translation and rotation. This is equivalent to considering the velocity of each of the individual particles comprising the body as was discussed in Chapter 6. The motion of a rigid body can be treated separately as translation of the center of mass (the position, velocity, and acceleration of the center of mass) and rotation of the body with respect to the center of mass. The position, velocity, and acceleration of the center of mass and their interrelations are as described above for particles. The position $_G\mathbf{r}$, velocity $_G\mathbf{v}$, and acceleration $_G\mathbf{a}$ of a rigid body, relative to the center of mass, and in terms of the angular velocity ($\boldsymbol{\omega}$), and angular acceleration ($\boldsymbol{\alpha}$) are summarized below (see chapter 6).

In chapters 5 and 6 we found that the position, velocity and acceleration of a rigid body, as a whole, can be described, for the purposes of linear momentum conservation, in terms of the position and velocity of the center of mass. Consequently:

$$\left(\frac{d(m\mathbf{v})_{sys}}{dt}\right) = m\left(\frac{d\mathbf{v}_G}{dt}\right) = m\mathbf{a}_G$$

Then, $\mathbf{v}_G$ and $\mathbf{a}_G$ can be expressed in terms of $\mathbf{r}_G$ for rectangular cartesian, cylindrical, or spherical coordinates according to Tables 10-1, 2, and 3. For example, in cylindrical coordinates, $\mathbf{v}_G = \dot{r}_G\mathbf{e}_r + r_G\dot\theta_G\mathbf{e}_\theta + \dot{z}_G\mathbf{e}_z$.

Then, to describe the motion of any point within the body (at $_G\mathbf{r}$, relative to the center of mass) we have that:

$$\boxed{\mathbf{r} = \mathbf{r}_G + {}_G\mathbf{r}}$$

(6 − 25)

The velocity is

$$\boxed{\begin{aligned}\mathbf{v} = \left(\frac{d\mathbf{r}}{dt}\right) &= \mathbf{v}_G + {}_G\mathbf{v} \\ &= \mathbf{v}_G + \boldsymbol{\omega} \times {}_G\mathbf{r}\end{aligned}}$$

(10 − 24)

and the acceleration is

$$
\begin{aligned}
\mathbf{a} = \left(\frac{d\mathbf{v}}{dt}\right) &= \mathbf{a}_G + \frac{d}{dt}\left(\boldsymbol{\omega} \times {}_G\mathbf{r}\right) \\
&= \mathbf{a}_G + \boldsymbol{\alpha} \times {}_G\mathbf{r} + \boldsymbol{\omega} \times {}_G\mathbf{v} \\
&= \mathbf{a}_G + \boldsymbol{\alpha} \times {}_G\mathbf{r} + \boldsymbol{\omega} \times (\boldsymbol{\omega} \times {}_G\mathbf{r})
\end{aligned}
$$

(10 − 25)

As an example, the motion of a rigid body is described in terms of a cylindrical coordinate system which is aligned with the axis of rotation. Figure 10-6 depicts the motion. In the cylindrical coordinate system with the z-axis aligned with the vector $\boldsymbol{\omega}$ at the center of mass, the angular velocity and angular acceleration vectors are expressed as:

$$\boldsymbol{\omega} = \dot{\theta}\mathbf{e}_z$$

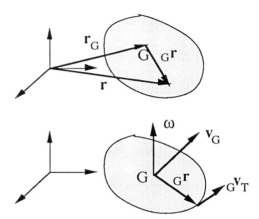

Figure 10-6: Rigid body kinematics in a cylindrical coordinate system

and

$$\boldsymbol{\alpha} = \ddot{\theta}\mathbf{e}_z$$

The position vector from the center of mass to any part of the rigid body is:

$$_G\mathbf{r} = {}_G r\,\mathbf{e}_r + {}_G z\,\mathbf{e}_z$$

The cross product of the angular velocity vector and the relative position vector for the cylindrical coordinate system is:

$$
\begin{aligned}
(\boldsymbol{\omega} \times {}_G\mathbf{r}) &= (\dot{\theta}\mathbf{e}_z) \times ({}_G r\,\mathbf{e}_r + {}_G z\,\mathbf{e}_z) \\
&= {}_G r\dot{\theta}\mathbf{e}_\theta
\end{aligned}
$$

The cross product of the angular acceleration vector and the relative position vector for the cylindrical coordinate system is calculated in the same way.

$$
\begin{aligned}
(\boldsymbol{\alpha} \times {}_G\mathbf{r}) &= (\ddot{\theta}\mathbf{e}_z) \times ({}_G r\,\mathbf{e}_r + {}_G z\,\mathbf{e}_z) \\
&= {}_G r\ddot{\theta}\mathbf{e}_\theta
\end{aligned}
$$

The final vector product in the acceleration expression is calculated for a cylindrical coordinate system:

$$
\begin{aligned}
\boldsymbol{\omega} \times (\boldsymbol{\omega} \times {}_G\mathbf{r}) &= (\dot{\theta}\mathbf{e}_z) \times ({}_G r\dot{\theta}\mathbf{e}_\theta) \\
&= -{}_G r\dot{\theta}^2\mathbf{e}_r
\end{aligned}
$$

Now the velocity and the acceleration of a point on the rigid body at $_G\mathbf{r}$ with respect to the center of mass can be expressed in the cylindrical coordinate system, with the axis of rotation defined in the z-direction:

$$\mathbf{v} = \mathbf{v}_G + \boldsymbol{\omega} \times {}_G\mathbf{r}$$
$$= \left(\dot{r}_G\mathbf{e}_r + r_G\dot{\theta}\mathbf{e}_\theta + \dot{z}_G\mathbf{e}_z \right) + {}_Gr\dot{\theta}\mathbf{e}_\theta \tag{10-26}$$

and

$$\mathbf{a} = \mathbf{a}_G + \boldsymbol{\alpha} \times {}_G\mathbf{r} + \boldsymbol{\omega} \times \boldsymbol{\omega} \times {}_G\mathbf{r}$$
$$= \left((\ddot{r}_G - r_G\dot{\theta}^2)\mathbf{e}_r + (r_G\ddot{\theta} + 2\dot{r}_G\dot{\theta})\mathbf{e}_\theta + \ddot{z}_G\mathbf{e}_z \right) + {}_Gr\ddot{\theta}\mathbf{e}_\theta - {}_Gr\dot{\theta}^2\mathbf{e}_r \tag{10-27}$$

Note that $_G\mathbf{v}$ and $_G\mathbf{a}$ can be obtained from Table 10-2 for $_G\dot{\mathbf{r}}$, because the body is rigid.

Example 10-5. A satellite is launched from the space shuttle with a constant angular velocity of 2 rad/s and translational velocity upward along the axis at a constant rate of 5 m/s (Figure 10-7). (What is the purpose of the rotation?) Express the velocity and acceleration of point A in terms of cylindrical components.

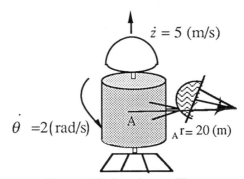

Figure 10-7: Rotating satellite

SOLUTION: We choose a cylindrical coordinate system with the coordinates along the axis of the satellite. For Point A, the velocity vector is given by equation 10-26 with motion of the center of mass only in the z-direction:

$$\mathbf{v} = {}_Gr\dot{\theta}\mathbf{e}_\theta + \dot{z}\mathbf{e}_z = (20)(2)\mathbf{e}_\theta + (5)\mathbf{e}_z = 40\mathbf{e}_\theta + 5\mathbf{e}_z \text{ (m / s)}$$

The acceleration of the center of mass is zero since $\mathbf{v}_G$ is constant, and the angular acceleration of the satellite is zero since the angular velocity is constant, $\boldsymbol{\alpha} = 0$. Hence equation (10-27) gives that the acceleration vector for the point at $(_Gr, \theta, _Gz)$ is:

$$\mathbf{a} = -{}_Gr\dot{\theta}^2\mathbf{e}_r = -(20)(2)^2\mathbf{e}_r = -80\mathbf{e}_r \text{ (m / s}^2)$$

Example 10-6. A disk is initially at rest. If a point at the outer radius of the disk undergoes a transverse acceleration as a function of time of $a_T = 2t$ (m / s^2) where t is in seconds, determine (a) the angular velocity of the disk as a function of time and (b) the angular position of the disk in radians as a function of time. A schematic of the disk is shown in Figure 10-8.

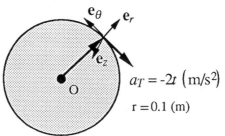

Figure 10-8: Rotating disk

SOLUTION: A cylindrical coordinate system is defined about the center of the disk. The velocity and hence the acceleration of the center of mass is zero so that the acceleration vector is given by (equation 10-27):

$$\mathbf{a} = {}_G r \ddot{\theta} \mathbf{e}_\theta - {}_G r \dot{\theta}^2 \mathbf{e}_r$$

where ${}_G r \ddot{\theta}$ $(= {}_G r \alpha)$ is the transverse (θ) acceleration component. Hence

$$-2t = {}_G r \alpha \qquad \text{or} \qquad \alpha = \frac{-2t}{{}_G r} = -20t \ (\text{rad}/\text{s}^2)$$

In vector form:

$$\boldsymbol{\alpha} = -20t \mathbf{e}_z$$

The kinematic relation between the angular acceleration and the angular velocity is:

$$\alpha \mathbf{e}_z = \left(\frac{d\omega}{dt}\right) \mathbf{e}_z = -20t \mathbf{e}_z$$

Rearranging and integrating gives:

$$\int_{\omega_o}^{\omega} d\omega = \int_{t_o}^{t} -20t \, dt$$

Since the body is initially at rest ($\omega_o = 0$ (rad/s)), and defining the beginning time period to be zero ($t_o = 0$ (s)) gives

$$\omega = -10t^2 \ (\text{rad}/\text{s}) \qquad \text{or} \qquad \omega \mathbf{e}_z = -10t^2 \mathbf{e}_z \ (\text{rad}/\text{s})$$

The kinematic relation between the angle θ and the angular velocity ω is:

$$\omega = \left(\frac{d\theta}{dt}\right) = -10t^2$$

Rearranging and integrating gives:

$$\int_{\theta_o}^{\theta} d\theta = \int_{t_o}^{t} -10t^2 \, dt$$

The initial angle (at $t_o = 0$) is assumed to be zero ($\theta_o = 0$ (rad)) giving the expression for the angular position:

$$\theta = \frac{-10}{3} t^3 \ (\text{rad})$$

Example 10-7. Disk A, as shown in Figure 10-9, starts from rest and rotates with a constant angular acceleration of $\alpha_A = -3 \ \text{rad}/\text{s}^2$. If disk A is in contact with disk B and no slipping occurs (the velocity and acceleration at point P is always equal to the velocity and acceleration at P'), calculate the angular velocity and angular acceleration of disk B after A has completed 5 revolutions.

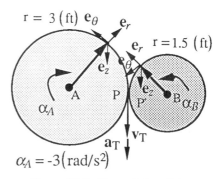

Figure 10-9: Two rotating disks

SOLUTION: First we will look at the motion of disk A and define a cylindrical coordinate system situated at the center of the disk. The angular acceleration vector is:

$$\alpha e_z = -3e_z \ (\text{rad} \ / \ \text{s}^2)$$

In the z-direction, the kinematic relationship is:

$$\alpha = \left(\frac{d\omega}{dt}\right) = -3$$

Rearranging and integrating gives:

$$\int_{\omega_o}^{\omega} d\omega = \int_{t_o}^{t} -3dt$$

The lower limits of both the integrals are zero because the body starts at rest which gives:

$$\omega_A = -3t \ (\text{rad} \ / \ \text{s})$$

Since we know that disk A completed 5 revolutions, let us derive the expression for the angular position as a function of time from the kinematic relationship in the z-direction :

$$\omega = \left(\frac{d\theta}{dt}\right) = -3t$$

Rearranging an integrating yields:

$$\int_{\theta_o}^{\theta} d\theta = \int_{t_o}^{t} -3t dt$$

The lower limit of the angular position is defined as zero giving:

$$\theta_A = \left(\frac{-3t^2}{2}\right) \ (\text{rad})$$

solving for the time yields:

$$t = \sqrt{\frac{-2\theta_A}{3}}$$

From the defined coordinate system the angular displacement is negative or:

$$t = \sqrt{\frac{-2(-5(2\pi))}{3}} = 4.58 \ (\text{s})$$

The angular velocity of disk A after 5 revolutions is:

$$\omega_A = (-3)(4.58) = -13.73 \ (\text{rad} \ / \ \text{s})$$

The velocity at point P is equal to the velocity at point P' and the acceleration at point P is equal to the acceleration at point P'. To express the velocity and acceleration at P' we define another cylindrical coordinate system, this time situated at the center of disk B. Notice that the coordinate systems are different. Furthermore, $v_\theta = r\dot\theta = r\omega$ so that

$$-_A r \omega_A = {_B}r\omega_B$$

The negative sign is necessary because of the different coordinate systems. Hence,

$$\omega_B = -\left(\frac{_Ar}{_Br}\right)\omega_A = -\left(\frac{3}{1.5}\right)(-13.74) = 27.46 \ (\text{rad / s})$$

In vector form:

$$\boldsymbol{\omega}_B = 27.46\mathbf{e}_z \ (\text{rad / s})$$

Likewise for the angular acceleration equation 10-27 with $\mathbf{a}_G = 0$ and $_G\ddot{\mathbf{r}} = 0$

$$-_Ar\alpha_A = {_Br}\alpha_B$$

Solving for the unknown angular acceleration:

$$\alpha_B = -\left(\frac{_Ar}{_Br}\right)\alpha_A = -\left(\frac{3}{1.5}\right)(-3) = 6 \ (\text{rad / s}^2)$$

In vector form:

$$\boldsymbol{\alpha}_B = 6\mathbf{e}_z \ (\text{rad / s}^2)$$

10.2.3 Relative Motion of Two Objects

Frequently it is useful and informative to consider the motion of one body relative to that of another moving body. To do so, the vector representations of each of the bodies separately with respect to a fixed coordinate system is an extremely convenient approach. Figure 10-10 shows the relative positions of two points A and B. If the position of body A is described by vector $\mathbf{r}_A$ and the position of body B is described by vector $\mathbf{r}_B$, then the position of point B with respect to A ($_A\mathbf{r}_B$) is:

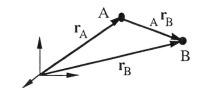

Figure 10-10: Relative position vectors

$$\boxed{_A\mathbf{r}_B = \mathbf{r}_B - \mathbf{r}_A}$$

$(10-28)$

Accordingly, their relative velocities can be described as the difference of their two velocities:

$$\boxed{_A\mathbf{v}_B = \left(\frac{d_A\mathbf{r}_B}{dt}\right) = \mathbf{v}_B - \mathbf{v}_A}$$

$(10-29)$

and similarly, their relative acceleration is simply the difference of their two individual accelerations:

$$_A\mathbf{a}_B = \left(\frac{d_A\mathbf{v}_B}{dt}\right) = \mathbf{a}_B - \mathbf{a}_A$$

(10 − 30)

In this way the relative motion of the two moving bodies can be described in a straightforward manner.

Example 10-8. Car A, traveling at a constant velocity of 50 (mph) in the positive $\mathbf{i}$ direction, crosses over car B traveling at a constant speed of 40 (mph) as shown in Figure 10-11. Find the relative velocity of car A with respect to car B.

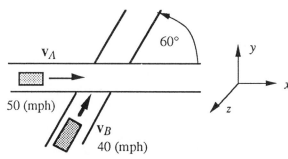

Figure 10-11: Two cars

SOLUTION: A cartesian coordinate system is used. The velocities of the cars are:

$$\mathbf{v}_A = 50\mathbf{i} + 0\mathbf{j} + 0\mathbf{k} \text{ (mph)}$$
$$\mathbf{v}_B = 40\cos(60°)\mathbf{i} + 40\sin(60°)\mathbf{j} + 0\mathbf{k} \text{ (mph)}$$

and the velocity of car A relative to car B is (equation 10-29):

$$_B\mathbf{v}_A = \mathbf{v}_A - \mathbf{v}_B$$
$$= (50\mathbf{i} + 0\mathbf{j} + 0\mathbf{k}) - \left(40\cos(60°)\mathbf{i} + 40\sin(60°)\mathbf{j} + 0\mathbf{k}\right)$$
$$= 30\mathbf{i} - 34.6\mathbf{j} + 0\mathbf{k}$$

and the distance between them is changing at a rate of 45.8 mph ($|_B\mathbf{v}_A|$).

10.2.4 Interdependent Motion of Two Particles

There are many problems in mechanics which involve the motion of two objects which are connected together by a rope or cable but which may move as part of a machine. For example, a block and tackle involves using a rope wound around a number of pulleys to suspend and move a heavy object. Mechanical leverage is gained in this kind of machine because the heavy object is moved a relatively short distance as the person or machine pulling at the other end of the rope moves through a longer distance. However, the motion of the two objects is related in a very definite way according to the motion of the movable pulleys which are part of the apparatus. Describing the relative displacements or the weights or multiple weights or bodies which are part of the configuration is the subject of kinematics and depends upon the arrangement of the apparatus.

Analyzing the relative motion of the objects related through a displacement and motion of the cord is facilitated by recognizing that the total length of cord as the displacements occur is constant, even though the lengths of the various portions of the cord may change due to the motion of the objects. Ultimately, as one section becomes shorter, another section must become longer by the same amount. Consequently, by adding up the length of all the sections of the cords which are changing in length, a constant total length is obtained. Differentiating this equation then gives the relative rates of change of the lengths of the various sections of the cord.

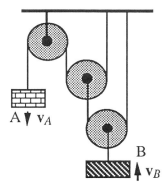

As an example, consider the cord and pulley arrangement of Figure 10-12. As block A moves, so does block B, although at a different speed, because of the cord and pulley arrangements. In dividing the cord into lengths which are changing, one need not consider any length which is constant such as the loop around each pulley, or any length of cord which is stationary such as that fixed to a support or point in space.

Figure 10-12: Pulley arrangement

By adding up all of the variable lengths which total a constant, one can then differentiate to obtain the relative speeds of the connected objects. This equation relating the speeds of the two bodies is used as a constraining relation, that is, a relation which provides the interdependency of the motion of the two objects for use in their two equations of motion. We can consider the two equations of motion separately; however, we must take into account this interdependence, or constraint, which relates their motions.

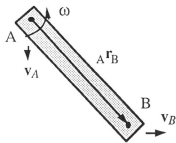

A second type of constraint relates any two points on a rigid body having angular velocity ω. The relative velocities for points of a rigid body can be expressed in terms of the relative position and the angular velocity with respect to that position as shown in Figure 10-13. Examples showing the determination of these interrelations follow.

Figure 10-13: Relative velocity
with angular velocity

$$_A\mathbf{v}_B = \mathbf{v}_B - \mathbf{v}_A = \omega \times {_A}\mathbf{r}_B$$

Example 10-9. Determine the velocity in the y-direction of block A as shown in Figure 10-14.

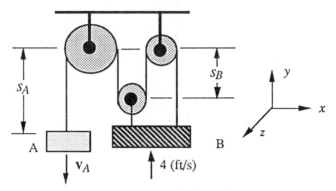

Figure 10-14: Pulleys and weight

SOLUTION: We will use a cartesian coordinate system as given in the figure. The velocity of block A is dependent on the velocity of block B. However, we know that the total length of the cord is constant. The total length of the cord is expressed as the sum of all the parts or:

$$l = s_A + 3s_B + k$$

where k is the total length of cord about the pulleys and also is a constant. Now we can differentiate the total length of the rope with respect to time or

$$\left(\frac{dl}{dt}\right) = \left(\frac{ds_A}{dt}\right) + \left(\frac{d3s_B}{dt}\right) + \left(\frac{dk}{dt}\right)$$

$$0 = v_A + 3v_B + 0$$

This equation relates the velocity of block B in the y-direction with the velocity of block A in the y-direction:

$$v_A\mathbf{j} = -3v_B\mathbf{j}$$

Substituting the known value gives:

$$v_A\mathbf{j} = -3(4)\mathbf{j}$$
$$= -12\mathbf{j} \text{ (ft / s)}$$

Example 10-10. The link shown in Figure 10-15 is guided by two blocks at A and B which move in the fixed slots. At the instant shown, the velocity of A is 4 m/s in the negative direction. Determine the velocity of B at this instant.

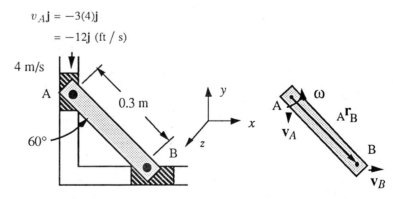

Figure 10-15: Motion of a link and free body diagram

SOLUTION: The cartesian coordinate is defined and the motion of the system is isolated as given in Figure 10-15. The velocities of block A and B are expressed as the vectors:

$$\mathbf{v}_A = 0\mathbf{i} - 4\mathbf{j} + 0\mathbf{k} \ (m/s)$$

$$\mathbf{v}_B = v_{xB}\mathbf{i} + 0\mathbf{j} + 0\mathbf{k} \ (m/s)$$

The unknown that must be determined is v_{xB} We can relate the velocity of A to the B by the relative velocity

$$\mathbf{v}_B = \mathbf{v}_A + {}_A\mathbf{v}_B$$

Furthermore, we can express the relative velocity of B with respect to A in terms of the cross product of the angular velocity vector and the relative position vector.

$$_A\mathbf{v}_B = \boldsymbol{\omega} \times {}_A\mathbf{r}_B$$

The relative position vector is:

$$_A\mathbf{r}_B = 0.3\sin(60°)\mathbf{i} - 0.3\cos(60°)\mathbf{j} + 0\mathbf{k} \ (m)$$

And the angular velocity vector is:

$$\boldsymbol{\omega} = 0\mathbf{i} + 0\mathbf{j} + \omega\mathbf{k} \ (rad/s)$$

The expression of the relative velocity is found after evaluating the cross products:

$$_A\mathbf{v}_B = (0\mathbf{i} + 0\mathbf{j} + \omega\mathbf{k}) \times (0.260\mathbf{i} - 0.150\mathbf{j} + 0\mathbf{k})$$

$$= 0.150\omega\mathbf{i} + 0.260\omega\mathbf{j} + 0\mathbf{k}$$

Substituting this expression into the equation relating the velocity of B to A gives:

$$\mathbf{v}_B = \mathbf{v}_A + {}_A\mathbf{v}_B$$

$$v_{xB}\mathbf{i} + 0\mathbf{j} + 0\mathbf{k} = 0\mathbf{i} - 4\mathbf{j} + 0\mathbf{k} + 0.150\omega\mathbf{i} + 0.260\omega\mathbf{j} + 0\mathbf{k}$$

From the **j** direction components,

$$0 = -4 + 0.260\omega \qquad or \qquad \omega = 15.4 \ (rad/s)$$

and from the **i** direction components,

$$v_{xB} = 0.150\omega = 0.150(15.4) = 2.3 \ (m/s)$$

The **k** direction equation is trivial.

In vector form, the velocity of block B is:

$$\mathbf{v}_B = 2.3\mathbf{i} + 0\mathbf{j} + 0\mathbf{k} \ (m/s)$$

Example 10-11. A car at point A is used to hoist a block B as shown in Figure 10-16. Determine the velocity and acceleration of the block when it reaches the elevation of 5 (m). The total length of the rope is 40 (m) and passes over a small pulley at D. The car is traveling in the x-direction at a velocity of 0.7 m/s

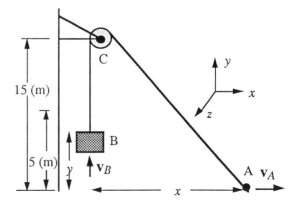

Figure 10-16: Motion of a block

SOLUTION: The cartesian coordinate system is defined in the figure. The total length of the rope is a constant of 40 (m) and can be expressed as:

$$l_{tot} = l_{AC} + l_{CB} = 40 \text{ (m)}$$

Furthermore, the length l_{CB} can be expressed in terms of the variable y by the following equation:

$$l_{CB} = 15 - y$$

The length l_{AC} can be expressed in terms of the variable x by the Pythagorean theorem.

$$l_{AC} = \sqrt{(15)^2 + (x)^2}$$

Now the total length of the rope can be related to the variables x and y as:

$$l_{tot} = \sqrt{(15)^2 + (x)^2} + 15 - y = 40 \text{ (m)}$$

Consequently,

$$y = \sqrt{225 + x^2} - 25 \text{ (m)}$$

The velocity of the block by the chain rule in the **j** direction is:

$$v_B = \left(\frac{dy}{dt}\right) = \left(\frac{dy}{dx}\right)\left(\frac{dx}{dt}\right)$$

The expression for the derivative of the variable y with respect to x is:

$$\left(\frac{dy}{dx}\right) = \left(\frac{x}{\sqrt{225 + x^2}}\right)$$

and the expression of the derivative of x with respect to t is the velocity in the **i** direction:

$$\left(\frac{dx}{dt}\right) = v_A = 0.7 \text{ (m / s)}$$

Substituting in the expression for the velocity of B gives:

$$v_B = \left(\frac{x}{\sqrt{225 + x^2}}\right) v_A$$

Evaluating this expression at y equal 5 (m) and x equal 26 (m) (from the relationship between y and x) gives:

$$v_B = \left(\frac{(26)}{\sqrt{225 + (26)^2}}\right)(0.7) = 0.607 \text{ (m / s)}$$

The expression for the acceleration of the block B in the **j** direction as given as the time derivative of the velocity:

$$a_B = \left(\frac{dv_B}{dt}\right) = \left(\frac{dv_B}{dx}\right)\left(\frac{dx}{dt}\right)$$

The derivative of the velocity of the block with respect to the variable x is:

$$\left(\frac{dv_B}{dx}\right) = \left(\left(\frac{1}{(225 + x^2)^{1/2}}\right) - \left(\frac{x^2}{(225 + x^2)^{3/2}}\right)\right) v_A = \left(\frac{225 v_A}{(225 + x^2)^{3/2}}\right)$$

And the acceleration is:

$$a_B = \left(\frac{225 v_A^2}{(225 + x^2)^{3/2}}\right)$$

Evaluating this expression at x equal 26 (m) gives:

$$a_B = \left(\frac{225(0.7)^2}{(225 + (26)^2)^{3/2}}\right) = 0.0058 \; (\text{m} / \text{s}^2)$$

10.3 PARTICLE DYNAMICS

10.3.1 Momentum Analysis

You now have the background required for stating the laws which govern the motion of a particle and describing this motion mathematically. The physics of the motion, for a single particle, is contained in the equation of linear momentum (Newton's Second Law of Motion) and the mathematics for describing the motion is contained in the kinematical relationships for position, velocity, and acceleration in terms of the coordinates of the particle in a coordinate system and their time derivatives and the base vectors for that coordinate system.

The problem now can be stated quite simply: the three scalar equations of linear momentum, embodied in the vector statement of conservation of linear momentum, provide equations relating forces to the acceleration (and through kinematic relations, to position and velocity) of the particle:

$$
\left(\frac{d\mathbf{p}_{sys}}{dt}\right) = \sum \mathbf{f}_{ext}
$$
$$
m_{sys}\left(\frac{d\mathbf{v}_G}{dt}\right) = \sum \mathbf{f}_{ext} \tag{5 – 9}
$$
$$
m_{sys}\mathbf{a}_G = \sum \mathbf{f}_{ext}
$$

If we know these forces that are exerted upon the particle as a function of time, then we have three ordinary differential equations to solve to obtain the position or velocity of that particle as a function of time. With the mathematical statement of the physics in hand, it then becomes a problem of applying your knowledge of mathematics, kinematics, and calculus to carry out that solution.

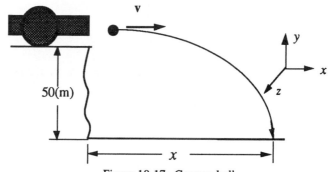

Figure 10-17: Cannon ball

Example 10-12. A cannon ball is fired from point A with a horizontal muzzle velocity of 100 (m/s), as shown in Figure 10-17. The ball has a mass of 60 (kg). The cannon is located at an elevation of 50 (m) above the ground.

a) If we can neglect the forces due to air resistance, determine the time for the cannon ball to strike the ground, the horizontal distance x and the magnitude of the velocity at impact.

b) If the force due to the air resistance is: $\mathbf{f}_D = -k\mathbf{v}$ (the drag force is proportional to the velocity of the body but acting in the opposite direction) where the drag coefficient is: $k = 30$ kg / m s determine the time for the cannon ball to strike the ground, the horizontal distance x and the magnitude of the velocity at the impact.

SOLUTION We choose the system to be the cannon ball and a cartesian coordinate system is defined with the origin at the initial position of the ball. This free body diagram is shown in Figure 10-17a. Since the system is the ball, the mass of the system is constant and the conservation of linear momentum for this system is given in Table 10-5.

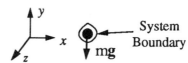

Figure 10-17a: Cannon ball
free body diagram

Table 10-5: Data for Example 10-12a (Linear Momentum, System = Cannon ball)

Description	$\left(\dfrac{d(m\mathbf{v})_{sys}}{dt}\right)$ (N)	=	$\sum \mathbf{f}_{ext}$ $m\mathbf{g}$ (N)
x-direction	$m\left(\dfrac{dv_x}{dt}\right)\mathbf{i}$	=	$0\mathbf{i}$
y-direction	$m\left(\dfrac{dv_y}{dt}\right)\mathbf{j}$	=	$(-mg)\mathbf{j}$
z-direction	$m\left(\dfrac{dv_z}{dt}\right)\mathbf{k}$	=	$0\mathbf{k}$

Each equation in Table 10-5 can be integrated separately and solved, in turn, for v_x, v_y, and v_z (given the velocity components at the muzzle as initial conditions) and then integrated a second time to give the position. From the x-direction equation

$$\int_{v_{ox}}^{v_x} dv_x = \int_{t_o}^{t} 0\,dt \qquad \text{or} \qquad v_x - v_{ox} = 0$$

(v_{ox} is the muzzle velocity). The horizontal component of velocity is constant.

From the y-direction equation,

$$m\int_{v_{oy}}^{v_y} dv_y = -\int_{t_o}^{t} mg\,dt \qquad \text{or} \qquad v_y - v_{oy} = -gt$$

by letting $t_o = 0$. The rate of fall increases linearly with time due to the gravitational force.

The z-direction equation integrates just like the x-direction equation and so $v_z = v_{oz}$. Using the initial velocity $\mathbf{v}_o = 100\mathbf{i} + 0\mathbf{j} + 0\mathbf{k}$ (m/s) gives the x, y, and z components of the velocity as

$$v_x = 100 \ (\text{m}/\text{s})$$
$$v_y = -9.8t \ (\text{m}/\text{s})$$
$$v_z = 0$$

or in vector form $\mathbf{v} = 100\mathbf{i} - 9.8t0\mathbf{j} + 0\mathbf{k}$ (m/s).

Position as a function of time can be found from the kinematic expressions for velocity: $v_x = \left(\dfrac{dx}{dt}\right)$, $v_y = \left(\dfrac{dy}{dt}\right)$, and $v_z = \left(\dfrac{dz}{dt}\right)$. Using the initial position vector $\mathbf{r}_o = 0\mathbf{i} + 0\mathbf{j} + 0\mathbf{k}$ (m) gives the position vector as a function of time as:

$$\mathbf{r} = 100t\mathbf{i} - \frac{9.8}{2}t^2\mathbf{j} + 0\mathbf{k} \ (\text{m})$$

Now, if we look at the y component of the position vector, we know that at the end of the time period $y = -50$ (m).

$$y\mathbf{j} = -4.9t^2\mathbf{j}$$

or

$$t = \sqrt{\left(\frac{y}{-4.9}\right)} = \sqrt{\left(\frac{-50}{-4.9}\right)} = 3.19 \ (\text{s})$$

Now, we can use the time to determine the x-component of the position vector or the horizontal distance travelled:

$$x = 100t = 100(3.19) = 319 \ (\text{m})$$

The final position vector is:

$$\mathbf{r}(3.19) = 319\mathbf{i} - 50\mathbf{j} + 0\mathbf{k} \ (\text{m})$$

The magnitude of the velocity at the impact is:

$$\mathbf{v} = \sqrt{(100)^2 + (9.8 \cdot 3.19)^2 + (0)^2} = 104.8 \ (\text{m}/\text{s})$$

Now we will consider the problem when there is a drag force acting on the cannon ball as it is traveling through the air. The free-body diagram for this situation with the system as the cannon ball is shown in Figure 10-17b. The conservation of linear momentum for this system is given in Table 10-6.

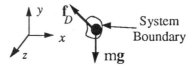

Figure 10-17b: Cannon ball free body diagram

Table 10-6: Data for Example 10-12b (Linear Momentum, System = Cannon ball)

Description	$\left(\dfrac{d(m\mathbf{v})_{sys}}{dt}\right)$ (N)	=	$\sum \mathbf{f}_{ext}$ mg (N)		$\mathbf{f}_D$ (N)
x-direction	$m\left(\dfrac{dv_x}{dt}\right)\mathbf{i}$	=	$0\mathbf{i}$	$-$	$kv_x\mathbf{i}$
y-direction	$m\left(\dfrac{dv_y}{dt}\right)\mathbf{j}$	=	$(-mg)\mathbf{j}$	$-$	$kv_y\mathbf{j}$
z-direction	$m\left(\dfrac{dv_z}{dt}\right)\mathbf{k}$	=	$0\mathbf{k}$	$-$	$kv_z\mathbf{k}$

Notice that the drag force acts in the opposite direction of the motion of the ball and hence the negative signs. Furthermore, just like the problem with no drag, each of the x, y, and z equations are independent. Each of the equations can be integrated independently. From the table, the x-direction equation gives:

$$m\left(\frac{dv_x}{dt}\right) = -kv_x$$

This equation can be separated and rearranged to yield:

$$\frac{dv_x}{v_x} = -\left(\frac{k}{m}\right)dt$$

Integrating both sides of the equation with the initial conditions $t_o = 0$ s, $v_{xo} = 100$ m / s gives:

$$\int_{v_{xo}}^{v_x} \frac{dv_x}{v_x} = \int_{t_o}^{t} -\left(\frac{k}{m}\right)dt$$

or:

$$\ln\left(\frac{v_x}{v_{xo}}\right) = -\left(\frac{k}{m}\right)t$$

Rearranging, the velocity in the x direction as a function of time is:

$$v_x = v_{xo}e^{-(k/m)t} \quad \text{(m / s)}$$

The position in the x-direction is determined from the definition of the velocity in the x-direction or:

$$v_x = \left(\frac{dx}{dt}\right)$$

Substituting the expression for the velocity in the x-direction gives:

$$v_{xo}e^{-(k/m)t} = \left(\frac{dx}{dt}\right)$$

This equation can also be separated and rearranged to give:

$$v_{xo}e^{-(k/m)t}dt = dx$$

Integrating both sides of the equation with the initial conditions $t_o = 0$ s, $x_o = 0$ m gives:

$$\int_{t_o}^{t}\left(v_{xo}e^{-(k/m)t}\right)dt = \int_{x_o}^{x}dx$$

Integrating and evaluating the limits of integration yields:

$$\left(\frac{v_{xo}m}{k}\right)\left(1 - e^{-(k/m)t}\right) = x - x_o$$

Rearranging the equation gives the position in the x-direction as a function of time:

$$x = \left(\frac{v_{xo}m}{k}\right)\left(1 - e^{-(k/m)t}\right) \text{ (m)}$$

  Now the velocity and position in the y-direction will be determined by the exact same procedure. The conservation of linear momentum in the y direction gives:

$$m\left(\frac{dv_y}{dt}\right) = -mg - kv_y$$

This equation can be separated and rearranged to:

$$-dt = \frac{dv_y}{g + (k/m)v_y}$$

This equation is integrated with the initial conditions $t_o = 0$ s, $v_{yo} = 0$ m / s as:

$$\int_{t_o}^{t} -dt = \int_{v_{yo}}^{v_y} \frac{dv_y}{g + (k/m)v_y}$$

Integrating and evaluating the integrals gives:

$$-t + t_o = \left(\frac{m}{k}\right)\ln\left(\frac{g + (k/m)v_y}{g + (k/m)v_{yo}}\right)$$

Since both of the initial conditions are zero the substitution yields:

$$-t = \left(\frac{m}{k}\right) \ln\left(\frac{g + (k/m)v_y}{g}\right)$$

Exponentiating both sides of the equation and rearranging gives the velocity in the y-direction as a function of time:

$$v_y = -\left(\frac{mg}{k}\right)\left(1 - e^{-(k/m)t}\right) \text{ (m/s)}$$

The position in the y-direction is calculated from the definition of the velocity in the y-direction as:

$$v_y = \left(\frac{dy}{dt}\right)$$

Substituting the expression for the velocity in the y direction into the equation gives:

$$\left(\frac{mg}{k}\right)\left(1 - e^{-(k/m)t}\right) = \left(\frac{dy}{dt}\right)$$

This equation can be separated and rearranged as:

$$\left(\frac{mg}{k}\right)\left(1 - e^{-(k/m)t}\right) dt = dy$$

Integrating both sides of the equation with the initial conditions $t_o = 0$ s, $y_o = 0$ m gives:

$$\int_{t_o}^{t} \left(\frac{mg}{k}\right)\left(1 - e^{-(k/m)t}\right) dt = \int_{y_o}^{y} dy$$

Integrating the expression and evaluating the expression at the limits of integration gives:

$$y = \left(\frac{m^2 g}{k^2}\right)\left(1 - e^{-(k/m)t}\right) - \left(\frac{mg}{k}\right)t \text{ (m)}$$

The velocity in the z-direction is trivial. As before, $v_{oz} = 0$ and there is no force in the z-direction and so $v_z = 0$ throughout the ball's flight. Now that we have derived all of the components of position and velocity as functions of time we need to solve for the time (t) when the position in the y-direction is -50 (m) or

$$-50 = \left(\frac{m^2 g}{k^2}\right)\left(1 - e^{-(k/m)t}\right) - \left(\frac{mg}{k}\right)t$$

This is a non-linear function of t and does not have an explicit solution. A trial and error process either using a bisection method, incremental search, or successive substitution is necessary to solve for the time. However, we know the time that is takes for the cannon ball to reach the ground with the drag force must be greater than the time required to reach the ground in the absence of air. So we know that the time must be greater than 3.19 (s). Evaluating the constants gives:

$$\left(\frac{m^2 g}{k^2}\right) = 39.2 \text{ (m)}$$

$$\left(\frac{k}{m}\right) = 0.5 \text{ (1 / s)}$$

$$\left(\frac{mg}{k}\right) = 19.6 \text{ (1 / s)}$$

And the root of the equation by trial and error is:

$$t = 4.3(s)$$

Now we can substitute the time into the position in the x-direction and calculate the horizontal distance that was traveled:

$$x(4.3) = 176 \text{ (m)}$$

And the components of the velocity at the impact are calculated as:

$$v_x(4.3) = 11.6 \text{ (m / s)} \qquad v_y(4.3) = -17.3 \text{ (m / s)}$$

and finally the magnitude of the velocity at the impact is:

$$|\mathbf{v}(4.3)| = \sqrt{(11.6)^2 + (-17.3)^2} = 20.8 \text{ (m / s)}$$

The following table compares the components of the velocity and position as functions of time for the case with no air resistance and with air resistance proportional to the velocity of the ball. The time, x-position at impact, components of the velocity at impact, and magnitude of the velocity at impact are also compared.

Description	No drag	Drag proportional to the velocity
v_x	v_{xo}	$v_{xo} e^{-(k/m)t}$
v_y	$-mgt$	$\left(\dfrac{mg}{k}\right)\left(1 - e^{-(k/m)t}\right)$
x	$v_{xo} t$	$\left(\dfrac{v_{xo} m}{k}\right)\left(1 - e^{-(k/m)t}\right)$
y	$-\left(\dfrac{mg}{2}\right) t^2$	$\left(\dfrac{m^2 g}{k^2}\right)\left(1 - e^{-(k/m)t}\right) - \left(\dfrac{mg}{k}\right) t$
Impact t	3.19 (s)	4.3 (s)
Impact x	319 (m)	176 (m)
Impact v_x	100 (m/s)	11.6 (m/s)
Impact v_y	−31.3 (m/s)	−17.3 (m/s)
Impact v	104.8 (m/s)	20.8 (m/s)

Even though the equations for the velocity and position look different, the method of arriving at the equations was the same. First the conservation of linear momentum equations were integrated with the initial conditions to obtain the velocities, and then from the definition of linear velocity, the position equations were integrated with the initial conditions. It does not matter how many forces are acting on the body or in what direction those forces are acting, the philosophy and solution strategy will be exactly the same for a rigid particle.

Example 10-13. From experimental measurements, the position of a 5 (lb_m) particle is defined in terms of its cylindrical coordinates by the parametric equations i.e., r, θ, z, expressed in terms of a single parameter, t): $r = (t^2+3t)$ ft, $\theta = (2t+3)$ rad, and $z = (t^3+4)$ ft where t is in seconds. Determine the magnitude of the resultant force acting on the particle at 2 seconds. (The forces cause the motion.) Figure 10-18 gives a schematic of the particle's motion.

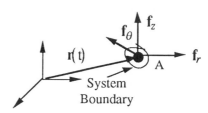

Figure 10-18: Motion of a particle

SOLUTION: Choose the particle to be the system and a cylindrical coordinate system is defined about the particle. The conservation of linear momentum equation is presented in Table 10-7.

Table 10-7: Data for Example 10-13 (Linear Momentum, System = Particle)

Description	$\left(\dfrac{d(m\mathbf{v})_{sys}}{dt}\right)$ (lb_f)	=	$\sum \mathbf{f}_{ext}$ $\mathbf{f}$ (lb_f)
r-direction	$m\left(\dfrac{dv_r}{dt}\right)\mathbf{e}_r$	=	$f_r \mathbf{e}_r$
θ-direction	$m\left(\dfrac{dv_\theta}{dt}\right)\mathbf{e}_\theta$	=	$f_\theta \mathbf{e}_\theta$
z-direction	$m\left(\dfrac{dv_z}{dt}\right)\mathbf{e}_z$	=	$f_z \mathbf{e}_z$

The components of the acceleration are given by the following kinematical relationships (Table 10-2):

$$a_r = \ddot{r} - \dot{r}\dot{\theta}^2$$
$$a_\theta = r\ddot{\theta} + 2\dot{r}\dot{\theta}$$
$$a_z = \ddot{z}$$

The function and time derivatives are evaluated from the given functions.

$$r = t^2 + 3t \qquad\qquad \theta = 2t + 3 \qquad\qquad z = t^3 + 4$$
$$\dot{r} = 2t + 3 \qquad\qquad \dot{\theta} = 2 \qquad\qquad \dot{z} = 3t^2$$
$$\ddot{r} = 2 \qquad\qquad\quad \ddot{\theta} = 0 \qquad\qquad \ddot{z} = 6t$$

Evaluating all the functions and derivative at 2 seconds gives:

$$r|_{t=2} = 10 \qquad\qquad \theta|_{t=2} = 7 \qquad\qquad z|_{t=2} = 12$$
$$\dot{r}|_{t=2} = 7 \qquad\qquad \dot{\theta}|_{t=2} = 2 \qquad\qquad \dot{z}|_{t=2} = 12$$
$$\ddot{r}|_{t=2} = 2 \qquad\qquad \ddot{\theta}|_{t=2} = 0 \qquad\qquad \ddot{z}|_{t=2} = 12$$

Substituting these expressions into the components of the acceleration gives:

$$a_r = -26 \ (\text{ft} / \text{s}^2)$$
$$a_\theta = 28 \ (\text{ft} / \text{s}^2)$$
$$a_z = 12 \ (\text{ft} / \text{s}^2)$$

Now we can solve the equations in Table 10-7 directly for the unknown component of force:

$$f_r = (5)(-26) \ \text{lb}_\text{m}\text{ft} / \text{s}^2 = -4.04 \ \text{lb}_\text{f}$$
$$f_\theta = (5)(28) \ \text{lb}_\text{m}\text{ft} / \text{s}^2 = 4.35 \ \text{lb}_\text{f}$$
$$f_z = (5)(12) \ \text{lb}_\text{m}\text{ft} / \text{s}^2 = 1.86 \ \text{lb}_\text{f}$$

The force vector at 2 seconds is:

$$\mathbf{f} = -4.04\mathbf{e}_r + 4.35\mathbf{e}_\theta + 1.86\mathbf{e}_z$$

And the magnitude of the force vector is:

$$f = \sqrt{f_r^2 + f_\theta^2 + f_z^2} = \sqrt{(-4.04)^2 + (4.35)^2 + (1.86)^2} = 6.22 \ \text{lb}_\text{f}$$

Example 10-14. Determine the banking angle θ of the circular track to the horizontal so that the wheels of the sports car shown in Figure 10-19 will not have to depend upon friction to prevent the car from sliding either up or down the curve. The car travels at a constant tangential velocity of 88 ft/s and the radius of curvature of the track is 500 ft.

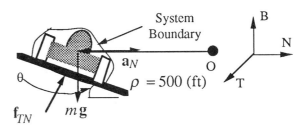

Figure 10-19: Motion of a car

SOLUTION: Choose the system to be the car. A path-oriented coordinate system is defined for the path of the car. For the defined coordinate system, the conservation of linear momentum is given in Table 10-8. The T, N, and B directions are the path tangential, normal, and binormal directions, respectively. The force normal to the track is denoted $\mathbf{f}_{TN}$.

Table 10-8: Data for Example 10-14 (Linear Momentum, System = Sports car)

Description	$\left(\dfrac{d(m\mathbf{v})_{\text{sys}}}{dt}\right)$ (lb$_f$)	=	$\sum \mathbf{f}_{\text{ext}}$		
			$m\mathbf{g}$ (lb$_f$)		$\mathbf{f}_{TN}$ (lb$_f$)
T-comp	$m\left(\dfrac{dv}{dt}\right)\mathbf{u}_T$	=	$0\mathbf{u}_T$	+	$0\mathbf{u}_T$
N-comp	$m = \dfrac{v_T^2}{\rho}\mathbf{u}_N$	=	$0\mathbf{u}_N$	+	$f_{TN}\sin\theta\,\mathbf{u}_N$
B-comp	$m0\mathbf{u}_B$	=	$(-mg)\mathbf{u}_B$	+	$f_{TN}\cos\theta\,\mathbf{u}_B$

The N-component and B-component equations give that:

$$N - \text{component}: \qquad f_{TN}\sin\theta = \frac{mv^2}{\rho}$$

$$B - \text{component}: \qquad f_{TN}\cos\theta = mg$$

Dividing the first equation by the second equation gives:

$$\tan\theta = \frac{v^2}{g\rho}$$

so that the angle is:

$$\theta = \tan^{-1}\left(\frac{v^2}{g\rho}\right) = \tan^{-1}\left(\frac{88}{(32.174)(500)}\right) = 25.7°$$

A banking angle of 25.7° is required so that the car will not have to depend upon friction to keep from sliding.

10.3.2 Energy Analysis

As has been discussed previously, the conservation of energy for a particle (in the absence of thermal effects) contains no additional information above and beyond that contained from a statement of linear momentum; in fact, it contains less information. This does not mean, however, that energy statements cannot be useful in analyzing problems or systems. In fact, it is entirely possible that a problem's solution using energy conservation is easy whereas it is very difficult using linear momentum. We should always be in the habit of taking a minute to assess the importance of each of the conservation laws and accounting statements which are available to us to determine whether these laws provide us with a useful alternative insight to the problem or additional information.

In particle dynamics problems, as for statics, we are not usually concerned with thermal energy effects or with electrical charge. Certainly, forces performing work on the particle and changes in kinetic energy may be important factors. Accordingly for a particle with no mass entering or leaving the system (the particle is the system), then the conservation of energy says that the work done on the particle results in changes in the kinetic and potential energy of the particle. Normally, we have no heat transfer and no changes in internal energy of the particle. An energy conservation equation, then for a single particle in the absence of charge and thermal effects is:

$$\frac{d}{dt}\left(E_K + E_P\right)_{sys} = \sum \dot{W}$$

(10 – 31)

where the terms E_K and E_P were defined in Chapter 7 to be:

$$E_K = \frac{mv_G^2}{2} \qquad E_P = mgh$$

From linear momentum for a particle

$$\left(\frac{d\mathbf{p}_{sys}}{dt}\right) = m\left(\frac{d\mathbf{v}}{dt}\right) = \sum \mathbf{f}_{ext}$$

(5 – 9)

Now, if the dot product of this equation with the particle velocity is formed

$$\mathbf{v} \cdot \left(\frac{d\mathbf{p}_{sys}}{dt}\right) = \frac{m}{2}\left(\frac{d(\mathbf{v} \cdot \mathbf{v})}{dt}\right) = \sum \mathbf{v} \cdot \mathbf{f}_{ext}$$

(10 – 32)

then the mechanical energy equation is obtained

$$\frac{d}{dt}\left(\frac{1}{2}mv_G^2\right) = \sum \mathbf{v} \cdot \mathbf{f}_{ext}$$

(10 – 33)

Each force $\mathbf{f}$ acting on the particle, which is moving with velocity $\mathbf{v}$, performs work on the particle at rate $\mathbf{f} \cdot \mathbf{v}$. This work rate is power. Furthermore, this rate of work is not always useful but often is an irreversible conversion of mechanical energy to thermal energy by frictional and drag forces. In this result, the rate of potential energy change appears as the rate of work done by gravity, $\mathbf{v} \cdot m\mathbf{g} = \left(\frac{d(-mgh)}{dt}\right)$, and so equation (10-33) can be written

$$\frac{d}{dt}\left(\frac{1}{2}mv_G^2\right) = \mathbf{v} \cdot m\mathbf{g} + \sum \mathbf{v} \cdot (\mathbf{f}_{contact})_{ext}$$

(10 – 34)

where $\mathbf{f}_{contact}$ represents all forces except gravity. This is the same as equation (10-31) and hence, we see that the total energy conservation statement for this problem, equation 10-31, is not independent of linear momentum conservation. It is seen in the next problem that it can be very useful and powerful, nonetheless.

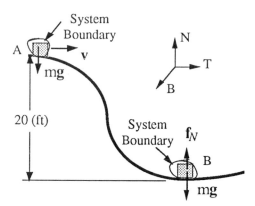

Figure 10-20: Block of ice

Example 10-15. A block of ice 10 lb_m with an initial velocity of 10 ft/s and on a frictionless surface starts out at the top of hill A. Determine the normal force on the cube when it arrives at B where the radius of curvature of the hill is $\rho = 30$ ft and the block is travelling horizontally (for an instant). Figure 10-20 shows the block of ice sliding down the hill.

SOLUTION: We choose the block of ice as the system and evaluate mechanical energy accounting in Table 10-9.

Table 10-9: Data for Example 10-15 (Mechanical energy, System = Block of ice)

$\dfrac{d}{dt}\left(E_K + E_P\right)_{sys}$	=	$\sum \dot{W}$
[energy/time]		[energy/time]
$\dfrac{d}{dt}\left(\dfrac{mv^2}{2} + mgh\right)_{sys}$	=	0

Since there is no friction, the only forces acting on the body are the force of gravity and the normal force exerted by the hill. Gravity is accounted for by the potential energy term and $\mathbf{v} \cdot \mathbf{f}_N$ for the normal force is always zero ($\mathbf{f}_N$ is always normal to $\mathbf{v}$, hence $\mathbf{v} \cdot \mathbf{f}_N = 0$). Hence, no work is done on the system or:

$$\left(\frac{mv^2}{2} + mgh\right)_{end} - \left(\frac{mv^2}{2} + mgh\right)_{beg} = 0$$

Dividing by the mass of the system gives:

$$\left(\frac{v^2}{2} + gh\right)_{end} - \left(\frac{v^2}{2} + gh\right)_{beg} = 0$$

The only unknown in this equation is the final velocity v_{end} which is found to be:

$$v_{end} = \sqrt{2g(h_{beg} - h_{end}) + v_{beg}^2} = \sqrt{2(32.\,174)(20) + (10)^2} = 37.\,24 \ (ft \,/\, s)$$

The normal force is calculated from the conservation of linear momentum at point B. The system is still the block of ice; however, a path-oriented coordinate system is defined at the position of the block. The conservation of linear momentum is given Table 10-10. Recall the kinematic relations for $\mathbf{a}_T$, $\mathbf{a}_N$, and $\mathbf{a}_B$ in Table 10-4.

Table 10-10: Data for Example 10-15b (Linear Momentum, System = Block of ice at point B)

Description	$\left(\dfrac{d(m\mathbf{v})_{sys}}{dt}\right)$ (lb$_f$)	=	$\sum \mathbf{f}_{ext}$		
			$m\mathbf{g}$ (lb$_f$)		$\mathbf{f}_N$ (lb$_f$)
T-component	$m\left(\dfrac{dv}{dt}\right)\mathbf{u}_T$	=	$0\mathbf{u}_T$	+	$0\mathbf{u}_T$
N-component	$ma_N\mathbf{u}_N$	=	$(-mg)\mathbf{u}_N$	+	$f_N\mathbf{u}_N$
B-component	$ma_B\mathbf{u}_B$	=	$0\mathbf{u}_B$	+	$0\mathbf{u}_B$

We now look at these three component equations. The T-component equation says that the velocity, at the very bottom of the hill, is constant. The B-component equation is satisfied trivially because $a_B = 0$ as a kinematic relation (Table 10-4). The N-component equation is more helpful, however. The kinematic relationship for the acceleration in the normal direction is (Table 10-4):

$$a_N = \frac{v^2}{\rho}$$

Substituting this value in the N-component of the linear momentum equation gives:

$$\frac{mv^2}{\rho} = -mg + f_N$$

Solving for the unknown force in the normal direction and substitution gives:

$$f_N = \frac{mv^2}{\rho} + mg = \left(\frac{(10)(37.24)^2}{(30)} + (10)(32.174)\right)\frac{\text{lb}_m\ \text{ft}}{\text{s}^2}\left|\frac{1\ \text{lb}_f}{32.174\ \text{lb}_m\ \text{ft}\ /\ \text{s}^2}\right. = 24.37\ (\text{lb}_f)$$

The force in the normal direction has a magnitude of 24.37 pounds force.

Example 10-16. Consider the cannon ball that was used in Example 10-12. The ball has a mass of 60 (kg) and an initial velocity in the x direction of 100 (m/s). Furthermore, the cannon ball was launched 50 (m) above the surface of impact. (a) For the case where there is no air resistance, use the mechanical energy equations to calculate the kinetic energy and the magnitude of the linear velocity of the ball at impact. (b) For the case where there is resistance, use the mechanical energy equations to calculate the work done on the ball by air resistance during the flight. Also, what happened to the work done by the air resistance.

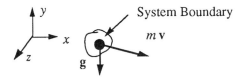

Figure 10-21: Cannon ball free body diagram

SOLUTION: Figure 10-21 shows the schematic of the the cannon ball in flight. For the case where there is no air resistance the accounting of mechanical energy is given in Table 10-11. There is no rate of work done on the ball for this case (except that due to gravity which is counted as a change in potential energy rather than work).

Table 10-11: Data for Example 10-16a (Mechanical energy, System = Cannon ball)

$\dfrac{d}{dt}\left(E_K + E_P\right)_{sys}$	=	$\sum \dot{W}$
[energy/time]		[energy/time]
$\dfrac{d}{dt}\left(\dfrac{mv^2}{2} + mgh\right)_{sys}$	=	0

This equation is integrated from the beginning time period to the ending time period:

$$\left(\frac{mv^2}{2} + mgh\right)_{end} - \left(\frac{mv^2}{2} + mgh\right)_{beg} = 0$$

The only unknown is the final kinetic energy of the ball which is:

$$\left(\frac{mv^2}{2}\right)_{end} = \left(\frac{mv^2}{2} + mgh\right)_{beg} - \left(mgh\right)_{end}$$

Substituting in the known values gives (let $h = 0$ at the cannon muzzle):

$$\left(\frac{mv^2}{2}\right)_{end} = (300,000 + 0) - (-29,400) = 329,400 \text{ (J)}$$

The final kinetic energy of the cannon ball at impact is 329,400 (J) and the magnitude of the velocity is:

$$v = \sqrt{\frac{2(329,400)}{m}} = \sqrt{\frac{2(329,400)}{60}} = 104.8 \text{ (m / s)}$$

The magnitude of the velocity of the cannon ball at impact is 104.8 (m/s) This is exactly the same answer that was obtained using linear momentum in Example 10-12. However, the solution using energy was much easier.

Now if we include the effects of the air resistance on the cannon ball, the accounting of mechanical energy must include the force that is doing work on the cannon ball through out its flight as given in Table 10-12

Table 10-12: Data for Example 10-16b (Mechanical energy, System = Cannon ball)

$\dfrac{d}{dt}\left(E_K + E_P\right)_{sys}$	=	$\sum \dot{W}$
		$\mathbf{f}_{ext} \cdot \mathbf{v}$
[energy/time]		[energy/time]
$\dfrac{d}{dt}\left(\dfrac{mv^2}{2} + mgh\right)_{sys}$	=	$(-k\mathbf{v}) \cdot \mathbf{v}$

This equation can be written as:

$$\frac{d}{dt}\left(\frac{mv^2}{2} + mgh\right) = -kv^2$$

In principle, this equation can be integrated over the time period of the flight of the cannon ball as:

$$\left(\frac{mv^2}{2} + mgh\right)_{end} - \left(\frac{mv^2}{2} + mgh\right)_{beg} = \int_{t_{beg}}^{t_{end}} (-kv^2)dt$$

In order to solve this problem for the final kinetic energy, we need to known the velocity as a function of time to perform the integration. Using only energy we would not be able to obtain a solution. However, because we know the velocity as a function of time from linear momentum in example 10-12 we can perform the integration and solve for the final kinetic energy. This integration is very tedious and will not be done here. More important is to view the integration over time of the rate of work done on the body as simply the total work due to that force during the time period or:

$$\left(\frac{mv^2}{2} + mgy\right)_{end} - \left(\frac{mv^2}{2} + mgy\right)_{beg} = W_D$$

Now from linear momentum, we know the final velocity (and hence the kinetic energy) and the only unknown is the work from the drag force. Substituting:

$$W_D = \left(13,000 - 29,400\right)_{end} - \left(300,000 + 0\right)_{beg}$$

$$= -316,000 \text{ (J)}$$

There was -316,000 (J) of work done on the cannon ball during the flight. The work was not useful but represents a conversion of mechanical energy to thermal energy. In fact the mechanical energy is lost and the cannon ball heats up during the flight. Remember, mechanical energy is not conserved but total energy is. The energy was converted to another form during the flight of the cannon ball.

10.4 RIGID BODY DYNAMICS

10.4.1 Momentum Analysis

Solving problems in rigid body dynamics requires using the linear and angular momentum conservation equations which were presented in chapters 5 and 6. (Remember that in this text, we restrict our attention to planar motion; any rotation of the rigid body is about a constant-direction axis. More general rotational motion is possible, of course, and the basis for a more general analysis which is within the conceptual and notational framework of this text is presented in Appendix E.) Obviously, there are a lot of situations which present simplifications to the equations, such as 1-dimensional translation without rotation, 1-dimensional translation with planar rotation, 2-dimensional translation without rotation, planar rotational motion without translation, planar rotational motion with constant translation, etc. The pertinent equations are given by equations (5-9) and (6-39)

$$m\left(\frac{d\mathbf{v}_G}{dt}\right) = \sum \mathbf{f}_{ext}$$

$$(5-9)$$

and

$$_G I_a \alpha = \sum (_G\mathbf{r} \times \mathbf{f})_{ext,a}$$

$$(6-39)$$

The latter result has alternate forms (Table 6-6) which may be used when convenient. Remember to try the different forms as one form may be considerably more easily applied than another to the problem at hand.

The linear momentum equations give information about the motion of the center of mass of the rigid body, while the angular momentum equations give information about angular motion and torques relative to a point. It is important to realize that the point about which the angular momentum is calculated can be moving and accelerating, and does not have to be stationary or moving at constant velocity. This makes the angular momentum equation extremely powerful in solving dynamic situations because we can pick the point of rotation such that unknown forces do not contribute to the torque and therefore do not appear in the equation.

Example 10-17. A telephone pole that is 50 ft in length and has a mass of 300 lb$_m$ falls from the vertical position. Determine the velocity of the tip of the pole as it strikes the earth using the conservation of momentum. A schematic of the figure is shown in Figure 10-22.

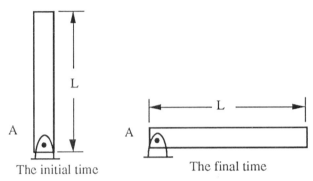

Figure 10-22: Telephone pole

SOLUTION: The system is the pole and it is a closed system. The point A is a pin. The conservation of mass and charge are trivial. Next, we will consider the conservation of linear momentum. Figure 10-22a is a free body diagram of the pole as it is falling. The conservation of linear momentum for this defined system is given in Table 10-13. It is important to recognize that the force at point A is not acting in the line of action of the pole and the components in the x-direction and y-direction cannot be related to the angle of the pole at this point.

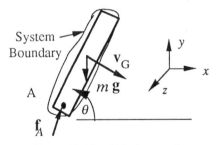

Figure 10-22a: Telephone pole
linear momentum free body diagram

Table 10-13: Data for Example 10-17a (Linear Momentum, System = Telephone pole)

Description	$\left(\dfrac{d(m\mathbf{v})_{\text{sys}}}{dt}\right)$ (lb$_f$)	=	$\sum \mathbf{f}_{\text{ext}}$		
			mg (lb$_f$)		$\mathbf{f}_A$ (lb$_f$)
x-direction	$m\left(\dfrac{d(v_{G,x})}{dt}\right)\mathbf{i}$	=	$0\mathbf{i}$	+	$f_{A,x}\mathbf{i}$
y-direction	$m\left(\dfrac{d(v_{G,y})}{dt}\right)\mathbf{j}$	=	$(-mg)\mathbf{j}$	+	$f_{A,y}\mathbf{j}$
z-direction	$m\left(\dfrac{d(v_{G,z})}{dt}\right)\mathbf{k}$	=	$0\mathbf{k}$	+	$0\mathbf{k}$

The velocity in the x-direction and the velocity in the y-direction are related to θ by the kinematical relationship:

$$v_{G,x} = |\mathbf{v}_G|\sin(\theta)$$
$$v_{G,y} = -|\mathbf{v}_G|\cos(\theta)$$

And the magnitude of the velocity of the center of mass is

$$|\mathbf{v}_G| = \frac{L}{2}|\omega|$$

so

$$v_{G,x} = \frac{L}{2}|\omega|\sin(\theta)$$

$$v_{G,y} = -\frac{L}{2}|\omega|\cos(\theta)$$

And the linear momentum equations that result are:

$$x\text{-dir}\quad f_{A,x} + 0 = m\frac{d}{dt}\left(\frac{L}{2}|\omega|\sin(\theta)\right)$$

$$y\text{-dir}\quad f_{A,y} - mg = m\frac{d}{dt}\left(-\frac{L}{2}|\omega|\cos(\theta)\right)$$

where

$$\omega = \left(\frac{d\theta}{dt}\right)$$

These are 2 independent equations and 3 unknowns. The unknowns are θ, $f_{A,x}$, and $f_{A,y}$ and there is no way to determine a solution by just using linear momentum.

Next, we will apply the conservation of angular momentum about point A which is an inertial point which is also fixed in the body. The convenience of using this point is that the external force $\mathbf{f}_A$ does not exert a torque about this point. Figure 10-22b is a free body diagram of the falling pole so we can account for angular momentum about the point A as the origin of our coordinate system. The only external force that exerts a torque is the force due to gravity; the force at the contact acts through the point A and does not contribute torque about point A. The conservation of angular momentum about point A is given in Table 10-14 (see equation (6-49) which is also the last equation in Table 6-6). The position vector from point A to the center of mass is:

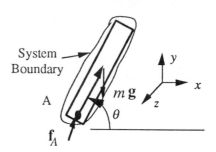

System
Boundary

A

mg

θ

$\mathbf{f}_A$

Figure 10-22b: Telephone pole
angular momentum free body diagram

$$_A\mathbf{r}_G = \left(\frac{L}{2}\right)\cos\theta\,\mathbf{i} + \left(\frac{L}{2}\right)\sin\theta\,\mathbf{j} + 0\mathbf{k} \text{ (ft)}$$

The cross product of $_A\mathbf{r}_G \times m\mathbf{g}$ is calculated:

$$_A\mathbf{r}_G \times m\mathbf{g} = \begin{vmatrix} \mathbf{i} & \mathbf{j} & \mathbf{k} \\ (L/2)\cos(\theta) & (L/2)\sin(\theta) & 0 \\ 0 & -mg & 0 \end{vmatrix} = 0\mathbf{i} + 0\mathbf{j} - (L/2)\cos(\theta)mg\mathbf{k} \text{ (ft}^2\text{ lb}_m/\text{s}^2)$$

Table 10-14: Data for Example 10-17b (Angular Momentum, System = Telephone pole)

Description	$_A I_z\left(\dfrac{d\omega}{dt}\right)\mathbf{k}$ (ft lb$_f$)	=	$\sum(_A\mathbf{r} \times \mathbf{f})_{ext}$ $(_A\mathbf{r} \times m\mathbf{g})$ (ft lb$_f$)
x-direction	$0\mathbf{i}$	=	$0\mathbf{i}$
y-direction	$0\mathbf{j}$	=	$0\mathbf{j}$
z-direction	$_A I_z\left(\dfrac{d\omega}{dt}\right)\mathbf{k}$	=	$(-mg)\left(\dfrac{L}{2}\right)\cos\theta\mathbf{k}$

Hence, the only torque and rotation are in the $\mathbf{k}$ direction giving .

$$_A I_z\left(\frac{d\omega}{dt}\right) = -mg\left(\frac{L}{2}\right)\cos\theta$$

From Table C-7, the moment of inertia about point A is:

$$_A I_z = \left(\frac{1}{3}\right)mL^2$$

and this moment of inertia about point A is not a function of time. Substituting back into the conservation of angular momentum gives

$$\left(\frac{1}{3}\right)mL^2\left(\frac{d\omega}{dt}\right) = -mg\left(\frac{L}{2}\right)\cos\theta$$

Rearranging gives:

$$\left(\frac{d\omega}{dt}\right) = -\left(\frac{3}{2}\right)\left(\frac{g}{L}\right)\cos\theta$$

where

$$\omega = \left(\frac{d\theta}{dt}\right)$$

so that

$$\left(\frac{d^2\theta}{dt^2}\right) = -\left(\frac{3}{2}\right)\left(\frac{g}{L}\right)\cos(\theta)$$

This equation is a nonlinear second order differential equation for θ which can be solved by numerical techniques and software packages. The solution of the equation is an expression or list of values for θ as a function of time t.

Initial conditions are required. We know that the initial velocity is zero or

$$\text{for } t = 0, \qquad \omega = 0 \qquad (i.e., \ d\theta/dt = 0)$$

and the initial position is slightly less than $(\pi/2)$ (If the angle were exactly $\pi/2$, the pole would never fall). Assume that at

$$t = 0 \qquad \theta = 1.56$$

From a numerical solution, the value of ω at $\theta = 0$ is found to be, approximately, -1.39 rad/sec.

Therefore, the linear velocity of the tip of the pole is:

$$v_{tip} = L\omega = (50)(-1.39) = -69.5 \text{ ft}/\text{s}$$

The magnitude of the linear velocity of the tip of the pole as it impacts the earth is 69.5 ft/s

Example 10-18. A cube and a cylinder are shown on the top on an inclined plane in Figure 10-23. The cylinder has a radius of 4 in. The incline has an angle of elevation of 30° from the horizontal, and the vertical distance is 3 ft. The mass of the block and the cylinder are both 5 lb$_m$. If the block and the cylinder initially start at rest, determine the linear and angular velocity of the block and the cylinder at the bottom of the incline for the following two cases using the conservation of momenta:

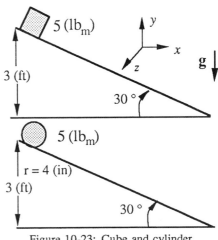

Figure 10-23: Cube and cylinder on inclined plane

a) The inclined plane is frictionless.

b) The inclined plane is not frictionless and has a coefficient of kinetic friction (μ_k) of 0.3.

SOLUTION: For the cube and the frictionless surface, Figure 10-23a shows the system or free body diagram of the cube. For convenience, the coordinate system is aligned with the motion of the block and the gravity vector is resolved into the components of the the new coordinate system. The only forces acting on the cube are the force due to gravity and the normal force exerted on the cube by the inclined plane. The conservation of the linear momentum for this system is given in Table 10-15. The angle β is the angle of inclination of the plane. Furthermore, the momentum of the cube is not changing in the y or z directions for the defined coordinate axis.

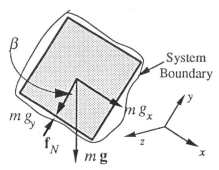

Figure 10-23a: Cube no friction
free body diagram

Table 10-15: Data for Example 10-18a (Linear Momentum, System = Block)

Description	$\left(\dfrac{d(mv)_{sys}}{dt}\right)$ (lb$_f$)	=	$\sum \mathbf{f}_{ext}$ mg (lb$_f$)		$\mathbf{f}_N$ (lb$_f$)
x-direction	$m\left(\dfrac{d(v_{G,x})}{dt}\right)\mathbf{i}$	=	$mg \sin\beta\mathbf{i}$	+	$0\mathbf{i}$
y-direction	$0\mathbf{j}$	=	$(-mg)\cos\beta\mathbf{j}$	+	$f_N\mathbf{j}$
z-direction	$0\mathbf{k}$	=	$0\mathbf{k}$	+	$0\mathbf{k}$

From Table 10-15, the y-direction equation says that the normal force is equal and opposite to the component of the force of gravity in the y-direction.

$$f_N = mg \cos(\beta)$$

The x-direction equation gives the the first order differential equation relating the change in velocity of the system to the forces acting on the system:

$$m\left(\frac{dv_x}{dt}\right) = mg \sin(\beta) \quad \text{or} \quad \left(\frac{dv_x}{dt}\right) = g \sin(\beta)$$

This equation can be separated and integrated since the angle and force due to gravity are constant.

$$\int_{v_o}^{v} dv_x = \int_{t_o}^{t} g \sin(\beta)dt$$

Because the cube starts from rest, at t_o, we have $v_o = 0$. Thus, integrating:

$$v_x = g \sin(\beta)t$$

Now if we know the time that the cube is at the bottom of the incline, we can substitute the time into the equation and solve the problem for the linear velocity. However, we do not know this time. The kinematical relationship between the position and the velocity allows us to solve for the time:

$$\left(\frac{dx}{dt}\right) = v_x = g\sin(\beta)t$$

This equation can be separated and integrated to give:

$$\int_{x_o}^{x} dx = \int_{t_o}^{t} g\sin(\beta)t\,dt$$

If the origin of the coordinate system is located at the top of the inclined plane then at $t = t_o = 0$, $x = 0$ and

$$x = \frac{g\sin(\beta)}{2}t^2$$

From the geometry of the block,

$$x = \frac{h}{\sin(\beta)}$$

where h is the verical height of 3 (ft). Substituting this into the previous equation gives:

$$\frac{h}{\sin(\beta)} = \frac{g\sin(\beta)}{2}t^2$$

which can be solved for t:

$$t = \sqrt{\frac{2h}{\sin^2(\beta)g}}$$

This expression for the time can be substituted back into the relationship for the linear velocity:

$$v_x = g\sin(\beta)t = g\sin(\beta)\sqrt{\frac{2h}{\sin^2(\beta)g}} = \sqrt{2hg}$$

Therefore, at the bottom of the ramp,

$$v_x = \sqrt{2hg} = \sqrt{2(3)(32.174)} = 13.9 \text{ (ft / s)}$$

The time required to reach the bottom is found to be:

$$t = 0.86 \text{ (s)}$$

Since all of the forces act through the center of mass of the cube, the angular momentum equations do not give any additional information.

We consider next the motion of the cylinder. Figure 10-23b shows the system or free body diagram. The coordinate system again is aligned with the motion of the cylinder and the gravity vector is resolved into the components of the new coordinate system. The only forces acting on the cylinder are the force due to gravity and the normal force exerted on the cube by the inclined plane. The conservation of the linear momentum for this system is given in Table 10-16. The angle β is the angle of inclination of the plane. Furthermore, the momentum of the cylinder is not changing in the y or z directions.

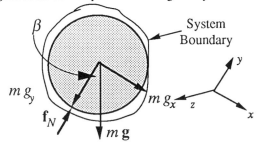

Figure 10-23b: Cylinder no friction free body diagram

Table 10-16: Data for Example 10-18a (Linear Momentum, System = Cylinder)

Description	$\left(\dfrac{d(m\mathbf{v})_{sys}}{dt}\right)$ (lb$_f$)	=	$\sum \mathbf{f}_{ext}$		
			mg (lb$_f$)		$\mathbf{f}_N$ (lb$_f$)
x-direction	$m\left(\dfrac{d(v_{G,x})}{dt}\right)\mathbf{i}$	=	$mg\sin\beta\,\mathbf{i}$	+	$0\mathbf{i}$
y-direction	$0\mathbf{j}$	=	$(-mg)\cos\beta\,\mathbf{j}$	+	$f_N\mathbf{j}$
z-direction	$0\mathbf{k}$	=	$0\mathbf{k}$	+	$0\mathbf{k}$

Comparing these equations to the equations for the cube, there is no difference. The conservation equations give the same math problem so we know that the final linear velocity of the cylinder is 13.9 (ft/s). Furthermore, the free body diagram shows that all of the forces act through the center of mass of the cylinder so the cylinder will not roll. The cylinder slides down the frictionless plane and the angular velocity of the cylinder is zero. Based on the previous discussion we can conclude that for any shaped object, for a frictionless surface, the same equation results.

Consider now the cube on the surface with friction, Figure 10-23c shows the system or free body diagram of the cube. The coordinate system is aligned with the motion of the block and the gravity vector is resolved into the components of the the new coordinate system. The only forces acting on the cube are the force due to gravity, normal force exerted on the cube by the inclined plane, and the force due to friction which is in the opposite direction of the motion. The conservation of linear momentum for this system is given in Table 10-17. The angle β is the angle of inclination of the plane. Furthermore, the momentum of the cube is not changing in the y or z directions for the defined coordinate axis.

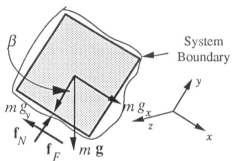

Figure 10-23c: Cube friction free body diagram

Table 10-17: Data for Example 10-18b (Linear Momentum, System = Block)

Description	$\left(\dfrac{d(m\mathbf{v})_{sys}}{dt}\right)$ (lb$_f$)	=	$\sum \mathbf{f}_{ext}$					
			mg (lb$_f$)		$\mathbf{f}_N$ (lb$_f$)		$\mathbf{f}_F$	
x-direction	$m\left(\dfrac{d(v_{G,x})}{dt}\right)\mathbf{i}$	=	$mg\sin\beta\,\mathbf{i}$	+	$0\mathbf{i}$	−	$\mu_K f_N\mathbf{i}$	
y-direction	$0\mathbf{j}$	=	$(-mg)\cos\beta\,\mathbf{j}$	+	$f_N\mathbf{j}$	+	$0\mathbf{i}$	
z-direction	$0\mathbf{k}$	=	$0\mathbf{k}$	+	$0\mathbf{k}$	+	$0\mathbf{k}$	

From Table 10-17, the y-direction equation, we can determine $\mathbf{f}_N$; the normal force is equal and opposite to the component of the force of gravity in the y-direction:

$$f_N = mg \cos(\beta)$$

This is the same result as that for the frictionless plane. The x-direction equation gives a first order differential equation relating the change in velocity of the system to the forces acting on the system:

$$m\left(\frac{dv_x}{dt}\right) = mg \sin(\beta) - \mu_K f_N$$

Substituting the normal force expression from the y-direction equation gives:

$$m\left(\frac{dv_x}{dt}\right) \quad \text{or} = mg \sin(\beta) - \mu_K mg \cos(\beta) \qquad \left(\frac{dv_x}{dt}\right) = g\left(\sin(\beta) - \mu_K \cos(\beta)\right)$$

This equation can be separated and integrated since the angle, the force due to gravity, and the kinetic friction coefficient all are constant.

$$\int_{v_o}^{v} dv_x = \int_{t_o}^{t} g\left(\sin(\beta) - \mu_K \cos(\beta)\right) dt$$

Using the initial condition that at $t = 0$, $v_o = 0$ gives

$$v_x = g\left(\sin(\beta) - \mu_K \cos(\beta)\right)t$$

Like the frictionless plane, we do not know the time that the block is at the bottom of the plane. We must use the kinematic relationship between velocity and position to solve for the time:

$$\left(\frac{dx}{dt}\right) = v_x = g\left(\sin(\beta) - \mu_K \cos(\beta)\right)t$$

To simplify this expression, let us define a new constant:

$$K = g\left(\sin(\beta) - \mu_K \cos(\beta)\right)$$

This equation can be separated and integrated to give:

$$\int_{x_o}^{x} dx = \int_{t_o}^{t} Kt\,dt$$

Defining $x_o = 0$ at $t_o = 0$ gives

$$x = \frac{K}{2}t^2$$

From before, $x = h \,/\, \sin(\beta)$ so that

$$\frac{h}{\sin(\beta)} = \frac{K}{2}t^2$$

The only unknown in this equation is the time t. Rearranging yields:

$$t = \sqrt{\frac{2h}{K \sin(\beta)}}$$

This expression for the time can be substituted back into the relationship for the velocity:

$$v_x = Kt = K\sqrt{\frac{2h}{K \sin(\beta)}} = \sqrt{\frac{2Kh}{\sin(\beta)}}$$

Evaluating K gives

$$K = g\left(\sin(\beta) - \mu_K \cos(\beta)\right) = 32.174\left(\sin(30) - 0.3\cos(30)\right) = 7.73 \text{ (ft / s}^2)$$

Then, the final velocity is:

$$v_x = \sqrt{\frac{2Kh}{\sin\beta}} = \sqrt{\frac{2(7.73)(3)}{\sin(30)}} = 9.6 \text{ (ft / s)}$$

and the time required to reach the bottom is

$$t = \frac{v_x}{K} = 1.25 \text{ (s)}$$

The velocity of the cube is smaller than for the frictionless case, and the time is longer to reach the bottom. The cube does not tip over so the angular velocity of the cube at the bottom of the incline is zero. However, since all of the forces do not act through the center of mass of the cube, the angular momentum equations do give additional information. The angular momentum equations will give the location of the normal force on the cube as was done in Example 9-7.

For the cylinder on the surface with friction we will assume that there is no slippage and that there is only rotation. Figure 10-23d shows the system or free body diagram of the cylinder. Again, the coordinate system is aligned with the motion of the cylinder and the gravity vector is resolved into the components of the the new coordinate system. The only forces acting on the cylinder are the force due to gravity, the normal force exerted on the cylinder by the inclined plane, and the frictional force causing the rotation. The conservation of linear momentum for this system is given in Table 10-18. The angle β is the angle of inclination of the plane. Furthermore, the momentum of the cylinder is not changing in the y or z directions for the defined coordinate axis.

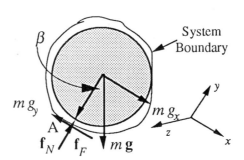

Figure 10-23d: Cylinder friction free body diagram

Table 10-18: Data for Example 10-18b (Linear Momentum, System = Cylinder)

Description	$\left(\dfrac{d(m\mathbf{v})_{sys}}{dt}\right)$ (lb$_f$)	=	$\sum \mathbf{f}_{ext}$ $m\mathbf{g}$ (lb$_f$)		$\mathbf{f}_N$ (lb$_f$)		$\mathbf{f}_F$
x-direction	$m\left(\dfrac{d(v_{G,x})}{dt}\right)\mathbf{i}$	=	$mg\sin\beta\mathbf{i}$	+	$0\mathbf{i}$	−	$f_F\mathbf{i}$
y-direction	$0\mathbf{j}$	=	$(-mg)\cos\beta\mathbf{j}$	+	$f_N\mathbf{j}$	+	$0\mathbf{i}$
z-direction	$0\mathbf{k}$	=	$0\mathbf{k}$	+	$0\mathbf{k}$	+	$0\mathbf{k}$

The y-direction equation gives the relationship of the normal force as:

$$f_N = mg\cos(\beta)$$

However, the x-direction equation is:

$$m\left(\frac{dv_x}{dt}\right) = mg\sin(\beta) - f_F$$

Unlike the sliding cube where we can relate the frictional force to the normal force through the coefficient of kinetic friction, for the rolling cylinder the magnitude of the frictional force is unknown. Because the (static-there is no slip) frictional force is unknown, we cannot solve this problem solely by linear momentum because there are 2 unknowns, the friction force and the velocity, and only 1 equation.

The conservation of angular momentum for planar motion about point (really a line) A (the contact of the cylinder with the plane), as stated in the fifth equation in Table 6-6, is given in Table 10-19. The axis of rotation is the z axis. Note that we have chosen point A (point A on the cylinder corresponds to point B in table 6-6) to be fixed in the cylinder on its surface and therefore moves away from the surface and then back towards it as the cylinder rolls down the plane (exercise: sketch a plot of y versus x for point A as the cylinder rolls down the plane). Also, note that we have chosen an instant in time at which point A is in contact with the cylinder and therefore, at that instant, its velocity is zero (the same as that of the surface). Then, although its acceleration is not zero at this (or any other time), its acceleration must be perpendicular to the plane of the surface because there is no slip between the cylinder and the surface; the only non-zero component of acceleration is in the y direction. This means that the acceleration of point A and the position vector of the cylinder's center of mass relative to point A must be collinear: $_B\mathbf{r}_G \times {}_A\mathbf{a} \equiv 0$. Also, the unknown friction force exerts zero torque about point A (this is why point A was chosen), leaving only gravity to exert a torque.

Table 10-19: Data for Example 10-18b (Angular Momentum, System = Cylinder)

Description	$_A I_z \alpha + m[_A\mathbf{r}_G \times \mathbf{a}_A]_z$ (ft lb$_f$)	=	$\sum {}_A T_z$ $(_A\mathbf{r} \times m\mathbf{g})$ (ft lb$_f$)
z-direction	$[_A I_z \alpha + m(0)]$	=	$(-Rmg\sin\beta)$

Using the kinematic relation that $\alpha = d\omega/dt$, the conservation of angular momentum reduces to:

$$_A I_z\left(\frac{d\omega}{dt}\right) = -Rmg\sin(\beta)$$

The moment of inertia about point A is calculated by adding mR^2 (according to the parallel axis theorem, eq 6-50) to $_GI_z$, given in Table C-7 of Appendix C:

$$_AI_z = {_G}I_z + mR^2 = \frac{1}{2}mR^2 + mR^2 = \frac{3}{2}mR^2$$

Substituting this expression into the previous equation:

$$-Rmg\sin(\beta) = \frac{3}{2}mR^2\left(\frac{d\omega}{dt}\right) \quad \text{or} \quad -\left(\frac{2g\sin(\beta)}{3R}\right) = \left(\frac{d\omega}{dt}\right)$$

This equation can be separated and integrated with the initial conditions that at $t = 0$, $\omega = 0$:

$$\omega = -\left(\frac{2g\sin(\beta)}{3R}\right)t$$

Like the linear momentum problems, we do not know the time that the cylinder is at the bottom of the inclined plane. Using the kinematical relationship between the angular velocity and the angular displacement together with the above result gives:

$$\left(\frac{d\theta}{dt}\right) = \omega = -\left(\frac{2g\sin(\beta)}{3R}\right)t$$

This equation can also be separated and integrated with the initial condition that the angular displacement is zero at the the initial time:

$$\theta = -\left(\frac{g\sin(\beta)}{3R}\right)t^2$$

Now the final angular displacement is related to the total distance the cylinder rolled and the radius of the cylinder by:

$$-\theta = \frac{\text{total distance}}{R} = \frac{h/\sin(\beta)}{R} = \frac{h}{R\sin(\beta)}$$

Substituting the expression for θ into the previous equation gives:

$$\frac{h}{R\sin(\beta)} = \left(\frac{g\sin(\beta)}{3R}\right)t^2$$

and solving for t gives:

$$t = \sqrt{\frac{3h}{g\sin^2(\beta)}} = \sqrt{\frac{3(3)\text{ ft}}{\sin^2(30°)}\left|\frac{s^2}{32.174\text{ ft}}\right.} = 1.06 \text{ s}$$

This expression for the time can now be substituted into the angular velocity equation.

$$\omega = -\left(\frac{2g\sin(\beta)}{3R}\right)t = -\left(\frac{2g\sin(\beta)}{3R}\right)\sqrt{\frac{3h}{g\sin^2(\beta)}} = -\left(\frac{2}{R}\right)\sqrt{\frac{gh}{3}}$$

Inserting values gives:

$$\omega = -\left(\frac{2}{4\text{ in}}\left|\frac{12\text{ in}}{\text{ft}}\right.\right)\sqrt{\frac{32.174\text{ ft}}{s^2}\left|\frac{3\text{ ft}}{3}\right.} = -34 \text{ (1/s)}$$

Note that the radius, 4 in, must be converted to feet (always include units and look for needed conversions). The angular velocity is 34 radians per second and the time required for the cylinder to reach the bottom of the plane is 1.06 seconds. The linear velocity of the center of the mass of the cylinder is calculated from the definition of the angular velocity:

$$v_G = \omega \times {}_A r_G$$

In the x-direction this equation reduces to:

$$v_x = -R\omega = -\frac{4 \text{ in}}{} \left| \frac{\text{ft}}{12 \text{ in}} \right| \frac{-34}{\text{s}} = 11.3 \text{ ft / s}$$

The velocity of the center of mass of the cylinder is 11.3 (ft/s). This velocity is faster than the velocity of the cube on the friction surface. We examine this problem further in the next section.

10.4.2 Energy Analysis

As was the case for particles, the dynamics of rigid bodies may be analyzed using energy concepts as well as the momentum laws. As you will see below, these concepts are not independent of linear momentum conservation, however and, in fact, do not contain as much information as the momentum equations. This does not mean, however, that the equations are not useful; in fact, they may be much more useful (or at least more easily applied) than the full momentum equations for the very reason that they are simplified relations.

In this section, we present two results. The first one is a true mechanical energy equation in that it has direct physical interpretation with respect to energy. The other result is an energy-like equation in that the result that is obtained carries the dimensions of energy, but its interpretation in terms of bonafide energy conepts is much less clear. Nevertheless, it is a correct result, and a useful one.

A Rigid Body Mechanical Energy Equation. Consider a rigid body undergoing both translational and rotational motion. Now, each differential element of the rigid body may be treated as its own system, having its own momentum and external forces acting upon it. Writing the conservation of linear momentum for a single such element (i) of differential mass dm, we have

$$\left(\frac{d v_i}{dt}\right) dm = (f_{\text{ext}})_i \tag{10-35}$$

The external forces in this equation are, of course, those which are external to, but acting on, this differential element system. Now, if we dot this differential element's velocity with its linear momentum equation, we obtain

$$\left(\frac{d v_i}{dt}\right) \cdot v_i dm = (f_{\text{ext}})_i \cdot v_i \tag{10-36}$$

and this is a mechanical energy equation for this differential element. This result can be summed over all such elements of the entire rigid body to give

$$\int v \cdot \left(\frac{dv}{dt}\right) dm = \int (g \cdot v) dm + \sum (f_{\text{ext}})_{i,\text{contact}} \cdot v_i \tag{10-37}$$

Note that we have separated the gravitational body force from the contact surface forces for each element and that the body forces are then summed by integration. Now, recall that $\mathbf{v} = \mathbf{v}_G + \mathbf{v}'$ and note also that

$$\mathbf{v} \cdot \left(\frac{d\mathbf{v}}{dt} \right) = \frac{d}{dt} \left(\frac{1}{2} \mathbf{v} \cdot \mathbf{v} \right)$$

so that

$$\int \frac{d}{dt} \left[\frac{1}{2} (\mathbf{v}_G + {}_G\mathbf{v}) \cdot (\mathbf{v}_G + {}_G\mathbf{v}) \right] dm = \int \left(\mathbf{g} \cdot (\mathbf{v}_G + {}_G\mathbf{v}) \right) dm + \sum \left((\mathbf{f}_{ext})_{i,\text{contact}} \cdot \mathbf{v}_i \right) \qquad (10-38)$$

The left-hand side of this equation previously was written in terms of the translational and rotational kinetic energy (equations 7-1 through 7-13). Also, the right-hand side becomes

$$\int \left(\mathbf{g} \cdot (\mathbf{v}_G + {}_G\mathbf{v}) \right) dm + \sum \left((\mathbf{f}_{ext})_{i,\text{contact}} \cdot \mathbf{v}_i \right) = m\mathbf{g} \cdot \mathbf{v}_G + \int {}_G\mathbf{v}\, dm \cdot \mathbf{g} + \sum \left((\mathbf{f}_{ext})_{i,\text{contact}} \cdot \mathbf{v}_i \right) \qquad (10-39)$$

in which the middle term is identically zero (Chapter 6).

Now the sum of the contact forces throughout the rigid body can be simplified. Consider two adjacent differential elements (Figure 10-24). At their contact surface they have the same velocity. Furthermore, the force which acts on the first element by the second is opposite and equal to that which acts on the second element by the first; forces represent an *exact exchange* of momentum between the system and surroundings. Consequently, when the sum over all such elements of the body is obtained, all of the contact force-velocity dot products cancel except those which are at the external surface of the entire rigid body. That is

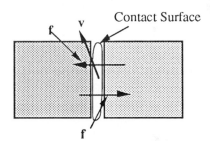

Figure 10-24: Two adjacent differential elements

$$\sum \left((\mathbf{f}_{ext})_{i,\text{contact}} \cdot \mathbf{v}_i \right) = \sum \left((\mathbf{f}_{ext})_{\text{surface}} \cdot \mathbf{v}_{\text{surface}} \right) \qquad (10-40)$$

Finally, then

$$\frac{d}{dt} \left(\frac{1}{2} m v_G^2 + \frac{1}{2} {}_G I_a \omega^2 \right) = m\mathbf{g} \cdot \mathbf{v}_G + \sum \left((\mathbf{f}_{ext})_{\text{surface}} \cdot \mathbf{v}_{\text{surface}} \right) \qquad (10-41)$$

This is a mechanical energy equation for the rigid body. The gravity term, in this form, represents work (actually work rate, i.e., power) done by gravity. The forces at the surface of the body also may do work but *only if the surface of the body is moving where the force acts on the body.* For example, if a frictional force acts on a body at its point of contact with a fixed (with respect to our inertial frame) surface *without slippage*, then there is no work. The term on the left-hand side, as has been mentioned before, represents changes in the body's translational and rotational kinetic energy. In this form, then, changes in the body's translational and rotational kinetic energy occur because of work done by gravity and other external forces. The potential energy form of this result is

$$\frac{d}{dt}\left(\frac{1}{2}mv_G^2 + \frac{1}{2}{}_GI_a\omega^2 + mgh_G\right) = \sum\left((\mathbf{f}_{ext})_{surface} \cdot \mathbf{v}_{surface}\right) \qquad (10-41b)$$

In this form, changes in the kinetic energy (both translational and rotational) and potential energy occur because of work done by external forces, except for gravity (which, again, is accounted for by potential energy). In the absence either of external forces (other than gravity) or of motion of the body at the point of contact of any such forces, this result says that there is an exact interchange between the body's kinetic (both translational and rotational) and potential energy (work done by gravity). Note that a zero velocity of the rigid body at the point of action of a contact force at the body's surface does *not* imply that the body has no motion at all. It simply implies that the motion *at the point of contact of the force* is zero with respect to the inertial frame of reference which is used to describe the motion.

A Rigid Body Energy-Like Equation ("Work-energy" theorem). Now a second useful result can be obtained much more easily by performing the dot product of the linear momentum equation for a rigid body (equation 5-9) with its center-of-mass velocity:

$$\frac{d}{dt}\left(\frac{mv_G^2}{2}\right) = m\mathbf{g} \cdot \mathbf{v}_G + \sum\left((\mathbf{f}_{ext})_{surface} \cdot \mathbf{v}_G\right) \qquad (10-42)$$

The gravity term has the same meaning as above, i.e., the work (rate) done by gravity but the other terms are different. The surface forces are dotted with the velocity of the center of mass instead of the velocity at their points of contact and therefore, have no clear interpretation in terms of work done by these forces. *Work is energy exchanged across the system boundary as the result of a force moving through a distance.* In this relation, the force is associated with a distance (velocity) which is different from that at its point of contact and therefore physical significance is lost. We might as well dot the force with the velocity of the moon, as far as a work interpretation is concerned! The left-hand side also is different in that only translational kinetic energy is represented. Rotational kinetic energy appears to be missing; it is included in the surface contact force term.

The advantage of this equation over the mechanical energy equation is that the only velocity which appears is the center-of-mass velocity. Furthermore, the result is not complicated by a rotational energy term. The tradeoff, however, is in the appearance of the surface force terms; these terms must be included even if the velocity at their point of action is zero which, as discussed above, would exclude them from the true mechanical energy equation.

Example 10-19. A telephone pole that is 50 (ft) in length and has a mass of 300 (lb$_m$) falls from the vertical position as given in problem 10-17. Use the mechanical energy accounting equation to determine the velocity of the tip of the pole as it strikes the earth. A schematic of the figure is shown in Figure 10-25.

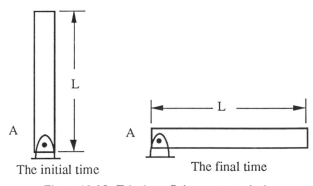

The initial time   The final time

Figure 10-25: Telephone Pole energy analysis

SOLUTION: The system is the pole and it is a closed system. The point A is a pin. Figure 10-25a gives a schematic of the pole as it is falling. The accounting of mechanical energy is given in Table 10-20. The force at the pin A does no work because the displacement and the velocity at that point are zero. Furthermore, the kinetic energy is distributed between the kinetic energy due to translation and rotation.

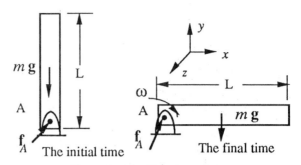

Figure 10-25a: Telephone Pole free body diagram

Table 10-20: Data for Example 10-19 (Mechanical energy, System = Telephone pole)

$\dfrac{d}{dt}\left(E_K + E_P\right)_{sys}$	$=$	$\sum(\dot{W})$
[energy/time]		[energy/time]
$\dfrac{d}{dt}\left(\dfrac{mv_G^2}{2} + \dfrac{{}_GI_z\omega^2}{2} + mgh_G\right)_{sys}$	$=$	0

This equation is integrated over the time period giving:

$$\left(\frac{mv_G^2}{2} + \frac{{}_GI_z\omega^2}{2} + mgh_G\right)_{end} - \left(\frac{mv_G^2}{2} + \frac{{}_GI_z\omega^2}{2} + mgh_G\right)_{beg} = 0$$

The potential energy at the end is zero and the kinetic energy at the beginning is zero reducing the equation:

$$\left(\frac{mv_G^2}{2} + \frac{{}_GI_z\omega^2}{2}\right)_{end} - \left(mgh_G\right)_{beg} = 0$$

The potential energy at the beginning is the height to the center of mass of the pole:

$$(mgh_G)_{beg} = \frac{L}{2}mg$$

To calculate the rotational kinetic energy, the moment of inertia *about the center of mass* is obtained from Table C-7 in Appendix C

$$_GI_z = \frac{1}{12}mL^2$$

To calculate the tranlational kinetic energy, the velocity of the center of mass and the angular velocity are related kinematically by

$$v_G = \frac{L}{2}\omega$$

Combining these results gives the kinetic energy at the end

$$(KE)_{end} = \frac{1}{2}m\left(\frac{L}{2}\omega\right)^2 + \frac{1}{2}\left(\frac{1}{12}\right)mL^2\omega^2 = \frac{1}{2}m\left(\frac{3}{12} + \frac{1}{12}\right)L^2\omega^2 = \frac{1}{2}m\left(\frac{1}{3}\right)L^2\omega^2$$

Substituting back into the mechanical energy equation gives

$$\frac{1}{2}m\left(\frac{1}{3}\right)L^2\omega^2 - \frac{L}{2}mg = 0$$

from which

$$\omega = \sqrt{\frac{3g}{L}}$$

Notice that the angular velocity of the pole is not dependent on the mass of the pole. Substituting the numerical values yields:

$$\omega = \sqrt{\frac{3(32.174)}{50}} = 1.389 \text{ (rad / s)}$$

The angular velocity of the pole when the pole strikes the earth is 1.389 radians per second
      The velocity at the tip is:
$$v_{tip} = L\omega = (50)(1.389) = 69.47 \text{ (ft / s)}$$

This is the magnitude of v_{tip}. The sign of the velocity of the tip is not known from the conservation of energy equation. From our understanding or our experience we know that the velocity is downward. It is important to note though that energy did not tell us the direction.
      Comparing this answer with the one obtained in Problem 10-17, we see that the answers are the same, as they should be. However the energy equation has the advantage of not requiring integration of the velocity to obtain information about the position.

Example 10-20. A cube and a cylinder are shown on the top on an inclined plane as shown in Figure 10-23. The cylinder has a radius of 4 (in). The incline has an angle of elevation of 30° from the horizontal, and the vertical distance in 3 (ft). The mass of the block and the cylinder are both 5 (lb$_m$). If the block and the cylinder initially start at rest, determine the magnitude of the linear and angular velocity of the block and the cylinder at the bottom of the incline for the two cases:

      a) The inclined plane is frictionless,
      b) The inclined plane is not frictionless and has a coefficient of kinetic friction (μ_k) of 0.3.
      Use the accounting of mechanical energy to determine the velocities in both cases. Also, for the surface that is not frictionless determine the work done by the friction on the cube and cylinder.

      SOLUTION: For the cube and the frictionless surface, Figure 10-23a (Example 10-18) shows the system or free body diagram of the cube. The accounting of mechanical energy for this system is given in Table 10-21. Notice that the normal force does no work because the dot product of the force and the motion is zero.

Table 10-21: Data for Example 10-20a (Mechanical energy, System = Block)

$\dfrac{d}{dt}\left(E_K + E_P\right)_{\text{sys}}$	=	$\sum \dot{W}$
[energy/time]		[energy/time]
$\dfrac{d}{dt}\left(\dfrac{mv^2}{2} + mgh\right)_{\text{sys}}$	=	0

This equation can be integrated over the time period giving:

$$\left(\frac{mv^2}{2} + mgh\right)_{\text{end}} - \left(\frac{mv^2}{2} + mgh\right)_{\text{beg}} = 0$$

There is no kinetic energy at the beginning and the potential energy terms are combined:

$$\left(\frac{mv^2}{2}\right) + mg(h_{\text{end}} - h_{\text{beg}}) = 0$$

The mass of the cube is divided from both sides of the equation:

$$\left(\frac{v^2}{2}\right) + g(h_{\text{end}} - h_{\text{beg}}) = 0$$

Rearranging gives the magnitude of the velocity of the center of mass:

$$v = \sqrt{2g(h_{\text{beg}} - h_{\text{end}})}$$

Substituting the known values gives:

$$v = \sqrt{2(32.\,174)(3 - 0)} = 13.\,9 \ (\text{ft} \ / \ \text{s})$$

The magnitude of the velocity of the center of mass of the cube is 13.9 (ft/s). The angular velocity of the cube is zero.

For the cylinder and the frictionless surface, Figure 10-23b shows the system or free body diagram of the cylinder. The accounting of mechanical energy for this system is given in Table 10-22. Notice that the normal force does no work because the dot product of the force and the motion is zero. Also, the cylinder does not roll but slips with no opposing force down the plane.

Table 10-22: Data for Example 10-20a (Mechanical energy, System = Cylinder)

$\dfrac{d}{dt}\left(E_K + E_P\right)_{\text{sys}}$	=	$\sum \dot{W}$
[energy]		[energy]
$\dfrac{d}{dt}\left(\dfrac{mv^2}{2} + mgh\right)_{\text{sys}}$	=	0

This equation is exactly the same as the cube for the frictionless surface. So we know that the magnitude of the velocity of the center of mass is 13.9 (ft/s). Since the cylinder is not rolling, the angular velocity is zero. Comparing these results with the answers obtained using linear and angular momentum we can see that they are the same.

For the cube and the surface with friction, Figure 10-23c shows the system or free body diagram of the cube. The accounting of mechanical energy for this system is given in Table 10-23. Notice that the normal force does no work but the frictional force does work and the rate of work is the force dotted with the velocity at the contact point.

Table 10-23: Data for Example 10-20b (Mechanical energy, System = Block)

$\dfrac{d}{dt}\left(E_K + E_P\right)_{sys}$	=	$\sum \dot{W}$
		$\mathbf{f}_{ext} \cdot \mathbf{v}$
[energy/time]		[energy/time]
$\dfrac{d}{dt}\left(\dfrac{mv^2}{2} + mgy\right)_{sys}$	=	$(-\mathbf{f}_F) \cdot \mathbf{v}$

We can integrate this equation over the time period of the cube slipping down the plane:

$$\left(\frac{mv^2}{2} + mgy\right)_{end} - \left(\frac{mv^2}{2} + mgy\right)_{beg} = \int_{t_{beg}}^{t_{end}} (-\mathbf{f}_F) \cdot \mathbf{v}\, dt$$

This equation cannot be solved unless we know the velocity as a function of time and the time period so we can perform the integration; we have no independent means of determining the work. In other words we cannot proceed any farther from just energy considerations; however, because we solved the linear momentum problem in example 10-17 we do know the velocity as a function of time. Moreover, we know the final velocity of system. Conceptually, the previous equation is:

$$\left(\frac{mv^2}{2} + mgy\right)_{end} - \left(\frac{mv^2}{2} + mgy\right)_{beg} = W_{friction}$$

The beginning kinetic energy is zero and the the potential energy terms are combined:

$$\left(\frac{mv^2}{2}\right) + mg(y_{end} - y_{beg}) = W_{friction}$$

Now from the conservation of linear momentum equations, we know the final linear velocity:

$$v_{end} = 9.6 \text{ (ft / s)}$$

The only unknown in the previous equation is the total work done by friction:

$$W_{friction} = \left(\frac{5 \text{ lbm} \mid 9.6^2 \text{ ft}^2}{2 \mid s^2} + \frac{5 \text{ lbm} \mid 32.174 \text{ ft} \mid (0-3) \text{ ft}}{s^2}\right) \frac{\text{lbf s}^2}{32.174 \text{ lbm ft}}$$

$$= -7.84 \text{ (ft lb}_f\text{)}$$

The friction did negative work on the block; the mechanical energy of the block was decreased.

For the cylinder and the surface with friction, Figure 10-23d shows the system or free body diagram of the cylinder. We are assuming that the cylinder does not slip but only rolls down the incline plane. The accounting of mechanical energy for this system is given in Table 10-24. Notice that the normal force does no work, because the dot product of the force and the motion is zero. The frictional force also does no work because the velocity at the point of contact is zero for the rolling cylinder.

Table 10-24: Data for Example 10-19b (Mechanical energy, System = Cylinder)

$\dfrac{d}{dt}\left(E_K + E_P\right)_{\text{sys}}$	$=$	$\sum \dot{W}$
[energy/time]		[energy/time]
$\dfrac{d}{dt}\left(\dfrac{mv_G^2}{2} + \dfrac{_GI_z\omega^2}{2} + mgh\right)_{\text{sys}}$	$=$	0

This equation can be integrated giving:

$$\left(\frac{mv_G^2}{2} + \frac{_GI_z\omega^2}{2} + mgh\right)_{\text{end}} - \left(\frac{mv_G^2}{2} + \frac{_GI_z\omega^2}{2} + mgh\right)_{\text{beg}} = 0$$

The kinetic energy at the beginning is zero and combining the potential energy terms gives:

$$0 = \left(\frac{mv_G^2}{2} + \frac{_GI_z\omega^2}{2}\right)_{\text{end}} + mg(h_{\text{end}} - h_{\text{beg}})$$

The moment of inertia of the cylinder about the center of mass of the body is:

$$_GI_z = \frac{1}{2}mR^2$$

The kinematical relationship between the angular velocity and the linear velocity is:

$$v_G = -R\omega$$

or

$$\omega^2 = \left(\frac{v_G}{r}\right)^2$$

Substituting these relationhips into the mechanical energy accounting equation gives:

$$0 = \left(\frac{mv_G^2}{2} + \frac{1}{2}\left(\frac{1}{2}mr^2\right)\left(\frac{v_G}{r}\right)^2\right)_{\text{end}} + mg(h_{\text{end}} - h_{\text{beg}})$$

The mass of the cylinder is divided from both sides of the equation and the equation is simplified:

$$0 = \left(\frac{3}{4}\right)v_G^2 + g(h_{\text{end}} - h_{\text{beg}})$$

Rearranging, the velocity of the center of mass of the cylinder is:

$$v_G = \sqrt{\frac{4}{3}g(h_{\text{beg}} - h_{\text{end}})}$$

which, when evaluated, is:

$$v_G = \sqrt{\frac{4}{3}(32.\,174)(3 - 0)} = 11.\,3 \text{ (ft / s)}$$

The velocity of the the center of mass of the cylinder is 11.3 (ft/s). Comparing this to the result using linear and angular momentum the answer is the same, as it should be.

10.5 RIGID-BODY DYNAMICS PROBLEM-SOLVING TOOLS AND STRATEGY

The tools which have been presented in this and earlier chapters and which are needed for addressing and understanding rigid-body dynamics problems are summarized below.

1) **The Laws.** This includes, for rigid-body dynamics, linear momentum, angular momentum, and the conservation of energy (along with an accounting of mechanical energy).

2) **Kinematic Relations.** It is important to deliberately realize the role of the kinematic relations presented in this chapter. These relations are mathematics relations (as opposed to physics) between position, velocity, and acceleration vectors and between angular position, angular velocity and angular acceleration. These are required because the system (rigid body) is changing momentum and/or angular momentum and descriptions of this change, in a mathematical sense, are crucial to being able to state the physical laws. Obviously, understanding the several coordinate systems is essential. Tables 10-1, 2, 3, and 4 are important summaries of kinematical relations for rectangular cartesian, cylindrical, spherical, and path coordinates.

4) **Descriptions of forces.** This may include concepts like static and kinetic friction, gravity, linear and non-linear springs, buoyant force (chapter 11), electrostatic forces, etc.

4) **Process Constraints.** This tool refers to descriptive terms of the nature of the motion of rigid bodies such as: rotation without translation (of the center of mass), translation without rotation, rotation about a fixed point in the body (which is not the center of mass so that there is translation of the center of mass), rotation about a point which is fixed with respect to each point in the body. You must be aware of the impact of these types of motions on the form of the conservation of angular momentum law which you use (see Table 6-6). Other isssues are frictionless surfaces (if they exist), static versus kinetic friction forces, slip versus no-slip at points of contact of the rigid body with its surroundings, initial conditions (such as positon, velocity, and acceleration), etc.

5) **Moments of Inertia.** Of course, you must have values for the moment of inertia of the rigid body of interest and about the appropriate axis to characterize the shape and mass distribution of the body with respect to the axis. Appendix Table C-7 lists some common values of center-of-mass moments of inertia. The parallel axis theorem is also a tool that can be used to calculate the moment of inertia with respect to an axis which is shifted parallel to the axis through the center of mass.

6) **Calculations of Work.** When considering the conservation of energy or an accounting of mechanical energy, any transfer of energy due to work must be included, of course. This requires understanding how work is calculated in terms of the forces associated with it: (work) $= \int \mathbf{f} \cdot d\mathbf{s}$. If a force acts at or across a system boundary, then there must be a displacement associated with that force which has a component collinear with the force in order for there to be work. Remember also that gravity is a force but that work due to the displacements associated with it are normally accounted for through changes in system potential energy. It would be wrong to count gravity as both producing changes in the system potential energy *and* transferring energy as work.

7) **Mathematics.** Mathematics of course, is an important tool in stating and solving problems. It plays two roles. One is as a language for expressing the physical concepts and laws (including the kinematic relations) and the other is as a technique for solving the mathematics problem which is thus obtained.

As with rigid-body statics, strategies for solving dynamics problems begins with assessing each of the laws in turn and, perhaps, for different systems. Note also that selection of the point (or points) about which to calculate torques (for angular momentum) can be an important decision with respect to ease of solving the problem, or even as to whether a useful result is obtained. The point which is used need not even be inertial, but great care must be exercised in selecting the correct form of the conservation of angular momentum result (see Table 6-6 for planar motion). Kinematic relations play a separate role and must not be confused with the physical laws. Once the laws are expressed mathematically, then the kinematic relations can be substituted to obtain the needed differential equations to be solved.

10.6 REVIEW

1) The motion of a rigid body is governed by the conservation of linear and angular momentum.

2) Each momentum law, being a vector equation, provides three scalar equations for analyzing the rigid body. Thus, there are a total of six simultaneous scalar equations which may be used to understand the motion of a finite sized rigid body.

3) These six equations may not all be nontrivial however. For two dimensional (planar) motion, three equations are nontrivial. For translational motion without rotation in one dimension only one equation is independent.

4) Because dynamics is the study of motion, we must understand how to mathematically describe the translation and rotational motion of a body. Kinematics deals with the mathematical description of motion without addressing the cause or physics of the motion.

5) In order to describe the position and orientation of a body and its changes (motion), its position in a coordinate system must be defined and followed with the motion. A number of different coordinate systems are commonly used (rectangular, cylindrical, spherical, or others). The position in the coordinate system is then described in terms of coordinates in the three directions and the unit base vectors for the three directions.

6) Because, in describing motion, we are describing changes in the position in a coordinate system, the changes in the base vectors with position in a coordinate system must be considered. A rectangular cartesian system has the same base vectors (in terms of both magnitude and direction) at all points in the coordinate system. A curvilinear coordinate system has base vectors which are functions of position. Cylindrical, spherical, and path coordinate systems are curvilinear systems.

7) In some cases, there are constraints placed on the motion which provide important kinematic information. One set of examples are the rope and pulley problems.

8) The translational and rotational equations of motion (linear and angular momentum conservation), which incorporate the kinematical expressions, provide a set of equations which can be solved for the body position, velocity and acceleration, provided the problem is suitably posed with sufficient given information.

9) A mechanical energy equation also can provide very useful information for analyzing dynamics in an efficient way. However, such an equation does not contain as much information as the momentum equations, nor is it unique information.

10.7 NOTATION

The following notation has been used throughout this chapter. The bold face lower case letters represent vector quantities and the upper case boldface letters represent tensor quantites.

<u>Scalar Variables and Descriptions</u>		<u>Dimensions</u>
E	energy	[energy]
f	magnitude of the vector $\mathbf{f}$	[mass length / time^2]
g	magnitude of the acceleration of gravity	[length / time^2]
h	vertical displacement	[length]
$_G I_z$	center-of-mass moment of inertia about axis a	[mass Length^2]
L	length	[length]
m	total mass	[mass]
r, θ, z	spatial cylindrical variables	
r, θ, ϕ	spatial spherical variables	
t	time	[time]
W	work	[energy]
$\dot{W}$	rate of work or power	[energy / time]
x, y, z	spatial rectangular cartesian variables	[length]
μ	coefficient of friction	[length]
ρ	radius of curvature	[1 / length]

<u>Vector Variables and Descriptions</u>		<u>Dimensions</u>
$\mathbf{a}$	linear acceleration vector	[length / time^2]
$\mathbf{e}_r, \mathbf{e}_\theta, \mathbf{e}_z$	cylindrical unit directional vectors	
$\mathbf{e}_r, \mathbf{e}_\theta, \mathbf{e}_\phi$	spherical unit directional vectors	
$\mathbf{f}$	force	[force]
$\mathbf{g}$	acceleration of gravity	[length / time^2]
$\mathbf{i}, \mathbf{j}, \mathbf{k}$	cartesian unit directional vectors	
$\mathbf{l}$	angular momentum	[mass length^2 / time]
$\mathbf{p}$	linear momentum	[mass length / time]
$\mathbf{r}$	position	[length]

u	unit directional vector	
v	velocity	[length / time] or
α	angular acceleration vector	[1 / time^2]
ω	angular velocity vector	[1 / time]

<u>Subscripts</u>

B	binormal component
contact	a force acting only at the system surface as opposed to a body force such as gravity
ext	external to the system
G	corresponding to the center of mass of the system
k	kinetic
K	kinetic energy
N	corresponding to the normal component
P	potential energy
sys	corresponding to the system
T	corresponding to the tangential component

10.8 SUGGESTED READING

Beer, F. P., and E. R. Johnston, Jr., *Mechanics for Engineers, Statics and Dynamics*, 4th edition, McGraw-Hill, New York, 1987

Halliday, D. and R. Resnick, *Physics for Students of Science and Engineering*, 2nd edition, John Wiley & Sons, 1963

Hibbeler, R. C., *Engineering Mechanics, Statics and Dynamics*, 3rd edition, MacMillan, New York, 1983

Sears, F. W., M. W. Zemansky, and H. D. Young, *University Physics*, Addison-Wesley, Reading, Massachussetts, 1982

QUESTIONS

1) Of the fundamental laws and accounting statements which are presented in the first eight chapters, which ones provide useful information for analyzing the motion of dynamic rigid body systems? Explain why those which you have omitted are trivial.

2) Give a verbal statement of the conservation of linear momentum for a dynamic system which is a point of zero volume.

3) Give a symbolic statement of the conservation of linear momentum for a dynamic system which is of zero volume and give verbal interpretations of each term in the symbolic equation.

4) Answer the two previous questions for a dynamic rigid body of finite size. What kind of equations govern the dynamics of rigid bodies (algebraic, O.D.E, etc.).

5) How many independent equations govern the linear (straight line, one dimensional) motion of a point particle?

6) How many independent equations govern the two dimensional motion of a point mass?

7) How many independent equations govern the three dimensional motion of a point mass?

8) How many independent equations govern the full three dimensional motion of a rigid finite-sized body?

9) Consider a baseball hit by a bat. Because of the compressibility of the ball it deforms when struck by the bat and the impact occurs over a finite period of time; the ball and the bat travel together for a finite period of time. During this time, the force (or forces distributed over the contact area) result in a change of momentum of the ball. Write a symbolic integral expression for this change in momentum over the time period of contact in terms of the forces acting upon the ball which are considered to be unknown funtions of time.

10) Consider two situations; 1) two cars collide head on each with a speed of 30 mph in such a way that as they collide they deform in exactly the same way, that is, as exact mirror images. This would be a perfect headon collision of identical vehicles. In this way as each vehicle deforms and tries to move forward it collides with the other vehicle which doing the same thing, 2) a single car colliding with an immovable and non-deformable solid wall.

Which of these two situations subjects the passengers to the most severe impact? Substantiate your explanation with both momentum and energy considerations.

11) Considering again the question of a baseball being hit by a bat, what might you as the batter do as far as the motion of the bat is concerned (after all, this is all that the ball knows about) so as to increase the distance that the ball ultimately travels (i.e., to hit the ball out of the ballpark)?

12) Consider a car driving downhill which is brought to a stop without skidding by applying the brakes. a) From the point of view of the entire car (including the wheels) what force is responsible for stopping the car? b) Does this force perform any work on the car? Explain. c) In light of your answer to part b, give a statement of the conservation of total energy for this situation and system. d) Now consider the stopping car in the context of accounting for mechanical energy, still with the total car as the system. What happens to the kinetic and potential energy as the car is brought to a stop (where does it go?)? e) Answer questions a) through d) for a skidding stop with the wheels locked.

13) Define kinematics and its importance to the analysis of dynamics problems.

14) Consider a tree that has been sawed off at its base and is now falling to the ground. What forces act upon this tree in order that it falls and what role does each of these forces play? Consider that the base of the tree acts like a frictionless pin.

SCALES

1) Practice your scales for chapters 5 and 6.

2) Do problems 6 and 7, Chapter 10, as SCALES.

3) A baseball of 5 1/8 ounces is struck by a 2 lb_m bat.

   a) Draw a free-body diagram for the ball at one instant while it is in contact with the bat. Of course, include all forces acting on the ball.
   b) Draw a free-body diagram for the bat at the same instant while it is contact with the ball. Of course, include all forces acting on the bat.
   c) How is the force of the bat on the ball related to the force of the ball on the bat? Is this always true? Why or why not?

4) After being hit by the bat in exercise (3), the ball is moving through the air.

   a) Draw a free-body diagram for the ball. Include all forces acting on the ball.
   b) Draw a free-body diagram for the air at the boundary between the ball and the air.
   c) How is the drag force of the air on the ball related to the force of the ball on the air?

5) A ball is dropped from the inside top of an evacuated chamber (evacuated to remove air friction) and falls to the bottom.

a) Draw a free-body diagram for the ball while it is falling.

b) Draw a free-body diagram for the surroundings while the ball is falling.

c) How is the force of the surroundings on the ball related to the force of the ball on the surroundings? Explain. What is the nature or origin of the force.

6) A particle of mass m is moving at constant velocity. What is the net force acting on it?

7) A particle of mass m is moving in a circle of radius R at constant velocity. What is the net force acting on it?

8) A particle of mass m is moving at constant acceleration. What is the net force acting on it?

9) You are pushing a box of mass m across a floor at constant velocity.

a) What is the net force acting on the box?

b) Draw a free-body diagram for the box.

c) Draw a free-body diagram for yourself. How are the force that you exert on the box and that which it exerts on you related? Explain.

10) You are pushing a box of mass m across a floor at constant acceleration.

a) What is the net force acting on the box?

b) Draw a free-body diagram for the box.

c) Draw a free-body diagram for yourself. How are the force that you exert on the box and that which it exerts on you related? Explain.

11) You are pushing against a box of mass m but it is not moving.

a) Draw a free-body diagram for the box.

b) Draw a free-body diagram for yourself. How are the force that you exert on the box and that which it exerts on you related? Explain.

12) In each of exercises 9 through 11 are you doing work on the box? Explain.

13) A cylinder is rolling down an incline plane without slip.

a) Draw a free-body diagram for the cylinder.

b) Draw a free-body diagram for the incline plane. How are the forces which exist at the point of contact between the cylinder and the plane and which act on the cylinder related to those which act on the plane?

c) How much work does the friction force which acts on the cylinder by the plane do on the cylinder? Is it zero or not zero? Explain.

14) A tree is chopped at its base until it starts to fall. The fact that it is still connected prevents the tree from moving away from its base even though there still is a force acting there. How much work is done by the force acting at its base?

15) A ball of radius R and mass m is rolling across a horzontal table without slip. The velocity of its center of mass is v_G.

a) What is the velocity of the point which is in contact with the table?

b) What is the velocity of the point which is at the top of the ball, relative to v_G?

c) If the ball's angular velocity is ω, how are ω and the velocity of the center of mass related?

d) How are ω and the velocity of a point on the top of the ball related?

e) Are the answers to (b) and (d) consistent with each other?

16) A ball of radius R and mass m is both rolling and slipping such that the velocity of the point which is in contact with the table (relative to a fixed coordinate system) is v_{slip} and the velocity of the center of mass (relative to a fixed coordinate system) is v_G.

a) What is the velocity of a point on top of the ball (again relative to a fixed coordinate system)?
b) What is the angular velocity ω in terms of v_G and v_{slip}?

17) A ball of mass m which is falling through the air is found to obey the differential relation

$$m\frac{dv_y}{dt} = -kv_y - mg$$

a) Explain where this relation comes from.
b) Obtain an expression for v_y as a function of time if, when $t = 0$, $v_y = 0$
c) If $m = 0.2$ kg and $k = 0.05$ kg/s, plot v_y versus time using a computer graphics package (plot the analytical equation) or spreadsheet to generate values of v_y over time from the analytical solution of the ODE.
d) Using a spreadsheet, generate a table of approximate values of v_y versus time by adding to the previous value the amount of change which occurs over a small time increment

$$\text{(change in } v_y) = (-kv_y / m - g)(\text{change in } t)$$

and then graph the results.
e) Compare the graphs produced in (c) and (d)

18) Do exercise (17d) if the ball's fall is described by the non-linear ODE

$$m\frac{dv_y}{dt} = -k\sqrt{v_y} - mg$$

PROBLEMS

1) A baseball with a mass of 5 1/8 ounces is struck by a bat 2 (ft) above the center of home plate. The magnitude of the velocity of the ball after being hit is 110 miles per hour and the velocity is directed at an angle $35\,^{\circ}$ above the horizontal.

a) For the time period from the instant the ball leaves the bat to the time the ball strikes the earth, if we neglect air resistance what is the force on the ball. From the conservation of linear momentum, determine the velocity in the horizontal and vertical direction as a function of time. From the definition of velocity, calculate the position of the ball in the horizontal and vertical direction as a function of time. Determine the time after being struck that the ball lands on the earth. How far in the horizontal direction did the ball travel? What was the velocity vector of the ball as it struck the earth? Determine the time when the ball was at it's maximum height? What was the height of the ball? Was the hit a home run in most major league ball parks?
b) Now, the effects of air resistance will be included. Assume that the force due to air resistance is proportional to the velocity of the ball but acting in the opposite direction, i.e.,

$$\mathbf{f}_{air} = -k\mathbf{v} \text{ (lb}_m \text{ ft} / \text{s}^2)$$

where $k = 0.1$ (lb$_m$ / s). From the conservation of linear momentum, determine the velocity in the horizontal and vertical direction as a function of time. From the definition of velocity, calculate the position of the ball in the horizontal and vertical directions as a function of time. Determine the time when the ball strikes the earth. How far in the horizontal direction did the ball travel? What was the velocity vector of the ball as it struck the earth? Determine the time when the ball is at its maximum height. What was the maximum height of the ball. Was the hit a home run in most major league ball parks?

2) A flat plate of mass 100 (g) and radius of 70 (cm) is rotating about the center of mass aligned with the z-axis as shown in the figure. The center of mass is stationary. The initial angular velocity and final angular velocity are given by the following vector

$$\omega_{beg} = 0\mathbf{i} + 0\mathbf{j} + 20\mathbf{k} \text{ (rpm)}$$

$$\omega_{end} = 0\mathbf{i} + 0\mathbf{j} + 35\mathbf{k} \text{ (rpm)}$$

The change in angular rotation of the body happened over a period of 5 (s).

a) Calculate the average torque that was exerted on the body to change the angular momentum of the disk.
b) Calculate the initial and final kinetic energy of the body. Consider the conservation of energy and determine the work that was done on the body over the time period.

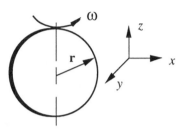

Problem 10-2: Rotating disk

3)

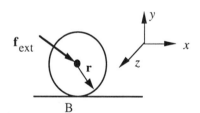

Problem 10-3: Rolling ball

A ball with a mass of 10 (lb$_m$) and a radius of 2 (ft) and a frictionless pin through the center is resting on a horizontal surface as shown in the figure. A variable force that is a continuous function of x acting through the center of mass of the ball acts on the ball for a distance of 4 (ft) in the x-direction as given in the table.

a) If the surface at point B is frictionless, calculate the linear velocity of the center of mass of the ball when it has moved 4 (ft).
b) If the surface at point B is not frictionless and the coefficient of friction is 0.4 and there is *no slip*, calculate the velocity of the center of mass and the angular velocity when the ball has moved 4 (ft).
c) Why or why isn't the value of the linear velocity the same in part a) and b)? Furthermore, is the total kinetic energy the same in a) and b) and why?

Problem 10-3 Data

x (ft)	f_x (lb$_f$)	f_y (lb$_f$)	f_z (lb$_f$)
0	1	0	0
1	2	-1	0
2	2	-2	0
3	1	-3	0
4	1	-4	0
5	2	-5	0
6	2	-6	0

4) A pulley system is shown in the figure where the pulleys may be approximated as frictionless and massless and the cord is massless. The mass of block A is 100 (lb$_m$) and the mass of block B is 50 (lb$_m$). The angle of the inclined plane is 40° and the angle that the cord makes with the wall is 50°. The coefficient of static friction is 0.3 and the coefficient of kinetic friction is 0.08. Determine the force in the cable and the motion of block A and B (Remember to check if the block is static.) In terms of energy for the system defined as both blocks, how is the energy of the system changing because of the rate of work being done on the system?

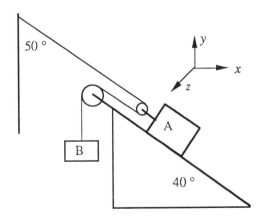

Problem 10-4: Sliding blocks

5) A bowling ball with mass m is released in the direction of the pins at velocity v_{Go} with no rotation. Initially, it slips against the surface of the lane. The friction force exerts a drag causing the ball to begin to rotate. The rotation is not immediately enough to match the translational velocity, however, and the ball still slips. After enough time, the rotation has increased (in magnitude) and the translational velocity has decreased (both due to the friction force) sufficiently for the ball to stop slipping.

a) Assess the motion of the ball using the conservation equations. You may want to consider both rate and finite time period forms in each case.

b) Show that when the ball has stopped slipping

$$v_G = \frac{5}{7}v_{Go}$$

(Note that this result does not involve a coefficient of friction.)

c) Calculate the total work done on the bowling ball due to the frictional force in terms of m and v_{Go}.
d) Make plots of velocity of the ball versus time and force acting on the ball versus time.

6) Given the following position vectors as a function of time, calculate the velocity and acceleration vectors. The coefficients a and b are constants.

a) $\mathbf{r} = \exp(t)\mathbf{i} + t^2\mathbf{j} + 3t\mathbf{k}$ (ft)
b) $\mathbf{r} = a\sin(t)\mathbf{i} + b\cos(t)\mathbf{j} + t^3\mathbf{k}$ (m)
c) $\mathbf{r} = a\mathbf{e}_r + t\mathbf{e}_z$ (m), $\theta = 2t$ (radians)   (Cylindrical)
d) $\mathbf{r} = \sin(t)\mathbf{e}_r + \exp(t)\mathbf{e}_z$ (m), $\theta = 3\sqrt{t}$ (radians)   (Cylindrical)
e) $\mathbf{r} = at\mathbf{e}_r$ (ft), $\theta = 4\sin(t)$, $\phi = 3\cos(t)$ (Spherical)

7) Given the following acceleration vectors and masses in the absence of a gravitational field, determine the net force causing the motion. (If the force is not constant, express it as a function of time.) For the initial velocity $\mathbf{v}_o = 3\mathbf{i} - 2\mathbf{j} + 1\mathbf{k}$ and position $\mathbf{r}_o = 2\mathbf{i} + 1\mathbf{j} - 3\mathbf{k}$, give the velocity and position as functions of time.

a) $\mathbf{a} = 5\mathbf{i} + 3\mathbf{j} - 2\mathbf{k}$ (ft / s^2) , $m = 30$(lb$_m$)   c) $\mathbf{a} = 0\mathbf{i} + 6\sin(t)\mathbf{j} + 0\mathbf{k}$ (ft / s^2) , $m = 5$(lb$_m$)
b) $\mathbf{a} = 5t\mathbf{i} + 0\mathbf{j} + 0\mathbf{k}$ (m / s^2) , $m = 2$(kg)

8) For Example 10-3, show that $\mathbf{u}_T(t) = (3\cos(3t)\mathbf{i} - 3\sin(3t)\mathbf{j} - 2\mathbf{k}) / \sqrt{13}$, $\mathbf{u}_N(t) = -(\sin(3t)\mathbf{i}+\cos(3t)\mathbf{j})$, $\mathbf{u}_B(t) = (-2\cos(3t)\mathbf{i}+ 2\sin(3t)\mathbf{j} - 3\mathbf{k}) / \sqrt{13}$, $\rho = 13 / 9$, and $1 / \sigma = 6 / \sqrt{13}$.

9) You are on a horizontal plane firing a projectile of 50 kg. The initial muzzle speed is 200 km/hr and the angle of inclination is 60 ° above the horizon. The only force acting on the projectile is the force of gravity.

a) Use the conservation of linear momentum to determine the position and velocity of the projectile as a function of time.
b) Determine the maximum height above the plane. Use both momentum and energy.
c) Determine the horizontal distance travelled when the projectile impacts the plane.
d) Determine the time of flight.
e) Determine the magnitude of the velocity (speed) at the instant of impact.

10) You are firing projectiles from a hill 200 ft above a valley floor. The projectile is 100 lb$_m$. Gravity is the only force acting on the projectile. Neglect the drag forces due to air resistance. The initial speed of the projectile is 100 mph and the direction is $\mathbf{u} = (1 / \sqrt{2})\mathbf{i} + (1 / \sqrt{2})\mathbf{j} + 0\mathbf{k}$. Gravity acts in the negative $\mathbf{j}$ direction.

a) Use the conservation of linear momentum to determine the position and velocity of the projectile as a function of time.
b) Determine the maximum height above the valley floor. Use both momentum and energy.
c) Determine the horizontal (x) distance travelled when the projectile impacts the valley floor.
d) Determine the time of flight.
e) Determine the magnitude of the velocity (speed) at the instant of impact.

11) A particle with mass 20 kg is subjected to the following constant force $2\mathbf{i} + 0\mathbf{j} - 3\mathbf{k}$ (N). Gravity does not act on the particle. The initial velocity is $0\mathbf{i} + 0\mathbf{j} + 4\mathbf{k}$ (m / s)

a) Determine the velocity of the particle as a function of time.

b) At what time is the speed minimum.

12) A particle of mass 50 lb_m has an initial velocity $\mathbf{v}_o = 50\mathbf{i} + 0\mathbf{j} + 0\mathbf{k}$ (ft / s), and position $\mathbf{r}_o = 0\mathbf{i} + 3000\mathbf{j} + 0\mathbf{k}$ (ft). Gravity acts on the particle in the negative $\mathbf{j}$ direction.

   a) Determine the velocity and the position of the particle as a function of time.
   b) Determine the time it iake to hit the surface $\mathbf{r} = 0\mathbf{j}$.
   c) Determine the velocity vector and the speed at impact.

13) A 30 kg particle is acted on by two (2) variable forces, $\mathbf{f}_1$ and $\mathbf{f}_2$. The initial velocity is $10\mathbf{i} + 12\mathbf{j} + 10\mathbf{k}$ (m / s).

$$\mathbf{f}_1 = 3\mathbf{i} - 5t\mathbf{j} + 0\mathbf{k} \text{ N}$$

$$\mathbf{f}_2 = 2t\mathbf{i} - 0\mathbf{j} - 4\mathbf{k} \text{ N}$$

where t is in seconds. Gravity does not act on the particle.

   a) For $\mathbf{f}_1$, what is the units on the 5 in the $\mathbf{j}$ component expression for the equation to be dimensionally consistant? For $\mathbf{f}_2$, what is the units on the 2 in the $\mathbf{i}$ component expression for the equation to be dimensionally consistant?
   b) Determine the velocity of the particle as a function of time.
   c) Determine the speed of the particle as a function of time.
   d) What is the velocity vector and speed of the particle at 5 s.

14) A 3 lb_m particle is revolving at a constant angular velocity of 5 rad/s. The particle is attached to a massless rope with a radius of 2 ft. The rope is attached to a spindle 1 ft above the ground. Gravity does not act on the particle. Neglect any drag forces due to air resistance. If the particle is spinning perpendidular to the acting gravity, calculate the tension in the rope.

15) A child has put 2 lb_m of water in a bucket. The childs arm is only 2 ft long. Neglect the mass of the bucket and the mass of the childs arms. The child wishes to rotate the bucket in a vertical plane without having the water spill. Determine the angular velocity to insure that no water spills on the child. At this constant angular velocity what is the tension in the childs arm when the bucket is closest to the ground.

16) A mass m is attached to a massless spring of force constant k. The mass and spring hang vertically. At some initial time, a person pulls down on the spring and lets go. Gravity acts on the mass but air resistance is negligible. The force due to gravity is $\mathbf{f}_g = m\mathbf{g}$. The force due to the spring is proportional to the position and acts in the opposite direction: $\mathbf{f}_{spring} = -k\mathbf{r}$. Define the system as the mass. Use the conservation of linear momentum to develop a relationship between the position, $\mathbf{r}$, and t, m, k, and $\mathbf{g}$. Do not integrate the equation.

17) As described in problem 10-16 a mass and a massless spring hang vetically. At some initial time, a person pulls down on the spring and lets go. Gravity acts on the mass and this time there is a drag force from the air resistance. The force due to gravity is $\mathbf{f}_g = m\mathbf{g}$. The force due to the spring is proportional to the position and acts in the opposite direction: $\mathbf{f}_{spring} = -k\mathbf{r}$. The drag force is proportional to the velocity and acts in the opposite direction. $\mathbf{f}_{drag} = -c\mathbf{v}$. Define the system as the mass. Use the conservation of linear momentum to develop a relationship between the position, $\mathbf{r}$, and t, m, k, c and $\mathbf{g}$. Do not integrate the equation.

18) A solid sphere of 5 kg and radius 0.5 m rests on the top on an inclined plane. The plane is 5 m high and the angle of inclination if 30 °. Determine the kinetic energy and the velocity of the center of mass of the sphere at the bottom of the inclined plane for the following cases.

   a) The plane is frictionless.
   b) The plane has friction. (The sphere does not slip.)

19) A lumberjack is chopping down a 75 ft tall tree. The mass of the tree is about 2 tons. Assume that the tree acts like a pin supported slender rod. Determine the kinetic energy and the velocity of the tip of the tree as it strikes the ground.

PART IV

MACROSCOPIC FLUID MECHANICS

In this section we consider the mechanics of fluids and the behavior of flow processes from a macroscopic viewpoint. That is, we do not consider the details of fluid flow to the point of being concerned with velocity profiles ($v = v(\mathbf{x})$), pressure distribution ($P = P(\mathbf{x})$), or other such details (with the exception of static fluids, for which we will obtain expressions for pressure variation with position). Instead we characterize fluid flow through pipes and channels in terms of average properties (average velocity, average pressure, etc.) across the cross-sectional area of the conduit. Then we calculate the mass flow rate through a conduit in terms of this average velocity. Also, we calculate the momentum flux (momentum/time/area) at a given cross section in terms of this average velocity and we calculate energy properties in terms of values of temperature, pressure, composition, elevation, velocity, etc., averaged over this cross-sectional area.

Accordingly, the equations which we use for analyzing such processes are the macroscopic equations presented in Chapter 1 through 8 representing the conservation and accounting of mass, the conservation and accounting energy equations (including the extended Bernoulli steady state mechanical energy equation), a macroscopic accounting equation for entropy, and a macroscopic equation for the conservation of linear and angular momentum. Usually we do not have to consider charge conservation. Furthermore, in designing fluid piping systems (i.e., sizing piping networks, pump requirements, piping support requirements etc.), we normally need only be concerned with mechanical energy so that we are able to use primarily the Bernoulli equation for flow (together with momentum). Neither the total energy conservation equation nor the thermal energy accounting equation is usually required.

This section consists of two chapters. The first deals with fluid statics. Here we begin with the equation of fluid statics which is obtained from considerations of the conservation of linear momentum (Chapter 5). Using this equation we are able to calculate the pressure in a fluid as a function of position when that fluid is static. If the system is multiphase, that is if there are layers of fluids which form different phases (e.g. water and oil), then, because of their different densities, pressure will vary differently with depth and we must take this into account. Additionally in Chapter 11 we use this pressure-depth relation to calculate the forces which are exerted upon submerged objects and express this as an equivalent net force acting at a single point (as we did in rigid body statics when we considered distributed loads acting on objects). The consideration of these distributed forces is useful for designing the support required for structures such as dams. We also use considerations of pressure-depth relations in evaluating buoyancy forces. Finally, in fluid statics we will consider the design of thin-walled pressure vessels and the pressure drop which occurs across curved static fluid interfaces.

In Chapter 12 we consider macroscopic fluid dynamics. Here we are concerned with the flow of fluid through conduits and with pumps and compressors for providing the energy to flow of fluids through pipes. Also, we will consider the flow of incompressible fluids in closed conduits, including a discussion of pressure-measurement devices. Finally we consider the forces which are associated with fluid flow and "rocket-type" problems.

11.1 INTRODUCTION

Fluid statics addresses the variation of pressure in static fluids and the consequent forces which are exerted on submerged surfaces. As with rigid body statics, the conservation of mass is of only trivial consequence as is the conservation of charge and the conservation of energy. In a static situation nothing is moving so that mass and energy are not crossing system boundaries. Likewise because a static situation is at equilibrium, entropy and the second law become a trivial consideration. We are left then with the conservation of linear and angular momentum.

In Chapter 5 the basic equation of fluid statics was presented. This result says that the pressure in a static fluid varies from point to point because of differences in the weight of the fluid which exists above each point. This weight depends upon the fluid density which in fact, may vary with position because of pressure differences. We will consider applications of pressure-depth relations in this chapter.

Additionally we will consider the pressure difference which exists across curved static interfaces between two fluid phases; if there is a static curved interface, then there is a pressure difference. This phenomenon is important in situations which involve the capillary flow of fluids such as the displacement of fluids through porous media and the wick action which is responsible for the rise of a fluid through a fabric material to a level above that of the bulk liquid.

This chapter will present, first the general situation of the variation of pressure throughout a single phase fluid. As mentioned above, this pressure varies because of elevation in the gravity field, i.e. it varies because of the weight of the fluid. Second, using this dependence of pressure on position, we will calculate the forces which are exerted on submerged areas or structures by the fluid pressure. The calculation of these forces is essential for designing objects to withstand pressure. Next we will discuss the phenomenon of buoyancy, why it exists and how to calculate the buoyant force. Then we will see how to design thin-walled pressure vessels to withstand a given fluid pressure. Finally, we will use this knowledge of thin-walled pressure vessels to address the question of the pressure drop which exists across a curved interface in a fluid. This is applicable to the calculation of capillary pressures and forces necessary to force a fluid through small openings.

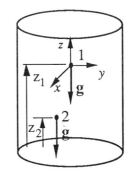

Figure 11-1: Pressure as a function of height

11.2 THE DEPENDENCE OF PRESSURE ON POSITION IN A STATIC FLUID

In Chapter 5 we obtained a relation for the dependence of pressure on elevation in a gravity field. If the coordinat system is aligned such that the z axis is aligned with the action of gravity and with the positive direction of the z axi representing increasing "height" as shown in Figure 11-1 then we saw that:

$$\left(\frac{dP}{dz}\right) = \rho g_z = -\rho(z)g$$

(5 − 16

That is, at any point in the fluid a pressure change is observed to occur upon displacement in the z direction. Thi result was a direct consequence of the conservation of linear momentum applied to static fluids; the sum of the forces at an point in the fluid must be balanced. Pressure acting on all sides around a fluid element must counteract or exactly balanc the weight of that element of fluid for it to be stationary. Furthermore, with the z axis oriented in the direction of gravit the variation of pressure in either the x or y directions is zero.

If the coordinate system is not aligned with gravity as shown in Figure 11-2, then the pressure will vary in each of the coordinate directions according to the component of the gravity vector which is in that same direction.

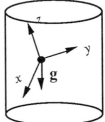

Figure 11-2: Pressure as a function of position with non-aligned coordinate axis

Accordingly:

$$\left(\frac{\partial P}{\partial x}\right)_{y,z} = \rho g_x$$

$$\left(\frac{\partial P}{\partial y}\right)_{x,z} = \rho g_y \qquad (11-1)$$

$$\left(\frac{\partial P}{\partial z}\right)_{x,y} = \rho g_z$$

This is the general fluid statics result but we will not consider it further at this time.

Returning now to the result where pressure varies in only one direction (say the z direction) we see that this can be integrated directly for *constant density* fluids such as liquids and constant magnitude of gravity. In this case then we have

$$P_2 - P_1 = -\rho g(z_2 - z_1) \qquad (11-2)$$

This expression can be rearranged to give:

$$P_2 = P_1 + \rho g(z_1 - z_2) \qquad (11-3)$$

where $(z_1 - z_2)$ is the depth of point 2 below point 1. In other words, as long as we are moving around within the same phase of constant density without crossing into or through another fluid (again, density does not change), then the pressure within that static fluid varies according to depth by the above equation. This result is totally independent of the route taken between two points. If we have layers of static fluids then we can move around in any one layer according to this relation provided that we use the density which is appropriate for that phase. Where two phases are in contact at an interface, then we have the constraint that the pressures in these two phases are equal (at least so long as the interface is flat, see Section 11.6).

If the density of the fluid is a function of position (i.e., of pressure) within a single phase (as it would be for a gas which is a compressible fluid) then the integration of equation (5-16) must be done while allowing for this variation of density with pressure. For example, if the fluid is an ideal gas at uniform temperature then we have the ideal gas law relation for density as a function of pressure and temperature:

$$\rho = \frac{n(MW)}{V}$$

$$= \frac{P(MW)}{RT} \qquad (11-4)$$

where n is the number of moles, MW is the molecular weight (mass per mole), V is the volume, R is the universal gas constant, and T is the absolute temperature. Using this relation in (5-16) gives:

$$\left(\frac{dP}{dz}\right) = -\rho(z)g$$

$$= -\left(\frac{P(MW)}{RT}\right)g$$

Separating variables and integrating from point 1 to point 2:

$$\frac{dP}{P} = -\left(\frac{(MW)}{RT}\right)g\,dz$$

$$\int_{P_1}^{P_2}\frac{dP}{P} = -\int_{z_1}^{z_2}\left(\frac{(MW)}{RT}g\right)dz$$

If the temperature is constant throughout the integration, the temperture can be taken out of the integral. The molecular weight, gas constant, and the force due to gravity are all constant, giving:

$$\ln\left(\frac{P_2}{P_1}\right) = -\left(\frac{(MW)g}{RT}\right)(z_2 - z_1)$$

This equation can be rearranged as:

$$P_2 = P_1 \exp\left(-\frac{(MW)g}{RT}(z_2 - z_1)\right) \tag{11-5}$$

This isothermal pressure equation for gases is reasonable so long as the elevation changes are not very great. Figure 11-3 shows a plot of the isothermal expansion of air at two different temperatures. However, on an atmospheric scale over distances of several thousand feet there certainly will be significant changes of temperature with elevation.

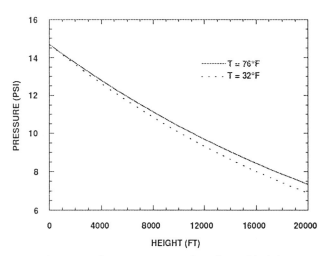

Figure 11-3: Pressure as a function of height
for an (obviously fictional) isothermal atmosphere

In this case we should account for the variation of density with pressure using a more realistic equation. We might use an equation which is obtained assuming that an element of the gas changes elevation following an adiabatic and reversible expansion. In this case we obtain (details are deferred to the chapters on thermodynamics):

$$P_2 = P_1\left(\frac{T_2}{T_1}\right)^{(\widetilde{C}_P/R)} \tag{11-6}$$

In reality, in atmospheric conditions the variation of temperature with elevation is not as great as would be indicated by this equation.

Example 11-1. For the dam shown in Figure 11-4, find the pressure acting on its face 15 (ft) below the surface of the lake.

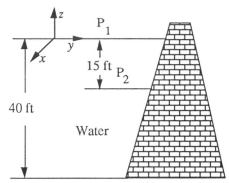

Figure 11-4: Pressure on the face of a dam

SOLUTION: The cartesian coordinate system is oriented in the direction of gravity with the origin at the lake level. The density of liquid water is assumed to be constant at a value of:

$$\rho_{H_2O} = 62.4 \ (\text{lb}_m \ / \ \text{ft}^3)$$

The pressure at point 2 is found by integrating equation (5-16), and for this case is:

$$P_2 = P_1 + \rho_{H_2O}g(z_1 - z_2)$$

The pressure at P_1 is atmospheric pressure. Substituting into the previous equation gives:

$$P_2 = (14.7 \ \text{psi}) + (62.4)(32.174)(0 - (-15)) \ \left(\frac{\text{lb}_m \ \text{ft}^2}{\text{ft}^3 \ \text{s}^2} \right)$$

The value for z_2 is negative because the coordinate system was defined at the level of the lake where P_1 was known. The position of z_2 was below the origin thus requiring the negative sign. Substituting in the proper conversion factors gives:

$$P_2 = 14.7 \ \text{psi} + 30,114.9 \frac{\text{lb}_m \ \text{ft}^2}{\text{ft}^3 \ \text{s}^2} \left| \frac{1 \ \text{lb}_f}{32.174 \ \text{lb}_m \ \text{ft} \ / \ \text{s}^2} \right| \frac{1 \ \text{ft}^2}{12^2 \ \text{in}^2} \left| \frac{1 \ \text{psi}}{1 \ \text{lb}_f \ / \ \text{in}^2} \right.$$

$$= (14.7 + 6.5) \ \text{psi} = 21.2 \ \text{psi}$$

The pressure at point 2 fifteen feet below the level of the dam is 21.2 (psi). *Pressure increases with depth; $P = P_o + \rho g$(depth).*

Example 11-2. For the tank containing acetone under a pressure (above atmospheric) of 25 kPa, calculate the pressure at the bottom of the tank if the liquid level is 3 m above the bottom of the tank. Figure 11-5 shows a schematic of the tank. The density of the acetone is $\rho_{Ace} = 792 \ \text{kg} \ / \ \text{m}^3$.

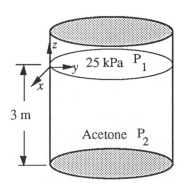

Figure 11-5: Pressure on the bottom of a tank

SOLUTION: A cartesian coordinate system is aligned with the direction of gravity with the origin at the top of the liquid level. As before,

$$P_2 = P_1 + \rho_{Ace}g(z_1 - z_2)$$

where point 1 is at the acetone surface and point 2 is at the bottom of the tank. Substituting numerical values gives:

$$P_2 = (25 \text{ kPa} + (792)(9.8)(0 - (-3))) \left(\frac{\text{kg m}^2}{\text{m}^3 \text{ s}^2} \right)$$

Again the value of z_2 was negative because the origin was defined at the top liquid level and z_2 was 3 (m) below the origin. With the required conversion factors, the answer is:

$$P_2 = 25\text{kPa} + 23284.8 \frac{\text{kg m}^2}{\text{m}^3 \text{ s}^2} \left| \frac{1 \text{ N}}{1 \text{ kg m} / \text{s}^2} \right| \frac{1 \text{ Pa}}{1 \text{ N} / \text{m}^2} \left| \frac{1 \text{ kPa}}{1000 \text{ Pa}} \right.$$

$$= (25 + 23.3) \text{ kPa} = 48.3 \text{ kPa}$$

The pressure at the bottom of the tank is 48.3 (kPa).

Example 11-3. The closed manometer shown in Figure 11-6 is at 20 °C. (A manometer is a device for measuring pressure differences. It consists of a bent tube filled with liquid and open at each end. Different pressures applied to the two ends result in a different liquid level in the two "legs" of the manometer. By measuring the difference in these levels, the pressure difference can be determined, provided the density of the manometer liquid is known.) If the pressure at point A is 70,000 (Pa), what is the pressure at point B? If we neglect the density of the air, what is the percent error in the calculation? The density of air and water are given $\rho_{air} = 1.2 \text{ kg} / \text{m}^3$ and $\rho_{H_2O} = 1000 \text{ kg} / \text{m}^3$.

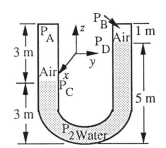

Figure 11-6: Pressure in manometer with two different liquids

SOLUTION: A cartesian coordinate system oriented with gravity is defined with the origin at the top of the inside of the manometer. The conservation of linear momentum in the z direction says:

$$\left(\frac{dP}{dz} \right) = -\rho(z)g$$

For this problem the density is a function of the position because we have two different types of fluids.

The pressure at the bottom of the manometer is the same whether obtained from the left or right arm of the manometer. For the left arm:

$$P_2 = P_A + \rho_{air}g(z_A - z_C) + \rho_{H_2O}g(z_C - z_2)$$

From the right arm

$$P_2 = P_B + \rho_{air}g(z_B - z_D) + \rho_{H_2O}g(z_D - z_2)$$

Equating these two results and solving for the unknown P_B gives:

$$P_B = P_A + \rho_{air}g(z_A - z_C) + \rho_{H_2O}g(z_C - z_2) - \rho_{H_2O}g(z_D - z_2) - \rho_{air}g(z_B - z_D)$$

Combining terms gives

$$P_B = P_A + \rho_{H_2O}g(z_2 - z_D - z_2 + z_C) + \rho_{air}g(z_D - z_B - z_C + z_A)$$

$$= P_A + \rho_{air}g(z_A - z_C) - \rho_{H_2O}g(z_D - z_C) - \rho_{air}g(z_B - z_D)$$

Substitution gives:

$$P_B = (70,000 \text{ Pa}) + \left[(1.2)(9.8)(3) - (1000)(9.8)(5 - 3) - (1.2)(9.8)(1)\right] \left(\frac{\text{kg m}^2}{\text{m}^3 \text{ s}^2}\right)$$

This last result is easily understood by realizing that the pressure in the water in the left leg at point C is the same as that at the same level in the right leg. Then, the pressure at point B can be obtained from that at point A by adding pressure due to moving deeper in the air in the left leg, and then subtracting pressure due to moving up through the water in the right leg and then through air, also in the right leg.

or

$$P_B = 70,000 \text{ Pa} + (-19600 + 23.52)\frac{\text{kg m}^2}{\text{m}^3 \text{ s}^2}\bigg|\frac{1 \text{ N}}{1 \text{ kg m} / \text{s}^2}\bigg|\frac{1 \text{ Pa}}{1 \text{ N} / \text{m}^2}$$

$$= (70,000 - 19576) \text{ Pa} = 50,424 \text{ Pa}$$

The pressure at point B in the manometer is 50,424 (Pa).

If we neglect the weight of the air we are assuming that the density of the air is close to zero as compared with the density of the water. This assumption reduces the expression for the presure as point B to:

$$P_B^* = P_A - \rho_{H_2O}g(z_D - z_C)$$

Substituting into this expression gives:

$$P_B^* = 70,000 \text{ Pa} - \left[(1000)(9.8)(5 - 3)\right] \left(\frac{\text{kg m}^2}{\text{m}^3 \text{ s}^2}\right)$$

$$= 70,000 \text{ Pa} - (19600)\frac{\text{kg m}^2}{\text{m}^3 \text{ s}^2}\bigg|\frac{1 \text{ N}}{1 \text{ kg m} / \text{s}^2}\bigg|\frac{1 \text{ Pa}}{1 \text{ N} / \text{m}^2}$$

$$= 50,400 \text{ Pa}$$

The percent error in this calculation is:

$$\%_{\text{err}} = \left|\frac{P_B - P_B^*}{P_B}\right| \times 100 = 0.046 \%$$

There is only 0.046 percent error in assuming that the density of air was approximately zero when compared to the density of water.

Example 11-4. The open manometer, contains two different immiscible liquids, A and B, as shown in Figure 11-7. Find (a) the elevation of the liquid surface in arm B, and (b) the total pressure at the bottom of the manometer (point 2). The density of the liquids A and B is given $\rho_A = 0.72$ g / mL and $\rho_B = 2.36$ g / mL.

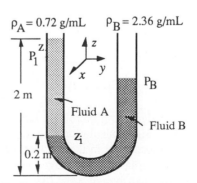

Figure 11-7: Pressure in open manometer with two different liquids

SOLUTION: A cartesian coordinate system aligned with gravity is defined with the origin at the top of liquid level A. Calculating as before, for the left arm of the manometer,

$$P_2 = P_1 + \rho_A g(z_1 - z_i) + \rho_B g(z_i - z_2)$$

where z_i is the position of the fluid interface.

For the right arm,

$$P_2 = P_B + \rho_B g(z_B - z_2)$$

Equating these two results and recognizing that P_B is equal to P_1 allows us to solve for z_B

$$\rho_B g(z_2 - z_B) = \rho_B g(z_2 - z_i) + \rho_A g(z_i - z_1) - P_1 + P_B$$
$$= \rho_B g(z_2 - z_i) + \rho_A g(z_i - z_1)$$

Dividing both sides of the equation by $\rho_B g$ gives:

$$(z_2 - z_B) = (z_2 - z_i) + \left(\frac{\rho_A}{\rho_B}\right)(z_i - z_1)$$

and finally:

$$z_B = z_i - \left(\frac{\rho_A}{\rho_B}\right)(z_i - z_1)$$

Substitution gives:

$$z_B = (-1.8) - \left(\frac{0.72}{2.63}\right)(-1.8 - 0) \text{ (m)} = -1.3 \text{ (m)}$$

The level of fluid B is 1.3 meters below the level of the defined coordinate system or 0.7 meters above the bottom of the manometer

The final part of the problem is calculated by evaluating the expression for the pressure at point 2 by any of the derived equations Since the expression from the right arm of the manometer contains only one term it will be the easiest to evaluate. Substituting values and conversion factors gives:

$$P_2 = 101 \text{ kPa} - (2.36)(9.8)\left(-2 - (-1.3)\right)\left(\frac{\text{g m}^2}{\text{mL s}^2}\right)$$

$$= 101 \text{ kPa} + 16.2\frac{\text{g m}^2}{\text{mL s}^2}\left|\frac{1 \text{ kg}}{1000 \text{ g}}\right|\frac{1 \text{ mL}}{1 \text{ cm}^3}\left|\frac{100^3 \text{ cm}^3}{1 \text{ m}^3}\right|\frac{1 \text{ N}}{1 \text{ kg m}/\text{s}^2}\left|\frac{1 \text{ Pa}}{1 \text{ N}/\text{m}^2}\right|\frac{1 \text{ kPa}}{1000 \text{ Pa}}$$

$$= (101 + 16.2) \text{ kPa} = 117.2 \text{ kPa}$$

11.3 FORCES ON SUBMERGED AREAS

With the knowledge of pressure as a function of depth we are now equipped to calculate the force which is exerted by a fluid on a surface. Such calculations are important for the design of storage tanks, pressure vessels, and structures such as dams. They are also essential in the calculation of buoyancy forces.

The pressure acting upon a submerged object by a fluid is a distributive force and can be treated accordingly, as was done in Chapter 9. The distributive force (per area) integrated over the entire area is used to calculate the total force acting by the pressure on the object:

$$
\boxed{
\begin{aligned}
\mathbf{f}_{eq} &= \int \mathbf{d}_A(\mathbf{r})\,dA \\
&= \int P(\mathbf{r})\mathbf{u}_N\,dA
\end{aligned}
}
\tag{11-7}
$$

Pressure acts perpendicular to a surface and so a unit vector which is normal to the surface establishes the direction of the force while the pressure is its magnitude. A similar integration of the torque exerted by the pressure can be used to calculate an equivalent point of application of this force:

$$
\boxed{
\begin{aligned}
\bar{\mathbf{r}} \times \mathbf{f}_{eq} &= \int \mathbf{r} \times \mathbf{d}_A(\mathbf{r})\,dA \\
&= \int \mathbf{r} \times P(\mathbf{r})\mathbf{u}_N\,dA
\end{aligned}
}
\tag{11-8}
$$

Knowledge of the total force and torque exerted by the fluid can be used to design the supporting members for a submerged object as well as the thickness (and other design variables) of the object itself.

The pressure-area product gives the *magnitude* of the force; the unit normal gives the *direction*. As an example, consider the pressure exerted by a fluid on one side of a curved surface as shown in Figure 11-8. For simplicity, the surface is taken to be curved in only one direction; it is flat in the direction which is parallel to the surface of the fluid. With the z axis aligned with this flat direction of the surface, the shape of the curve can be defined by some function of x, that is, $y = h(x)$.

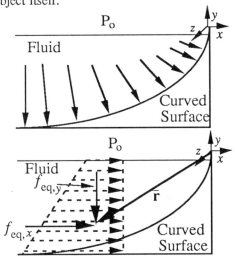

Figure 11-8: Pressure on a curved surface and equivalent force

Equivalently, we can say that the surface is defined by the relation:

$$
h(x) - y = 0
\tag{11-9}
$$

which fits the general form:

$$
f(x, y, z) = h(x) - y = \text{constant}
\tag{11-10}
$$

where the constant in this case is zero. Considering this general form we can see that as long as we move around within the surface f does not change, $(df)_A = 0$. Because f is a function of the independent variables x, y, and z we write this total differential as

$$df = \left(\frac{\partial f}{\partial x}\right)_{y,z} dx + \left(\frac{\partial f}{\partial y}\right)_{x,z} dy + \left(\frac{\partial f}{\partial z}\right)_{x,y} dz \qquad (11-11)$$

which can be written in terms of the del operator and a differential change in the position vector tangent to the surface as:

$$df = \nabla f \cdot (d\mathbf{r})_{\text{tangent}} = 0 \qquad (11-12)$$

or

$$df = \left(\left(\frac{\partial f}{\partial x}\right)_{y,z} \mathbf{i} + \left(\frac{\partial f}{\partial y}\right)_{x,z} \mathbf{j} + \left(\frac{\partial f}{\partial z}\right)_{x,y} \mathbf{k}\right) \cdot (dx\mathbf{i} + dy\mathbf{j} + dz\mathbf{k})_{\text{tan}}$$

$$= \left(\frac{\partial f}{\partial x}\right)_{y,z} dx + \left(\frac{\partial f}{\partial y}\right)_{x,z} dy + \left(\frac{\partial f}{\partial z}\right)_{x,y} dz \qquad (11-13)$$

This is a general result for any surface defined by equation (11-10). In this result, ∇f is a vector which, when dotted with a vector which is tangent to the surface $(d\mathbf{r})_{\text{tan}}$ gives the scalar zero; ∇f is normal to the surface. From this we can obtain a unit normal vector by normalizing ∇f.

$$\mathbf{u}_N = \frac{\nabla f}{|\nabla f|} \qquad (11-14)$$

Applying this to the problem at hand we see that:

$$\nabla f(x,y,z) = \left(\frac{\partial f(x,y,z)}{\partial x}\right)\mathbf{i} + \left(\frac{\partial f(x,y,z)}{\partial y}\right)\mathbf{j} + \left(\frac{\partial f(x,y,z)}{\partial z}\right)\mathbf{k}$$

$$= \left(\frac{dh(x)}{dx}\right)\mathbf{i} + \left(\frac{d(-y)}{dy}\right)\mathbf{j} + 0\mathbf{k}$$

$$= \left(\frac{dh(x)}{dx}\right)\mathbf{i} - \mathbf{j} + 0\mathbf{k}$$

And the magnitude of ∇f is:

$$|\nabla f| = \sqrt{\left(\frac{dh(x)}{dx}\right)^2 + 1}$$

Therefore, a unit normal vector to the surface of interest is given by:

$$\mathbf{u}_N = \frac{\left(\frac{dh(x)}{dx}\right)\mathbf{i} - \mathbf{j}}{\sqrt{\left(\frac{dh(x)}{dx}\right)^2 + 1}} = \frac{\left(\frac{dy}{dx}\right)\mathbf{i} - \mathbf{j}}{\sqrt{\left(\frac{dy}{dx}\right)^2 + 1}} \qquad (11-15)$$

We now wish to obtain the force exerted by pressure upon this area. The pressure is isotropic, that is, at a point it is exerted equally in all directions so that on a surface, the force which is exerted by pressure is normal to the surface. Consequently, the total force exerted at a point on the surface is calculated by integrating $P(\mathbf{r})\mathbf{u}_N$ over the area:

$$\mathbf{f}(\mathbf{r}) = \int_A P(\mathbf{r})\mathbf{u}_N dA \tag{11 - 16}$$

Now, the pressure, P, is a function of position, $P = P(\mathbf{r})$, as discussed previously. For example, if the cartesian coordinate system is defined such that the y axis is aligned with the force per mass due to gravity, and the density of the fluid is constant, then the pressure as a function of position is:

$$P(y) = P_o + \rho g(y_o - y)$$

where y_o and P_o are the values at the defined origin.

The normal vector is calculated as above and the differential area can be calculated according to

$$dA = wds \tag{11 - 17}$$

where w is the width of the surface taken parallel to the surface of the fluid and ds is the arc length along the curve. For given dx and dy along the curve,

$$ds = \sqrt{(dx)^2 + (dy)^2} \tag{11 - 18}$$

or

$$ds = \sqrt{1 + \left(\frac{dy}{dx}\right)^2}\, dx \tag{11 - 19}$$

so that we can write

$$dA = wds$$
$$= w\sqrt{1 + \left(\frac{dy}{dx}\right)^2}\, dx \tag{11 - 20}$$

Putting equations (11-15) and (11-20) into (11-16) and integrating along the entire curve which defines the area, we obtain the total force exerted by this pressure.

$$\mathbf{f}_{eq} = \int_C P(\mathbf{r}) \left(\frac{\frac{dy}{dx}\mathbf{i} - \mathbf{j}}{\sqrt{1 + \left(\frac{dy}{dx}\right)^2}} \right) w\sqrt{1 + \left(\frac{dy}{dx}\right)^2}\, dx \tag{11 - 21}$$

$$\mathbf{f}_{eq} = \int_C P(\mathbf{r}) w \left(\frac{dy}{dx}\mathbf{i} - \mathbf{j} \right) dx$$

where the integral is calculated along the path C. From this result we see that the x and y components of this force are given by:

$$
\begin{aligned}
f_{\text{eq},x} &= \int_C P(\mathbf{r})w\left(\frac{dy}{dx}\right)dx \\
&= \int_C P(\mathbf{r})w\,dy
\end{aligned}
$$

(11 − 22)

and

$$
f_{\text{eq},y} = -\int_C P(\mathbf{r})w\,dx
$$

(11 − 23)

Note that if $P = P(y)$ (i.e., if P is independent of x), then $f_{\text{eq},x}$ is independent of $y(x)$; the force depends upon the pressure-depth relation but not upon the shape of the curve that defines the surface. However, $f_{\text{eq},y}$ does depend upon $y(x)$, i.e., upon the shape of the surface.

These results have a rather simple geometrical interpretation which is quite general and useful. The force in the y direction is simply the pressure times the projected area of the surface onto the horizontal (x-z) plane; the force in the x direction is simply the pressure times the projected area of the surface onto the vertical (y-z) plane. More generally, the force exerted by a pressure against a surface in a given direction is the integral of that pressure over the projected area of that surface onto a plane normal to the direction desired. This fairly simple statement is not true for all kinds of forces and is a direct consequence of the fact that pressure is isotropic, that is, at a point it is the same in every direction. Note however that the pressure times this projected area must be integrated along the curve so that the pressure assumes the value that it has along the curve in spite of the fact that we are projecting the curve onto a flat plane. This result is true regardless of the source of the pressure which exists in the fluid.

Another physical interpretation is useful for the case where the object is submerged in a fluid below the free surface of that fluid. In this case the pressure along the surface of the object is the weight per projected area of the column of fluid above it. This weight per area integrated over the object's projected surface is equal to the force exerted by the pressure in the direction of gravity. If the pressure is exerted from below the surface, then the same interpretation exists, at least quantitatively, although in this case the force would be exerted upwards against gravity rather than down. The amount of force, however, would still be that equal to the total weight of fluid which would be lying on top of the (lower) surface.

This result also has implications on the buoyant force exerted by a fluid on a body. If a submerged body is divided into a top surface and a bottom surface, then the force exerted downward (in the direction of gravity) on the top surface, according to the above discussion, is the weight of fluid which exists directly above it up to the surface of the fluid. Likewise, the force exerted from below the object is equal to that of the weight of the fluid that would exist above its bottom surface including the volume occupied by the object, although, this force is exerted upward. The difference of these two forces (one up and one down) is the difference in weight of the two columns of fluid, i.e. it is the weight of the fluid which has been displaced by the object.

Still another way of viewing this is that if instead of the object, the fluid itself resided in that volume, then the integration of the pressure over the same area, exactly equals the weight of the fluid which exists in place of the object. This is true because the fluid is static, i.e. perfectly at equilibrium with respect to the surface and gravity forces. If this region of fluid is now removed and replaced with the object, then the pressure integrated over the area still produces exactly the same force and must then be exactly equal to the weight of the fluid which was removed to make room for the object.

This result can be stated succinctly as Archimedes Principle: the buoyant force exerted on an object by a surrounding fluid is equal to the weight of the fluid which is displaced by the object. To the extent that fluid density varies with depth, this buoyant force will also vary with depth.

In addition to the amount of force exerted against a surface submerged in a fluid, we should consider the equivalent point of application of that force in accordance with the discussions in Chapter 9. To obtain the x and y coordinates of the center of pressure (c.p.) we write

$$\bar{\mathbf{r}} \times \mathbf{f}_{eq} = \int_{C} \mathbf{r} \times P(\mathbf{r}) \mathbf{u}_{N} \, dA \qquad (11-24)$$

and from these we obtain the two results

$$\bar{r}_x = \frac{\displaystyle\int P(\mathbf{r}) w x \, dx}{\displaystyle\int P(\mathbf{r}) w \, dx} \qquad \text{dependent on } y(x) \qquad (11-25)$$

and

$$\bar{r}_y = \frac{\displaystyle\int P(\mathbf{r}) w y \, dy}{\displaystyle\int P(\mathbf{r}) w \, dy} \qquad \text{independent of } y(x) \qquad (11-26)$$

Now if we view Pw as a load per unit length along the curve, then $\bar{r}_x$ and $\bar{r}_y$ are the x and y coordinates of the centroid of the Pw versus x and Pw versus y load curves, in accordance with the discussion of distributed loads in Chapter 9. Note that if the submerged area is planar, then this center of pressure lies on the plane. If it is non-planar then the center of pressure is above or below the curve. Figure 11-8 shows the force due to pressure on a submerged object in the coordinate directions and with the location $(\bar{r}_x, \bar{r}_y)$.

Example 11-5. Determine the magnitude and location of the resultant hydrostatic force acting on the submerged plate AB as shown in Figure 11-9. The width of the plate (in the x direction) is 2 m.

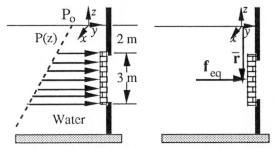

Figure 11-9: Forces on a submerged object and equivalent force

SOLUTION: The system is the plate AB and a cartesian coordinate system aligned with gravity is defined so that the origin is located at the level of the water right above the plate as shown in Figure 11-9.

The equivalent force in the horizontal direction is calculated from equation 11-22 (note the change of axes).

$$f_{eq,y} = \int_{z_1}^{z_2} P(z)w\,dz$$

where the pressure as a function of position for the constant density fluid is:

$$P(z) = P_o + \rho g(z_o - z)$$

The values of z_o and P_o are constants and given as:

$$z_o = 0 \text{ (m)} \qquad P_o = 101 \text{ (kPa)} = 1 \text{ (atm)}$$

Substituting this expression into the equation for the total force in the horizontal direction gives:

$$f_{eq,y} = \int_{z_1}^{z_2} \left(P_o + \rho g(z_o - z) \right) w\,dz$$

We will always integrate in the positive coordinate direction so the limits of integration are:

$$z_2 = -2 \text{ (m)} \qquad z_1 = -5 \text{ (m)}$$

Integrating the equation gives:

$$f_{eq,y} = w(P_o + \rho g z_o)(z_2 - z_1) + \left(\frac{-\rho g w}{2} \right)(z_2^2 - z_1^2)$$

Because z_o is defined as zero at the origin of the coordinate system, the equation reduces to:

$$f_{eq,y} = (w P_o)(z_2 - z_1) + \left(\frac{-\rho g w}{2} \right)(z_2^2 - z_1^2)$$

Substituting the numerical values with conversion factors gives:

$$f_{eq,y} = (2)(101)\left((-2) - (-5) \right) \text{ (kPa m}^2)+$$

$$\left(\frac{-(1000)(9.8)(2)}{2} \right)\left((-2)^2 - (-5)^2 \right) \left(\frac{\text{kg m}^4}{\text{m}^3 \text{ s}^2} \right)$$

$$= \frac{606 \text{ kPa m}^2}{} \left| \frac{1 \text{ kN}}{1 \text{ kPa m}^2} + \frac{205800 \text{ kg m}}{\text{s}^2} \right| \frac{1 \text{ N}}{1 \text{ kg m / s}^2} \left| \frac{1 \text{ kN}}{1000 \text{ N}} \right.$$

$$= (606 + 206) \text{ kN} = 812 \text{ kN}$$

The location of the force is found from the conservation of angular momentum for the equivalent force:

$$\bar{\mathbf{r}} \times \mathbf{f}_{eq} = \int \mathbf{r} \times P(\mathbf{r})\mathbf{u}_N\,dA$$

The position to the point of application of the force is only in the negative z direction so:

$$-\bar{r}_z \mathbf{k} \times f_{eq,y}\mathbf{j} = \int_{z_1}^{z_2} -z\mathbf{k} \times P(z)\mathbf{j}wdz$$

Evaluating the cross product gives:

$$\bar{r}_z f_{eq,y}\mathbf{i} = \int_{z_1}^{z_2} zP(z)wdz\mathbf{i}$$

Looking at the $\mathbf{i}$ component gives the scalar equation:

$$\bar{r}_z f_{eq,y} = \int_{z_1}^{z_2} zP(z)wdz$$

Substituting the expression for the pressure as a function of position:

$$\bar{r}_z f_{eq,y} = \int_{z_1}^{z_2} z\left(P_o + \rho g(z_o - z)\right)wdz$$

Integrating yields:

$$\bar{r}_z f_{eq,y} = \left(\frac{w}{2}\right)(P_o + \rho g z_o)(z_2^2 - z_1^2) + \left(\frac{-\rho g w}{3}\right)(z_2^3 - z_1^3)$$

Since the value of z_o was defined to be zero at the origin of the defined coordinate system the equation reduces to:

$$\bar{r}_z f_{eq,y} = \left(\frac{wP_o}{2}\right)(z_2^2 - z_1^2) + \left(\frac{-\rho g w}{3}\right)(z_2^3 - z_1^3)$$

Substituting numerical values gives:

$$\bar{r}_z f_{eq,y} = \left(\frac{(2)(101)}{2}\right)\left((-2)^2 - (-5)^2\right) \text{ (kPa m}^3)+$$

$$\left(\frac{-(1000)(9.8)(2)}{3}\right)\left((-2)^3 - (-5)^3\right)\left(\frac{\text{kg m}^5}{\text{m}^3 \text{ s}^2}\right)$$

$$= \frac{-2121 \text{ kPa m}^3}{} \left|\frac{1 \text{ kN}}{1 \text{ kPa m}^2}\right. + \frac{-764,400 \text{ kg m}^2}{\text{s}^2} \left|\frac{1 \text{ N}}{1 \text{ kg m}/\text{s}^2}\right|\frac{1 \text{ kN}}{1000 \text{ N}}$$

$$= (-2121 - 764) \text{ kN m} = -2885 \text{ kN m}$$

Solving for the position gives:

$$\bar{r}_z = \left(\frac{-2885}{812}\right)$$

$$= -3.55 \text{ (m)}$$

The equivalent force vector is:

$$\mathbf{f}_{eq} = 0\mathbf{i} + 812\mathbf{j} + 0\mathbf{k} \text{ (kN)}$$

and the point of application of the force from the defined coordinate system is:

$$\bar{r} = 0\mathbf{i} + 0\mathbf{j} - 3.55\mathbf{k} \text{ (m)}$$

Example 11-6. Determine the magnitude of the resultant hydrostatic force acting on the surface of the sea wall shaped in the form of a parabola as shown in Figure 11-10. The wall is 5 (m) in width and the equation for the curve defining the wall (for a cartesian coordinate system origin at the top of the liquid level and the y axis aligned with the gravitational force is: $y = -4x^2$ where the coefficient 4 has the units of (1/m) for the equation to be dimensionally consistent. The density of the sea water is 1025 (kg / m^3) and the pressure at the water's surface is 1 atmoshphere (101 kPa).

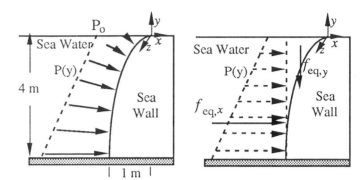

Figure 11-10: Forces on a sea wall and equivalent force

SOLUTION: The system is the sea wall and the coordinate system is defined in the problem statement. From the figure the forces are only acting the y and x direction and the magnitude of the total force is simply:

$$|\mathbf{f}_{eq}| = \sqrt{f_{eq,x}^2 + f_{eq,y}^2}$$

The forces in the x and y directions are found by integrating the pressure over the curved surface (see equation (11-22) and the previous problem). For the x-direction,

$$f_{eq,x} = \int_{y_1}^{y_2} P(y) w \, dy \qquad \text{or} \qquad \int_{y_1}^{y_2} \left(P_o + \rho g (y_o - y) \right) w \, dy$$

where the limits of integration (from integrating in the positive direction) are:

$$y_2 = 0 \text{ (m)} \qquad\qquad y_1 = -4 \text{ (m)}$$

Integrating the equation gives:

$$f_{eq,x} = w(P_o + \rho g y_o)(y_2 - y_1) + \left(\frac{-\rho g w}{2} \right)(y_2^2 - y_1^2)$$

Since the value of y_o is defined at the origin of the coordinate system, the equation reduces to:

$$f_{eq,x} = (w P_o)(y_2 - y_1) + \left(\frac{-\rho g w}{2} \right)(y_2^2 - y_1^2)$$

Substituting numerical values and conversion factors yields:

$$f_{eq,x} = (5)(101)\left((0) - (-4) \right) \text{ (kPa m)} + \left(\frac{-(1025)(9.8)(5)}{2} \right)\left((0)^2 - (-4)^2 \right)\left(\frac{\text{kg m}^4}{\text{m}^3 \text{ s}^2} \right)$$

$$= \frac{2020 \text{ kPa m}^2}{} \left| \frac{1 \text{ kN}}{1 \text{ kPa m}^2} \right. + \frac{401800 \text{ kg m}}{\text{s}^2} \left| \frac{1 \text{ N}}{1 \text{ kg m} / \text{s}^2} \right| \frac{1 \text{ kN}}{1000 \text{ N}}$$

$$= (2020 + 401) \text{ kN} = 2421 \text{ kN}$$

To calculate the component of the total force in the y direction we need to solve the following equation:

$$f_{eq,y} = -\int_{x_1}^{x_2} P(y)wdx$$

To begin with, the pressure is expressed as a function of y and we must integrate this distributed force along the curve. Therefore we can write the pressure along the curve as:

$$P(y) = P_o + \rho g(y_o - y) + P$$
$$P(x) = P_o + \rho g(-4x_o^2 - (-4x^2))$$

where at the origin, $x_o = 0$ and $P_o = 101$ (kPa). Substituting for the pressure as a function of x along the curve gives:

$$f_{eq,y} = -\int_{x_1}^{x_2} \left(P_o + \rho g(-4x_o^2 - (-4x^2)) \right) wdx$$

Now points 1 and 2 are at the bottom $(y = -4)$ and top $(y = 0)$ of the dam for which the limits of integration for the integration variable, x, are:

$$x_2 = 0 \text{ (m)} \qquad x_1 = -1 \text{ (m)}$$

Integrating the equation gives:

$$f_{eq,y} = -w(P_o - \rho g 4x_o^2)(x_2 - x_1) - \left(\frac{4\rho g w}{3} \right)(x_2^3 - x_1^3)$$

Since the value of x_o is defined at the origin of the coordinate system, the equation reduces to:

$$f_{eq,y} = -(wP_o)(x_2 - x_1) - \left(\frac{4\rho g w}{3} \right)(x_2^3 - x_1^3)$$

Substituting the known values with the conversion factors gives:

$$f_{eq,y} = -(5)(101)\left((0) - (-1) \right) \text{ (kPa m}^2) - \left(\frac{4(1025)(9.8)(5)}{3} \right) \left((0)^3 - (-1)^3 \right) \left(\frac{\text{kg m}^4}{\text{m}^3 \text{ s}^2} \right)$$

$$= \frac{-505 \text{ kPa m}^2}{} \left| \frac{1 \text{ kN}}{1 \text{ kPa m}^2} - \frac{66966.7 \text{ kg m}}{\text{s}^2} \right| \frac{1 \text{ N}}{1 \text{ kg m} / \text{s}^2} \left| \frac{1 \text{ kN}}{1000 \text{ N}} \right.$$

$$= (-505 - 66.97) \text{ kN} = -572 \text{ kN}$$

Now the magnitude of the resultant force is:

$$f_{eq} = \sqrt{(2421)^2 + (-572)^2} = 2488 \text{ (kN)}$$

The components of the force are shown in the Figure 11-10.

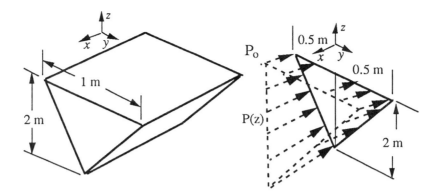

Figure 11-11: Example 11-7 Forces on a wall and equivalent force

Example 11-7. Determine the magnitude and location of the resultant force acting on the ends of the water trough as shown in Figure 11-11. The density of water is $\rho = 1000$ (kg / m^3).

SOLUTION: We choose the right end plate of the trough as the system (we could choose the other end just as well) and define a cartesian coordinate system at the liquid surface as shown in Figure 11-11.

To determine the total force acting on the plate we need to integrate the pressure over the area:

$$\mathbf{f}_{eq} = \int_A P(\mathbf{r})\mathbf{u}_N dA$$

The force acting on the plate is only in the horizontal direction, so that the only non-zero force component is $f_{eq,x}$:

$$f_{eq,x} = \int_A P(\mathbf{r})dA$$

The pressure as a function of the position and (constant density fluid) is:

$$P(z) = P_o + \rho g(z_o - z)$$

where the values of $z_o = 0$ (m) and $P_o = 101$ (kPa). Also, $dA = wdz$ where the width is not constant over the integration and can be expressed as a function of z:

$$w = \frac{z + 2}{2}$$

In this way, w varies linearly with z and gives the correct values at the top ($z = 0$, $w = 1$) and bottom ($z = -2$, $w = 0$) of the tank. Making these substitutions into the equation for the total force gives:

$$f_{eq,x} = \int_{z_1}^{z_2} \left(P_o + \rho g(z_o - z) \right) \left(\frac{z + 2}{2} \right) dz$$

where the limits of integration, (integrating in the positive z-direction) are:

$$z_1 = -2 \text{ (m)} \qquad z_2 = 0 \text{ (m)}$$

Integrating the equation gives:

$$f_{eq,x} = \left(\frac{1}{2}\right)\left[\left(\frac{-\rho g}{3}\right)(z_2^3 - z_1^3) + \left(\frac{P_o + \rho g z_o - 2\rho g}{2}\right)(z_2^2 - z_1^2) + 2(P_o + \rho g z_o)(z_2 - z_1)\right]$$

and because $z_o = 0$ this reduces to:

$$f_{eq,x} = \left(\frac{1}{2}\right)\left[\left(\frac{-\rho g}{3}\right)(z_2^3 - z_1^3) + \left(\frac{P_o - 2\rho g}{2}\right)(z_2^2 - z_1^2) + 2P_o(z_2 - z_1)\right]$$

Substituting numerical values into the equation with the proper conversion factors gives:

$$f_{eq,x} = \left(\frac{1}{2}\right)\left[\left(\frac{-(1)(9.8)}{3}\right)\left((0)^3 - (-2)^3\right)\right.$$
$$\left. + \left(\frac{(101) - 2(1)(9.8)}{2}\right)\left((0)^2 - (-2)^2\right) + 2(101)\left((0) - (-2)\right)\right]$$

The total force is:

$$f_{eq,x} = 107.5 \text{ (kN)}$$

The magnitude of the force is positive, but from the figure the direction of the force is in the negative x direction.

$$\mathbf{f}_{eq} = -107.5\mathbf{i} + 0\mathbf{j} + 0\mathbf{k} \text{ (kN)}$$

The location of the equivalent force is determined from the conservation of angular momentum through:

$$\bar{\mathbf{r}} \times \mathbf{f}_{eq} = \int_A \mathbf{r} \times P(\mathbf{r})\mathbf{u}_N dA$$

Values for the x and y position coordinates are zero giving that

$$\bar{r}_z\mathbf{k} \times f_{eq,x}\mathbf{i} = \int_A z\mathbf{k} \times P(\mathbf{r})\mathbf{i}\,dA \quad \text{or} \quad \bar{r}_z f_{eq,x}\mathbf{j} = \int_{z_1}^{z_2} zP(\mathbf{r})dA\mathbf{j}$$

Substituting for pressure and w as functions of position gives:

$$\bar{r}_z f_{eq,x} = \int_{z_1}^{z_2} z\left(P_o + \rho g(z_o - z)\right)\left(\frac{z+2}{2}\right)dz$$

and with $z_o = 0$ this is

$$\bar{r}_z f_{eq,x} = \int_{z_1}^{z_2} z\left(P_o - \rho g z\right)\left(\frac{z+2}{2}\right)dz$$

Integrating this equation gives:

$$\bar{r}_z f_{eq,x} = \left(\frac{1}{2}\right)\left[\left(\frac{-\rho g}{4}\right)(z_2^4 - z_1^4) + \left(\frac{P_o - 2\rho g}{3}\right)(z_2^3 - z_1^3) + P_o(z_2^2 - z_1^2)\right]$$

Substituting numerical values and conversion factors gives:

$$\bar{r}_z f_{eq,x} = \left(\frac{1}{2}\right)\left(\left(\frac{-(1)(9.8)}{4}\right)\left((0)^4 - (-2)^4\right) + \right.$$

$$\left.\left(\frac{(101) - 2(1)(9.8)}{3}\right)\left((0)^3 - (-2)^3\right) + (101)\left((0)^2 - (-2)^2\right)\right)$$

$$= -73.9 (kN\ m)$$

Solving for $\bar{r}_z$ gives:

$$\bar{r}_z = \left(\frac{-73.9}{107.5}\right) = -0.69\ (m)$$

The position vector to the point of application of the equivalent force is:

$$\bar{r} = 0\mathbf{i} + 0\mathbf{j} - 0.69\mathbf{k}\ (m)$$

11.4 BUOYANCY

The variation of pressure with depth in a fluid provides an interesting phenomenon called buoyancy. Because the pressure is greater on a submerged object at its bottom than near its top, the pressure, integrated over the entire surface of the body provides a force which is upward, relative to gravity:

$$\mathbf{f}_B = -\rho_{fluid} V \mathbf{g} \qquad\qquad (11-27)$$

where V is the volume of the body. If this net upward force (called the buoyant force, or simply buoyancy) is sufficient, i.e. equal to the weight of the object itself, then the object will float (be suspended) in the fluid:

Figure 11-12 shows an object immersed in a fluid with all of the forces acting on the body. Even if buoyancy is not enough to support the entire body's weight, it will serve to unweight the object to some extent. Consequently, when submerged in fluids, objects will appear to weigh less. If an object is hollow, or its density is small compared to that of the fluid, then its weight can be quite small (relative to buoyancy) and a considerable net upward force can be achieved if immersed totally in the fluid.

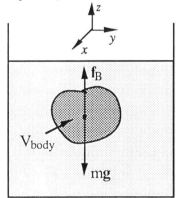

Figure 11-12: Buoyancy

If unrestrained, then, the object will rise and achieve equilibrium at the interface with a less dense fluid (say air). At this equilibrium position, the object displaces an amount of fluid equal to its own weight – it floats on the surface. This phenomenon has been known (if not understood) for centuries, of course, and has enabled man to navigate the rivers and oceans. We also are naturally familiar with the ability of ourselves to float in water if we expand our lungs enough with air. The human body is very nearly neutrally buoyant and with some very small adjustments in the expansion of the chest cavity we can make ourselves float on the surface or sink well below it. Finally, it should be noted that buoyancy acts through the center of mass of the displaced fluid while the weight of the object acts through the center of mass of the submerged or floating object. This is an important consideration in the stability of a body acted upon by buoyant forces.

Example 11-8. A cube of cork 2 (ft) on each side floats in water as shown in Figure 11-13. The density of cork is 15 (lb_m / ft^3). Find the depth z of the cube in the water.

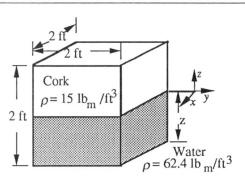

Figure 11-13: Buoyant force acting on a cork cube

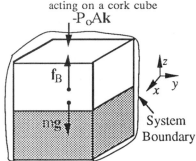

Figure 11-13a: Free body diagram

SOLUTION: The system is the cube, and a cartesian coordinate system aligned with the acceleration of gravity with the origin defined at the water level is shown in the figure. The conservation of linear momentum for this static system says: $\sum \mathbf{f}_{ext} = \mathbf{0}$ A free body diagram of the cube with all the external forces acting on the cube are shown in Figure 11-13a.

From the figure, there are no forces in the x or y direction, so the vector equation in the $\mathbf{k}$ direction gives:

$$-P_o A\mathbf{k} - mg\mathbf{k} + f_B\mathbf{k} = 0\mathbf{k}$$

The buoyant force is equal to the pressure at the depth z integrated over the area of the cube. The pressure as a function of depth for the constant density fluid is:

$$P(z) = -\rho g(z - z_o) + P_o$$

And the buoyant force is calculated as:

$$f_B = \int_A P(z)dA$$

$$= \text{weight of fluid displaced}$$

Since the pressure is constant over the area of the bottom face the equation reduces to:

$$f_B = P(z)A$$

$$= \left(-\rho_{\text{fluid}}g(z - z_o) + P_o\right)A$$

Substituting this back into the vector equation yields:

$$-P_o A\mathbf{k} - mg\mathbf{k} + \left(-\rho_{\text{fluid}}g(z - z_o) + P_o\right)A\mathbf{k} = 0\mathbf{k}$$

Because of the defined coordinate system where z_o is equal to zero, the equation reduces to:

$$-mg - \rho_{\text{fluid}}gzA = 0$$

Rearranging this equation gives:

$$z = -\left(\frac{m}{\rho_{\text{fluid}}A}\right)$$

$$= -\left(\frac{\rho_{\text{cube}}V}{\rho_{\text{fluid}}A}\right)$$

$$= -\left(\frac{(15)(8)}{(62.4)(4)}\right)\left(\frac{\text{ft}^3}{\text{ft}}\right)$$

$$= -0.48 \text{ (ft)}$$

The bottom of the cube is 0.48 (ft) below the level of the water.

Example 11-9. An object 6 (in) thick by 8 (in) wide by 12 (in) long is weighed in water at a depth of 20 (in) and found to weigh 11.0 (lb$_f$) as shown in Figure 11-14. What is the weight of the object in the air (or in the absence of a buoyant force) and what is the density of the object?

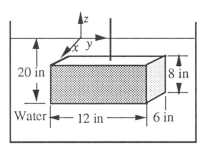

Figure 11-14: Buoyant force acting on an object

SOLUTION: The system is the object and a cartesian coordinate system aligned with the acceleration of gravity with the origin defined at the level of the liquid. A free body diagram is shown in Figure 11-14a.

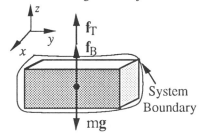

Figure 11-14a: Free body diagram or system

The conservation of linear momentum for this system is given by the vector equation:

$$\sum \mathbf{f}_{\text{ext}} = 0$$

From the figure, the only forces acting on the body are in the **k** direction, so the x and y equations only give trivial solutions. In the **k** direction the equation is:

$$-mg\mathbf{k} + f_{B,z}\mathbf{k} + f_{T,z}\mathbf{k} = 0\mathbf{k}$$

The buoyant force is calculated from the equation:

$$f_{B,z} = \rho_{\text{fluid}} V g$$

so the only unknown in the equation is the product of the mass times the force of gravity or the weight. Substituting and rearranging gives:

$$mg\mathbf{k} = \rho_{\text{fluid}} V g\mathbf{k} + f_{T,z}\mathbf{k}$$

Substituting the known values into the equation with the correct conversion factors gives:

$$mg = (62.4)(576)(32.174)\left(\frac{\text{lb}_\text{m}\ \text{in}^3\ \text{ft}}{\text{ft}^3\ \text{s}^2}\right) + (11\ \text{lb}_\text{f})$$

$$= \frac{115641\ \text{lb}_\text{m}\ \text{in}^3\ \text{ft}}{\text{ft}^3\ \text{s}^2}\left|\frac{1\ \text{lb}_\text{f}}{32.174\ \text{lb}_\text{m}\ \text{ft}/\text{s}^2}\right|\frac{1\ \text{ft}^3}{12^3\ \text{in}^3} + 11\ \text{lb}_\text{f}$$

$$= (20.8 + 11)\ \text{lb}_\text{f} = 31.8\ \text{lb}_\text{f}$$

The density of the solid is calculated from the definition of density:

$$\rho_{\text{object}} = \left(\frac{m}{V}\right)$$

Where the mass is equal to :

$$m = \left(\frac{31.8}{g}\right)$$

$$= 31.8\ (\text{lb}_\text{m})$$

The density of the object is:

$$\rho_{\text{object}} = \left(\frac{31.8}{576}\right)\left(\frac{\text{lb}_\text{m}}{\text{in}^3}\right)\left(\frac{12\ \text{in}}{1\ \text{ft}}\right)^3$$

$$= 95.4\ (\text{lb}_\text{m}/\text{ft}^3)$$

11.5 THIN-WALLED PRESSURE VESSELS

Another example of the importance of understanding pressure as a function of depth is the design of thin-walled pressure vessels. Storage and pressure vessels must be able to sustain the forces exerted by the pressure of the contained fluid. This pressure can vary significantly with depth and must be taken into account in the design of large (tall) storage vessels. In the case of true pressure vessels the concern is less with the increase of pressure with depth than with the biasing pressure.

The design of the wall thickness of a pressure vessel is a fairly straightforward problem in statics. The wall material is capable of sustaining a certain maximum (allowable) stress which is the total force tending to pull the wall apart across its thickness divided by the cross sectional area across this thickness. There is a pressure difference from one side of the wall to another and to the extent the wall is curved this pressure difference leads to a force which must be counteracted by the stress in the wall.

As an example, consider a cylindrical vessel with pressure P_i on the inside and P_o on the outside. The cross section through the cylinder is shown in Figure 11-15. The pressure on the inside must be counteracted by the force in the walls which act parallel to the surface of the wall. This circular cross section is shown cut in half. With this half circular section viewed as the system, a static analysis of the forces shows clearly how the pressure integrated over the surfaces of the vessel must be balanced by the force in the walls. This leads to the equation (assuming that the wall thickness is small compared to the radius)

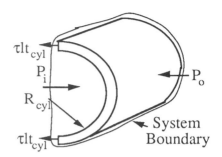

Figure 11-15: Thin walled pressure vessel

$$(P_i - P_o)2R_{cyl}l = 2(\tau t l) \qquad (11-28)$$

where τ is the stress in the wall (force per area), t is the wall thickness, and l is the wall length. This can be arranged to give

$$t_{cyl} = \left(\frac{(P_i - P_o)R_{cyl}}{\tau} \right) \qquad (11-29)$$

For a given material (τ), desired vessel size (R_{cyl}), and design pressure difference, the required wall thickness can be calculated. Alternatively, for a given vessel, as the inside pressure changes, the stress in the walls changes accordingly to accommodate; so long as the yield stress of the material is not exceeded, the vessel will remain intact.

This analysis was for the walls of a cylindrical vessel. Other shapes can be used (spherical, ellipsoidal) and different relations are obtained relating the difference in pressure between inside and outside, the vessel parameters, and the stress in the wall. The principles in obtaining these relations, however, are of course the same. For a spherical surface:

$$t_{sph} = \left(\frac{(P_i - P_o)R_{sph}}{2\tau} \right) \qquad (11-30)$$

and for an ellipsoidal surface for which the minor diameter equals one-half the major diameter (D_{ell}) is:

$$t_{ell} = \left(\frac{(P_i - P_o)D_{ell}}{2\tau} \right) \qquad (11-31)$$

From these results we see that a spherical shape is more effective in sustaining a pressure difference in the sense that for a given pressure difference, vessel size, and wall thickness, the stress in the wall will be less for a spherical shape. More to the point of economic design, for a given size, pressure, and allowable stress the wall thickness may be less. For this reason for high pressure gas storage, spherical tanks are frequently used. For moderate pressures they are not used, however, because the fabrication cost is greater than for a simple cylindrical tank. There is an optimization between the cost of the shape and the cost of materials (wall thickness) required to sustain a given pressure difference.

Furthermore, the shapes are frequently combined to achieve an optimum balance between effectiveness of pressure containment and cost. A vessel which is cylindrical in shape may be fabricated with either hemispherical or elliptical ends to provide adequate strength in the ends. If the ends were flat, then for pressure vessels the thicknesses required would be considerable in order to achieve adequate strength.

Example 11-10. If 1000 (lb$_m$) of steam at P_{in} = 370 (psia) and specific volume $\widehat{V}$ = 1.402 (ft^3 / lb$_m$) is stored in a cylindrical tank that is 20 (ft) in length, calculate the thickness of the steel necessary in the tank. If the steam is stored in a spherical tank, calculate the thickness for this tank. Figure 11-16 gives a schematic of the two tanks. The allowable stress in the steel is ($\tau = 13,700$ (psi)).

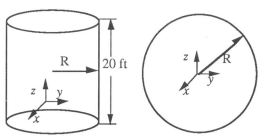

Figure 11-16: Cylindrical and spherical pressure vessels

SOLUTION: The thickness of the steel in the tank is calculated from the conservation of linear momentum for a section of the tank. This equation was derived as:

$$t_{cyl} = \left(\frac{(P_i - P_o)R_{cyl}}{\tau} \right)$$

and

$$t_{sph} = \left(\frac{(P_i - P_o)R_{sph}}{2\tau} \right)$$

In order to use these equations we must determine the radius of the cylindrical and spherical tanks. We know that the total volume of the steam is:

$$V_{tot} = m\widehat{V}$$
$$= (1000)(1.402)$$
$$= 1402 \ (ft^3)$$

The volume of a cylinder and a sphere are calculated by the following geometrical relationships:

$$V_{cyl} = l\pi R_{cyl}^2 \qquad V_{sph} = \left(\frac{4}{3}\right)\pi R_{sph}^3$$

From the known volume the radius for the cylinder and the sphere are calculated:

$$R_{cyl} = 4.72 \ (ft) \qquad R_{sph} = 6.94 \ (ft)$$

The thickness of the cylindrical tank is:

$$t_{cyl} = \left(\frac{(370 - 14.7)(4.72)}{13,700} \right) \left(\frac{psi \ ft}{psi} \right) = 0.12 \ (ft)$$

For the spherical tank the thickness is:

$$t_{sph} = \left(\frac{(370 - 14.7)(6.94)}{2(13.700)} \right) \left(\frac{psi\ ft}{psi} \right)$$

$$= 0.09 \ (ft)$$

The thickness required of the spherical tank is less than that of the cylindrical tank.

11.6 PRESSURE DROP ACROSS STATIC CURVED FLUID INTERFACES

In the above discussion we saw that a pressure difference across a curved solid wall resulted in stress within the wall. If the pressure changed, then the stress changed accordingly.

At the interface between liquids and gases or between two immiscible liquids, a phenomenon occurs which can be quantified in a completely analogous manner to that of pressure vessels. In this case, however, the stresses in the interface are established by the properties of the materials involved and not by the pressure drop-curvature combination. The stress at the interface is the surface tension and does not change according to the situation. Instead, if an interface is curved then this implies that a pressure drop exists across the interface in accordance with calculations analogous to those above. If the stress is fixed (as surface tension) then the amount of curvature indicates the pressure difference across the interface.

For example, for a spherical interface the pressure difference across the curved surface at equilibrium is given by

$$(P_i - P_o) = \frac{2\sigma}{R} \tag{11-32}$$

In this situation, the stress in the wall (interfacial tension-actually an energy per area) is a fixed property of the materials (fluid phases) and if the shape of the interface is dictated, then the pressure difference is established as a consequence. Values of surface tension for a number of fluids are given in Table C-11 in Appendix C. Equivalently, if the pressure difference is fixed for a static interface, then the shape of the interface is established as a consequence.

Before considering this question further, we must first address the property of wettability as determined by the contact angle. The contact angle is the angle of intersection of three phases: one solid surface and two fluid phases. If a drop of water is placed on a clean glass slide, then the drop will spread as shown in Figure 11-17(a). At the point of contact of the water with the slide we see a well-defined angle. The angle is normally measured through the more-dense fluid phase which in this case is the water (as opposed to the air). If the angle is zero degrees then we say that the liquid perfectly wets the surface (Figure 11-17(b)). If on the other hand the angle is 180 degrees (Figure 11-17(c)) then the fluid is perfectly nonwetting. Any angle between these values implies both wetting and nonwetting character and a contact angle of $90°$ indicates exactly balanced or neutral wetting.

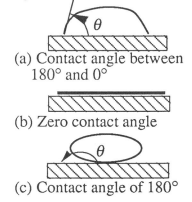

(a) Contact angle between $180°$ and $0°$

(b) Zero contact angle

(c) Contact angle of $180°$

Figure 11-17: Contact angles and wettabilities

The contact angle is a property of the three phases involved: the solid material, which the fluids contact, and the two fluid phases. Furthermore, it is a property that is fundamentally different from interfacial tension. The latter depends upon the properties of two phases at their surface of contact whereas the former depends on the properties of three phases along their line of contact. However, as we will see, both the contact angle and the interfacial (or surface) tension are important properties in establishing the effect of surface forces in a number of situations.

To return to the example of a water, if the drop is placed on an oily film on the glass slide, then we see that the drop now is essentially nonwetting and forms a very distinct, nearly spherical bead (as in water beading on a newly polished car). The surface that the water is in contact with has changed and therefore the contact angle has changed although the surface tension of the water in the air is still the same.

Returning to the question of pressure drop across fluid interfaces, we consider the situation of capillary rise in a thin tube with which we are all familiar. If a small tube (say a small soda straw) is placed vertically in a glass of water then we see that the water rises in the tube to an equilibrium position well above the water level in the glass, Figure 11-18(a).

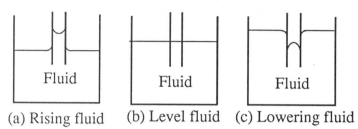

(a) Rising fluid   (b) Level fluid   (c) Lowering fluid

Figure 11-18: Fluid rising, stationary or falling in a capillary tube

If you believe fluid statics and the results obtained from the variation of pressure in a static fluid with depth (and you should) , then you believe that the pressure in the water just beneath the meniscus in the straw is less than that in the glass at the water surface outside the straw. This difference in pressure associated with the capillary rise is calculated according to the fluid statics equation presented above and is simply equal to (for a constant density fluid) $\rho g h$ where h is the amount of capillary rise.

We now need to reconcile this with our discussion of pressure drop across curved interfaces. Where the fluid makes contact with the capillary there is a well-defined contact angle which is established in accordance with the discussion above. This angle then forces the existence of a curved interface near the walls of the tube which in turn, together with the interfacial tension, establishes a pressure difference across the interface. To the extent this difference is not consistent with the pressure which is established in the water at the interface by considerations of fluid statics, the meniscus will be pushed up or down until agreement is reached. For example, if the contact angle is zero and the tube is small enough, then the shape of the meniscus will be very nearly spherical. In this case the pressure drop across the interface will be given by the equation for a hemispherical surface

$$(P_i - P_o) = \left(\frac{2\sigma}{R_{cap}} \right)$$

and the fluid will rise in the capillary until this pressure difference agrees with that calculated by fluid statics

$$\left(\frac{2\sigma}{R_{cap}} \right) = \rho g h \qquad (11-33)$$

from which we can establish a direct relation between the amount of capillary rise and interfacial tension.

$$\boxed{\sigma = \left(\frac{\rho g R_{cap}}{2} \right) h} \qquad (11-34)$$

f the surface tension is very small, then the amount of this rise will also be small; if the surface tension is large, then the capillary rise will be larger. This relation has obvious implications on measuring interfacial tension.

Note again the role of the contact angle in establishing this capillary rise. If the contact angle is 90° so that there s no curvature established at the wall of the tube, then the interface is flat and/or equivalently the radius of curvature is infinity so that there is no pressure drop across the interface and hence no capillary rise (Figure 11-18b) regardless of the interfacial tension. Going to the opposite extreme of complete wettability, if the contact angle is 180 degrees so that the fluid is totally nonwetting then the curvature established is in the opposite direction so that at equilibrium the pressure is greater on the liquid side of the interface than on a gas side resulting in a depression of the meniscus at equilibrium. This is the case that we observe when a glass tube is placed in a reservoir of mercury (Figure 11-18c).

The direction of capillary rise (or fall) is established by the contact angle. The quantitative amount of capillary rise is established by both the contact angle (through its effect on the shape of the interface, i.e. the radius of curvature) and by the amount of stress which exists within the interface which is set by the interfacial or surface tension.

Apart from capillary rise in a soda straw this phenomenon has application to a number of practical situations. In distillation columns, for example, contact between a liquid and vapor phase is usually made by bubbling gas up through a pool of liquid on a tray. The tray has small holes perforating it and the breakthrough of gas bubbles through these holes, up through the liquid, is dependent upon sufficient pressure to force the bubbles through the hole. The amount of pressure required depends upon the size of the hole and the surface tension. These then become important parameters in the design of sieve tray columns.

As a second example, if we are trying to flow a liquid through a porous medium (for example, if we are trying to displace oil through the pores of a consolidated, but still porous, sandstone) then we can think of this as the passage of the liquid through channels of varying thicknesses. If a drop is being displaced from a tube of one diameter to a tube of another diameter (Figure 11-19) then different radii of curvature exist at the two ends of the drop.

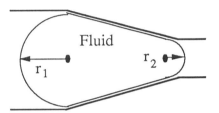

Figure 11-19: Tube with different radii

As a result (assuming the same contact angle at the front and the rear) the pressure drop across the two ends is different. The small end exhibits a larger pressure drop than does the larger end. The net result from one end of the bubble to the other is that in order to displace the drop through the porous medium, a finite pressure drop is required. If a drop is moving from a small diameter to a large, then the opposite effect is observed–the drop is "pulled" through the opening. This pressure drop, in the case of oil recovery, can be prohibitive when multiplied over the distances involved from an injection to a production well (when the natural reservoir drive pressure has diminished it is common practice to inject water in one set of wells to provide adequate pressure to displace the oil and move it to another set of production wells).

In such displacements the tortuosity or configuration of the channels can make a difference as well. In the above discussion the assumption was made that the tube walls are parallel so that for a given contact angle a curved surface is established. If, however, the walls are not parallel but rather conical, then the curvature of the fluid is affected and consequently the pressure drop across the interface is affected. For example, consider the fortuitous situation of having the walls of the tube shaped in such a way that they just cancel out the curvature induced by the contact angle (Figure 11-20).

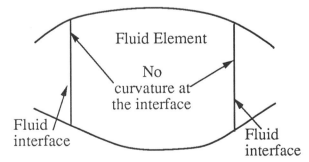

Figure 11-20: Canceling out of the contact angle

In this case, there would be no curvature established at the interface and therefore no pressure drop in spite of the fluid interfacial tension being finite. Equivalently, if the wettability of the fluid which is being mobilized can be changed by some kind of a treatment, then this can affect the pressures required to displace the drop. Pretreatments of reservoirs with wetting agents have been considered as a means of enhancing oil recovery. Obviously another way of enhancing the displacement is to lower the interfacial tension and the use of surfactants for this purpose have also been extensively considered.

Example 11-11. A glass tube with diameter of 3 (mm) is placed in water as shown in Figure 11-21. Determine the height of the column of water above the liquid level, assuming the water completely wets the glass tube. The physical property data for water are $\rho = 1$ (g / cm^3) and $\sigma = 73.05$ (dyne / cm).

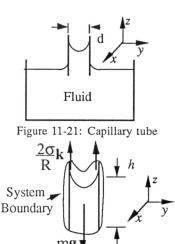

Figure 11-21: Capillary tube

Figure 11-21a: Free body diagram

SOLUTION: A free body diagram of the column of fluid is given in Figure 11-21a with a cartesian coordinate system aligned with the gravitational force.

The conservation of linear momentum in the z direction is given by equation (11-33):

$$\left(\frac{2\sigma}{R_{cap}}\right) = \rho g h$$

Rearranging and solving for the height of the fluid gives:

$$h = \left(\frac{2\sigma}{R_{cap}\rho g}\right)$$

Substituting the known values in with the proper conversion factors gives:

$$h = \left(\frac{2(73.05)}{(0.15)(1)(980)}\right)\left(\frac{\text{dyne cm}^3 \text{ s}^2}{\text{cm}^3 \text{ g}}\right)\left(\frac{1 \text{ g cm / s}^2}{1 \text{ dyne}}\right) = 0.99 \text{ (cm)}$$

The height of the column of water above the level of the fluid is 0.99 (cm).

11.7 FLUID STATICS PROBLEM-SOLVING TOOLS

The tools which have been presented in this and earlier chapters and which are needed for addressing and understanding fluid statics are summarized below.

1) **The Laws.** The non-trivial conservation laws for fluid statics are linear and angular momentum. The basic relation between fluid pressure and depth results from steady-state (static) linear momentum. The stress in a pressure vessel wall is obtained directly from linear momentum conservation.

2) **Descriptions of Forces.** Force concepts tools are gravity, pressure (an isotropic force per area), buoyancy (resulting from pressure varying with depth), wall stress in pressure vessels (also force per area but not isotropic, necessarily), and distributed forces.

3) **Properties of Matter.** These include fluid density, interfacial (or surface) tension, and allowable stress.

4) **Mathematics.** Of course, mathematics tools are always present. In fluid statics, the required tools are integration of density over depth, integration of pressure over a surface to obtain an equivalent force and torque, and solutions of algebraic equations.

11.8 REVIEW

1) The pressure in a static fluid varies with depth (i.e. with the height of a column of fluid) due to the additional weight of the column of fluid. The calculations of pressure versus depth are linear for an incompressible fluid and may exhibit a variety of nonlinearities for compressible fluids depending upon whether the fluid is considered to be isothermal, adiabatic, or some combination of the two.

2) A knowledge of the pressure versus depth relationship can be used to calculate the forces that are exerted upon submerged areas and objects due to this pressure. Calculations of such forces are important in designing structures such as dams, tunnels and storage or pressure vessels.

3) The fact that pressure varies with depth implies that there exists a buoyant force upon a submerged object which serves to unweight it when it is immersed in a fluid. If this buoyant force is large enough, then objects will float at the surface of the fluid.

4) From a straight forward static force analysis, it is concluded that a pressure difference exists across a static curved fluid interface.

5) A property of three phases which are in simultaneous contact produces a contact angle which exists along the line of contact, also referred to as wettability. This contact angle has importance in that it establishes a curved interface between the two fluid phases and in accordance with statement four, there therefore exists a pressure difference across this curved interface.

6) The amount of pressure difference which exists across a curved interface depends upon the shape of the surface (interface) and the interfacial tension of that surface.

7) Some important consequences of the existence of a pressure drop across curved interfaces are capillary rise (wick action), two-phase flow through small tubes or pores such as the displacement of oil during enhanced oil recovery, bubble flow through distillation column sieve trays, and percolation two-phase flow.

8) The phenomenon of capillary rise in a small tube can be used as a method of measuring interfacial tension.

11.9 NOTATION

The following notation has been used throughout this chapter. The bold face lower case letters represent vector quantities.

Scalar Variables and Descriptions   Dimensions

A	area	[length^2]
$\tilde{C}_P$	molar constant pressure heat capacity	[energy / moletemperature]
D	diameter	[length]
f	magnitude of the vector $\mathbf{f}$	[mass length / time^2]
g	magnitude of the acceleration of gravity	[length / time^2]
h	vertical distance above the fluid level	[length]
l	length	[length]
m	total mass	[mass]
n	number of moles	
P	pressure	[force / length^2]
R	universal gas constant	[energy / mole temperature]
R	radius	[length]
t	thickness	[length]
T	temperature	[temperature]
V	volume	[length^3]
w	width	[length]
x, y, z	spatial rectangular cartesian variables	[length]
ρ	density	[mass / length^3]
σ	surface tension	[force / length]
τ	wall stress	[force / length^2]

Vector Variables and Descriptions   Dimensions

$\mathbf{d}_A$	distibuted force per area	[force / length^2]
$\mathbf{f}$	force	[force]
$\mathbf{g}$	acceleration of gravity	[length / time^2]
$\mathbf{i}, \mathbf{j}, \mathbf{k}$	cartesian unit directional vectors	
$\mathbf{r}$	position	[length]
$\mathbf{u}$	unit direction vector	

Subscripts

eq	equivalent or equal to
ext	external to the system or acting at the system boundary
i	internal or inside the system
N	normal to the surface
o	external or outside the system

Superscripts

<blockquote>
¯   position of equivalent force
</blockquote>

Vector Operator		Dimensions
∇	gradient operator of a scalar or vector quantity	[1 / length]

11.10 SUGGESTED READING

Currie, I. G., *Fundamental Mechanics of Fluids*, McGraw- Hill, New York, 1974
Fox, R. W., and A. T. McDonald, *Introduction to Fluid Mechanics*, 3rd edition, John Wiley & Sons, New York, 1985
Halliday, D. and R. Resnick, *Physics for Students of Science and Engineering, Part I*, John Wiley & Sons, 1960
LeMehaute, B., *An Introduction to Hydrodynamics and Water Waves*, Springer-Verlag, New York, 1976
Vennard, J. K., and R. L. Street, *Elementary Fluid Mechanics*, 6th edition, John Wiley & Sons, New York, 1982
Whitaker, S., *Introduction to Fluid Mechanics*, Prentice-Hall, Englewood Cliffs, N. J., 1968
White, F. M., *Fluid Mechanics*, McGraw-Hill, New York, 1986

QUESTIONS

1) Of the fundamental equations and laws discussed in the first eight chapters, which ones provide non-trivial information in the analysis of isothermal static fluids?

2) Which forces act upon a static fluid and therefore play a role in the conservation of linear momentum?

3) The pressure in a fluid is an isotropic stress. Explain what is meant by isotropic.

4) What are the dimensions of pressure?

5) How do you calculate the force exerted by a pressure against an area if the pressure is constant at all points of the area?

6) How do you calculate the force exerted against an area by a pressure if the pressure varies across the area?

7) List three circumstances in which the pressure exerted by a static fluid plays a role in the design of a structure. In each case describe the way in which the pressure affects the design of the structure.

8) Apart from providing a supply of water, explain the function of an elevated water storage tank.

9) Explain in physical terms why a liquid rises to a different level in different sized soda straws. Which straw has the larger rise for a given drink, the smaller tube or the bigger one?

10) In what ways are a pressure vessel and a liquid drop similar with respect to pressure difference and wall stresses? In what ways are they different?

11) What is Archimedes' principle?

12) What are the units of pressure in SI and in the American Engineering System? How much is one atmosphere of pressure in SI units and how much is it in the engineering system of units? Define one bar of pressure and how many bars are there in one atmosphere?

13) A hydrometer is used to test the electrolyte in a car battery. Explain what the hydrometer measures and how it works.

14) Compare the equations given in this chapter for designing pressure vessels to those equations which appear in engineering handbooks. In what ways are they the same and in what ways are they different?

15) Explain the concepts of wettability and interfacial or surface tension and the roles that each plays in capillary action (i.e. the rise of a fluid in a small diameter tube).

16) Give two situations in which interfacial phenomena are important.

SCALES

1) The origin of a coordinate system is located 30 feet below the surface of a lake which is at 70°F. The vertical direction is the z direction. The air above the lake is at 1 atmosphere. One end of a flat metal plate 5 ft wide extends from the point (0,0,1) to the point (0,3,4). The 5-foot width is in the x direction; the bottom, horizontal, edge of the plate lies on the x axis.

    a) Sketch the plate beneath the surface of the lake.
    b) Obtain a mathematical expression for the position of the edge of the plate in terms of $z = f(y)$.
    c) Calculate the horizontal force exerted on one side of the plate by the water in the y direction.
    e) Calculate the downward force exerted on the plate by the water.

2) The origin of a coordinate system is located 30 feet below the surface of a lake which is at 70°F. The vertical direction is the z direction. The air above the lake is at 1 atmosphere. One end of a curved metal plate 5 ft wide extends from the point (0,0,1) to the point (0,3,4). The curve is exponential of the form $z = ae^{by}$ where $a = 1$ and $b = 0.462$). The 5-foot width is in the x direction.

    a) Sketch the plate beneath the surface of the lake.
    b) Calculate the horizontal force exerted on one side of the plate by the water in the y direction.
    c) Calculate the downward force exerted on the plate by the water.

3) The origin of a coordinate system is located 30 feet below the surface of a lake which is at 70°F. The vertical direction is the z direction. The air above the lake is at 1 atmosphere. One end of a curved metal plate 5 ft wide extends from the point (0,0,1) to the point (0,3,4). The curve is exponential of the form $z = 5 - 4e^{-0.462y}$. The 5-foot width is in the x direction.

    a) Sketch the plate beneath the surface of the lake.
    b) Calculate the horizontal force exerted on one side of the plate by the water in the y direction.
    c) Calculate the downward force exerted on the plate by the water.

4) Compare the answers to exercises (1), (2), and (3). How do the horizontal forces compare to each other? How do the vertical forces compare to each other?

5) Convert 30 psi to bars, N/m^2, and atmospheres.

6) Calculate, for each object in parts (a) through (d), its volume, its weight in a vacuum if it is made of quartz (see Table C-5), the buoyant force exerted on it if totally immersed in water, and its weight in water.

    a) A sphere of diameter 6 inches.
    b) A right circular cone whose base is diameter 6 inches and whose height is 8 inches,
    b) A rectangular parallelepiped with edge lengths of 5 cm, 8 cm, and 10 cm.
    c) A pyramid whose base is a square with edge lengths of 6 cm and height 9 cm.
    d) A truncated cone whose base is 12 cm in diameter, top is 8 cm in diameter, and whose height is 5 cm.

7) The normal stress in the wall of a cylindrical pressure vessel is 5,000 psi. What is this in N/m^2?

8) The interfacial tension in a fluid can be thought of either as a force/length or as energy/area. As anyone who has cut firewood knows, it takes energy to create a new surface. The interfacial tension of water is 73.05 dyne/cm. What is this in N/m and J/m^2?

PROBLEMS

1) A diver's suit can only withstand a finite amount of stress (force per unit area) before the suit might fail, possibly killing the diver. If the suit can be used safely at an absolute external pressure of 300 (psi) what is the maximum depth that the diver can descend in,

    a) Sea water.

    b) A fluid having density as a function of depth given by:

$$\rho(z) = ke^{(-z/a)} \ (lb_m \ / \ ft^3)$$

    where $k = 2 \ lb_m \ / \ ft^3$ and $a = 3$ ft with $z = 0$ at the fluid surface and z positive in the upward direction.

2) A vertical cylindrical tank contains water and benzene. The level from the bottom of the tank to the top liquid surface is 20 ft. The diameter of the tank is 10 ft. Additionally, the tank is pressurized to 5 atm at the benzene surface. The presssure at the benzene-water interface is 0.2 atm greater than that at the surface of the benzene. Determine (a) the position of the benzene-water interface, (b) the pressure at the bottom of the tank, and (c) the minimum thickness of the tank required at the bottom if the allowable tensile stress for the tank material is 10,000 psi. The density of benzene and water may be taken as 56.1 and 62.2 lbm/ft^3.

3) The mass of a balloonist is 150 lb$_m$; the mass of the balloon and the basket is 100 lb$_m$, and the mass of the air inside the balloon when it is inflated is 1700 lb$_m$. The total volume of the balloon can be approximated as a sphere having a diameter of 40 ft. If the balloon is static (neutrally buoyant) at an elevation of 2000 ft, calculate the density of the surrounding air.

4) Figure problem 11-4 shows forces on a submerged wall:

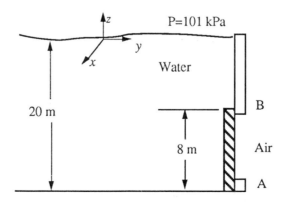

Problem 11-4: Forces on a wall

    a) Calculate the total force per unit width acting on the plate AB and the position of that force per unit width due to the water. Sketch the equivalent force acting of the plate.

b) Calculate the total force per unit width acting on the plate AB and the position of that force per unit width due to the air. Sketch the equivalent force acting on the plate.

c) If the plate is not attached at AB but is simply pressed up against points A and B, determine the reaction forces per unit width acting at A and B on the plate.

5) Figure problem 11-5 shows forces on an inclined wall:

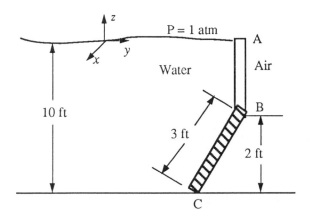

Problem 11-5: Forces on an inclined wall

a) Calculate the total force per unit width acting on the plate BC and the position of that force per unit width due to the water. Sketch the equivalent force acting of the plate.

b) Calculate the total force per unit width acting on the plate BC and the position of that force per unit width due to the air. Sketch the equivalent force acting on the plate.

c) If the plate is attached at B by a hinge or pin support and rests on the bottom at point C as a frictionless contact, determine the forces per unit width acting at B and C.

6) Figure problem 11-6 shows forces on a submerged barrel:

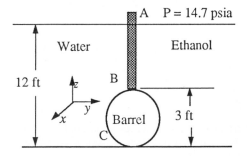

Problem 11-6: Forces on a submerged barrel

a) Calculate the total force per unit width acting on the barrel and the position of that force per unit width due to the water. Sketch the equivalent force acting on the barrel.

b) Calculate the total force per unit width acting on the barrel and the position of that force per unit width due to the ethanol. Sketch the equivalent force acting on the barrel.

c) If the barrel is free to move horizontally at point B, will the barrel move and why. Also in, what direction will the barrel move. Based on your answers here, if you wanted the barrel to stay exactly where it is by adjusting the water level, would you increase or decrease the height of the water and why?

7) Natural gas is stored in high pressure spherical tanks at an inside pressure of 1,000 (psi). If the tank is fabricated from steel with an allowable stress of $\tau = 12,000$ (psi), determine the thickness of the steel and the total mass of the steel for a tank with a 40 (ft) diameter. If the tank is cylindrical with a 20 (ft) diameter and spherical ends and capable of holding the same volume of pressurized natural gas, determine the thickness and the total mass of steel for fabricating the tank from the same material.

8) A 4 (mm) capillary tube is placed in a liquid with a known density of 0.83 (g/ml). After a sufficient time has passed, you notice that the fluid in the column is 0.70 (cm) above the liquid level. The surface of the liquid in the capillary is spherical in nature. Use the conservation of linear momentum to determine the surface tension of the unknown fluid.

9) Hoover dam on the Colorado River outside of Las Vegas, Nevada was constructed to provide water and generate electricity to the states of Arizona, Nevada, and California. The total capacity of water is 30.5×10^6 (acre ft) (1acre = $43,560\text{ft}^2$ and 640acre = 1mi^2). At flood stage, the level of the water is 1229 (ft) above sea level. The level of the river below is 503 (ft) above sea level. The average temperature of the water is about 40 °F. The thickness of the dam at the roadway atop the dam is about 45 (ft) while that of the base is 660 (ft). Assume that the upstream face of the dam is vertical. The dam is constructed from reinforced concrete.

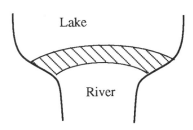

Problem 11-9: Hoover dam

a) At flood stage, the dam can handle a maximum volumetric flow rate of 90,000 (ft^3 / s) through the turbines and generators. If the total capacity of power that the dam can generate is 1.4×10^9 (W), estimate the overall efficiency of the hydroelectric plant, η, where $\eta = \left(\dfrac{\text{electric power generated}}{\text{power input}} \right)$.

b) The projected area of the face of the dam approximates a trapezoid with a base of 800 (ft) and a top of 2000 (ft). Determine the total force and the position of that total force below the surface of the water, due to the lake acting on the dam.

c) In the previous answer, we assumed that the dam was a flat plate. In reality, an aerial photograph reveals the curvature of the structure is as shown in the figure. Why do you think the dam is shaped this way?

10) A plastic beach ball with a diameter of 1 (ft) and a mass of 3 (oz) is placed at the bottom of an 8 (ft) deep swimming pool. The temperature of the water in the pool is 60 °F. Neglect the resistance of the water.

    a) Use the conservation of energy to determine the velocity of the ball just as it breaks the surface of the water.

    b) Use the conservation of momentum to determine the velocity of the ball just as it breaks the surface of the water.

    c) Determine the time that it takes for the ball to reach the surface. Can you use both momentum and energy to determine the time and why?

11) After releasing the ball from the bottom of the pool as described in the previous problem, you actually measure the velocity at the surface to be 1/10 the previously calculated value. This discrepancy is because we neglected the effect of the water resistance. We propose a model for the force due to the water resistance which is proportional to the velocity of the ball but acting in the opposite direction: $\mathbf{f}_{water} = -k\mathbf{v}$ where k has units of lb_m / s.

    a) Use the conservation of the energy to determine the total work that this force due to water resistance did on the ball.

    b) Use the conservation of momentum to determine the proportionality constant, and determine the velocity of the ball as a function of time.

    c) Determine the time that it takes for the ball to reach the surface. Can you use both momentum and energy to determine the time and why?

    d) We discover that our original proposed model for the force due to the water resistance is not correct and we propose another model, that the force is proportional to the magnitude of the velocity squared and acts in the opposite direction: $\mathbf{f}_{water} = -cv^2\mathbf{u}_v$ where c has units of lb_m / ft and $\mathbf{u}_v$ is a unit vector in the direction of $\mathbf{v}$. Outline a strategy that you would take to determine the proportionality constant and the velocity as a function of time.

12) The atmosphere surrounding the earth can be thought of as a deep pool of fluid (air) and we live at the bottom of the pool. However, the fluid in the pool does not have a constant density. Also, from measurement, the gravitational force per unit mass (g) varies as a function of position in the pool as shown in the table.

Problem 11-12 Data)

Height above sea level h (m)	Density ρ (kg / m³)	Magnitude of g g (m / s²)
0	1.225	9.807
10,000	0.414	9.776
20,000	0.089	9.745
30,000	0.012	9.715
40,000	0.004	9.684
50,000	0.001	9.654
60,000	0.000	9.624

    a) If the pressure measured at 60,000 (m) above the earth's surface is 20 (N / m²), calculate the pressure at sea level from the given data.

b) If the pressure at sea level is measured to be 1.01×10^5 (N / m^2), determine the percent error between the calculation of part (a) and the measurement.

c) Discuss sources of error in the calculation, i.e., what would it take to obtain a more accurate comparison. (More data? Different calculation procedure? Explain.)

13) Mercury is used in a U-tube manometer to measure pressure differences. The mercury interface in one leg of the manometer is 20 cm higher than that in the other leg. The fluid above the mercury is air. The manometer is 25 °C. Calculate the measured pressure difference in kPa.

14) A manometer is filled with carbon tetrachloride at 20 °C and used to measure a pressure difference. The liquid interface in one of the manometer legs is 4 in higher than that in the other leg. The fluid above the carbon tetrachloride is air. Calculate the measured pressure difference in psi.

15) A manometer is attached at two points to a pipe containing flowing water. The manometer fluid is mercury and the temperature is 25 °C. The level of the mercury in the upstream leg is 6 inches lower than that in the downstream leg. The fluid in the pipe and above the mercury in the manometer is water. Calculate the measured pressure difference between the two points in the pipe, in psi.

16) To measure the presure difference in a horizontal pipe with fluid flow, a manometer is attached to two points in the pipe. The manometer fluid is mercury at 10 °C. The fluid in the pipe and in the manometer above the mercury is water. The mercury-water interface in the upstream leg of the manometer is 30 cm below the pipe and the interface in the downstream leg is 15 cm below the pipe. Calculate the pressure difference between the two points in the pipe, in kPa.

17) A manometer open at both ends contains both water and mercury. The vertical column of water (in the right leg) is 5 in above the water/mercury interface, which is also in the right leg. The temperature of manometer is 40 °C. What is the height of the mercury level (in the left leg), above the water/mercury interface? There is no water in the left leg. Sketch the manometer, indicating the liquid levels and the position of the interface.

18) Water and an unknown liquid (less dense than water, and immiscible with it) are put in the two legs of an open manometer (open to the atmosphere). In the right leg of the manometer, the vertical column of the unknown liquid is 45 cm above the water/liquid interface. In the left leg of the manometer, the unknown liquid is 30 cm above the water-liquid interface. The water-liquid interface is 12 cm lower in the right leg than it is in the left leg. The temperature is 20 °C. Determine the density of the unknown liquid.

19) One end of a 2 mm diameter capillary tube open at both ends is submerged in ethanol at 20 °C. The tube is vertical. Assuming the ethanol completely wets the capillary tube, calculate the height in mm of the ethanol in the capillary column above the liquid surface.

20) A 1 mm capillary tube is placed in mercury at 20 °C. The mercury does not wet the surface of the capillary tube. Calculate the distance in mm of the mercury in the capillary columm below the liquid surface.

CHAPTER
TWELVE
FLUID DYNAMICS

12.1 INTRODUCTION

Fluid dynamics deals with the motion (dynamics) of fluids. As such, it is concerned with a broad range of problems covering the details of fluid motion on a differential-scale (velocity profiles, i.e. velocity expressed as a function of position) as well as on a macroscopic-scale such as the flow of fluid through pipes and pipe networks. In these latter cases we are concerned with average flow rates rather than velocity profiles, with pressure drops which occur over finite distances in pipes, and with the energy required from pumps or compressors to achieve the flows (or energy delivered by turbines).

It is these macroscopic-scale problems which we are concerned with in this course. Such considerations suggest extensive use of the extended Bernoulli (mechanical energy) accounting equation and this is certainly the case. Additionally, however we will make considerable use of mass, linear momentum, and angular momentum conservation. Total energy and entropy may play roles as well, although these are less frequently used directly in these applications.

We will begin this chapter with a review of these fundamental laws from the perspective of their applicability to macroscopic fluid dynamics problems. Part of this discussion will involve reconsideration of the velocities which appear in those equations. Second we will consider the most basic and frequently-encountered macroscopic fluid mechanics problem: the flow of fluid in pipes. We will consider this for constant density flows, i.e. for flows of incompressible fluids but also for flows of compressible fluids for which the pressure and temperature changes are small enough that the densities are nearly constant. Analysis of these types of problems will allow us, as two examples, to size pumps for a given piping system, or to determine the most economical pipe diameter as a design problem. Third, we will consider the forces which exist on macroscopic systems because of the fluid flow. These forces are analyzed by the conservation of linear and angular momentum and arise because of the weight of the fluid, because of changes in the fluid's velocity either in direction or magnitude, and because of pressure changes.

12.2 REVIEW OF THE BASIC EQUATIONS

In macroscopic-scale fluid dynamics, normally we are not concerned with individual species in multicomponent systems but rather with the fluid taken as a whole in terms of average fluid physical properties and average velocities. Likewise, we are not normally concerned with charge effects (although this could be a factor in some situations). Finally, apart from the total energy equation, the equation which is usually of greatest concern is that for steady state mechanical energy (the extended Bernoulli equation). Direct considerations of entropy may be of interest in some situations but this is the exception, rather than the rule.

Considering these equations, then, we have, first, the conservation of total mass which we may write as:

$$\left(\frac{dm_{\text{sys}}}{dt}\right) = \sum \dot{m}_{\text{in/out}} \tag{3-15}$$

Now, normally in pipe flow situations we are concerned with an average flow of fluid in one direction only, normal to the cross section of the pipe. In terms of this velocity and the cross sectional area of the pipe we can rewrite the above equation as:

$$\left(\frac{dm_{\text{sys}}}{dt}\right) = \sum (\rho v A)_{\text{in/out}} \tag{12-1}$$

where the sumations are taken over all the inlet streams and outlet streams. If the process is steady state, then of course the right hand side, representing the accummulation of mass within the system, is zero. Again the density and velocity are averaged over the cross-sectional area for each inlet or outlet conduit. This averaging process will be discussed more fully below in section 12.2.1.

The second equation to be considered is that of linear momentum. The general macroscopic form which we have used is:

$$\left(\frac{d\mathbf{p}_{\text{sys}}}{dt}\right) = \sum (\dot{m}\mathbf{v})_{\text{in/out}} + \sum \mathbf{f}_{\text{ext}} \tag{5-1}$$

and describes the relations between the forces which act on a piping (or other) system. For example, the forces which must be provided by the supporting members of a pipe are affected by the weight of the fluid, by the changes in momentum of the fluid between inlet and outlet streams, and by pressure changes along the pipe length. Piping systems must be adequately designed, not only from the view point of pipe diameter and pump size, but also with respect to the mechanical strength of the members which hold the piping system in place.

The third primary equation which is considered in fluid dynamics is the steady state extended Bernoulli (steady state mechanical energy accounting) equation:

$$0 = \dot{m}(\widehat{PE}_{\text{in}} - \widehat{PE}_{\text{out}}) + \dot{m}(\widehat{KE}_{\text{in}} - \widehat{KE}_{\text{out}}) + \dot{m}\int_{P_{\text{out}}}^{P_{\text{in}}} \frac{dP}{\rho} + \sum \dot{W}_{\text{shaft}} - \sum \dot{F} \tag{7-50}$$

or, in alternate form (let $\widehat{V}dP = d(P\widehat{V}) - Pd\widehat{V}$, where $\widehat{V} = 1/\rho$)

$$0 = \left[\dot{m}\left(\widehat{KE} + \widehat{PE} + \frac{P}{\rho} \right) \right]_{\text{in/out}} - \dot{m}\int_{\widehat{V}_{\text{out}}}^{\widehat{V}_{\text{in}}} Pd\widehat{V} + \sum \dot{W}_{\text{shaft}} - \sum \dot{F}$$

In this form we see that mechanical energy enters and leaves the system with mass in the form of kinetic energy, potential energy, and flow work. Furthermore, it may enter due to shaft work. Inside the system, mechanical energy is generated or consumed by conversion. It may be converted to internal energy through friction ($\dot{F}$, where $\dot{F}$ is always positive) or though expansion or compression ($-Pd\widehat{V}$) of the fluid. Expansion ($\widehat{V}_{\text{out}} > \widehat{V}_{\text{in}}$) results in a gain or generation of mechanical energy inside the system while compression ($\widehat{V}_{\text{out}} < \widehat{V}_{\text{in}}$) results in a loss. For an incompressible fluid (and liquids under most circumstances are nearly so), this expansion/contraction term is zero.

Several difficulties arise in using this equation and should be appreciated. First, as described above, the extended Bernoulli equation is simplified for incompressible fluids but for compressible flows we must calculate an integral over a specific path through the process and this can be a very difficult, if not impossible, calculation. Another complication of this equation concerns the kinetic energy, which involves the velocity of the fluid. As with mass and momentum input rates, this varies across the pipe cross section. Consequently, the kinetic energy must be integrated over the area. This will be considered further below. Still another complication is the term representing the frictional conversion of mechanical energy to internal energy of the fluid. This term represents a loss of mechanical energy due to friction within the fluid. This conversion process occurs both near the walls of the pipe, where a drag force is exerted on the fluid by the walls, thereby resulting in a loss of mechanical energy, and it also occurs inside the pipe due to viscous drag that occurs between adjacent elements of the fluid. To the extent these elements are moving at different velocities and to the extent the fluid is viscous (i.e. to the extent that it has a nonzero viscosity), this difference in velocity results in a drag between the two elements and consequently in a conversion of mechanical energy to internal energy in the same way that your hands get warmer when you rub them together. In a few situations this loss of mechanical energy can be calculated *a priori* but in general this cannot be done and we must resort to empirical determinations and correlations of these losses in terms of the flow situation. However, observation, formalized by the second law of thermodynamics says that the terms representing the conversion from mechanical energy are positive:

$$\dot{F} \geq 0 \qquad\qquad (12-2)$$

Fortunately for us this has been done by those who preceded us and the data are readily available. As a result, however, a significant portion of the material in this chapter deals with evaluating these friction losses for specific pipe elements (straight pipe, valves, expansions, contractions, elbows, etc.).

12.2.1 Velocity Profiles and Average Velocities

As a consequence of dealing with fluids, we are confronted with the fact that adjacent elements of the materials may be moving relative to each other. This was not the situation for rigid bodies; in fact the definition of a rigid body was that adjacent elements of material did not experience relative motion.

Because of this ability of different elements to have motion relative to each other, it is a fact of fluid flow, and one which we must recognize in these discussions, that the velocity of a fluid through a pipe is not everywhere the same across the pipe cross section. Experimentally it has been observed that almost universally the fluid "sticks" to the wall and consequently has a *zero velocity at the wall* or near other nonmoving or stationary solid surfaces (if a solid surface is moving, then almost always the fluid in immediate contact with that surface is moving with exactly the same velocity as the surface). Now, if the fluid is experiencing no slip at the wall, then in order for the fluid to have a non-zero average velocity, there must be a region of the fluid which has a flow velocity greater than the average in order to make up for this deficit near the wall. Obviously, then, the velocity in a pipe is more properly characterized by a velocity profile across the pipe; the velocity of the fluid varies from point-to-point across the pipe.

While it is not appropriate subject matter for this course to derive the shape of these profiles (although rest assured that the calculations involve using conservation of mass and conservation of linear and angular momentum), we will nevertheless, make use of these results or, at the very least, we must be aware of the fact that these velocity variations exist.

However, before we proceed we must have an understanding of the terms laminar and turbulent with respect to flow. If a fluid flows through a pipe sufficiently slowly, then its flow will be in one direction only, the axial direction, and it will be well organized in that it will be zero at the pipe wall (no slip) and will increase towards the center of the pipe where it will have a maximum velocity. This will be the preferred stable structure of the flow; as long as there are no disturbances to upset this flow, it can be maintained indefinitely. It's as though each cylindrical shell of fluid within the pipe is flowing downsteam as a single coherant shell independent of the others with the shells near the center moving faster than the shells near the wall; there is no mixing from one shell to another. We call this *laminar flow*. It has been observed however, that if the flow velocity becomes high enough, then the shells cannot persist and this kind of laminar, nicely ordered, velocity profile becomes disturbed. If a high enough flow rate is imposed then the flow becomes very turbulent and the velocity profile is nearly flat. This flat profile we will refer to as plug flow, that is, it is as though the fluid is proceeding as a uniform plug through the pipe. This is only an approximation, of course, in that first, very near the wall, we still have the no slip condition which holds, and second, the fluid is turbulent with almost random fluctuations imposed on the average flow. On average, however, over a period of time the fluid velocity will be the same at one point of the pipe cross section as it is at another (except very near the wall).

Consider now as one example, the laminar profile of a newtonian fluid. It has been observed that the velocity varies in a parabolic manner; the velocity as a function of position in the pipe forms a parabaloid, Figure 12-1a. A profile which might exist for turbulent flow for the same fluid at higher flow rates is also shown in Figure 12-1b. At low flow rates the order which is imposed on the fluid by the no slip condition at the wall is able to propagate all the way to the center of the tube by means of the stresses in the fluid. At high flow rates however, the energy in the fluid is too great for this order to be maintained.

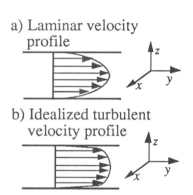

a) Laminar velocity profile

b) Idealized turbulent velocity profile

Figure 12-1: Velocity profiles

In this context you can imagine that a viscous fluid is better able to sustain and propagate the order than is a less viscous fluid, which has lower stresses, in the same flow situation. In fact, this is the case and we will observe later how this order is preserved as a function of flow rate and fluid stresses (viscosity).

In the present discussion we are interested in the effect of these velocities on the forms of the equations as given above. Considering the conservation of mass we can write that for a small shell of fluid the mass flow rate across shell area dA is given by (see eq 3-22d):

$$\left(\begin{array}{c} \text{mass flow rate} \\ \text{across area } dA \end{array} \right) = \rho v dA \qquad (12-3)$$

and this can be integrated over the entire cross sectional area to give the total mass flow rate across that area. Accordingly, equation 3-19 becomes

$$\left(\frac{dm_{\text{sys}}}{dt} \right) = \left(\int_A (\rho v) dA \right)_{\text{in}} - \left(\int_A (\rho v) dA \right)_{\text{out}} \qquad (12-4)$$

In these equations v *is the magnitude of the velocity in the axial direction.* That is, it is assumed that we can neglect the other components of the velocity in that they do not transport mass across the system boundary. This mass flow rate, integrated over the entire area, can be used to define an average velocity through the area. Accordingly we write:

$$\int_A (\rho v) dA = \rho \langle v \rangle A \qquad (12-5)$$

and this serves to define the average velocity. In this equation it is assumed that the density of the fluid is constant across the area. This implies constant temperature, pressure, and composition across the area, generally a valid assumption in the absence of heat transfer. The average velocity, then, is defined as:

$$\langle v \rangle = \frac{\int v dA}{A} \qquad (12-6)$$

In terms of average velocities, then, we can write the total mass conservation equation as:

$$\left(\frac{dm_{\text{sys}}}{dt} \right) = \sum (\rho \langle v \rangle A)_{\text{in/out}} \qquad (12-7)$$

This is generally the form of the equation that will be used for fluid dynamics.

We must also consider a similar approach to the conservation of linear momentum for fluid systems. Again, considering the velocity to be uni-directional normal to the cross sectional area, we can write the rate at which momentum enters the system across a boundary as the velocity (momentum per mass) times the mass flow rate at that boundary and times the differential area:

$$\left(\begin{array}{c} \text{momentum flow rate} \\ \text{across area } dA \end{array} \right) = (v\mathbf{u}_N \rho v) dA = (\rho v^2 \mathbf{u}_N) dA \qquad (12-8)$$

In this equation we write the velocity as the magnitude of the velocity times its unit vector in the direction which is normal to the area and the mass flow rate is $\rho v dA$. This must be integrated over the entire cross sectional area, then, and summed over all of the areas across which flow occurs. Accordingly,

$$\left(\begin{array}{c} \text{total momentum flow} \\ \text{rate across area } A \end{array} \right) = \sum \left(\int_A (\rho v^2 \mathbf{u}_N) dA \right) \qquad (12-9)$$

For any given surface we can define an average of the squared velocities in a manner completely analogous to defining the average velocity. Accordingly, we write:

$$\langle v^2 \rangle = \frac{\int v^2 dA}{A} \qquad (12-10)$$

and the conservation of linear momentum can be written in terms of these averages , again assuming density is constant across the area, according to:

$$\left(\frac{d\mathbf{p}_{\text{sys}}}{dt} \right) = \sum (\rho \langle v^2 \rangle \mathbf{u}_N A)_{\text{in/out}} + \sum \mathbf{f}_{\text{ext}} \qquad (12-11)$$

which, in terms of the mass flow rate across each area, can be rewritten

$$\left(\frac{d\mathbf{p}_{\text{sys}}}{dt}\right) = \sum\left(\frac{\langle v^2\rangle}{\langle v\rangle}\rho\langle v\rangle A\mathbf{u}_N\right)_{\text{in/out}} + \sum\mathbf{f}_{\text{ext}}$$

$$\left(\frac{d\mathbf{p}_{\text{sys}}}{dt}\right) = \sum\left(\frac{\langle v^2\rangle}{\langle v\rangle}\dot{m}\mathbf{u}_N\right)_{\text{in/out}} + \sum\mathbf{f}_{\text{ext}}$$

$$(12-12)$$

Evidently, $\langle v^2\rangle / \langle v\rangle\mathbf{u}_N$ is the average momentum per mass (**v**) written in terms of $\langle v^2\rangle$ and $\langle v\rangle$. Note that in these equations, the average of the square of the velocity over a cross section is not the same as the square of the average velocity. Consequently, the ratios do not cancel out. Furthermore, remember that this is a vector equation and that the direction of each of these mass flow rates is important in considering the forces which are involved. This will become clear in problems that are worked later in this chapter.

Finally a similar consideration must be made for the mechanical energy accounting equation. In this equation we must consider the kinetic energy of each flow stream which is written in terms of the velocity squared i.e. we have

$$\left(\begin{array}{c}\text{kinetic energy flow}\\\text{rate across area } dA\end{array}\right) = \left(\frac{1}{2}v^2\rho v\right)dA$$

$$(12-13)$$

This specific kinetic energy times the differential mass flow rate can be integrated over the area to give the total rate of kinetic energy crossing the system boundary at this surface:

$$\left(\begin{array}{c}\text{total kinetic energy flow}\\\text{rate across area } A\end{array}\right) = \int_A\left(\frac{1}{2}v^3\rho\right)dA$$

$$(12-14)$$

Again, as for the mass and momentum flow rates, we can define an average velocity cubed by the relation:

$$\langle v^3\rangle = \frac{\int v^3 dA}{A}$$

$$(12-15)$$

from which we can write the mechanical energy accounting equation as

$$0 = \sum(\widehat{PE}\rho\langle v\rangle A)_{\text{in/out}} + \sum\left(\frac{\rho\langle v^3\rangle A}{2}\right)_{\text{in/out}} + (\rho\langle v\rangle A)\left(\int_{P_{\text{out}}}^{P_{\text{in}}}\frac{dP}{\rho}\right) + \sum\dot{W} - \sum\dot{F}$$

$$(12-16)$$

which can be rewritten in terms of kinetic energy per unit mass and mass flow rate as:

$$0 = \sum(\widehat{PE}\rho\langle v\rangle A)_{\text{in/out}} + \sum\left(\frac{\langle v^3\rangle}{2\langle v\rangle}\rho\langle v\rangle A\right)_{\text{in/out}} + \left(\int_{P_{\text{out}}}^{P_{\text{in}}}\frac{dP}{\rho}\right)(\rho\langle v\rangle A) + \sum\dot{W} - \sum\dot{F}$$

$$(12-17)$$

or as:

$$0 = \sum\dot{m}(\widehat{PE}_{\text{in/out}} + \sum\dot{m}\left(\frac{\langle v^3\rangle}{2\langle v\rangle}\right)_{\text{in/out}} + \dot{m}\left(\int_{P_{\text{out}}}^{P_{\text{in}}}\frac{dP}{\rho}\right) + \sum\dot{W} - \sum\dot{F}$$

$$(12-18)$$

These equations, (12-7), (12-12), and (12-18) will be the basic starting equations to be considered for the fluid dynamics applications which follow. In these equations we have used the average velocity, the average of the velocity squared and the average of the velocity cubed across each of the cross sectional areas through which fluid is flowing into or out of the system.

In the momentum equation the ratio of the average of the square of the velocity to the average velocity appeared and in the Bernoulli equation the ratios of the average of the velocity cubed to the average velocity appeared. These velocity-squared and velocity-cubed averages differ from the square and cube of the average velocity if the velocity profiles are not flat. In some references you will see a factor which represents the extent of this deviation. Accordingly it is common to define a factor, α, which, when used for the calculation of kinetic energy, will allow us to use the square of the average velocity just as we would like to use it in order to calculate kinetic energy. Accordingly, we will write:

$$\left(\frac{\langle v \rangle^2}{2\alpha_{\text{KE}}}\right) = \left(\frac{\langle v^3 \rangle}{2\langle v \rangle}\right) \tag{12-19}$$

Accordingly, then,

$$\alpha_{\text{KE}} \equiv \left(\frac{\langle v \rangle^3}{\langle v^3 \rangle}\right) \tag{12-20}$$

We could define a similar relation for the momentum (equation 12-10) in which case:

$$\left(\frac{\langle v \rangle}{\alpha_{\text{Mom}}}\right) \equiv \left(\frac{\langle v^2 \rangle}{\langle v \rangle}\right) \tag{12-21}$$

so that:

$$\alpha_{\text{Mom}} = \left(\frac{\langle v \rangle^2}{\langle v^2 \rangle}\right) \tag{12-22}$$

In terms of the average velocity and these alpha correction factors, we can sumarize the equations for the conservation of mass and linear momentum and accounting of mechanical energy as:

$$\left(\frac{dm_{\text{sys}}}{dt}\right) = \sum (\rho \langle v \rangle A)_{\text{in/out}} \tag{12-1}$$

$$\left(\frac{d\mathbf{p}_{\text{sys}}}{dt}\right) = \sum \left(\frac{\langle v \rangle}{\alpha_{\text{Mom}}} \rho \langle v \rangle \mathbf{u}_N A\right)_{\text{in/out}} + \sum \mathbf{f}_{\text{ext}} \tag{12-23}$$

$$0 = \sum (gh\rho \langle v \rangle A)_{\text{in/out}} + \sum \left(\frac{\langle v \rangle^2}{2\alpha_{\text{KE}}} \rho \langle v \rangle A\right)_{\text{in/out}} + \left(\int_{P_{\text{out}}}^{P_{\text{in}}} \frac{dP}{\rho}\right)(\rho \langle v \rangle A) + \sum \dot{W} - \sum \dot{F} \tag{12-24}$$

Note that in each of these equations we have, for each area through which mass is flowing, a property, such as momentum per unit mass or energy per unit mass, times a mass flow rate where the mass flow rate is written as $\rho\langle v\rangle A$. The mass per unit mass for the conservation of mass equation is of course equal to unity. Remember also that the Bernoulli equation is only applicable for steady-state situations.

Example 12-1. The conduit shown in Figure 12-2 has inside diameters of 9 in and 12 in at points 1 and 2 respectively. If water is flowing through the conduit at a velocity of 18 ft/s at point 2 and the velocity profile is flat or fully turbulent, find (a) the velocity of the water at point 1, (b) the volumetric flow rate at point 1, and (c) the mass flow rate at point 1.

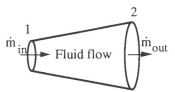

Figure 12-2: Steady state flow through a conduit

SOLUTION: The system is defined as the section of conduit between point 1 and point 2 as described in Figure 12-2a: The conservation of mass for this system is:

$$\dot{m}_{in} - \dot{m}_{out} = \left(\frac{dm_{sys}}{dt}\right)$$

$$= 0$$

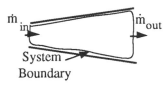

Figure 12-2a: System for conduit

The mass flow rate out of the system equals the mass flow into the system because the system is operating at steady state.

Expressing the mass flow rates in terms of the density, average velocity, and cross-sectional areas gives (eq 3-22d):

$$0 = (\rho\langle v\rangle A)_{in} - (\rho\langle v\rangle A)_{out}$$

Because the density of water is constant for this problem, $\rho = 62.4 \; (lb_m / ft^3)$, the equation reduces to:

$$0 = (\langle v\rangle A)_{in} - (\langle v\rangle A)_{out}$$

The only unknown in this equation is the average velocity at point 1 the entering average velocity. Rearranging gives:

$$\langle v\rangle_{in} = \left(\frac{A_{out}}{A_{in}}\right)\langle v\rangle_{out} = \left(\frac{D_{out}}{D_{in}}\right)^2\langle v\rangle_{out}$$

Substituting numerical values yields:

$$\langle v\rangle_{in} = \left(\frac{12}{9}\right)^2(18) = 32 \text{ ft} / \text{s}$$

The average velocity at point 1 is 32 ft/s.

For the second part of the problem, the volumetric flow at point 1 is simply the cross sectional area at point 1 times the average linear velocity at that section.

$$\dot{V}_{in} = (\langle v\rangle A)_{in} = \left(\frac{\langle v\rangle_{in}\pi D_{in}^2}{4}\right)$$

ubstituting values with the proper conversion factors gives:

$$\dot{V}_{in} = \left(\frac{(32)\pi(9)^2}{4} \right) \frac{ft\ in^2}{s} \left| \frac{1\ ft^2}{144\ in^2} = 14.1\ ft^3 / s$$

he volumetric flow rate at point 1 is 14.1 ft^3 / s.

The final part of the question is calculated from the definition of density

$$\dot{m}_{in} = \dot{V}_{in}\rho$$

ubstituting into this equation yields:

$$\dot{m}_{in} = (14.1)(62.4) \left(\frac{ft^3\ lb_m}{s\ ft^3} \right) = 882\ lb_m / s$$

The mass flow rate at point 1 is 882 lb_m / s.

Example 12-2. Water is forced into an emulsifier s shown in Figure 12-3 at a volumetric flow rate of 0.2 1^3 / s through the pipe at point B. Oil with a density of 0.6 /ml is pumped into the process at a volumetric flow rate of .01 m^3/s through pipe A. If the liquids are incompressible (the densities remain constant), the volume change due to nixing is zero (there is no generation or consumption of olume), and the mixture is a homogeneous mixture of lobules, what is the average velocity of the mixture leaving he process through pipe C having an inside diameter of .5 m. Also, what is the density of the mixture leaving the rocess?

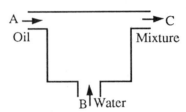

Figure 12-3: Mixing of two liquids

SOLUTION: The system is defined as the emulsi- er. The system is at steady state, furthermore there are o reactions occuring. The defined system is shown in igure 12-3a. The conservation of total mass for this system perating at steady state is given by the following equation. This equation is presented in tabular form in Table 12-1.

$$0 = \sum (\dot{m}_i)_{in/out}$$

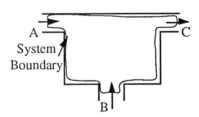

Figure 12-3a: System for mixing of liquids

Table 12-1: Data for Example 12-2 (Mass, System = Mixing Process)

Species	$\left(\dfrac{d(m_i)_{\text{sys}}}{dt}\right)$ (kg/s)	=	Stream A (kg/s)		Stream B (kg/s)		Stream C (kg/s)
Oil	0	=	$\dot{m}_{\text{oil}}$	+	0	−	?
Water	0	=	0	+	$\dot{m}_{\text{water}}$	−	?
Total	0	=	$\dot{m}_{\text{oil}}$	+	$\dot{m}_{\text{water}}$	−	?

If we look at the total conservation of mass equation it has only one unknown, the total mass flow rate leaving the system. The mass flow rates entering the system can be expressed as the product of the volumetric flow rate of each entering stream time the density of each stream (eq 3-22c):

$$0 = (\dot{V}\rho)_{\text{oil}} + (\dot{V}\rho)_{\text{H}_2\text{O}} - \dot{m}_{\text{out}}$$

Rearranging for the unknown gives:

$$\dot{m}_{\text{out}} = (\dot{V}\rho)_{\text{oil}} + (\dot{V}\rho)_{\text{H}_2\text{O}}$$

Substitution yields:

$$\dot{m}_{\text{out}} = (0.01)(600) + (0.2)(1000) \left(\frac{\text{kg m}^3}{\text{s m}^3}\right) = 206 \text{ kg / s}$$

The mass flow rate leaving the system at point C is 206 kg/s.

Because there is no volume change on mixing, the volumetric flow rates of the entering streams must be equal to the volumetric flow rates of the leaving streams.

$$0 = \dot{V}_{\text{oil}} + \dot{V}_{\text{H}_2\text{O}} - \dot{V}_{\text{mix}}$$

Rearranging this equation gives:

$$\dot{V}_{\text{mix}} = \dot{V}_{\text{oil}} + \dot{V}_{\text{H}_2\text{O}}$$

Substituting values allows us to calculate the unknown volumetric flow rate of the mixture:

$$\dot{V}_{\text{mix}} = (0.01) + (0.2) = 0.21 \text{ m}^3 / \text{s}$$

Now, the average velocity leaving the system at point C is the volumetric flow rate leaving the system divided by the cross-sectional area.

$$\langle v \rangle_{\text{out}} = \left(\frac{\dot{V}_{\text{out}}}{A_{\text{out}}}\right) = \left(\frac{4\dot{V}_{\text{out}}}{\pi D_{\text{out}}^2}\right)$$

Substitution gives:

$$\langle v \rangle_{\text{out}} = \left(\frac{4(0.21)}{\pi(0.5)^2}\right)\left(\frac{\text{m}^3}{\text{s m}^2}\right) = 1.07 \text{ m / s}$$

The average velocity of stream C leaving the system is 1.07 m/s

Finally, the density of the stream leaving the system is calculated from the definition of density (see eq 3-22, 3-22c):

$$\rho_{\text{out}} = \left(\frac{\dot{m}}{\dot{V}}\right)_{\text{out}}$$

Substituting values gives:

$$\rho_{\text{out}} = \left(\frac{206}{0.21}\right)\left(\frac{\text{kg s}}{\text{s m}^3}\right) = 981 \text{ kg} / \text{m}^3$$

The density of the stream at point C is 981 kg / m^3.

12.3 FLOW IN PIPES AND OTHER CLOSED CONDUITS

12.3.1 Steady State Flow of Constant Density Materials

In this section we are concerned only with the steady state flow of constant density materials. Furthermore, if we consider a situation which has one inlet (point 1) and one outlet (point 2), such as flow through a pipe then the extended Bernoulli (mechanical energy accounting) equation is:

$$0 = \dot{m}g\sum z_{\text{in/out}} + \dot{m}\sum \frac{\langle v\rangle^2_{\text{in/out}}}{2\alpha_{\text{KE}}} + \dot{m}\sum \frac{P_{\text{in/out}}}{\rho} + \sum \dot{W}_i - \sum \dot{F}_i \qquad (12-25)$$

Again, this is for a constant density fluid. Such a piping process might involve the flow of a fluid from a tank through a number of elbows and lengths of pipe, a valve, changes in pipe size (expansions or contractions), and then finally discharge from the end of the pipe. This general process in shown in Figure 12-4.

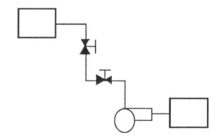

Figure 12-4: General piping process

The problem is to describe the various factors involved, changes in elevation of the fluid, velocity changes, pump energy input and mechanical energy losses, in terms of each other. The Bernoulli equation does this for us by accounting for the mechanical energy of the system (fluid between the input and output). The fluid entering the system has a certain potential and kinetic energy and it leaves with different potential energy and kinetic energy. Between input and output, it has energy added to it through the pump and it has mechanical energy converted to thermal energy by way of frictional dissipation within the pipe, at the walls, around the elbows (bends) in the pipe, and through other restrictions such as valves. This is one equation and if we know all terms except one in this equation, then we can solve for that one in terms of the others. As one example, if we know the fluid condition (potential energy, kinetic energy, and pressure) at the input and output and if we know the desired flow rate and piping system, then (as described below) we can calculate the friction losses and then, from the equation, determine the size pump that is required to provide that flow rate. As a second example, if we know the fluid conditions at the input and output in the form of kinetic and potential energy and pressure, and if we know the work delivered by the pump, then we can calculate the frictional losses which exist.

Along with the Bernoulli equation, the conservation of mass for the constant density fluid allows us to relate the average velocity of the entering stream to the average velocity of the leaving stream if we know the cross sectional areas of the entering and leaving streams:

$$0 = \dot{m}_{\text{in}} - \dot{m}_{\text{out}}$$
$$0 = (\rho \langle v \rangle A)_{\text{in}} - (\rho \langle v \rangle A)_{\text{out}} \qquad (12-26)$$

or:

$$\langle v \rangle_{\text{out}} = \left(\frac{A_{\text{in}}}{A_{\text{out}}} \right) \langle v \rangle_{\text{in}} \qquad (12-27)$$

Flow areas, inside and outside diameters for various nominal pipe sizes and schedule numbers are provided in Appendix C Table C-12. In this equation, as discussed previously in this section, the velocities that are used are the average velocity across the flow cross-sectional area. For turbulent flow, we may assume that the velocity profile is essentially flat in which case:

$$\boxed{\quad \text{Turbulent Flow} \qquad \alpha_{\text{KE}} = 1 \quad} \qquad (12-28)$$

For laminar flow, however, the profile is definitely not flat; α_{KE} deviates considerably from unity. For Newtonian fluid in perfect laminar flow, the parabolic velocity profile gives

$$\boxed{\quad \text{Laminar Flow} \qquad \alpha_{\text{KE}} = \frac{1}{2} \quad} \qquad (12-29)$$

Before we can proceed further with specific applications and examples of this equation we must understand better the frictional losses which occur. These losses occur because fluids undergo deformation and flow (they are not rigid) and they exhibit viscosity, that is, they resist flow. This resistance is completely analogous to the resistance we feel when we slide two solid objects together. For example, if we rub our hands together, they become warmer. When adjacent elements of a fluid move at different velocities, that is when they move with a non-zero relative velocity, then this relative displacement can be a sliding (shear) of the elements past each other (other kinds of deformation are possible, such as elongation). When this is the case, the finite viscosity creates a loss of mechanical energy, that is, a conversion of mechanical energy to thermal energy. This occurs when the fluid is in contact with the wall at which point the fluid velocity is zero due to the no-slip condition described earlier in this chapter. This fluid at the wall is in contact with a moving fluid next to the wall and this relative displacement creates viscous forces and consequent dissipation of mechanical energy to thermal energy. Likewise, throughout the fluid and across the pipe cross-section, wherever there is a velocity gradient in the fluid, there are stresses which are generated and therefore there is viscous dissipation.

Now if we look back at the Bernoulli equation for the steady-state flow of a constant density fluid, with one inlet and one outlet, the mass flow rate in the equation is the same for each term. Therefore, we can divide each of the terms by the mass flow rate yielding:

$$0 = g(z_{\text{in}} - z_{\text{out}}) + \left(\frac{\langle v \rangle_{\text{in}}^2}{2\alpha_{\text{KE}}} - \frac{\langle v \rangle_{\text{out}}^2}{2\alpha_{\text{KE}}} \right) + \left(\frac{P_{\text{in}} - P_{\text{out}}}{\rho} \right) + \sum \left(\frac{\dot{W}_i}{\dot{m}} \right) - \sum \left(\frac{\dot{F}_i}{\dot{m}} \right) \qquad (12-30)$$

or:

$$0 = g(z_{in} - z_{out}) + \left(\frac{\langle v \rangle_{in}^2}{2\alpha_{KE}} - \frac{\langle v \rangle_{out}^2}{2\alpha_{KE}} \right) + \left(\frac{P_{in} - P_{out}}{\rho} \right) + \sum \widehat{W}_i - \sum \widehat{F}_i \qquad (12-31)$$

where the term $\widehat{W}_i$ represents the work done on the system per unit mass and $\widehat{F}_i$ the irreversible conversion of mechanical energy to thermal energy per unit mass (friction losses).Then, the rate of work of mode i done on the system and the rate of irreversible conversion of mode i is calculated as the the mass flow rate times the work or conversion of mechanical to thermal energy per unit mass:

$$\dot{W}_i = \dot{m}\widehat{W}_i$$
$$\dot{F}_i = \dot{m}\widehat{F}_i \qquad (12-32)$$

In general the conversion of mechanical energy to thermal energy per unit mass between the inlet and outlet is dependent on the density and vicosity of the fluid (ρ, μ), the size, length and material (or roughness) of the conduit, (D, l, ϵ), the magnitude of the average velocity in the conduit $(\langle v \rangle)$, and any geometrical factors including contractions, enlargements, fittings, and valves (K_i) that the fluid must flow through. This can be expressed in an equation as:

$$\widehat{F}_i = \widehat{F}_i(\rho, \ \mu, \ D, \ l, \ \epsilon, \ \langle v \rangle, \ K_i) \qquad (12-33)$$

This equation can be also be expressed in terms of a dimensionless friction factor $\mathfrak{f}$ as

$$\widehat{F}_i = \widehat{F}_i(\mathfrak{f}, \ l, \ D, \ \langle v \rangle, \ K_i) \qquad (12-34)$$

where the friction factor is a function of the fluid properties, the conduit characteristics, and average velocity.

$$\mathfrak{f} = \mathfrak{f}(\rho, \ \mu, \ D, \ \epsilon, \ \langle v \rangle) \qquad (12-35)$$

This friction factor is explained further in the next paragraphs.

The quantitative characterization of this conversion of mechanical to thermal energy is one of the most significant developments of fluid mechanics. Osborne Reynolds established that these friction losses (in terms of a dimensionless friction factor, $\mathfrak{f}$) can be expressed as a function of a dimensionless group of numbers which has come to be called the Reynolds number:

$$\mathfrak{f} = \mathfrak{f}\left(N_{Re}, \frac{\epsilon}{D} \right) \qquad (12-36)$$

This Reynolds number is given by:

$$N_{Re} = \left(\frac{D\langle v \rangle \rho}{\mu} \right) \qquad (12-37)$$

Reynolds found that the flow of all Newtonian fluids in smooth pipes can be characterized by the same curve when this friction factor is plotted versus Reynolds number. Or for ϵ equal to zero for a *smooth* pipe, and K_i equal to unity for a straight pipe (no valves or fittings).

$$\mathfrak{f} = \mathfrak{f}(N_{Re}) \qquad (12-38)$$

He found this to be true for all viscosities, densities, flow rates, and pipe sizes. This correlation of these widely varying experiments by means of a single non-dimensional relationship has been termed the most celebrated application of dimensional analysis in fluid mechanics.

The plot of this friction factor versus Reynolds number is shown in Appendix C, Figure C-1. In this figure, Reynolds observed a linear relationship (for a log-log scale) for laminar flow as long as the Reynolds number was below 2,000.

$$\text{Laminar Flow} \qquad N_{Re} < 2000 \qquad\qquad (12-39)$$

Above about 4,000, he observed that a different relationship was followed because of the existence of turbulent flow.

$$\text{Turbulent Flow} \qquad N_{Re} > 4000 \qquad\qquad (12-40)$$

Notice that the onset of turbulence causes an increase in the friction factor. Evidently the turbulent flow is a lot more dissipative of mechanical energy than is the laminar flow. For example, for the same Reynolds number of 4,000, laminar flow, if it were obtained, would have a factor of less than 0.005. Turbulent flow, on the other hand for the same Reynolds number, has a friction factor for flow through smooth pipe of about 0.01, twice as great. At this Reynolds number the friction losses for smooth pipe are twice as great for turbulent flow as they would be for laminar flow. Further experiments showed that a roughening of the surface of the pipe further increased the effect of turbulence so that added roughness created a greater resistance to flow or a greater conversion of mechanical energy to thermal energy, a greater dissipation of mechanical energy. Notice further that for each amount of roughness (expressed as the *relative roughness*, that is, the amount of surface roughness relative to the diameter of the pipe) the friction factor eventually becomes constant at a high enough Reynolds number. The values of roughness for a number of materials are given in Table C-14 in Appendix C.

This friction factor plot can be used to calculate the friction losses, i.e. the dissipation of mechanical energy, for a given flow situation. The friction factor that is read from the chart is known as the Fanning Friction factor. There are other friction factor definitions and it is important to realize that all of the equations in this and following sections use the Fanning Friction factor. These friction losses represent the losses due to flow through sections of straight pipe and do not include additional losses due to flow around bends or through valves or restrictions. For a given flow rate, pipe size (D) and fluid viscosity and density, one can calculate the Reynolds number. From this Reynolds number and a roughness of the pipe (if the Reynolds number is in a turbulent regime) the friction factor chart gives the friction factor. Knowing the friction factor and flow parameters, the conversion of mechanical energy to thermal energy from flow in straight pipes is given as:

$$\text{Straight Pipe:} \qquad \widehat{F}_i = \frac{4fl}{D}\left(\frac{\langle v \rangle^2}{2}\right) \qquad\qquad (12-41)$$

where l is the total length of pipe and D is the inside diameter of the pipe. Using this friction factor the mechanical energy losses can be calculated by equation (12-41) and then included in the Bernoulli equation. If the objective is to determine the pumping requirements for this particular flow situation, then the pressure difference, the potential energy change, and a kinetic energy change will be known and the Bernoulli equation can be solved directly to give this pump requirement. Now there are other applications of the Bernoulli equation which are not as direct, but this is certainly the most straightforward and the one that must be mastered before all others.

Equation for the Friction Factor. Churchill (1977) has obtained an expression for the Fanning Friction factor which can be used for computer calculations of pipe flow problems. This equation is:

$$f = 2\left(\left(\frac{8}{N_{Re}}\right)^{12} + \left(\frac{1}{(A+B)^{3/2}}\right)\right)^{(1/12)} \tag{12-42}$$

where the parameters A and B are given by the following equations:

$$A = \left[2.457 \ln\left(\frac{1}{(7/N_{Re})^{0.9} + (0.27\epsilon/D)}\right)\right]^{16}$$

$$B = \left(\frac{37,530}{N_{Re}}\right)^{16} \tag{12-43}$$

Note that all of the parameters and constants in this equation are dimensionless so that as long as a consistent set of units is used for the various quantities used to calculate the Reynolds number, and ϵ/D, the same result for the friction factor is obtained.

Example 12-3. Ethanol at 20 °C is flowing through a 2 (in) Nominal Schedule 40 steel pipe with a mass flow rate of 10 lb_m/s and 100 lb_m/s in two different cases. If the pipe is horizontal and the length of the pipe is 20 ft as shown in Figure 12-5, calculate the difference in pressure between the entrance of the pipe and the exit of the pipe (the pressure drop) for the two different cases.

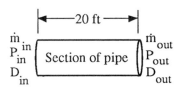

Figure 12-5: Flow in pipe

SOLUTION: Choose the system as the volume of the pipe between the entrance and the exit as shown in Figure 12-5a. A cartesian coordinate system is defined at the entrance of the pipe. Furthermore this system is operating at steady state.

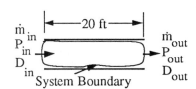

Figure 12-5a: System for flow in pipe

From Appendix C the following data about the liquid ethanol and the pipe are obtained:

$$\rho = 49.4 \; lb_m/ft^3 \qquad \mu = 8.1 \times 10^{-4} \; lb_m/ft \; s$$

$$D_{in} = 2.067 \; in \qquad A_{in} = 3.35 \; in^2$$

For this system, the conservation of mass says:

$$\dot{m}_{in} - \dot{m}_{out} = 0$$

For the constant density fluid and the constant cross sectional area, because the dimensions of the pipe do not change, the entering and leaving average linear velocities are the same:

$$\langle v \rangle_{in} = \langle v \rangle_{out}$$

Furthermore the velocity of the entering stream is calculated from the mass flow rate as:

$$\langle v \rangle_{in} = \left(\frac{\dot{m}}{A\rho} \right)_{in}$$

Substituting numerical values with the proper conversion factors for case (a) gives:

$$\langle v \rangle_{in} = \left(\frac{10}{(3.35)(49.2)} \right) \frac{lb_m\ ft^3}{s\ in^2\ lb_m} \left| \frac{12^2\ in^2}{1\ ft^2} \right. = 8.7\ ft\ /\ s$$

For case (b) the average velocity at the entrance is:

$$\langle v \rangle_{in} = 87\ ft\ /\ s$$

The equation for the accounting of mechanical energy for this system at steady state is given by:

$$0 = g(z_{in} - z_{out}) + \left(\frac{\langle v \rangle_{in}^2}{2\alpha_{KE}} - \frac{\langle v \rangle_{out}^2}{2\alpha_{KE}} \right) + \left(\frac{P_{in} - P_{out}}{\rho} \right) + \sum \widehat{W}_i - \sum \widehat{F}_i$$

For this system, there is no change in elevation. There is also no change in kinetic energy because the average velocity at the entrance is the same as the average velocity at the exit. Furthermore there is no work done on the system. However, there is irreversible conversion of mechanical energy to thermal energy. The resulting equation is:

$$0 = \left(\frac{P_{in} - P_{out}}{\rho} \right) - \sum \widehat{F}_i$$

This equation is valid for both cases, and the only unknown in the equation is the difference between the pressure at the entering stream and the pressure at the exit stream. The irreversible conversion of mechanical to thermal energy can be calculated. Rearranging yields:

$$(P_{in} - P_{out}) = \rho \sum \widehat{F}_i$$

Since the system is defined as the section of straight pipe, the only irreversible conversion of mechanical to thermal energy occurs due to the flow in the straight pipe.

$$\widehat{F}_i = \frac{4fl}{D} \left(\frac{\langle v \rangle^2}{2} \right)$$

From the fluid properties, velocity, and the pipe characteristics, we can calculate the Reynolds number:

$$N_{Re} = \left(\frac{D \langle v \rangle \rho}{\mu} \right)$$

$$= \left(\frac{(49.4)(2.067)(8.7)}{8.1 \times 10^{-4}} \right) \frac{lb_m\ in\ ft^2\ s}{ft^3\ s\ lb_m} \left| \frac{1\ ft}{12\ in} \right. = 91,000$$

Note that the Reynolds number is rounded to two significant figures because this is the accuracy of the viscosity. In a calculation which multiplies and divides numbers, the accuracy of the answer can be no greater than that of the least precise number. The Reynolds number is above 4000 so the flow is fully turbulent. Because the flow is fully turbulent, we need to know the relative roughness along with the Reynolds number to calculate the Fanning friction factor. From Table C-14 in Appendix C, the roughness of commercial steel is 0.0018 in and the relative roughness is:

$$\frac{\epsilon}{D} = \frac{0.0018}{2.067} = 0.001$$

Then, from Figure C-1 in Appendix C, the Fanning friction factor is:

$$\mathfrak{f} = 0.006$$

The Fanning friction factor is substituted into the equation for the friction losses to give:

$$\widehat{F}_i = \frac{4(0.006)(20)}{2.067}\left(\frac{8.7^2}{2}\right)\frac{\mathrm{ft}^3}{\mathrm{in\ s}^2}\left|\frac{12\ \mathrm{in}}{1\ \mathrm{ft}}\right.\left|\frac{1\ \mathrm{lb_f\ s}^2}{32.174\ \mathrm{lb_m\ ft}}\right. = 3.3\ \mathrm{lb_f\ ft\ /\ lb_m}$$

Now using this value to calculate the change in pressure gives:

$$(P_{\mathrm{in}} - P_{\mathrm{out}}) = \rho\sum\widehat{F}_i$$

$$= (49.4)(3.3)\frac{\mathrm{lb_m\ lb_f\ ft}}{\mathrm{ft}^3\ \mathrm{lb_m}}\left|\frac{1\ \mathrm{ft}^2}{12^2\ \mathrm{in}^2}\right.\left|\frac{1\ \mathrm{psi\ in}^2}{1\ \mathrm{lb_f}}\right. = 1.12\ \mathrm{psi}$$

The pressure leaving the system is 1.12 (psi) lower than the pressure at the entrance.

For case (b) where the mass flow rate is 100 $\mathrm{lb_m}$ / s the velocity was calculated from the conservation of mass as:

$$\langle v\rangle_{\mathrm{in}} = 87\ \mathrm{ft\ /\ s}$$

The same accounting equation for mechanical energy is true:

$$(P_{\mathrm{in}} - P_{\mathrm{out}}) = \rho\sum\widehat{F}_i$$

However, the friction losses will be different because of the different velocity. First we will calculate the Reynolds number for this flow:

$$N_{\mathrm{Re}} = \left(\frac{D\langle v\rangle\rho}{\mu}\right)$$

$$= \left(\frac{(49.4)(2.067)(87)}{8.1\times10^{-4}}\right)\frac{\mathrm{lb_m\ in\ ft}^2\ \mathrm{s}}{\mathrm{ft}^3\ \mathrm{s\ lb_m}}\left|\frac{1\ \mathrm{ft}}{12\ \mathrm{in}}\right. = 910,000$$

Here, the Reynolds number is above 4000 also and the flow is fully turbulent. With the calculated relative roughness of 0.001 and Figure C-1 in Appendix C, the Fanning friction factor is:

$$\mathfrak{f} = 0.005$$

Substituting the Fanning friction factor into the equation for the irreversible conversion of mechanical energy to thermal energy gives:

$$\widehat{F}_i = \frac{4(0.005)(20)}{2.067}\left(\frac{87^2}{2}\right)\frac{\mathrm{ft}^3}{\mathrm{in\ s}^2}\left|\frac{12\ \mathrm{in}}{1\ \mathrm{ft}}\right.\left|\frac{1\ \mathrm{lb_f\ s}^2}{32.174\ \mathrm{lb_m\ ft}}\right. = 273.1\ \mathrm{lb_f\ ft\ /\ lb_m}$$

Finally, the frictional losses are substituted in the equation to calculate the change in pressure as:

$$(P_{\mathrm{in}} - P_{\mathrm{out}}) = \rho\sum\widehat{F}_i$$

$$= (49.4)(273.1)\frac{\mathrm{lb_m\ lb_f\ ft}}{\mathrm{ft}^3\ \mathrm{lb_m}}\left|\frac{1\ \mathrm{ft}^2}{12^2\ \mathrm{in}^2}\right.\left|\frac{1\ \mathrm{psi\ in}^2}{1\ \mathrm{lb_f}}\right. = 93.7\ \mathrm{psi}$$

The pressure at the outlet of the system is 93.7 psi lower than the pressure at the entrance.

By increasing the mass flow rate by 10 times through the conduit we increased the irreversible conversion of mechanical energy to thermal energy by 30 times.

Losses in Pipe Fittings. The friction factors which are obtained from the Moody chart are for straight runs of pipe. Where there are bends in the pipe or fittings, additional friction losses need to be added (K_i). These are normally expressed as equivalent lengths of pipe, that is, a certain kind of fitting is equivalent to a certain number of feet of straight pipe under the same circumstances.

$$l_{eq} = \text{Equivalent length of straight pipe} \qquad (12-44)$$

Table C-13 in Appendix C contains these equivalences for a variety of the common fittings such as 45° scale and 90° scale elbows, tees, and various kinds of valves. The irreversible conversion of the mechanical energy to thermal energy per unit mass for flow through fittings is given by:

$$\text{Fittings:} \qquad \hat{F}_i = \frac{4 f l_{eq}}{D}\left(\frac{\langle v \rangle^2}{2}\right) \qquad (12-45)$$

Specialized fluid flow manuals, such as engineering handbooks and the Crane technical manual number 401 contain more detailed tables.

Enlargements and Contractions. Mechanical energy losses above and beyond those experienced in straight-run pipe also exist in enlargements and contractions (K_i). A sudden expansion or a sudden contraction induces a disturbance to the flow field which results in additional losses of mechanical energy. With these expansions and contractions, the usual method of calculating friction losses is to write

$$\hat{F}_i = K_i \left(\frac{\langle v \rangle^2}{2 \alpha_{KE}}\right) \qquad (12-46)$$

The velocity used in this equation is the larger of the two average velocities, that is, it is the upstream velocity for sudden expansion and the downstream velocity for a contraction. Furthermore, the larger the change in diameter between the two sections, the greater the losses. Also, for a given diameter ratio, flow rate expansions cause a greater loss of mechanical energy than do contractions. The constant K_i varies from zero to one. A sudden expansion and sudden contraction are shown in Figure 12-6.

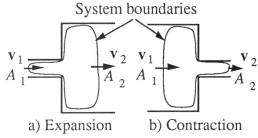

a) Expansion   b) Contraction

Figure 12-6: Sudden expansion or contraction

With the expansion, the irreversible conversion of mechanical energy to thermal energy per unit mass is only a function of the velocity of the fluid upstream of the expansion:

Expansions:
$$\widehat{F}_i = K_{\text{Exp}}\left(\frac{\langle v\rangle_1^2}{2\alpha_{\text{KE}}}\right)$$

(12 − 47)

Where the coefficient is given as a function of the cross sectional areas,

$$K_{\text{Exp}} = \left[1 - \left(\frac{A_1}{A_2}\right)\right]^2$$

(12 − 48)

(Note, however, that when a fluid flows from a pipe to the open atmosphere, no friction loss should be counted. Hence, if a pipe discharges to a tank above the liquid level in the tank, then there is no expansion friction loss.)

The irreversible conversion of mechanical energy to thermal energy per unit mass is expressed in term of the loss coefficient due to contractions as:

Contractions:
$$\widehat{F}_i = K_{\text{Con}}\left(\frac{\langle v\rangle_2^2}{2\alpha_{\text{KE}}}\right)$$

(12 − 49)

Where the coefficient is given as a function of the cross-sectional areas,

$$\text{for} \qquad \left(\frac{A_2}{A_1}\right) < 0.715, \qquad K_{\text{Con}} = 0.4\left(1.25 - \frac{A_2}{A_1}\right)$$

$$\text{for} \qquad \left(\frac{A_2}{A_1}\right) > 0.715, \qquad K_{\text{Con}} = 0.75\left(1 - \frac{A_2}{A_1}\right)$$

(12 − 50)

Types of Pipe Flow Problems. Now that we know how to calculate the irreversible conversions of mechanical energy to thermal energy let us again look at the extended Bernoulli equation,

$$0 = g(z_{\text{in}} - z_{\text{out}}) + \left(\frac{\langle v\rangle_{\text{in}}^2}{2\alpha_{\text{KE}}} - \frac{\langle v\rangle_{\text{out}}^2}{2\alpha_{\text{KE}}}\right) + \left(\frac{P_{\text{in}} - P_{\text{out}}}{\rho}\right) + \sum\widehat{W}_i - \sum\widehat{F}_i$$

(12 − 31)

This equation allows us to solve several different types of problems involving the system variables. The different types of problems are discussed below.

As discussed above, the most straight-forward problem is to be given the specific piping system, fluid design flow rate, pressure drop, and elevation change and then to find the pump requirements in order to meet these specifications.

Unknown $\qquad (\widehat{W})$

Known $\qquad (h_{\text{in}}, h_{\text{out}}, \rho, \mu, l, P_{\text{in}}, P_{\text{out}}, A_{\text{in}}, A_{\text{out}}, \dot{m})$

This can be done directly by calculating a Reynolds number from the given flow rate, fluid viscosity, and density which is then used to calculate the friction factor which is then used, together with the other information in Bernoulli's equation to calculate the pump requirements.

Another straight-forward problem is to calculate the maximum height or the change in elevation achievable for a given piping system, pressure drop, pump work, and mass flow rate along with the type of fluid and diameter of the pipe:

Unknown $(h_{in} - h_{out})$

Known $(\rho, \mu, l, P_{in}, P_{out}, \widehat{W}, A_{in}, A_{out}, \dot{m})$

Finally, if the only unknown is the pressure drop with all of the other parameters about the piping system known, this calculation is also straightforward:

Unknown $(P_{in} - P_{out})$

Known $(h_{in}, h_{out}, \rho, \mu, l, \widehat{W}, A_{in}, A_{out}, \dot{m})$

Other problems involving the same parameters may not be as straightforward, however. For example, one problem is that for a given pipeline system, fluid and changes in pressure, elevation, and kinetic energy to find the maximum achievable flow rate:

Unknown $(\dot{m})$

Known $(h_{in}, h_{out}, \rho, \mu, l, P_{in}, P_{out}, \widehat{W}, A_{in}, A_{out})$

With the flow rate unknown at the beginning, the changes in kinetic energy and the Reynolds number cannot be calculated. Consequently, the friction losses cannot be calculated directly. However, for an assumed flow rate, the problem becomes one of finding the pump work which can then be compared with the existing pump work, that is the problem can be treated iteratively as a trial and error problem. Choose a flow rate, calculate the pumping requirements as for the first kind of problem, then adjust the flow rate as necessary so that the calculated pumping requirement matches that which is available.

Another kind of problem is to have a flow rate specified, a pump work, fluid and pipe roughness specified and then find the pipe diameter or the cross sectional area.

Unknown (A)

Known $(h_{in}, h_{out}, \rho, \mu, l, P_{in}, P_{out}, \widehat{W}, \dot{m})$

Again as an iterative approach, one could select the diameter and use all of the other information to calculate the pump requirements. These pump requirements can then be compared to the available pump and to the extent the two differ, the pipe diameter can be adjusted.

Example 12-4. Liquid water at 100 °C is flowing through a horizontal straight pipe at a flow rate of 3 lb_m / s in a commercial steel schedule 40 1 in nominal diameter as shown in Figure 12-7. If the water is pumped through a 37.3 m section of pipe from tank 1 at atmospheric pressure to tank 2, also at atmospheric pressure, determine the rate of work (power) that needs to be *done on the fluid*, i.e., energy added to the fluid, to transfer the fluid from one tank to another. The liquid level is the same in each tank (i.e., the water surface in each tank is at the same elevation).

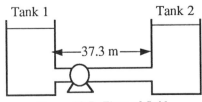

Figure 12-7: Flow of fluid
through pipe

SOLUTION: The system is defined as all of the water in the two tanks plus that in the pipe connecting them, as shown in Figure 12-7a. Hence, the inlet to the system is at the surface of the water in tank 1 and the outlet is at the surface of the water in tank 2. The system is operating at steady state. Hence, we must assume that water enters tank 1 at the same rate at which it leaves. Likewise, water is removed from tank 2 at the same rate at which it enters.

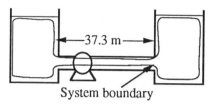

System boundary

Figure 12-7a: System for flow of fluid through pipe

The general accounting of mechanical energy for this system at steady state is:

$$0 = g(z_{in} - z_{out}) + \left(\frac{\langle v \rangle_{in}^2}{2\alpha_{KE}} - \frac{\langle v \rangle_{out}^2}{2\alpha_{KE}} \right) + \left(\frac{P_{in} - P_{out}}{\rho} \right) + \sum \widehat{W}_i - \sum \widehat{F}_i$$

From the figure shown, the entering and leaving streams are at the same elevation (the liquid levels in the two tanks), and the changes in kinetic energy are zero (the velocity of the water at the surface of the water in each tank is nearly zero because the cross-sectional area for flow (the area of each tank) is large ($v = \dot{m} / [\rho A]$) and hence their differences are zero (or very nearly so). Furthermore, the pressures at the entrance and exit of the system (the liquid surfaces in the two tanks) are both atmospheric. These observations reduce the accounting of mechanical energy to:

$$0 = \sum \widehat{W}_i - \sum \widehat{F}_i$$

or the total rate of work done on the system per unit mass is equal to the total irreversible conversions of mechanical energy to thermal energy. This equation can be rearranged to give:

$$\sum \widehat{W}_i = \sum \widehat{F}_i$$

For flow through the pipe the friction losses are calculated according to

$$\widehat{F}_{pipe} = \frac{4 f l}{D} \left(\frac{\langle v \rangle_{pipe}^2}{2} \right)$$

where $\langle v \rangle_{pipe}$ is the velocity of water through the pipe, averaged over the pipe cross-section. The length of the pipe is converted to feet

$$l = \frac{37.3 \text{ m} \mid 3.28 \text{ ft}}{\mid 1 \text{ m}} = 122.3 \text{ ft}$$

The physical properties of the water at the conditions stated are given in Appendix C:

$$\rho = 59.8 \text{ lb}_m / \text{ft}^3 \qquad \mu = 1.9 \times 10^{-4} \text{ lb}_m / \text{ft s}$$

Furthermore, from the pipe specifications, the inside diameters and areas are:

$$D = 1.049 \text{ in} \qquad A = 0.864 \text{ in}^2$$

Now the average velocity of water through the pipe can be calculated as:

$$\langle v \rangle_{pipe} = \left(\frac{\dot{m}}{A\rho} \right)_{in} = \left(\frac{3}{(0.864)(59.8)} \right) \frac{\text{lb}_m \text{ ft}^3 \mid 12^2 \text{ in}^2}{\text{s in}^2 \text{ lb}_m \mid 1 \text{ ft}^2} = 8.36 \text{ ft} / \text{s}$$

From the given physical property data, and the characteristics of the pipe and average velocity, the Reynolds number is:

$$N_{Re} = \left(\frac{D\langle v \rangle_{pipe} \rho}{\mu} \right) = \left(\frac{(59.8)(1.049)(8.36)}{1.9 \times 10^{-4}} \right) \frac{lb_m \; in \; ft^2 \; s}{ft^3 \; s \; lb_m} \left| \frac{1 \; ft}{12 \; in} \right. = 230,000$$

Because of the magnitude of the Reynolds number, the flow is assumed to be fully turbulent. Since the flow is fully turbulent, the Fanning friction factor is both a function of the Reynolds number and a function of the relative roughness. From Table C-14 in Appendix C, the relative roughness of the pipe is calculated:

$$\left(\frac{\epsilon}{D} \right) = \left(\frac{0.0018}{1.049} \right) = 0.002$$

For the calculated Reynolds number and relative roughness, the Fanning friction factor from Figure C-1 is:

$$\mathfrak{f} = 0.0059$$

Substituting the Fanning friction factor and the other known quanties into the equation to calculate the pipe frictional losses gives:

$$\widehat{F}_{pipe} = \frac{4(0.0059)(122.3)}{1.049} \left(\frac{8.36^2}{2} \right) \frac{ft^3}{in \; s^2} \left| \frac{12 \; in}{1 \; ft} \right| \frac{1 \; lb_f \; s^2}{32.174 \; lb_m \; ft} = 35.8 \; lb_f \; ft \, / \, lb_m$$

For the contraction from tank 1 to the pipe,

$$\widehat{F}_{Con} = K_{Con} \left(\frac{\langle v \rangle_2^2}{2 \alpha_{KE}} \right)$$

and because $A_2 \, / \, A_1 \approx 0$ (the area of tank 1 is much larger than that of the pipe), $K_{Con} = 1 \, / \, 2$. Therefore

$$\widehat{F}_{Con} = \frac{1}{2} \left(\frac{8.36^2}{2(1)} \right) \frac{ft^2}{s^2} \left| \frac{1 \; lb_f \; s}{32.174 \; lb_m \; ft} \right. = 0.54 \; lb_f \; ft \, / \, lb_m$$

Similarly, for the expansion from the pipe to tank 2

$$\widehat{F}_{Exp} = K_{Exp} \left(\frac{\langle v \rangle_1^2}{2 \alpha_{KE}} \right) = \frac{\langle v \rangle_1^2}{2 \alpha_{KE}}$$

because $A_1 \, / \, A_2 \approx 0$ and hence $K_{Exp} = 1$. Therefore,

$$\widehat{F}_{Exp} = \left(\frac{8.36^2}{2(1)} \right) \frac{ft^2}{s^2} \left| \frac{1 \; lb_f \; s^2}{32.174 \; lb_m \; ft} \right. = 1.09 \; lb_f \; ft \, / \, lb_m$$

Combining the pipe, contraction, and expansion losses gives the total work per mass required of the pump:

$$\widehat{W} = (35.8 + 0.5 + 1.1) = 37.4 \; lb_f \; ft \, / \, lb_m$$

To calculate the total rate of work done on the system, we multiply the mass flow rate times the work per unit mass:

$$\dot{W} = \dot{m}\widehat{W} = (3)(37.4) \frac{lb_f \; ft}{s} \left| \frac{1 \; hp \; s}{550 \; lb_f \; ft} \right. = 0.20 \; hp$$

A rate of work of $1 \, / \, 5$ horsepower needs to be provided by the pump to move the fluid from one tank to another.

Example 12-5. An elevated storage tank containing methanol at 0 °C is shown in Figure 12-8. If it is necessary to have a volumetric flow rate of 0.02 ft³ / s at point 2, determine the level of the liquid in the tank. The figure shows that the fluid travels through 15 ft of schedule 40, 1 in nominal commercial steel pipe and one standard 90° elbow followed by 65 ft of 3 / 4 in schedule 40 commercial steel pipe, with two standard 90 °elbows and 2 open globe valves.

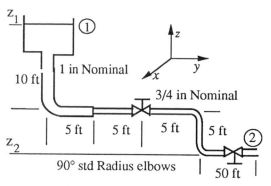

Figure 12-8: Piping network

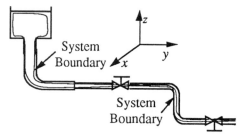

Figure 12-8a: System for
piping network

SOLUTION: For this problem, the system is defined as the volume from the level of the tank through the piping network to the exit of the pipe as shown in Figure 12-8a. A cartesian coordinate system is defined at the exit of the system with the z-direction aligned with the acceleration of gravity. The system is operating at steady state.

The conservation of mass for this system says:

$$\dot{m}_{\text{in}} - \dot{m}_{\text{out}} = 0$$

or for the constant density material:

$$\langle v \rangle_{\text{in}} = \left(\frac{A_{\text{out}}}{A_{\text{in}}} \right) \langle v \rangle_{\text{out}}$$

The accounting of mechanical energy for the steady state system is:

$$0 = g(z_{\text{in}} - z_{\text{out}}) + \left(\frac{\langle v \rangle_{\text{in}}^2}{2\alpha_{\text{KE}}} - \frac{\langle v \rangle_{\text{out}}^2}{2\alpha_{\text{KE}}} \right) + \left(\frac{P_{\text{in}} - P_{\text{out}}}{\rho} \right) + \sum \widehat{W}_i - \sum \widehat{F}_i$$

The pressure at the entrance and the exit are the same because both are atmospheric, and no work is being done on the system yielding:

$$0 = g(z_{\text{in}} - z_{\text{out}}) + \left(\frac{\langle v \rangle_{\text{in}}^2}{2\alpha_{\text{KE}}} - \frac{\langle v \rangle_{\text{out}}^2}{2\alpha_{\text{KE}}} \right) - \sum \widehat{F}_i$$

Because the system is defined to include the water in the tank, the entering average velocity is approximately zero and can be assumed to be negligible. Hence:

$$0 = g(z_{\text{in}} - z_{\text{out}}) + \left(0 - \frac{\langle v \rangle_{\text{out}}^2}{2\alpha_{\text{KE}}} \right) - \sum \widehat{F}_i$$

The only unknown in the equation is the difference between the height of the entering stream and that of the leaving stream so rearranging gives:

$$(z_{\text{in}} - z_{\text{out}}) = \left(\frac{1}{g} \right) \sum \widehat{F}_i + \frac{\langle v \rangle_{\text{out}}^2}{2g\,\alpha_{\text{KE}}}$$

The change in height is a function of the irreversible conversion of mechanical to thermal energy and the fluid velocity. The irreversible conversions are due to contractions, valves and fittings, and flow through straight pipe of different diameters.

The following physical properties for methanol at 0 °C are provided in Appendix C

$$\rho = 50.5 \text{ lb}_m / \text{ft}^3 \qquad \mu = 0.551 \times 10^{-3} \text{ lb}_m / \text{ft s}$$

Furthermore, data for the two pipe diameters are obtained from Table C-12 (the subscripts 3/4 and 1 indicate the pipe diameters):

$$D_{3/4} = 0.824 \text{ in} \qquad A_{3/4} = 0.534 \text{ in}^2$$
$$D_1 = 1.049 \text{ in} \qquad A_1 = 0.864 \text{ in}^2$$

To begin, the linear velocity in the 3 / 4 in and 2 in diameter pipes are calculated from the given volumetric flow rate:

$$\langle v \rangle_{3/4} = \left(\frac{\dot{V}}{A_{3/4}} \right) = \left(\frac{0.02}{0.534} \right) \frac{\text{ft}^3}{\text{in}^2} \bigg| \frac{12^2 \text{ in}^2}{1 \text{ ft}^2} = 5.39 \text{ ft} / \text{s}$$

$$\langle v \rangle_{1} = \left(\frac{\dot{V}}{A_{1}} \right) = \left(\frac{0.02}{0.864} \right) \frac{\text{ft}^3}{\text{in}^2} \bigg| \frac{12^2 \text{ in}^2}{1 \text{ ft}^2} = 3.33 \text{ ft} / \text{s}$$

Now let us calculate the Reynolds number for each of the flow conditions in the the different diameter pipes (check the units for complete cancellation):

$$N_{Re3/4} = \left(\frac{D_{3/4} \langle v \rangle_{3/4} \rho}{\mu} \right) = \left(\frac{(50.5)(0.824)(5.39)}{0.551 \times 10^{-3}} \right) \frac{\text{lb}_m \text{ in ft}^2 \text{ s}}{\text{ft}^3 \text{ s lb}_m} \bigg| \frac{1 \text{ ft}}{12 \text{ in}} = 33,900$$

$$N_{Re1} = \left(\frac{D_{1} \langle v \rangle_{1} \rho}{\mu} \right) = \left(\frac{(50.5)(1.049)(3.33)}{0.551 \times 10^{-3}} \right) \frac{\text{lb}_m \text{ in ft}^2 \text{ s}}{\text{ft}^3 \text{ s lb}_m} \bigg| \frac{1 \text{ ft}}{12 \text{ in}} = 26,680$$

The Reynolds numbers are greater than 4000 and the flow is turbulent in both pipes.

Now that all the preliminary calculations are done let us now begin to calculate all of the frictional losses of mechanical energy. First let us look at the straight run of pipe as shown in Figure 12-8b.

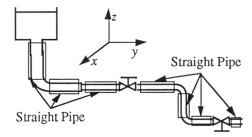

Figure 12-8b: Straight pipe

The frictional losses from the straight run of pipe are calculated as:

Straight Pipe:
$$\widehat{F}_{pipe} = \frac{4 f_1 l_1}{D_1} \left(\frac{\langle v \rangle_1^2}{2} \right) + \frac{4 f_{3/4} l_{3/4}}{D_{3/4}} \left(\frac{\langle v \rangle_{3/4}^2}{2} \right)$$

The total length of the 1 and 3 / 4 in pipes are 15 and 65 ft, respectively. Because the flow in each of the sections of pipe is turbulent, the Fanning friction factor in each of the sections is not only a function of the Reynolds number but also the relative roughness. From Table C-14 in Appendix C, the relative roughness for each of the pipes is:

$$\left(\frac{\epsilon}{D_1}\right) = \left(\frac{0.0018}{1.049}\right) = 0.00171$$

$$\left(\frac{\epsilon}{D_{3/4}}\right) = \left(\frac{0.0018}{0.824}\right) = 0.00218$$

For the Reynolds numbers and relative roughnesses for the two pipe diameters, the Fanning friction factors from Figure C-1 in Appendix C are found to be

$$f_1 = 0.0071 \qquad f_{3/4} = 0.0071$$

Substituting the calculated and known values into the equation to calculate the frictional losses gives:

$$\text{Straight Pipe: } \widehat{F}_{\text{pipe}} = \left[\frac{4(0.0071)(15)}{1.049}\left(\frac{3.33^2}{2}\right) + \frac{4(0.0071)(65)}{0.824}\left(\frac{5.39^2}{2}\right)\right] \frac{\text{ft}^3}{\text{in s}^2}\left|\frac{12 \text{ in}}{1 \text{ ft}}\right|\frac{1 \text{ lb}_f \text{ s}^2}{32.174 \text{ lb}_m \text{ ft}}$$

$$= 13.0 \text{ lb}_f \text{ ft} / \text{lb}_m$$

Now let us calculate the losses due to all of the fittings and valves as given in Figure 12-8c.

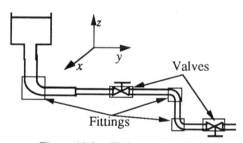

Figure 12-8c: Fittings and valves

$$\text{Fittings and Valves: } \widehat{F}_i = \sum \frac{4f_1 l_{\text{eq}}}{D_1}\left(\frac{\langle v\rangle_1^2}{2}\right) + \sum \frac{4f_{3/4} l_{\text{eq}}}{D_{3/4}}\left(\frac{\langle v\rangle_{3/4}^2}{2}\right)$$

For the 1 in diameter pipe there is only one fitting, a standard 90° elbow. From Appendix C, the dimensionless equivalent length is:

$$\left(\frac{l_{\text{eq}}}{D_1}\right) = 32$$

For the 3 / 4 in diameter pipe there are 2 standard 90° elbows and 2 open globe valves. The total dimensionless equivalent length is the sum of all the individual equivalent lengths or:

$$\left(\frac{l_{\text{eq}}}{D_{3/4}}\right) = 32 + 32 + 300 + 300 = 664$$

Substituting the calculated dimensionless equivalent lengths and the appropriate Fanning friction factor and linear average velocity is:

Fittings and Valves:
$$\widehat{F}_i = \left[4(0.0071)(32) \left(\frac{3.33^2}{2} \right) + 4(0.0071)(664) \left(\frac{5.39^2}{2} \right) \right] \frac{ft^2}{s^2} \left| \frac{1 \; lb_f \; s^2}{32.174 \; lb_m \; ft} \right.$$

$$= 8.7 \; lb_f \; ft \; / \; lb_m$$

Finally the last frictional losses due to expansions and contractions are calculated. This losses are shown in Figure 12-8d.

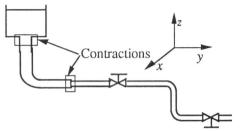

Figure 12-8d: Contractions

The following equation allows us to calculate the frictional losses due to contractions:

Contractions:
$$\widehat{F}_{Con} = K_{Con} \left(\frac{\langle v \rangle^2}{2 \alpha_{KE}} \right)$$

The α_{KE} is always equal to 1 in this problem. There are 2 contractions; one contraction is from the tank to the 1 in diameter pipe and the other is from the 1 in diameter pipe to the 3 / 4 in diameter pipe. For the first contraction, the ratio of the areas is approximately zero:

$$\left(\frac{A_2}{A_1} \right) = 0$$

so the loss coefficient for this contraction as given in Equation 12-50 is

$$K_{Con} = 0.5$$

For the second contraction, from 1 in diameter pipe to the 3 / 4 in diameter pipe the ratio of the areas is:

$$\left(\frac{A_2}{A_1} \right) = \left(\frac{0.534}{0.864} \right) = 0.618$$

Since this value is less than 0.715 as given in Equation 12-50, the loss coefficient is:

$$K_{Con} = 0.4 \left(1.25 - \frac{A_2}{A_1} \right) = 0.4(1.25 - 0.618) = 0.253$$

The total loss due to the contractions is calculated as:

$$\widehat{F}_{Con} = K_{Con} \left(\frac{\langle v \rangle_1^2}{2 \alpha_{KE}} \right) + K_{Con} \left(\frac{\langle v \rangle_{3/4}^2}{2 \alpha_{KE}} \right)$$

$$= \left[(0.5) \left(\frac{3.33^2}{2} \right) + (0.253) \left(\frac{5.39^2}{2} \right) \right] \frac{ft^2}{s^2} \left| \frac{1 \; lb_f \; s^2}{32.174 \; lb_m \; ft} \right.$$

$$= 0.20 \; lb_f \; ft \; / \; lb_m$$

Now, the total losses are given as the sum or:

$$\sum \widehat{F_i} = 12.0 + 8.7 + 0.2 = 20.9 \text{ lb}_f \text{ ft} / \text{lb}_m$$

The original equation to determine height of the liquid level in the tank is presented again:

$$(z_{in} - z_{out}) = \left(\frac{1}{g}\right) \sum \widehat{F_i} + \frac{\langle v \rangle^2_{out}}{2g\,\alpha_{KE}}$$

Substituting yields:

$$(z_{in} - z_{out}) = \frac{20.9 \text{ lb}_f \text{ ft}}{\text{lb}_m} \left|\frac{s^2}{32.174 \text{ ft}}\right|\frac{32.174 \text{ lb}_m \text{ ft}}{\text{lb}_f \text{ s}^2} + \left(\frac{1}{2(1)}\right) \frac{5.39^2 \text{ ft}^2}{s^2}\left|\frac{s^2}{32.174 \text{ ft}}\right.$$

$$= 21.4 \text{ ft}$$

The level of the methanol in the tank is 21.4 (ft) above the exit of the pipe, i.e. the methanol is 6.4 feet deep.

Piping Networks. In the above discussion we have considered only flow through a single piping system. This system might involve a variety of elements such as pumps, elbows, pipe bends, constrictions, elevation change, velocity changes, etc. but still there is only one flow path through the system.

More complex systems are possible and even common such as those involving multiple flow paths. There may be branch points where the flow can take one, two, or even more additional paths with each path having its own selection of pipe bends, valves, pumps, etc.

While the analysis of such systems becomes considerably more complicated in terms of variables and parameters, the fundamental principles upon which the analysis is based remain the same. Mass is conserved and the Bernoulli equation is applicable as an accounting of mechanical energy for flow through any parts of the process. Frequently for analysis, multiple systems must be considered simultaneously, that is, a number of systems must be defined and expressed using the available principles, then the systems' equations must be solved simultaneously to obtain the information desired. We will not discuss these kinds of more complex problems further in this text. Remember, however, that the Bernoulli equation, conservation of mass, and your own ingenuity in applying these equations will enable you to analyze the situations at hand.

Non-Newtonian Fluids. The previous discussion of flow through pipes has been restricted to incompressible Newtonian fluids, however, we must realize that not all fluids are Newtonian, that is, they do not all obey the Newtonian relationship for stress as a function of shear rate (velocity gradient). Approaches for non-Newtonian fluids are completely analogous to those for Newtonian fluids except that they have their own friction factor chart and Reynolds number. Some of these results for power law fluids and Bingham plastics have been presented by Darby (1988). We will not address these further here.

Forms of the Bernoulli Equation. As presented previously, all of the terms in the Bernoulli equation carry the units of energy (per time or per mass). This is termed the *energy form of the Bernoulli equation*:

$$0 = g(z_{in} - z_{out}) + \left(\frac{\langle v \rangle^2_{in}}{2\alpha_{KE}} - \frac{\langle v \rangle^2_{out}}{2\alpha_{KE}}\right) + \left(\frac{P_{in} - P_{out}}{\rho}\right) + \sum \widehat{W_i} - \sum \widehat{F_i} \qquad (12-31)$$

Other forms are possible. If the energy form is multiplied by the density of the fluid, then the following equation is obtained:

$$0 = \rho g(z_{in} - z_{out}) + \rho\left(\frac{\langle v \rangle_{in}^2}{2\alpha_{KE}} - \frac{\langle v \rangle_{out}^2}{2\alpha_{KE}}\right) + (P_{in} - P_{out}) + \rho\sum\widehat{W}_i - \rho\sum\widehat{F}_i \qquad (12-51)$$

From the pressure term it is evident that all of these terms carry units of pressure and this is referred to as the *pressure form of the Bernoulli equation*.

Likewise, the energy form could be divided by the gravitational force per unit mass g. If this is done, then from the potential energy term it is obvious that all of these terms carry the dimensions of length which is referred to in engineering jargon, as head. Consequently, this is referred to as the *head form of the Bernoulli equation*:

$$0 = (z_{in} - z_{out}) + \left(\frac{\langle v \rangle_{in}^2}{2g\alpha_{KE}} - \frac{\langle v \rangle_{out}^2}{2g\alpha_{KE}}\right) + \left(\frac{P_{in} - P_{out}}{g\rho}\right) + \left(\frac{1}{g}\right)\sum\widehat{W}_i - \left(\frac{1}{g}\right)\sum\widehat{F}_i \qquad (12-52)$$

From a physical principles viewpoint, the Bernoulli equation is an energy equation. Therefore we should view the terms as energy or energy conversion. The pressure form and the head form are simply different ways of describing these energy terms which arise from multiplying or dividing the original equation by a constant; pressure and head are not properties we can account for, mechanical energy is. You should be aware of these different forms, however, because they are frequently used and it is not uncommon, for example, to talk about a pressure head or a velocity head or a pump head or friction head. You should understand that when this is done, the terms refer to certain terms in the Bernoulli equation. The adjective such as pressure, velocity, pump or friction refers to the type of energy and the head part of the term refers to the fact that the equation has been converted to the dimensions of length.

Flow Through Non-circular Conduits. In the previous discussion, the pipes that have been considered were of a circular cross-section. Other cross-sections can be described using these friction factor charts provided that instead of using the pipe diameter, a hydraulic mean diameter is used. This hydraulic diameter is defined by the equation:

$$D_{Hyd} = 4R_{Hyd}$$
$$= 4\left(\frac{\text{cross-sectional area of flow stream}}{\text{wetted perimeter}}\right) \qquad (12-53)$$

and then the frictional loss for flow in a non-circular conduit is calculated in an analogous way as for the circular pipe according for the equation

$$\text{Non-Circular:} \qquad \widehat{F}_i = \frac{4fl}{D_{Hyd}}\left(\frac{\langle v \rangle^2}{2}\right) \qquad (12-54)$$

where the standard friction factor chart is used to calculate the friction factor, with the exception that the hydraulic diameter is used to calculate the Reynolds number and the relative roughness rather than the pipe diameter. The Reynolds number becomes

$$\text{Non-Circular:} \qquad N_{Re} = \frac{D_{Hyd}\langle v \rangle \rho}{\mu} \qquad (12-55)$$

Note that according to this definition of hydraulic diameter, for a circular cross-section the diameter equals the hydraulic diameter, as we would hope:

$$\text{Circular:} \qquad D_{\text{Hyd}} = 4\left(\frac{(\pi D^2 / 4)}{\pi D}\right)$$
$$= D \qquad\qquad (12-56)$$

This hydraulic diameter calculation is applicable only for turbulent flow. Calculations for the laminar flow are also available.

As an example of the calculation of a hydraulic diameter, consider the annulus area between an outer square cross-section and an inner circular cross-section as shown in Figure 12-9.

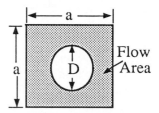

Figure 12-9: Flow through an annulus

According to the definition of hydraulic diameter, we have:

$$D_{\text{Hyd}} = 4\left(\frac{a^2 - (\pi D^2 / 4)}{\pi D + 4a}\right)$$
$$= \left(\frac{4a^2 - \pi D^2}{4a + \pi D}\right)$$

Different Definitions of Friction Factor. In different disciplines of engineering, it has been common to use different definitions of the friction factor which have correspondingly different friction factor charts. Perhaps the most common definitions of friction factor are the Fanning friction factor, given above (used by mechanical and chemical engineers) and a friction factor which is four times as great (used commonly by civil engineers). As defined above, the Fanning friction factor is used to calculate the frictional losses due to laminar flow according to

$$\text{Laminar:} \qquad f = \left(\frac{16}{N_{\text{Re}}}\right) \qquad\qquad (12-57)$$

The civil engineering friction factor or Darcy friction factor defines a friction factor which is four times that of the above equation. In that case, the friction loss (F) would be calculated as described previously but without the constant four. The differences are of no consequence whatsoever, so long as one uses a consistent definition throughout the calculations in both the equations and figures. Thus, the civil engineering friction factor f_{Darcy} in the laminar region would be

$$\text{Laminar} \qquad f_{\text{Darcy}} = 4f = \left(\frac{64}{N_{\text{Re}}}\right) \qquad\qquad (12-58)$$

Likewise, the Churchill equation gives the Fanning friction factor as defined in this chapter.

Measurement of Flow Through Pipes. The Bernoulli equation can be used to advantage for the design of inline flow measurement devices in pipes. One common technique is to use a narrowing of the pipe either by means of a tapered pipe section or by means of an inline pipe orifice. The fact that the flow area is narrower requires that the velocity increase, and simultaneously, that the pressure decrease. This change in pressure as the cross-sectional area is changed by a known amount gives a measurement of the flow velocity. The same result can be achieved with simple inline orifice plate which obviously would be a much cheaper design but not without introducing considerably more friction loss across the measurement device.

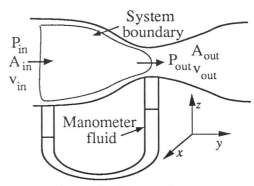

For example, consider the venturi meter in some more detail. As shown in Figure 12-10, the pipeflow cross-sectional area changes from A_{in} to A_{out} while, for an incompressible fluid, the volumetric flow rate past each area remains the same.

Figure 12-10: Venturi meter

According to the Bernoulli equation, the pressure and velocity are related as

$$\left(\frac{\langle v \rangle_{in}^2}{2} - \frac{\langle v \rangle_{out}^2}{2} \right) + \left(\frac{P_{in} - P_{out}}{\rho} \right) = 0 \qquad (12-59)$$

In this equation we have assumed that friction losses are negligible and the flow is fully turbulent. Solving the Bernoulli equation for the downstream velocity which is leaving the system by taking into account the conservation of mass (which gives the ratio between upstream and downstream velocities in terms of cross-sectional areas),

$$\langle v \rangle_{in} = \left(\frac{A_{out}}{A_{in}} \right) \langle v \rangle_{out}$$

we obtain

$$\langle v \rangle_{out} = \sqrt{\frac{2(P_{in} - P_{out})}{\rho \left(1 - (A_{out} / A_{in})^2 \right)}} \qquad (12-60)$$

Now in reality, there is some frictional loss through a meter of this sort and, traditionally, this is taken into account by introducing a correction factor, referred to as the coefficient of discharge. Accordingly, this last equation is written:

$$\langle v \rangle_{throat} = K_V \sqrt{\frac{2(P_{in} - P_{out})}{\rho \left(1 - (A_{out} / A_{in})^2 \right)}} \qquad (12-61)$$

where the discharge coefficient for the venturi meter is between zero and one:

$$0 < K_V \leq 1$$

The coefficient of discharge is determined by calibrating the meter and is included as documentation when the meter is purchased. For such a meter then, a measurement of the pressure decrease, along with the density of the fluid and the known cross-sectional areas, is sufficient to allow calculation of the velocity. In short, measuring a pressure drop allows calculation of the fluid flow rate. Note that if the meter is not horizontal, then the $(P_{in} - P_{out})$ term must include the gravity head, $\rho g(h_{in} - h_{out})$. See example 12-7.

The orifice flow meter works on exactly the same principle except that a flat plate with a center hole is used in place of the converging section of pipe. An orifice meter is shown in Figure 12-11. This makes the cost of the meter considerably less, although it does introduce more disturbance to the fluid flow.

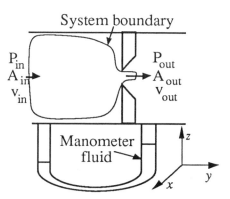

Figure 12-11: Orifice meter

Exactly the same equation is obtained for the orifice meter as for the venturi meter except that a different value for the coefficient of discharge is required.

$$\langle v \rangle_{throat} = K_O \sqrt{\frac{2(P_{in} - P_{out})}{\rho\left(1 - (A_{out} / A_{in})^2\right)}} \qquad (12-62)$$

where the discharge coefficient of the orifice meter also is between zero and one:

$$0 < K_O \leq 1$$

The orifice, although costing much less than the venturi, introduces more turbulence into the fluid flow creating a larger loss of mechanical energy:

$$K_O < K_V \qquad (12-63)$$

Again, a calibration is required to determine K_0.

Pitot Tube.   Another device for measuring fluid flow velocities is the pitot tube. This is a tube which is inserted into the flow stream and aligned pointing upstream and colinear with the direction of flow as shown in Figure 12-12.

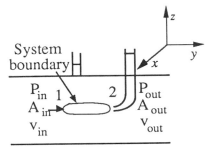

Figure 12-12: Pitot tube meter

As the fluid impacts the open end of the tube, it becomes stagnant and, in the context of the Bernoulli equation, the kinetic energy is converted to pressure energy. A reading of the pressure then is used to obtain a direct indication of the upstream velocity. Because the velocity at the opening of the pitot tube is zero (point 2), the Bernoulli equation written between an upstream velocity (point 1) and the entrance to the tube is:

$$\left(\frac{\langle v \rangle_{in}^2}{2} - 0 \right) + \left(\frac{P_{in} - P_{out}}{\rho} \right) = 0$$

Rearranging and solving for the entering velocity gives:

$$\langle v \rangle_{in} = \sqrt{\frac{2(P_{out} - P_{in})}{\rho}} \qquad\qquad (12 - 64)$$

In this result we have ignored the friction losses, which is normally a valid assumption for a pitot tube. Consequently, we can calculate the flow velocity in terms of the difference between the pressure in the opening of the tube and the upstream pressure. Again, a measurement of pressure is used to measure flow velocity.

Example 12-6.   Water flows through a venturi meter with diameters of 8 cm and 4 cm at a volumetric flow rate of 0.02 m³ / s. The manometer pressure gauge across the meter is deflected by 1 m as shown in Figure 12-13. The density of the liquid in the gauge is 13.0 g / cm³. If the pressure gauge is aligned with the gravity field, and the density of liquid water is constant during the process, calculate the coefficient of discharge of the meter K_V

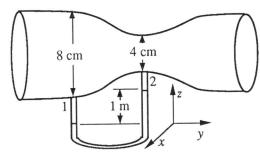

Figure 12-13: Venturi meter
(horizontal)

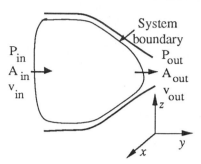

Figure 12-13a: System for
horizontal venturi meter

SOLUTION: The system is defined as the fluid in the venturi meter between points 1 and 2. Furthermore, the system is operating at steady state. The system definition is shown in Figure 12-13a.

The coefficient of discharge was obtained from the steady state accounting of mechanical energy by equation (12-61). This equation has been rearranged in terms of *outlet* velocities yielding

$$\langle v \rangle_{\text{out}} = K_V \sqrt{\frac{2(P_{\text{in}} - P_{\text{out}})}{\rho\left(1 - (A_{\text{out}} / A_{\text{in}})^2\right)}}$$

Since the diameters of the venturi meters have been given and the cross-sectional areas are circular, the equation becomes

$$\langle v \rangle_{\text{out}} = K_V \sqrt{\frac{2(P_{\text{in}} - P_{\text{out}})}{\rho\left(1 - (D_{\text{out}} / D_{\text{in}})^4\right)}}$$

Because the fluid is incompressible, the volumetric flow rate through the system is constant and the average velocity leaving the system can be expressed as the volumetric flow rate divided by the cross-sectional area.

$$\langle v \rangle_{\text{out}} = \left(\frac{\dot{V}_{\text{out}}}{A_{\text{out}}}\right) = \left(\frac{4\dot{V}_{\text{out}}}{\pi D_{\text{out}}^2}\right)$$

Substituting this expression into the previous equation yields:

$$\left(\frac{4\dot{V}_{\text{out}}}{\pi D_{\text{out}}^2}\right) = K_V \sqrt{\frac{2(P_{\text{in}} - P_{\text{out}})}{\rho\left(1 - (D_{\text{out}} / D_{\text{in}})^4\right)}}$$

The only unknown in this equation is the coefficient of discharge K_V. Solving for K_V yields:

$$K_V = \left(\frac{4\dot{V}_{\text{out}}}{\pi D_{\text{out}}^2}\right)\left(\sqrt{\frac{2(P_{\text{in}} - P_{\text{out}})}{\rho\left(1 - (D_{\text{out}} / D_{\text{in}})^4\right)}}\right)^{-1}$$

The pressure at the entrance and outlet points is expressed in terms of the manometer fluid density, gravitational force per unit mass and the change in elevation:

$$K_V = \left(\frac{4\dot{V}_{\text{out}}}{\pi D_{\text{out}}^2}\right)\left(\sqrt{\frac{-2\rho_{\text{man}}g(z_{\text{in}} - z_{\text{out}})}{\rho\left(1 - (D_{\text{out}}/D_{\text{in}})^4\right)}}\right)^{-1}$$

The numerical values are substituted into the equation and evaluated with the proper conversion factors to give:

$$K_V = \left(\frac{4(0.02)}{\pi(4)^2}\right)\left(\frac{m^3}{s\,cm^2}\right)\left(\frac{100\,cm}{1\,m}\right)^2\left(\sqrt{\frac{-2(13.0)(9.8)(-1)}{(1.0)\left(1 - (4/8)^4\right)}}\right)^{-1}\left(\frac{s}{m}\right) = 0.965$$

The coefficient of discharge for the venturi meter is 0.965.

Example 12-7. For the venturi meter shown in Figure 12-14 the deflection of mercury manometer is 14.3 in. If the manometer is aligned with gravity, determine the volumetric flow rate of water through the meter. The density of water is assumed to be constant throughout the process. Furthermore, assume that there are no friction losses.

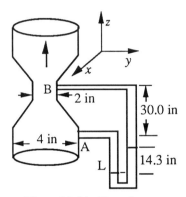

Figure 12-14: Venturi meter
(vertical)

SOLUTION: Choose the sytem as the fluid in the meter between points A and B with a cartesian coordinate system defined at the lower level of the manometer fluid as shown in Figure 12-14a. The system is operating at steady state.

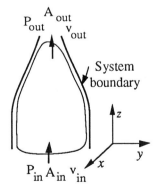

Figure 12-14a: System for
vertical venturi meter

For the defined system, the steady state conservation of mass equation is:

$$0 = \dot{m}_{\text{in}} - \dot{m}_{\text{out}}$$

For the constant density fluid, water, we can relate the average velocity of the entering stream to the average velocity of the exit streams through the entering and leaving cross sectional areas

$$\langle v \rangle_{\text{out}} = \left(\frac{A_{\text{in}}}{A_{\text{out}}} \right) \langle v \rangle_{\text{in}}$$

or for the circular cross sectional area:

$$\langle v \rangle_{\text{out}} = \left(\frac{D_{\text{in}}}{D_{\text{out}}} \right)^2 \langle v \rangle_{\text{in}}$$

Substitution yields:

$$\langle v \rangle_{\text{out}} = \left(\frac{4}{2} \right)^2 \langle v \rangle_{\text{in}} = 4 \langle v \rangle_{\text{in}}$$

The average velocity of the exit stream is four times the average velocity of the entering stream.

For the defined system the steady-state accounting of mechanical energy is:

$$0 = g(z_{\text{in}} - z_{\text{out}}) + \left(\frac{\langle v \rangle_{\text{in}}^2}{2\alpha_{\text{KE}}} - \frac{\langle v \rangle_{\text{out}}^2}{2\alpha_{\text{KE}}} \right) + \left(\frac{P_{\text{in}} - P_{\text{out}}}{\rho} \right) + \sum \widehat{W}_i - \sum \widehat{F}_i$$

This equation can be simplified by recognizing that work is zero, there are no mechanical energy losses, and the flow is assumed to be fully turbulent.

$$0 = g(z_{\text{in}} - z_{\text{out}}) + \left(\frac{\langle v \rangle_{\text{in}}^2}{2} - \frac{\langle v \rangle_{\text{out}}^2}{2} \right) + \left(\frac{P_{\text{in}} - P_{\text{out}}}{\rho} \right)$$

Substituting the relationship for the average velocity leaving the system yields:

$$0 = g(z_{\text{in}} - z_{\text{out}}) + \left(\frac{\langle v \rangle_{\text{in}}^2}{2} - \frac{4^2 \langle v \rangle_{\text{in}}^2}{2} \right) + \left(\frac{P_{\text{in}} - P_{\text{out}}}{\rho} \right)$$

or:

$$0 = g(z_{\text{in}} - z_{\text{out}}) - 7.5 \langle v \rangle_{\text{in}}^2 + \left(\frac{P_{\text{in}} - P_{\text{out}}}{\rho} \right)$$

The only unknown in this equation is the average linear velocity of the entering stream. Rearranging yields:

$$7.5 \langle v \rangle_{\text{in}}^2 = g(z_{\text{in}} - z_{\text{out}}) + \left(\frac{P_{\text{in}} - P_{\text{out}}}{\rho} \right) = \frac{(P_{\text{in}} + \rho g z_{\text{in}}) - (P_{\text{out}} + \rho g z_{\text{out}})}{\rho}$$

The pressure at the entrance and the pressure at the exit are related to g, the density of the manometer fluid, and the height of the mercury column:

$$7.5 \langle v \rangle_{\text{in}}^2 = g(h_{\text{in}} - h_{\text{out}}) + \left(\frac{-\rho_{\text{Hg}} g(z_{\text{in}} - z_{\text{out}})}{\rho_{\text{H}_2\text{O}}} \right)$$

The density of mercury and the density of water are found in Appendix C

$$\rho_{H_2O} = 62.4 \ (\text{lb}_{\text{m}} / \text{ft}^3) \qquad \rho_{Hg} = 848 \ (\text{lb}_{\text{m}} / \text{ft}^3)$$

Substituting values into the equation gives:

$$7.5\langle v\rangle_{in}^2 = \left[(32.174)(0-30) + \left(\frac{-(848)(32.174)(0-14.3)}{62.4}\right)\right]\left(\frac{1}{12}\right) = 440.6 \text{ ft}^2/\text{s}^2$$

Solving for the average velocity entering the system gives:

$$\langle v\rangle_{in} = \sqrt{\frac{440.6}{7.5}} = 7.7 \text{ ft}/\text{s}$$

The average velocity entering the meter is 7.7 ft/s.

Now, the volumetric flow rate through the meter can be calculated as the product of the average velocity at the entrance times the cross sectional area:

$$\dot{V}_{in} = \langle v\rangle_{in} A_{in} = \left(\frac{\langle v\rangle_{in}\pi D_{in}^2}{4}\right)$$

Substituting the known values gives:

$$\dot{V}_{in} = \left(\frac{(7.7)\pi(4)^2}{4}\right)\left(\frac{1}{144}\right) = 0.67 \text{ ft}^3/\text{s}$$

The volumetric flow rate through the meter is $0.67 \text{ ft}^3/\text{s}$.

Example 12-8. A pitot tube is used to measure the linear velocity of a stream of water at the center of the pipe as shown in Figure 12-15. The stagnation pressure head is 15 ft and the static pressure head in the pipe is 10 ft. Determine the velocity at the center of the pipe.

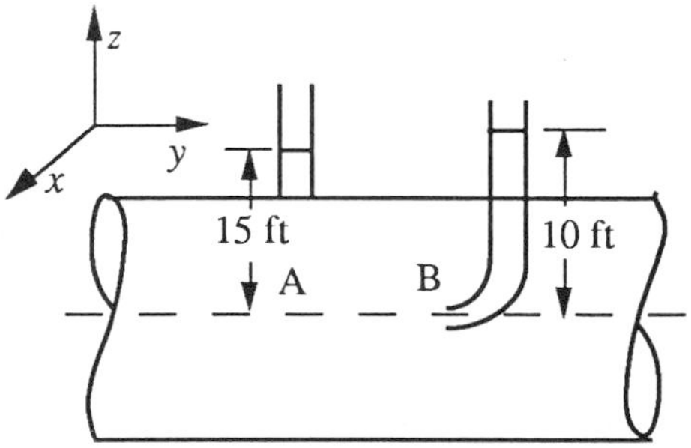

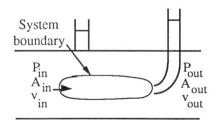

Figure 12-15: Pitot tube

Figure 12-15a: System for pitot tube

SOLUTION: The system is defined as the volume between the points A and B as shown in Figure 12-15a. A cartesian coordinate system with the axis aligned with gravity is also shown in the figure. The system is operating at steady state.

From the above, a steady state mechanical energy accounting equation gave

$$\langle v\rangle_{in} = \sqrt{\frac{2(P_{out} - P_{in})}{\rho}}$$

The only unknown in this equation is the velocity of the entering stream. The pressure is expressed in terms of the density, gravitational force per unit mass, and the difference between static and stagnation liquid heads to give:

$$\langle v \rangle_{\text{in}} = \sqrt{\frac{2g\rho(h_{\text{out}} - h_{\text{in}})}{\rho}}$$

The density of water cancels out and the final equation is:

$$\langle v \rangle_{\text{in}} = \sqrt{2g(h_{\text{out}} - h_{\text{in}})}$$

Substituting numerical values into the equation gives:

$$\langle v \rangle_{\text{in}} = \sqrt{2(32.174)(15 - 10)} = 17.94 \text{ ft} / \text{s}$$

12.4 FORCES ASSOCIATED WITH FLUID SYSTEMS

In the previous section we were concerned with the flow of fluids and the design of piping systems and networks. This has required using the conservation of mass and accounting of mechanical energy in the form of the extended Bernoulli equation to calculate information such as the pumping requirements for a given fluid system. This has also involved calculating the mechanical energy losses due to frictional effects and we have also discussed the use of the Bernoulli equation and the interconversion between pressure and kinetic energy as a means of calculating fluid flowrates. All of these topics have addressed the question of fluid flow inside pipes.

In this section we address the forces on piping systems and the forces which fluids exert on their surroundings as a result of momentum considerations. Accordingly, here we are concerned with the conservation of linear and angular momentum. As developed in Part II the equation for the conservation of linear momentum in terms of the average fluid velocities at the various cross sections where fluid enters and leaves the system is:

$$\left(\frac{d\mathbf{p}_{\text{sys}}}{dt} \right) = \sum \left(\frac{\langle v \rangle}{\alpha_{\text{Mom}}} \rho \langle v \rangle \mathbf{u}_N A \right)_{\text{in/out}} + \sum \mathbf{f}_{\text{ext}} \qquad (12-23)$$

This is an unsteady state result which can be simplified for steady state for which the change in the system momentum with time is zero. Accordingly, at steady state:

$$0 = \sum \left(\frac{\langle v \rangle}{\alpha_{\text{Mom}}} \rho \langle v \rangle \mathbf{u}_N A \right)_{\text{in/out}} + \sum \mathbf{f}_{\text{ext}}$$

In the context of a static piping system, this equation says that the external forces acting on the system must support both the weight of the piping system (including fluid) and changes in the momentum of the fluid between inlet and outlet areas. For example, if a fluid enters a pipe flowing in one direction and makes a 90 degree turn in going through the pipe to leave in another direction, then the momentum of the fluid has changed (due to the change in direction of flow but not due to a change in the magnitude of the flow velocity). This change in momentum requires that a force act on the fluid. This is shown in

Figure 12-16. Forces that must be counted in the momentum anaylsis are the pressure-area force at the entrance and exit of the defined system, and any drag or frictional forces that the wall exerts on the flowing fluid. The drag stresses at the wall integrated over the wall area act opposite to the direction of flow. Of course, in the context of a non-steady state situation, fluid leaving a system with a given velocity brings about a change in the momentum of the system thereby causing an acceleration; this is a rocket-type problem.

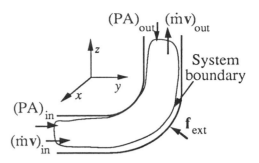

Figure 12-16: Forces on a pipe due to the change in momentum

Consider the macroscopic statement of linear momentum:

$$\left(\frac{d\mathbf{p}_{sys}}{dt}\right) = \left(\frac{d(m\mathbf{v})_{sys}}{dt}\right) = \sum(\dot{m}\mathbf{v})_{in/out} + \sum \mathbf{f}_{ext} \qquad (5-1)$$

Here, each $\mathbf{v}$ is expressed with respect to the same inertial reference frame while the $\dot{m}_{in}$ and $\dot{m}_{out}$ terms are with respect to a defined system boundary. The frame of reference must be constant velocity but the system boundary may accelerate. To illustrate the previous points, consider (1) the above statement for a second reference frame, moving at constant velocity $\mathbf{v}_{Ref}$ with respect to the first and (2) a system boundary fixed in space versus one moving at velocity $\mathbf{v}_{sys}$

Reference Frame Velocity. For a constant reference frame velocity of $\mathbf{v}_{Ref}$, each velocity in and out ($\mathbf{v}_{in}$ and $\mathbf{v}_{out}$) is expressed with respect to this moving frame. The resulting linear momentum equation becomes:

$$\frac{d}{dt}\left(m_{sys}(\mathbf{v}_{sys} - \mathbf{v}_{Ref})\right) = \sum\left(\dot{m}(\mathbf{v} - \mathbf{v}_{Ref})\right)_{in/out} + \sum \mathbf{f}_{ext} \qquad (12-65)$$

Rearranging the terms:

$$\left(\frac{d(m\mathbf{v})_{sys}}{dt}\right) - \mathbf{v}_{Ref}\left(\frac{dm_{sys}}{dt}\right) = \sum(\dot{m}\mathbf{v})_{in/out} - \mathbf{v}_{Ref}\left(\sum \dot{m}_{in/out}\right) + \sum \mathbf{f}_{ext} \qquad (12-66)$$

Note that the forces must be frame indifferent; they are the same for different inertial frames. Now, by conservation of total mass,

$$\left(\frac{dm_{sys}}{dt}\right) = \sum \dot{m}_{in/out} \qquad (3-19)$$

we see that equation (12-66) is in fact identical to (12-23). The inertial frame that is choosen does not alter the physical situation. Note also that for a constant-mass system

$$\left(\frac{dm_{sys}}{dt}\right) = 0$$

one inertial reference frame may be used for the system and a different frame (which need not even be inertial) may be used for the $\mathbf{v}_{in}$ and $\mathbf{v}_{out}$ velocities.

System Boundary Fixed versus System Boundary Moving. A distinction must now be made between the motion of the *system boundary* and the the motion of the *contents* of the system. The boundary velocity, relative to the velocity of the mass crossing the boundary, establishes the mass flowrate crossing the boundary:

$$\dot{m}_{in} = \rho(\mathbf{v}_{in} - \mathbf{v}_{bound,in}) \cdot \mathbf{n}$$

where $\mathbf{n}$ is a unit normal to A_{in}. But the boundary has no momentum because it is massless and therefore does not contribute to the system momentum. The motion of the contents establishes the system momentum; what is inside the system boundary has momentum and this momentum must be expressed in terms of an inertial frame for Newton's laws of motion to hold. However, the boundary of the system can be defined to accelerate without violating the conservation of momentum.

So, choosing different system boundary velocities will result in different mass flow rates $\dot{m}_{in}$ and $\dot{m}_{out}$, which in turn affects the mass and momentum accumulation in the system, and consequently (through equation (12-65)) affects calculating the external forces acting on the system.

As always with momentum equations it should be remembered that these are vector equations and therefore consist in general of three (directional) scalar equations, each of which is an independent equation providing different information. Examples illustrating the application of the conservation of linear momentum to fluid dynamics problems follow.

Example 12-9. A stream of water with constant density strikes a cart traveling at constant velocity $\mathbf{u}$ in the horizontal (y) direction as shown in Figure 12-17. The cross-sectional area of the entering and leaving streams are equal. The velocity of the entering stream with respect to an inertial reference frame is

$$\mathbf{v}_{in} = 0\mathbf{i} + v_1\mathbf{j} + 0\mathbf{k} \ (m \, / \, s)$$

and the angle of turning measured with respect to the cart is θ. Obtain equations for f_y and f_z, the horizontal and vertical forces required to keep the cart from accelerating due to the change in the water's direction (i.e., excluding the weight of the cart and water).

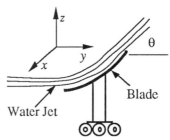

Figure 12-17: Moving cart problem

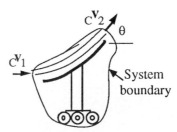

Figure 12-17a: Moving cart problem with moving system

SOLUTION: We choose the reference frame to be moving with the cart (because u is constant this is still an inertial frame), and the system boundary also is moving with the cart, as shown in Figure 12-17a. There is no accumulation of mass in this system and the conservation of mass for this system is given in Table 12-2. Make note of the notation. The pre-subscript for the velocity indicates that the moving cart is the reference frame.

Table 12-2: Data for Example 12-9 (Mass, System = Moving cart)

$\left(\dfrac{dm_{sys}}{dt}\right)$		$\sum \dot{m}_{in/out}$		
		Stream 1		Stream 2
[mass / time]		[mass / time]		[mass / time]
Total	0	$=$	$(_cv\rho A)_1$	$-$	$(_cv\rho A)_2$

Table 12-3: Data for Example 12-9 (Momentum, System = Moving cart)

	$\left(\dfrac{d\mathbf{p}_{sys}}{dt}\right)$		$\sum (\dot{m}\mathbf{v})_{in/out}$				$+$	$\sum \mathbf{f}_{ext}$
			Stream 1		Stream 2			
	[LM/time]		[LM/time]		[LM/time]			[LM/time]
x-direction	0	$=$	0	$-$	0		$+$	f_x
y-direction	$\left(\dfrac{d(mv)_{sys}}{dt}\right)$	$=$	$\dot{m}_1(_cv_1)$	$-$	$\dot{m}_2(_cv_2)\cos\theta$		$+$	f_y
z-direction	0	$=$	0	$-$	$\dot{m}_2(_cv_2)\sin\theta$		$+$	f_z

The two magnitudes of the velocities are related through mass conservation:

$$_cv_1 = {}_cv_2$$

There is no accumulation of linear momentum for this system and the conservation statements are given in Table 12-3. Recognizing that $\dot{m}_1 = \dot{m}_2 = \rho(_cv_1)A$ and that $_cv_1 = {}_cv_2$, the y-direction result is

$$f_y = -\dot{m}(_cv_1)(1 - \cos\theta) = -\rho A(_cv_1)^2(1 - \cos\theta)$$

and for the z direction

$$f_z = \dot{m}(_cv_1)\sin\theta = \rho A(_cv_1)^2 \sin\theta$$

Alternatively, in terms of the fixed frame:

$$_cv_1 = v_1 - u$$

and so

$$f_y = -\rho A(v_1 - u)^2(1 - \cos\theta)$$
$$f_z = \rho A(v_1 - u)^2 \sin\theta$$

Example 12-10. The piping connection shown divides a jet of water so that 2.00 ft³ / s goes in each direction as shown in Figure 12-18. The magnitude of the velocity entering the system is 50 ft/s and the cross-sectional areas of the entering and leaving streams are equal and the magnitude of the linear velocities leaving the connection are equal. Determine the x and y components of the net force f required to split the water and change its direction. The density of the water is constant. There are no frictional forces and the pressure-area forces are negligible.

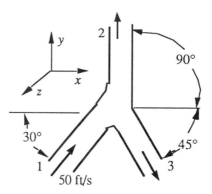

Figure 12-18: Forces
due to moving fluids

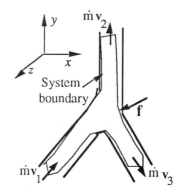

Figure 12-18a: System for
forces due to moving fluids

SOLUTION: The system is defined as the connection. The system is not moving and is at steady state with coordinate axis shown. This system is shown in Figure 12-18a.

Table 12-4: Data for Example 12-10 (Mass, System = Volume impacting plate)

Species	$\left(\dfrac{dm_{sys}}{dt}\right)$	=	$\sum \dot{m}_{in/out}$				
			Stream 1		Stream 2		Stream 3
	(lb_m / s)		(lb_m / s)		(lb_m / s)		(lb_m / s)
Total	0	=	$\rho \dot{V}_1$	−	$\rho \dot{V}_2$	−	$\rho \dot{V}_3$

The conservation of mass for this system is given in Table 12-4. From the table the total volumetric flow rate into the defined system is:

$$\dot{V}_1 = 2\dot{V}_2$$
$$= 2(2) = 4 \text{ ft}^3 / s$$

The mass flow rates in and out of the system are simple the volumetric flow rate times the density of water:

$$\dot{m}_1 = \dot{V}_1\rho = (4)(62.4) = 250 \; (\text{lb}_\text{m} \, / \, \text{s})$$

The mass flow rates of the streams leaving at points 2 and 3 are:

$$\dot{m}_2 = 125 \; \text{lb}_\text{m} \, / \, \text{s} \qquad \dot{m}_3 = 125 \; \text{lb}_\text{m} \, / \, \text{s}$$

Finally, the conservation of mass tells us the relationship between the magnitudes of the entering and leaving velocities because we know that the cross-sectional areas are equal.

$$0 = (\rho A v)_1 - (\rho A v)_2 - (\rho A v)_3$$

Because density is constant, the cross-sectional areas are the same, and $\dot{V}_2 = \dot{V}_3$:

$$v_1 = v_2 + v_3 = 2v_2$$

Solving for v_2 gives:

$$v_2 = v_3 = \left(\frac{50}{2}\right) = 25 \; (\text{ft} \, / \, \text{s})$$

The velocity vectors for points 1, 2, and 3 are:

$$\mathbf{v}_1 = 50\cos(30)\mathbf{i} + 50\sin(30)\mathbf{j} + 0\mathbf{k} \; \text{ft} \, / \, \text{s}$$
$$\mathbf{v}_2 = 0\mathbf{i} + 25\mathbf{j} + 0\mathbf{k} \; \text{ft} \, / \, \text{s}$$
$$\mathbf{v}_3 = 25\cos(45)\mathbf{i} - 25\sin(45)\mathbf{j} + 0\mathbf{k} \; \text{ft} \, / \, \text{s}$$

The conservation of linear momentum for this system is given in Table 12-5.

Table 12-5: Data for Example 12-10 (Momentum, System = Impacting volume)

$\left(\dfrac{d\mathbf{p}_\text{sys}}{dt}\right)$	=	$\sum(\dot{m}\mathbf{v})_{\text{in/out}}$			+	$\sum \mathbf{f}_\text{ext}$
(lb$_\text{m}$ ft / s^2)		Stream 1 (lb$_\text{m}$ ft / s^2)	Stream 2 (lb$_\text{m}$ ft / s^2)	Stream 3 (lb$_\text{m}$ ft / s^2)		(lb$_\text{m}$ ft / s^2)
x-direction 0	=	$\dot{m}_1 v_{1x}$ −	$\dot{m}_2 v_{2x}$ −	$\dot{m}_3 v_{3x}$ +		f_x
y-direction 0	=	$\dot{m}_1 v_{1y}$ −	$\dot{m}_2 v_{2y}$ −	$\dot{m}_3 v_{3y}$ +		f_y
z-direction 0	=	0 −	0 −	0 +		f_z

The z-direction equation gives that the force in the z-direction must be zero. The x-direction is:

$$0 = \dot{m}_1 v_{1x} - \dot{m}_2 v_{2x} - \dot{m}_3 v_{3x} + f_x$$

The only unknown in this equation is the force in the x-direction. Rearranging and substituting gives:

$$f_x = \dot{m}_2 v_{2x} + \dot{m}_3 v_{3x} - \dot{m}_1 v_{1x} = \dot{m}_2(v_{2x} + v_{3x}) - \dot{m}_1 v_{1x}$$

$$= (125)\left(0 + 25\cos(45)\right) - (250)(50\cos(30)) = (2210 - 10{,}825)\left(\frac{1}{32.\,174}\right)$$

$$= -268 \text{ (lb}_f)$$

The equation in the y-direction gives:

$$0 = \dot{m}_1 v_{1y} - \dot{m}_2 v_{2y} - \dot{m}_3 v_{3y} + f_y$$

The only unknown in this equation is the force on the system in the y-direction. Rearranging gives:

$$f_y = \dot{m}_2 v_{2y} + \dot{m}_3 v_{3y} - \dot{m}_1 v_{1y} = \dot{m}_2(v_{2y} + v_{3y}) - \dot{m}_i v_{1y}$$

Substituting values gives:

$$f_y = (125)\left(25 - 25\sin(45)\right) - (250)(50\sin(30)) = (915 - 6{,}250)\left(\frac{1}{32.\,174}\right) = -166 \text{ lb}_f$$

The force on the system represented in vector notation is $\mathbf{f} = -268\mathbf{i} - 166\mathbf{j} + 0\mathbf{k}$ (lb$_f$)

Example 12-11. A 4 in diameter jet of water has a velocity of 120 ft/s and strikes a blade moving the same direction at a constant speed of 60 ft/s as shown in Figure 12-19. The angle of the blade is 145° from the defined coordinate axis. Calculate the x and y components of the the net force $\mathbf{f}$ exterted by the blade on the water in order to change its direction.

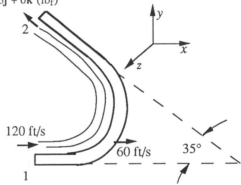

Figure 12-19: Forces on a blade

SOLUTION: Choose the system as the volume occupied by the water between points 1 and 2. Even though the system is moving at 60 ft/s, the system is still at steady state. Figure 12-19a shows the system. Furthermore, the coordinate system is defined to move with the system at 60 ft/s and is an inertial frame.

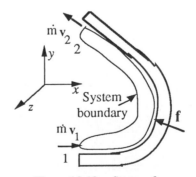

Figure 12-19a: System for forces on a blade

Table 12-6: Data for Example 12-11 (Mass, System = Volume between 1 and 2)

$\left(\dfrac{dm_{sys}}{dt}\right)$	=	$\sum \dot{m}_{in/out}$		
		Stream 1		Stream 2
[mass / time]		[mass / time]		[mass / time]
Total 0	=	$\rho_B v_1 A_1$	–	$\rho_B v_2 A_2$

The conservation of mass for this system is given in Table 12-6. Note that v_1 and v_2 are with respect to the moving blade and hence the pre-subscript B. Because the density of the fluid is constant and the cross-sectional areas are the same, the magnitudes of the entering and leaving velocities are equal:

$$_B v_1 = {}_B v_2$$

The conservation of linear momentum for this system is given in Table 12-7.

Table 12-7: Data for Example 12-11 (Momentum, System = Volume between 1 and 2)

$\left(\dfrac{d\mathbf{p}_{sys}}{dt}\right)$	=	$\sum (\dot{m}\mathbf{v})_{in/out}$			+	$\sum \mathbf{f}_{ext}$
		Stream 1		Stream 2		
$(lb_m\ ft\ /\ s^2)$		$(lb_m\ ft\ /\ s^2)$		$(lb_m\ ft\ /\ s^2)$		$(lb_m\ ft\ /\ s^2)$
x-direction 0	=	$_B v_1 \dot{m}_1$	–	$_B v_2 \cos\theta\, \dot{m}_2$	+	f_x
y-direction 0	=	0	–	$_B v_2 \sin\theta\, \dot{m}_2$	+	f_y
z-direction 0	=	0	–	0	+	f_z

Substituting in the mass flow rate expression $\dot{m}_1 = \dot{m}_2 = \rho A ({}_B v_1)$, the x-direction equation gives:

$$x\text{-dir}\qquad 0 = \rho A ({}_B v_1)^2 - \rho A ({}_B v_1)^2 \cos\theta + f_x$$

Rearranging, this equation gives:

$$x\text{-dir}\qquad 0 = ({}_B v_1)^2 \rho A - ({}_B v_1)^2 \cos\theta\, \rho A + f_x$$

In terms of the fixed coordinates, the force in the x direction is:

$$f_x = (v_1 - u)^2 \rho A (\cos\theta - 1)$$

Substituting values gives:

$$f_x = (120 - 60)^2 (62.4)\left(\frac{\pi 4^2}{4}\right)(\cos(35) - 1) = -510{,}520 \frac{ft^2\ lb_m\ in^2}{ft^3\ s^2}\left|\frac{1\ ft^2}{144\ in^2}\right|\frac{1\ lb_f\ s^2}{32.174\ lb_m\ ft} = -110\ lb_f$$

The y-direction equation gives, with the substitution of the mass flow rate:

$$y\text{-dir}\qquad 0 = -({}_B v_2)({}_B v_1)\rho A + f_y$$

Rearranging and recognizing, through mass conservation, that $(_Bv_2) = (_Bv_1)$ gives:

$$y\text{-dir} \qquad 0 = -(_Bv_1)^2 \sin\theta \rho A + f_y$$

With respect to the fixed frame $(_Bv_1) = v_1 - u$ and hence the force in the y-direction is:

$$f_y = (v_1 - u)^2 \rho A \sin\theta$$

Substituting values gives:

$$f_y = (120 - 60)^2(62.4)\left(\frac{\pi 4^2}{4}\right)\sin(35) = 1.62 \times 10^6 \frac{\text{ft}^2 \ \text{lb}_m \ \text{in}^2}{\text{ft}^3 \ \text{s}^2}\left|\frac{1 \ \text{ft}^2}{144 \ \text{in}^2}\right|\frac{1 \ \text{lb}_f \ \text{s}^2}{32.174 \ \text{lb}_m \ \text{ft}} = 350 \ \text{lb}_f$$

The force exerted by the blade on the water, represented in vector notation is $\mathbf{f} = -110\mathbf{i} + 350\mathbf{j} + 0\mathbf{k} \ \text{lb}_f$.

12.5 FLUID DYNAMICS PROBLEM-SOLVING TOOLS

The various tools which have been presented in this and earlier chapters for evaluating and understanding fluid dynamics are summarized below.

1) **The Laws.** Especially useful for fluid mechanics are the conservation laws of mass (e.g. to relate $\rho v A$ products of inlet and outlet streams), linear and angular momentum (e.g. to calculate support forces). Mechanical energy accounting also is essential (to calculate pump requirements).

2) **Properties of Matter.** This tool includes fluid density, viscosity, and friction factors (really a combination of fluid and system properties).

3) **Descriptions of Forces.** Most notable is the fluid pressure (actually a force per area) which actually plays the role of a form of energy (flow work) in the mechanical energy equation.

4) **System Characteristics.** For fluid pipe flow problems, this includes information such as pipe diameter and length, valves, expansions and contractions, pump size, elevation change between inlet and outlet, etc.

5) **Mathematics.** The mathematics required for the fluid dynamics of this chapter consists of manipulating and solving algebraic equations or simultaneous algebraic equations (or an algebraic equation in concert with graphical data as in the friction factor chart). Solutions sometimes may be trial and error. Software packages or spreadsheets can be especially useful mathematics tools for these problems.

Strategic methods include considering the various laws and other tools in a deliberate way; be conscious of what the tools are and the roles that they play. Also, consider the value of using different system boundaries. As has been evident in other applications, different system boundaries can provide different information. Be open minded and creative!

12.6 REVIEW

1) When dealing with the macroscopic flow of isothermal fluids, we are concerned with the conservation of mass, the conservation of momentum, and an accounting of mechanical energy. The conservation of total energy does not provide any additional non-trivial information in the absence of thermal (i.e., heat transfer and temperature) effects.

2) Because we are dealing with macroscopic flow of a fluid rather than point-by-point flow, we calculate the average flow of a fluid across a macroscopic cross-sectional area in terms of an average velocity across that area. If the fluid velocity is not uniform across the cross section, then this average does not necessarily represent the fluid velocity at any particular point, but rather is descriptive of the total (or average) mass flow rate of the fluid.

3) Likewise, an average momentum flowrate and an average kinetic energy flowrate can be calculated. For convenience, these momentum and kinetic energy flowrates are expressed in terms of the average fluid velocity but must then be adjusted using an appropriate correction factor (α_{Mom} or α_{KE}).

4) For the steady-state flow of an incompressible fluid through a single conduit (i.e., there is only one entrance and one exit for the system) the extended (or engineering) Bernoulli equation is extremely valuable. This equation represents an accounting of mechanical energy for the system and says that mechanical energy entering the system in the form of kinetic, potential, or pressure energy, or work either leaves the system in these same forms or is lost due to frictional dissipation.

5) A key to applying the engineering Bernoulli equation to specific flow problems is to be able to calculate the friction losses which occur throughout the system. A friction factor chart or equation (the Churchill equation) is used to calculate these losses for any sections of straight pipe. Tables of losses for pipe fittings (valves, tees, elbows, etc.) in terms of equivalent sections of straight pipe are available for these sections. The losses in sudden enlargements and contractions can be included by using equations for these cases. When these kinds of losses are all added together, the combined loss is used in the Bernoulli equation.

6) This basic approach used for newtonian fluids in a single pipe can has been extended to networks of pipes, the flow of non-newtonian fluids, and to non-circular conduits.

7) The extended Bernoulli equation also provides the basis for many devices used to measure the flow of a fluid in a pipe. By changing the area of a pipe cross-section in a measurement device, the pressure in the device is necessarily changed with respect to that upstream of the device, in accordance with the accounting of mechanical energy provided by the Bernoulli equation. Therefore, a simple measurement of pressure change in flowing through the device provides a direct measurement of the fluid flowrate. Example devices using this principle are the venturi meter, the orifice meter and the pitot tube.

8) A macroscopic momentum conservation law is applied to piping systems to determine the forces associated with the flow of fluid through a pipe. Any time a fluid changes momentum (due to a change in flowrate or due to a change in direction) a force is involved. Also, differences in upstream and downstream pressure provide a net force on the system which must be accounted for in structural design.

9) The macroscopic momentum equation also is used in the design of fluid motive devices with one example being rocket propulsion systems.

12.7 NOTATION

The following notation has been used throughout this chapter. The bold face lower case letters represent vector quantities.

Scalar Variables and Descriptions

		Dimensions
A	area	[length^2]
D	diameter	[length]
D_{Hyd}	hydraulic diameter	[length]
$\dot{F}$	rate of conversion of mechanical to thermal energy	[energy / mass]
f	Fanning friction factor	
f_D	Darcy friction factor	
g	magnitude of the acceleration of gravity	[length / time^2]
h	elevation	[length]
K	loss coefficient	
KE	kinetic energy	[energy]
l	length	[length]
m	total mass	[mass]
$\dot{m}$	total mass flow rate	[mass / time]
N_{Re}	Reynolds number	
P	pressure	[force / length^2]
PE	potential energy	[energy]
R_{hyd}	hydraulic radius	[length]
t	time	[time]
$\langle v \rangle$	average velocity	[length / time]
$\langle v^2 \rangle$	average squared velocity	[length^2 / time^2]
$\langle v^3 \rangle$	average cubed velocity	[length^3 / time^3]
$\dot{V}$	volumetric flow rate	[length^3 / time]
$\dot{W}$	rate of work	[energy / time]
x, y, z	spatial rectangular cartesian variables	[length]
α	velocity profile correction	
ϵ	conduit roughness	[length]
μ	viscosity	[mass / length time]
ρ	density	[mass / length^3]

Vector Variables and Descriptions

		Dimensions
f	force	[force]
g	acceleration of gravity	[length / time^2]
i, j, k	cartesian unit directional vectors	
v	velocity	[length / time]
u	unit direction vector	

Subscripts

Con	contraction
eq	equivalent or equal to

Exp	expansion
ext	external to the system or acting at the system boundary
KE	related to the kinetic energy
Mom	related to the linear momentum
N	normal to the surface
O	orifice meter
P	pitot tube
Ref	reference frame
V	venturi meter
out	external or outside the system

<u>Superscripts</u>

^   per unit mass

12.8 SUGGESTED READING

Churchill, S. W., *Chemical Engineering*, November 7, p. 91, 1977

Currie, I. G., *Fundamental Mechanics of Fluids*, McGraw- Hill, New York, 1974

Darby, R., "Pipe Flow in Non-Newtonian Fluids," in *Encyclopedia of Fluid Mechanics*, N. P. Cheremisinoff (ed), Vol 7, Ch 2, pp. 19-54, Gulf Publishing Co., 1988.

Fox, R. W., and A. T. McDonald, *Introduction to Fluid Mechanics*, 3rd edition, John Wiley & Sons, New York, 1985

Halliday, D. and R. Resnick, *Physics for Students of Science and Engineering, Part I*, John Wiley & Sons, 1960

LeMehaute, B., *An Introduction to Hydrodynamics and Water Waves*, Springer-Verlag, New York, 1976

Vennard, J. K., and R. L. Street, *Elementary Fluid Mechanics*, 6th edition, John Wiley & Sons, New York, 1982

Whitaker, S., *Introduction to Fluid Mechanics*, Prentice-Hall, Englewood Cliffs, N. J., 1968

White, F. M., *Fluid Mechanics*, McGraw-Hill, New York, 1986

QUESTIONS

1) Of the fundamental laws and usual accounting equations, which are most applicable for fluid dynamics problems?

2) For the flow of a fluid through a pipe, explain the difference between the average fluid velocity and the fluid velocity profile. How are the two related? Why are the alpha factors used in the energy and momentum equation?

3) What kind of equation is the extended Bernoulli equation (i.e. where does it come from and what property does it account for?) Give a physical (verbal) interpretation of each term in the Bernoulli equation.

4) What does the friction factor in the Bernoulli equation represent, in terms of energy, and how is it obtained for

   a) the flow of a fluid in a straight pipe,

   b) the flow of a fluid through valves and pipe fittings, and

   c) the flow of a fluid through expansion or contractions in pipes

5) What is the Churchill equation and what are its advantages?

6) List five kinds of pipe flow problems (as determined by the quantities which are known and those which are unknown) and explain a strategy or solution procedure for each kind.

7) Explain the meaning of "energy," "pressure," and "head," forms of the Bernoulli equation. Which form connotes a proper physical meaning of this equation?

8) What is the hydraulic radius and how is it used for pipe flow problems?

9) Explain the physical difference between laminar flow and turbulent flow in a pipe. What is the significance of the kind of flow with respect to calculating pressure drop, required pumpwork, or other terms from the Bernoulli equation?

10) What is the relative roughness and what is its importance?

11) What is the value of the α_{KE} for turbulent flow? What is its value for a Newtonian fluid in laminar flow?

SCALES

1) Fill in the blanks in each row of the following table and indicate whether the flow is laminar or turbulent. Use Figure C-1 to determine f, assuming smooth pipe.

v_{avg}	ρ	μ	I.D.	$\dot{m}$	N_{Re}	f	Laminar or Turbulent
2 ft/s	62.3 lbm/ft^3	1 cp	2 in				
1 m/s	800 kg/m^3	2 mPa· s	7 cm				
1 cm/s	50 lbm/ft^3	5 cp	1 mm				
		2 cp	1 cm	0.5 lbm/s			

2) Using a spreadsheet and the Churchill equation, calculate f for each case in problem 1, for relative roughnesses of 0, 0.01, and 0.001.

3) Convert the following dimensional quantities to SI or convert from SI to Engineering units, as appropriate. Show the appropriate conversion factor(s) and cancellation of units. (Note: centipoise [cp] is not SI so convert it to SI.)

a) 0.5 hp

b) 75 psi

c) 3.5 cp

d) 82 lbm/ft^3

e) 700 kg/m^3

f) 2.3 m/s

g) 2 lbm/s

h) 25 (dimensionless)

i) 95 lbf

j) 400 ft-lbf

k) 1.5 atm

l) 75 kPa

4) What are the density and viscosity of each fluid at the indicated temperature (at 1 atm pressure)?

a) water at 30°C
b) ethanol at 20°C
c) sea water at 15°C
d) acetone at 20°C

5) At an instant in time, a mass of 5 kg is moving with velocity $\mathbf{v} = 3\mathbf{i} + 2\mathbf{j} - \mathbf{k}$ m/s at an elevation of 20 m. At this time, what are its $\widehat{KE}, \widehat{PE}, KE$, and PE, all in SI units of energy/mass.

6) Water at a point in a pipe is flowing at a pressure of 100 psi. What is the flow work (energy per mass) at this point? If the mass flowrate is 5 lbm/s, what is the rate of flow work at this point?

7) What is the equivalent length (in feet) of 2 in I.D. pipe (for friction loss calculation) for each fitting in the table?

Fitting or Valve	L_{equiv}
45°elbow	
90°elbow	
Gate valve, open	
Globe valve, open	

8) For the following upstream and downstream pipe sizes or situation, calculate the appropriate K_{exp} or K_{con}.

I.D.		
Upstream	Downstream	K_{exp} or K_{con}
1.049 in	2.067	
2.067	1.049	
2.067	tank (below the (liquid surface)	
2.067	tank (above the) liquid surface	

PROBLEMS

1) You are designing a piping system to pump water from a lake to a tank for storage. The pipe is 1 in schedule 40 commercial steel with a relative roughness of 0.005 and is to provide water at a flow rate of 10 gal/min. If the level of water in the storage tank is 20 feet above the water level of the lake and there are 4 90°standard radius elbows, 1 open globe valve, and 80 feet of straight pipe, what size pump is required? The pipe exit is beneath the water level in the tank.

2) You want to size a pump that will provide water at a flow rate of 5 ft/s through 500 feet of 2-in schedule 40 PVC pipe (assume that it is smooth and that the I.D. is 2.00 in). The pipe inlet is at 1 atm pressure and an elevation of 30 feet and the outlet is at 40 psia and an elevation of 130 feet. In addition to the pipe, there are 4 90°standard radius elbows, 1 open globe valve, and 1 open gate valve. You may assume that there are no expansions or contractions.

    a) Assuming that the pump operates at 100% efficiency, determine the size (hp) of pump required.
    a) Assuming that the pump operates at 75% efficiency, determine the size of pump required.

3) You are designing a piping system to pump water from one tank to another. The pipe is 1 in schedule 40 commercial steel with a relative roughness of 0.005 and is to provide water at a flow rate of 10 gal/min. The pipe exit is 1 ft above the water level in the second tank so that the water exits to the atmosphere straight from the pipe. If the level of water in the first tank is 20 feet above the water level of the second and there are 4 90°standard radius elbows, 1 open globe valve, and 80 feet of straight pipe, what size pump is required?

4) Methanol at 0 °C is being pumped at steady-state through a 3 in, Schedule 40, commercial steel pipe over a total distance of 1000 ft including equivalent lengths of pipe for valves, elbows, etc. The relative roughness of the pipe is 0.01. Included in the total distance is an increase in height of 50 ft. A 0.5 hp motor is available to transfer the fluid, the mass flow rate is 40 lb_m / min, and the pressure at the entrance is atmospheric. The exit pressure is unknown.

a) Calculate the Reynolds number for the flow and determine if the flow is laminar or turbulent.
b) From the graph of friction factor, determine the Fanning friction factor.
c) Calculate the friction losses in the piping system.
d) Use the steady-state mechanical energy accounting equation to calculate the unknown exit pressure.

5) Water at 20 °C is flowing at steady state through a 4 in Schedule 40 pipe at a mass flow rate of 30 lb_m / s. The water flows through one (1) gate valve and a 10 hp motor is available to transfer the liquid. If the pressure at the entrance is atmospheric and the pressure at the exit is twice atmospheric and the pipe is laid at a 30 ° incline up the side of a hill, what is the maximum height that we can move the water? What is the maximum height if the pipe is vertical?

   a) Calculate the Reynolds number for the flow and determine if the flow is laminar or turbulent.
   b) From the graph of friction factor, determine the Fanning friction factor.
   c) Write the mathematical expression to calculate the frictional losses in the piping system.
   d) Use the steady-state mechanical energy accounting equation to determine the frictional losses and the length of the pipe.

6) Benzene at 0 °C is pumped at steady-state through a 2 in Schedule 40 straight pipe for a distance of 2000 ft. The entrance pressure is 20 psia and the exit pressure is 100 psia. The relative roughness of the pipe is 0.001. If a 7 hp motor is available to move the fluid, what is the mass flow rate that is moving through the pipe? Hint: A trial and error solution is required when the unknown cannot be solved for explicitly. First, assume that the flow is completely turbulent and use this as the first guess.

7) Liquid mercury at 30 °C is flowing at steady state down a vertical glass tube with a 0.1 cm diameter. If the entrance pressure is 1.3 bar and the exit pressure is 1 bar, determine the mass flow rate and the velocity of the mercury if the tube is 1 m in length. You may assume that $\epsilon = 0$.

8) Water is flowing through a 2 in Schedule 40 pipe at a mass flow rate of 10 lb_m / s. The pressure at the entrance of the 90° square elbow is 30 psi

   a) Determine pressure at the exit of the elbow and neglect any changes in potential energy.

   b) For the system defined as the inside volume of the 90 ° square elbow, determine the force required to change the water's direction.

9) Benzene is flowing through a 3 in Schedule 40 commercial steel 90 °elbow, followed be a sudden expansion to a 4 in Schedule 40 pipe at a mass flow rate of 0.5 (lb_m / min). The pressure at the entrance of the elbow is 25 psi.

   a) Estimate the pressure at the exit of the expansion and neglect any changes in potential energy.
   b) For the system defined as the inside volume of the elbow and expansion, determine the net force that is acting on this system.

10) Water is flowing through a 6 in I.D. pipe. To determine the mass flow rate through the pipe you install a venturi meter for which $K_V = 0.98$. The diameter of the throat of the venturi is 3 in. If the manometer fluid of the venturi is mercury and the manometer reading is 5 cm, calculate the mass flow rate. Also, estimate the rate of mechanical energy loss from flowing through the venturi meter. If the venturi meter is replaced by a 3 in orifice meter and the mass flow rate is equal to that calculated for the venturi, determine the manometer reading if the loss coefficient is $K_O = 0.63$. Estimate, the rate of mechanical energy loss from flowing through the orifice meter.

11) A cylindrical tank of radius 10 ft and height 20 ft with its axis vertical is full of water but has a leak at the bottom. Assuming the water escapes at a rate proportional to the depth of water in the tank and that 10 percent of the original amount escapes during the first hour, derive the function that defines the volume of water in the tank as a function of time.

12) Water at 70 °F flows at 12 ft/s in a 150 ft length 2 inch schedule 40 steel pipe through an open globe valve and one 90° standard radius elbow. What is the pressure drop across the pipe? State any and all assumptions.

13) A piping system is proposed to move water from a holding pond to an elevated storage tank at atmospheric pressure. Water with density 62.4 lb_m / ft^3 and viscosity 1 cp is flowing through a steel pipe system at a mass flow rate of 80 lb_m / s. The steel pipe has an inside diameter of 4.026 in and a roughness of 0.0018 in. The outlet of the pipe is 100 ft higher than the entrance of the pipe. A 20 (hp) motor is available to move the water. In the piping system there are five (5) gate valve and four (4) $90°$ square elbows. Determine the maximum total length of straight pipe that can be used.

14) Liquid methanol at 20 °C is flowing through a horizontal 2.5 in inside diameter steel pipe at a mass flow rate of 100 lb_m / min. The density of the liquid methanol is 49.4 lb_m / ft^3. The pipe system has a pump which is capable of delivering 5 hp and the exit of the pump is a 2 in inside diameter pipe. If we assume steady-state and neglect any frictional losses, calculate the pressure a the exit of the pump relative to the pressure at the entrance of the pump.

15) You are a young engineer in a process plant faced with the following situation. A piping system is in place to pump a fluid at 20 °C from a storage tank at 1 atmosphere of pressure to a reactor which is at 2 atmospheres. The piping system is 2-in schedule 40 commercial steel. There is a total straight run of 40 ft of pipe and two (2) $90°$ standard radius elbows. The pump is 2 hp and the elevation increase from inlet to outlet is 20 feet. The fluid density is 93.6 lb_m / ft^3 and viscosity is 5 cp. You suspect that the flowrate is not what it should be. To test this hypothesis, you insert a venturi meter in the line (for which $K_V = 1$, i.e., friction losses in the meter are negligible) and measure the pressure drop using a mercury manometer. The manometer reading is 250 mm and the area ratio for the meter is A_{in} / A_{out} = 4.

    a) What is the measured mass flow rate and the average velocity through the pipe?
    b) Does the answer to part a) agree with your design expectation? Explain.
    c) Why might the answers to a) and b) differ?
    d) How else might you measure the fluid flow rate (besides with an in-line meter)?

16) Water is flowing at a constant mass flow rate of 50 kg/s through a horizontal diverging section of pipe. The inlet inside diameter is 0.25 m and exit inside diameter is 0.5 m. If we neglect the losses due to friction, and the inlet pressure is 505 (kPa), calculate the outlet pressure. Why does this answer seem unusual. Using the conservation of linear momentum for this diverging section what terms must be included and what is the direction and magnitude of the force at the wall.

ELECTRICAL CIRCUITS

13.1 INTRODUCTION

In this chapter and the one following it, we apply the conservation of charge and conservation of energy statements in the form of Kirchhoff's current and voltage laws to study the behavior of electrical circuits. The focus in this chapter is on resistive circuits (no capacitors or inductors), and we will develop a set of tools and general analysis techniques for such circuits. In the later chapter, these tools and techniques will be generalized so that they can be applied to circuits containing inductors and capacitors as well. First, we review.

13.1.1 Current and Voltage

The instantaneous rate at which charge (i) passes some point is called the electric *current*. The symbol for current is the lower case i and in terms of a system and system boundary the current is the net charge flow rate into or out of the system. For a given system, the conservation of charge states:

$$\boxed{\left(\frac{dq_{sys}}{dt}\right) = \sum i_{in/out}}$$

(4 – 9)

The unit of electric current is the *ampere*.

The potential difference (voltage) between two points is the amount of work per unit charge it would take to move a charge from one point to the other. The unit of voltage is the *volt*. Hence, a charge at voltage V above a reference required qV work to bring it to this voltage; it possesses electrical energy qV; voltage is electrical energy per charge.

13.1.2 Kirchhoff's Laws

The steady state conservation of net charge and the accounting of electrical energy are Kirchhoff's current law (KCL) and Kirchhoff's voltage law (KVL) respectively.

> KCL — At any connection point (node) in a circuit, the sum of the currents flowing toward that point is equal to the sum of the currents flowing away from that point.

> KVL — For any loop in an electrical circuit the algebraic sum of the potential differences across each element encountered in traversing the loop must be zero.

$$\sum_{\text{Loop}} V = 0 \qquad (7-60)$$

Futhermore, the accounting of electrical energy for steady-state operation is also an important relationship to remember.

$$\left(\frac{d(E_E)_{\text{sys}}}{dt}\right) = \sum (iV)_{\text{in/out}} + \sum \dot{W}_{\text{ele}} - \sum \dot{F} \qquad (7-56)$$

However, this equation is not independent of the voltage law, and provides a way to check for errors in the solution of circuit problems.

13.1.3 Resistor

Resistance is defined as the ratio between an applied voltage (voltage difference across the resistor) and a measured current through the resistor. Devices for which the measured current is directly proportional to the applied voltage are called resistors and these satisfy the relation

$$i = \left(\frac{V}{R}\right) \qquad \text{or} \qquad V = Ri \qquad (7-61)$$

which is called Ohm's Law.

13.1.4 Definitions and Notation

An *independent voltage source* is a circuit element that maintains a prescribed potential difference across its terminals regardless of the current through the element. That is, the voltage is not a function of the current. A voltage source can be constant (in which case the circuit is called a DC-direct current circuit), or time varying. The schematic symbol for a voltage source is shown in Figure 13-1.

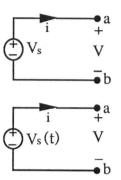

Figure 13-1: Independent voltage source

An *independent current source* is a circuit element that maintains a prescribed current through it regardless of the voltage that is across its terminals. That is, the current is not a function of voltage. As with a voltage source, a current source may be constant (DC) or time-varying. The schematic symbol for a current source is shown in Figure 13-2.

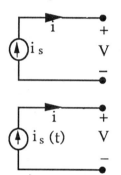

Figure 13-2: Independent current source

A *short circuit* is a voltage source of zero volts. That is, there are zero volts across it, no matter what the current is. In effect, a short circuit is like an ideal wire; there will be no volts across it even if there is current through it. Since the voltage across a resistor is the resistence times the current flowing through the resistor, one can think of a short circuit as a zero ohm resistor. A short circuit is depicted as a straight connection between two points (Figure 13-3a).

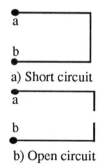

a) Short circuit

b) Open circuit

Figure 13-3: Circuit elements

An *open circuit* is a current source of zero amperes. That is, there is zero current through it, no matter how many volts are across it. In effect, an open circuit is a broken wire; there will be no current through it. Since the current through a resistor is the voltage across the resistor divided by the resistance, one can think of an open circuit as a resistor of infinity ohms. An open circuit is depicted as a broken connection between two points shown in Figure 13-3b.

13.1.5 Power or Rate of Work

The power used or supplied by any element of a circuit is given by

$$\dot{W} = Vi$$

where V is the *voltage across* the element and i is the *current through* it. If a positive current enters the higher potential side of an element that has a potential difference across it, then power is consumed by that element; electrical energy leaves the circuit. On the other hand if the positive current enters the lower end, that element is a supplier of power; electrical energy enters the circuit.

Voltage and current sources are usually suppliers of power (they force charge to move through a potential difference and so supply a power $\dot{W} = Vi$). Resistors are always consumers of power by converting electrical energy to thermal (internal) energy and the power consumption for resistors is given by the following equations:

$$\text{Resistors} \qquad \dot{F} = i^2 R = Vi = \left(\frac{V^2}{R}\right)$$

Capacitors and inductors are reversible users of power when they are strengthening their fields (storing more energy), and are reversible suppliers of power when their fields are weakening (returning some of the stored energy).

The combination of Kirchhoff's current law, Kirchhoff's voltage law, and the voltage/current relations for resistors, capacitors, and inductors allows one to write equations that can then be solved to obtain all the voltages and currents in a circuit. For circuits that contain only sources and resistors these are linear algebraic equations, while for circuits consisting of sources, resistors, inductors, and capacitors we obtain sets of linear *differential-integro* equations.

13.2 EQUIVALENT RESISTANCE

13.2.1 Resistors in Series

The circuit shown consists of a voltage source and three resistors "in series." The currents through each resistor and voltages across each are labeled in Figure 13-4. Application of Kirchhoff's Current Law for the system defined as each of the resistors and the voltage source is given in Table 13-1.

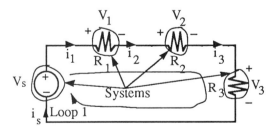

Figure 13-4: Resistors in series

Table 13-1: Resistors in Series (Charge, System = Resistors)

	0	=	$\sum i_{\text{in/out}}$			
System	(C/s)		i_s (A)	i_1 (A)	i_2 (A)	i_3 (A)
Resistor 1	0	=		i_1 −	i_2	
Resistor 2	0	=			i_2 −	i_3
Resistor 3	0	=	$-i_s$		+	i_3
Voltage Source	0	=	i_s −	i_1		

The equations of the table give:

$$i_s = i_1 = i_2 = i_3$$

That is, the resisitive elements in series all have the same current through them because there is no accumulation of charge in any of the resistors and there is only one loop.

Kirchhoff's voltage law for the loop is given in Table 13-2 and gives the result that

Table 13-2: Resistors in Series (Electrical Energy-Voltage)

	$\sum V$				=	0
Loop	V_s (V)	V_1 (V)	V_2 (V)	V_3 (V)		
Loop 1	V_s −	V_1 −	V_2 −	V_3	=	0

$$V_s = V_1 + V_2 + V_3$$

Ohm's Law relates each resistor's voltage to the current through each resistor as:

$$V_1 = (Ri)_1 = R_1 i_s$$
$$V_2 = (Ri)_2 = R_2 i_s$$
$$V_3 = (Ri)_3 = R_3 i_s$$

Substituting these relationships into Kirchhoff's Voltage Law gives:

$$
\begin{aligned}
V_s &= V_1 + V_2 + V_3 \\
&= R_1 i_s + R_2 i_s + R_3 i_s \\
&= (R_1 + R_2 + R_3) i_s
\end{aligned}
$$

and *as far as the total voltage and total current is concerned*, the circuit acts like the simplified circuit in Figure 13-5. The source cannot tell whether it is connected to three resistors, or to one resistor of size:

$$R_{eq} = \left(\frac{V_s}{i_s}\right)$$

$$= (R_1 + R_2 + R_3)$$

(13 – 1)

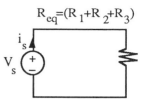

$$R_{eq} = (R_1 + R_2 + R_3)$$

Figure 13-5: Simplified Series Circuit

Note also that the equivalent resistance of several resistors in series is always larger than the *largest* of the resistors.

For a circuit with a number of resistors in series, the equivalent resistance is simply the sum of the resistances:

$$R_{eq} = \left(\frac{V_s}{i_s}\right) = \sum R_i$$

(13 – 2)

Returning to the original circuit, since all resistors have the same current through them, and since the voltage across each resistor is the current times resistance, one can see that the largest resistor has the largest voltage drop, while the smallest resistor has the smallest voltage drop.

In fact, the voltage across each resistor is a percentage of the total voltage V_s, with the largest resistor having the largest percentage of the total voltage, while the smallest resistor has the smallest percentage.

$$V_1 = R_1 i_s$$

$$= R_1 \left(\frac{V_s}{R_1 + R_2 + R_3}\right)$$

$$= \left(\frac{R_1}{R_1 + R_2 + R_3}\right) V_s$$

and for the voltage across the second resistor

$$V_2 = R_2 i_s$$

$$= R_2 \left(\frac{V_s}{R_1 + R_2 + R_3}\right)$$

$$= \left(\frac{R_2}{R_1 + R_2 + R_3}\right) V_s$$

For several resistors in series, the voltage across an individual resistor is related to the voltage source, the equivalent resistance and the particular resistance by:

$$V_i = R_i i_s$$

$$= \left(\frac{R_i}{R_{eq}}\right) V_s$$

(13 – 3)

The series resistance circuit is sometimes called a voltage divider circuit since the total voltage is apportioned among the resistors. In terms of power, the energy supplied by V_s is dissipated in the three resistors in direct proportion to their size.

Example 13-1. Find the current through and the voltage across each resistor in the circuit shown in Figure 13-6. Also for the circuit, calculate the power dissipated as heat in the resistors.

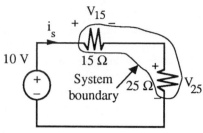

Figure 13-6: Resistors in Series

SOLUTION: For the circuit the equivalent resistance is calculated from the given information.

$$R_{eq} = 15 + 25$$
$$= 40 \ (\Omega)$$

and thus the current through the circuit is:

$$i_s = i_{15} = i_{25}$$
$$= \left(\frac{V_s}{R_{eq}} \right)$$
$$= \left(\frac{10}{40} \right)$$
$$= 0.25 \ (A)$$

The voltage across the 15 (Ω) resistor is

$$V_{15} = iR_{15}$$
$$= (0.25)(15)$$
$$= 3.75 \ (V)$$

and the voltage across the 25 (Ω) resistor is

$$V_{25} = iR_{25}$$
$$= (0.25)(25)$$
$$= 6.25 \ (V)$$

Note that the voltages across the resistors add to the total voltage of the source because of Kirchhoff's Voltage Law and the fact that the electrical field is a potential field.

The electrical energy dissipated or lost in the circuit is calculated:

$$\dot{F} = i^2 R_{eq}$$

Substituting in the known quantities gives:

$$\dot{F} = (0.25)^2 40$$
$$= 2.5 \ (W)$$

The resistors dissipate electrical energy in the form of heat at a rate of 2.5 (W).

13.2.2 Resistors in Parallel

The circuit shown in Figure 13-7 consists of a voltage source and three resistors "in parallel." The currents through each resistor and voltages across each are labeled. Application of Kirchhoff's current law for System 1 is shown in Table 13-3. Six loops can be found (the left, the middle, the right, the left two, the right two, and around the outside). Kirchhoff's voltage law for each loop is shown in Table 13-4.

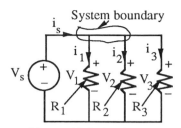

Figure 13-7: Resistors in parallel

Table 13-3: Resistors in Parallel (Charge)

	0	=	$\sum i_{in/out}$			
System			i_s	i_1	i_2	i_3
	(C/s)		(A)	(A)	(A)	(A)
System 1	0	=	i_s —	i_1 —	i_2 —	i_3

Table 13-4: Resistors in Parallel (Electrical Energy-Voltage)

	$\sum V$				=	
Loop	V_s	V_1	V_2	V_3		
	(V)	(V)	(V)	(V)		
Loop 1	V_s —	V_1			=	0
Loop 2	V_s	—	V_2		=	0
Loop 3	V_s		—	V_3	=	0
Loop 4		V_1 —	V_2		=	0
Loop 5			V_2 —	V_3	=	0
Loop 6		V_1	—	V_3	=	0

Even though there are six loops, only three (3) give independent information. These relate the different voltage: across the resistors as:

$$V_s = V_1 = V_2 = V_3$$

or elements in parallel all have the same voltage across them. Ohm's Law relates each resistor's current to its voltage (and hence to V_s)

$$i_1 = \left(\frac{V}{R}\right)_1 = \left(\frac{V_s}{R_1}\right)$$

$$i_2 = \left(\frac{V}{R}\right)_2 = \left(\frac{V_s}{R_2}\right)$$

$$i_3 = \left(\frac{V}{R}\right)_3 = \left(\frac{V_s}{R_3}\right)$$

Thus the KCL equation for system 1 can be written

$$i_s = i_1 + i_2 + i_3$$

$$= \left(\frac{V}{R}\right)_1 + \left(\frac{V}{R}\right)_2 + \left(\frac{V}{R}\right)_3$$

$$= V_s\left(\frac{1}{R_1} + \frac{1}{R_2} + \frac{1}{R_3}\right)$$

and *as far as the total voltage and total current are concerned*, the circuit acts like the simplified circuit in Figure 13-8.

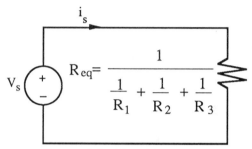

Figure 13-8: Simplified circuit for parallel resistors

The source cannot tell whether it is connected to three resistors, or to one resistor of size

$$R_{eq} = \left(\frac{V_s}{i_s}\right) = \left[\frac{1}{\left(\dfrac{1}{R_1} + \dfrac{1}{R_2} + \dfrac{1}{R_3}\right)}\right] \qquad (13-4)$$

Note also that the equivalent resistance of several resistors in parallel is always smaller than the *smallest* of the resistors. For a circuit with a number of resistors in parallel, the equivalent resistance is:

$$R_{eq} = \left(\frac{V_s}{i_s}\right) = \left[\frac{1}{\sum\left(\dfrac{1}{R_i}\right)}\right] \qquad (13-5)$$

Returning to the original circuit, because all resistors have the same voltage across them, and because the current through each resistor is the voltage divided by resistance, one can see that the largest resistor has the *least* current, while the smallest resistor has the largest current. In fact, the current in each branch is a percentage of the total current i_s, with the the largest resistor having the smallest percentage of the total current, while the smallest resistor has the largest percentage.

Example 13-2. Find the currents in each branch and the total current in the circuit shown in Figure 13-9. Also, calculate the rate of electrical energy dissipation in the circuit.

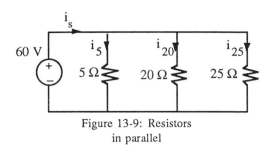

Figure 13-9: Resistors
in parallel

SOLUTION: Because the voltage across each resistor is 60 volts (V_s = 60 V), the currents are given by:

$$i_5 = \left(\frac{60}{5}\right) = 12 \text{ (A)}$$

$$i_{20} = \left(\frac{60}{20}\right) = 3 \text{ (A)}$$

$$i_{25} = \left(\frac{60}{25}\right) = 2.4 \text{ (A)}$$

Thus the total current is the sum of all the branch currents as

$$i_s = 12 + 3 + 2.4 = 17.4 \text{ (A)}$$

Note that the equivalent resistance is

$$R_{eq} = \left(\frac{1}{\left(\frac{1}{5} + \frac{1}{20} + \frac{1}{25}\right)}\right) = \left(\frac{1}{(0.2 + 0.05 + 0.04)}\right) = 3.448 \text{ }(\Omega)$$

and that it is smaller than any of the resistors. Also note that

$$i_s = \left(\frac{V_s}{R_{eq}}\right) = \left(\frac{60}{3.448}\right) = 17.4 \text{ (A)}$$

The power dissipated in the resistors is calculated from:

$$\dot{F} = i^2 R_{eq}$$

Substituting values gives:

$$\dot{F} = (17.4)^2(3.448) = 1.044 \text{ (kW)}$$

The resistors dissipate the the electrical energy at a rate of 1.044 (kW).

Current Divider Circuit. For the special case of two resistors in parallel as shown in Figure 13-10, the formula for R_{eq} takes on the simple form.

$$R_{eq} = \cfrac{1}{\left(\cfrac{1}{R_1} + \cfrac{1}{R_2}\right)}$$

$$= \left(\frac{R_1 R_2}{R_1 + R_2}\right) \qquad (13-6)$$

$$= \frac{\text{product}}{\text{sum}}$$

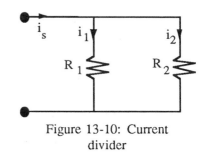

Figure 13-10: Current divider

The currents in each branch are easily calculated as

$$i_1 = \left(\frac{R_2}{R_1 + R_2}\right) i_s \qquad i_2 = \left(\frac{R_1}{R_1 + R_2}\right) i_s$$

Note that for the current divider circuit, the percentage of the total current down each branch is given by the *other* resistor over the sum, and also notice that the larger resistor has the smaller current.

The pipe analogy may be helpful when thinking of resistors in parallel. It is clear that, for a given pressure (voltage), two parallel pipes will be able to carry more water (current) than just one pipe. However, the larger diameter pipe (smaller resistance) will carry the larger amount of water.

Example 13-3. Find the current i_{tot} and the voltage V_5 in the circuit shown in Figure 13-11.

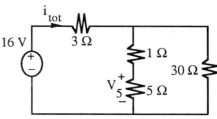

Figure 13-11: Example 13-3

SOLUTION: An approach is to first simplify the circuit by resistance combinations, then find the total current, and after this build the circuit back up. We see that the 1 (Ω) is in series with the 5 (Ω), and so the circuit behaves as simplified in Figure 13-11a.

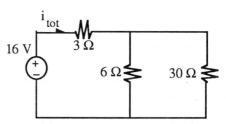

Figure 13-11a: Simplification in series

Next, we see that the 6 Ω is in parallel with the 30 Ω, and so their combination is equivalent to:

$$R_{eq} = \frac{(6)(30)}{(6 + 30)} = 5 \ (\Omega)$$

and this is shown in Figure 13-11b.

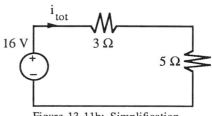

Figure 13-11b: Simplification in parallel

Therefore, the total current through the equivalent circuit is calculated as:

$$i_{tot} = \left(\frac{16}{(3 + 5)} \right) = 2 \text{ A}$$

Going back to the original picture with the total current labelled as shown in Figure 13-11c, we have and we see that the right side of the circuit is a current divider.

$$i_5 = \left(\frac{30}{(30 + 6)} \right) 2$$

$$= 1.667 \text{ A}$$

Now the voltage across the 5 Ohm resistor is calculated as:

$$V_5 = 5 i_5$$

$$= 8.3333 \text{ V}$$

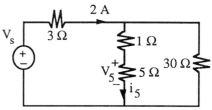

Figure 13-11c: Partially labeled circuit

The fully reconstructed circuit is shown in Figure 13-11d. The reader should check to see that all KCL and KVL equations are satisfied. Also, determine the rate of electrical energy dissipation in the resistors.

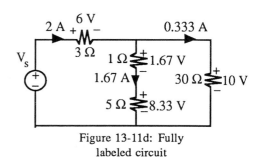

Figure 13-11d: Fully labeled circuit

13.3 SOURCE TRANSFORMATION

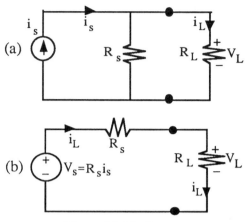

In Figure 13-12, as far as the resistor named R_L is concerned, it cannot tell the difference between circuit (a) (current source with R_s in parallel) and circuit (b) (voltage source with R_s in series such that $V_s = R_s i_s$). That is, as far as outside effects are concerned, a current source in parallel with a resistor has the same effect as a properly chosen voltage source in series with the resistor. To prove this statement, let us calculate the voltage across R_L and the current through R_L for each of the circuits.

Figure 13-12: Source transformations

The circuit on the left is a current divider, and so

$$i_L = \left(\frac{R_s}{R_s + R_L}\right) i_s \qquad (13-7)$$

and

$$V_L = R_L i_L$$
$$= \left(\frac{R_L R_s}{R_s + R_L}\right) i_s \qquad (13-8)$$

The circuit on the right is a single loop, and so

$$i_L = \left(\frac{V_{\text{tot}}}{R_{\text{tot}}}\right)$$
$$= \left(\frac{R_s}{R_s + R_L}\right) i_s \qquad (13-9)$$

and

$$V_L = R_L i_L$$
$$= \left(\frac{R_L R_s}{R_s + R_L}\right) i_s \qquad (13-10)$$

Since the resulting voltage and current are identical for any value of R_L, circuit (a) and circuit (b) are equivalent as far as R_L is concerned.

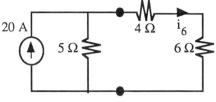

Example 13-4. Find i_6 in the circuit shown in Figure 13-13.

Figure 13-13: Example 13-4
Source Transformations

SOLUTION: We can perform a source transformation to simplify the problem as shown in Figure 13a. The equivalent voltage source is:

$$V_s = iR = (20)(5) = 100 \text{ V}$$

and, from the equivalent circuit it is clear that

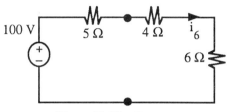

$$i_6 = \left(\frac{100}{15}\right) = 6.67 \text{ A}$$

Figure 13-13a: Source
Transformations

Note that the problem could have also been done as a current divider problem.

Example 13-5. Find the power used or supplied by the 6 volt source as shown in Figure 13-14.

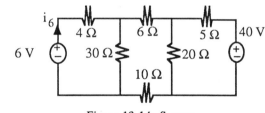

Figure 13-14: Source
Transformations

SOLUTION: Since the power for the 6 volt source is the product of the voltage across it and the current through it, if we can find the current we are done because the voltage across it is definitely 6 V. We can do a succession of source transformations as illustrated below in Figure 13-14a.

From these transformations, the (clockwise) current i_6 can be calculated as:

$$i_6 = \left(\frac{\text{total volts}}{\text{total R}}\right) = \frac{(-19.2 + 6)}{16} = -0.825 \text{ A}$$

and so (because the 0.825 A is entering the + side of the 6 V source) the 6 V source is a *user* and the power consumed is

$$\dot{F} = Vi = (0.825)(6) = 4.95 \text{ W}$$

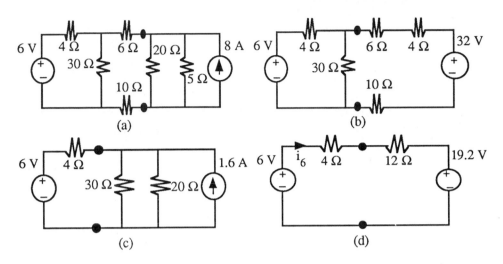

Figure 13-14a: Source Transformation

13.4 THE SUPERPOSITION PRINCIPLE

The circuits we have been looking at are made up of only independent sources and resistors. It can be shown that such circuits, as well as circuits that also contain inductors and capacitors, are *linear* systems. A distinguishing feature of a linear system is that, if it is driven by several independent sources, then the total response to all the sources can be calculated by finding the responses to each source acting alone (all other sources off), and adding these responses. Note that an "off" voltage source is a zero volt voltage source or *short circuit*, while turning off a current source makes it a zero ampere current source or *open circuit*.

Example 13-6. Find the values of i and V_4 in the circuit shown in Figure 13-15.

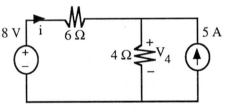

Figure 13-15: Example 13-6
Superposition Principle

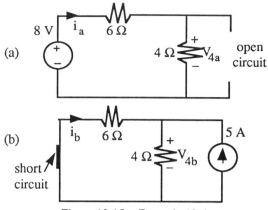

(a)

SOLUTION: The superposition principle allows us to treat the original problem as two separate ones as shown in Figure 13-15a.

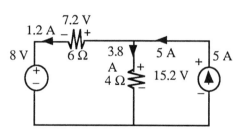

(b)

Figure 13-15a: Example 13-6
Superposition Principle

Circuit (a) is a single loop voltage divider.

$$i_a = \left(\frac{8}{10}\right) = 0.8 \text{ (A)}$$

$$V_{4a} = \left(\frac{4}{6+4}\right)8 = 3.2 \text{ (V)}$$

Circuit (b) is a current divider. The negative sign is required because of the assumed direction of the current i_b. The result can be checked by performing KVL and KCL on the circuit and showing that the results are consistent as shown in Figure 13-15b.

Figure 13-15b: Example 13-6
Completed circuit

$$i_b = -\left(\frac{4}{6+4}\right)5 = -2 \text{ A}$$

$$V_{4b} = (5-2)(4) = 12 \text{ V}$$

Thus

$$i = i_a + i_b = 0.8 + (-2.0) = -1.2 \text{ A}$$

$$V_4 = V_{4a} + V_{4b} = 3.2 + 12 = 15.2 \text{ V}$$

13.5 NODAL ANALYSIS

Application of Kirchhoff's current law at each connection point of a circuit, Kirchhoff's voltage law around each loop of the circuit, and Ohm's law to relate branch voltages and branch currents leads to a large number of equations with a large number of unknowns. The node voltage and the mesh current methods are systematic procedures that arrive at a minimum number of simultaneous equations. These may then be solved, and from the solution one can calculate any voltage or current in the circuit.

Nodal analysis is based on writing KCL equations at the connection points, but doing so in terms of *node voltages*. Ohm's law and KVL is automatically built into the procedure. Nodal analysis consists of the following steps:

1. Find all the connection points (nodes) of the circuit.

2. Choose one of these nodes as "reference node." While any node may be chosen as reference node, it will be seen convenient to pick either the node with the most connections to it, or a node that appears at the bottom of the diagram.

3. At each of the other nodes, define and label each *node voltage* as the voltage rise from the reference node to that node.

4. For each node (except the reference node) write a Kirchhoff current law equation, but do so in terms of the node voltages and resistor values. That is, when referring to the current i below, note that

$$i = \left(\frac{V_{ab}}{R}\right)$$

$$= \left(\frac{(V_{an} - V_{bn})}{R}\right)$$

$$= \left(\frac{(V_a - V_b)}{R}\right)$$

where V_a and V_b are the node voltages of node A and node B, with respect to reference node n as shown in Figure 13-16.

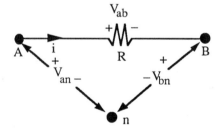

Figure 13-16: Reference node

5. Solve the simultaneous equations for the node voltages

6. Each branch voltage can then be calculated as the difference between two node voltages, and Ohm's law can be applied to find each branch current.

Example 13-7. Use nodal analysis to find the current labeled i_x and the voltage labelled V_y in Figure 13-17.

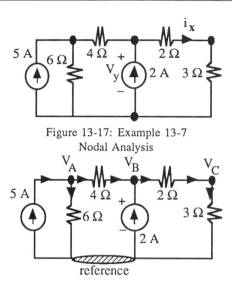

Figure 13-17: Example 13-7
Nodal Analysis

Figure 13-17a: Example 13-7
Node Voltage

SOLUTION: There are four connection points; let us choose the bottom point as reference node, and label the node voltages as V_A, V_B, and V_C as shown in Figure 17a. We assume currents as shown in the figure. If the actual directions are different, we will find out from the values of the voltages which are determined.

At node A, 5 amperes comes into the node, and current can leave the node by going *down* the 6 (Ω) resistor or to the *right* through the 4 (Ω) resistor. At node B there is 2 amperes into the node from the current source and additional current from the 4 (Ω) resistor, while current can leave to the *right* through the 2 (Ω) resistor. At node C we can state that the current entering from the 2 (Ω) resistor minus current leaving through the 3 Ω resistor sum to zero. The equations are presented in Table 13-5. Note that current will flow from high to low potential so that the assumed direction of current flow implies the order of the voltages.

Table 13-5: Data for Example 13-7 (Nodal Analysis)

	0	=			$\sum i_{in/out}$			
	(C/s)		(A)	(A)	(A)	(A)	(A)	(A)
Node A	0	=	5	$-\left(\dfrac{V_A - 0}{6}\right)$	$-\left(\dfrac{V_A - V_B}{4}\right)$			
Node B	0	=			$\left(\dfrac{V_A - V_B}{4}\right)$	$+\ \ 2$	$-\left(\dfrac{V_B - V_C}{2}\right)$	
Node C	0	=					$\left(\dfrac{V_B - V_C}{2}\right)$	$-\left(\dfrac{V_C - 0}{3}\right)$

These three independent equations can be solved for the three independent unknowns

$$V_A = 22 \text{ (V)}$$
$$V_B = 16.667 \text{ (V)}$$
$$V_C = 10 \text{ (V)}$$

Thus, the voltage across the 6 (Ω) resistor is $V_A = 22$ (V), as is the voltage across the 5 (A) source. The current *down* the 6 (Ω) resistor is:

$$i_6 = \left(\frac{22}{6}\right) = 3.667 \text{ (A)}$$

The voltage across the 2 (A) source is $V_B = 16.667$ (V), and the voltage across the 3 (Ω) is $V_C = 10$ (V). The current *down* the 3 (Ω) resistor is therefore:

$$i_3 = \left(\frac{10}{3}\right) = 3.333 \text{ (A)}$$

The voltage across the 4 Ω is

$$V_A - V_B = 22 - 16.667 = 5.333 \text{ (V)}$$

and the current to the *right* through the 4 (Ω) is therefore

$$i_4 = \left(\frac{5.333}{4}\right) = 1.333 \text{ (A)}$$

The voltage across the 2 (Ω) is given by

$$V_B - V_C = 16.667 - 10 = 6.667 \text{ (V)}$$

and so the current to the *right* through the 2 (Ω) is

$$i_2 = \left(\frac{6.667}{2}\right) = 3.333 \text{ (A)}$$

As it turns out, our assumed directions for current were all correct. The full diagram is shown in Figure 13-7b. The voltage V_y is seen to be 16.667 V, and the current x is 3.333 A.

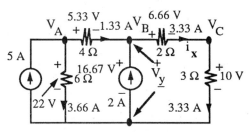

Figure 13-17b: Example 13-7
Completed Circuit

Example 13-8. Use Nodal Analysis to find all voltages and currents for the circuit shown in Figure 13-18.

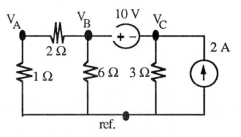

Figure 13-18: Example 13-8
Nodal Analysis

SOLUTION: The bottom node is chosen as reference, and the other nodes are labelled as shown in Figure 13-18a. We assume current directions are as indicated in the figure. At node A the picture and corresponding KCL equation are

$$-\left(\frac{(V_A - 0)}{1}\right) + \left(\frac{(V_B - V_A)}{2}\right) = 0$$

At node B one would like to say that the current going *left* through the 2 Ω resistor plus current going *down* the 6 Ω resistor plus current going *right* through the 10 volt source add to zero. However, the current through the source is independent of the voltage across it, and so an Ohm's Law statement does not apply. However, what may be written is that node B is definitely 10 (V) higher in potential than node C

$$V_B - V_C = 10$$

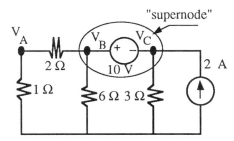

At node C one would like to write that the current *down* the 3 (Ω) resistor plus current to the *left* through the 10 volt source equals the 2 A coming into node C. Again, it is not obvious how the write the voltage source current in terms of node voltages. Consider the situation depicted in Figure 13-18a.

Figure 13-18a: Example 13-8
Super node

Clearly, the sum of the currents leaving this "supernode" must equal zero, and so we can write that current going *left* through the 2 Ω plus current going *down* the 6 Ω plus current *down* the 3 Ω equals the 2 (A) coming in from the source.

$$-\left(\frac{(V_B - V_A)}{2}\right) - \left(\frac{(V_B - 0)}{6}\right) - \left(\frac{(V_C - 0)}{3}\right) + 2 = 0$$

Table 13-6: Data for Example 13-8 (Nodal Analysis)

	0	=			$\sum i_{in/out}$		
	(C/s)		(A)	(A)	(A)	(A)	(A)
Node A	0	=	$-\left(\dfrac{V_A - 0}{1}\right)$	$+ \left(\dfrac{V_B - V_A}{2}\right)$			
Node B–C	0	=		$- \left(\dfrac{V_B - V_A}{2}\right)$	$- \left(\dfrac{V_B - 0}{6}\right)$	$- \left(\dfrac{V_C - 0}{3}\right)$	$+ \quad 2$

Table 13-6 shows the equations for node A and supernode B-C. A third equation is that for voltage across the independent source, $V_B - V_C = 10$. These three independent equations can be solved for V_A, V_B, and V_C

$$V_A = 2.133 \text{ (V)} \qquad V_B = 6.4 \text{ (V)} \qquad V_C = -3.6 \text{ (V)}$$

All voltages and currents are displayed in the diagram below in Figure 13-18b. Note that there is 1.2 amperes of current going up the 3 Ω resistor since $V_C < 0$. To recap, when a voltage source is encountered using the nodal analysis method, first write down the relation that is known to exist between the two nodes it is connected to, and then obtain a second equation by forming a "supernode" and summing all currents leaving the supernode.

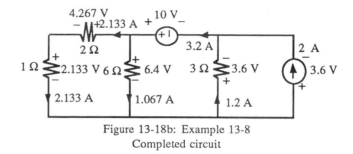

Figure 13-18b: Example 13-8
Completed circuit

13.6 MESH ANALYSIS

Mesh analysis involves the writing of a minimal set of KVL equations, but doing so in terms of "mesh currents." The steps for mesh analysis are

1. Find all the meshes. A mesh is a loop that does not contain any other loop. Each "window pane" of the circuit comprises a mesh as shown in Figure 13-19.

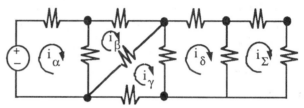

Figure 13-19: Mesh analysis

2. For each mesh, define a *mesh current* as a current that exists in the perimeter of the mesh. Note that the mesh currents automatically satisfy KCL since at any given node any mesh current that touches it goes into *and* out of the node.

3. For each mesh write a Kirchhoff voltage law equation, but do so in terms of the mesh currents and resistor values. That is, when referring to the voltage V in Figure 13-20, note that:

$$V = Ri$$
$$= R(i_\alpha - i_\beta)$$

where i_α and i_β are mesh currents, and so $i_\alpha - i_\beta$ describes the current *down* the branch as shown in Figure 13-20.

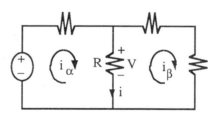

Figure 13-20: Mesh currents

4. Solve the simultaneous equations for the mesh currents

5. Each branch current can then be calculated as either a mesh current or the difference between two mesh currents, and Ohm's Law can be applied to find each branch voltage.

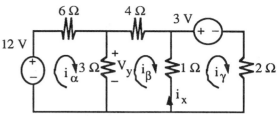

Example 13-9. Use mesh analysis to find the current i_x and the voltage V_y as shown in Figure 13-21.

Figure 13-21: Example 13-9
Mesh analysis

SOLUTION: The three mesh currents are labeled. For the first mesh, one can write that the voltage across the 6 (Ω), plus voltage across the 3 (Ω), minus the 12 (V) of the source, must add to zero. That is: six times the current going to the *right* through the 6 (Ω), plus three times the current going *down* the 3 (Ω), minus 12 volts, must equal zero.

For the second mesh we can state that four times the current going *right* through the 4 (Ω), plus one times the current going *down* the 1 (Ω), *plus* three times the current going *up* the 3 (Ω), adds to zero.

Table 13-7: Data for Example 13-9 (Mesh Analysis)

			$\sum V$					=	0
(V)	(V)	(V)	(V)	(V)	(V)	(V)		(V)
Mesh α	12	$-$	$6i_\alpha$	$-$	$3(i_\alpha - i_\beta)$					=	0
Mesh β			$-$	$3(i_\beta - i_\alpha)$	$-$	$4i_\beta$	$-$	$1(i_\beta - i_\gamma)$			=	0
Mesh γ						$-$	$1(i_\gamma - i_\beta)$	$-$	$2i_\gamma$	$-$	3	=	0

The three mesh equations are given in Table 13-7. Note that each resistor results in a voltage drop for the assumed direction of current flow because resistors are users of power. The voltage sources are positive or negative depending on the polarity indicated.

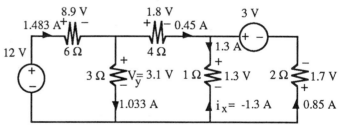

Figure 13-21a: Example 13-9
Completed circuit element

The simultaneous solution of the three equations gives:

$$i_\alpha = 1.483 \text{ A} \qquad i_\beta = 0.45 \text{ A} \qquad i_\gamma = -0.85 \text{ A}$$

The full diagram is constructed in Figure 13-21a. The voltage labeled V_y has value

$$V_y = 3(i_\alpha - i_\beta)$$
$$= 3.1 \text{ (A)}$$

nd the current labeled i_x is

$$i_x = (i_\gamma - i_\beta)$$
$$= -1.3 \ (A)$$

neaning that there is 1.3 A downward through the resistor.

Example 13-10. Perform mesh analysis on the cir-
uit shown in Figure 13-22.

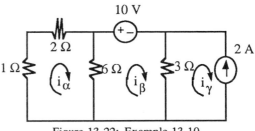

Figure 13-22: Example 13-10
Mesh Analysis

SOLUTION: There are three meshes and the mesh currents are defined in the diagram. Around the first mesh we can write the
quations as given in Table 13-8. For the second mesh, we have chosen to think of the current *down* the 3 (Ω) and current *up* the 6 (Ω).
For the third mesh the voltage across the current source is independent of the current through it and so cannot be written in terms of mesh
urrents. However, it is definitely true that the current i_γ as named has value -2 (A). Therefore a third equation is

$$i_\gamma = -2 \ (A)$$

Table 13-8: Data for Example 13-10 (Mesh Analysis)

		$\sum V$			=	0
(V)	(V)	(V)	(V)	(V)		(V)
Mesh α	$-1i_\alpha$	$- \quad 2i_\alpha$	$- \quad 6(i_\alpha - i_\beta)$			=	0
Mesh β			$- \quad 6(i_\beta - i_\alpha)$	$- \quad 10$	$- \quad 3(i_\beta - i_\gamma)$	=	0

At this point we can solve for the mesh currents (we have three equations and three unknowns) as

$$i_\alpha = -2.\,133 \ (A) \qquad i_\beta = -3.\,2 \ (A) \qquad i_\gamma = -2 \ (A)$$

All branch currents and branch voltages can now be calculated. Note that the current *down* the 6 (Ω) is

$$(i_\alpha - i_\beta) = \left(-2.\,133 - (-3.\,2) \right) = 1.\,067 \ (A)$$

nd the curent *down* the 3 (Ω) is

$$(i_\beta - i_\gamma) = \left((-3.\,2) - (-2) \right) = -1.\,2 \ (A)$$

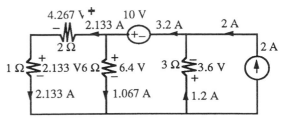

That is, a current of 1.2 A is going *up* that branch. The complete diagram is shown in Figure 13-22a.

Figure 13-22a: Example 13-10
Completed circuit

Example 13-11.   Write the set of mesh analysis equations for the circuit shown in Figure 13-23.

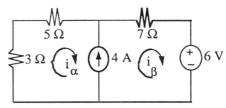

Figure 13-23: Example 13-11
Mesh Analysis

SOLUTION: The two mesh currents are described. However, due to the presence of the current source in the first mesh a KVL equation for that mesh is not easily written in terms of the mesh currents. Instead we write that

$$(i_\beta - i_\alpha) = 4$$

At this point we can go around the current source, forming an equation out of two meshes as

$$-5i_\alpha - 7i_\beta - 6 - 3i_\alpha = 0$$

These two equations can then be solved to obtain $i_\alpha = -2.27$ and $i_\beta = 1.73$.

Thus, when there are current sources in mesh analysis, write down the relation that describes the current through the source, and then write a KVL equation around the outside of the combination of the two meshes.

13.7 THEVENIN EQUIVALENT CIRCUIT

Suppose an electrical circuit can be broken into two parts as depicted in Figure 13-24. In addition, suppose that we are concerned with the effect of Circuit A on Circuit B. That is, we are concerned with the pair of terminals ab that connect Circuit A and Circuit B.

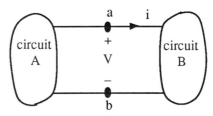

Figure 13-24: Thevenin
equivalent circuit

Thevenin Theorem — As far as Circuit B is concerned, no matter how complicated Circuit A really is, it behaves as if it were the simple circuit consisting of a voltage source in series with a resistor.

The simple circuit is the Thevenin Equivalent Circuit for Circuit A, and the value of the voltage source and resistor can be determined by performing two tests on Circuit A. The Thevenin Equivalent Circuit is shown in Figure 13-25.

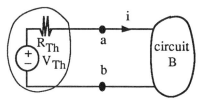

Figure 13-25: Thevenin equivalent

1. The value of the voltage source, V_{Th}, can be found by measuring the *open circuit voltage* V_{oc} of Circuit A. That is: find the voltage across terminal ab when no current flows out of Circuit A.

2. The value of the resistance, R_{Th}, can be found by looking at the total resistance seen at terminals ab, *when all sources in Circuit A are turned off*. That is all voltage sources are made *short circuits*, and all current sources are made *open circuits*.

Example 13-12. Find the simplest equivalent (Thevenin equivalent) circuit as far as the resistor R_L is concerned as shown in Figure 13-26.

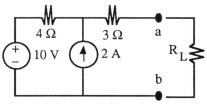

Figure 13-26: Example 13-12 Thevenin Equivalent

SOLUTION: First we must calculate the open circuit voltage V_{oc} as shown in Figure 13-26a. Since $i_3 = 0$, we have that $V_3 = 0$ and therefore $V_{oc} = V_{CS}$. Around the first loop, the voltage across the 4 (Ω), plus V_{CS}, must equal 10 volts. However, since $i_3 = 0$, there is 2 (A) going to the *left* through the 4 (Ω). Therefore

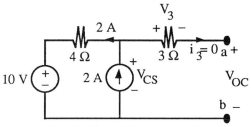

Figure 13-26a: Example 13-12 Open circuit

$$-8 + V_{CS} - 10 = 0$$

or

$$V_{CS} = V_{oc} = 18 \ (V)$$

and thus the Thevenin equivalent circuit has a voltage source of 18 volts.

Next, to find the Thevenin resistance, turn off all sources and find total resistance seen at ab in Figure 13-26b. Thus

$$R_{Th} = 3 + 4 = 7 \ (\Omega)$$

The equivalent circuit is given in Figure 13-26c.

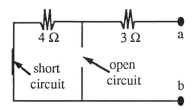

Figure 13-26b: Example 13-12
Thevenin Resistance

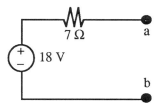

Figure 13-26c: Example 13-12
Equivalent circuit

13.7.1 Norton Form

Note that a source transformation applied to the Thevenin equivalent circuit leads to another form of the simplest equivalent circuit. This form, involving a current source in parallel with a resistance is called the *Norton* form as shown in Figure 13-27.

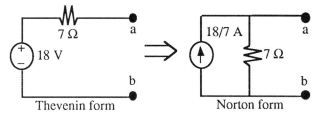

Figure 13-27: Norton form

13.8 MAXIMUM POWER TRANSFER

The problem of maximum power transfer can be described with reference to Figure 13-28. The problem is to determine the value of R_L so that the largest possible number of watts that *can* be delivered to R_L from Circuit A *is* delivered to R_L. Immediately we can simplify the problem by application of Thevenin's Theorem to Circuit A. That is, the problem now can be cast as "for the circuit shown below what value of R_L will get maximum power?"

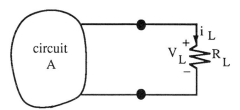

Figure 13-28: Maximum
Power Transfer

This is displayed in Figure 13-29. This is now a one-loop circuit. The power dissipated in R_L is

$$\dot{F}_L = i_L^2 R_L$$

and the current is

$$i_L = \left(\frac{V_{Th}}{R_L + R_{Th}}\right)$$

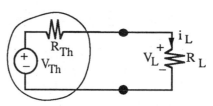

Figure 13-29: Thevenin
Equivalent

Thus

$$\dot{F}_L = \left(\frac{V_{Th}}{R_L + R_{Th}}\right)^2 R_L$$

$$= \left(\frac{V_{Th}^2 R_L}{(R_L + R_{Th})^2}\right)$$

Since V_{Th} and R_{Th} are fixed, $\dot{F}_L$ is a function of R_L, and this function is to be maximized by choice of R_L. To find this maximum, we can differentiate the expression for $\dot{F}_L$ with respect to R_L and set it equal to zero, and solve for R_L.

$$\left(\frac{d\dot{F}_L}{dR_L}\right) = V_{Th}^2 \left(\frac{(R_L + R_{Th})^2 - 2R_L(R_L + R_{Th})}{(R_L + R_{Th})^4}\right)$$

$$= 0$$

This expression is equal zero when

$$(R_L + R_{Th})^2 = 2R_L(R_L + R_{Th})$$

or

$$R_L = R_{Th}$$

Thus, maximum power transfer occurs when R_L is chosen to be equal to the Thevenin equivalent resistance of the circuit connected to R_L. To find this maximum power, substitute $R_L = R_{Th}$ in the defining equation for power

$$\dot{F}_L = \left(\frac{V_{Th}}{(R_L + R_{Th})}\right)^2 R_L$$

$$= \frac{V_{Th}^2 R_{Th}}{(R_{Th} + R_{Th})^2}$$

$$= \left(\frac{V_{Th}^2}{4R_{Th}}\right)$$

Example 13-13. Find the value of R_L so that it obtains the maximum power possible. Also find the value of this maximum power for the circuit shown in Figure 13-30.

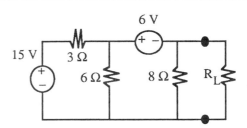

Figure 13-30: Example 13-13
Maximum Power

SOLUTION: First find the Thevenin equivalent circuit. To do this we find the open circuit voltage as shown in Figure 13-30a. It is clear that

$$V_{oc} = 8i_\beta$$

and

$$-3i_\alpha - 6(i_\alpha - i_\beta) + 15 = 0$$

$$-6 - 8i_\beta - 6(i_\beta - i_\alpha) = 0$$

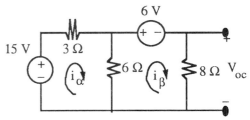

Figure 13-30a: Example 13-13
Open Circuit

Solving gives

$$i_\alpha = \left(\frac{29}{15}\right) = 1.933 \text{ (A)}$$

$$i_\beta = 0.4 \text{ (A)}$$

. Thus

$$V_{oc} = (8)(0.4) = 3.2 \text{ (V)}$$

$$= V_{Th}$$

Next, find R_{Th} by turning off the sources and calculating R_{eq} as shown in Figure 13-30b. Obtaining that

$$R_{Th} = R_{eq}$$

$$= 1.6 \text{ } (\Omega)$$

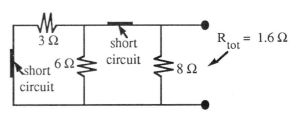

Figure 13-30b: Example 13-13
Thevenin Equivalent Resistance

Thus, the problem reduces to the circuit shown in Figure 13-30c.

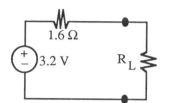

Figure 13-30c: Example 13-13
Circuit Reduction

It is clear that R_L should be chosen as $R_L = 1.6$ (Ω), and that the resulting circuit takes on the values shown in Figure 13-30d. The power delivered to R_L is

$$P_L = i_L^2 R_L = 1^2 (1.6) = 1.6 \text{ W}$$

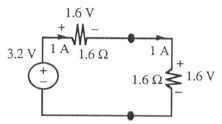

Figure 13-30d: Example 13-13
Completed Circuit

13.9 RESISTIVE CIRCUIT PROBLEM SOLVING TOOLS

1) **The Laws.** The laws which are of primary use for electrical circuits are the conservation of charge and the conservation of energy (and/or an accounting of electrical energy) in the form of Kirchhoff's current and voltage laws.

2) **Circuit Element Definitions.** A resistor can be defined as a linear device which converts electrical energy to internal energy, energy which ultimately is dissipated from the circuit as heat. Mathematically, it is defined according to equation (7-61). In addition to the basic definitions of circuit elements, certain relationships also are important for simplifying segments of circuits, equivalent resistances for resistors in series or parallel, for example.

3) **Circuit Description.** Of course, the description of the circuit of interest, what its elements are and how they are positioned relative to each other, is essential. Related concepts are the voltage divider circuit (resistors in series) and the current divider circuit (resistors in parallel), and source transformations and Thevenin equivalent circuits (including the Norton form) and the concept of maximum power transfer.

4) **Mathematics.** As always, mathematics is a keystone tool for problem solving. Included in this chapter are all the tools of solving simultaneous linear equations, including the superposition principle.

Strategies of circuit analysis include (of course) using the various tools, in order as appropriate, and selecting perhaps several systems for analysis. The use of mesh analysis versus nodal analysis is simply an issue of which part or parts of the circuit are selected as the system and is analogous to the statics structure analysis techniques of section versus joint analysis.

13.10 REVIEW

1) Circuit analysis is the problem of calculating the voltages across and the currents through each of the elements in an electric circuit.

2) A resistor is a circuit element which has the property that the voltage across it is proportional to the current through it. That is, resistors obey Ohm's law, $V = Ri$.

3) Voltage sources are circuit elements which maintain a prescribed voltage across their terminals, independent of the current through them, while current sources maintain a prescribed current, independent of the voltage across them.

4) The power used or supplied by a circuit element is the product of the voltage across it and the current through it.

5) Kirchhoff's current law (steady-state charge flow rate equation) and Kirchhoff's voltage law (steady-state accounting of electrical energy) are the basic circuit analysis tools. By writing KCL equations for each connection point and KVL equations for each loop, one obtains a set of constraint equations that must hold. When combined with the properties of the circuit elements (Ohm's Law, properties of sources, etc.), these equations can be solved for the voltages and currents of each element.

6) Since resistors in series each have the same current, several resistors in series produce the same effect as one resistor whose value is the sum of the resistances.

7) Total voltage is divided proportionately across resistors in series, the largest resistor getting the largest percentage of the total volts.

8) Resistors in parallel have the same voltage across them and produce the same effect as one resistor of appropriate value. Since the total current due to several resistors in parallel is the sum of the currents in each resistor, the equivalent resistance is always smaller than any of the parts. Its value is calculated as the reciprocal of the sum of the reciprocals of the resistors in parallel.

9) For resistors in parallel the largest percentage of total current goes through the smallest resistance.

10) A current source (i_s) in parallel with a resistance (R) has the same effect on the rest of the circuit as a properly chosen voltage source ($V_s = Ri_s$) in series with the resistance (R). This source transformation property may be useful in simplifying circuits for analysis.

11) Electrical circuits made up of voltage sources, current sources, resistors, capacitors, and inductors are linear systems. The superposition principle for linear systems states that the response due to several inputs can be calculated by finding the responses due to each input acting alone, and then summing these responses. Setting a voltage source to zero means replacing it with a short circuit, while setting a current source to zero means replacing it with an open circuit.

12) Nodal analysis builds KVL and Ohm's law into the writing of KCL equations so that a smaller set of equations is obtained. This set can be solved to find the node voltages, and from these each voltage and current can be obtained.

13) Mesh analysis builds KCL and Ohm's law into the writing of KVL equations so that a smaller set of equations is obtained. From these we can solve for the mesh currents, and obtain all voltages and currents.

14) The Thevenin Theorem states that the effect of any two-terminal network is equivalent to that of a simple network consisting of a voltage source in series with a resistor. Thus, in analyzing the voltages and currents of a television plugged into a wall outlet, it is not necessary to include all the detail of the power distribution network. Instead we model the outlet by its Thevenin equivalent, which would be a voltage source of approximately $110\sqrt{2}\cos{(2\pi 60t)}$ volts in series with a very small resistance.

15) Application of the source transformation property to the Thevenin equivalent circuit leads to the Norton form.

16) Maximum power is delivered by a circuit to a resistor when the value of the resistor is chosen so that it matches the value of the Thevenin equivalent resistance of the circuit it is connected to. This concept of "impedance matching" is why, for example, audio speakers are usually listed as 8 (Ω), and amplifiers are designed so that their Thevenin equivalent resistance is 8 (Ω).

13.11 NOTATION

The following notation was used in this chapter.

	Scalar Variables	Dimensions
$\dot{F}$	rate of electrical energy dissipation i	[energy / time]
i	current	[charge / time]
q	charge	[charge]
R	resistance	[energy time / charge^2]
t	time	[time]
V	voltage	[energy / charge]
$\dot{W}$	rate of work	[energy / time]

	Subscripts
ele	electrical
in	input or entering
out	output or leaving
eq	equivalent
Th	Thevenin equivalent

13.12 SUGGESTED READING

Fitzgerald, A., D. Higginbotham, and A. Grabel, *Basic Electrical Engineering*, 5th edition, McGraw-Hill, New York, 1981
Halliday, D. and R. Resnick, *Physics for Students of Science and Engineering, Part II*, 2nd edition, John Wiley & Sons, New York, 1962
Nilsson, J., *Electrical Circuits*, 3rd edition, Addison-Wesley, Reading, Massachussetts, 1989

Paul, C., S. Nasar, and L. Unnewehr, *Introduction to Electrical Engineering*, McGraw-Hill, New York, 1986
Roadstrum, W. H., and D. H. Wolaver, *Electrical Engineering for All Engineers*, Harper and Row, New York, 1987

QUESTIONS

1) State under what conditions a circuit element is a consumer of power, and under what conditions it is a supplier of power.

2) Why must resistors always be consumers of power?

3) Can you think of a situation wherein an automobile battery is a supplier of power?

4) Can you think of a situation wherein an automobile battery is a consumer of power?

5) When the engine is being cranked, the voltage across a "12 volt" car battery will often be as small as 8 volts. Why?

6) Conversely, once the car is running the terminal voltage of the battery may measure 14 volts. Why?

7) A 100 (Ω), 300 (Ω), and 500 (Ω) resistor are in series. Which one gets the largest current? Which one gets the largest voltage?

8) A 100 (Ω), 300 (Ω), and 500 (Ω) resistor are in parallel. Which one gets the largest current? Which one gets the largest voltage?

9) The superposition principle states that one can calculate the voltages and currents due to several sources as the sum of the responses due to each source acting alone. Can the power used by an element due to several sources be calculated by adding the powers due to each source acting alone?

10) When is nodal analysis probably preferable to mesh analysis? When is mesh analysis probably preferable to nodal analysis?

11) What is the difference between the heating element in a 1000 watt hair dryer and that in an 1800 watt hair dryer?

12) In some apartments, why do the lights dim momentarily when the refrigerator motor comes on? Hint: Thevenin equivalent circuit and voltage divider circuit.

13) What would you expect to happen if an 8 (Ω) speaker is connected to a terminal on the amplifier that says 4 (Ω)? How about if it is connected to a terminal that says 16 (Ω)?

SCALES

1) A battery provides 6 V on open circuit and it provides 5.4 V when delivering 8 A. What is the internal resistance of the battery?

2) A short circuit across a non-ideal current source draws 20 A. If a 10 Ohm resistor across the source draws 18 A, what is the internal resistance of the source?

3) A series circuit consists of a 240 V source and 12 Ohm, 16 Ohm, and 20 Ohm resistors. Find the voltages across each of the resistors.

4) A resistor R is in series with an 8 Ohm resistor. It absorbs 100 Watts when the two are connected across a 60 V line. Find the resistance R.

5) Resistors R_1, R_2, and R_3 are in series with a 100 V source. The measured voltage drop across R_1 and R_2 is 50 V, and across R_2 and R_3 it is 80 V. The total resistance is 50 Ohms. Find the values R_1, R_2, and R_3.

6) A series circuit consists of a 120 V source and two resistors. One resistor is 30 Ohms while the other has 45 V across it. Find the value of the second resistor.

7) Five resistors of values 4, 5, 6, 7, and 8 Ohms, respectively are connected in series with a 240 V source. Find the voltage across the 8 Ohm resistor.

8) A resistor is to be connected in parallel with a 30 Ohm resistor in order to produce a total resistance of 12 Ohms. Find the value of resistor needed.

9) A current of 120 A flows into the parallel combination of a 60 Ohm and 90 Ohm resistor. Find the current in the 90 Ohm.

10) Resistors of values 5, 6, 12, and 20 Ohms are in parallel. Find the voltage across the combination if 40 A flows into the circuit.

11) A voltage source is connected in series with 4, 5, and 6 Ohm resistors. The total current is 9 A. Find the source voltage.

12) An automobile battery is modelled as a 12 V ideal voltage source in series with a 0.3 Ohm internal resistance. It is to be charged from a 15 V source. The charging current is to be limited to 2 A. What series resistor will limit the charging current to no more than 2 A?

13) Two resistors are in series across a 240 V line. One is 300 Ohm and absorbs 27 W. What is the value of the other resistor?

14) A 12 V source has an internal resistance of 0.3 Ohms. What current source and resistance combination is equivalent to this?

15) A 3 A current source in parallel with a 9 kOhm resistor is equivalent to what circuit involving a resistor and voltage source?

16) A car battery has an open circuit voltage of 12 V. The terminal voltage drops to 9.6 V when the battery supplies 80 A to a starter motor. Find an equivalent circuit for the battery.

17) A current source of 4 A is connected in parallel with resistors of values 2, 6, and 30 Ohms. What resistor connected in parallel with these will receive maximum power possible, and what is the power?

PROBLEMS

1) Determine the current i_6 in the figure shown.

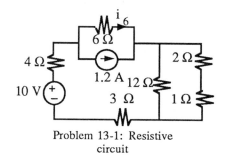

Problem 13-1: Resistive circuit

2) Find the power used or supplied by each of the elements in the circuit shown.

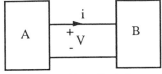

Problem 13-2: Resistive circuit

3) The circuits are represented as A and B, and connected as shown. The voltage and current in the interconnection is labeled in the figure.

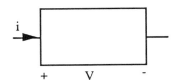

Problem 13-3: Resistive circuit

For each case below, find the power and determine if A is supplying B or vice-versa

   a) $i = 15$ A, $V = 20$ V
   b) $i = 4$ A, $V = -50$ V
   c) $i = -5$ A, $V = 100$ V
   d) $i = -16$ A, $V = -25$ V

4) For the given circuit in the figure where $V(t) = 170\cos(100\pi t)$ V and $i(t) = 7\cos(100\pi t + 40^o)$ A,

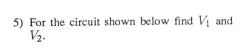

Problem 13-4: Resistive circuit

   a) Find the power as a function of time and sketch the graph.
   b) Find the maximum power that is ever used by the element.
   c) Find the maximum power ever supplied by the element.

HINT: $\cos x \cos y = \dfrac{1}{2}\cos(x - y) + \dfrac{1}{2}\cos(x + y)$

5) For the circuit shown below find V_1 and V_2.

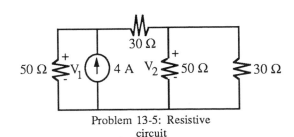

Problem 13-5: Resistive circuit

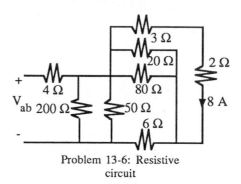

6) For the given circuit, if 8 A is down the
2 (Ω) resistor, what is V_{ab}

Problem 13-6: Resistive
circuit

7) For the given circuit where R_1 = 1000 (Ω) and
R_L = 500 (Ω), find the power dissipated in each
of the resistors.

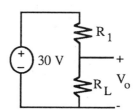

Problem 13-7: Resistive
circuit

a) Find R_1 and R_L so the V_o is the same but $\dot{F}_1$ and $\dot{F}_L$ are both less than 0.25 (W).

8) For the same circuit used in problem 7, R_L = ∞ (Ω) and V_o = 6 (V). If the smallest R_L that will ever be connected is 3000
(Ω), but for any resistor R_L that is greater the 3000 (Ω) to infinite resistance we must guarantee the $V_o \geq 5$ (V).

a) Find R_1 and R_L so that V_o = 6 (V) for R_L = ∞ (Ω) and $V_o \geq 6$ (V) for any $R_L \geq 3000$ (Ω)

b) What worst case (maximum) power would be dissipated when R_1 for your design in part (a).

CHAPTER FOURTEEN

INDUCTANCE AND CAPACITANCE

14.1 INTRODUCTION

In this chapter the techniques for analysis of resistive circuits are applied to circuits which contain inductance and capacitance. It will be seen that the equations now become sets of differential-integro equations. For the special case of circuits containing only one energy storage element (inductor or capacitor), analysis leads to a first order ordinary differential equation which may be solved in a straightforward mannner. We examine the behavior of such circuits.

While, in general, more complicated examples cannot be easily solved, for the special case wherein all the sources are sinusoidal, there is a generalization of the concept of resistance that will allow the methods developed for resistive circuits to be applied to AC problems, and the steady-state solution found. The properties of circuits which contain inductance and capacitance will be seen to depend on the frequency of the sources, and so such circuits can be used as frequency selective *filters*.

14.2 REVIEW OF THE BASIC EQUATIONS

The analysis of dynamic circuits requires the understanding of the concepts of conservation of charge, through Kirchhoff's current law, and electrical energy accounting in the form of Kirchhoff's voltage law. The conservation of net charge for any circuit says:

$$\left(\frac{dq_{\text{sys}}}{dt}\right) = \sum i_{\text{in/out}} \qquad (4-9)$$

And Kirchhoff's voltage law says that the summation of the voltage rises and drops around a closed loop must sum to zero:

$$\sum_{\text{Loop}} V = 0 \tag{7 - 60}$$

The direct-form accounting of electrical energy equation states that energy added to the circuit due to current, together with conversions between electrical and other forms, contribute to changes in electrical energy stored in the circuit:

$$\left(\frac{d(E_E)_{\text{sys}}}{dt} \right) = \sum (iV)_{\text{in/out}} + \sum \dot{W}_{\text{ele}} - \sum \dot{F} \tag{7 - 56}$$

Note that the electrical energy equation is not independent of the voltage and current laws but provides a way to check that the solution is correct. The electrical energy is stored in the system in the form of an electric field in a capacitor and a magnetic field in an inductor. A resistor cannot store electrical energy; in fact, it converts it to thermal energy.

The conservation of mass and momentum give trivial results when applied to circuit problems. However, the second law of thermodynamics places contraints on the direction of conversions between electrical, thermal, mechanical and chemical energy. Losses must always be greater than zero:

$$\dot{F} \geq 0$$

14.2.1 Circuit Elements

Resistor. The defining relationship for a resistor is given by Ohm's Law. The voltage across a resistor is proportional to the current flowing through the resistor and the constant of proportionality is the resistance:

$$\boxed{V = iR \qquad i = \left(\frac{V}{R} \right)} \tag{7 - 61}$$

Chapter 13 gives a detailed explanation of resistors in circuits and methods of solving for the unknown voltages and currents. Figure 14-1 shows the schematic of a resistor.

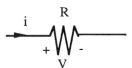

Figure 14-1: Resistor

Inductor. A coil of wire produces a directed magnetic field whose strength is proportional to the size of the current in the coil. Such a coil of wire is called an *inductor* and is shown in Figure 14-2.

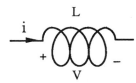

Figure 14-2: Inductor

Faraday's Law states that a changing magnetic field induces a voltage proportional to the time rate of change of the magnetic flux. Since the strength of the magnetic field is proportional to the current, if the current changes with time, then the magnetic field (and magnetic flux) will be changing with time. Thus a voltage is induced across the ends of an inductor whenever the current through it is *changing*. The size of this voltage is given by (Chapter 7)

$$V = L\left(\frac{di}{dt}\right)$$

(7 − 64)

where L is called the inductance of the coil, and the unit of inductance is the Henry. Note that, in contrast to the resistor (whose voltage is proportional to $i(t)$), the inductor has its voltage proportional to $\left(\frac{di}{dt}\right)$. In particular, if the current is constant, then the voltage is zero.

If both sides of the voltage-current relation for the inductor are integrated over some interval of time, we obtain that:

$$\int_{t_o}^{t} \frac{V}{L}\,dt = \int_{i_o}^{i} di$$

and integrating yields:

$$i = i_o + \left(\frac{1}{L}\right)\int_{t_o}^{t} V\,dt$$

(14 − 1)

where i_o is the initial current in the inductor at time t_o.

The time rate of change of electrical energy storage in the magnetic field of an inductor is (Chapter 7)

Inductor: $$\left(\frac{dE_L}{dt}\right) = Li\left(\frac{di}{dt}\right)$$

(7 − 66)

and the amount of energy stored (E_L) when the current is i is seen to be

$$E_L = \frac{1}{2}Li^2$$

Note that the amount of energy does not depend on whether the current is positive or negative.

Capacitor: A capacitor is formed by a pair of charged parallel plates separated by some distance. One plate has charge $+q$ and the other $-q$. For such an element, the *voltage across* it is directly proportional to the amount of *positive charge on one of the plates*

$$q_+ = CV$$

(7 − 67)

where C is the capacitance, and the unit of capacitance is the Farad. A schematic for a capacitor is shown in Figure 14-3.

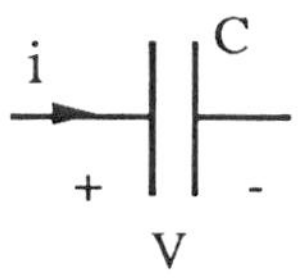

Figure 14-3: Capacitor

By differentiating both sides of the relation above with respect to time we obtain

$$\left(\frac{dq_+}{dt}\right) = C\left(\frac{dV}{dt}\right)$$

which says that if the amount of positive charge on one of the plates changes with time, then the voltage between the plates will also change. The relation between the voltage across (V) and the current (i) "through" a capacitor is (see equations 7-67 through 7-73)

$$i = C\left(\frac{dV}{dt}\right)$$

Note that although no charges jump the gap between the plates, charges do move onto and off of the plates, and this charge movement constitutes the current "through" the capacitor. Note also that if the voltage is constant, then the current is zero for a capacitor.

By integrating both sides of the voltage-to-current relation with respect to time over some time interval we obtain

$$V = V_o + \left(\frac{1}{C}\right)\int_{t_o}^{t} i\,dt \tag{14 - 2}$$

where V_o is the voltage on the capacitor at time t_o.

The rate of storage of energy in the electric field of a capacitor is (Chapter 7)

Capacitor:
$$\left(\frac{dE_C}{dt}\right) = \left(\frac{q}{C}\right)\left(\frac{dq}{dt}\right) = CV\left(\frac{dV}{dt}\right) \tag{7 - 72}$$

and the energy stored in the electric field when a voltage V exists between the plates is

$$E_C = \frac{1}{2}CV^2 = \frac{1}{2}\frac{q_+^2}{C}$$

Note that the *amount* of energy stored in the field does not depend on which of the plates is positive and which is negative.

The defined characteristic equations for all of the circuit elements are shown in Table 14-1:

Table 14-1: Circuit Element Equations

Element	Voltage*	Current†	Electrical Energy Storage
Resistor	$V = iR$	$i = \left(\dfrac{V}{R}\right)$	0
Inductor	$V = L\left(\dfrac{di}{dt}\right)$	$i = i_o + \left(\dfrac{1}{L}\right)\int_{t_o}^{t} V\,dt$	$\dfrac{1}{2}Li^2$
Capacitor	$V = V_o + \left(\dfrac{1}{C}\right)\int_{t_o}^{t} i\,dt$	$i = C\left(\dfrac{dV}{dt}\right)$	$\dfrac{1}{2}CV^2$

* The voltage across the circuit element: $V_{in} - V_{out} = V$
† The current through the circuit element: $i_{in} = i_{out} = i$

14.2.2 Nodal and Mesh Anaysis

The ideas of nodal and mesh analysis carry over to circuits which contain inductors and capacitors as well as resistors, provided that instead of Ohm's law, we use the voltage-current relations for inductors and capacitors in writing the equations. The voltage-current relations are summarized in Table 14-1.

1) Since the voltage across an inductor depends on the time rate of *change* of the current, if the current is *constant* then the voltage is *zero*.

2) Since the current through the inductor depends on the integral (area under the graph) of the voltage function, the inductor current must be *continuous*. That is, no matter how messy the voltage function is, the current through an inductor cannot have any jumps or discontinuities.

3) Since the current through a capacitor depends on the time rate of *change* of the voltage across it, if the voltage is *constant* then the current is *zero*. That is, if the voltage is constant, then no charge is flowing onto or off the plates.

4) Since the voltage across a capacitor depends on the integral (area under the graph) of the current function, the capacitor voltage must be *continuous*. That is: no matter how messy the current function is, the voltage across a capacitor cannot have any jumps or discontinuities.

5) For circuits in which all the sources are constant (DC), then *after a long time* all voltages and currents will be constant also. Items 1 and 3 then imply that for circuits whose sources are DC, after a long time, all inductors look like short circuits (no voltage across, but possibly current through them), and all capacitors look like open circuits (no current through them, but possibly voltage across them).

6) For circuits in which all the sources are sines and cosines (i.e. sinusoidal), then since derivatives, integrals, multiplication by constants, and sums of sinusoids of the same frequency, produce sinusoids, then, *even after a long time*, every voltage and current in such a circuit will be sinusoidal (AC).

For writing nodal or mesh analysis equations the same overall procedures apply as in resistive circuits. However, the relations between voltage and current for capacitors and inductors involve derivative and integral expressions instead of Ohm's law. Nonetheless, the idea of writing the KCL equations in nodal analysis in terms of differences of node voltages, or the KVL equations of mesh analysis in terms of mesh currents and differences of mesh currents is the same for circuits which contain inductors and capacitors as well as resistors.

Example 14-1. For the circuit shown in Figure 14-4, the initial time of interest is $t_o = 0$ and the initial values of inductor currents and capacitor voltages are as labelled. Write first a set of nodal analysis equations that describe the circuit for $t > 0$, and second, a set of mesh analysis equations that describe the circuit for $t > 0$.

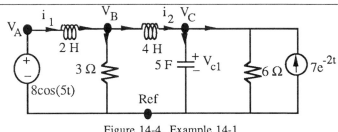

Figure 14-4. Example 14-1 circuit

The initial conditions at t equal zero are given below:

$$V_{C1}(0) = 14 \text{ V} \qquad i_{L1}(0) = 0.6 \text{ A} \qquad i_{L2}(0) = 1.8 \text{ A}$$

SOLUTION: Nodal Analysis – the node voltages are labelled, and one node chosen as reference node. At node A we can write

$$V_A - 0 = 8 \cos(5t)$$

At node B, the current going *right* through the inductor, minus current going *down* through the resistor, minus current going *right* through the other inductor must equal zero. Note that the initial value of the current going *right* through the 2 H inductor is 0.6 A, while the initial value of the current going to the *right* through the 4 H inductor is 1.8 A. The nodal equation is

$$\left[0.6 + \frac{1}{2}\int_0^t (V_A - V_B)dt\right] - \left(\frac{V_B - 0}{3}\right) - \left[1.8 + \frac{1}{4}\int_0^t (V_B - V_C)dt\right] = 0$$

Note how we speak of $(V_A - V_B)$ when talking of the current *right* through the 2 (H) inductor, and of $(V_B - V_C)$ when talking of the current going *right* through the 4 (H) inductor. At node C, the current going *right* through the 4 (H) inductor, minus current going *down* the 5 F capacitor, minus current going *down* the 6 Ω resistor, plus the current into the node from the current source must sum to zero. That is:

$$\left[1.8 + \frac{1}{4}\int_0^t (V_B - V_C)dt\right] - 5\left(\frac{d(V_C - 0)}{dt}\right) - \frac{(V_C - 0)}{6} + 7e^{-2t} = 0$$

These three equations have three unknowns in terms of *functions* of time V_A, V_B, V_C. Furthermore, they are differential-integro equations but they can be solved with some effort giving the unknown voltages as functions of time. Once these voltages are known as functions of time one could then calculate each branch voltage and current.

SOLUTION: Mesh Analysis – the circuit is redrawn and the mesh currents are shown in Figure 14-4a.

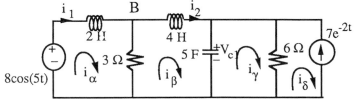

Figure 14-4a. Example 14-1 Mesh Currents

For the first mesh, the voltage across the 2 (H), plus voltage across the 3 (Ω), minus the source voltage must equal zero.

$$8\cos(5t) - 2\left(\frac{di_\alpha}{dt}\right) - 3(i_\alpha - i_\beta) = 0$$

Around the second mesh, the voltage across the 4 (H), plus voltage across the capacitor, plus voltage (from bottom to top) across the 3 Ω must equal zero.

$$-4\left(\frac{di_\beta}{dt}\right) - \left[14 + \frac{1}{5}\int_0^t (i_\beta - i_\gamma)dt\right] - 3(i_\beta - i_\alpha) = 0$$

Around the third mesh

$$-6(i_\gamma - i_\delta) - \left[-14 + \frac{1}{5}\int_0^t (i_\gamma - i_\beta)dt\right] = 0$$

where note that we talk of current going up the capacitor here, and treat the initial voltage of the capacitor as (-14) as we go from bottom to top.

The last mesh equation is

$$i_\delta = -7e^{-2t}$$

There are 4 equations and 4 unknowns i_α, i_β, i_γ, and i_δ. Each of the unknowns is expressed as a function of time. These equation are also differential-integro, and can be solved with some effort. From the solution of the mesh currents as a function of time one could obtain expressions for each voltage and current in the circuit as a function of time.

14.3 FIRST ORDER RESPONSE

The solution of high order differential equations is beyond the scope of this text, but substantial insight into the behavior of circuits containing inductance or capacitance can be obtained by looking at the special case of circuits having only one inductor or one capacitor. That is: circuits which contain *one energy storage device*. These circuits were explained in Chapter 7 and only a short review of the problem will be given here.

First, consider the case shown in Figure 14-5 with the *inductor and resistor in series*. At $t = 0$ the inductor current has a current i_0, and the switch goes down from a to b at time $t = 0$.

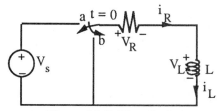

Figure 14-5. First Order
Response – Inductor

For $t > 0$ we can write for the one loop

$$V_R + V_L = 0$$

and the current flowing through the resistor is equal to the current flowing through the inductor from Kirchhoff's voltage law:

$$i_R = i_L = i$$

The voltage law can be written as:

$$Ri + L\left(\frac{di}{dt}\right) = 0$$

From an electrical energy viewpoint, this result (multiplied by the current, i) says that at any instant in time the rate of energy depletion in the inductor is equal to the rate of electrical energy dissipation in the resistor.

The solution of this equation with the initial condition $t = 0$, $i = i_o$, gives the current through the inductor and resistor as a function of time:

$$i = i_o e^{\frac{-t}{(L/R)}}$$

Note that the current decreases from its initial value to zero at a rate determined by (L/R). The voltages across the resistor and inductor are calculated from the defining relationships:

$$V_R = Ri_o e^{\frac{-t}{(L/R)}} \qquad V_L = -Ri_o e^{\frac{-t}{(L/R)}}$$

Graphs of current as a function of time and the voltages across the resistor and inductor as functions of time are shown in Figure 14-6.

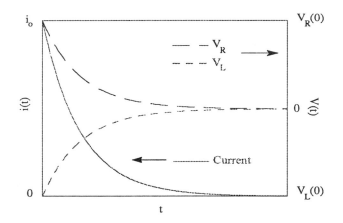

Figure 14-6: Current through and Voltage across the RL circuit elements

An accounting for energy in the circuit would show that initially the inductor stored energy in the amount $(L\,/\,2)i_o^2$, but for $t > 0$ this energy is being dissipated as heat in the resistor and, consequently, the amount of energy stored decreases with time towards zero. As time goes to infinity, all of the energy originally stored in the magnetic field of the inductor will have been dissipated as heat in the resistor.

Consider the case for the *resistor and capacitor in series* in Figure 14-7. At $t = 0$ a plate of the capacitor has a charge q_o or equivalently a voltage across it of $V_o = q_o\,/\,C$ volts. The switch moves down from a to b at $t = 0$ and we are concerned with circuit behavior for $t > 0$.

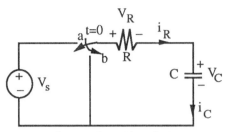

Figure 14-7: First Order
Response – Capacitance

For $t > 0$ we can write for the one loop:
$$V_R + V_C = 0$$

Since the current through the resistor is equal to the current through the capacitor from Kirchhoff's current law:
$$i_C = i_R = i$$

The voltage equation is now:
$$iR + V_C = 0$$

where the current (through the capacitor) is:
$$i = C\left(\frac{dV_C}{dt}\right)$$

So the voltage equation becomes:
$$RC\left(\frac{dV_C}{dt}\right) + V_C = 0$$

From an electrical energy viewpoint, this result (multiplied by the current) says that the rate of energy depletion in the capacitor is equal to the rate of electrical energy dissipation by the resistor.

The solution of this equation with the initial condition $t = 0$, $V_C = V_o$, gives the voltage across the capacitor as a function of time:
$$V_C = V_o e^{\frac{-t}{RC}}$$

Note that the voltage decreases from its initial value to zero at a rate determined by RC. The voltage across the resistor is

$$V_R = -V_o e^{\frac{-t}{RC}}$$

The current through the resistor and capacitor are calculated from the defining relationships.

$$i = -\left(\frac{V_o}{R}\right) e^{\frac{-t}{RC}}$$

Graphs of current as a function of time and the voltages across the resistor and capacitor as functions of time are shown in Figure 14-8. Note that the current through the capacitor is shown as negative, meaning that it actually flows from the capacitor to the resistor, left to right in Figure 14-7.

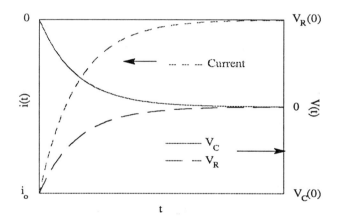

Figure 14-8: Current through and Voltage across the RC circuit elements

An accounting for energy in the circuit would show that initially the capacitor stored energy in the amount $(C/2)V_o^2$, but for $t > 0$ this energy is being dissipated as heat in the resistor and, consequently, the amount of energy stored decreases with time towards zero. As time goes to infinity, all of the energy originally stored in the electric field of the capacitor will have been dissipated as heat in the resistor.

An accounting for charge in the circuit would show that the initial positive charge on the top plate of the capacitor will have left that plate and circulated around to the bottom plate cancelling negative charge there. Note that the current i_C (positive charge going onto the top plate) is *negative*. Over any interval of time, the integral of the current describes the net change of charge on the plates during that interval.

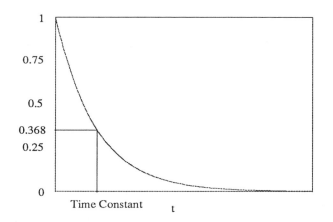

Based on the similarity of the functions for both of the circuits a parameter is defined called the time constant which describes the rate of decay of the functions as shown in Figure 14-9.

Figure 14-9: Time constant

The time constant for a first order circuit is the value of time it will take for the exponential term to equal $0.368 (= e^{-1})$.

The time constant of a circuit determines the rate of decay of the voltages and currents. Circuits with small time constants decay faster towards zero than circuits with larger time constants. Since $e^{-4} = 0.0183$ and $e^{-5} = 0.0067$, the initial values have, for all practical purposes, decayed to zero after about four or five time constants of time have passed. Note that the R-L circuit has a time constant given by L / R seconds, while the R-C circuit has a time constant RC seconds. The different decay rates are shown in Figure 14-10.

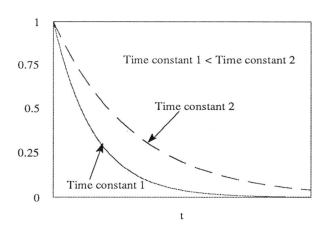

Figure 14-10: Different time constants

Other physical situations also behave similarly to these RC and RL circuits. The canon ball problem with air resistance gives that exact same mathematical problem, as well as the discharging plate and tank emptying problem. These problems also have characteristic time constants but they may be much larger in magnitude. This analogy between the electrical circuits and other physical phenoma is very powerful and arises because of the conservation laws. We can use the behavior of certain devices such as circuits to model or predict the behavior of other physical phenonoma such as throwing a ball through the air.

14.3.1 First Order Circuits with Sources

We will consider the two cases illustrated in Figure 14-11. One circuit is a resistor and inductor in series with a constant voltage source and the other is a resistor and capacitor in series with a constant voltage source. Assume that in the first circuit there is no initial current in the inductor. That is $i_o = 0$ meaning that the inductor has no stored energy. Likewise for the second circuit assume the capacitor is initially uncharged, or $V_O = 0$, meaning that there is no initital stored energy.

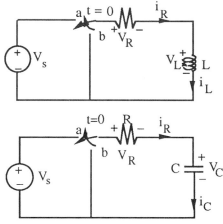

Figure 14-11: Inductive and Capacitive Circuits

The discussions for both circuits will be done together to further emphasize the similarity of the equations developed and the single math problem that is solved. At $t = 0$ the switches come up, connecting the constant voltage sources to the rest of the circuit. For $t > 0$ we can write Kirchhoff's voltage laws for each of the circuits.

$$V_s - V_R - V_L = 0 \qquad\qquad V_s - V_R - V_C = 0$$

Kirchhoff's current law for each of the circuits says:

$$i_s = i_R = i_L = i \qquad\qquad i_s = i_R = i_C = i$$

From the defining relationships for the circuit elements, the voltage laws become:

$$Ri + L\left(\frac{di}{dt}\right) = V_s \qquad\qquad RC\left(\frac{dV_C}{dt}\right) + V_C = V_s$$

In terms of electrical energy accounting, the electrical energy supplied by the source is either stored in the inductor (or capacitor) or dissipated in the resistor.

These equations can be separated and integrated:

$$\frac{di}{(i - (V_s / R))} = -\frac{R}{L}dt \qquad\qquad \frac{dV_C}{(V_C - V_s)} = -\frac{1}{RC}dt$$

Integrating the equations with the prescribed initial conditions and rearranging gives:

$$i = \left(\frac{V_s}{R}\right)\left(1 - e^{\frac{-t}{(L/R)}}\right) \qquad\qquad V_C = V_s(1 - e^{\frac{-t}{RC}})$$

Table 14-2: First Order Circuits with Sources

Element	Voltage	Current
RL-Circuit		
Resistor	$V_R = V_s(1 - e^{\frac{-t}{(L/R)}})$	$i_R = \left(\frac{V_s}{R}\right)\left(1 - e^{\frac{-t}{(L/R)}}\right)$
Inductor	$V_L = V_s e^{\frac{-t}{(L/R)}}$	$i_L = \left(\frac{V_s}{R}\right)\left(1 - e^{\frac{-t}{(L/R)}}\right)$
RC-Circuit		
Resistor	$V_R = V_s e^{\frac{-t}{RC}}$	$i_R = \left(\frac{V_s}{R}\right) e^{\frac{-t}{RC}}$
Capacitor	$V_C = V_s(1 - e^{\frac{-t}{RC}})$	$i_C = \left(\frac{V_s}{R}\right) e^{\frac{-t}{RC}}$

Note that the inductor current and capacitor voltage increase from their initial values of zero, towards final values $i(\infty) = (V_s / R)$ and $V_C(\infty) = V_s$, respectively. The rate of increase depends on the time constants L / R and RC. The voltages and currents of the elements in the two different circuits are shown in Table 14-2. The graphs are sketched in Figure 14-12 and Figure 14-13.

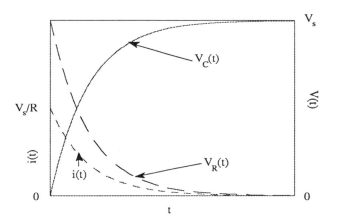

Figure 14-12: Currents through and Voltages across the RC circuit elements

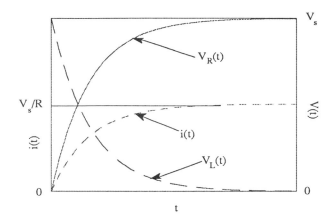

Figure 14-13: Currents through and Voltages across the RL circuit elements

As before, one can account for energy and charge for any interval of time. It is seen that in the R-L case, the battery supplies energy, some of which is dissipated as heat in the resistor, while the rest becomes stored in the magnetic field of the inductor. For the R-C case, the battery forces charge onto the plates of the capacitor, and energy is dissipated as heat in the resistor and stored in the electric field of the capacitor. Note that in the steady state (that is, after a long time) the R-L circuit still has a current, and so energy is still being supplied by the source and is being dissipated as heat in the resistor, while for the R-C case, no current flows at steady state and thus there is no further energy exchange once the capacitor is "fully charged".

14.3.2 General Procedure for First Order Circuits

For circuits containing only one inductor or one capacitor and constant-valued sources, the following general procedure can be used to find the voltages and currents whether or not there are any sources in the circuit for the time period of interest.

1. Find the initial value of the inductor current or the capacitor voltage. Either this value is given, or it can be inferred from the circuit's operation previous to the period of interest.

2. Find the value that the inductor current or capacitor voltage will attain at steady state. This can be readily done for the case at hand since after a long time, if all sources are constant, inductors will act like short circuits and capacitors will act like open circuits.

3. Find the time constant of the circuit. This is the value L / R_{eq} for R-L circuits, and $R_{eq}C$ for R-C circuits, where R_{eq} stands for the equivalent resistance seen by the inductor or capacitor with all sources turned off.

4. Then we can write immediately:

$$i_L(t) = i_L(\infty) + \left(i_L(0) - i_L(\infty)\right) e^{\frac{-t}{(L/R)}}$$

$$V_C(t) = V_C(\infty) + \left(V_C(0) - V_C(\infty)\right) e^{\frac{-t}{RC}}$$

5. From $i_L(t)$ or $V_C(t)$ any other voltage or current can be found.

14.4 SECOND ORDER RESPONSE

Responses of higher order are also present in physical systems. The first order response is the exponential decay and is found most frequently; however, observed physical phenomona such as continuous oscillation or damped (decaying) oscillation as observed in a swinging pendulum are not described by a first order response. It is a second order response. A second order natural response does predict oscillatory behavior either continuous, decaying or increasing resulting in instability. In electrical circuits, if both types of electrical energy storage elements, capacitors and inductors, are present in the circuit, the circuit will behave with a second order response.

14.4.1 Inductive-Capacitive Circuit

The LC circuit is shown in Figure 14-14. For $t > 0$ (after the switch is thrown from a to b), the inductor and the capacitor are in series. The initial voltage across the capacitor at $t = 0$ is $V_C(0) = V_s$ which is the voltage of the constant voltage source.

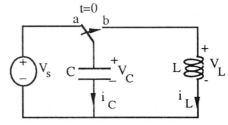

Figure 14-14: LC circuit

After the switch has been thrown, Kirchhoff's voltage law for this circuit says:

$$V_C - V_L = 0$$

Kirchhoff's current law says:

$$i_L = i = -i_C$$

Substituting the defining relationships between the current through the element and the voltage across the element reduces Kirchhoff's voltage law to:

$$-L\left(\frac{di}{dt}\right) + \left[V_s + \left(\frac{1}{C}\right) \int_0^t (-i)dt\right] = 0$$

If the inductance L, and capacitance C are constant, the time derivative of Kirchhoff's voltage law is:

$$-L\left(\frac{d^2 i}{dt^2}\right) - \left(\frac{1}{C}\right)i = 0$$

The voltage source is constant so its time derivative is zero. This equation is rearranged as:

$$LC\left(\frac{d^2i}{dt^2}\right) + i = 0$$

In terms of electrical energy accounting, this result says that electrical energy stored in the circuit oscillates between the inductor and capacitor forever.

This equation is a second order differential equation. It can be solved, but we will stop here. The important point in this problem was setting up the equations using both the conservation of charge through Kirchhoff's current law and electrical energy accounting plus KCL through Kirchhoff's voltage law. Conceptually, this equation gives the current as a continuous function of time which is sinusiodal in nature. This oscillation occurs because energy in the capacitor is transferred to the inductor and vice versa. Furthermore, the oscillations do not decay because (in this ideal case) there are no dissipative elements in the circuit.

14.4.2 Inductive-Capacitive-Resistive Circuit

The LRC circuit is shown in Figure 14-15. We have simply added a resistor in series with the LC circuit from the previous discussion. The initial voltage across the capacitor at $t = 0$ is V_s, the voltage of the constant voltage source.

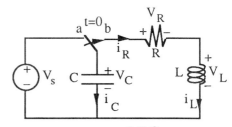

Figure 14-15: LRC circuit

After the switch is thrown, Kirchhoff's voltage law for this circuit says:

$$V_C - V_R - V_L = 0$$

Kirchhoff's current law says:

$$i_R = i_L = i = -i_C$$

Substituting the defining relationships between the current through the element and the voltage across the element into Kirchhoff's current law gives:

$$\left[V_s + \left(\frac{1}{C}\right)\int_0^t (-i)dt\right] - iR - L\left(\frac{di}{dt}\right)$$

If the inductance L, capacitance C, and resistance R are constant, the time derivative of Kirchhoff's voltage law is:

$$-\left(\frac{1}{C}\right)i - R\left(\frac{di}{dt}\right) - L\left(\frac{d^2i}{dt^2}\right) = 0$$

This equation is rearranged as:

$$L\left(\frac{d^2i}{dt^2}\right) + R\left(\frac{di}{dt}\right) + \left(\frac{1}{C}\right)i = 0$$

In terms of electrical energy accounting, this says that electrical energy storage oscillates between the capacitor and inductor but with dissipation in the resistor; the oscillations die out (decay) over a period of time due to electrical energy losses from the circuit in the resistor. This is a damped oscillation.

This equation is a second order differential equation. This equation also has a solution but we will stop here. Again the essence of this excercise was to present the concepts of electrical energy storage and dissipation in the circuit and how they are expressed quantitatively. This was done by using the conservation of charge and accounting of electrical energy through Kirchhoff's voltage law. Conceptually, this equation gives an oscillatory function of time for the current. However, the oscillations must decay because the circuit contains a dissipative element, the resistor. Eventually the function becomes zero and all the electrical energy initially stored in the capacitor has been lost in the form of heat through the resistor.

14.5 INDUCTIVE, CAPACITIVE CIRCUIT PROBLEM SOLVING TOOLS

1) **The Laws.** The laws which are of primary use for inductive and capacitive electrical circuits are the conservation of charge and the conservation of energy (and/or an accounting of electrical energy) in the form of Kirchhoff's current and voltage laws.

2) **Circuit Element Definitions.** This includes the definitions of the resistor, capacitor, and inductor. The resistor converts electrical energy to internal energy, the capacitor stores electrical energy in an electric field, and an inductor stores electrical energy in a magnetic field.

3) **Circuit Description.** Again, the description of the circuit of interest, what its elements are and how they are positioned relative to each other, is essential. In this chapter, the nature of first-order circuits, such as the RL and RC circuits was discussed, as well as second-order RLC circuits.

4) **Mathematics.** And, again, mathematics is an essential tool for problem solving. Circuits involving inductors and capacitors are dynamic circuits and hence, in addition to simultaneous algebraic equations, ordinary differential equations are encountered and solution skills for these problems are essential.

Strategies of circuit analysis include (of course) using the various tools, in order as appropriate, and selecting perhaps several systems for analysis. The use of mesh analysis versus nodal analysis appears again as a system selection strategy.

14.6 REVIEW

1) A current in a coil of wire will set up a magnetic field.

2) For such a device, a voltage is induced across it, proportional to the time rate of change of the current in the coil. The proportionality constant is called the inductance of the coil, and is measured in Henries.

3) Anytime a current exists through an inductor, energy is being stored in a magnetic field.

4) The only time there is a voltage across an inductor is when the current is changing. That is there will be no voltage across an inductor whose current is constant.

5) A capacitor is formed by parallel plates separated by space.

6) If charge is on the plates, then an electric field is set up in the region between the plates, and a potential difference is created. The voltage is proportional to the amount of charge. The proportionality constant is called capacitance ($CV = q$), measured in the unit Farads.

7) The voltage-current relation for a capacitor is obtained by noting that the current (charge flow rate) is the capacitance times the time rate of change of voltage.

8) Anytime there is a voltage across a capacitor, energy is being stored in an electric field.

9) The only time there is a capacitor current is when the voltage is changing. That is there will be no current for a capacitor whose voltage is constant.

10) Circuit analysis for circuits involving inductors and capacitors can be accomplished by nodal and mesh analysis. However, since the voltage-current relations for inductors and capacitors involve derivatives and integrals, we obtain sets of differential-integro equations which would need to be solved.

11) For the special case of circuits containing only one inductor or one capacitor (that is, one energy storage element) the equation is a first order linear, differential equation which can be readily solved.

12) The current through an inductor cannot instantaneously jump.

13) After a long time, if all sources are constant, inductors will have a constant current, and therefore zero voltage (act as short circuits).

14) The voltage across a capacitor cannot instantaneously jump.

15) After a long time, if all sources are constant, the voltage across capacitors will be constant, and therefore their current will be zero (act as open circuits).

16) The rate at which an inductor current proceeds from its initial value to its final value is determined by the time constant of the circuit. Likewise, the rate at which a capacitor voltage goes from its initial to final value is determined by the circuit time constant.

14.7 NOTATION

The following notation was used in this chapter.

Scalar Variables		Dimensions
C	capacitance	[charge / volt]
$\dot{F}$	rate of electrical energy dissipation i	[energy / time]
i	current	[charge / time]
L	inductance	[volt time^2 / charge]
q	charge	[charge]
R	resistance	[volt time / charge]
t	time	[time]
V	voltage	[energy / charge]
$\dot{W}$	rate of work	[energy / time]

Subscripts	
ele	electrical
in	entering the system
out	leaving the system
sys	system

14.8 SUGGESTED READING

Fitzgerald, A., D. Higginbotham, and A. Grabel, *Basic Electrical Engineering*, 5th edition, McGraw-Hill, New York, 1981

Halliday, D. and R. Resnick, *Physics for Students of Science and Engineering, Part II*, 2nd edition, John Wiley & Sons, New York, 1962

Nilsson, J., *Electrical Circuits*, 3rd edition, Addison-Wesley, Reading, Massachussetts, 1989

Paul, C., S. Nasar, and L. Unnewehr, *Introduction to Electrical Engineering*, McGraw-Hill, New York, 1986

Roadstrum, W. H., and D. H. Wolaver, *Electrical Engineering for All Engineers*, Harper and Row, New York, 1987

QUESTIONS

1) Under what conditions can a capacitor have 100 volts across it, but no current? Under what conditions can a capacitor have 10 amperes of current, but no voltage?

2) Under what conditions can an inductor have 100 volts across it, but no current? Under what conditions can an inductor have 10 amperes of current, but no voltage?

3) Why can an inductor current not instantaneously jump?

4) Why can a capacitor voltage not instantaneously jump?

5) For switched circuits with constant sources, what properties allow one to find the initial conditions for inductor current or capacitor voltage? Assume that before $t = 0$, the circuit was operating that way for a "long time."

6) For switched circuits with constant sources, what properties allow one to find the final conditions for inductor current or capacitor voltage?

7) How does one determine "a long time?"

8) Explain in words why it makes physical sense that when the resistance of an RC circuit is increased, the time constant should increase.

9) Explain in words why it makes physical sense that when the capacitance of an RC circuit is increased, the time constant should increase.

10) Explain in words why it makes physical sense that when the resistance of an RL circuit is increased, the time constant should decrease.

11) Explain in words why it makes physical sense that when the inductance of an RL circuit is increased, the time constant should increase.

SCALES

1) A capacitor charges at a steady rate to 10 mC in 0.02 ms, and it discharges in 1 microsecond. What are the magnitudes of the charging and discharging currents?

2) If a steady current flows to a capacitor, find the time required for the capacitor to charge to 2.5 mC if the current is 30mA.

3) A 4 μF, 5 μF, and 6 μF capacitor are in parallel across a voltage source of 250 V. Find the total stored energy.

4) A 4 μF, 5 μF, and 6 μF capacitor are in series with a 250 V source. Find the total stored energy.

5) Two capacitors are in series with a 24 V source. One capacitor is 20 μF and it has 8 V across it. What is the value of the other capacitor?

6) A 5 μF capacitor charged to 100 V is connected across an uncharged 10 μF capacitor that is in series with a 2000 Ohm resistor. Find the final stored energy of each capacitor.

7) If the voltage across a 2 μF capacitor is $200t$ V for $t < 1$ s, 200 V for 1 s $< t < 5$ s, and $3200 - 600t$ V for $t > 5$ s. Find the capacitor current.

8) How long does it take for a 20 μF capacitor, initially charged to 100 V and discharging through a 3 MOhm resistor, to have only 2 V across it?

9) A 1000 Ohm resistor in series with an initially uncharged 2 μF capacitor are switched into series with a 100 V source. What are the initial current, final current, final capacitor voltage, and time required to reach 95the final voltage?

10) A 5 μF capacitor is initially charged to 200 V. It is discharged through a 250 kOhm resistor. What is the voltage across the capacitor at $t = 0.2$ s? What is the current at this time?

11) At $t = 0$, the closing of a switch connects in series a 250 V source, a 1 kOhm resistor, and the parallel combination of a 1.2 kOhm resistor and uncharged 0.2 μF capacitor. Find the initial and final values of the capacitor voltage, initial and final values of the capacitor current, and the time constant. Also find the capacitor voltage and current at $t = 1.5$ ms.

12) A 150 mH inductor has a constant current of 4 A. Find the voltage.

13) A current that is uniformly increasing from 50 mA to 90 mA in the time 5 ms to 7 ms is through a 90 mH inductor. Find the induced voltage during this time.

14) A current given by $i = 0.35t$ A flows through a 40 mH inductor. Find the energy stored at $t = 0.5$ s.

15) A switch is closed so that a 20 V source, a 5 Ohm resistor and a 4 mH inductor are in series. What is the value of the current at time $t = 1.2$ ms? Also find the inductor voltage at t = 1.2 ms.

16) The coil for a relay has resistance of 30 Ohms and an inductance of 2 H. If 12 V is applied to the coil, how long will it be before the current is less than 250 mA?

17) A 2 H inductor, 450 Ohm resistor and 50 V source have been connected in series for a long time. Find the energy stored in the inductor.

PROBLEMS

1) Consider the circuit shown in the figure. At $t = 0$ the switch comes up. The components have the values: $R_1 = 1000\ \Omega$, $R_2 = 9000\ \Omega$, and $C = 1 \times 10^{-6}$ F.

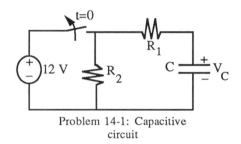

Problem 14-1: Capacitive circuit

a) Find the initial positive charge on one of the capacitor plates at $t = 0$.

b) Find the initial energy stored in the electric field of the capacitor.

c) Determine the voltage across each of the elements as a function of time and the current through each of the elements as a function of time.

d) Find the charge that will remain on the plates as $t \rightarrow \infty$.

e) Find the energy that is stored in the electric field of the capacitor as $t \rightarrow \infty$.

f) For the time period $t = 0$ to $t = 0.03$ seconds write electrical energy accounting equations. Show that the change in energy stored in the electric field of the capacitor is equal to the energy dissipated by the resistors during the time period.

2) Consider the circuit shown in the figure. At $t = 0$ the switch comes up. The components have the values: $R_1 = 500\ \Omega$, $R_2 = 300\ \Omega$, and $L = 6$ H.

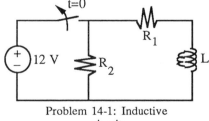

Problem 14-1: Inductive circuit

a) Determine the initial current through the inductor.

b) Find the initial energy stored in the magnetic field of the inductor.

c) Determine the voltage across each of the elements as a function of time and the current through each of the elements as a function of time.

d) Find the current through the inductor as $t \rightarrow \infty$.

e) Find the energy that is stored in the magnetic field of the inductor as $t \rightarrow \infty$.

f) For the time period $t = 0$ to $t = 0.03$ seconds write an electrical energy accounting equation for the circuit. Show that the change in energy stored in the magnetic field of the inductor is equal to the energy dissipated by the resistors during the time period.

THERMODYNAMICS

In earlier chapters, a handful of conservation principles plus some additional accounting equations were presented as the foundation structure for engineering and science. Because of their extreme importance to our problem-solving abilities and to our understanding of science and engineering, these principles bear repeating here.

First, we have five conservation principles which count those unique properties which are conserved, i.e., which are neither generated nor consumed within a system but which may exchange with the surroundings: conservation of total mass (including conservation of the elements), conservation of total charge, conservation of total energy, and conservation of linear and angular momentum. These principles establish nine independent scalar equations, plus a scalar equation for each element, which describe the system's behavior. The equations of mass, charge, and energy are scalar equations and therefore provide three independent equations and the conservation of linear and angular momentum equations are vector equations, each consisting of three more scalar equations.*

Second, we have a number of accounting relations which count subsets of mass, charge, or energy which are not necessarily conserved. Individual species of mass (compounds, ions, e.g.) may undergo chemical reaction and therefore may be generated or consumed, subject to the constraint of conservation of total mass and of the elements. Positive and negative charges may be generated or consumed locally, subject to the conservation of total or net charge. Conversions between mechanical, electrical, and thermal energy may occur so that each form, considered individually may not be conserved,

---

* In the absence of nuclear conversions or relativistic effects, mass and energy are conserved separately. When nuclear conversions are important, mass and energy together are conserved.

even though total energy is. Separate species mass, positive or negative charge, mechanical, electrical, and thermal energy accounting equations are used when appropriate to increase our understanding of processes, and machines.

Finally, in chapter eight, another property was introduced which is used to quantify the observation that certain events happen in only one direction. During processes, changes in the entropy of the system and surroundings together do not necessarily total zero so that entropy is not a conserved property. In fact, not only is it not conserved, it (the entropy of the surroundings and system together) must always either increase or remain the same during any process, it never decreases. This second law of thermodynamics provides a constraint upon the processes and a way to verify whether the processes are theoretically feasible.

With these foundations in hand, you have a structure from which to analyze a great many situations in thermodynamics as well as other areas such as solid and fluid mechanics and electrical circuits. In order to complete the framework, however, two more things need to be done at this time.

First, is to describe the manner in which the properties which appear within the fundamental equations (internal energy, enthalpy, entropy, density) can be calculated. That is, internal energy, as one example, can be expressed as a function of a set of independent variables. How these variables are chosen and how they are used to guide experimental measurements which can be used for calculating internal energy and the other properties is conveniently presented within the context of a discussion of thermodynamics. Second, we need to formalize the definition and discussion of entropy. This requires a discussion of ideal cycles known as Carnot (car-nō) cycles. The discussion of these cycles leads to a definition of a thermodynamic temperature scale. Then, the combination of the Carnot cycle discussion and thermodynamic temperature scale discussions leads to the observation that a state property, entropy, can be defined which has the desired properties for quantifying the direction in which things happen and the concept of irreversible processes.

This section on thermodynamics provides this additional information on the second law and the properties of materials. Additionally, it provides applications to a large number of situations which can be analyzed using these thermodynamic principles.

The term thermodynamics refers, as a literal translation, to heat (thermo) and changes (dynamics). This term goes back to the original motivation for thermodynamics which was to describe and understand heat engines, that is machines or processes for converting heat to work such as steam engines, internal combustion engines, etc. Much of thermodynamics, however, is concerned more directly with static or equilibrium situations and, therefore, may be more properly referred to as thermostatics. The entire body of knowledge dealing with properties of materials refers rigorously to equilibrium situations. The study of thermodynamic properties of materials is a very major part of the study of thermodynamics. These properties include energy properties such as internal energy, enthalpy, the Gibbs free energy, the Helmholtz free energy, entropy, pressure-volume-temperature relationships and, phase equilibrium phenomena. This latter study is largely beyond the scope of this course but is an extremely important part of chemical engineering and other areas that involve chemical processing and equilibrium phenomena.

CHAPTER FIFTEEN

CALCULATION OF
THE THERMAL PROPERTIES OF MATERIALS

5.1 INTRODUCTION

As discussed in the introduction to this section, the objective of thermodynamic property relationships is to describe properties that are required for the conservation laws and accounting equations (internal energy, enthalpy, entropy) in terms of proper set of independent variables. This set must consist of the right number of intensive variables and, practically speaking, the variables must be *observable* or *experimentally measurable* quantities such as temperature, pressure, composition, volume, or heat capacity. For example, entropy itself, required for using the second law of thermodynamics, is not directly measurable; neither are internal energy and enthalpy which are required for energy calculations. Consequently we must have relationships for these properties as functions of the various observable quantities.

Figure 15-1 outlines the interrelationships and structure of that part of thermodynamics dealing with the calculation of the thermodynamic properties ($\widehat{U}$, $\widehat{H}$, $\widehat{A}$, $\widehat{G}$, and $\widehat{S}$). The rationale of these calculations evolves from the conservation laws themselves (Boxes 1-4) plus the zeroth and second laws of thermodynamics which lead to definitions of temperature and entropy (Box 5 - Chapter 16). With this basis, fundamental relationships between the thermodynamic properties of a single phase, single component material are obtained (e.g., $d\widehat{U} = Td\widehat{S} - Pd\widehat{V}$) (Box 6). Then, combining these property relations with appropriate mathematical concepts, relationships for the changes in internal energy, enthalpy, and entropy can be obtained which are entirely in terms of measurable properties: thermal data (heat capacities) and volumetric (P-$\widehat{V}$-T) data (Box 7). (Obtaining these relationships and the mathematical formalism required to do so is covered more extensively later in this chapter.) Obviously, P-$\widehat{V}$-T and heat capacity relationships and their determination from data play an important role in thermodynamics (Boxes 8-11). Furthermore, if *reference states* and corresponding reference values for enthalpy, entropy, and internal energy are defined (Box 12), then we can, as a matter of convenience, calculate absolute values of enthalpy, internal energy, and entropy (Box 13). Additional thermodynamic properties such as the *Gibbs free energy* $(\widehat{G})$ and the *Hemholtz free energy* $(\widehat{A})$, also have special value to thermodynamics and calculations of these properties can be made with no more additional data or definitions (Box 14). Some special forms of the thermal property relations also are useful (Box 15). Understanding these interrelationships and the underlying logic of these thermodynamic calculations is extremely important to understanding and applying the conservation laws and the second law of thermodynamics.

With this structure in mind, then, this chapter focusses first on the details of the relationships which are outlined in Figure 15-1. In learning this material, keep in mind the bottom-line objective: to be able to calculate, for use in the first (total energy conservation) and second laws of thermodynamics, the thermal properties which are not directly measurable (internal energy, enthalpy, entropy) from properties which are measurable (P-$\widehat{V}$-T and heat capacity data). Then, motivated by this need for heat capacity and volumetric properties of materials, sections follow which deal with these topics. Section 15.4 presents volumetric properties and relationships (equations of state) for both ideal and nonideal gases and for liquids. Section 15.5 discusses the special forms of the fundamental property relations which are useful for the various equations of state. Then, section 15.6 deals with phase equilibrium and latent heat (enthalpy change) associated with phase transitions and section 15.7 addresses chemical reaction equilibrium and heats of reaction (enthalpy change) associated with reactions.

This is a long chapter, but not meant to be intimidating. The purpose of combining thermal properties, volumetric properties and other information in this single long chapter is to provide a focus on the objective of calculating $\widehat{U}$, $\widehat{H}$, (and $\widehat{A}$ and $\widehat{G}$) which are required for applying the fundamental laws.

15.2 APPLICABILITY OF THE FUNDAMENTAL LAWS

From the above introduction, it is apparent that the foundation laws play a vitally important role in the structure of thermodynamics and that more than one or two of the laws are important. In fact, with the exception of charge, all of the conservation laws plus the second law of thermodynamics comprise the foundation. How these are used in various contexts will become more apparent from working specific problems, but to summarize, it is this: thermodynamics deals with thermal energy changes; mass changes (conservation of mass) affect thermal energy changes through the thermal properties of matter (internal energy, enthalpy, and entropy) and entropy changes place a constraint on the manner in which thermal changes occur; furthermore, by considering mechanical energy changes (which we can do by applying momentum conservation) we can simplify (by eliminating certain mechanical energy quantities) these thermal energy calculations. Interwoven with the physics of thermodynamics are a number of mathematical concepts and proofs which complicate the directness of the relationships and procedures.

So, in contrast to isothermal rigid body mechanics which relies on linear and angular momentum, isothermal fluid mechanics which is governed by mass conservation and linear and angular momentum, and electrical circuits which are understood through charge conservation and electrical energy accounting, classical thermodynamics is based upon nearly all of the fundamental laws (there are situations where charge and electrical energy must be considered as well, further broadening the applicability of thermodynamics). Is it any wonder that the study of thermodynamics can be a very confusing and frustrating experience or that it typically requires more than one attempt at understanding its foundation concepts, structure, and details of calculations?

This chapter presents the bottom-line details of thermodynamic calculations and problem solving in the context of the foundation laws and with a view to providing an appreciative picture of the beauty and wonder of the structure of thermodynamics. It is with great respect to those who have gone before us and whose efforts, imagination, and creativity contributed to the development of this picture that we attempt to convey this structure.

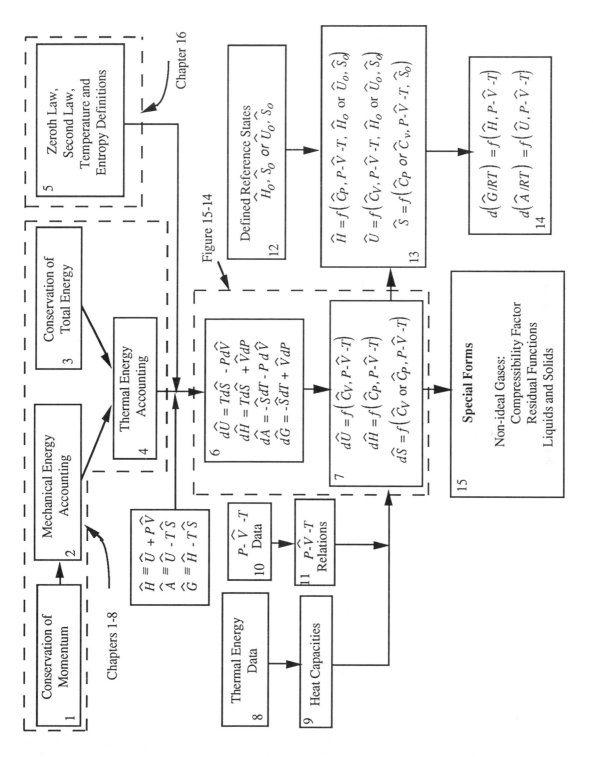

Figure 15-1: Interrelations of thermodynamic and volumetric properties and data and the conservation laws and accounting equations.

15.3 THERMODYNAMIC PROPERTIES OF HOMOGENEOUS MATERIALS OF CONSTANT COMPOSITION

In this section we present the methods for calculating the thermodynamic properties in terms of heat capacity and volumetric properties which are depicted in Figure 15-1. First, in sections 15.3.1 through 15.3.5, we provide a more detailed discussion of Figure 15-1 with an objective of at arriving at the results which are used for the bottom-line calculations of thermodynamic properties. Specifically, we discuss Boxes 6 through 15 but without the details of going from Box 6 to 7. This is intended to provide 1) a summary of the important relations and results, along with an overview of the structure of the calculations but without the complications of some of the more-involved details and 2) a perspective from which to approach the further study of the additional details of thought, mathematics, and physics which are required to obtain these results; with an appreciation and knowledge of the specific objective, the details are more interesting and comprehensible. Then, in Section 15.3.6 we present the missing details which are required to obtain the expressions for changes in the thermal properties in terms of the observables (those required to go from Box 6 to Box 7 in Figure 15-1).

This section is restricted to homogeneous materials, materials which, within our region or volume of interest, are everywhere the same. We are talking about materials of uniform composition, temperature, and pressure. Practically speaking, for this text, we are interested in single-component systems, although the term homogeneous certainly does not imply that restriction.

15.3.1 State Functions, Internal energy, and Thermal energy Accounting

We begin with a discussion of the concepts of the *state of existence* of a material, *state variables*, and *state functions*. You already have a notion of these concepts but now should develop a more complete understanding of their meaning and importance to thermodynamics and to the calculation of changes in $\widehat{U}$, $\widehat{H}$, and $\widehat{S}$.

For example, you know of the importance of temperature and pressure to the state of existence of water. At 1 atm (14.696 lb_f / in^2, 101.325 kPa or 1.01325 bar) of pressure and 0°C (32°F, 273.15 K), liquid water exists in equilibrium with ice; they coexist together. Below 0°C, it is ice; above 0°C, it is water (still at 1 atm). At 1 atm and 100°C (212°F, 373.15 K), liquid water exists in equilibrium with steam; Above 100°C at 1 atm, water exists as steam. Furthermore, you believe or readily accept, although you may not realize its importance, that water exists in these states at these conditions of temperature and pressure regardless of its history, i.e., regardless of what has happened to it previously, regardless of the path of previous states which it has traversed. Figure 15-2 shows three possible paths from state 1 to state 2 on a $P - V$ diagram: path a–b, path c, and path d–e. The change in state properties from State 1 to State 2 will not depend upon which of these three paths is taken.

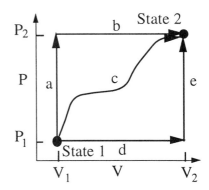

Figure 15-2: Changes in state properties

In fact, this concept of path or history independence is extremely important to our ability to understand and apply the conservation laws and the second law of thermodynamics. Total energy conservation is formulated in the context of the state functions of internal energy ($\widehat{U}$) and specific volume ($\widehat{V}$) along with the intensive state property of pressure (or alternatively in terms of internal energy and enthalpy, per mass, $\widehat{U}$ and $\widehat{H}$). Furthermore, entropy achieves its importance and, in fact, was defined originally, because of the conclusion that the observations of directionality and irreversibility leading to the second law implied the existence of a new state function, an additional state intensive property of matter (entropy per mass, $\widehat{S}$) which is path or history independent. Further discussion of the arguments leading to entropy are postponed until Chapter 16, after better familiarity is reached with the meaning of entropy and its relation to temperature and pressure.

At this point, you may feel that there is nothing unique or special about state functions, that all physical states or properties of matter are path or history independent. In fact, this is not true. For example, the state of stress of a material, which is related to the material's deformation and/or rate of deformation, may well depend on the history of deformation and not just the current deformation.

> A state function is an intensive property of matter which depends on the current state of existence of the matter and not on the "path" of states taken to reach it.

Thermodynamic state functions are internal energy, enthalpy, Helmholtz free energy, Gibbs free energy, and entropy. Volume also is a state function and temperature and pressure are intensive state variables, variables which, along with the state extensive functions, can be used to define the state of existence of a pure material (composition is another variable, or set of variables, which is required to fully define the state for a multicomponent material). Figure 15-3 shows how the volume is a function of the state variables pressure and temperature.

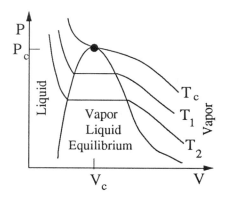

Figure 15-3: Volume as a function of the state variables P and T

In Chapter 7, we discussed the very important concept of the conservation of total energy. In this discussion, we introduced internal energy as a measure of the state of energy of materials, representing its internal molecular motion and position.

Energy is stored within matter in the form of vibration, translation, and rotation of all the various elements of mass within the material. For example, a gas in a closed container contains a very large number of molecules and each of these molecules is moving around within the container. The average speed of the motion of the molecules and their vibrations and rotations reflects the energy of the gas. Now, if this gas is contained within a stationary vessel, then, when the translational motion of these molecules is added together in the vectorial sense, the total or net motion is zero. The average *magnitudes* of the translational motions of the molecules is not zero, however, and the greater the internal energy of the gas, the greater this average velocity.

Now in addition to translational, rotational, and vibrational motion of the molecules and submolecular particles, energy is manifested in the force fields which exist between these various bodies of mass. To the extent force fields exist between two bodies, the energy of the two bodies changes as the separation between them is changed. That is, the internal energy of a material is affected not just by the motion of the molecules and submolecular particles but also by the extent to which the molecules or other particles are separated from each other.

This picture of the internal energy of matter has developed as a model which describes physical observations. Total energy is conserved when an internal energy function is defined. Stated somewhat differently, whenever the state of matter is changed from one condition to another by two different processes such that as a result of both processes its change of state, in terms of temperature, pressure, and all other measured properties is exactly the same, then the same amount of energy was required in order to bring about this change of state. This amount of energy is the change of internal energy of that mass, and it is a "state function;" the same amount is required, independent of how that change of state was achieved, that is, regardless of the distribution between heat and work in order to bring about that change in internal energy.

Implied by the above statements is the notion that in changing the internal energy by a required amount, the distribution of energy between the heat and work required is not unique. In fact, this is true and leads to the notion that *heat and work are not state functions*. In fact, they are not properties of matter at all but represent energy in transit from the surroundings to the system and vice versa; *heat and work are path functions*. That is, if a piece of mass is taken as a closed system, and its state is changed by the transfer of heat and work, then the relative amounts of heat and work required to bring about a fixed or predefined change in energy is not unique. The amount of heat transferred and the amount of work, do indeed depend upon the specific path followed. Taken together, they result in a change in internal energy which defines a specific change in the state of the matter.

The thermal energy equation which was obtained in Chapter 7 provides the necessary relationship for beginning an understanding of thermodynamic changes of state. Remember that this result was obtained by considering the conservation of total energy together with an accounting of mechanical energy (obtained by integrating the conservation of linear momentum for a differential volume element, a process which is well beyond the scope of this text). When these two results were applied to a steady-state, single-input, single-output, macroscopic, open-flow process the following thermal energy accounting statement was obtained:

$$0 = \widehat{H}_{in} - \widehat{H}_{out} - \int_{P_{out}}^{P_{in}} \widehat{V} dP + \sum \left(\frac{\dot{Q}_i}{\dot{m}} \right) + \sum \left(\frac{\dot{F}_i}{\dot{m}} \right) \qquad (7-55)$$

which can be rearranged as:

$$\widehat{H}_{out} - \widehat{H}_{in} = \widehat{Q} + \widehat{F} + \int_{P_{in}}^{P_{out}} \widehat{V} dP \qquad (15-1)$$

where the terms $\widehat{Q}$ and $\widehat{F}$ represent the total heat transfer per unit mass and the total irreversible conversion of mechanical to thermal energy per unit mass. It is important to note that these quantities are not state properties but rather are dependent on the path.

The change in enthalpy between the inlet and outlet conditions is the result of heat transferred to the material across the system boundary plus the conversion of mechanical energy to thermal energy within the system boundary due to mechanical irreversibilities, plus an integral of $\widehat{V} dP$, *along the specific process path which the material follows between the inlet and outlet states*. If the system which is chosen is a very small or differential system, then this result is written as

$$d\widehat{H} = \widehat{V} dP + \delta \widehat{Q} + \delta \widehat{F} \qquad (15-2)$$

Now, two physical views of this thermal energy equation are possible, each of which has its own value and appropriateness. On the one hand, the macroscopic equation says that for a steady-state process as defined and restricted above, the relation between entering and leaving enthalpies is given by Equation 15-1. In other words, it relates the current states of existence of two different pieces of mass, that which is entering the system and that which is leaving. The two enthalpies are related by the heat transfer (per unit mass) to the system, the irreversible conversion of mechanical energy to thermal energy (per unit mass) which occurs within the system, as well as the integral of $\widehat{V} dP$ along the path which is followed by the fluid in flowing through the system.

The alternative view can be understood in light of the differential volume element relation, Equation 15-2. This view says that as we integrate along the path from entrance to exit, we are following a specific piece of mass and watching its properties change along the path. According to this interpretation, the change in enthalpy (per unit mass) between inlet and outlet states, represented by the macroscopic equation (Eq. 15-1), are changes in enthalpy of the same piece of mass brought about by heat transfer per unit mass, irreversible mechanical energy conversion per unit mass, and the pressure integral along the path which is followed by this specific piece of mass. From this perspective, we are accounting for the thermal energy changes of a closed system (the mass which we are following) as it moves along its flow path.

These are important concepts and their differences should be well noted. Again, the thermal energy result may be viewed either as describing a steady state, open system with the inlet mass at state 1 and the outlet at state 2 (Eq. 15-1), or it may be viewed as describing an unsteady-state, closed system (which is a differential piece of mass) as it moves through a process and changes from state 1 to state 2 (Eq. 15-2). In either case, the same result for changes in enthalpy holds. From a problem-solving viewpoint, either interpretation may be useful, although in describing process changes, the steady-state interpretation is most often appropriate. However, for the purposes of this chapter which are to relate thermodynamic property changes, the closed-system interpretation is most appropriate.

Consider now another form of this result. Because of the relationship between enthalpy and internal energy ($\widehat{H} = \widehat{U} + P\widehat{V}$), the differential form of the thermal energy equation is written alternatively as:

$$d\widehat{U} = -Pd\widehat{V} + \delta\widehat{Q} + \delta\widehat{F} \qquad (15-3)$$

and we can say that the internal energy of an element of mass changes as the result of heat transfer, the irreversible conversion of mechanical energy to thermal energy, and $Pd\widehat{V}$.

Now these two differential forms of the thermal energy accounting equation for a differential size element of mass are relations which describe changes in state functions (enthalpy and internal energy) in terms of path (process) properties (heat transfer, mechanical energy losses, and $-Pd\widehat{V}$ or $\widehat{V}dP$ integrals). So, on the one hand we have changes in state functions and on the other hand we have changes in path functions. Now the $\widehat{V}dP$ and $Pd\widehat{V}$, even though they represent path quantities, are at least expressed in terms of state properties; they are represented in terms of properties which can be associated uniquely with a state of existence. Heat transfer and mechanical energy losses, however, are not properties which are associated with a state of existence, but rather are process properties. They do not represent changes in the matter but rather are parameters of the process which is being used to effect changes in matter. Consequently, $\widehat{Q}$ and $\widehat{F}$ are of little value in describing or allowing us to calculate property changes independent of a given process.

However, because, for the purposes of this discussion, we are only interested in calculating changes in internal energy or enthalpy, given changes in state properties, we can arbitrarily select a specific process to go from one state to another. Consequently, we can choose a particular path which will make the calculation of $\widehat{Q}$ and $\widehat{F}$ most convenient. The most convenient such path is a reversible path, for in this case the irreversible conversion of mechanical energy to thermal energy is exactly zero (there are no mechanical energy losses) and the heat transfer can be expressed exactly in terms of changes in entropy (for a closed system) according to:

$$\delta\widehat{Q}_{\text{rev}} = Td\widehat{S} \qquad (15-4)$$

where T is *absolute* temperature. Hence, for a reversible path we may write (for a closed system consisting of one unit of mass)

$$\boxed{d\widehat{U} = Td\widehat{S} - Pd\widehat{V}} \qquad (15-5)$$

Now this qualifier, that this result is for a reversible path is not as restrictive as it might at first seem. Because $\widehat{S}$ is a state function and temperature and pressure are state properties, we now have the entire right hand side written in terms of state functions and state properties. The left-hand side is also a state function. Consequently, for any defined change of state the left-hand side will be uniquely defined and likewise, the right-hand side will be uniquely defined, *regardless of the path taken to achieve that change of state*. This statement requires further clarification and emphasis. The result was obtained by recognizing that it is true for a reversible path. However, because it involves only state properties and state functions, the result still holds regardless of whether a reversible path is followed or not. The corresponding result for enthalpy is:

$$\boxed{d\widehat{H} = T\,d\widehat{S} + \widehat{V}\,dP}$$

$$(15-6)$$

These last two results are extremely important to the foundations of thermodynamics. They have their roots in the fundamental conservation laws and in the second law of thermodynamics and in that sense represent an accounting of thermal energy for a process. Yet, they are more fundamental than that. *They are process independent.* They relate changes in internal energy or enthalpy to other changes in state variables and state functions, regardless of the process used to bring about those changes. *They are property relations and are not restricted to specific processes.* A thermal energy accounting equation we may properly associate with a specific process whereas these relations are property relations independent of path, i.e. independent of the process under consideration.

A few comments on the physical meanings of $P\,d\widehat{V}$ and $\widehat{V}\,dP$ are appropriate. It is common to refer to $-P\,d\widehat{V}$ as reversible work which occurs as a fluid changes volume by a differential amount at pressure P. A pressure is exerted at the system boundary equal to P and this pressure exerted over volume change $d\widehat{V}$ represents the force-distance product (per unit mass), i.e., the work (per unit mass) as this volume change is carried out.

While this is a valid interpretation for this kind of process, we prefer to think in somewhat different terms. In accordance with the preceeding paragraphs, both $T\,d\widehat{S}$ and $-P\,d\widehat{V}$ are quantities which are related to changes in state of matter and are not necessarily representative of the process by which those changes are occurring. On the other hand, when we talk about $-P\,d\widehat{V}$ representing reversible work, such an interpretation only has meaning for a particular process, that is, one which is reversible and one for which those changes are brought about by work.

Furthermore, it is important to realize that the work shown in Figure 15-4 in raising the cylinder from the floor to the table, while maintaining constant temperature and pressure within the cylinder, is not included in the thermal accounting equation. This mechanical energy has been subtracted from the total energy to obtain the statement of thermal energy.

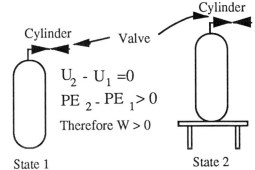

Figure 15-4: Raising a cylinder
to a table top

A physical interpretation of $\widehat{V}\,dP$ is not so appropriate, however. In fact, it arises simply as a result of the definition of $\widehat{H}$ in terms of $\widehat{U}$,

$$d\widehat{H} = d\widehat{U} + d(P\widehat{V}) = d\widehat{U} + P\,d\widehat{V} + \widehat{V}\,dP$$

$$(15-7)$$

which was done to account for (isotropic, i.e., pressure) flow work at the entrance and exit to the system.

15.3.2 The Single Phase Fundamental Property Relations

From the previous section, we have two fundamental property relations:

$$d\widehat{U} = T d\widehat{S} - P d\widehat{V} \tag{15 - 5}$$

$$d\widehat{H} = T d\widehat{S} + \widehat{V} dP \tag{15 - 6}$$

These relate changes in internal energy (per unit mass), or enthalpy, to entropy, volume, temperature and pressure.

Now, also it is convenient for some situations to define two other energy functions called the Helmholtz free energy ($\widehat{A}$), and the Gibbs free energy ($\widehat{G}$) according to:

$$\boxed{\widehat{A} \equiv \widehat{U} - T\widehat{S}} \tag{15 - 8}$$

$$\boxed{\widehat{G} \equiv \widehat{H} - T\widehat{S}} \tag{15 - 9}$$

Then, the total change in the Helmholtz and Gibbs free energies are given by:

$$d\widehat{A} = d\widehat{U} - T d\widehat{S} - \widehat{S} dT \tag{15 - 10}$$

$$d\widehat{G} = d\widehat{H} - T d\widehat{S} - \widehat{S} dT \tag{15 - 11}$$

and substituting for $d\widehat{U}$ and $d\widehat{H}$ gives two additional property relations:

$$d\widehat{A} = -\widehat{S} dT - P d\widehat{V} \tag{15 - 12}$$

$$d\widehat{G} = -\widehat{S} dT + \widehat{V} dP \tag{15 - 13}$$

These four relations for $d\widehat{U}$, $d\widehat{H}$, $d\widehat{A}$, and $d\widehat{G}$ are the *four fundamental thermodynamic property relationships*

$$\boxed{d\widehat{U} = T d\widehat{S} - P d\widehat{V}} \tag{15 - 5}$$

$$\boxed{d\widehat{H} = T d\widehat{S} + \widehat{V} dP} \tag{15 - 6}$$

$$d\widehat{A} = -\widehat{S}\,dT - P\,d\widehat{V}$$

$$(15 - 12)$$

$$d\widehat{G} = -\widehat{S}\,dT + \widehat{V}\,dP$$

$$(15 - 13)$$

They are so important that they should be committed to memory. Remember that all of these equations say the same thing, they just do it in different ways using different variables. Although having roots in an accounting of thermal energy, these results transcend such a statement for any particular process. They represent relations between the thermodynamic state functions ($\widehat{U}$, $\widehat{H}$, $\widehat{A}$, $\widehat{G}$, $\widehat{S}$) and state properties (P, $\widehat{V}$, T) and hold true independent of the actual process which causes these changes. Note again that T *must* be an absolute scale (Kelvin or Rankine).

15.3.3 Calculation of Changes in $\widehat{U}$, $\widehat{H}$, and $\widehat{S}$ in Terms of Observables

In the preceding section we presented the four fundamental thermodynamic relationships:

$$d\widehat{U} = T\,d\widehat{S} - P\,d\widehat{V} \qquad\qquad (15 - 5)$$

$$d\widehat{H} = T\,d\widehat{S} + \widehat{V}\,dP \qquad\qquad (15 - 6)$$

$$d\widehat{A} = -\widehat{S}\,dT - P\,d\widehat{V} \qquad\qquad (15 - 12)$$

$$d\widehat{G} = -\widehat{S}\,dT + \widehat{V}\,dP \qquad\qquad (15 - 13)$$

These results show how changes in the internal energy, enthalpy, Helmholtz free energy, and the Gibbs free energy are related to temperature, pressure, volume, and entropy. However, as has been mentioned above, these equations are not in a readily usable form, practically speaking, because entropy is not directly measurable. That is, the energy functions which are not directly measurable are expressed in terms of another quantity which also is not directly measurable, entropy, and as a result, it is not immediately clear that anything has been accomplished by arriving at these relationships.

It turns out, however, that these four relations for the energy functions can be written in terms of observable quantities, heat capacities and P-$\widehat{V}$-T data and relationships (equations of state). These four thermodynamic property relations then have a tremendous impact upon our ability to apply the fundamental laws.

In this section we present these practical forms of the fundamental thermodynamic property relations for internal energy and enthalpy. We also give related practical results for calculating entropy and the difference between constant pressure and constant volume heat capacities. The details of obtaining these results are deferred until section 15.3.6. In presenting the results at this point in the chapter without giving the details of the derivation, we certainly do not want to de-emphasize the importance of understanding the logic of arriving at them. We feel that how these results are obtained is important and that to appreciate the results more fully, one should be able to derive them unassisted. In fact, it is to encourage this ability that that we first present the bottom-line results and how they are used in thermodynamic calculations. Then after this familiarity with the results is developed, we show how they are derived.

So, without further ado, the bottom-line result for internal energy, i.e., the fundamental property relation for internal energy (per unit mass) *in terms of observable quantities* is

$$d\widehat{U} = \widehat{C}_V \, dT + \left[T \left(\frac{\partial P}{\partial T} \right)_{\widehat{V}} - P \right] d\widehat{V}$$

(15 – 14)

and the corresponding result for enthalpy (per unit mass) is

$$d\widehat{H} = \widehat{C}_P \, dT + \left[\widehat{V} - T \left(\frac{\partial \widehat{V}}{\partial T} \right)_{P} \right] dP$$

(15 – 15)

Again, note that temperature must be on an absolute scale. These are perfectly general results for a single-phase, single-component (or homogeneous of constant composition) material. Changes in the energy state functions can be calculated directly from changes in temperature using the material's heat capacity ($\widehat{C}_V$ or $\widehat{C}_P$) and changes in volume or pressure using its equation of state (P-$\widehat{V}$-T relationships). Changes in temperature reflect the kinetic motion of the molecular and submolecular particles and therefore relate directly to the energy functions; temperature has a major effect on internal energy and enthalpy. The equation of state plays a role because as a material, say a gas, is compressed to a smaller and smaller volume, even if done at constant temperature by removing heat, energy changes may occur. Moving molecules closer and closer together in the presence of force fields between the molecules requires energy and therefore introduces some changes in internal energy and enthalpy. In most processes and with most materials, these equation-of-state effects are small compared to temperature effects. Some familiarity with the relative sizes of temperature versus equation-of-state terms will be obtained from working problems.

Now the previous equations were written in differential form. To calculate changes in internal energy or enthalpy due to finite changes in temperature and volume (or pressure) and without changes in phase, we can simply integrate these equations over the appropriate temperature and volume (or pressure) ranges. Accordingly, we have

$$\widehat{U}_2 - \widehat{U}_1 = \int_{T_1}^{T_2} \widehat{C}_V \, dT + \int_{\widehat{V}_1}^{\widehat{V}_2} \left[T \left(\frac{\partial P}{\partial T} \right)_{\widehat{V}} - P \right] d\widehat{V}$$

(15 – 16)

and

$$\widehat{H}_2 - \widehat{H}_1 = \int_{T_1}^{T_2} \widehat{C}_P \, dT + \int_{P_1}^{P_2} \left[\widehat{V} - T \left(\frac{\partial \widehat{V}}{\partial T} \right)_{P} \right] dP$$

(15 – 17)

In carrying out the integration for internal energy, the first term, the integral over temperature, is calculated following a constant volume path and the second term, the equation-of-state term, is calculated over a constant temperature path. We could pick any other path to go from state 1 to state 2 and we would get exactly the same result because internal energy is a state function and therefore path independent. However, to pick an arbitrary path is not convenient because then we would have to allow for changes in heat capacity along the path which we have chosen in a way which we may not, in fact, know.

Simarly, for changes in enthalpy, the temperature integral is chosen over a constant pressure path and the equation-of-state term is integrated over a constant temperature path.

These paths for these calculations are illustrated in Figure 15-5. The internal energy change for what may be a typical process involving changes in temperature and volume is shown at state 1 and state 2. In three dimensions, a two-step process is shown whereby the first part is at constant volume (V_1) with the temperature changing between the initial and final temperature (and the internal energy changing from U_1 to U'). The second part is at constant temperature (T_2) while the volume changes from the initial volume to the final volume. A similar plot of enthalpy as a function of temperature and pressure could be made to show the changes which occur along a two-step process of which the first part is at constant pressure and the second part at constant temperature.

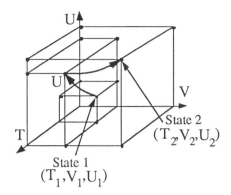

Figure 15-5: Schematic showing the changes in internal energy with T and V

In these relationships, the heat capacities are a measure of the amount of heat necessary to change the temperature of a material. In general, materials which have heavy molecules and more degrees of freedom in their motion will have higher heat capacities. Small molecules with limited degrees of freedom, e.g. the monotomic gases, have low heat capacities. Note that these heat capacities are expressed in terms of energy per unit mass per degree of temperature change. Typical heat capacities are shown in Appendix C Tables 15-18 for a number of materials and will vary with temperature, in general.

The equation-of-state contribution to changes in internal energy and enthalpy generally are not large, as stated above. In the presence of temperature effects, these equation-of-state effects are relatively small. In fact, in one particular case, that of ideal gases (gases which follow the ideal gas equation of state) the contribution to changes in internal energy and enthalpy by the equation of state is identically zero. This is a direct result of the fact that for an ideal gas, there are no force fields between molecules.

One further note: if the heat capacity of a material is essentially constant over the temperature range under consideration and if the equation-of-state effect is negligible, then the change in internal energy or enthalpy is calculated very simply as the product of the heat capacity and the change in temperature:

$$\widehat{C}_V = \text{constant and} \left(\frac{\partial \widehat{U}}{\partial \widehat{V}}\right)_T \approx 0: \qquad \widehat{U}_2 - \widehat{U}_1 = \widehat{C}_V(T_2 - T_1) \tag{15-18}$$

or

$$\widehat{C}_P = \text{constant and} \left(\frac{\partial \widehat{H}}{\partial P}\right)_T \approx 0: \qquad \widehat{H}_2 - \widehat{H}_1 = \widehat{C}_P(T_2 - T_1) \tag{15-19}$$

Changes in entropy also can be calculated in terms of the heat capacities and equation-of-state functions. The result in terms of the constant-volume heat capacity is:

$$d\widehat{S} = \left(\frac{\widehat{C}_V dT}{T}\right) + \left(\frac{\partial P}{\partial T}\right)_{\widehat{V}} d\widehat{V} \qquad (15-20)$$

and that in terms of constant-pressure heat capacity is:

$$d\widehat{S} = \left(\frac{\widehat{C}_P dT}{T}\right) - \left(\frac{\partial \widehat{V}}{\partial T}\right)_P dP \qquad (15-21)$$

Although we give all of these functions in dual form, that is, in terms of either constant volume or constant pressure heat capacities, the constant pressure forms are most often the more appropriate simply because it is usually the constant pressure heat capacities which are measured experimentally. As we will show later in section 15.3.6, if we know one of the heat capacities, we can calculate the other from equation-of-state relations. One form of the result is:

$$\widehat{C}_P = \widehat{C}_V + T\left(\frac{\partial P}{\partial T}\right)_{\widehat{V}}\left(\frac{\partial \widehat{V}}{\partial T}\right)_P \qquad (15-22)$$

Furthermore, note that changes in internal energy can be calculated directly from changes in enthalpy (and hence constant pressure heat capacity) by using the definition of enthalpy in terms of internal energy : $\widehat{H} = \widehat{U} + P\widehat{V}$. Hence

$$\begin{aligned} d\widehat{U} &= d\widehat{H} - d(P\widehat{V}) \\ &= d\widehat{H} - P d\widehat{V} - \widehat{V} dP \end{aligned} \qquad (15-23)$$

and

$$d\widehat{U} = \widehat{C}_P dT - \left[P d\widehat{V} + T\left(\frac{\partial \widehat{V}}{\partial T}\right)_P dP\right] \qquad (15-24)$$

Equations 15-14 (or 15-24), 15-15, and 15-21 are the fundamental results for calculating finite changes in internal energy, enthalpy, and entropy in terms of observable heat capacities and equations of state. These results may appear in alternate forms, as discussed later in section 15.5 but it should always be remembered that these results are the fundamental working relationships. Again, note that T is *absolute* temperature.

One further note should be made concerning terminology. Processes which hold one variable constant are named accordingly. For example, a *constant-temperature process* moves along an *isotherm* and is *isothermal*. A *constant-pressure process* moves along an *isobar* and is *isobaric*. A *constant-volume process* moves along an *isochore* and is *isochoric* (also referred to as *isometric*). A *constant-entropy process* moves along an *isentrope* and is *isentropic* (note "isen-" not "iso-").

Example 15-1. Calculate the change in $\widehat{U}$, $\widehat{H}$, and $\widehat{S}$ for nitrogen that undergoes a change in state from $T_1 = 25\ °C$ to $T_2 = 200$ °C at 1 atm as shown in Figure 15-6. Assume that nitrogen behaves according to the ideal gas equation of state $P\widehat{V} = RT\,/\,(MW)$ and the constant-pressure specific heat capacity (for a diatomic molecule) is: $\widehat{C}_P = \dfrac{7R}{2(MW)}$

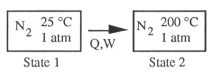

State 1　　　　　　State 2

Figure 15-6: Change in State

SOLUTION: The change in state is shown on the P-$\widehat{V}$ and P-T diagrams in Figure 15-7. To calculate the change in specific internal energy, specific enthalpy, and specific entropy, we have available the following equations.

$$\widehat{U}_2 - \widehat{U}_1 = \int_{T_1}^{T_2} \widehat{C}_V\, dT + \int_{\widehat{V}_1}^{\widehat{V}_2} \left[T\left(\frac{\partial P}{\partial T}\right)_{\widehat{V}} - P \right] d\widehat{V}$$

$$\widehat{H}_2 - \widehat{H}_1 = \int_{T_1}^{T_2} \widehat{C}_P\, dT + \int_{P_1}^{P_2} \left[\widehat{V} - T\left(\frac{\partial \widehat{V}}{\partial T}\right)_P \right] dP$$

$$\widehat{S}_2 - \widehat{S}_1 = \int_{T_1}^{T_2} \frac{\widehat{C}_P\, dT}{T} - \int_{P_1}^{P_2} \left(\frac{\partial \widehat{V}}{\partial T}\right)_P dP$$

However, the most convenient one to use is enthalpy because we are given expressions for the constant pressure heat capacities.

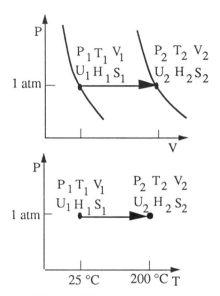

Figure 15-7: Changes in State
for Example 15-1

After the change in enthalpy is calculated, the change in internal energy is calculated as:

$$\widehat{U}_2 - \widehat{U}_1 = (\widehat{H}_2 - \widehat{H}_1) - (P_2\widehat{V}_2 - P_1\widehat{V}_1)$$

The change in enthalpy for a constant pressure process is:

$$\widehat{H}_2 - \widehat{H}_1 = \int_{T_1}^{T_2} \widehat{C}_P\, dT$$

Substituting the expression for $\widehat{C}_P$ yields:

$$\widehat{H}_2 - \widehat{H}_1 = \int_{T_1}^{T_2} \frac{7R}{2(MW)} dT$$

Since the gas constant and molecular weight are constant the integration is:

$$\widehat{H}_2 - \widehat{H}_1 = \frac{7R}{2(MW)}(T_2 - T_1)$$

The value for the gas constant is found in Table A-4 in Appendix A to be $R = 8.314$ J / gmol K and the molecular weight of nitrogen is 28 g / gmol. Furthermore, the final and initial absolute temperatures are:

$$T_1 = 273.15 + 25 = 298 \text{ K} \qquad T_2 = 273.15 + 200 = 473 \text{ K}$$

Substituting values gives:

$$\widehat{H}_2 - \widehat{H}_1 = \left(\frac{7(8.314)}{2(28)}\right)\left(473 - 298\right) = 181.9 \text{ J / g}$$

The change in enthalpy of the nitrogen from going from state 1 to state 2 was 181.9 (J/g).

Now the change in internal energy of the gas is determined from:

$$\widehat{U}_2 - \widehat{U}_1 = (\widehat{H}_2 - \widehat{H}_1) - (P_2\widehat{V}_2 - P_1\widehat{V}_1)$$

The product of the pressure and the volume at the different states can be related to the temperature at the different states by the equation of state.

$$\widehat{U}_2 - \widehat{U}_1 = (\widehat{H}_2 - \widehat{H}_1) - (P_2\widehat{V}_2 - P_1\widehat{V}_1)$$

$$= (\widehat{H}_2 - \widehat{H}_1) - \left(\frac{RT_2}{(MW)} - \frac{RT_1}{(MW)}\right)$$

$$= (\widehat{H}_2 - \widehat{H}_1) - \left(\frac{R}{(MW)}\right)(T_2 - T_1)$$

Substituting values into this equation gives:

$$\widehat{U}_2 - \widehat{U}_1 = 181.9 - \left(\frac{8.314}{28}\right)(473 - 298) = 129.9 \text{ J / g}$$

The change in specific internal energy of the gas going from state 1 to state 2 is 129.9 (J/g).

Finally, the change in entropy of the gas is calculated. Since the process is a constant pressure process, the change in entropy is calculated as:

$$\widehat{S}_2 - \widehat{S}_1 = \int_{T_1}^{T_2} \frac{\widehat{C}_P dT}{T}$$

Substituting in the expression for the constant pressure specific heat capacity gives:

$$\widehat{S}_2 - \widehat{S}_1 = \int_{T_1}^{T_2} \frac{7R dT}{2(MW)T}$$

Factoring out the constants R and MW and integrating gives:

$$\widehat{S}_2 - \widehat{S}_1 = \frac{7R}{2(MW)} \ln\left(\frac{T_2}{T_1}\right)$$

Substituting values gives:

$$\widehat{S}_2 - \widehat{S}_1 = \frac{7(8.314)}{2(28)} \ln\left(\frac{473}{298}\right) = 0.480 \text{ J / g K}$$

The change in entropy for the gas going from state 1 to state 2 is 0.480 J / g K. You should verify the units in all of these calculations.

Example 15-2. Air (MW = 29) expands from P_1 = 10 bar to P_2 = 3 bar and an initial temperature of T_1 = 75 °C to a final temperature of T_2 = 95 °C as shown in Figure 15-8. If the air obeys the ideal gas equation of state $P\widehat{V} = RT / (MW)$ and the dimensionless constant-pressure heat capacity in this ideal gas state is given by:

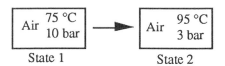

State 1   State 2

Figure 15-8: Change in State

$$\frac{\widetilde{C}_P(T)}{R} = A + BT + CT^2 + DT^{-2} + ET^3$$

where:

$$A = 3.355$$
$$B = 0.575 \times 10^{-3}$$
$$C = 0$$
$$D = -0.016 \times 10^5$$
$$E = 0$$

as shown in Table C-16 in Appendix C, calculate the change in $\widehat{U}$, $\widehat{H}$, and $\widehat{S}$ for this process.

SOLUTION: P-$\widehat{V}$ and P-T diagrams show the change in state of the air (Figure 15-9). To calculate the change in specific internal energy, specifc enthalpy, and specific entropy, we have

$$\widehat{U}_2 - \widehat{U}_1 = \int_{T_1}^{T_2} \widehat{C}_V \, dT + \int_{\widehat{V}_1}^{\widehat{V}_2} \left[T\left(\frac{\partial P}{\partial T}\right)_{\widehat{V}} - P \right] d\widehat{V}$$

$$\widehat{H}_2 - \widehat{H}_1 = \int_{T_1}^{T_2} \widehat{C}_P \, dT + \int_{P_1}^{P_2} \left[\widehat{V} - T\left(\frac{\partial \widehat{V}}{\partial T}\right)_P \right] dP$$

$$\widehat{S}_2 - \widehat{S}_1 = \int_{T_1}^{T_2} \frac{\widehat{C}_P \, dT}{T} - \int_{P_1}^{P_2} \left(\frac{\partial \widehat{V}}{\partial T}\right)_P dP$$

Because we are given the constant-pressure heat capacity data, it is more convenient to calculate first the change in enthalpy and then calculate the change in internal energy from the definition of enthalpy.

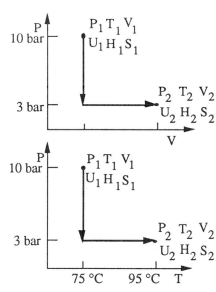

Figure 15-9: Changes in state for Example 15-2

The equation for the change in enthalpy, says that the enthalpy changes due to changes in temperature and changes in pressure. Then, because the pressure is not constant for this process, we need to determine the expression for

$$\left[\widehat{V} - T\left(\frac{\partial \widehat{V}}{\partial T}\right)_P \right]$$

from the provided equation of state.

First the partial derivative is determined. From the ideal gas equation of state, we can write the volume as a function of the pressure by the following equation:

$$\widehat{V} = \frac{RT}{(MW)P}$$

Differentiating this function with respect to the variable T at constant P (treating P as a constant) yields:

$$\left(\frac{\partial \widehat{V}}{\partial T}\right)_P = \frac{R}{(MW)P}$$

and substituting back gives:

$$\left[\widehat{V} - T\left(\frac{\partial \widehat{V}}{\partial T}\right)_P\right] = \left[\widehat{V} - T\frac{R}{(MW)P}\right]$$

again using the equation of state, this relationship reduces to:

$$\text{Ideal Gas}: \quad \left[\widehat{V} - T\left(\frac{\partial \widehat{V}}{\partial T}\right)_P\right] = \left[\widehat{V} - \widehat{V}\right] \equiv 0$$

We have shown that if the gas behaves as an ideal gas, then the change in enthalpy is only a function of the change in temperature or:

$$\widehat{H}_2 - \widehat{H}_1 = \int_{T_1}^{T_2} \widehat{C}_P dT$$

even though the gas changed pressure.

In terms of the given heat capacity relation, the specific constant-pressure heat capacity is:

$$\widehat{C}_P = \left(\frac{R}{(MW)}\right)\left(\frac{\widetilde{C}_P(T)}{R}\right)$$

Because constant pressure heat capacity data are given as a power function of temperature as shown in Tables C-15 through C-18 in Appendix C, a general expression for integrating the heat capacity is shown:

$$\int_{T_1}^{T_2} \frac{\widetilde{C}_P(T)}{R} dT = \int_{T_1}^{T_2} (A + BT + CT^2 + DT^{-2} + ET^3) dT$$

$$= \left[A(T_2 - T_1) + \frac{B}{2}(T_2^2 - T_1^2) + \frac{C}{3}(T_2^3 - T_1^3) - D\left(\frac{1}{T_2} - \frac{1}{T_1}\right) + \frac{E}{4}(T_2^4 - T_1^4)\right]$$

where the temperature is in Kelvin. It is important to remember that the previous expression is valid for all the materials whose heat capacities are given in the tables for the specified temperature range. The change in enthalpy from state 1 to state 2, then is:

$$\widehat{H}_2 - \widehat{H}_1 = \int_{T_1}^{T_2} \widehat{C}_P dT$$

$$= \left(\frac{R}{MW}\right)\left[A(T_2 - T_1) + \frac{B}{2}(T_2^2 - T_1^2) + \frac{C}{3}(T_2^3 - T_1^3) - D\left(\frac{1}{T_2} - \frac{1}{T_1}\right)\right]$$

Substituting the numerical values for the constants ($E = 0$), the gas constant ($R = 8.314$ J / mol K), $T_1 = 348$ K, and $T_2 = 368$ K gives:

$$\widehat{H}_2 - \widehat{H}_1 = \left(\frac{8.314}{29}\right)\left[(3.355)(368 - 348) + \frac{0.575 \times 10^{-3}}{2}(368^2 - 348^2)\right.$$

$$\left. + \frac{0}{3}(368^3 - 348^3) - (-0.016 \times 10^5)\left(\frac{1}{368} - \frac{1}{348}\right)\right]$$

$$= 20.4 \text{ J} / \text{g}$$

The change in $\widehat{H}$ of the air from state 1 to state is 20.4 J/g.

The change in internal energy then is calculated from the definition of the enthalpy as:

$$\widehat{U}_2 - \widehat{U}_1 = (\widehat{H}_2 - \widehat{H}_1) - (P_2\widehat{V}_2 - P_1\widehat{V}_1)$$

The product of the pressure and the volume at the different states can be related to the temperature at the different states by the ideal gas equation of state, and hence

$$\widehat{U}_2 - \widehat{U}_1 = (\widehat{H}_2 - \widehat{H}_1) - \left(\frac{RT_2}{MW} - \frac{RT_1}{MW}\right)$$

$$= (\widehat{H}_2 - \widehat{H}_1) - \left(\frac{R}{MW}\right)(T_2 - T_1)$$

Substituting values into this equation allows us to calculate the change in internal energy as:

$$\widehat{U}_2 - \widehat{U}_1 = 20.4 - \left(\frac{8.314}{29}\right)(368 - 348) = 14.6 \text{ J} / \text{g}$$

The change in $\widehat{U}$ of the air going from state 1 to state 2 is 14.6 J/g.

Finally, the change in specific entropy is calculated from:

$$\widehat{S}_2 - \widehat{S}_1 = \int_{T_1}^{T_2} \frac{\widehat{C}_P dT}{T} - \int_{P_1}^{P_2} \left(\frac{\partial \widehat{V}}{\partial T}\right)_P dP$$

This equation says that the change in $\widehat{S}$ is a function of both the change in the temperature and the change in pressure. Furthermore, the partial derivative of volume with respect to the temperature at constant pressure is:

$$\left(\frac{\partial \widehat{V}}{\partial T}\right)_P = \frac{R}{(MW)P}$$

Substituting this expression into the equation gives:

$$\widehat{S}_2 - \widehat{S}_1 = \int_{T_1}^{T_2} \frac{\widehat{C}_P dT}{T} - \int_{P_1}^{P_2} \frac{R}{(MW)P} dP$$

Because the integration of the constant pressure heat capacity divided by the temperature over the temperature range is required to calculate the change in entropy, a general expression will be derived as was done in the enthalpy calculation.

For the given dimensionless constant pressure heat capacities in Appendix C tables C-15 through C-18, the integral is:

$$\int_{T_1}^{T_2} \frac{\tilde{C}_P(T)}{RT}dT = \int_{T_1}^{T_2}\left(\frac{A}{T} + B + CT + \frac{D}{T^3} + ET^2\right)dT$$

$$= \left[A\ln\left(\frac{T_2}{T_1}\right) + B(T_2 - T_1) + \frac{C}{2}(T_2^2 - T_1^2) - \frac{D}{2}\left(\frac{1}{T_2^2} - \frac{1}{T_1^2}\right) + \frac{E}{3}(T_2^3 - T_1^3)\right]$$

In terms of $\widehat{C}_P$ the integral is:

$$\int_{T_1}^{T_2} \widehat{C}_P \frac{dT}{T} = \left(\frac{R}{MW}\right)\left[A\ln\left(\frac{T_2}{T_1}\right) + B(T_2 - T_1) + \frac{C}{2}(T_2^2 - T_1^2) - \frac{D}{2}\left(\frac{1}{T_2^2} - \frac{1}{T_1^2}\right) + \frac{E}{3}(T_2^3 - T_1^3)\right]$$

Again, these relationships are completely general for any of the materials whose constant-pressure heat capacities are given in Tables C-15 through C-18. Substituting the known values into the expression, allows use to calculate the integral:

$$\int_{T_1}^{T_2} \widehat{C}_P \frac{dT}{T} = \left(\frac{8.314}{29}\right)\Big[(3.355)\ln\left(\frac{368}{348}\right) + (0.575 \times 10^{-3})(368 - 348) +$$

$$\frac{0}{2}(368^2 - 348^2) - \frac{-0.016 \times 10^5}{2}\left(\frac{1}{368^2} - \frac{1}{348^2}\right)\Big]$$

The integral is:

$$\int_{T_1}^{T_2} \widehat{C}_P \frac{dT}{T} = 0.057 \text{ J / g K}$$

The effect of the change in pressure on the change in entropy is calculated:

$$\int_{P_1}^{P_2} \frac{R}{(MW)P}dP = \frac{R}{(MW)}\ln\left(\frac{P_2}{P_1}\right)$$

Substituting values gives

$$\int_{P_1}^{P_2} \frac{R}{(MW)P}dP = \frac{8.314}{29}\ln\left(\frac{3}{10}\right) = -0.345 \text{ J / g K}$$

Substituting these values allows us to calculate the change in entropy for the air from state 1 to state 2:

$$\widehat{S}_2 - \widehat{S}_1 = (0.057) - (-0.345) = 0.402 \text{ J / g K}$$

The change in $\widehat{S}$ of the air going from state 1 to state 2 is 0.402 J/g K.

Example 15-3. Calculate the change in $\widehat{U}$, $\widehat{H}$, and $\widehat{S}$ for hydrogen gas that is initially at $T_1 = 100$ °C and $V_1 = 100$ m³, and reaches a final state of $T_2 = 100$ °C and $V_2 = 200$ m³ as shown in Figure 15-10. The gas obeys the ideal gas equation of state and the dimensionless constant-pressure heat capacity is:

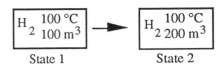

Figure 15-10: Change in State

$$\frac{\tilde{C}_P}{R} = A + BT + CT^2 + DT^{-2} + ET^3$$

where the constants are:

$$A = 3.249$$
$$B = 0.422 \times 10^{-3}$$
$$C = 0$$
$$D = 0.083 \times 10^5$$
$$E = 0$$

from Table C-16 in Appendix C.

SOLUTION: The change of state is shown in the P-T and P-$\widehat{V}$ diagrams (Figure 15-11). The molecular weight of hydrogen is 2 g/gmol, and the gas constant R is 8.314 J/mol K. The equations to calculate the changes in $\widehat{U}$, $\widehat{H}$, and $\widehat{S}$ are given below:

$$\widehat{U}_2 - \widehat{U}_1 = \int_{T_1}^{T_2} \widehat{C}_V\, dT + \int_{\widehat{V}_1}^{\widehat{V}_2} \left[T\left(\frac{\partial P}{\partial T}\right)_{\widehat{V}} - P \right] d\widehat{V}$$

$$\widehat{H}_2 - \widehat{H}_1 = \int_{T_1}^{T_2} \widehat{C}_P\, dT + \int_{P_1}^{P_2} \left[\widehat{V} - T\left(\frac{\partial \widehat{V}}{\partial T}\right)_P \right] dP$$

$$\widehat{S}_2 - \widehat{S}_1 = \int_{T_1}^{T_2} \frac{\widehat{C}_P\, dT}{T} - \int_{P_1}^{P_2} \left(\frac{\partial \widehat{V}}{\partial T}\right)_P dP$$

Because we have data about the constant pressure heat capacities, is it convenient to calculate the change in enthalpy first, and then, calculate the change in internal energy from the definition of enthalpy.

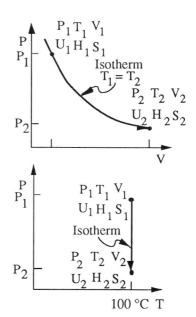

Figure 15-11: Changes in States for Example 15-3

The change in enthalpy for the ideal gas behavior is (see example 15-2):

$$\widehat{H}_2 - \widehat{H}_1 = \int_{T_1}^{T_2} \widehat{C}_P\, dT$$

Then, since there is no change in temperature ($T_1 = T_2$) the heat capacity integral is zero:

$$\widehat{H}_2 - \widehat{H}_1 = \int_{T_1}^{T_2} \widehat{C}_P\, dT$$
$$= 0 \text{ J} / \text{g}$$

The change in $\widehat{H}$ of an ideal gas for a constant temperature (isothermal) process is 0 J/g.

Now, let us calculate the change in $\widehat{U}$. From the definition of enthalpy,

$$\widehat{U}_2 - \widehat{U}_1 = (\widehat{H}_2 - \widehat{H}_1) - (P_2\widehat{V}_2 - P_1\widehat{V}_1)$$

which, from the equation of state is (example 15-2)

$$\widehat{U}_2 - \widehat{U}_1 = (\widehat{H}_2 - \widehat{H}_1) - \left(\frac{R}{(MW)}\right)(T_2 - T_1)$$

Since the process is isothermal ($T_1 = T_2$), the change in $\widehat{U}$ is:

$$\widehat{U}_2 - \widehat{U}_1 = (\widehat{H}_2 - \widehat{H}_1)$$
$$= 0 \, \text{J}/\text{g}$$

In fact, the change in $\widehat{U}$ for any isothermal process of an ideal gas is zero.

Finally, the change in $\widehat{S}$ is calculated by:

$$\widehat{S}_2 - \widehat{S}_1 = \int_{T_1}^{T_2} \frac{\widehat{C}_P dT}{T} - \int_{P_1}^{P_2} \left(\frac{\partial \widehat{V}}{\partial T}\right)_P dP$$

which, by using the equation of state (example 15-2) gives:

$$\widehat{S}_2 - \widehat{S}_1 = \int_{T_1}^{T_2} \frac{\widehat{C}_P dT}{T} - \int_{P_1}^{P_2} \frac{R}{(MW)P} dP$$

Since this process is an isothermal process, the temperature difference is zero and the change in entropy reduces to:

$$\widehat{S}_2 - \widehat{S}_1 = -\int_{P_1}^{P_2} \frac{R}{(MW)P} dP$$

Integrating this equation and evaluating the integral at the initial and final pressures gives:

$$\widehat{S}_2 - \widehat{S}_1 = -\frac{R}{(MW)} \ln\left(\frac{P_2}{P_1}\right)$$

We can relate the initial and final pressures to the initial and final volumes and temperatures through the equation of state to give:

$$\widehat{S}_2 - \widehat{S}_1 = -\frac{R}{(MW)} \ln\left(\frac{V_1}{V_2}\right)$$

Substituting values gives:

$$\widehat{S}_2 - \widehat{S}_1 = -\frac{8.314}{2} \ln\left(\frac{100}{200}\right) = 2.88 \, \text{J}/\text{g K}$$

The change in specific entropy of the nitrogen gas is 2.88 J/g K.

15.3.4 Reference States

The preceding equations allow us to calculate *changes* in enthalpy, internal energy, and entropy from one state to another; they do not allow us to calculate enthalpy, internal energy, or entropy in an absolute sense and in fact it is not possible to do so uniquely, nor is it necessary. All we really need are changes in enthalpy, internal energy, or entropy for use in the fundamental laws.

Nevertheless, we do find it convenient to assign values to these properties in a reference state which can then be used to calculate values in other states. This allows values to be tabulated or plotted over ranges of temperature and pressure from which differences between states can be readily calculated for use with the fundamental laws. When these differences are obtained, the reference-state value cancels, making its actual value immaterial.

Consequently, the selection of states is arbitrary and so long as we are consistent between calculations we may use whatever reference we wish. However, because the energy functions are related to each other (for example, $\widehat{H} = \widehat{U} + P\widehat{V}$, $\widehat{G} = \widehat{H} - T\widehat{S}$, etc.) we need only to define reference states for two of the functions and then from these, reference states for all of the others can be immediately calculated. Reference values usually are chosen either for enthalpy and entropy or for internal energy and entropy. Given a *reference state*, (with measured values T_o, P_o, V_o, etc.) if we assign *reference values* for *enthalpy and entropy*, then the reference values for all of the thermodynamic energy functions can be calculated:

$$\widehat{U}_o = \widehat{H}_o - (P\widehat{V})_o = \widehat{H}_o - P_o\widehat{V}_o \qquad (15-25)$$

$$\widehat{A}_o = \widehat{H}_o - (P\widehat{V})_o - (T\widehat{S})_o = \widehat{H}_o - P_o\widehat{V}_o - T_o\widehat{S}_o \qquad (15-26)$$

and

$$\widehat{G}_o = \widehat{H}_o - (T\widehat{S})_o = \widehat{H}_o - T_o\widehat{S}_o \qquad (15-27)$$

Likewise, if we define reference values for *internal energy and entropy* for the reference state then the reference values for the complete set are calculated as:

$$\widehat{H}_o = \widehat{U}_o + (P\widehat{V})_o = \widehat{U}_o + P_o\widehat{V}_o \qquad (15-28)$$

$$\widehat{A}_o = \widehat{U}_o - (T\widehat{S})_o = \widehat{U}_o - T_o\widehat{S}_o \qquad (15-29)$$

and

$$\widehat{G}_o = \widehat{U}_o + (P\widehat{V})_o - (T\widehat{S})_o = \widehat{U}_o + P_o\widehat{V}_o - T_o\widehat{S}_o \qquad (15-30)$$

As example reference states we see that in some steam tables (tables C-22 and C-23), the triple point of water (liquid, ice, and vapor are in simultaneous equilibrium) is selected as the reference state at which the internal energy and entropy of the liquid are arbitrarily defined to be zero. Again these reference state values, and indeed the reference state itself, are arbitrary. The point is that once the reference state is established and two of the thermodynamic functions are arbitrarily assigned values, then all the other thermodynamic functions can be calculated from the relationships between the energy functions and heat capacity and equation-of-state data.

Reference states which have been used for other systems and other situations are given below. Note that, consistent with the arbitrariness of reference states, the choices have not been unique.

Table 15-1: Reference States

Material	Reference State	Assigned Reference Value	
Water[a]	Triple Point	$\widehat{U}_{\text{liq}} = 0$	$\widehat{S}_{\text{liq}} = 0$
Water[b]	Saturated Liquid at 32° F	$\widehat{H}_{\text{liq}} = 0$	$\widehat{S}_{\text{liq}} = 0$
Ammonia[c]	Saturated Liquid at -40° F	$\widehat{H}_{\text{liq}} = 0$	$\widehat{S}_{\text{liq}} = 0$
Carbon dioxide[d]	Saturated Liquid at -40° F	$\widehat{H}_{\text{liq}} = 0$	$\widehat{S}_{\text{liq}} = 0$
Gases	0° C, 1 atm (Ideal Gas)	$\widehat{H} = 0$	$\widehat{S} = 0$
Gases	0° F, 1 atm (Ideal Gas)	$\widehat{H} = 0$	$\widehat{S} = 0$
Gases	60° F, 1 atm (Ideal Gas)	$\widehat{H} = 0$	$\widehat{S} = 0$
Gases	25° C, 1 atm (Ideal Gas)	$\widehat{H} = 0$	$\widehat{S} = 0$

a) Reference 0.01° C, 0.611 kPa
b) Reference 32° F, 0.08854 psia
c) Reference -40° F, 10.41 psia
d) Reference -40° F, 140 psia

5.3.5 Calculations of the Gibbs and Helmholtz Free Energies

The above equations allow us to calculate three of the five $UHAGS$ relations: $\widehat{U}, \widehat{H}, \widehat{S}$. We have not yet discussed calculations of the Helmholtz and Gibbs free energies, $\widehat{A}$ and $\widehat{G}$. With reference states included in the calculations of $\widehat{U}$, $\widehat{H}, \widehat{S}$ we can now obtain values for $\widehat{G}$ and $\widehat{A}$. For the Gibbs free energy it is convenient to work with $(\widehat{G} / RT)$ and recognize that:

$$d\left(\frac{\widehat{G}}{RT}\right) = \left(\frac{1}{RT}\right)d\widehat{G} - \left(\frac{\widehat{G}}{RT^2}\right)dT \qquad (15-31)$$

Then because of the $UHAG$ relations for $\widehat{G}$, i.e. that $d\widehat{G} = -\widehat{S}dT + \widehat{V}dP$ and the definition of $\widehat{G}$, i.e. $\widehat{G} = \widehat{H} - T\widehat{S}$ we obtain:

$$d\left(\frac{\widehat{G}}{RT}\right) = \left(-\frac{\widehat{S}}{RT}dT + \frac{\widehat{V}}{RT}dP\right) - \left(\frac{\widehat{H}}{RT^2} - \frac{T\widehat{S}}{RT^2}\right)dT \qquad (15-32)$$

which reduces to:

$$d\left(\frac{\widehat{G}}{RT}\right) = \left(\frac{\widehat{V}}{RT}\right)dP - \left(\frac{\widehat{H}}{RT^2}\right)dT \qquad (15-33)$$

As a result, changes in $(\widehat{G} / RT)$ (rather than $\widehat{G}$ itself) can be conveniently calculated in terms of equation of state data and enthalpy relations (including a reference state, necessary for calculating actual values for $\widehat{H}$).

Similarly, for the Helmholtz free energy we can write:

$$d\left(\frac{\widehat{A}}{RT}\right) = \left(\frac{1}{RT}\right)d\widehat{A} - \left(\frac{\widehat{A}}{RT^2}\right)dT \qquad (15-34)$$

and using the relation that $d\widehat{A} = -\widehat{S}dT - Pd\widehat{V}$ and that $\widehat{A} = \widehat{U} - T\widehat{S}$ this reduces to:

$$d\left(\frac{\widehat{A}}{RT}\right) = -\left(\frac{P}{RT}\right)d\widehat{V} - \left(\frac{\widehat{U}}{RT^2}\right)dT \qquad (15-35)$$

so that changes in $(\widehat{A}/RT)$ can be calculated from equation of state data and the internal energy.

All of the $UHAGS$ properties now are related directly to heat capacity and equation-of-state data and arbitrary reference states.

Example 15-4. From the data in example 15-1 and a reference state defined as:

$$T_o = 0\ ^\circ C \quad \text{and} \quad P_o = 1\ \text{atm} \quad \widehat{H}_o = 0\ \text{J}/\text{g}$$
$$\widehat{S}_o = 0\ \text{J}/\text{g K}$$

calculate the change in $\widehat{G}$ and $\widehat{A}$ for nitrogen gas ($MW = 28$) going from state 1 to state 2 as shown in Figure 15-12.

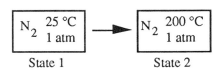

Figure 15-12: Changes in State

SOLUTION: The change in state is shown in P-T and P-$\widehat{V}$ diagrams below. The values for the changes in $\widehat{H}$ and $\widehat{U}$ are (from example 15-1):

$$\widehat{H}_2 - \widehat{H}_1 = 181.9\ \text{J}/\text{g}$$
$$\widehat{U}_2 - \widehat{U}_1 = 129.9\ \text{J}/\text{g}$$

The constant pressure heat capacity is: $\widehat{C}_P = \dfrac{7R}{2(MW)}$. There are several ways to do this calculation but the easiest and most straight forward simply comes from the definition of the Helmholtz and Gibbs free energies.

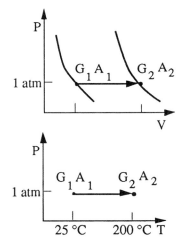

Figure 15-13: Change in State
for Example 15-4

The changes in the Helmholtz (A) and Gibbs (G) free energy are:

$$\widehat{G}_2 - \widehat{G}_1 = (\widehat{H}_2 - T_2\widehat{S}_2) - (\widehat{H}_1 - T_1\widehat{S}_1)$$
$$= (\widehat{H}_2 - \widehat{H}_1) - (T_2\widehat{S}_2 - T_1\widehat{S}_1)$$

and:

$$\widehat{A}_2 - \widehat{A}_1 = (\widehat{U}_2 - T_2\widehat{S}_2) - (\widehat{U}_1 - T_1\widehat{S}_1)$$
$$= (\widehat{U}_2 - \widehat{U}_1) - (T_2\widehat{S}_2 - T_1\widehat{S}_1)$$

The changes in $\widehat{U}$ and $\widehat{H}$ have already been calculated in example 15-1. However, the entropy at state 1 and the entropy at state 2 must be calculated with respect to the defined reference.

The entropy of the gas with respect to the reference is calculated as:

$$\widehat{S} - \widehat{S}_o = \int_{T_o}^{T} \frac{\widehat{C}_P dT}{T} - \int_{P_o}^{P} \left(\frac{\partial \widehat{V}}{\partial T} \right)_P dP$$

Where the values of T_o and P_o are the reference temperature and pressure. If the value of $\widehat{S}_o$ is defined, this equation is rearranged to calculate the value of $\widehat{S}$ with respect to the defined reference:

$$\widehat{S} = \int_{T_o}^{T} \frac{\widehat{C}_P dT}{T} - \int_{P_o}^{P} \left(\frac{\partial \widehat{V}}{\partial T} \right)_P dP + \widehat{S}_o$$

For this problem, the actual pressure and the reference pressure are the same (both states are at 1 atm) so the equation reduces to:

$$\widehat{S} = \int_{T_o}^{T} \frac{\widehat{C}_P dT}{T} + \widehat{S}_o$$

Substituting the known heat capacity into the equation and integrating yields:

$$\widehat{S} = \frac{7R}{2(MW)} \ln \left(\frac{T}{T_o} \right) + \widehat{S}_o$$

Substituting values gives

$$\widehat{S}_1 = \frac{7(8.314)}{2(28)} \ln \left(\frac{298}{273} \right) + 0 = 0.091 \text{ J} / \text{g K}$$

The entropy at state 2 is:

$$\widehat{S}_2 = \frac{7R}{2(MW)} \ln \left(\frac{T_2}{T_o} \right) + \widehat{S}_o$$

$$\widehat{S}_2 = \frac{7(8.314)}{2(28)} \ln \left(\frac{473}{273} \right) + 0 = 0.571 \text{ J} / \text{g K}$$

Substituting these values into the equations for the change in Gibbs and Helmholts free energies along with the other known quantities gives:

$$\widehat{G}_2 - \widehat{G}_1 = (\widehat{H}_2 - \widehat{H}_1) - (T_2\widehat{S}_2 - T_1\widehat{S}_1)$$

$$= (181.9) - \left((473)(0.571) - (298)(0.091) \right) = -61.1 \text{ J} / \text{g}$$

and:

$$\widehat{A}_2 - \widehat{A}_1 = (\widehat{U}_2 - \widehat{U}_1) - (T_2\widehat{S}_2 - T_1\widehat{S}_1)$$

$$= (129.9) - \left((473)(0.571) - (298)(0.091) \right) = -113 \text{ J} / \text{g}$$

The Gibbs free energy decreases by 61.1 J/g in going from state 1 to state 2. and Helmholtz free energy decrease by -113 J/g.

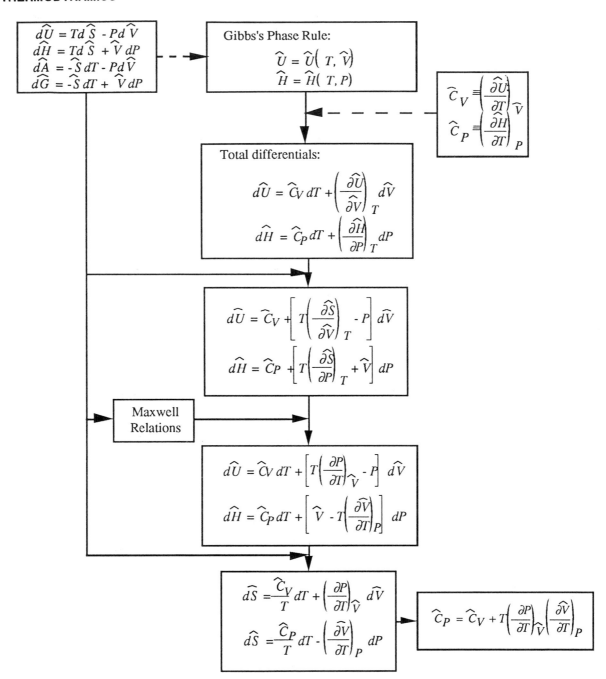

Figure 15-14: Expression of the thermodynamic properties in terms of thermal (heat capacity) and volumetric (P-V-T) observable quantities.

15.3.6 More Detail on Thermodynamic Property Calculations

We now provide details on obtaining the bottom-line property relations (Equations 15-14 and 15-15) and related equations for entropy (Equations 15-20 and 15-21) in terms of observable heat capacity and $P\text{-}\widehat{V}\text{-}T$) (equation of state) relations. These are the details relating boxes 6 and 7 of Figure 15-1.

The steps involved in the derivation are outlined schematically in Figure 15-14. The Gibbs phase rule, the fundamental property relations, and mathematics relating total and partial derivatives (which includes the Maxwell relations, to be discussed later) are used to obtain the results.

The Gibbs Phase Rule. The phase rule of J. Willard Gibbs establishes the number of independent variables which are required to define a given system with respect to equilibrium. The rule applies only to equilibrium and says nothing whatsoever about nonequilibrium situations. It begins with the premise (based on physical observation and supported by the fundamental property relations) that a single-phase, single-component system requires that two independent intensive variables be specified in order to fully define the state; there are *two degrees of freedom*. Example intesive variables are T, P, $\widehat{V}$, $\widehat{U}$, $\widehat{H}$, $\widehat{A}$, $\widehat{G}$, and $\widehat{S}$.

For example, if we have a single component, homogeneous (everywhere the same) gas phase, then specifying the temperature and pressure of that gas phase (as well as the material) is sufficient to completely specify other properties such as volume, internal energy, entropy, etc. Again, we are talking about intensive properties, that is, temperature, pressure, the amount per unit mass (energy per unit mass, enthalpy per mass, entropy per mass, volume per mass etc.) If, in addition to the property values per unit mass, we also know the total amount of mass, then we also know the total properties, total internal energy, enthalpy, entropy, volume, etc.

Now if instead of a gas phase, we have a single component as a vapor in equilibrium with its own liquid, then, *by specifying that equilibrium exists, we lose one degree of freedom*, that is, we have an additional constraint placed upon the condition of the system, resulting in one less independent variable. Now instead of two independent intensive variables, we would have to specify only one independent variable (one degree of freedom) along with the constraint that vapor-liquid equilibrium exists in order to fully define the state of the system.

For example, if we have water at one atmosphere of total pressure and equilibrium between the vapor and liquid, then the temperature of the water must be 100 °C (212 °F).

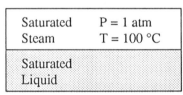

Figure 15-15: Vapor Liquid Equilibrium

Similarly, if we had liquid water in equilibrium with its own solid (ice) still at one atmosphere of pressure, then that is sufficient to specify the exact state of the system and we would know that the temperature must be 0 °C (32 °F). In addition to knowing the temperature, we could just as well state that the internal energy per unit mass of liquid water and of the solid ice are now established (relative to the defined reference value, of course).

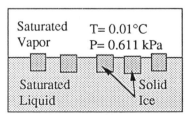

Figure 15-16: Vapor, liquid, and solid equilibrium

Now to look at another situation, if the original gas phase consisted of two components instead of one, then there would be one additional degree of freedom, that is, one more independent variable would have to be specified in order to fully establish the state of the system. Consequently for a single phase gas consisting of two components, the number of independent variables that would have to be specified is three (e.g., T, P, and one fractional composition such as mole or mass fraction).

Gibbs's phase rule can be generalized to the following equation.

$$\boxed{F = C + 2 - \phi}$$

$$(15-36)$$

where:

F = the number of degrees of freedom (independent variables)

C = the number of components.

ϕ = the number of phases.

Each additional component adds a degree of freedom and each additional phase subtracts a degree of freedom. The 2 comes from the initial postulate that two degrees of freedom exist for one component and one phase. So for a single phase, single component system we would say that the internal energy per unit mass (an intensive state property) can be specified as a function of two independent variables, say temperature and volume (per unit mass):

$$\widehat{U} = \widehat{U}(T, \widehat{V})$$

$$(15-37)$$

If T and $\widehat{V}$ are specified, then $\widehat{U}$ is fully defined for a given material. If we add a second phase in equilibrium, then we would have only one independent variable

$$\widehat{U} = \widehat{U}(T)$$

$$(15-38)$$

If we have a single phase and two components, then we would have a third independent variable required to fully specify the state so that we could write:

$$\widehat{U} = \widehat{U}(T, \widehat{V}, x)$$

$$(15-39)$$

In the last equation, the third variable chosen is the mole fraction of one of the components. An alternative way of expressing this would be to say that the total internal energy of the system in this case would be a function of temperature, pressure and the number of moles of one species along with the total number of moles, i.e.,

$$n\widetilde{U} = U(T, P, n_1, n)$$

$$(15-40)$$

The last equation looks like there are four independent variables. However, here we are writing the total system internal energy and not the internal energy per unit mass, so that in addition to specifying the composition we must also specify the total number of moles present in the system.

Example 15-5. For the given systems at equilibrium, use the Gibb's Phase Rule to determine the number of independent variables that are necessary to fix or define completely the state of the system.

a. Pure liquid water not in equilibrium with vapor or solid.

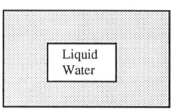

Figure 15-17: Pure liquid water

b. A solution of liquid salt water in equilibrium with the vapor

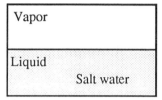

Figure 15-18: Salt solution in equilibrium with the vapor

c. Solid sulphur in two different equilbrium crystal structures.

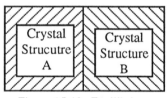

Figure 15-19: Equilibrium crystal structures of sulphur

d. A solution of liquid alcohol and water in equilibrium with solid

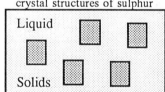

Figure 15-20: Alcohol-water solid-liquid equilibrium

SOLUTION: The Gibbs phase rule is:

$$F = C + 2 - \phi$$

where F is the number of independent variables, C is the number of components, and ϕ is the number of phases in equilibrium.

a. For this system, there is only one component and only one phase, so Gibbs phase rule says that 2 independent intensive variables must be specified to fix the state of the system:

$$F = 1 + 2 - 1$$
$$= 2$$

b. For this system, there are two components and two phases, so Gibbs phase rule says that 2 independent intensive variables must be specified to fix the state of the system:

$$F = 2 + 2 - 2$$
$$= 2$$

c. For this system, there is only one component and two phases, so Gibbs phase rule says that 1 independent intensive variable must be specified to fix the state of the system:

$$F = 1 + 2 - 2$$
$$= 1$$

d. For this system, there are two components and two phases, so the Gibbs phase rule says that 2 independent intensive variables must be specified to fix the state of the system:

$$F = 2 + 2 - 2$$
$$= 2$$

An Intermediate Relation for $d\widehat{U}$. With Gibbs phase rule now at our disposal, we know how many independent intensive variables are required to fully specify a state and hence define all of the other independent variables. This is crucial information for converting the fundamental property relations to a form that involve only observable properties. We write internal energy for a single component single phase as a function of two independent variables:

$$\widehat{U} = \widehat{U}(T, \widehat{V}) \tag{15 - 41}$$

We could pick temperature and pressure as the two independent variables, however, for reasons which are not immediately obvious to the first-time observer, we will pick temperature and volume. Now with internal energy written as a function of temperature and volume, mathematically we can expand this in terms of partial derivatives and differentials in the two independent variables as:

$$d\widehat{U} = \left(\frac{\partial \widehat{U}}{\partial T}\right)_{\widehat{V}} dT + \left(\frac{\partial \widehat{U}}{\partial \widehat{V}}\right)_{T} d\widehat{V} \tag{15 - 42}$$

In this result, $(\partial \widehat{U} / \partial T)_{\widehat{V}}$ is defined to be the constant-volume heat capacity:

$$\boxed{\widehat{C}_V \equiv \left(\frac{\partial \widehat{U}}{\partial T}\right)_{\widehat{V}}} \tag{15 - 43}$$

This is the constant-volume heat capacity and, it turns out, is a property which we can measure experimentally. If we have a system and maintain constant volume and carry out heat transfer experiments, we can very precisely measure the heat transfer to the closed system at constant volume. We can also measure the temperature change of the body, and from the thermal energy equation, we have that:

$$d\widehat{U} = \delta\widehat{Q} + \delta\widehat{W} \tag{15 - 44}$$

The change in volume is zero so reversible work $-Pd\widehat{V}$, is also zero and the equation reduces to (provided there is no conversions of mechanical energy to $\widehat{U}$ through irreversibilities, i.e., $\delta\widehat{F} = 0$ in eqaution 15-3):

$$d\widehat{U} = \delta\widehat{Q} \qquad (15-45)$$

The change in temperature dT is measured so equation (15-43) can be evaluated directly from experimental data:

$$\left(\frac{\partial\widehat{U}}{\partial T}\right)_{\widehat{V}} \approx \left(\frac{\Delta\widehat{U}}{\Delta T}\right)_{\widehat{V}} \approx \left(\frac{\delta\widehat{Q}}{dT}\right) \qquad (15-46)$$

Hence, by measuring the temperature change in response to a given heat transfer, we can calculate the change in internal energy. The measurement is made at constant volume simply by conducting it in a rigid constant volume cell.*

Although the first coefficient $\widehat{C}_V$ is one that we can measure directly, the second coefficient $(\partial\widehat{U} / \partial\widehat{V})_T$ would seem to pose more of a challenge but is still manageable. This term must be converted to another form in terms of properties which are measurable. This we do through the fundamental property and Maxwell relations. From the fundamental property relation for internal energy (equation 15-5) we can obtain an expression for this partial derivative by dividing by $d\widehat{V}$ at constant temperature:

$$\left(\frac{d\widehat{U}}{d\widehat{V}}\right)_T = T\left(\frac{d\widehat{S}}{d\widehat{V}}\right)_T - P\left(\frac{d\widehat{V}}{d\widehat{V}}\right)_T \qquad (15-47)$$

or

$$\left(\frac{\partial\widehat{U}}{\partial\widehat{V}}\right)_T = T\left(\frac{\partial\widehat{S}}{\partial\widehat{V}}\right)_T - P \qquad (15-48)$$

which gives:

$$d\widehat{U} = \widehat{C}_V\, dT + \left[T\left(\frac{\partial\widehat{S}}{\partial\widehat{V}}\right)_T - P\right]d\widehat{V} \qquad (15-49)$$

Now we have the needed partial derivative in terms of temperature and pressure which looks like an improvement because these are observable quantities. However, the partial derivative still is not directly measurable experimentally due to the dependence on entropy and must be converted further. This conversion is done through the use of the Maxwell relations.

The Maxwell Relations. A minor digression is necessary at this point to explain the existence of the Maxwell relations. If we have a function f which is written in terms of two independent variables,

$$f = f(x, y) \qquad (15-50)$$

then we can write an equation for the total differential change in f which arises from differential changes in x and y

$$df = \left(\frac{\partial f}{\partial x}\right)_y dx + \left(\frac{\partial f}{\partial y}\right)_x dy \qquad (15-51)$$

Now if we take the partial derivative multiplying dx and differentiate it with respect to y at constant x, we have:

$$\left[\frac{\partial}{\partial y}\left(\frac{\partial f}{\partial x}\right)_y\right]_x = \frac{\partial^2 f}{\partial y\,\partial x} \qquad (15-52)$$

---

* However, constant volume is not really as simple as it sounds. This is not a constant pressure process and as the pressure changes, the cell is not perfectly rigid and will adjust its volume somewhat, compromising the accuracy of the measurement.

and if we take the partial derivative multiplying dy and differentiate it with respect to x while holding y constant, we have:

$$\left[\frac{\partial}{\partial x} \left(\frac{\partial f}{\partial y} \right)_x \right]_y = \frac{\partial^2 f}{\partial x \, \partial y} \tag{15-53}$$

Now it is a result of mathematics and differential calculus that these two second partial derivatives are the same even though they are differentiated with respect to the two independent variables in opposite order, therefore,

$$\frac{\partial^2 f}{\partial y \, \partial x} = \frac{\partial^2 f}{\partial x \, \partial y} \tag{15-54}$$

If we look at each of the fundamental property relationships, we see that in each case we have a differential in a thermodynamic state property written in terms of differential changes in two other properties which are independent variables. For example, for the property relationship for $d\widehat{A}$, $\widehat{A}$ evidently is written as a function of temperature and volume $(\widehat{A} = \widehat{A}(T, \widehat{V}))$ so that $d\widehat{A}$ is:

$$d\widehat{A} = \left(\frac{\partial \widehat{A}}{\partial T} \right)_{\widehat{V}} dT + \left(\frac{\partial \widehat{A}}{\partial \widehat{V}} \right)_T d\widehat{V} \tag{15-55}$$

Then, comparing this equation to the relation $d\widehat{A}$, we have that:

$$-\widehat{S} = \left(\frac{\partial \widehat{A}}{\partial T} \right)_{\widehat{V}} \tag{15-56}$$

and

$$-P = \left(\frac{\partial \widehat{A}}{\partial \widehat{V}} \right)_T \tag{15-57}$$

We can write the other three fundamental property relationships in terms of their independent variables and show that the following relationships are true:

$$T = \left(\frac{\partial \widehat{U}}{\partial \widehat{S}} \right)_{\widehat{V}} = \left(\frac{\partial \widehat{H}}{\partial \widehat{S}} \right)_P \tag{15-58}$$

$$-P = \left(\frac{\partial \widehat{U}}{\partial \widehat{V}} \right)_{\widehat{S}} = \left(\frac{\partial \widehat{A}}{\partial \widehat{V}} \right)_T \tag{15-59}$$

$$\widehat{V} = \left(\frac{\partial \widehat{H}}{\partial P} \right)_{\widehat{S}} = \left(\frac{\partial \widehat{G}}{\partial P} \right)_T \tag{15-60}$$

$$-\widehat{S} = \left(\frac{\partial \widehat{A}}{\partial T} \right)_{\widehat{V}} = \left(\frac{\partial \widehat{G}}{\partial T} \right)_P \tag{15-61}$$

Furthermore, carrying out the cross partial derivatives and equating the terms gives

$$-\left(\frac{\partial \widehat{S}}{\partial \widehat{V}}\right)_T = \frac{\partial^2 \widehat{A}}{\partial \widehat{V}\,\partial T}$$

$$= \frac{\partial^2 \widehat{A}}{\partial T\,\partial \widehat{V}} \tag{15 – 62}$$

$$= -\left(\frac{\partial P}{\partial T}\right)_{\widehat{V}}$$

or

$$-\left(\frac{\partial \widehat{S}}{\partial \widehat{V}}\right)_T = -\left(\frac{\partial P}{\partial T}\right)_{\widehat{V}} \tag{15 – 63}$$

The entire set of Maxwell's relationships are calculated by carrying out these cross partial derivatives from equations 15-5, 5, 12, and 13. The Maxwell relationships are listed below:

$$\left(\frac{\partial T}{\partial \widehat{V}}\right)_{\widehat{S}} = -\left(\frac{\partial P}{\partial \widehat{S}}\right)_{\widehat{V}} \tag{15 – 64}$$

$$\left(\frac{\partial T}{\partial P}\right)_{\widehat{S}} = \left(\frac{\partial \widehat{V}}{\partial \widehat{S}}\right)_P \tag{15 – 65}$$

$$\left(\frac{\partial \widehat{S}}{\partial \widehat{V}}\right)_T = \left(\frac{\partial P}{\partial T}\right)_{\widehat{V}} \tag{15 – 66}$$

$$\left(\frac{\partial \widehat{S}}{\partial P}\right)_T = -\left(\frac{\partial \widehat{V}}{\partial T}\right)_P \tag{15 – 67}$$

The Bottom-Line Results. This result for $(\partial \widehat{S}/\partial \widehat{V})_T$ (equation 15-63), inserted into equation 15-49 gives the bottom-line result stated previously:

$$d\widehat{U} = \widehat{C}_V\,dT + \left[T\left(\frac{\partial P}{\partial T}\right)_{\widehat{V}} - P\right]d\widehat{V} \tag{15 – 14}$$

which is the desired relationship for changes in internal energy (per unit mass) of a single component in a single phase as a result of changes in temperature and volume and in terms of totally measurable properties, namely the constant-volume heat capacity and P-V-T relationships (equation of state). A result for enthalpy (given previously) can be obtained in a similar way and is:

$$d\widehat{H} = \widehat{C}_P dT + \left[\widehat{V} - T\left(\frac{\partial \widehat{V}}{\partial T}\right)_P\right] dP \qquad (15-15)$$

In the equations for $d\widehat{H}$, $\widehat{C}_P$ is the constant-pressure heat capacity and arises in a completely analogous way to the constant-volume heat capacity and is defined according to

$$\boxed{\widehat{C}_P \equiv \left(\frac{\partial \widehat{H}}{\partial T}\right)_P} \qquad (15-68)$$

In obtaining equation 15-15 enthalpy is written first as a function of temperature and pressure rather than temperature and volume. Again, as with internal energy, this is a set of independent variables which is convenient but its convenience is not immediately obvious. The convenience is seen through the second fundamental property relation (equation 15-6) which says that at constant pressure, reversible heat transfer ($Td\widehat{S} = \delta \widehat{Q}_{rev}$) causes changes in enthalpy. (See also equation 15-2 applied to a reversible constant pressure process.) Compare this to equation 15-45 which says that at constant volume heat transfer causes changes in internal energy. Consequently, at constant pressure, if we measure changes in the system temperature which result from known heat added to the system, we obtain the change in enthalpy with respect to temperature at constant pressure. Therefore, *when concerned with enthalpy*, the constant pressure heat capacity is *the appropriate observable property* and the corresponding logical dependence is $\widehat{H} = \widehat{H}(T, P)$.

For second-law calculations, we must also be able to calculate changes in entropy (per mass). Entropy is also a state property and it exists as a function of two independent intensive variables for a single component single phase situation. Equations for differential changes in entropy are easily obtained by combining the first two property relationships (equations 15-5 and 15-6) with equations (15-14) or (15-15) respectively. Thus, we obtain:

$$d\widehat{S} = \left(\frac{\widehat{C}_V}{T}\right) dT + \left(\frac{\partial P}{\partial T}\right)_{\widehat{V}} d\widehat{V} \qquad (15-20)$$

and

$$d\widehat{S} = \left(\frac{\widehat{C}_P}{T}\right) dT - \left(\frac{\partial \widehat{V}}{\partial T}\right)_P dP \qquad (15-21)$$

Incidentally, from there last two equations we can obtain a relationship between the constant-pressure and constant-volume heat capacities in terms of the P-$\widehat{V}$-T relations. That is,

$$\widehat{C}_P = \widehat{C}_V + T\left(\frac{\partial P}{\partial T}\right)_{\widehat{V}} \frac{d\widehat{V}}{dT} + T\left(\frac{\partial \widehat{V}}{\partial T}\right)_P \frac{dP}{dT} \qquad (15-69)$$

Now, we can use more facts of multivariable calculus to manipulate equation (15-69) to a more convenient form. $f = f(x, y)$ then

$$df = \left(\frac{\partial f}{\partial x}\right)_y dx + \left(\frac{\partial f}{\partial y}\right)_x dy \qquad (15-70)$$

from which

$$\left(\frac{\partial f}{\partial x}\right)_f = \left(\frac{\partial f}{\partial x}\right)_y \left(\frac{\partial x}{\partial x}\right)_f + \left(\frac{\partial f}{\partial y}\right)_x \left(\frac{\partial y}{\partial x}\right)_f \qquad (15-71)$$

or

$$0 = \left(\frac{\partial f}{\partial x}\right)_y + \left(\frac{\partial f}{\partial y}\right)_x \left(\frac{\partial y}{\partial x}\right)_f \qquad (15-72)$$

i.e.,

$$\left(\frac{\partial f}{\partial x}\right)_y = -\left(\frac{\partial f}{\partial y}\right)_x \left(\frac{\partial y}{\partial x}\right)_f \qquad (15-73)$$

Using this relationship, the constant-pressure heat capacity can be related to the constant-volume heat capacity in the following ways.

$$\widehat{C}_P = \widehat{C}_V + T\left(\frac{\partial P}{\partial T}\right)_{\widehat{V}} \left(\frac{\partial \widehat{V}}{\partial T}\right)_P \qquad (15-22)$$

$$\widehat{C}_P = \widehat{C}_V - T\left(\frac{\partial P}{\partial T}\right)_{\widehat{V}}^2 \left(\frac{\partial \widehat{V}}{\partial P}\right)_T \qquad (15-74)$$

Consequently, if we have an equation of state for a material, then we can calculate the difference between $\widehat{C}_P$ and $\widehat{C}_V$. Certain specific cases will be addressed later, but one immediate observation which is quite evident is that for an incompressible material where $(\partial \widehat{V} / \partial P)_T$ and $(\partial \widehat{V} / \partial T)_P$ are zero, $\widehat{C}_P$ and $\widehat{C}_V$ are the same. Because solids and liquids are very nearly incompressible, it follows that constant pressure and constant volume heat capacities for solids and liquids are essentially the same. For less dense materials, however, such as gases and supercritical fluids, compressibilities are significant and there are significant differences between the two heat capacities. Exactly what those differences are depends, as is indicated by equation (15-22) or (15-74), upon the P-$\widehat{V}$-T relations.

With this relationship between $\widehat{C}_V$ and $\widehat{C}_P$, we need only one kind of heat capacity plus P-$\widehat{V}$-T relation. Usually, constant-pressure heat capacities are more often reported in the literature.

In summary, the above relations for $d\widehat{U}$, $d\widehat{H}$, and $d\widehat{S}$ (Equations 15-14 or 15-24, 15-15, and 15-20 or 15-21) can be used to calculate changes in $\widehat{U}$, $\widehat{H}$, and $\widehat{S}$ from one state to another and in terms of experimentally measurable properties. All we need are heat capacity data and information on the manner in which temperature, pressure, and volume are interrelated. Again, these relationships are for single component and single phase systems which are of uniform properties. The P-$\widehat{V}$-T relationships are referred to as volumetric properties and are the subject of the next section.

Example 15-6. Show that,

$$\left(\frac{\partial \widehat{H}}{\partial P}\right)_T = \widehat{V} - T\left(\frac{\partial \widehat{V}}{\partial T}\right)_P$$

SOLUTION: First we will start with the UHAG relationship defining $\widehat{H}$ as:

$$d\widehat{H} = Td\widehat{S} + \widehat{V}dP$$

Now we divide both sides of the equation by dP and hold the temperature constant or:

$$\left(\frac{\partial \widehat{H}}{\partial P}\right)_T = T\left(\frac{\partial \widehat{S}}{\partial P}\right)_T + \widehat{V}\left(\frac{\partial P}{\partial P}\right)_T$$

or

$$\left(\frac{\partial \widehat{H}}{\partial P}\right)_T = T\left(\frac{\partial \widehat{S}}{\partial P}\right)_T + \widehat{V}$$

From the Maxwell relationships we can relate the partial derivative of entropy with respect to pressure at constant temperature to the partial derivative of the volume with respect to the temperature at constant pressure or:

$$\left(\frac{\partial \widehat{V}}{\partial T}\right)_P = -\left(\frac{\partial \widehat{S}}{\partial P}\right)_T$$

Substituting this expression back into the previous equation gives:

$$\left(\frac{\partial \widehat{H}}{\partial P}\right)_T = \widehat{V} - T\left(\frac{\partial \widehat{V}}{\partial T}\right)_P$$

15.4 VOLUMETRIC PROPERTIES OF MATERIALS

As was pointed out, in the above relations the interrelationships between the pressure, volume, and temperature of a material must be known in order to calculate the desired changes in $\widehat{U}$, $\widehat{H}$, or $\widehat{S}$. These P-$\widehat{V}$-T relationships are referred to as equations-of-state and an example is the ideal gas equation of state. Again, for a single component in a single phase we can express any one of these three state variables as a function of the other two. For example, we can express volume as the function of temperature and pressure (for an ideal gas, $\widehat{V} = RT / (P(MW))$). Stated somewhat differently, we can think of any two of these variables as independent variables and the third as a dependent variable which means that we must specify two in order to have the third defined. Mathematically, however, the relationships are not usually that explicit. In fact, the P-$\widehat{V}$-T relationships may be quite complex.

Typical volumetric properties in the form of P versus $\widehat{V}$ and P versus T are shown in Figures 15-21 and 15-22. From a pressure versus temperature viewpoint, typical behavior is characterized by three equilibrium boundaries between two phases (2ϕ): solid-vapor (sublimation curve, a two-phase or 2ϕ boundary), solid-liquid (fusion curve), and liquid-vapor (vaporization curve) regions. These curves intersect at a point of simultaneous equilibrium between all three phases (3ϕ), the triple point.

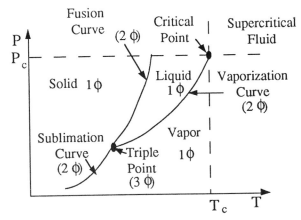

Figure 15-21: P-T diagram

An additional feature is that the vaporization curve terminates at a point where the difference between vapor and liquid ceases to exist (they have the same density and other properties at the temperature and pressure of this point); this point is called the critical point with temperature T_c and pressure P_c. Beyond this point, generally taken to mean the region where temperature and pressure both exceed the critical values, the material is referred to simply as a fluid.

For most materials, the fusion curve slopes as indicated; increasing pressure at constant temperature from a point near the fusion curve but in the liquid region causes freezing to the solid state and the solid is more dense than the liquid (the deviation of this fusion curve from the vertical is greatly exaggerated). For water, however, the reverse slope holds so that increasing pressure of the solid at constant temperature near the fusion curve causes melting and the liquid is more dense than the solid. This fact enables us to iceskate; the pressure exerted by the blade of the skates on the ice surface creates a thin film of water lubricant, making a very slippery surface. More importantly, when water freezes at the surface of our lakes and rivers, it remains at the surface to insulate the rest of the water rather than dropping to the bottom and allowing more water (and fish) to freeze.

From a pressure versus volume viewpoint, the equilibrium curves are opened up to two-phase regions (areas) and the triple point is opened up to a triple line. These regions represent points at which a homogeneous material does not exist; rather two phases (or three phases) of different density (specific volume) exist in equilibrium. Being in equilibrium, they are in pressure and thermal equilibrium; they are at the same pressure and temperature. Therefore, two phases which are in equilibrium lie at opposite ends of a constant pressure line which spans or traverses the two-phase region of interest.

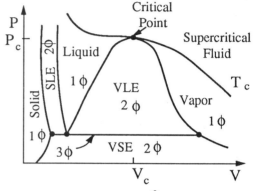

Figure 15-22: P-$\widehat{V}$ diagram

In the case of the triple line, there are points of the solid, liquid, and vapor regions which touch this line; solid, liquid, and vapor are in equilibrium. Note that within a single-phase region, pressure may be represented as a function of two independent variables, say $\widehat{V}$ and T, in accordance with the Gibbs phase rule. If two phases are in equilibrium, say liquid and vapor, then for a given temperature, there is only one pressure at which the behavior is two-phase, also in accordance with the Gibbs phase rule. Finally, if three phases are in equilibrium, then there is only one temperature-pressure pair, the triple line, at which this occurs.

To represent all of these rather complex behaviors in a single analytical equation of state obviously is a considerable challenge. The primary complication is accurately describing the widely different behavior of the different phases which occur at various conditions of temperature and pressure. Generally, at low pressure-high temperature regions, materials exist as gases for which the intermolecular distances, and therefore the intermolecular forces, are relatively small and do not dominate material behavior. At higher pressure-lower temperature regions, materials exist in a condensed state as liquids or solids with drastically smaller intermolecular distances so that collisions are much more frequent and forces are much larger and much more important to the volumetric behavior. Exactly what high temperature-low pressure or low temperature-high pressure means depends upon the particular material at hand. To describe these drastically different regions of behavior with a single analytical equation is almost more than can be expected, although the effort has been and continues to be made.

For this reason, equations-of-state normally are considered to be primarily for gases, or for liquids or solids; a single equation spanning two regions of phase behavior will not give highly accurate results over the entire domain. The equations considered below are divided into equations of state for gases and equations for liquids (and solids).

15.4.1 Equations of State for Gases: Ideal Gases

The simplest equation of state for gases, and one with which you are already familiar, is that for ideal gases (note: molar volume = $\tilde{V} = V/n$, specific volume = $\hat{V} = \tilde{V}/(MW)$):

$$PV = nRT \quad \text{or} \quad P\tilde{V} = RT \quad \text{or} \quad P\hat{V} = RT/(MW)$$

$(15-75)$

This model is applicable to gases when the size of the molecules is small compared to intermolecular distances and when the force fields between molecules are negligible. This is the situation for all gases approaching zero pressures at high temperatures. For this model the equation of state derivatives which are needed for enthalpy and internal energy calculations are ($\tilde{V} = V/n$):

$$\left(\frac{\partial \hat{V}}{\partial T}\right)_P = \frac{R}{P(MW)}$$

$(15-76)$

and

$$\left(\frac{\partial P}{\partial T}\right)_{\hat{V}} = \frac{R}{\hat{V}(MW)}$$

$(15-77)$

from which it is seen (combine these results with equations 15-14, 15-15, and 15-75) that for an ideal gas both enthalpy and internal energy do not change with pressure; the internal energy and enthalpy of an ideal gas are functions of temperature only and not of pressure or volume (see example 15-2). This is a direct consequence of the fact that ideal behavior insists that there are no forces between molecules so that bringing molecules closer together by changing pressure or volume at constant temperature results in no changes in energy. Likewise, an ideal gas undergoing a free expansion, that is, expanding against zero external force (i.e., into a vacuum) experiences no change in internal energy and therefore no change in temperature; there are no intermolecular forces for molecules to work against in moving farther apart.

The use of molar properties is convenient for the equations of state, and the conversion between the property per unit mole and property per unit mass is simply the molecular weight of the particular species, as noted above and repeated here:

$$\hat{V}(MW) = \tilde{V} \qquad \hat{U}(MW) = \tilde{U} \qquad \hat{C}_P(MW) = \tilde{C}_P$$

15.4.2 Equations of State for Gases: Virial Equations

These results which are observed for ideal gases are not true in general for real gases. For real gases there will be changes in internal energy or enthalpy due to compression at constant temperature and real gases undergoing a free expansion will change their temperature. While these effects generally are relatively small in the presence of other energy effects, such as heat transfer and work, they nevertheless do indicate deviations from ideal behavior.

Consequently, one of the first corrections that can be made to the ideal gas law as an alternative equation of state is to postulate a form which can be used to represent small deviations from the ideal behavior. For this representation it is convenient to express the ideal gas behavior in the form:

$$\frac{P\tilde{V}}{RT} = 1$$

$(15-78)$

Then, to the extent that ideal behavior is not obeyed at a particular set of conditions, the right hand side of this equation will deviate from unity. Consequently, for real gases an analogous expression is

$$\frac{P\tilde{V}}{RT} = Z \qquad (15-79)$$

where Z is called the *compressibility factor*.

Now, the deviation of this compressibility factor from unity can be expressed in terms of a series expansion about one of the variables. Two such expansions are conventionally done. One is termed the *pressure virial expansion* and is expressed in terms of powers of pressure. The other is termed the *density virial expansion* and is written in terms of powers of the molar density, i.e. powers of reciprocal molar volume. Hence, we may write the pressure and density virial equations

$$\boxed{\text{Pressure Virial Equation:} \qquad \frac{P\tilde{V}}{RT} = 1 + B'P + C'P^2 + D'P^3 + \cdots} \qquad (15-80)$$

and

$$\boxed{\text{Density Virial Equation:} \qquad \frac{P\tilde{V}}{RT} = 1 + \frac{B}{\tilde{V}} + \frac{C}{\tilde{V}^2} + \frac{D}{\tilde{V}^3} \cdots} \qquad (15-81)$$

These expansions, when carried to enough terms for a given gas for a given situation must give the same value of the compressibility factor. However, because the expansions are in terms of different variables, (P or $\tilde{V}^{-1}$) the coefficients from corresponding forms of the expansions are not equal. Relations between the two can be obtained, however, and are:

$$B' = \frac{B}{RT}$$

$$C' = \frac{C - B^2}{(RT)^2} \qquad (15-82)$$

$$D' = \frac{D - 3BC + 2B^3}{(RT)^3}$$

etc..

In these equations the coefficient of the first-order term is called the second virial coefficient and that for the second-order term is the third virial coefficient and so forth. (The first term of the expansion which would be the first virial coefficient is, of course, unity).

Because these virial expansions are written from the viewpoint of deviations of a gas from ideal behavior, they are both most applicable for situations where the deviations are not excessive. Consequently, they are intended to apply to gases which may exhibit moderate deviations from ideal behavior but not so extreme as dense gas or liquid behavior. Also, because the two equations are written in terms of different expansion variables, they are not equally representative of the deviations; that is, one of them represents the deviations better than the other when these deviations become appreciable. Specifically, *for moderate deviations, i.e. for the pressure less than about 15 bar at subcritical temperatures, the pressure virial equation, truncated after the second virial coefficient, is generally used.* In this region this pressure expansion is quite adequate for accurately representing the behavior of real gases and at the same time is computationally very convenient because both volume and pressure can be explicitly evaluated. However, at higher pressures from 15 bar to about 50 bar, the third virial coefficient is needed and in this case the density virial expansion gives more accurate calculations with essentially no more difficulty than the comparable pressure virial expansion. Consequently, *in the range of 15 to 50 bar the density virial expansion using terms through the third virial coefficient is the form of choice.*

Now, in principle, one could express deviations which occur at any temperature and pressure in terms of a virial expansion. However, to do so requires a greater and greater number of coefficients. Consequently, the virial expansions are not generally used for excessive deviations from ideal behavior.

The parameters of the density virial equation, i.e. the second, third, and higher virial coefficients carry physical meanings which are derived from the interactions which occur between molecules. The second virial coefficient accounts for two-body (i.e. molecule-molecule) interactions, the third virial coefficient accounts for three-body interactions, the fourth coefficient accounts for four-body interactions, etc. Because the likelihood of a greater number of molecules colliding simultaneously becomes increasingly small, these higher-order virial coefficient terms (for the density virial equation) contribute progressively less to the compressibility factor (unless the density is too large), resulting in the fact that the density virial equation is more accurate than the pressure virial for appreciable deviations from ideal behavior.

In these equations the virial coefficients are functions of temperature so that the compressibility factor for a gas is a function of two independent variables, either temperature and pressure (for the pressure virial expansion) or temperature and density (for the density virial expansion). In section 15.6.2 we look further at the use of Z in calculating changes in internal energy, enthalpy, and entropy.

15.4.3 Equations of State for Gases: Cubic Equations

A class of equations of state termed (cubic equations of state) exhibit cubic polynomial behavior and as a result, with appropriate parameters, may reasonably describe both vapor and liquid P-$\widehat{V}$-T behavior. For this reason they are capable of expressing P-$\widehat{V}$-T behavior over a very broad range (with a reasonably small number of parameters). The most well-known cubic equation of state, and the original one, is the *Van der Waals equation*, a two parameter model, given by

$$\text{Van der Waals EOS:} \quad P = \frac{RT}{\widetilde{V} - b} - \frac{a}{\widetilde{V}^2} \qquad (15-83)$$

In this equation a corrects for attractive interactions between the molecules and b corrects for finite molecular volumes.

A second cubic equation of state which is used rather widely is the *Redlich-Kwong equation* given by:

$$\text{Redlich-Kwong EOS:} \quad P = \frac{RT}{\widetilde{V} - b} - \frac{a}{T^{1/2}\widetilde{V}(\widetilde{V} + b)} \qquad (15-84)$$

Another cubic equation of state is the *Peng-Robinson equation* and, like the Redlich-Kwong equation, is widely used in the natural gas industry:

$$\text{Peng-Robinson EOS:} \qquad P = \frac{RT}{\widetilde{V} - b} - \frac{a}{\widetilde{V}(\widetilde{V} + b) + (\widetilde{V} - b)b} \qquad (15-85)$$

Still another cubic equation of state is the *Soave modification* of the Redlich-Kwong equation.

$$P = \frac{RT}{\widetilde{V} - b} - \frac{a}{\widetilde{V}(\widetilde{V} + b)} \qquad (15-86)$$

Typical cubic equation-of-state behavior is given in Figure 15-23 and demonstrates that the cubic behavior over the two-phase region is incorrect, physically, in that pressure is constant in this region at a given temperature, as the volume changes between liquid and vapor. Consequently, in the two-phase region the equation of state must be replaced by a constant-pressure isotherm. If this is done, then we see that, at least qualitatively, the cubic form matches real behavior.

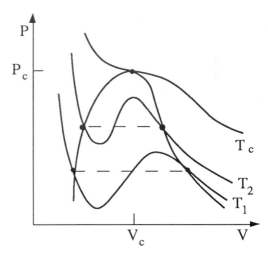

Figure 15-23: Cubic equations
of state

For these equations of state one may estimate parameters by a regression procedure using P-$\widehat{V}$-T data over a wide range of conditions. When this is done for the Van der Waals equation, it is observed that both a and b are functions of temperature. The Redlich-Kwong equation and the Peng-Robinson and Soave modifications, however, are successful at incorporating more of the P-$\widehat{V}$-T behavior into the functional relation itself (rather than in the parameters) in that the a parameter for these equation is found to be essentially temperature-independent. The b parameters, however, are temperature dependent.

As an alternative procedure for determining the equation of state parameters, one may use theoretical constraints which exist at the critical point. These constraints can be especially useful for the cubic equations of state which involve two parameters. At the critical point two partial derivatives are known, on theoretical principles, to be equal to zero:

$$\left(\frac{\partial P}{\partial \widetilde{V}}\right)_{T,cr} = 0 \qquad (15-87)$$

and

$$\left(\frac{\partial^2 P}{\partial \widetilde{V}^2}\right)_{T,cr} = 0 \qquad (15-88)$$

Differentiating the equation of state to obtain these derivatives and setting the results to zero allows calculation of these two parameters *in terms of the critical properties*. Although these may not be the optimum parameters for calculations over a wider range (the parameters obtained this way are constants and therefore do not follow the temperature dependencies described above), they nevertheless do give this correct derivative behavior at the critical point and, furthermore, they give reasonable P-$\widehat{V}$-T behavior at the critical point and elsewhere and can normally be obtained for most materials because the critical constants are normally known. This advantage of availability and reasonable accuracy presents a considerable argument in favor of using values of these two parameters calculated from the critical properties. Table C-19 in Appendix C contains critical properties of various compounds. Values of a and b based on this critical point behavior for the cubic equations discussed above and in terms of the critical constants are given in Table 15-2.

Table 15-2: Cubic EOS Parameters Based on Critical Point Derivatives

EOS	a	b	Z_c
Van der Waals	$\dfrac{9}{8}RT_c\widetilde{V}_c$	$\dfrac{1}{3}\widetilde{V}_c$	0.375
Van der Waals	$\dfrac{27\,RT_c^2}{64\,P_c}$	$\dfrac{RT_c}{8P_c}$	0.375
Redlich-Kwong	$0.42748\dfrac{R^2T_c^{2.5}}{P_c}$	$0.08664\dfrac{RT_c}{P_c}$	0.34
Soave/Redlich/Kwong	$0.42748\dfrac{R^2T_c^2}{P_c}$	$0.08664\dfrac{RT_c}{P_c}$	0.34
Peng-Robinson	$0.4274\dfrac{R^2T_c^2}{P_c}$	$0.0778\dfrac{RT_c}{P_c}$	0.32

Still higher-order equations have been used to provide more accurate representations of P-$\widehat{V}$-T behavior. One widely used example is the eight-parameter density virial expansion of Benedict-Webb-Rubin equation (Benedict, et al., 1940, 1942, 1951)

$$P = \frac{RT}{\widetilde{V}} + \frac{B_oRT - (A_oC_o/T^2)}{\widetilde{V}^2} + \frac{bRT - a}{\widetilde{V}^3} +$$
$$\frac{a\alpha}{\widetilde{V}^6} + \left(\frac{c}{\widetilde{V}^3T^2}\right)\left(1 + \frac{\gamma}{\widetilde{V}^2}\right)\exp\left(-\frac{\gamma}{\widetilde{V}^2}\right) \qquad (15-89)$$

where A_o, B_o, a, b, c, α, and γ are constant parameters for a specific material.

In general, the cubic equations of state are more widely used for calculations because of the speed with which the calculations can be performed on computers. This ease of calculation outweighs some sacrifice of accuracy.

15.4.4 Equations of State for Gases: Generalized Corresponding States Correlations

Two-parameter Corresponding States. All of the above equations of state may be converted to a two-parameter generalized corresponding states form based upon the observation that materials obey a corresponding states principle. This principle says that all materials, when expressed in terms of reduced temperature (T/T_c) and pressure (P/P_c) (or volume (V/V_c)), obey similar or corresponding behavior. That is, if we plot the compressibility factor for all materials as a function of the reduced pressure and temperature, then the same family of curves are observed for all materials; the functions for all materials overlay. For a given material, the two parameters for a given equation of state are the material's critical temperature and pressure (or volume). If these are known, then the P-V-T behavior can be calculated. In fact, using the parameters of Table 15-2 in the equations of state is equivalent to a corresponding states assumption. The parameters a and b are replaced by functions of T_c and P_c (or $\tilde{V}_c$). While this principle is not obeyed exactly, it is obeyed quite well for non-ideal gases and leads to generalized compressibility factor equations for materials in terms of the reduced variables.

The critical temperature, pressure, and volume for each of the materials may vary considerably (see table C-19 in Appendix C) but when the curves are normalized with respect to these constants for each material, that is, when the state of the material is expressed relative to its distance away from critical conditions, similar behavior is observed. For example, the Van der Waals equation of state can be expressed in terms of reduced quantities $(P_r = P/P_c, T_r = T/T_c, \tilde{V}_r = \tilde{V}/\tilde{V}_c)$ by the equation (using the values of a, b, and Z_c from Table 15-2)

$$\text{VDW, CS Form:}\qquad P_r = \frac{8T_r}{3\tilde{V}_r - 1} - \frac{3}{\tilde{V}_r^2} \tag{15-90}$$

An alternative form of this equation written for the compressibility factor in terms of the reduced variables is:

$$\text{VDW, CS Z Form:}\qquad Z = \frac{\tilde{V}_r}{\tilde{V}_r - 1/3} - \frac{9}{8T_r\tilde{V}_r} \tag{15-91}$$

Three-parameter Corresponding States. Although the two-parameter assumption works quite well, it can be improved by adding a third parameter. The most popular three-parameter approach is that of Pitzer and coworkers who introduced the *acentric factor* (ω) to account for non-spherically shaped molecules. Methods for calculating the compressibility factor from T_c, P_c and ω, either directly or in pressure virial form, are presented in Smith and Van Ness (1987).

15.4.5 Equations of State for Liquids

Some equations of state have been presented specifically for liquids. These include the Rackett (1970) equation for *saturated liquids* (liquids in equilibrium with vapor). The Tait equation has been presented for *compressed liquids* (liquids which are below saturation) (Smith and Van Ness, 1975). The Rackett equation is:

$$\hat{V}^{\text{sat}} = \hat{V}_c^{(1-T_r)^{0.2857}} \tag{15-92}$$

where $\widehat{V}^{\,\mathrm{sat}}$ is the specific volume of the saturated liquid at the given reduced temperature. The Tait equation is

$$\widetilde{V} = \widetilde{V}_o - D \ln\left(\frac{P + E}{P_o + E}\right)$$

(15 – 93)

where $\widetilde{V}$ and $\widetilde{V}_o$ are molar volumes at the pressure of interest (P) and the reference pressure P_o respectively. D and E are constants for a given pressure and must be determined by sufficient experimental data. A generalized correlation also may be used for liquids as developed by Lydersen, Greenkorn, and Hougen and given by Smith and Van Ness (1987).

15.5 THERMODYNAMICS PROBLEM-SOLVING TOOLS AND STRATEGY

At this point we are ready to summarize eight tools which are available for solving thermodynamics problems. This summary should serve as a strategy checklist. If you are having difficulty solving a thermodynamics problem, or if a solution procedure is not immediately obvious, then reconsider this list. The chances are that there is something on it which you are not including or which you do not fully understand.

So, the tools with which you are now equipped are:

1) **The Laws.** This includes the five conservation laws (mass, charge, linear momentum, angular momentum, energy) and the second law of thermodynamics. It also includes accountings of mechanical, thermal, and electrical energy, if necessary. Mass, total and thermal energy, and entropy equations are especially important for thermodynamics problems. Momentum (both linear and angular) and charge equations are less frequently required.

2) **Properties of Matter: The Gibbs Phase Rule.** This is materials information that guides us in establishing what and how much we need to know in order to have a state of existence fully specified.

3) **Properties of Matter: Fundamental Property Relations.** The fundamental property relations, introduced in section 15.3.2, are extremely valuable thermodynamic tools. They are the basis for the observable–non-observable relations (tool 4) but also are frequently directly useful on their own.

4) **Properties of Matter: Relations for $\widehat{U}$, $\widehat{H}$, $\widehat{A}$, $\widehat{G}$, and $\widehat{S}$ in terms of Observables.** These property relations are essential to using the laws. The laws require the use of energy and entropy properties, we cannot measure these properties directly, and so, we must be able to calculate them in terms of properties which can be observed, namely heat capacity, pressure, volume (per mass), and temperature.

5) **Properties of Matter: Heat Capacities and Equations of State Relations or Data.** As stated in (4), data for the observable properties are essential. These properties may exist in the form of analytical equations (e.g., $\widehat{C}_P$ as a function of temperature or the ideal gas equation of state which relate T, P, and $\widehat{V}$) or they may exist in the form of tables of values (e.g., steam tables which list values of $\widehat{H}$, $\widehat{V}$, and $\widehat{S}$ for given values of T and P). Note that tables like those for steam provide both item (4) and (5) information. Steam tables are discussed in section 15.6.3.

6) **Definitions and Calculations of Work.** Clearly, in order to use the First Law of Thermodynamics (conservation of energy), work must be understood. Frequently, this means calculating the work in terms of other process information

and especially from the fundamental concept of work being produced by a force acting over a distance. This will be discussed more fully in section 15.5.1, below.

7) **Process Constraints.** Of course, we also must know about the actual process with which we are dealing. For example, possible characteristics which must be noted are process geometry, size, mass, and whether it is isochoric, isobaric, isothermal, isentropic, isenthalpic, adiabatic, reversible, closed, open, or isolated.

8) **Mathematics.** Of course, underlying all of the above tools is mathematics and this includes algebra, partial derivatives, total derivatives, and some integral and differential calculus. Fortunately, most of thermodynamics is not difficult mathematically. The challenge is largely conceptual, requiring recognizing the pieces listed in items 1 through 5 and their interrelations.

Note that the tools list for thermodynamics is longer than for the previous sections. This is why thermodynamics can seem so puzzling and difficult to grasp. It is an intricate combination of several fundamental laws, a wide range of properties information, concepts such as work and heat, and mathematics. To feel confident with thermodynamics, you must have a clear organization of this information in mind as well as experience with using it.

The sections which follow discuss tool six (work calculations) first, followed by the forms of the property relations for some specific materials existing as a single phase (ideal gases, non-ideal gases, solids and liquids) and give some example problems which use them. Also, calculations of property changes with phase change and chemical reaction are presented. Keep in mind the tools which are available as these problems are presented.

15.5.1 Work Calculations.

The caculations of work for specific cases can seem tricky and difficult. Perhaps the best way to improve your understanding is to consider specific examples within the context of the fundamental work definition

$$(\text{Work}) = \int \mathbf{f} \cdot d\mathbf{s}$$

As noted previously, 1) work exists if there is an external force *and* an accompanying collinear displacement, and (according to the definition of this text) 2) work is positive if it adds energy to the system and negative if it removes it.

Example 15-7. An ideal gas is initially contained in the left chamber of a two-chamber, perfectly-rigid container (Figure 15-24). The right chamber is a perfect vacuum. The partition separating the two chambers is also perfectly rigid. Suppose that the partition now ruptures, allowing the gas to expand into the right chamber so that, after a time, both chambers are completely filled to the same pressure. If the gas is the system, what is the work associated with this process?

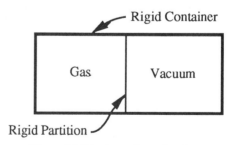

Figure 15-24: A gas in a chamber

SOLUTION: Figure 15-25 shows this expansion process at three times during the expansion: 1) before the gas has expanded, 2) while the gas is expanding but before any of its molecules have reached the walls of the right chamber, and, finally, after the expansion is complete. Note the system boundary at each time which completely contains every molecule of the gas. With no gas molecules outside the system boundary, there is, by definition, a perfect vacuum outside the boundary. Note that this boundary must move (very rapidly!) with the gas in order to stay ahead of it. (Because the boundary is a massless creation of our imagination, this motion is possible.)

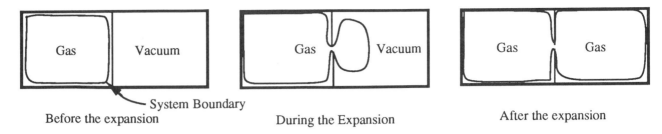

Figure 15-25: Free expansion of a gas into a vacuum

The work associated with this process is easily determined. Because of the vacuum which exists outside the system boundary, the external force exerted on the moving boundary is always zero. Where the system boundary is moving, the external force is zero and hence the work is zero. However, when the gas molecules impact the wall, a force is required to change their momentum and keep the gas from continuing to expand. But now the motion of the system boundary is stopped so that the work is still zero; the external force is not zero but motion associated with it is (the walls are perfectly rigid) and hence the work is zero. So, at no point during the expansion is a non-zero external force accompanied by a collinear motion. Hence, (work) = 0.

Example 15-8. A gas is contained in the left chamber of a two-chamber vessel by a piston held in place by stops. The piston is massless and frictionless. The right chamber is completely evacuated. At an instant in time, the stops are removed so that the piston accelerates (rapidly) to the right due to the unopposed pressure of the gas. Considering the gas as the system, what is the work associated with this process?

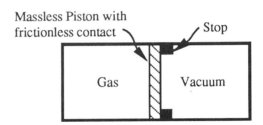

Figure 15-26: Gas contained in a cylinder by a piston

SOLUTION: The situation is, in essence, no different from the previous example. With the gas in the left chamber as the system, the external force, as the gas expands, is zero. After the piston hits the stop, it now exerts a force against the gas but motion is zero, hence, there is never an external force on the gas which is accompanied by motion so as to result in work: (work) = 0.

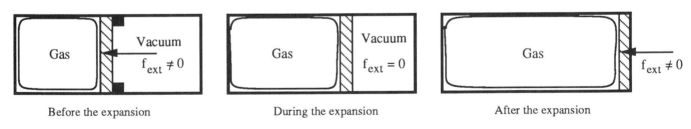

Figure 15-27: Free expansion of the gas against the massless and frictionless piston (work = 0)

Note that the external force is not zero before the stops are removed, drops to zero when the stops are removed (zero force is required to accelerate a *massless* piston and friction is zero) and becomes non-zero again after the expansion.
Also, note that in Figure 15-27, this very rapid gas expansion is irreversible, resulting in entropy generation.

Example 15-9. Suppose that the piston in the previous problem has mass m. How does this affect the work?

SOLUTION: This time, the work (for the system = gas) will not be zero because the external force against the gas during the expansion is not zero. This must be true because a finite force is required to accelerate the piston away from the gas and that force on the piston is accompanied by an opposite and equal-in-magnitude reaction force on the gas where the piston contacts the gas. This latter force is an external force on the gas during the expansion. We can't say exactly what the value of the work is, however, unless more detail is known about the nature of the expansion or unless the kinetic energy of the piston is specified at the end of the expansion, just before it strikes the end of the cylinder (explain).

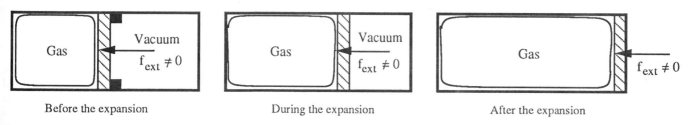

| Before the expansion | During the expansion | After the expansion |

Figure 15-28: Expansion of the gas against the frictionless piston of mass m (work $\neq 0$)

Example 15-10. A gas is contained in a cylinder of constant cross-sectonal area by a piston held in place by stops (Figure 15-29). The piston is massless and frictionless. To the right of the piston, the cylinder is open to the atmosphere which has a constant pressure of 1 bar. At an instant in time, the stops are removed so that the piston accelerates to the right. Considering the gas as the system, what is the work associated with this process? The cross-sectional area of the piston (A) is 50 cm^2 and the piston travels a distance L of 5 cm before it impacts a second set of stops. The expansion process is shown in Figure 15-30.

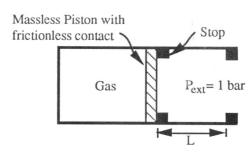

Figure 15-29: Gas contained in a cylinder by a piston

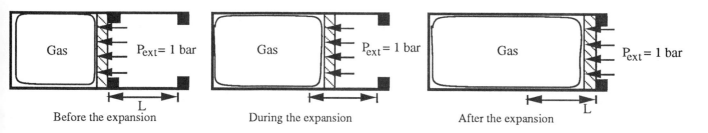

| Before the expansion | During the expansion | After the expansion |

Figure 15-30: Expansion of a gas against a constant atmospheric pressure

SOLUTION: In this case, the work is not zero because there is an external force accompanying the motion of the piston. The pressure acting against the piston (and therefore acting against the gas) is constant at 1 bar. The force exerted by this pressure is given by

$$\text{Force} = PA$$

where $A = 40$ cm^2, the cross-sectional area of the cylinder. Both the pressure and the area are constant throughout the expansion and so the force is constant. The force acts over distance L. Consequently, the work is given by

$$(\text{work}) = -(PA)L = -P(V_{\text{end}} - V_{\text{beg}})$$

Note that the work is negative because the force and displacement are counter to each other; hence energy leaves the gas because of the work. Numerically, the work is found to be

$$(\text{work}) \; = \; - \frac{1 \; \text{bar}}{} \frac{\left| 50 \; \text{cm}^2 \right.}{} \frac{\left| 5 \; \text{cm} \right.}{} \frac{\left| 10^5 \; \text{N} \right.}{\left| \; \text{m}^2 \text{bar} \right.} \frac{\left| \; \text{m}^2 \right.}{\left| 10^4 \; \text{cm}^2 \right.} \frac{\left| \; \text{m} \right.}{\left| 100 \; \text{cm} \right.} \frac{\left| 1 \; \text{J} \right.}{\left| \; \text{Nm} \right.} \; = \; -25 \; \text{J}$$

Also note that before the stops are removed and after impact of the piston with the second set of stops, work ceases because there is no longer any displacement (the force acting against the gas remains the same, however).

Example 15-11. As one final example, consider a vertical cylinder with a piston which contains a gas at an initial pressure P_{beg} contained in volume V_{beg}. The piston is not constrained by any stops but rather has a weight of sand on top which just balances the PA force of the gas. The piston again is frictionless but may possess mass. Also, there may be an external, non-zero, atmospheric pressure. The only issue is that the weight of the piston plus the external force of the atmosphere against the piston plus the weight of sand exactly balances the upward force of the gas in the cylinder. Now, over time, the sand is removed, one grain at a time. This allows the gas to expand (slightly each time) until its pressure is reduced and again it just balances the weight of the sand. This process in continued until the final pressure is P_{end} and the final volume is V_{end}. Determine an expression for the value of the work associated with this expansion.

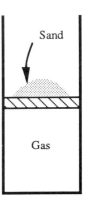

Figure 15-31: Gas contained in a cylinder by a piston

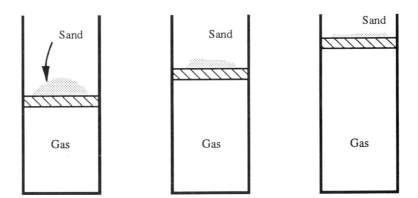

Figure 15-32: Expansion of a gas against a balancing external pressure

SOLUTION: The key to this calculation is to realize that the (total) external force always balances (or nearly so) the force exerted by the pressure of the gas. Hence, for the value of the magnitude of the external force we can *substitute* the PA product calculated using the pressure of the gas and the work becomes:

$$(\text{Work}) \; = \; - \int_{V_{\text{beg}}}^{V_{\text{end}}} P \, dV$$

Note well that the pressure P varies as the gas expands and hence the requirement for the integration. Again, this calculation of work using the system pressure is special and applies only because the external force is expressed in terms of the system internal pressure. Also, note that the work is negative if the gas volume is increasing against the external pressure, in which case energy is removed from the gas as the result of the expansion. If the reverse process occurred (add sand grains back to decrease the volume of the gas), then the force and displacement would occur in the same direction and hence the work (calculated using the same expression) would automatically be positive, indicating that the energy of the gas increases because of the compression.

One further note: this last example is a *reversible* expansion (enropy generation is zero) because the external pressure always nearly balances the system pressure. Hence, the expansion is slow. The other examples, discussed previously, are characterized by a definite imbalance between the initial external force and the system pressure. Hence, the expansion occurs very rapidly and is accompanied by turbulence of the gas during the expansion. These processes are inherently irreversible (generate entropy).

15.6 THERMODYNAMICS PROBLEMS AND SPECIAL FORMS OF THE PROPERTY RELATIONS

The bottom-line relationships for the fundamental thermodynamic property relations initially were presented in Section 15.3.3 and then derived in Section 15.3.6. These relations allowed calculating changes in internal energy, enthalpy, and entropy in terms of heat capacities and equations of state. They involve integrals of heat capacity over temperature and equation-of-state functions over either pressure or volume. These relationships are essential for using the first and second laws of thermodynamics.

Now, all calculations of internal energy, enthalpy, and entropy could be made using the property relations in these forms. However, because of the forms of the equations of state which are used for the various states of substances (ideal gas, non-ideal gas, dense gas, liquid, solid), forms of the property relations which are compatible with these special forms of the equations of state are more convenient. In this section we present these forms for these special cases.

15.6.1 Ideal Gases

An ideal gas obeys the familiar equation of state

$$PV = nRT$$

$$(15 - 75)$$

or, in terms of volume per mass

$$P\widehat{V} = \frac{RT}{MW}$$

$$(15 - 94)$$

Now from equations (15-14) and (15-94), it is easily shown that

$$d\widehat{U}^{ig} = \widehat{C}_V \, dT$$

$$(15 - 95)$$

and from equations (15-15) and (15-94) (see example 15-2),

$$d\widehat{H}^{ig} = \widehat{C}_P \, dT$$

$$(15 - 96)$$

and from equations (15-20), (15-21), and (15-94),

$$d\widehat{S}^{ig} = \frac{\widehat{C}_V^{ig}}{T}dT + \frac{R}{MW}\frac{d\widehat{V}}{\widehat{V}}$$

$$d\widehat{S}^{ig} = \frac{\widehat{C}_P^{ig}}{T}dT - \frac{R}{MW}\frac{dP}{P}$$

(15 − 97)

Furthermore, the difference between the two heat capacities is

$$\widehat{C}_P^{ig} - \widehat{C}_V^{ig} = \frac{R}{MW}$$

(15 − 98)

For an ideal gas, both internal energy and enthalpy are functions of temperature only and not of pressure or volume, a result of the fact that an ideal gas has no intermolecular force fields. Entropy, however, is a function of both temperature and volume (or pressure), as indicated by the above relation. Note that in the above relations, the heat capacities, even though for ideal gases, may vary from one gas to another because of the differing degrees of freedom that molecules of different size and structure may have. In general these heat capacities are functions of temperature, even for an ideal gas. Nevertheless, finite changes in internal energy which occur as a result of finite changes in temperature can be determined simply by integrating the above expression, once the dependence of the heat capacity on temperature is known. Likewise, enthalpy and entropy can be calculated. In fact, the ideal gas heat capacities for a great many materials are available (Appendix C) and play a role even for calculations of highly non-ideal materials, as we shall see in section 15.6.2. These integrations can be expressed in terms of reference values to obtain absolute values of internal energy, enthalpy, and entropy, as discussed in Section 15.3.4.

Example 15-12. One mole of an ideal gas, confined in a cylinder by a piston at 3 atm and 100 °C, expands isothermally such that the final temperature is also 100 °C against a constant resisting pressure of 1 atm as shown in Figure 15-33. The final volume is twice the initial volume. Furthermore, the piston is massless and frictionless. For the process calculate, ΔU, ΔH, ΔS and the work W and the heat transfer Q. Furthermore, *if* the process were reversible, then $-\int P dV$ would be the work and $\int T dS$ would be the heat transfer. Calculate these values for such a reversible, isothermal process. Note that the "delta" notation is defined to mean that $\Delta U \equiv U_{end} - U_{beg}$.

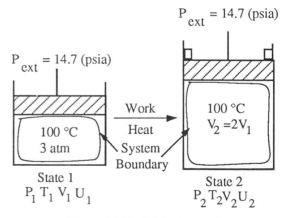

Figure 15-33: Frictionless and massless piston

SOLUTION: First, we assess the laws (Tool 1). Choose the gas in the cylinder as the system. **Conservation of mass:** the system is closed so that the mass of the system is constant. **Conservation of charge:** trivial. **Conservation of linear momentum:** The piston is moving but is *massless* and therefore is not changing momentum, even though it may accelerate. Zero force is required to accelerate zero mass. Furthermore, the piston is frictionless. Hence, the force (pressure) exerted by the gas on the piston is the same as that exerted by the surrounding atmosphere on the piston. (Note that if the piston were not massless and frictionless, then the pressure of the gas would have to be greater than the surrounding atmosphere in order to accelerate the piston to the stops.) **Conservation of angular momentum:** trivial. **Conservation of energy:** For the closed system the only way the energy of the gas (system) can change is by heat and work. There is an external force applied by the surroundings on the system (1 atm pressure at the gas/piston boundary - see conclusions of conservation of linear momentum) and this force acts over a distance and so there must be work. The energy of the gas exists as internal energy only (neglect changes in kinetic energy - the gas is low density - and in potential energy - the gas is low density and the distance involved is small). Consequently, conservation of energy (first law of thermodynamics) says

$$\Delta U \equiv U_{\text{end}} - U_{\text{beg}} = Q + W$$

The second law of thermodynamics: The entropy of the system can change only by heat transfer or generation. With no mass exchange, entropy cannot change by this mode. This is not a "slow" process so that there are almost certainly irreversibilities (in this case, conversion of mechanical energy to internal energy due to the viscosity of the gas). Consequently, we can only say, at this point, that the entropy of the system does, in fact, change and that the entropy of the universe increases due to the process.

Consider now the Gibbs Phase Rule (Tool 2). For the *initial state* we know both the temperature and pressure. These are two, independent, intensive variables for this single phase system and hence are enough to pin down the intensive state of the system (but not its actual size, i.e., the total amount of the gas or the total volume). However, we also know that there is one mole so that, in fact, the size of the system is specified, as well. For the *final state*, we know only the temperature but we also know that its volume is twice the initial volume. Therefore, its specific volume is twice the initial specific volume and, because for the initial state we can determine all intensive variables, we also will then be able to detemine the final state specific volume. Hence, enough information is stated about the final state to fully pin it down; we just need to deduce the values.

Consider next the UHAGS in terms of observables (Tool 4). For an ideal gas, internal energy and enthalpy depend only on temperature according to

$$d\widehat{U}^{\text{ig}} = \widehat{C}_V^{\text{ig}} dT$$

$$d\widehat{H}^{\text{ig}} = \widehat{C}_P^{\text{ig}} dT$$

But we know from the problem statement that the final-state temperature is the same as the initial-state temperature and hence,

$$\widehat{U}_{\text{end}} - \widehat{U}_{\text{beg}} = 0 \qquad \text{and} \qquad \widehat{H}_{\text{end}} - \widehat{H}_{\text{beg}} = 0$$

regardless of the actual temperture relation for the heat capacities. Back to the first law, this tells us that

$$Q = -W$$

which we will come back to.

Looking at entropy, for an ideal gas,

$$d\widehat{S}^{\text{ig}} = \widehat{C}_P^{\text{ig}} dT - \frac{R}{MW} \frac{dP}{P}$$

which, for an isothermal process is

$$d\widehat{S}^{\text{ig}} = -\frac{R}{MW} \frac{dP}{P} \qquad \text{or} \qquad \Delta S = -nR \ln\left(\frac{P_2}{P_1}\right)$$

If we knew the final pressure, we could calculate this change in entropy from beginning state to final state.

Next, consider Heat capacity and equation of state relations (Tool 5). For an ideal gas,

$$PV = nRT \qquad \text{or} \qquad P\widehat{V} = \frac{RT}{MW}$$

For a constant-mass system, this equation gives

$$\frac{P_1 V_1}{T_1} = \frac{P_2 V_2}{T_2}$$

and hence, because $V_2 = 2V_1$ and $T_1 = T_2$

$$P_1 = 2P_2$$

so that $P_2 = 1.5$ atm. This lets us immediately calculate the change in entropy:

$$S_{\text{end}} - S_{\text{beg}} == -nR \ln \left(\frac{P_2}{P_1} \right)$$

$$= -(1)(8.314) \ln \left(\frac{1}{2} \right) = 5.763 \text{ J} / \text{K}$$

The change in entropy of the gas from state 1 to state 2 is 5.763 J/K.

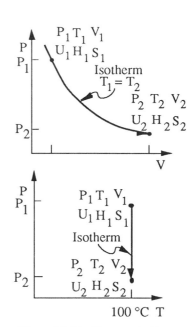

The change in state from state 1 to state 2 is approximated on the P-T and P-V diagrams (Figure 15-34). Actually, because the expansion is rapid, the notion that the process is constant pressure may be incorrect; in fact, as the gas expands, there is certainly a state of non-uniform pressure within the cylinder and hence, the concept of "pressure of the gas" is rather meaningless. In spite of this, however, we can certainly calculate the end state after the system has reached a new equilibrium state.

Figure 15-34: Changes in State
for Example 15-12

To calculate the work, and heat transfer we need to look at the process and the fundamental definition of work (tool 6). By definition, the work that is done on the system is defined as:

$$W = -\int_{V_1}^{V_2} P_{\text{ext}} dV$$

where the external pressure is that exerted by the piston on the gas inside the cylinder. From momentum considerations above, we established that this pressure (for this massless piston) is equal to the external atmospheric pressure, a constant during the expansion. Hence:

$$W = -P_{ext}(V_2 - V_1)$$
$$= -P_{ext}(2V_1 - V_1)$$
$$= -P_{ext}V_1$$

The initial volume can be related to the initial pressure and temperature by the ideal gas law:

$$W = -P_{ext}\frac{nRT_1}{P_1}$$

Here the value of the gas constant R is 8.314 J/mol K as given in Appendix A. Substituting values yields:

$$W = -(1)\frac{(1)(8.314)(373)}{(3)} = -1,033.7 \text{ J}$$

Energy left the system in the form of work at an amount of 1,033.7 J.

Then, using the first law result, the heat transfer to the system is:

$$Q = -W$$
$$= 1,033.7 \text{ J}$$

The heat transferred to the system is 1,033.7 J.

The integral of the state variables is calculated next but we should remember that this is a hypothetical calculation only; because of the irreversibilities, we cannot say exactly what path is followed from state 1 to state 2 and hence how P is related to V during the expansion. The first integral is the PdV expression. Using the equation of state gives:

$$-\int_{V_1}^{V_2} PdV = -\int_{V_1}^{V_2} \frac{nRT}{V}dV$$

Then, assuming that the process is isothermal and that the temperature is uniform throughout the cylinder during the process (not necessarily true, as discussed above), the temperature, gas constant, and the number of moles can be removed from the integral sign.

$$-\int_{V_1}^{V_2} PdV = -nRT \int_{V_1}^{V_2} \frac{dV}{V}$$

Integrating gives:

$$-\int_{V_1}^{V_2} PdV = -nRT \ln\left(\frac{V_2}{V_1}\right)$$

Substituting values with the temperature in Kelvin as 373 K gives:

$$-\int_{V_1}^{V_2} PdV = -(1)(8.314)(373)\ln(2) = -2,149.5 \text{ (J)}$$

The integral of $-PdV$ for an isothermal process yielded a value of -2,149.5 (J). *If* this had been a reversible process, this would have been the work.

The integral of TdS is calculated

$$\int_{S_1}^{S_2} TdS = T\int_{S_1}^{S_2} dS$$
$$= T(S_2 - S_1)$$

again assuming that the temperature is constant and uniform. From the previous calculation, the change in entropy for the process is known. Substituting values gives:

$$\int_{S_1}^{S_2} TdS = (373)(5.763) = 2,149.5 \ (J)$$

The integral of TdS gives a value of 2149.5 (J) for this process and *would* have been the heat transfer, *if* the process were reversible.

Example 15-13: The Tank Filling Problem. A tank containing an ideal gas at P_1 and T_1 is connected to a line that fills the tank as shown in Figure 15-35. The line is always at constant pressure and constant temperature (P_{in} and T_{in}). The tank is initially isolated from the line by a closed valve. After the valve is open, the tank begins to fill with the gas from the line. If $\widehat{C}_P$ and $\widehat{C}_V$ for the gas are independent of temperature and if the process is adiabatic, derive an expression relating the temperature of the tank T_T to the temperature of the line T_{in}.

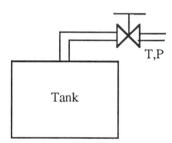

Figure 15-35: Tank filling problem with ideal gas

*SOLUTION:*The system is defined to be the tank, the valve, and the pipe between them. The system is not at steady state, however; there is only one entrance. Figure 15-36 shows the defined system boundary.

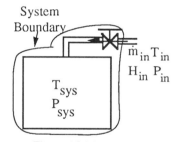

Figure 15-36: System for example 15-13

First we look at the Laws (Tool 1). For this system, the **conservation of mass** states:

$$\left(\frac{dm_{sys}}{dt}\right) = \dot{m}_{in} - \dot{m}_{out}$$

There is no mass leaving the system so the time rate of change of the mass in the system is equal to the rate of mass entering the system:

$$\left(\frac{dm_{sys}}{dt}\right) = \dot{m}_{in}$$

The **conservation of total energy** states:

$$\left(\frac{dE_{sys}}{dt}\right) = \left(\dot{m}(\widehat{H} + \widehat{PE} + \widehat{KE})\right)_{in} - \left(\dot{m}(\widehat{H} + \widehat{PE} + \widehat{KE})\right)_{out} + \dot{Q} + \dot{W}$$

for which a number of simplifications are possible. First, there is no mass leaving the system, so that term is zero. Second, there is no work being done on the system except the flow work of the entering stream and this is included in its enthalpy. Also, the process is adiabatic. If we neglect the changes in potential and kinetic energy of the system, and assume that the potential and kinetic energy contributions due to the entering mass are negligible, the following equation results.

$$\left(\frac{d(m\widehat{U})_{\text{sys}}}{dt} \right) = (\dot{m}\widehat{H})_{\text{in}}$$

This equation can be integrated over some finite time period from t_o to t

$$\int_{t_o}^{t} (\dot{m}\widehat{H})_{\text{in}} dt + = \int_{t_o}^{t} d(m\widehat{U})_{\text{sys}}$$

For the first integral the specific enthalpy of the mass entering is always at the same conditions and therefore is a constant. Also, note that $\dot{m}\,dt = dm$. Integrating yields:

$$\widehat{H}_{\text{in}}(m - m_o) = (m\widehat{U}_T) - (m\widehat{U}_T)_o$$

where m is the total amount of gas in the tank. Note that if the tank is initially empty, then the internal energy of the contents of the tank equals the enthalpy of the entering stream. This equation can be rearranged to give:

$$m(\widehat{H}_{\text{in}} - \widehat{U}_T) - m_o(\widehat{H}_{\text{in}} - \widehat{U}_{T_o}) = 0$$

From the definition fo enthalpy,

$$\widehat{H} = \widehat{U} + P\widehat{V}$$

Moreover, because the gas is ideal, $P\widehat{V}$ can be replaced by $RT / (MW)$ (tool 5, EOS)

$$\widehat{H} = \widehat{U} + \frac{RT}{(MW)}$$

Consequently,

$$\widehat{H}_{\text{in}} = \widehat{U}_{\text{in}} + \frac{RT_{\text{in}}}{(MW)}$$

Substituting this expression for the enthalpy into the previous equations and rearranging gives:

$$m\left(\widehat{U}_{\text{in}} - \widehat{U}_T + \frac{RT_{\text{in}}}{(MW)} \right) - m_o\left(\widehat{U}_{\text{in}} - \widehat{U}_{T_o} + \frac{RT_{\text{in}}}{(MW)} \right) = 0$$

Now, for an ideal gas *with temperature-independent heat capacities*, (tool 5, heat capacities)

$$\widehat{U}_{\text{in}} - \widehat{U}_T = \widehat{C}_V(T_{\text{in}} - T_T)$$
$$\widehat{U}_{\text{in}} - \widehat{U}_{T_o} = \widehat{C}_V(T_{\text{in}} - T_o)$$

So that

$$m\left(\widehat{C}_V(T_{\text{in}} - T_T) + \frac{RT_{\text{in}}}{(MW)} \right) - m_o\left(\widehat{C}_V(T_{\text{in}} - T_o) + \frac{RT_{\text{in}}}{(MW)} \right) = 0$$

Now, for an ideal gas (tool 5),

$$\widehat{C}_V + \frac{R}{(MW)} = \widehat{C}_P$$

So that

$$m(\widehat{C}_P T_{\text{in}} - \widehat{C}_V T_T) - m_o(\widehat{C}_P T_{\text{in}} - \widehat{C}_V T_{To}) = 0$$

Rearranging gives:

$$T_T = \frac{\widehat{C}_P}{\widehat{C}_V} T_{\text{in}} - \frac{m_o}{m}\left(\frac{\widehat{C}_P}{\widehat{C}_V} T_{\text{in}} - T_{To}\right)$$

so the temperature of the system is a function the heat capacities of the gas, the initial temperature, the temperature of the line, the initial amount of mass, and the amount of mass present in the tank at the time of interest.

For the special case in which the tank is initially empty ($m_o = 0$),

$$T_T = \frac{\widehat{C}_P}{\widehat{C}_V} T_{\text{in}}$$

In this case, the temperature of the tank is only a function of the heat capacities and the temperature of the line. Interestingly, it is independent of the amount of mass which has entered the tank.

Example 15-14: The Tank Emptying Problem. An ideal gas with constant heat capacities is stored in an insulated tank. If the valve on the tank is opened slowly and allows the gas to leave the tank, the state of the gas that remains in the tank will change. Derive expressions that relate the state of the system to the amount left in the tank. Figure 15-37 shows the tank and valve.

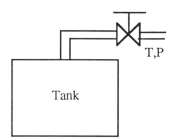

Figure 15-37: Tank emptying problem

SOLUTION: For this system definition, let us choose as the system the gas that will be left in the tank after some has left. This system is not at steady state. Furthermore, the system boundary increases (the remaining gas expands) as gas is removed from the tank. However, this system is closed by definition. Figure 15-38 shows the system at both the initial state and final states. Remember, for this system no mass crosses the system boundary.

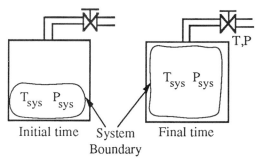

Figure 15-38: System for Example 15-14

Tool 1, The Laws. The **conservation of mass** for this system says that the mass of the system is constant. However, the **conservation of total energy** for this system says:

$$\left(\frac{dE_{\text{sys}}}{dt}\right) = \dot{Q} + \dot{W}$$

If we neglect the changes in potential and kinetic energy of the system, the following equation results:

$$\left(\frac{d(m\widehat{U})_{\text{sys}}}{dt}\right) = \dot{Q} + \dot{W}$$

Insteady of looking at the time derivative, now let us look at differential changes or:

$$md\widehat{U} = \delta Q + \delta W$$

Because the system always expands against a pressure that matches its own (the surroundings, i.e., the rest of the gas in the tank, is at the same pressure) and the expansion is slow, the process may be assumed to be reversible (tool 7). Hence, the heat and work are for a reversible process,

$$md\widehat{U} = \delta Q_{\text{rev}} + \delta W_{\text{rev}}$$

and furthermore, for a well-insulated tank (tool 7), $\delta Q_{\text{rev}} = 0$:

$$md\widehat{U} = \delta W_{\text{rev}}$$

Now, the work is that external force acting over a distance. The force in the pressure acting against the system but is this case, this is the same as the system pressure. Hence, $\delta W_{\text{rev}} = -PdV$. Also, for an ideal gas (tool 5), $d\widehat{U} = \widehat{C}_V dT$ so that the previous result becomes:

$$-PdV = m\widehat{C}_V dT$$

Since the gas is ideal (tool 5) this equation becomes:

$$-\frac{nRT}{V}dV = m\widehat{C}_V dT$$

Separating the variables gives:

$$\frac{dT}{T} = -\frac{nR}{m\widehat{C}_V}\frac{dV}{V}$$

Noting that for an ideal gas $R = \widetilde{C}_P - \widetilde{C}_V$ and that $m\widehat{C}_V = n\widetilde{C}_V$ gives

$$\frac{dT}{T} = -(\gamma - 1)\frac{dV}{V}$$

where $\gamma \equiv \widetilde{C}_P / \widetilde{C}_V$. Integrating this expression with constant γ gives:

$$\ln\left(\frac{T_2}{T_1}\right) = (\gamma - 1)\ln\left(\frac{V_1}{V_2}\right)$$

or:

$$\frac{T_2}{T_1} = \left(\frac{V_1}{V_2}\right)^{(\gamma-1)}$$

This result says that as this adiabatic, reversible (isentropic) expansion proceeds from one state to the next, the product $TV^{(\gamma-1)}$ will always be the same value; $TV^{(\gamma-1)} = $ constant. Furthermore, by eliminating T by using the ideal gas law, it can be shown that $PV^\gamma = $ constant. Finally, by eliminating V, it is seen that $PT^{\gamma/(1-\gamma)} = $ constant

To summarize, for any isentropic expansion (or compression) of an ideal gas with temperature-independent heat capacity ratio,

Isentropic expansion or compression of an ideal gas	$PV^\gamma = $ constant $TV^{(\gamma-1)} = $ constant $PT^{\gamma/(1-\gamma)} = $ constant

Hence, if the initial state is known, then the constants can be calculated. Then given the state 2 pressure, the volume and temperature of state 2 can be easily calculated.

The Ideal Gas and the Ideal Gas Temperature.

Ideal gas behavior can be viewed as a means of arriving at a definition of absolute temperature. From the virial equation of state we see that even real gases as they approach zero pressure or zero density approach a compressibility factor of unity. Stated differently, $P\tilde{V}$ in the limit of low pressure is a function of temperature only and that fact may be used to define temperature. This has, in fact, been done by arbitrarily setting this function of temperature equal to a constant times temperature according to the equation

$$P\tilde{V} = f(T)$$
$$= RT \qquad (15-99)$$

Any other functional relationship could have been chosen but this simple relationship is obviously the easiest to work with. The proportionality constant is called R and termed the ideal gas constant. Then to further define temperature the temperature of the triple point of water is arbitrarily assigned a value 273.16 K.

Once these two criteria are arbitrarily set, the ideal gas temperature scale is fully fixed. Precise measurements of $P\tilde{V}$ as a function of pressure for a variety of gases can be extrapolated to zero pressure to obtain the zero-pressure value of $P\tilde{V}$. From this zero-pressure value measured at the temperature of the triple point of water, the ideal gas constant can be determined:

$$R \equiv \lim_{P \to 0} \frac{(P\tilde{V})_{T=273.15\ K}}{273.15\ K} \qquad (15-100)$$

Values of R in a number of unit systems are given in Appendix A Table A-4. Then, from the zero-pressure value of $P\tilde{V}$ measured at other temperatures, numerical values of temperature are obtained:

$$T \equiv \frac{(P\tilde{V})_T}{R} \qquad (15-101)$$

This ideal gas law temperature actually is not totally arbitrary. As will be seen later, it is entirely consistent with an alternative definition of temperature which is made through arguments concerning the definition of entropy.

15.6.2 Non-Ideal Gases

Now because of the availability of ideal gas heat capacities, it is convenient to express calculations of changes in the thermodynamic properties in terms of these ideal gas heat capacities, even for non-ideal gases. This is done by selecting a conveniently choosen calculational path for changes in enthalpy, e.g., where we want to go from (T_1, P_1) to (T_2, P_2) (Figure 15-39). Instead of a direct or arbitrary path, we choose a special three step path. First, a constant-temperature path is followed to zero pressure (where all real gases behave as ideal gases). Second, followed by a constant-pressure ideal gas path (at zero pressure) is followed from T_1 to T_2. Third, and finally, another constant temperature path is followed from zero pressure to P_2.

Remember that because the property functions are state functions, we may choose *any* convenient path for calculating property changes and will obtain the same property changes as we would by any other path. So, for example, a change in enthalpy is calculated according to:

$$\widehat{H}_2 - \widehat{H}_1 = \int_{P_1}^0 \left[\widehat{V} - T\left(\frac{\partial \widehat{V}}{\partial T}\right)_P\right]_{T_1} dP + \int_{T_1}^{T_2} \widehat{C}_P^{\mathrm{ig}} dT + \int_0^{P_2} \left[\widehat{V} - T\left(\frac{\partial \widehat{V}}{\partial T}\right)_P\right]_{T_2} dP$$

$$= -\widehat{H}^R(T_1, P_1) + \int_{T_1}^{T_2} \widehat{C}_P^{\mathrm{ig}} dT + \widehat{H}^R(T_2, P_2) \qquad (15-102)$$

In this result, the two pressure integrals are referred to as residual enthalpies $(\widehat{H}^R)$ and represent differences in enthalpy between the non-ideal gas (at non-zero pressure) and the ideal gas (at zero pressure):

$$\widehat{H}^R(P, T) \equiv \widehat{H}(P, T) - \widehat{H}^{\mathrm{ig}}$$

$$= \int_0^P \left[\widehat{V} - T\left(\frac{\partial \widehat{V}}{\partial T}\right)_P\right]_{T_2} dP \qquad (15-103)$$

The temperature integral gives the change in enthalpy which would occur at zero pressure where the real gas behaves as an ideal gas. This result is the general equation for calculating changes in enthalpy of a real (non-ideal) gas from one state to another. Internal energy, of course, can be calculated directly from enthalpy, pressure, and volume ($\widehat{U} = \widehat{H} - P\widehat{V}$).

For entropy, in terms of T and P an analogous procedure is followed:

$$\widehat{S}_2 - \widehat{S}_1 = -\int_{P_1}^0 \left[\left(\frac{\partial \widehat{V}}{\partial T}\right)_P\right]_{T_1} dP + \int_{T_1}^{T_2} \frac{\widehat{C}_P^{\mathrm{ig}}}{T} dT - \int_0^{P_2} \left[\left(\frac{\partial \widehat{V}}{\partial T}\right)_P\right]_{T_2} dP$$

$$= -\widehat{S}^R(T_1, P_1) + \int_{T_1}^{T_2} \frac{\widehat{C}_P^{\mathrm{ig}}}{T} dT + \widehat{S}^R(T_2, P_2) \qquad (15-104)$$

where:

$$\widehat{S}^R(T, P) \equiv \widehat{S}(T, P) - \widehat{S}^{\mathrm{ig}}$$

$$= -\int_0^P \left(\frac{\partial \widehat{V}}{\partial T}\right)_P dP \qquad (15-105)$$

These residual functions can be calculated directly from the appropriate equation of state or they can be obtained from graphs based upon the generalized corresponding states correlation of Pitzer and coworkers (Smith and Van Ness, 1987).

If state 1 is the reference state, then $\widehat{H}_2 - \widehat{H}_1$ is simply the "absolute" enthalpy of state 2. Hence

$$\widehat{H}(T,P) = \widehat{H}_o^{ig} + \int_{T_o}^{T} \widehat{C}_P^{ig} \, dT + \widehat{H}^R(T,P) \tag{15-106}$$

and for entropy:

$$\widehat{S}(T,P) = \widehat{S}_o^{ig} + \int_{T_o}^{T} \frac{\widehat{C}_P^{ig}}{T} \, dT - R \ln\left(\frac{P}{P_o}\right) + \widehat{S}^R(T,P) \tag{15-107}$$

The residual forms of the equations are useful because the major changes in the thermodynamic properties with temperature and pressure are related to the non-residual terms whereas the residual terms amount to corrections (and therefore usually are fairly small) for deviations from ideal behavior.

Compressibility Factor Equations for U, H, A, G, S. The equations developed for the thermodynamic properties so far have been written directly in terms of heat capacities and equation-of-state variables $P, \widetilde{V}$, and T. However, it is frequently convenient for gases, because of the way equations-of-state are sometimes written, to express these equation-of-state properties in terms of compressibility factors. For an ideal gas we have that

$$\frac{P\widetilde{V}}{RT} = 1 \tag{15-108}$$

and to the extent real gases are non-ideal this same ratio may be expressed as the variable Z which differs from unity:

$$\frac{P\widetilde{V}}{RT} = Z \tag{15-109}$$

This compressibility factor (Z) is a function of two independent variables for a single phase material and so we may think of it as a function of temperature and pressure, temperature and volume, volume and pressure, or other combination of two state functions.

The above equations for calculating the thermodynamic functions can then be written in terms of the P-$\widehat{V}$-T variables, heat capacities, and this compressibility factor. The heat capacities are as expressed above while the residual functions (deviations from ideal behavior due to pressure at constant temperature) are expressed in terms of the compressibility factors in the following ways:

$$\widehat{H}^R(T,P) = -\frac{RT^2}{MW} \int_0^P \left(\frac{\partial Z}{\partial T}\right)_P \frac{dP}{P} \tag{15-110}$$

$$\widehat{S}^R(T,P) = -\frac{RT}{MW} \int_0^P \left(\frac{\partial Z}{\partial T}\right)_P \frac{dP}{P} - \frac{R}{MW} \int_0^P (Z-1)\frac{dP}{P} \tag{15-111}$$

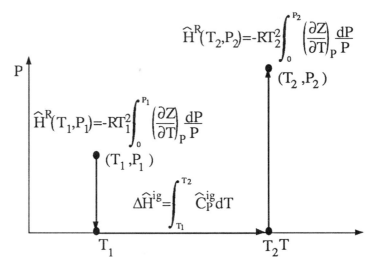

Note that with equations for calculating $\widehat{H}$ and $\widehat{S}$, the other properties ($\widehat{U}$, $\widehat{A}$, and $\widehat{G}$) can be determined directly from their definitions (equations 7-44, 15-8, and 15-9).

$$\widehat{H}^R(T_2,P_2)=-RT_2^2\int_0^{P_2}\left(\frac{\partial Z}{\partial T}\right)_P\frac{dP}{P}$$

(T_2,P_2)

$$\widehat{H}^R(T_1,P_1)=-RT_1^2\int_0^{P_1}\left(\frac{\partial Z}{\partial T}\right)_P\frac{dP}{P}$$

(T_1,P_1)

$$\Delta\widehat{H}^{ig}=\int_{T_1}^{T_2}\widehat{C}_P^{ig}dT$$

Figure 15-39: Residual enthalpy calculations using residual functions

Example 15-15. Before entering a reactor, ethylene gas at approximately atmospheric pressure is heated from 100 °F to 500 °F in a flow-through heat exchanger as shown in Figure 15-40. If the flow rate is 100 lb$_m$ / min calculate the rate of heat transfer in Btu/min.

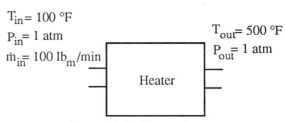

Figure 15-40: Heat exchanger

SOLUTION: For this problem we will define the system as the heat exchanger as shown in Figure 15-41. The system is operating at steady state.

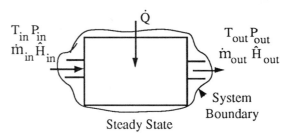

Figure 15-41: System for example 15-15

Tool 1, The Laws:

The **conservation of mass** for this defined system as steady state says:

$$\left(\frac{dm_{sys}}{dt}\right) = \dot{m}_{in} - \dot{m}_{out}$$

$$= 0$$

so that the mass flow rate entering the system is equal to the mass flow rate leaving the system called the single mass flow rate $\dot{m}$.

For the system with one entering stream and one leaving stream, **the conservation of energy (the first law of thermodynamics)** says:

$$\left(\frac{dE_{\text{sys}}}{dt}\right) = \left(\dot{m}(\widehat{H} + \widehat{PE} + \widehat{KE})\right)_{\text{in}} - \left(\dot{m}(\widehat{H} + \widehat{PE} + \widehat{KE})\right)_{\text{out}} + \dot{Q} + \dot{W}$$

$$= 0$$

There is no work for this process. Furthermore, the changes in kinetic and potential energy from the entering to the leaving stream is assumed to be negligible. This reduces the first law for this system to:

$$\dot{Q} = \dot{m}(\widehat{H}_{\text{out}} - \widehat{H}_{\text{in}})$$

We see that the enthalpy of the entering ethylene changes due to the heat transfer.

Tool 2, Gibb's Phase Rule. With two intensive variables specified for both the entering and leaving streams, it is clear that both states are fully specified.

Tool 4, Property Relations. We know how to calculate the change in enthalpy in terms of observables from the property relation:

$$\widehat{H}_{\text{out}} - \widehat{H}_{\text{in}} = \int_{T_{\text{in}}}^{T_{\text{out}}} \widehat{C}_P dT + \int_{P_{\text{in}}}^{P_{\text{out}}} \left[\widehat{V} - T\left(\frac{\partial \widehat{V}}{\partial T}\right)_P\right] dP$$

Because the pressure at the entrance is equal to the pressure at the exit, the change in enthalpy is:

$$\widehat{H}_{\text{out}} - \widehat{H}_{\text{in}} = \int_{T_{\text{in}}}^{T_{\text{out}}} \widehat{C}_P dT$$

Table C-15 in Appendix C gives the dimensionless constant pressure heat capacity for enthylene as:

$$\frac{\widetilde{C}_P(T)}{R} = A + BT + CT^2 + DT^{-2}$$

where the constants are:

$$A = 1.424$$
$$B = 14.394 \times 10^{-3}$$
$$C = -43.92 \times 10^{-7}$$
$$D = 0$$

and the temperatures are in Kelvin. Converting the temperatures to Kelvin gives $T_{\text{in}} = 311$ K and $T_{\text{out}} = 533$ K. The heat capacity example 15-2 as:

$$\int_{T_{\text{in}}}^{T_{\text{out}}} \widehat{C}_p dT = \left(\frac{R}{MW}\right)\left[A(T_{\text{out}} - T_{\text{in}}) + \frac{B}{2}(T_{\text{out}}^2 - T_{\text{in}}^2) + \frac{C}{3}(T_{\text{out}}^3 - T_{\text{in}}^3) - D\left(\frac{1}{T_{\text{out}}} - \frac{1}{T_{\text{in}}}\right)\right]$$

The value for the gas constant here is 8.314 J/mol K and the molecular weight of ethylene is 28 g/gmol. Substituting in the values and evaluating the expression gives:

$$\int_{T_{\text{in}}}^{T_{\text{out}}} \widehat{C}_p dT = \left(\frac{8.314}{28}\right)\left[(1.424)((533) - (311)) + \frac{14.394 \times 10^{-3}}{2}(533^2 - 311^2) +\right.$$

$$\left. \frac{-43.92 \times 10^{-7}}{3}(533^3 - 311^3) - (0)\left(\frac{1}{533} - \frac{1}{311}\right)\right]$$

The integral gives:

$$\widehat{H}_{\text{out}} - \widehat{H}_{\text{in}} = \int_{T_{\text{in}}}^{T_{\text{out}}} \widehat{C}_P dT = 441.5 \text{ J} / \text{g}$$

Substituting this value into the equation to calculate the heat transfer with the proper conversion factors yields:

$$\dot{Q} = \dot{m}(\widehat{H}_{\text{out}} - \widehat{H}_{\text{in}})$$

$$= (100)(441.5)\frac{\text{J}}{\text{g}}\frac{\text{lb}_{\text{m}}}{\text{min}}\left|\frac{454 \text{ g}}{1 \text{ lb}_{\text{m}}}\right|\frac{1 \text{ Btu}}{1055 \text{ J}} = 19,000 \text{ Btu} / \text{min}$$

The heat transfer rate must be 19,000 (Btu/min) to increase the temperature of the ethylene from 100 °F to 500 °F before entering the reactor.

15.6.3 Calculations of U, H, A, G, and S for Solids and Liquids

For solids and liquids it is usually adequate to represent P-$\widehat{V}$-T behavior in terms of the volume expansivity due to the nearly incompressible behavior of these materials. Consequently, by writing $\widehat{H} = \widehat{H}(T, P)$ we have:

$$d\widehat{H} = \left(\frac{\partial \widehat{H}}{\partial T}\right)_P dT + \left(\frac{\partial \widehat{H}}{\partial P}\right)_T dP \qquad (15-112)$$

which may be written in terms of heat capacities and the volume expansivity (β)

$$\boxed{d\widehat{H} = \widehat{C}_P dT + \widehat{V}(1 - T\beta)dP} \qquad (15-113)$$

where β is defined as:

$$\boxed{\beta \equiv \frac{1}{\widehat{V}}\left(\frac{\partial \widehat{V}}{\partial T}\right)_P} \qquad (15-114)$$

Likewise, changes in entropy may be calculated as

$$\boxed{d\widehat{S} = \widehat{C}_P \frac{dT}{T} - \beta \widehat{V} dP} \qquad (15-115)$$

Practically speaking, β and $\widehat{V}$ often are constant, or nearly so, over the temperature ranges, simplifying the integrations.

As described before, changes in $\widehat{U}$, $\widehat{A}$, and $\widehat{G}$ can also be determined once changes in $\widehat{H}$ and $\widehat{S}$ are known.

Thermodynamic Diagrams, Tables, and Computer Software. Thermodynamic diagrams and tables provide a convenient way to present the state properties of solids liquids and gases. An example of a pressure versus enthalpy diagram for a pure substance is provided in Figure 15-42. Additionally, tabulations of data are available.

Examples are steam tables (Tables C-22 or 23). These figures and tables present values for the thermodynamic properties and when available eliminate the need to calculate values ourselves. The calculations have already been done, following the procedures outlined above, and then given in the figure or table. Whenever available, we should take advantage of these resources rather than doing the calculations ourselves. Note that values of all properties are not given; values not listed can be calculated from those that are by using the property definitions. For example, a table may include only values of $\widehat{H}, \widehat{S}, T, P$, and $\widehat{V}$ from which $\widehat{U} \ (= \widehat{H} - P\widehat{V})$, $\widehat{A} \ (= \widehat{U} - T\widehat{S})$, and $\widehat{G} \ (= \widehat{H} - T\widehat{S})$ can be readily calculated.

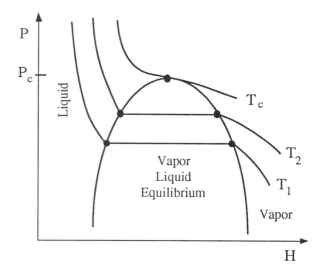

Figure 15-42: Pressure enthalpy diagram

Example 15-16. In a power plant, saturated water at 200 kPa is pumped from a condenser to the boiler pressure of 2000 kPa as shown in Figure 15-43. If the pump operates reversibly and adiabatically, how much work is required per unit mass to pump the water to this pressure?

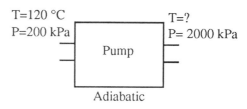

Figure 15-43: Pump

Also, what is the approximate temperature rise of the liquid water if the value of β is

$$\beta = \frac{1}{\widehat{V}} \left(\frac{\partial \widehat{V}}{\partial T} \right)_P$$

$$= 457.59 \times 10^{-6} \ (1 \ / \ K)$$

SOLUTION: The system is defined as the pump as shown in the schematic in Figure 15-44. Furthermore the system is operating at steady state.

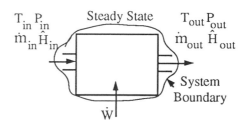

Figure 15-44: System for example 15-16

Tool 1, The Laws: The **conservation of mass** for this steady-state system states:

$$\left(\frac{dm_{sys}}{dt}\right) = \dot{m}_{in} - \dot{m}_{out}$$

$$= 0$$

Or the mass flow rate entering the system is equal to the mass flow rate leaving the system.

The **conservation of energy** for this steady-state flow process states:

$$\left(\frac{d(E_{sys})}{dt}\right) = \left(\dot{m}(\widehat{H} + \widehat{PE} + \widehat{KE})\right)_{in} - \left(\dot{m}(\widehat{H} + \widehat{PE} + \widehat{KE})\right)_{out} + \dot{Q} + \dot{W}$$

$$= 0$$

For this system, that is operating adiabatically, the heat transfer is zero. Furthermore, the changes in potential and kinetic energy of the water entering and leaving the system are approximately zero and the operation is steady state. These observations reduce the previous equation to:

$$0 = (\dot{m}\widehat{H})_{in} - (\dot{m}\widehat{H})_{out} + \dot{W}$$

Since $\dot{m}_{in} = \dot{m}_{out} (= \dot{m}$, say) it can be factored out as:

$$0 = \dot{m}(\widehat{H}_{in} - \widehat{H}_{out}) + \dot{W}$$

Rearranging and solving for the unknown work per unit mass gives:

$$\widehat{W} = (\widehat{H}_{out} - \widehat{H}_{in})$$

Obviously, if we can determine the change in enthalpy, we can find the work.

Tool 2, (Gibb's Phase Rule): Entering the pump the water is saturated liquid at 200 kPa ($C = 1$, $\phi = 2$, because is is saturated, so $F = 1 + 2 - 2 = 1$) and hence specifying the pressure is sufficient to fully define the state. All other intensive state properties can be determined. For the exit state, we know the pressure (2000 kPa) but that is all. We cannot assume saturation. bu we do know that the water is compressed adiabatically (no entropy change due to heat transfer) and reversibly (no entropy increase due to generation) and consequently, as each unit of water passes through the pump (as a closed system, no entropy change due to mass exchange) its entropy does not change (isentropic process). So, in addition to knowing the exit pressure, it turns out we know the exit entropy (the same as the entering entropy) and the exit state then is fully specified. So at least in principle we can determing the exit temperature, etc. At this point, if we had complete table of properties for subcooled (below saturation) water, we could just look up the values in the tables and be done. In the absence of the tables, we will proceed to Tool 3.

To help us calculate the change in enthalpy for the entering states and leaving states, a $P\text{-}\widehat{V}$ diagram is shown in Figure 15-45.

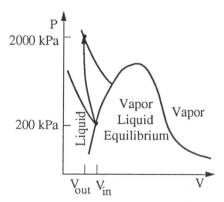

Figure 15-45: P-V diagram
for example 15-16

Tool 3, Fundamental Property Relations: Ordinarily at this point, we would use a UHAG relation wich is formulated in terms of observables. For enthalpy change at constant entropy, however, it is convenient to use the relation

$$d\widehat{H} = \widehat{V}dP + \delta\widehat{Q} + \delta\widehat{F} \qquad \text{or} \qquad d\widehat{H} = Td\widehat{S} + \widehat{V}dP$$

which for isentropic (adiabatic, $\delta Q = 0$, and reversible, no conversion of mechanical energy to thermal energy by friction, $\delta F = 0$) compression in the pump (tool 7) is

$$d\widehat{H} = \widehat{V}dP$$

Integrating this equation from the entering state to the leaving state gives

$$\widehat{H}_{out} - \widehat{H}_{in} = \int_{P_{in}}^{P_{out}} \widehat{V}dP$$

In order to calculate this integral, we need to know how the specific volume of the liquid varies with the pressure (tool 5, liquid EOS). A good approximation for liquids is that the volume of the liquid is constant for changes in pressure; the fluid is incompressible. With this assumption, the integration becomes

$$\widehat{H}_{out} - \widehat{H}_{in} = \widehat{V}(P_{out} - P_{in})$$

Where $\widehat{V}$ is the value of the specific volume of the saturated liquid at the conditions stated. From the Steam Tables in Appendix C (Table C-23) the value of the specific volume is:

$$\widehat{V} = 0.0010601 \text{ m}^3 / \text{kg}$$

and the value of the entering stream temperature is:

$$T_{in} = 393 \text{ K}$$

The change in enthalpy is calculated by substituting the known values into the previous equation

$$\widehat{H}_{out} - \widehat{H}_{in} = (0.0010601)(2000 - 200)\frac{\text{m}^3 \text{ kPa}}{\text{kg}}\left|\frac{1 \text{ J}}{1 \text{ m}^3 \text{ Pa}}\right. = 1.908 \text{ J} / \text{g}$$

Then, from the first law results, $\widehat{W}$ is

$$\widehat{W} = (\widehat{H}_{out} - \widehat{H}_{in}) = 1.908 \text{ J} / \text{g}$$

To determine the temperature change, we again return to enthalpy relations (tools 3 and 4). There are two forms of the thermal energy accounting which help us relate temperature change to the process. One equation is the thermal energy equation

$$d\widehat{H} = \widehat{V}dP + \delta\widehat{Q} + \delta\widehat{F}$$

which says that enthalpy increases because of pressure increases, heat transfer, and irreversible conversions of mechanical energy thermal energy (friction). Then, in terms of observables, we have a second result, equation (15-113)

$$d\widehat{H} = \widehat{C}_P dT + \widehat{V}(1 - T\beta)dP$$

which, combined with the previous relation, tells us that temperature increases because of compression (just like pumping air in a bicycle pump), heat transfer, and friction:

$$\widehat{C}dT = \widehat{V}T\beta dP + \delta\widehat{Q} + \delta\widehat{F}$$

If the process is adiabatic ($\delta Q = 0$) and reversible ($\delta F = 0$) then

$$\frac{\widehat{C}_P dT}{T} = \widehat{V} \beta dP$$

and the temperature increase is readily calculated.

The heat capacity of the liquid water is $\widehat{C}_P = 4.180 \text{ J} / \text{g K}$ and is assumed to be constant over the integration. Furthermore, the value of β and $\widehat{V}$ are assumed to be independent of the pressure and temperature over the range of integration. Integration yields:

$$\ln\left(\frac{T_{\text{out}}}{T_{\text{in}}}\right) = \frac{\widehat{V}\beta}{\widehat{C}_P}(P_{\text{out}} - P_{\text{in}})$$

Solving for the temperature of the water leaving the pump gives:

$$T_{\text{out}} = T_{\text{in}} \exp\left(\frac{\widehat{V}\beta}{\widehat{C}_P}(P_{\text{out}} - P_{\text{in}})\right)$$

Substituting in the known values and the proper conversion factors gives:

$$T_{\text{out}} = 393 \exp\left(\frac{(0.0010601)(457.59 \times 10^{-6})}{4.180}(2000 - 200)\frac{\text{m}^3 \text{ kPa K}^2 \text{ g}}{\text{kg K J}}\left|\frac{1 \text{ J}}{1 \text{ m}^3 \text{ Pa}}\right.\right) = 393.08 \text{ K}$$

The temperature of the water increased by 0.08 K a very small amount. Friction due to viscosity will add to this value.

15.7 PHASE TRANSITIONS

So far we have considered property changes only for single-phase changes of state, that is for changes in temperature or pressure of a single phase material. Of even greater magnitude, when they exist, are the property changes associated with phase transitions. The energy associated with vaporizing water, for example, at constant temperature and pressure is three orders of magnitude greater than that required to change its temperature by one degree, in a single phase (per unit mass).

We may evaluate changes in the thermodynamic properties associated with phase transitions by first looking at the $U H A G$ for changes in the Gibbs free energy:

$$d\widehat{G} = -\widehat{S}dT + \widehat{V}dP$$

Now because phase transitions occur at constant temperature and pressure (that is two phases that are in equilibrium exist at the same temperature and pressure) we see that the change in the Gibbs free energy (per mole or mass of material) must be zero as a material transfers between two phases which are in equilibrium. Stated differently, we see that the Gibbs free energies of two phases which are in equilibrium must be equal. If we are referring to the two phases as the alpha and the beta phases (as is done by Smith and Van Ness (1987)) then we write that:

$$\widehat{G}^{\alpha} = \widehat{G}^{\beta} \tag{15 - 116}$$

Here $\widehat{G}^{\alpha}$ and $\widehat{G}^{\beta}$ represent the Gibbs free energies (per mass) for each of the two phases.

Now from this observation, the change in enthalpy associated with the change in phase at constant temperature and pressure may be deduced. As two phases in equilibrium change temperature and pressure (simultaneously in order to maintain equilibrium) it must be true that:

$$d\widehat{G}^{\alpha} = d\widehat{G}^{\beta} \tag{15 - 117}$$

Accordingly, by writing the $d\widehat{G}$ relation for each phase we have:

$$-\widehat{S}^{\alpha} dT + \widehat{V}^{\alpha} dP^{\text{sat}} = -\widehat{S}^{\beta} dT + \widehat{V}^{\beta} dP^{\text{sat}} \tag{15 - 118}$$

In this result, because the two phases are being maintained in equilibrium for a given change in temperature, the pressure must change in such a way as to maintain saturation and hence the dP^{sat} notation. This result can be rearranged to give:

$$\left(\frac{dP^{\text{sat}}}{dT}\right) = \frac{\widehat{S}^{\alpha} - \widehat{S}^{\beta}}{\widehat{V}^{\alpha} - \widehat{V}^{\beta}} \tag{15 - 119}$$

which relates the changes in entropy and volume associated with a phase change at constant temperature and pressure to the slope of the saturation pressure versus temperature curve. Now the change in entropy associated with the phase change at constant temperature and pressure is related to the change in enthalpy according to:

$$d\widehat{H} = T d\widehat{S} \tag{15 - 120}$$

because $dP = 0$ in moving from one phase to the other. Consequently the above result can be expressed in terms of enthalpy changes associated with the phase change according to:

$$\boxed{\left(\frac{dP^{\text{sat}}}{dT}\right) = \frac{\widehat{H}^{\alpha} - \widehat{H}^{\beta}}{T(\widehat{V}^{\alpha} - \widehat{V}^{\beta})}} \tag{15 - 121}$$

This result is the *Clapeyron equation* and is very important for relating the thermodynamic properties of phase change to the P-T saturation curve. Note that no restriction on the phases being considered exists in this derivation; it holds for vapor-liquid, and solid-liquid, solid-vapor, solid (I)-solid (II), etc. phase transitions equally well.

For *vapor-liquid equilibrium*, we may write the phase transitions in terms of liquid and vapor transitions for which we have:

$$\left(\frac{dP^{\text{sat}}}{dT}\right) = \frac{\widehat{H}^{v} - \widehat{H}^{l}}{T(\widehat{V}^{v} - \widehat{V}^{l})} \tag{15 - 122}$$

Now because gases are much less dense than liquids, we may approximate the change in volume due to the phase transition as the vapor volume; the volume of the liquid phase is negligible compared to that of the vapor phase (per mole or per mass). If this vapor volume is then adequately represented by the ideal gas relation then the Clapeyron equation becomes:

$$\left(\frac{dP^{\text{sat}}}{dT}\right) = \frac{(\widehat{H}^{v} - \widehat{H}^{l})P^{\text{sat}}}{RT^2} \tag{15 - 123}$$

which may be rearranged to give an expression for the enthalpy change associated with the liquid-to-vapor (heat of vaporization) transition as:

$$\lambda_{\text{Vap}} = \widehat{H}^v - \widehat{H}^l = -R\left(\frac{d \ln P^{\text{sat}}}{d(1/T)}\right)$$

(15 – 124)

This result is termed the *Clausius-Clapeyron equation* and is a very important result for establishing approximate values for enthalpy changes due to liquid-vapor transition in terms of saturation pressures and temperatures. The heat of vaporization is λ_{Vap} Experimentally, it is observed that a plot of $\ln P^{\text{sat}}$ versus $1/T$ is nearly linear suggesting that the enthalpy change due to vaporization is nearly constant, i.e. independent of temperature. In fact, heats of vaporization are not nearly as temperature independent as the slope of $\ln P^{\text{sat}}$ versus $(1/T)$ would seem to imply due to the imperfections of the assumptions associated with the equation over a broad range of temperature; errors in the assumptions over a range of temperature tend to counteract the actual dependence of the enthalpy change with temperature.

If the vapor pressure versus temperature data are available for a material, then the heats of vaporization may be estimated directly from this equation. Alternatively, they may be measured calorimetrically although the availability of such data are rather hit and miss.

As a third alternative for vapor-liquid phase transitions, the heats of vaporization may be estimated from normal boiling point data or from a known heat of vaporization at a given temperature. The method of Riedel (for heats of vaporization) can be used to estimate the heat of vaporization at the normal boiling point (ΔH_n) of a material in terms of the normal boiling point temperature (T_n), reduced normal boiling point temperature (T_{rn}), and the material's critical pressure:

$$\frac{\Delta \widehat{H}_n}{RT_n} = \frac{1.092(\ln P_c - 1.013)}{0.930 - T_{rn}}$$

(15 – 125)

where P_c has units of bar. For the heats of vaporization at other temperatures, the Watson equation may be used:

$$\frac{\Delta \widehat{H}_2}{\Delta \widehat{H}_1} = \left(\frac{1 - T_{r2}}{1 - T_{r1}}\right)^{0.38}$$

(15 – 126)

The heat of vaporization at one temperature may be used to estimate the heat of vaporization at a second temperature in terms of the two reduced temperatures.

15.7.1 Saturation Vapor Pressures

The Clausius-Clapeyron equation, which is an approximate relationship between the heat of vaporization and the relationship between saturation pressure and temperature, suggests that the saturation vapor pressure be viewed in terms of an exponential relationship with $1/T$. That is, it suggests plotting $\ln P^{\text{sat}}$ versus $1/T$. The relationship, however, says nothing whatsoever about what this dependency should be. As a matter of experimental observation, however, and as mentioned above, we see that for a great many substances this relationship is quite linear and might be represented by an equation of the form:

$$\ln P^{\text{sat}} = A - \frac{B}{T}$$

(15 – 127)

The parameters A and B would be constants for each species. However, because the data are not perfectly linear a three parameter model is generally more widely used and it is known as the *Antoine equation*:

$$\ln P^{\text{sat}} = A - \frac{B}{T + C}$$

$(15 - 128)$

This equation is very similar in form to the linear relationship but has a correction to the temperature. The Antoine constants A, B, and C are readily available for a great many compounds in references such as *The Properties of Gases and Liquids* by Reid, Prausnitz, and Sherwood (1977). For general applicability in calculations of liquid vapor pressures, the Antoine equation should be used as a first choice. For interpolation between data points that are reasonably close together, the simpler two-parameter equation may be used instead, although for computer calculations this results in little savings.

For even more accurate calculations of vapor pressure over a wider temperature range, the Riedel (1954) equation (for vapor pressures), a four parameter equation, is available (Smith and Van Ness, 1987).

Example 15-17. For the given data for water, calculate the change in enthalpy of the water if the water goes from a liquid to a vapor at 200 °C and 1554.9 kPa. Also, compare, the calculated value with a value obtained from using the Riedel equation and the Watson equation.

Example 15-17 Data

T(K)	P (kPA)	$\widehat{V}^{\text{liq}}$ (ml/g)	$\widehat{V}^{\text{vap}}$ (ml/g)
471.15	1490.9	1.153	132.4
473.15	1554.9	1.156	127.2
475.15	1621.9	1.160	122.1

SOLUTION: The following P-T and P-$\widehat{V}$ diagrams show the calculation of the change in enthalpy of the water as it goes from the liquid state to the vapor state.

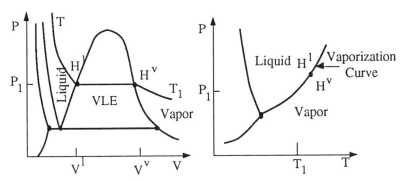

Figure 15-46: Changes in State for Example 15-17

The exact thermodynamic relationships for calculating the change in enthalpy for a change in phase is given as:

$$\widehat{H}^{v} - \widehat{H}^{l} = T(\widehat{V}^{v} - \widehat{V}^{l})\left(\frac{dP}{dT}\right)$$

From the given information in the problem the temperature is:

$$T = 473.15 \text{ K}$$

and the change in specific volume is:

$$\widehat{V}^v - \widehat{V}^l = 127.2 - 1.156 = 126.0 \text{ cm}^3 / \text{g}$$

The slope of the vapor pressure curve at 473.15 K is approximated from the given data:

$$\left(\frac{dP}{dT}\right) \approx \left(\frac{\Delta P}{\Delta T}\right)$$

We will approximate the slope of the curve at 473.15 K by a central difference formula or we will choose the values of the and temperatures that bracket the point at $T = 473.15$ as:

$$\left(\frac{\Delta P}{\Delta T}\right) = \left(\frac{P_2 - P_1}{T_2 - T_1}\right)$$
$$= \left(\frac{1621.9 - 1490.9}{475.15 - 471.15}\right)$$
$$= 32.75 \text{ kPa} / \text{K}$$

Now these values can be substituted into the equation to calculate the change in specific enthalpy as:

$$\widehat{H}^v - \widehat{H}^l = T(\widehat{V}^v - \widehat{V}^l)\left(\frac{dP}{dT}\right)$$
$$= (473.15)(126)(32.75)\left(\frac{1}{1000}\right)$$
$$= 1,952.4 \text{ J} / \text{g}$$

Now the change in specific enthalpy is converted to the change in molar enthalpy by dividing by the molecular weight.

$$\widetilde{H}^v - \widetilde{H}^l = (\widehat{H}^v - \widehat{H}^l)(MW)$$
$$= (1,952.4)(18)$$
$$= 35,144 \text{ J} / \text{mol}$$

The change in molar enthalpy due to the vaporization of water at 473.15 K as calculated from the thermodynamic data is 35,144 J/mol.

Now we will calculate the enthalpy change due to vaporization by the empirical equations of Riedel and Watson. The Reidel equation allows us to calculate the change in molar enthalpy at the normal boiling point from the data at the critical point. After that, the Watson equation enables us the calculate the change in enthalpy due to vaporization at any temperature. The Riedel equation is:

$$\frac{\Delta \widetilde{H}_n / T_n}{R} = \frac{1.029(\ln P_c - 1.03)}{0.930 - T_{rn}}$$

where the pressure is in bars.

The critical values for water are found in Table C-19 in Appendix C and are listed below:

$$P_c = (217.6)(1.013)$$
$$= 220.43 \text{ bar}$$
$$T_c = 647.3 \text{ K}$$

The normal boiling point for water at 1 atm is 100 °C or 373 °K so the reduced normal boiling point is:

$$T_{rn} = \frac{T_n}{T_c}$$
$$= \frac{373}{647.3}$$
$$= 0.576$$

Substituting these known values in the Riedel equations yields:

$$\frac{\Delta \widetilde{H}_n / T_n}{R} = \frac{1.029(\ln(220.43) - 1.03)}{0.930 - 0.576}$$
$$= 13.52$$

The change in molar enthalpy due to the phase change is give by:

$$\Delta \widetilde{H}_n = RT_n 13.52$$
$$= (8.314)(373)(13.52)$$
$$= 41,927 \text{ J / mol}$$

The change in molar enthalpy from the vaporization of water at the normal boiling point is 41,927 J/mol.

We can use the Watson equation to adjust the change in molar enthalpy to the temperature of 473.15 K. The Watson equation is:

$$\frac{\Delta \widetilde{H}_2}{\Delta \widetilde{H}_1} = \left(\frac{1 - T_{r2}}{1 - T_{r1}} \right)^{0.38}$$

If we choose the point 1 as the normal boiling point and the point 2 as the temperature of interest, we can calculate the reduced temperature as point 2 as:

$$T_{r2} = \frac{T_2}{T_c}$$
$$= \frac{473.15}{647.3}$$
$$= 0.731$$

Now substituting the known values into the equation gives:

$$\frac{\Delta \widetilde{H}_2}{\Delta \widetilde{H}_1} = \left(\frac{1 - 0.731}{1 - 0.576} \right)^{0.38}$$
$$= 0.841$$

Now the change in molar enthalpy can by calculated as:

$$\Delta \widetilde{H}_2 = 0.841 \Delta \widetilde{H}_1$$
$$= (0.841)(41,927)$$
$$= 35,269.6 \text{ J} / \text{mol}$$

The change in molar enthalpy due to the vaporization of water at 473 as calculated from the empircal equations is 35,269.5 J/mol. This value compares favorably with the value determined from the volumetric data.

15.8 ENTHALPY CHANGES ASSOCIATED WITH CHEMICAL REACTIONS

As reactions occur between compounds, extensive energy conversions occur. Changes in chemical bonding results in the release or absorption of thermal energy. If a process involves chemical reactions, then in order to calculate the total energy conversions one must be able to calculate the energies associated with the entering reactants and the exiting products (the enthalpies of different compounds) or, equivalently, the heats of reactions, the enthalpy differences between reactants and products.

15.8.1 Heats of Reaction

If a set of reactants is allowed to react to produce a set of products, then the *heat of reaction is defined to be the enthalpy of the products (in their final state) minus that of the reactants in their initial state.* If we furthermore define a standard state for reactants and products, (say 25 °C and one atmosphere) and standard states of physical existence (gas, liquid, solid) then if the reactants all exist in the standard state initially and the products are returned to the standard state by heat transfer, then this heat of reaction is called the *standard heat of reaction*. For example, if a hypothetical process brings reactants in at their standard state, reacts these materials to produce the products, and adds or removes heat to produce the products at their standard states when they leave the process, then (if the process operates at steady state, if there is no work exchange between the process and the surroundings, and if kinetic and potential energy effects are negligible) the heat transferred is this standard heat of reaction. An exothermic reaction is one for which the required heat transfer is negative (heat is evolved), and an endothermic reaction is one for which the required heat transfer is positive (heat is absorbed). Note that the term "standard," when used with respect to heats of reaction, refers to a standard state of existence of a material and not a standard temperature. In fact, standard heats of reaction may be defined at any temperature. The standard states which are chosen for materials for standard heats of reactions are

1) All gases exist as ideal gases

2) Solids and liquids exist in the state in which they are naturally found at the given temperature and the standard pressure (usually chosen to be one atmosphere).

3) The standard state of all materials is taken to be the state of pure materials.

Note again that the standard heat of reaction does not imply any particular reaction temperature or even any particular temperature for the initial and final products. It does imply that the reactants and products are at the same temperature and that the pressure of the reactants and products is one atmosphere and that the gases exist as pure gases in the ideal state and liquids and solids exist as pure liquids and solids in the physical state (liquid or solid) in which they exist at the specified temperature and one atmosphere of pressure.

If the heat of reaction is known at a given temperature for reactions involving the standard states of the reactants and products, then the heat of reaction can be determined between nonstandard states by making adjustments at the given temperature to nonstandard states. For example, ideal gas properties can be converted to nonideal values using the calculations given earlier for changes in enthalpy with pressure. Also, adjustments may be made for nonstandard states of existence. For example, heats of vaporization or heats of fusion may be used to account for a different phase such as solid versus liquid or liquid versus vapor.

15.8.2 Standard Heats of Formation

Now because of the large number of possible reactions which could be considered (some of which may not even be discovered or well defined yet and also due to the infinite number of temperatures which may be of interest) a tabulation of standard heats of reaction would be totally out of hand. That is, far too many reactions and reaction temperatures would have to be included. To circumvent this difficulty, the necessary information is presented in the form of tables of *standard heats of formation*. These heats of formation, in contrast to the standard heats of reaction, are for a single standard temperature (25 °C) and for a particular kind of reaction: the formation of the various compounds from their respective elements. Table C-20 and 20 gives standard heats of formation of several compounds. These formation reactions may be purely hypothetical in that a particular compound may not be able to be formed directly from its elements at the standard temperature and pressure. However, that fact is not a restriction on the standard heats of formation of compounds.

The standard heats of reaction at 25 °C can be calculated by combining the heats of formation (which also are 25 °C) in the right proportions (as established by the reaction stoichiometry) to give the standard heat of reaction. The fact that we might be combining a set of reactions which in fact do not occur to produce the desired products is immaterial. This is true because the heats of reaction represent enthalpy changes which are functions of state and not path. We can choose any path that we like to go from the reactants to the products at the given conditions and whatever path we choose we will get the same answer. Choosing a path by using the standard heats of formation is a convenient path because of the availability of the heats of formation data.

Once the heats of reaction at 25 °C have been determined, the standard heat of reaction can be determined at other temperatures by using the heat capacities of reactants and products in their standard states. These concepts are demonstrated in several problems which follow.

Example 15-18. Gaseous methane is oxidized completely in pure oxygen at 100 °C and 1 atm as shown in Figure 15-47. The mass flow rate of methane (MW = 16 g/gmol) is 1 g/min and the product gases contain only gaseous carbon dioxide and water. Calculate the amount of heat that must be transferred from the reactor if the product gases leave the reactor at 100 °C and 1 atm. The reactor is operating at steady state.

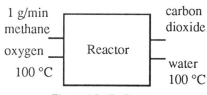

Figure 15-47: Reactor

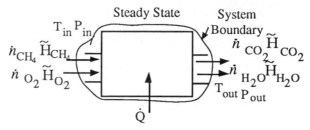

Figure 15-48: System
for Example 15-18

SOLUTION: The reactor is defined as the system and is shown in Figure 15-48. Furthermore, the system is operating at steady state.

The complete balanced combustion reaction of methane to carbon dioxide and water is:

$$CH_4 + 2O_2 \longrightarrow CO_2 + 2H_2O$$

That is, 1 mole of methane combines with 2 moles of oxygens and forms 1 mole of carbon dioxide and 2 moles of water.

Since the system is defined as the reactor and the reactor is operating the steady state, the total conservation of elemental mass is given by:

$$\sum (\dot{n}_i)_{in} - \sum (\dot{n}_i)_{out} = 0$$

or the total number of carbon atoms entering the system must equal the total number of carbon atoms leaving the system as discussed in Chapter 3. We can determine the number of moles of methane entering the system by dividing the mass flow rate by the molecular weight of methane:

$$\dot{n}_{CH_4} = \frac{\dot{m}_{CH_4}}{(MW)_{CH_4}}$$

$$= \frac{1}{16}$$

$$= 0.063 \text{ gmol / min}$$

This also can be thought of as 0.063 gram moles per minute of carbon enters the system with the methane and 0.252 gram moles per minute of elemental hydrogen enters with the methane.

If we know that 0.063 gram moles of methane enters the system per minute and the combustion reaction is complete, than we know from the stoichiometric balanced equation that for every molecule of methane that enters the system 2 moles of molecular oxygen must also enter. Thus we can calculate the number of moles of molecular oxygen entering the system as:

$$\dot{n}_{O_2} = 2\dot{n}_{CH_4}$$

$$= 2(0.063)$$

$$= 0.126 \text{ (gmol / min)}$$

Furthermore, from the balanced stoichiometric equation we can also determine the number of moles of carbon dioxide and water leaving the reactor. The number of moles of carbon dioxide is equal to the number of moles of methane entering the system.

$$\dot{n}_{CO_2} = \dot{n}_{CH_4}$$

$$= (0.063) \text{ gmol / min}$$

The number of moles of water leaving the reactor is 2 times the number of moles of methane entering the system or

$$\dot{n}_{H_2O} = 2\dot{n}_{CH_4}$$
$$= 2(0.063)$$
$$= 0.126 \text{ gmol} / \text{min}$$

Now we need to apply the conservation of energy to this system operating at steady state. The conservation of energy says:

$$\sum \left(\dot{n}(\widetilde{H} + \widetilde{PE} + \widetilde{KE}) \right)_{in} - \sum \left(\dot{n}(\widetilde{H} + \widetilde{PE} + \widetilde{KE}) \right)_{out} + \sum Q + \sum \dot{W} = \left(\frac{dE_{sys}}{dt} \right)$$
$$= 0$$

There is no work being done on the defined system, furthermore the change in kinetic and potential energies of the entering streams will be assumed to be negligible and this reduces the equation to:

$$\sum (\dot{n}_i \widetilde{H}_i)_{in} - \sum (\dot{n}_i \widetilde{H}_i)_{out} + \dot{Q} = 0$$

Rearranging the equation allows us to solve for the unknown rate of heat transfer.

$$\dot{Q} = \sum (\dot{n}_i \widetilde{H}_i)_{out} - \sum (\dot{n}_i \widetilde{H}_i)_{in}$$

Since there is no change in pressure from the entering and leaving states the enthalpies of the entering the leaving stream are functions of temperature only at 1 atm.

$$Q = \sum (\dot{n}_i \widetilde{H}_i(T))_{out} - \sum (\dot{n}_i \widetilde{H}_i(T))_{in}$$

In general, to calculate a value of the molar enthalpy at a certain temperature a reference state must be defined. A reference state that is defined to use for problems involving reactions is that at the temperature of 298 K and a pressure of 1 atm, the molar enthalpies of the elements are zero.

$$T_o = 293 \text{ K} \quad \text{and} \quad P_o = 1 \text{ atm} \qquad \widetilde{H}_o = 0 \text{ for elements}$$

Now, if we are dealing with compounds, molar enthalpies of those compounds at 298 K and 1 atm do not equal zero but equal the molar heat of formation of the compound. An extensive table is provided in Appendix C Table C-20 and 21. The molar heats of formation at 298 for methane as a gas, oxygen, carbon dioxide and water as a gas are given in the following table. The units for the molar heats of formation are joules per gmol.

Molar Heats of Formation

Compound	$\Delta \widetilde{H}_{298}^{f}$ (J/gmol)
CH_4	-74,520
O_2	0
CO_2	-393,509
H_2O	-241,818

Now, if the compound is not at 298 K then the enthalpy is affected by the difference in temperature between the reference temperature and the temperture that the compound is at. So for a defined reference state, the molar enthalpy of a compound at a temperature T with respect to that reference state is:

$$\widetilde{H}_i(T) = \int_{T_o}^{T} \widetilde{C}_{Pi} dT + \Delta \widetilde{H}_{298i}^{f} + \widetilde{H}_o(T_o)$$

At 1 atm, the enthalpy of the compound i at temperature T is calculated as the change in enthalpy from initially starting out at elements at 298 K and forming the compound, then changing the temperature from 298 to some temperature T. The defined reference states for all the initial elements at 298 and 1 atm is zero so the previous equation reduces to:

$$\widetilde{H}_i(T) = \int_{T_o}^{T} \widetilde{C}_{Pi} dT + \Delta \widetilde{H}^f_{298i}$$

This equation is valid for both the entering and leaving compounds. If there is more than one compound entering or leaving compound than we must multiply the molar enthalpy times the moles of the individual compound to calculate the total enthalpy of the entering or leaving streams. For the entering stream the equation is:

$$\sum (\dot{n}_i \widetilde{H}_i(T_{\text{in}})) = \int_{T_o}^{T_{\text{in}}} \sum (\dot{n}_i \widetilde{C}_{Pi}) dT + \sum (\dot{n}_i \Delta \widetilde{H}^f_{298i})$$

If the constant pressure heat capacity is expressed as a power series in temperature as given in Appendix C or:

$$\widetilde{C}_{Pi} = R(A_i + B_i T + C_i T^2 + D_i T^{-2})$$

then the molar flow rate of each compound times each the molar heat capacity is:

$$\dot{n}_i \widetilde{C}_{Pi} = R(\dot{n}_i A_i + \dot{n}_i B_i T + \dot{n}_i C_i T^2 + \dot{n}_i D_i T^{-2})$$

Now if we sum over all the compounds then:

$$\sum (\dot{n}_i \widetilde{C}_{Pi}) = R\left(\sum (\dot{n}_i A_i) + \sum (\dot{n}_i B_i) T + \sum (\dot{n}_i C_i) T^2 + \sum (\dot{n}_i D_i) T^{-2}\right)$$
$$= R\left(\mathcal{A} + \mathcal{B}T + \mathcal{C}T^2 + \mathcal{D}T^{-2}\right)$$

where the values of $\mathcal{A}, \mathcal{B}, \mathcal{C},$ and $\mathcal{D}$ are defined as

$$\mathcal{A} = \sum (\dot{n}_i A_i)$$
$$\mathcal{B} = \sum (\dot{n}_i B_i)$$
$$\mathcal{C} = \sum (\dot{n}_i C_i)$$
$$\mathcal{D} = \sum (\dot{n}_i D_i)$$

Now the integral can be evaluated and the total enthalpy is:

$$\sum (\dot{n}_i \widetilde{H}_i(T_{\text{in}})) = R\left[\mathcal{A}(T - T_o) + \frac{\mathcal{B}}{2}(T^2 - T_o^2) + \frac{\mathcal{C}}{3}(T^3 - T_o^3) - \mathcal{D}\left(\frac{1}{T} - \frac{1}{T_o}\right)\right] +$$
$$\sum (\dot{n}_i \Delta \widetilde{H}^f_{298i})$$

For the entering stream, the following table presents the molar flow rates and heat capacity data for methane and oxygen.

Entering Data

Compound	n_i gmol / min	A_i	B_i	C_i	D_i
CH$_4$	0.063	1.702	9.08×10^{-3}	-3.450×10^{-7}	0
O$_2$	0.126	3.639	0.506×10^{-3}	0	-0.227×10^5

From the entering data, the heat capacity values for the entering streams are:

$$\mathcal{A} = 0.566$$
$$\mathcal{B} = 0.636 \times 10^{-3}$$
$$\mathcal{C} = -1.363 \times 10^{-7}$$
$$\mathcal{D} = -0.029 \times 10^5$$

The sumation of the heats of formations for the entering stream is:

$$\sum (\dot{n}_i \Delta \tilde{H}^f_{298i}) = (0.063)(-74,520) + (0.126)(0)$$
$$= -4695 J \,/\, min$$

The integration over the heat capacity is calculated as:

$$\int_{T_o}^{T} \sum (\dot{n}_i \tilde{C}_{Pi}) dT = R \left[\mathcal{A}(T - T_o) + \frac{\mathcal{B}}{2}(T^2 - T_o^2) + \frac{\mathcal{C}}{3}(T^3 - T_o^3) - \mathcal{D}\left(\frac{1}{T} - \frac{1}{T_o}\right) \right]$$

Substituting the known values in yields:

$$\int_{T_o}^{T} \sum (\dot{n}_i \tilde{C}_{Pi}) dT = (8.314) \left[(0.566)(373 - 298) + \frac{0.636 \times 10^{-3}}{2}(373^2 - 298^2) + \right.$$
$$\left. \frac{-1.363 \times 10^{-7}}{3}(373^3 - 298^3) - (-0.029 \times 10^5)\left(\frac{1}{373} - \frac{1}{298}\right) \right]$$

Evaluating gives:

$$\int_{T_o}^{T} \sum (\dot{n}_i \tilde{C}_{Pi}) dT = 460 \text{ J} / \text{min}$$

The total entering enthalpy rate to the system is:

$$\sum (\dot{n}_i \tilde{H}_i(T_{in})) = \int_{T_o}^{T_{in}} \sum (n_i \tilde{C}_{Pi}) dT' + \sum (n_i \Delta \tilde{H}^f_{298i})$$
$$= 460 + (-4694)$$
$$= -4,234 \text{ J} / \text{min}$$

For the leaving stream, the following heat capacity data and molar flow rates are presented in the following table.

Leaving Data

Compound	n_i gmol / min	A_i	B_i	C_i	D_i
CO_2	0.063	5.457	1.045×10^{-3}	0	-1.157×10^5
H_2O	0.126	3.470	1.450×10^{-3}	0	0.121×10^5

From the leaving data, the heat capacities data are calculated as:

$$\mathcal{A} = 0.781$$
$$\mathcal{B} = 0.249 \times 10^{-3}$$
$$\mathcal{C} = 0$$
$$\mathcal{D} = -0.058 \times 10^5$$

Following the same procedure as used with the entrance stream, the total enthalpy leaving the system is calculated as:

$$\sum (\dot{n}_i \tilde{H}_i(T_{\text{out}})) = -54,753 \text{ J / min}$$

Finally, we can calculate the rate of heat transfer from the system as:

$$\dot{Q} = \sum (\dot{n}_i \tilde{H}_i(T))_{\text{out}} - \sum (\dot{n}_i \tilde{H}_i(T))_{\text{in}}$$
$$= (-54,753) - (-4,234)$$
$$= -50,519 \text{ (J / min)}$$

The rate of heat transfer from the defined system is 50,519 J/min for the product combustion gases to leave at 100 °C.

15.9 REVIEW

Thermodynamic Properties

1) The primary objective of this chapter is to relate changes in the thermodynamic energy properties and entropy ($\widehat{U}$, $\widehat{H}$, $\widehat{A}$, $\widehat{G}$, $\widehat{S}$) to observable properties (P-$\widehat{V}$-T, $\widehat{C}_V$, and $\widehat{C}_P$).

2) The Gibbs phase rule is used to establish the number of independent intensive variables which are required for representing the thermodynamic properties (expressed per mass) of a homogeneous material. For a single component, single phase system, two independent variables are required. Each additional component increases the number of independent variables by one and each additional equilibrium phase decreases the number by one.

3) The internal energy of matter is observed to be a state function, meaning that its value depends only upon the current state of existence of the matter and not on the path taken to reach that state.

4) Heat and work are path functions, even for reversible processes; for a given change in internal energy of the system, the distribution of this energy change between heat and work is not unique.

5) The thermal energy accounting equation can be expressed in terms of changes in entropy, a state function, and volume, a state variable, of the system. This version of the thermal energy equation, although derived for a reversible process, is true as a general statement for single phase, closed systems of one component and is of fundamental importance to thermodynamics.

6) Definitions of $\widehat{U}$, $\widehat{A}$, and $\widehat{G}$, and the fundamental property relation for $d\widehat{H}$ leads to relations for $d\widehat{U}$, $d\widehat{A}$, and $d\widehat{G}$, all equivalent representations of the thermal energy accounting equation for a closed, single phase, single component system.

7) Heat capacities arise as natural properties which are required to relate changes in $\widehat{U}$ or $\widehat{H}$ in the absence of a phase change (transition) to temperature changes.

8) The Maxwell relations are required in order to relate certain thermodynamic partial derivatives to observables using the P-$\widehat{V}$-T equations of state.

9) Heat capacities and P-$\widehat{V}$-T equations of state allow calculation of total changes in $\widehat{U}$, $\widehat{H}$, and $\widehat{S}$ for a single component, single phase system.

10) Reference states for two of the $UHAGS$ functions are required in order to calculate any of them in an absolute sense. These references may be chosen arbitrarily and once done for two, references for the others are also established. Usually either $\widehat{U}$ and $\widehat{S}$ or $\widehat{H}$ and $\widehat{S}$ are assigned arbitrary reference values in an arbitrary state.

11) With reference states and values defined for either internal energy and entropy or enthalpy and entropy, relations for calculating changes in the Gibbs and Helmholtz free energies can be written.

Volumetric Properties

12) In accordance with item 9, P-$\widehat{V}$-T equations of state are important elements to calculating the energy and entropy functions. For gases, the ideal gas, virial (both pressure and density), and cubic (such as Van der Waals, Peng-Robinson, and Redlich-Kwong) relations are commonly used. For liquids, although the cubic equations may work reasonable well in some cases, special equations such as the Rackett equation for saturated liquids and the Tait equation for compressed liquids are especially appropriate.

13) The virial equations (and others) may be written in terms of the compressibility factor, $Z = P\widetilde{V} / RT$. Accordingly, the equations for calculating the thermodynamic energy and entropy functions may be rewritten in terms of this compressibility factor.

14) An important observation of thermodynamics is the principle of corresponding states: all materials behave very similarly when the independent and dependent variables are expressed relative to the critical properties, i.e., when the critical point is used to normalize or benchmark each materials behavior. The various equations of state can then be rewritten in a corresponding-states form, i.e., in terms of reduced pressure, volume, and temperature.

15) Calculations of the energy and entropy functions for solids and liquids is conveniently done using an equation of state expressed in terms of the volume expansivity (β) and the isothermal compressibility (κ). These properties are nearly constant over moderate changes in temperature and pressure which facilitates calculations of $P\text{-}\widehat{V}\text{-}T$ behavior. Furthermore, their values are quite small and frequently may be considered zero in the energy and entropy function calculations. Hence, for example, there is practically no difference between the constant pressure and constant volume heat capacities for solids and liquids and changes in enthalpy may be calculated as the heat capacity integrated over the temperature change.

Phase Transitions

16) Property differences which exist between two phases which are in equilibrium can be determined by recognizing that the Gibbs free energy of two phases which are in equilibrium must be equal. This leads to the Clapeyron equation for the change in enthalpy accompanying *any* phase change and the Clausius-Clapeyron as an approximate relation for enthalpy changes for *vapor-liquid* phase changes.

17) Other methods can be used to determine heats of vaporization. They may be measured calorimetrically; they may be estimated from normal boiling point data using the method of Riedel; or they may be estimated from a known heat of vaporization at another temperature using the Watson relation.

Chemical Reactions

18) Changes in the energy functions also occur as the result of chemical reactions. Changes in enthalpy accoompanying reactions are termed heats of reaction; the heats of reaction accompanying the formation of compounds from the elements are termed heats of formation; heats of formation associated with the hypothetical formation of compounds from the elements in their arbitrarily defined standard states at 25 °C and 1 atm and to the compounds in their defined standard states at 25 °C and 1 atm are termed standard heats of formation.

19) Reactions which involve combustion with oxygen have special importance and these heats of reaction are termed heats of combustion.

20) Using calorimetric measurements, heats of reaction have been determined for a great many reactions (many of which are combustion reactions) and from them standard heats of formation have been calculated and are available in tabular form.

21) Calculations of changes in enthalpy due to reactions requires a) calculating the enthalpy associated with changing the reactants to heat-of-formation standard conditions (both temperature and pressure changes, as appropriate) in accordance with the relations presented for single components, b) calculating the heats of reaction at standard conditions, and c) calculating the changes in enthalpy associated with changing the products to the final conditions (again using the single component equations for each compound). Other calculational paths are possible between the reactant and product states, but this one takes advantage of the heats of reaction at heat-of-formation standard conditions (which can be calculated from the readily available standard heats of formation)

15.10 NOTATION

The following notation is used in this chapter.

Scalar Variables and Descriptions		Dimensions
A	Helmholtz free energy	[energy]
C_P	constant pressure heat capacity	[energy / temperature]
C_V	constant volume heat capacity	[energy / temperature]
E	energy	[energy]
F	number of degrees of freedom	
$\dot{F}$	rate of mech/elec energy consumption	[energy / time]
G	Gibbs free energy	[energy]
H	enthalpy	[energy]
KE	kinetic energy	[energy]
m	total mass	[mass]
$\dot{m}$	total mass flow rate	[mass / time]
MW	molecular weight	[mass / mole]
$\dot{n}$	total molar flow rate	[moles / time]
P	pressure	[force / length^2]
PE	potential energy	[energy]
Q	heat	[energy]
$\dot{Q}$	heat flow rate	[energy / time]
R	gas constant	[energy / temperature mass]
S	Entropy	[energy / temperature]
t	time	[time]
T	absolute temperature	[temperature]
U	internal energy	[energy]
V	volume	[length^3]
W	work	[energy]
$\dot{W}$	rate of work or power	[energy / time]
x	mole fraction	[mole$_i$ / moles]
Z	compressibiltiy factor	
β	volume expansivity	
ϕ	number of phases in equilibrium	

Subscripts	
beg	evaluated at the beginning of the time period
c	corresponding to the critical point
in	input or entering
n	corresponding to the normal boiling point
out	output or leaving

r	reduced variable
shaft	work associated with rotating equipment
sys	system or within the system boundary

<u>Superscripts</u>

sat	corresponding to the saturation
R	residual property
$\hat{}$	property per mass
$\sim$	property per mole

15.11 SUGGESTED READING

Abbott, M. M., and H. C. Van Ness, *Theory and Problems of Thermodynamics*, 2nd edition, McGraw-Hill, New York, 1989

Benedict, M., G. B. Webb, and L. C. Rubin, *J. Chem. Phys.*, **8**, 334 (1940).

Benedict, M., G. B. Webb, and L. C. Rubin, *J. Chem. Phys.*, **10**, 747 (1942).

Benedict, M., G. B. Webb, and L. C. Rubin, *Chem. Eng. Prog.*, **47**, 419 (1951).

Çengel, Y. A., and M. A. Boles, *Thermodynamics, An Engineering Approach*, McGraw-Hill New York 1989

Hougen, O. A., K. M. Watson, and R. A. Ragatz, *Chemical Process Principles, Part II Thermodynamics*, 2nd edition, John Wiley & Sons, New York, 1959

Rackett, H. G., *J. Chem. Eng. Data*, **15**, 514 (1970).

Redlich, O., and J. N. S. Kwong, *Chem. Rev.*, **44**, 233 (1949).

Reid, R. C., J. M. Prausnitz, and T. K. Sherwood, *The Properties of Gases and Liquids*, 3rd edition, McGraw-Hill, New York, 1977

Riedel, *Chem. Ing. Tech.*, **26**, 679 (1954).

Sandler, S. I., *Chemical and Engineering Thermodynamics*, John Wiley & Sons New York 1989

Smith, J. M., and H. C. Van Ness, *Introduction to Chemical Engineering Thermodynamics*, 3rd edition, McGraw-Hill, New York, 1975

Smith, J. M., and H. C. Van Ness, *Introduction to Chemical Engineering Thermodynamics*, 4th edition, McGraw-Hill, New York, 1987

Soave, G., *Chem. Eng. Sci.*, **27**, 1197 (1972).

Van Wylen, G. J., and R. E. Sonntag, *Fundamentals of Classical Thermodynamics*, 3rd edition, John Wiley & Sons, New York, 1986

QUESTIONS

1) Explain the role of the various fundamental laws and accounting relations to the science of thermodynamics.

2) What is a state property? What is a state function? Give examples of state properties and state functions.

3) What is the physical meaning of the thermal energy accounting equation with respect to the law of the conservation of total energy?

4) What dual meaning do the fundamental property relations have? What is the significance of these relationships with respect to each of these interpretations?

5) What is the basic difficulty related to the use of the fundamental property relations for calculating changes in the thermodynamic properties? How does this difficulty motivate our objective or purpose in obtaining equations such as 15-16 and 15-17?

6) What are reference states and why are they important to thermodynamics?

7) State the Gibbs Phase rule and its use in thermodynamics.

8) What are the Maxwell relations and what role do they play in thermodynamics?

9) Apart from telling us how to calculate a material's density given its temperature and pressure, what role does a thermodynamic equation-of-state play in thermodynamics?

10) What is a Virial equation? What's the difference between a pressure virial equation and a density virial equation? When might one be used as opposed to the other?

11) What is the principle of corresponding states? What is a two-parameter corresponding states model and what is a three-parameter corresponding states model?

12) How does the enthalpy of an ideal gas change with temperature at constant pressure? How does it change with pressure at constant temperature? How does the internal energy of an ideal gas change with temperature at constant volume? How does it change with volume at constant temperature?

13) Explain in physical terms why moving the molecules of an ideal gas closer together at constant temperature does not increase their energy.

14) The entropy change of an ideal gas depends upon temperature change at constant volume and upon volume change at constant temperature. Explain in physical terms why both of these effects are important when considering entropy.

15) What is the definition of compressibility factor for a gas?

16) What is the definition of a residual function?

17) What is the physical meaning of the Clapeyron equation? What situation does the Clapeyron equation describe?

18) What restrictions were made for the Clapeyron equation in obtaining the Clausius-Clapeyron equation, i.e. what kind of situation is the Clausius-Clapeyron equation restricted to as opposed to the Clapeyron equation? What is the usefulness of a Clausius-Clapeyron equation?

19) Define heat of reaction and heat of formation.

20) From memory and your understanding of thermodynamics, sketch Figures 15-1 and 15-14.

SCALES

1) Look up values of $\widehat{U}$, $\widehat{H}$, $\widehat{S}$, and $\widehat{V}$ for $T = 400°C$ an $P = 600$ kPa in the steam tables.

2) Given that $P = 200$ kPa and $\widehat{V} = 2.0129$ m^3/kg, look up T, $\widehat{U}$, $\widehat{H}$, and $\widehat{S}$ from the steam tables. Discuss Gibbs's phase rule for this condition.

3) For the condition in exercise (2), show that the value of $\widehat{H}$ calculated from $\widehat{U}$, P, and $\widehat{V}$ is the same as that given in the table. Try this for some other conditions in the table. Is it necessary to give both $\widehat{U}$ and $\widehat{H}$ in a table?

4) How are saturated liquid and saturated vapor properties related at the critical point?

5) Given 1 kg of a mixture of saturated steam and saturated water which is 10% steam, at 170°C, determine U, H, S, and V for the entire mixture. What is the pressure?

6) Given a 5 kg mixture of steam and water at 200 kPa which is 75% steam, what are the values of U, H, S, and V for the total mixture and per unit mass of the mixture?

7) The constant pressure heat capacity of air in the ideal gas state is given by the relation (see Table C-16):

$$\tilde{C}_P / R = A + BT + DT^{-2}$$

where $A = 3.355$, $B = 0.575 \times 10^{-3}$, and $D = -0.016 \times 10^5$, where T must be in Kelvin. What is the change in enthalpy at constant pressure for 5 moles of air for a change in temperature from 0°C to 100°C?

8) For the constant pressure heat capacity of air in the ideal gas state (exercise (7)) calculate $\hat{C}_P$ in J/kg/K at temperature intervals of 20 K from 300 K to 400 K. Note that the molecular mass of air must be used to convert from $\tilde{C}_P$ to $\hat{C}_P$.

9) For the constant pressure heat capacity of air in the ideal gas state (exercise (7)) calculate $\hat{C}_P$ in Btu/lbm/°F at temperature intervals of 20 °F from 500 °R to 600 °R. (Careful. T must still be in Kelvin to calculate the heat capacity. Then the units of $\tilde{C}_P$ are determined by the value (and corresponding units) of R that is used.

10) Using Table C-17, calculate $\hat{C}_P$ for liquid water from 300 K to 600 K at temperature intervals of 50 K.

11) Using the results of exercise (10), what is the change in enthalpy (per mass) of liquid water in going from 300 K to 600 K? (Note that the pressure must be greater than 1 atmosphere in order to maintain water in the liquid state.)

12) Using Table C-17, calculate $\hat{C}_P$ for liquid water from 0°C to 100°C at temperature intervals of 10°C.

13) From the definition of enthalpy in terms of internal energy, the results of exercise (12), and the density of water as a function of temperature, calculate the change in both enthalpy and internal energy as water is heated from 0°C to 100°C at 1 atmosphere. How do the changes in H and U compare? How do $\hat{C}_P$ and $\hat{C}_V$ compare for water?

14) What is the ideal-gas-state constant-pressure heat capacity of SO_2 at 500 K?

15) What is the ideal-gas-state constant-volume heat capacity of SO_2 at 500 K? (Hint: See the result from exercise (14) and adjust accordingly.)

16) What is the heat capacity of ethanol at room temperature and pressure?

17) By how much does the enthalpy of liquid water change as it is heated from 100°C to 200°C?

18) By how much does the enthalpy of liquid water change as it is heated from 200 F to 400 F?

19) Water exits a power cycle pump at 100 °C and 1000 kPa. It then enters a boiler where it is heated to saturation, still at 1000 kPa. What is its change in enthalpy due to heating in the boiler.

20) The van der Waals equation of state (one of many non-ideal gas equations) is:

$$P = \frac{RT}{\tilde{V} - b} - \frac{a}{\tilde{V}^2}$$

where a and b are parameters that tailor the general equation for a specific gas. These parameters can be estimated from critical P, V, and T values, as discussed in the text. For water, $T_c = 647.3$ K, $P_c = 217.6$ atm, and $V_c = 56 \times 10^{-6}$ m^3/mol (see Table C-19). Estimate the van der Waals constants for water using these critical properties.

21) Estimate the volume occupied by 1 mole (g-mole) of steam at 20 bar and 400°C by three methods:

    a) ideal gas law
    b) van der Waals equation of state (use the parameters calculated in exercise (20), and
    c) Steam tables.

You may use a root-finder program on your calculator or computer for (a) and (b). Which method is most accurate for these conditions? Least accurate? What is the percentage deviation of the two poorest from the best? What is the value of the compressibility factor at these conditions? What is its value for an ideal gas?

22) What are the van der Waals constants, a and b, for ammonia estimated from its critical properties?

23) Using the results of exercise (22) and the van der Waals equation of state, what is the pressure of 1 kg of ammonia which occupies 0.2600 m^3 at 60°C? What would be the pressure if it were an ideal gas? (You may use a root-finder program on your calculator or computer.) Compare these results with Table C-25. What is the value of the compressibility factor at these conditions? What is its value for an ideal gas?

24) Two moles of a gas, which you assume is ideal, is in a rigid container at 2 bar and 300 K. What is its volume? What is its pressure if the temperature is raised to 400K by adding heat? 500K? What is its temperature if the pressure is reduced to 1 bar by removing heat?

25) Two moles of a gas, which you assume is ideal, is in a container at 2 bar and 300 K. What is its volume? If the temperature is maintained at 300K, what is its volume if the pressure is changed to 1 bar? 3 bar? If the pressure is now reduced to 1 bar by removing heat and maintaining the volume, what is its temperature?

26) Determine values of $\widehat{U}$, $\widehat{H}$, $\widehat{S}$, and $\widehat{V}$ for $T = 400$°C an $P = 650$ kPa from the steam tables.

27) Determine values of $\widehat{U}$, $\widehat{H}$, $\widehat{S}$, and $\widehat{V}$ for $T = 425$°C an $P = 650$ kPa from the steam tables.

28) Given that $P = 200$ kPa and $\widehat{V} = 1.6000$ m^3/kg, determine T, $\widehat{U}$, $\widehat{H}$, and $\widehat{S}$ from the steam tables.

29) Given that $P = 475$ kPa and $\widehat{V} = 0.7000$ m^3/kg, determine T, $\widehat{U}$, $\widehat{H}$, and $\widehat{S}$ from the steam tables.

30) Estimate the volume occupied by 1 mole (g-mole) of steam at 20 atm and 400°C by using the Peng-Robinson equation of state using parameters estimated from critical properties. Compare the result with that obtained in exercise (21).

31) How many Joules is 5 Btu?

32) Give the values of R, the ideal gas constant, for the following sets of units:

    a) Btu/lbmole R
    b) psi/lbmole R
    c) ft lb$_f$/lbmole R
    d) J/mol K (i.e., J/gmol K)
    e) kJ/mol K

(Hint: see Table A-4.)

33) What is the reference state used for the data of Table C-22?

34) What are the reference values of H and S in the reference state for Table C-22?

35) What is the reference state used for the data of Table C-25?

36) What are the reference values of H and S in the reference state for Table C-25?

PROBLEMS

1) A rigid closed tank contains 50 lb$_m$ of oxygen at 20 psia. At these conditions, the gas can be assumed to obey the ideal gas law:

$$PV = nRT$$

Before placing the tank outside in the sun, you record the temperature of the gas inside the tank to be 25 °C. You set the tank out in the bright sunlight for the entire day and return before sunset and note that the temperature has changed to 60 °C. If the $\widehat{C}_P$ and $\widehat{C}_V$ for oxygen are given by the following functions:

$$\widehat{C}_V = \frac{5R}{2(MW)} \qquad \widehat{C}_P = \frac{7R}{2(MW)}$$

a) Make a sketch of the process showing the initial and final states.

b) How many intensive properties are given for state 1? How many are needed to fully define its state? How many are known for state 2 (without doing state 2 calculations)?

c) Calculate the change in pressure of the gas.

d) Calculate the change in internal energy of the gas.

e) Calculate the change in enthalpy of the gas.

f) Calculate the change in entropy of the gas.

g) From the conservation of total energy, why did the internal energy of the gas change? Quantify the result.

h) From the second law of thermodynamics, can we make any conclusions about this process from the given information and why? What is the possible change in the entropy of the surroundings and why?

2) In your house you have a refrigerator which keeps the food and perishables cold. Assume that there is no heat transfer from the inside of the house to the outside and no mass transfer from the inside of the house to the outside. Explain what will happen to the temperature of your house if the refrigerator door is left open. Use the basic laws that you know to argue your case.

3) For a reversible adiabatic expansion or compression of one mole of an ideal gas, derive the relationship between the initial temperature T_1 and the final temperature T_2 in terms of the initial volume V_1 and the final volume V_2. Assume that $\widehat{C}_P$ and $\widehat{C}_V$ are independent of temperature. (Hint: Treating the gas as a closed system, how can its entropy change, in general? Does the

entropy of this gas change during this process? Explain.) Also, derive an expression for the pressure of the gas as only a function of the initial pressure P_1 the initial volume V_1 and the volume V. This result is:

$$P_1 V_1^\gamma = P_2 V_2^\gamma$$

Also, show the following results:

$$P_1 T_1^{\gamma/(1-\gamma)} = P_2 T_2^{\gamma/(1-\gamma)}$$

$$V_1 T_1^{1/(\gamma-1)} = V_2 T_2^{1/(\gamma-1)}$$

where:

$$\gamma = \frac{\widehat{C}_P}{\widehat{C}_V}$$

For the following situations (problems 4 through 8), involving 0.5 (lb$_m$) of nitrogen ($MW = 28$) starting from an initial state of 25 °C and 1 (atm) and ending at 25 °C and 20 (atm), calculate the change in all of the thermal properties ($\Delta U, \Delta H, \Delta S, \Delta P, \Delta T, \Delta V$) of the nitrogen and $-\int P dV$ and $+\int T dS$. If the process is a two-step process, calculate the changes of all the thermal properties for each of the two steps and the total change for the entire process. Also, calculate $-\int P dV$ and $\int T dS$ for each of the steps and the total process. On different $P - V$ and $P - T$ diagrams, show the path of the process. Also, clearly label the reversible work on the $P - V$ diagram. $\widehat{C}_P$ and $\widehat{C}_V$ for nitrogen are given by the following functions:

$$\widehat{C}_V = \frac{5R}{2(MW)} \qquad \widehat{C}_P = \frac{7R}{2(MW)}$$

The gas behaves according to the ideal gas law:

$$PV = nRT$$

4) A reversible isothermal compression.

5) A two-step process which accomplishes the same change in state via:

   a) A reversible decrease in volume to the final volume at constant pressure followed by:
   b) A reversible increase in pressure to the final pressure at constant volume.

6) A two-step process which accomplishes the same change in state via:

   a) A reversible increase in pressure to the final pressure at constant volume followed by:
   b) A reversible decrease in volume to the final volume at constant pressure.

7) A two-step process which accomplishes the same change in state via:

   a) A reversible adiabatic compression to the final pressure followed by:
   b) A reversible decrease in volume to the final volume at constant pressure.

8) A two-step process which accomplishes the same change in state via:

   a) A reversible adiabatic compression to the final volume followed by:
   b) A reversible decrease in pressure to the final pressure at constant volume.

9) Liquid benzene is flowing at steady state through a horizontal 4 in Schedule 40 pipe. The mass flow rate is 70 (lb_m / s). The inlet pressure is twice atmospheric and the length of the tube is 50 (ft). The inlet temperature is 20 °C and the exit temperature is 70 °C. Assume that the density and the viscosity of the the benzene are not functions of temperature and pressure and assume that the heat capacity is only a function of temperature.

    a) Calculate the rate of mechanical energy loss and the outlet pressure.

    b) From the definition for the change in enthalpy in terms of observable quantities, calculate the change in enthalpy from entering the length of pipe to leaving the length of pipe.

    c) From the thermal energy accounting equation, determine the rate of heat transfer to the length of pipe.

10) Two (2) kilograms of neon gas with constant pressure and constant volume heat capacities:

$$\widehat{C}_P = \frac{5}{2} \frac{R}{(MW)}$$

$$\widehat{C}_V = \frac{3}{2} \frac{R}{(MW)}$$

is compressed *isothermally* at 25 °C from 1 (atm) to 4 (atm).

    a) If the process occurs reversibly, calculate ΔT, ΔP, ΔV and ΔU, ΔH, ΔS, and Q and W. For this process, the temperature of the surroundings is always 25 °C. Calculate the change in entropy of the surroundings and the change in entropy of the universe. How does a reversible process change the entropy of the universe?

    b) The process really requires 25 % more work than the reversible isothermal compression (because of friction etc.) to reach the final state of 4 (atm) and 25 °C. For this process, calculate ΔT, ΔP, ΔV and ΔU, ΔH, ΔS, and Q and W. (Note: The work represents the total amount of work, including the reversible and irreversible work.) For this process, the temperature of the surroundings is always 25 °C. Calculate the change in entropy of the surroundings and the change in entropy of the universe. How does an irreversible process change the entropy of the universe?

11) Two (2) kilograms of neon gas with constant pressure and constant volume heat capacities

$$\widehat{C}_P = \frac{5}{2} \frac{R}{(MW)}$$

$$\widehat{C}_V = \frac{3}{2} \frac{R}{(MW)}$$

are compressed *adiabatically* starting at 25 °C and 1 (atm) to 4 (atm).

    a) If the process occurs reversibly, calculate ΔT, ΔP, ΔV and ΔU, ΔH, ΔS, and Q and W. For this process, the temperature of the surroundings is always 25 °C. Calculate the change in entropy of the surroundings and the change in entropy of the universe. How does a reversible adiabatic process change the entropy of the universe?

    b) The process really requires 25 % more work than the reversible adiabatic compression (because of friction etc.) to reach the final state at 4 (atm). For this process, calculate ΔT, ΔP, ΔV and ΔU, ΔH, ΔS, and Q and W. (Note: The work, represents the total amount of work. Also, if the process is truly adiabatic then the heat transfer is zero regardless if the process is reversible or not.) For this process, the temperature of the surroundings is always 25 °C. Calculate the change in entropy of the surroundings and the change in entropy of the universe, assuming no irreversibilities in the surroundings. How does an irreversible process change the entropy of the universe?

12) Consider an ideal gas contained in an insulated cylinder with the axis of symmetry aligned with gravity. A frictionless piston of mass 20 (kg) rests on top of the gas and the ambient pressure on the other side of the piston is constant at 1 bar. Furthermore, the temperature of the gas is 300 K, and the volume is 0.5 (L) and its diameter is 5 (cm). An electrical heater in the gas cylinder is used to heat the gas slowly. After a time, the piston has been raised by 0.2 (m).

a) What is the pressure in the cylinder i) at the beginning, ii) after the piston has risen by 0.1 m and iii) at the end of the expansion.

b) How much heat was transferred to the gas in the cylinder to raise the piston 0.2 (m)?

The heat capacities of the gas are $\widetilde{C}_P = 7R / 2$ and $\widetilde{C}_V = 5R / 2$.

13) Consider the expansion of an ideal gas, contained in an insulated cylinder, against a piston. Before the expansion, the piston is held in place by a set of stops. The stops are then removed and the piston is free to move until retained by a second set of stops. As a result of the expansion, the volume of the cylinder is increased by a factor of two. For the cases below, determine the change in internal energy and the final temperature and pressure of the gas. The initial pressure is 10 (bar), and the volume is 2 (L) and the temperature is 300 K.

a) The piston expands against a vacuum, the piston is massless, and its motion is frictionless.

b) The piston expands against a pressure of 2 bar, the piston is massless, and its motion is frictionless.

c) How would your answer to part b) be affected by the presence of i) friction and ii) a piston of non-zero mass.

The heat capacities of the gas are independent of the temperature and equal to

$$\widetilde{C}_P = \frac{5}{2}R \qquad \widetilde{C}_V = \frac{3}{2}R$$

14) One pound mole of a gas is initially at 450 R and 10 (psia) in a rigid container. After some time, the temperature of the gas was increased to 600 R. The gas obeys the following equation of state.

$$PV = nRT$$

where

$$R = 10.73 \ (\text{ft}^3 \ \text{psia} / \text{lb}_{\text{mol}} \ {}^\circ\text{R})$$

The constant-volume and constant-pressure heat capacities of the gas are continuous functions of temperature as given by the data.

Problem 15-14 Data

T (R)	$\tilde{C}_V$ (Btu / lb$_{mol}$ R)	$\tilde{C}_P$ (Btu / lb$_{mol}$ R)
400	2.0	4.0
450	2.5	4.5
500	3.0	5.0
550	3.5	5.5
600	4.0	6.0
650	4.5	6.5
700	5.0	7.0

a) For the given change in state, calculate the change in the state variables ΔT, ΔP, ΔU, ΔH and ΔS.

b) Calculate Q and W for this process.

15) Steam is flowing at steady state through a reversible adiabatic turbine. The mass flow rate is 5 (lb$_m$ / s), the entering temperature is 900 °F, and the entering pressure is 300 (psia). The steam leaves the turbine at 100 (psia). Neglect the changes in potential energy and kinetic energy of the entering and leaving streams. Thermodynamic data are provided in the table.

Problem 15-15 Data

		700 °F	800 °F	900 °F	1000 °F
P(psia) = 300	$\hat{H}$ (Btu / lb$_m$)	1367.4	1419.7	1471.8	1524.4
	$\hat{S}$ (Btu / lb$_m$ R)	1.6742	1.7175	1.7570	1.7945

		500 °F	600 °F	700 °F	800 °F
P(psia) = 100	$\hat{H}$ (Btu / lb$_m$)	1278.6	1327.9	1377.5	1427.5
	$\hat{S}$ (Btu / lb$_m$ R)	1.7080	1.7570	1.8015	1.8428

a) How is the entropy of the entering steam related to the entropy of the leaving steam? Explain.
b) Determine the temperature of the leaving steam and explain.
c) Determine the rate of work or power done on the turbine and explain.

16) Methane at 570 R and 3 (atm) is expanded isothermally and reversibly to 1 (atm). Calculate the heat transfer per unit mass $(\hat{Q})$ and work per unit mass $(\hat{W})$ associated with this process for two cases:

a) a nonflow closed process, and
b) a steady-state flow process.

You may assume the methane ($MW = 16$) behaves as an ideal gas at these conditions. For each case, carefully define your system and consider the applicability or usefulness of each of the conservation laws and accounting relations – and use them, if appropriate.

17) The process of firing a 1/8 (lb_m) bullet from a gun is very complex. The overall objective of the process is to convert the thermal/chemical energy stored in the molecules of the gun powder into mechanical energy of the bullet. Even though the process is complex, we have the knowledge to make some approximations that simplify the problem. For simplification, consider the entire process as a two (2) step process that is much easier to analyze. For the first step:

> The 1 (in^3) volume of gun powder is ignited behind the bullet resulting in the rapid formation of a high pressure gas. For simplification, assume that the combustion occurs instantaneously at a constant volume producing a gas at high pressure and temperature. (In reality neither assumption is exactly true.)

For the second step:

> Assume the gas mixture expands adiabatically and reversibly down the barrel of the gun. (In reality, the expansion is definitely irreversible. We are also assuming that there is no heat transfer to the bullet or the gun in this approximation.) This expansion results in propelling the bullet out of the gun barrel. The pressure of the gases as function of the expansion volume down the gun barrel is given in the table.

Problem 15-17 Data

V (in^3)	1	2	3	4	5	6
P (psia)	400	150	85	58	42	32

a) Calculate the work that the hot gases did on the bullet during the reversible adiabatic expansion if the total volume of the inside of the gun barrel is 4 (in^3).

b) Neglect the air pressure from the outside (i.e. assume that the work done by this force is small compared to the work that is done by the hot gases). If the bullet initially starts at rest, and the total inside volume of the gun barrel is 4 (in^3), determine the magnitude of the velocity of the bullet just as the bullet leaves the gun barrel. (This is known as the muzzle velocity.)

c) If the gun were pointed straight up and fired, what is the maximum height that bullet will travel with respect to the end of the barrel of the gun? State and explain any assumptions in this calculation.

d) If the length of the barrel of the gun were greater (with the diameter of the barrel the same), what would you conclude about the muzzle velocity and the maximum height of the bullet if fired straight up? Explain?

SIXTEEN

MORE ON ENTROPY
AND THE SECOND LAW OF THERMODYNAMICS

16.1 INTRODUCTION

Up to now, we have not considered any of the formal arguments which led to the definition of entropy and the development of the second law of thermodynamics. Instead we have simply made statements on how to count entropy for a system. For example, we observe that the increase (or decrease) in the entropy of a *closed system* for a *reversible process* is equal to the heat transferred to (or from) the system divided by the temperature of the system. More generally, the second law of thermodynamics was stated for an open process which may be either reversible or irreversible according to a statement for the system

$$\left(\frac{dS_{\text{sys}}}{dt}\right) = \sum(\dot{m}\widehat{S})_{\text{in/out,sys}} + \sum(\dot{S}_Q)_{\text{in/out,sys}} + \sum(\dot{S})_{\text{gen(non-Q),sys}} \qquad (8-6)$$

and for the surroundings

$$\left(\frac{dS_{\text{sur}}}{dt}\right) = -\sum(\dot{m}\widehat{S})_{\text{in/out,sys}} + \sum(\dot{S}_Q)_{\text{in/out,sur}} + \sum(\dot{S})_{\text{gen(non-Q),sur}} \qquad (8-7)$$

together with the fact that these two sum to give a total change of entropy associated with the process *that cannot be negative*:

$$\left(\frac{d(S_{sys})}{dt}\right) + \left(\frac{d(S_{sur})}{dt}\right) \geq 0$$

$$\left(\frac{d(S_{uni})}{dt}\right) \geq 0 \qquad (8-2)$$

In these equations the entropy of a system may change (either increase or decrease) as a result of *entropy entering or leaving* the system (or surroundings) because of *mass entering or leaving* the system (or surroundings); the mass contains an amount of entropy according to its state. The entropy also may increase or decrease as a result of heat transfer, or it may increase (but not decrease) as a result of irreversibilities which occur within the system. Irreversibilities occur as a result of an imbalance in driving forces. The equilibration of the system so as to remove compositional gradients is an irreversible process; the equilibration of the system so as to remove temperature gradients is an irreversible process; and the equilibration of the system so as to remove velocity gradients is also an irreversible process. To the extent such processes occur, entropy increases. These are the facts of entropy and the second law of thermodynamics as we have so far considered them.

As a matter of completeness and to obtain a more satisfying understanding of entropy and the second law, we must now further consider the definition of entropy and its use to understand power cycles. This definition and its development were motivated by a need to understand some of the observations that were made concerning heat engines (power cycles), devices which convert heat to work. This discussion follows a series of classical observations and resulting conclusions culminating in the definition of entropy as a state function and in the discovery of the second law. Preceding our discussion of these classical arguments is an introduction to the concept of power cycles and refrigeration cycles and the implications of the first and second laws on these cycles. In most treatments of this material, the second law is not yet established at this point. In this discussion, however, we will take advantage of our understanding of the second law early on to facilitate our understanding of these processes. Remember, however, that historically the second law was established through an understanding of the processes and not vice versa.

16.2 ANALYSIS OF CYCLIC PROCESSES

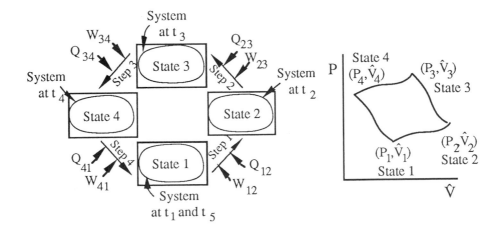

Figure 16-1: Arbitrary sequence of processes

A *cyclic process* can be defined as one for which a given element of a material (called the *working fluid*) passes through a sequence of states which return it to its original state. The sequence is then continued over and over again for this same element of the working fluid. The changes in state for such an element are brought about by heat transfer and work with no exchange of mass between the working element and its surroundings. The working element is chosen as a closed system and changes its state as a result of heat and work. Additionally, in the cycles we will consider, there are no changes in the gravitational energy or in the kinetic energy of the working fluid.

Consider, e.g., Figure 16-1 where we show an arbitrary set of processes (steps) which involve four states of existence for the working fluid. For each of these processes (steps), heat transfer and work take place in order to effect the change of state and they are indicated with a subscript which signifies the beginning and ending states for that step. Each state also is indicated by a point on the $P - \widehat{V}$ diagram (remember that for a pure component existing as a single phase, specifying two state variables is sufficient for specifying its state, e.g., $P, \widehat{V}$). Exactly how we go from state 1 to state 2 is not specified and therefore the corresponding path on the $P - \widehat{V}$ diagram is qualitative only.

Now considering a specific element of the working fluid as the (closed) system, a first law analysis of this fluid in going around a complete cycle, i.e., from state 1 to state 2 to state 3, etc., until it returns back to state 1, we have that

$$0 = Q_{12} + Q_{23} + Q_{34} + Q_{41} + W_{12} + W_{23} + W_{34} + W_{41} \qquad (16-1)$$

The heat transfer for the total cycle and work are the sums of the individual heat and work for the four steps in going from state to state. Also, the change in energy of the system from beginning to end is zero because we have returned the material back to its original state; its change in energy, with its initial and final states being identical, is zero.

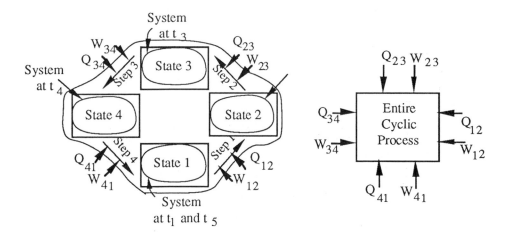

Figure 16-2: System definition for cyclic process

Note that if the system were considered at times t_1 and t_2 (corresponding to states 1 and 2) as the beginning and ending times, then there would be a change in the energy of the system and this would have to be included in the first law statement. For any *complete* cycle, however, there is no change of state of the working fluid.

Now we could consider a number of such elements of the working fluid, passing through the steps one after the other. At any specific time, each element is at a different point in the cycle. However, each one eventually goes through the entire cycle and it is for this reason that it is this per mass view of a cyclic process that we are interested in. Whether for the entire cycle or for any one step within the cycle we will be interested in the changes to a single piece of mass which are brought about by that cycle or process.

In considering the entire cyclic process, then, we will consider the changes in state of a single piece of mass in going from one state to another and associate with that piece of mass the appropriate amount of heat and work. The schematic of such an entire cycle will not necessarily show each of the individual steps within the cyclic process but it will show the heat and work, separately for the various steps, at least as is appropriate (Figure 16-2). The work terms for all of the steps may be combined into a single work term.

16.2.1 The Power Cycle (Heat Engine)

The objective of power cycles is to convert heat transferred to the system to useful work in accordance with Figure 16-3. Note that the process can be shown with the arrows for heat transfer and work indicating *additions* to the energy of the system or process, in accordance with our sign convention for heat and work which is that they are positive when energy is added to the system. The process also could be depicted in an equivalent way by using *absolute values* of the heat and work.

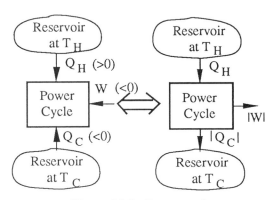

Figure 16-3: Power cycle

Now, considering the entire (closed) cycle, the first law (the conservation of total energy) says that

$$0 = \widehat{Q}_C + \widehat{Q}_H + \widehat{W} \qquad (16-2)$$

while the second law for the process and surroundings together for one complete cycle, says

$$\Delta \widehat{S}_{\text{tot}} \geq 0 \qquad (16-3)$$

Now from equation 16-2 we see that *if* $\widehat{Q}_C = 0$, then (because $T_C < T_H$ and $\Delta \widehat{S}_{\text{tot}} = \Delta \widehat{S}_{\text{sys}} + \Delta \widehat{S}_{\text{sur}} = 0 - \widehat{Q}_H / T_H$) $\Delta \widehat{S} < 0$, contradicting the second law; a finite amount of heat must be rejected to a lower temperature. Stated differently, we say that heat cannot be converted to work with 100 percent efficiency as the sole result of a process ($\widehat{Q}_C \neq 0$). Of course, this was one of the early observations that led to the development of the second law in the first place.

16.2.2 The Refrigeration Cycle

The purpose of a refrigeration cycle is not to deliver work but rather to remove heat from a low temperature and exhaust it at a higher temperature. This is depicted schematically in Figure 16-4. Now again, one of the original observations which led to the development of the second law is that heat cannot be transferred from a low temperature to a high temperature as the sole event of a process.

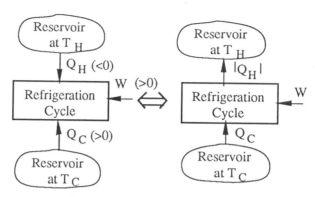

Figure 16-4: Refrigeration cycle

For Figure 16-4, then, if, for the complete cycle (for which $\Delta \widehat{S}_{sys} = 0$) heat is transferred through the process from the low temperature to the high temperature *without work*, i.e., $\widehat{W} = 0$, then (because $\widehat{Q}_C = -\widehat{Q}_H$ and because $T_C < T_H$, $\Delta \widehat{S}_{tot} = \Delta \widehat{S}_{sys} + \Delta \widehat{S}_{sur} = 0 - \widehat{Q}_C / T_C - \widehat{Q}_H / T_H$) the entropy change of the system and surroundings together must be negative, in direct conflict with the second law of thermodynamics.

Thus, the assumption that $\widehat{W}_C = 0$ is impossible. Accordingly, if we add enough work to the process, then $\widehat{Q}_H$ will be increased in absolute value relative to the value of $\widehat{Q}_C$ so that the total change of entropy is positive. This is exactly what is done in a refrigeration cycle. Energy is added to the process through the work of a compressor and heat is rejected to the surroundings (high temperature reservoir).

In the above examples the quantitative statement of the second law of thermodynamics which was presented earlier in the course was used to arrive at conclusions concerning the practical operation of heat engines and refrigerators. As was mentioned, this is actually in the reverse order of events, historically. What actually happened was that in an effort to invent these engines, careful experiments and physical observations led to a statement of the second law of thermodynamics.

In the rest of this chapter we present the pertinent physical observations and the classical results which derive from these observations concerning definitions of temperature, entropy, and the second law of thermodynamics. First, however, we present the concept of a Carnot engine.

16.2.3 The Carnot Heat Engine and Refrigerator

A crucial element to the discussion of entropy and the second law is the concept of a Carnot process. It is not a process which is physically achievable; it is a hypothetical machine. Nevertheless, the concept is important, as will be seen below, in establishing the theoretical limits of our ability to extract work from heat.

A *Carnot heat engine* is the most efficient reversible process which converts heat extracted *from a high temperature reservoir* $|\widehat{Q}_H|$, to work delivered to the surroundings $|\widehat{W}|$, while rejecting heat $|\widehat{Q}_C|$ *to a low-temperature reservoir*. By *reversible*, we mean that for a given $\widehat{Q}_H$, $\widehat{W}$, and $\widehat{Q}_C$ the process can be driven in the *opposite direction*, thereby removing the same (absolute) value of heat ($|\widehat{Q}_C|$) *from the low-temperature reservoir*, rejecting the same value of heat ($|\widehat{Q}_H|$) *to the high-temperature reservoir*, while exchanging the same value of work with the surroundings ($|\widehat{W}|$) (this time *from* the surroundings). Note that this definition says nothing about the kind of process or steps which might make up a Carnot engine. The Carnot engine, when driven in reverse, is a *Carnot refrigerator*.

16.3 PHYSICAL OBSERVATIONS RELATING TO TEMPERATURE AND THE SECOND LAW OF THERMODYNAMICS

16.3.1 The Observations

1) If two bodies are each, separately, in thermal equilibrium with a third, then they are in thermal equilibrium with each other.

2) Total mass is conserved.

3) Total energy is conserved.

4) No apparatus can convert heat completely into work with no other effect on the system and surroundings.

5) The transfer of heat from cold to hot as the sole result of a process is not possible.

We are now ready to discuss the conclusions which may be made from the above observations.

16.3.2 Conclusion 1

> There is a property of matter which quantitatively and reproducibly measures the "degree of hotness" of a body; one body can be used to sense the hotness of another (thermometer).

Observation 1 is called the *zeroth law of thermodynamics* and provides the basis for defining the state property of temperature, essential to the development of the science of thermodynamics. By itself, however, it is not sufficient to provide an actual definition of temperature.

16.3.3 Conclusion 2

The efficiency of a heat engine is given by

$$\eta_{th} = 1 - \frac{|\hat{Q}_C|}{\hat{Q}_H} \qquad (16-4)$$

Observations 3 (the first law of thermodynamics) and 4 motivate a definition of thermal efficiency in terms of $\hat{Q}_C$ and $\hat{Q}_H$. Because all of the heat transferred from the higher temperature ($\hat{Q}_H$) cannot be converted into useful work we are obviously interested in how much of the heat transferred actually is converted to work and this we express as a function of $\hat{Q}_H$. Accordingly, we define the thermal efficiency of a heat engine (power cycle) as

$$\eta_{th} \equiv \frac{|\hat{W}|}{\hat{Q}_H} \qquad (16-5)$$

which, from equation 16-2, gives equation 16-4.

16.3.4 Conclusion 3

> The thermal efficiency of a heat engine is less than unity.

Observation 4, which says that the heat transferred from a high temperature reservoir cannot be converted totally into work, says that $|\widehat{Q}_C| \neq 0$ and hence that the thermal efficiency must be less than 100 percent:

$$\eta_{th} < 1$$

$$(16-6)$$

16.3.5 Conclusion 4

This conclusion consists of two parts:

> 1) A Carnot cycle is the most efficient cycle or process for converting heat to work, and

> 2) The Carnot efficiency is a function of the hot and cold reservoir temperatures and not of the specific working fluid or mechanism for achieving a Carnot cycle.

The *first part* is reached by considering two processes, with one driving the other as shown in Figure 16-5. If two Carnot processes are used (which could be done because they are reversible) then the $|\widehat{Q}_H|$ rejected by the refrigerator is equal to the $|\widehat{Q}_H|$ absorbed by the power cycle. Likewise, the $|\widehat{Q}_C|$ absorbed by the refrigerator is the same as the $|\widehat{Q}_C|$ rejected by the power cycle (heat engine). Finally, the work transfers between the two are also identical (in absolute value).

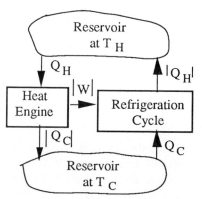

Figure 16-5: Two processes with one driving the other

Consequently, considering the two processes together as a single system, the net effect is that no heat has been transferred to or from the high temperature reservoir and also, no net heat has been transferred to or from the low temperature reservoir. Furthermore, no work has left the system. There are no changes in the surroundings.

Consider, now, another situation with respect to the same figure. Suppose that the heat engine is operating *more efficiently* than the reversible Carnot refrigerator but producing exactly the same amount of work $|\widehat{W}|$ used by the refrigerator. Because the heat engine is more efficient, the heat transferred from the high temperature reservoir to produce this amount of work is *less* than the amount of heat rejected by the refrigerator to the high temperature reservoir. There is a net transfer of heat for the two processes together *to the high temperature reservoir*. Also, considering the heat engine, because $|\widehat{Q}_H|$ delivered to it is reduced compared to what it was for the Carnot engine, it must be true that $|\widehat{Q}_C|$ for this more efficient engine is also less than it was for the Carnot heat engine (and for the Carnot refrigerator). Consequently, there is a net transfer of heat, considering the engine-refrigerator combination together, from the low temperature reservoir to the two-engine process.

Therefore, from the low temperature side, heat has been transferred from the low temperature to the engine-refrigerator process and at the high temperature side, heat has been transferred from the engine-refrigerator process to the high temperature reservoir. *Heat has been transferred from the low temperature reservoir to the high temperature reservoir with no net work from the surroundings.* This is a direct contradiction of observation 5. Therefore, the assumption that there can be a more efficient heat engine than the reversible Carnot engine must be false.

Now, this conclusion was obtained without any discussion, whatsoever, of the actual nature of the Carnot cycle or of its working fluid. Therefore, we are led to the conclusion that the efficiency of a Carnot cycle (that is, of a reversible heat engine or refrigerator) is independent of the actual process or material and therefore dependent only upon the temperatures of the hot and cold reservoirs. This was the *second part* of Conclusion 4.

16.3.6 Conclusion 5

> The existence of a temperature function which can be used as a rigorous basis for defining a temperature scale is implied by the Carnot cycle.

Consider the two separate processes for taking heat $\widehat{Q}_H$, delivering work $\widehat{W}$ and rejecting $\widehat{Q}_C$ shown in Figure 16-6. One process is simply a single Carnot engine operating between the two reservoirs at T_H and T_C. The other process, however, consists of two Carnot engines, operating in series, through an intermediate reservoir at temperature T_I. The intermediate heat transferred is called $\widehat{Q}_I$. Each of these two processes in series produces work and the two together produce an amount of work which is equal to that produced by the single Carnot engine (as required by the first law, for the same $\widehat{Q}_H$ and $\widehat{Q}_C$). Now, from the previous conclusion, we know that the efficiencies of each of these engines depends only upon the temperatures between which each operates.

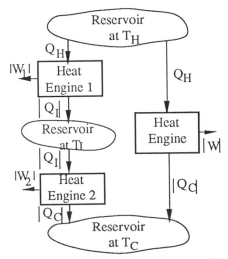

Figure 16-6: Two processes between the same T_H and T_C

For the single heat engine, this dependency can be expressed as

$$\eta_{th} = 1 - \frac{|\widehat{Q}_C|}{|\widehat{Q}_H|} = f(T_H, T_C) \tag{16-7}$$

which can be stated in terms of a different function as

$$\frac{|\widehat{Q}_H|}{|\widehat{Q}_C|} = g(T_H, T_C) \tag{16-8}$$

where $g = 1 / (1 - f)$. Now for the two engines in series, the corresponding results are

$$\frac{|\widehat{Q}_H|}{|\widehat{Q}_I|} = g(T_H, T_I) \tag{16-9}$$

and

$$\frac{|\widehat{Q}_I|}{|\widehat{Q}_C|} = g(T_I, T_C) \tag{16-10}$$

and each of these efficiencies is represented using the (same) function of the two temperatures between which they operate. Now considering the product of these ratios of the heats transferred for the two engines which are in series, we have

$$\frac{|\widehat{Q}_H|}{|\widehat{Q}_I|} \frac{|\widehat{Q}_I|}{|\widehat{Q}_C|} = \frac{|\widehat{Q}_H|}{|\widehat{Q}_C|} = g(T_H, T_I)g(T_I, T_C) \tag{16-11}$$

Now this is compared to equation 16-8 which gives the same ratio of $|\widehat{Q}_H| / |\widehat{Q}_C|$ as independent of T_I. The intermediate temperature, which is quite arbitrary, cancels from the product in equation 16-11, further implying that

$$g(T_H, T_C) = \frac{h(T_H)}{h(T_C)} \tag{16-12}$$

Consequently, from equations 16-7, 16-8, and 16-12, the thermal efficiency of a Carnot cycle can be expressed as

$$\eta_{th} = 1 - \frac{h(T_H)}{h(T_C)} \tag{16-13}$$

The temperature function $h(T)$ establishes the existence of a *thermodynamic temperature*.

16.3.7 Conclusion 6

The ideal gas temperature is a valid choice for a thermodynamic temperature scale.

Consider a cyclic process using an ideal gas which consists of the following four steps,

    1) Adiabatic compression from T_C to T_H

    2) Isothermal expansion with heat transfer $\widehat{Q}_H$

    3) Adiabatic expansion from T_H to T_C

    4) Isothermal compression with heat transfer $\widehat{Q}_C$.

Now if the working fluid of this process is an ideal gas and therefore is one for which $P\widetilde{V} = RT$, then it can be shown that for a complete cycle

$$\frac{|\widehat{Q}_H|}{|\widehat{Q}_C|} = \frac{T_H}{T_C} \qquad (16-14)$$

and therefore the temperature scale defined by the ideal gas relationship is a valid thermodynamic temperature, i.e, it follows the form of equation 16-12. In fact, it is the simplest relation which satisfies equation 16-12. This can be shown in a straightforward way using the relationships and procedures for analyzing processes of chapters 7, 8 and 15.

16.3.8 Conclusion 7

The thermal efficiency of a Carnot cycle is given by

$$\eta_{th} = 1 - \frac{T_C}{T_H} \qquad (16-15)$$

where T is an ideal gas absolute temperature scale.

This result follows directly from equations 16-7 and 16-14. The efficiency of a Carnot cycle can be calculated directly from the temperatures of the two reservoirs.

16.3.9 Conclusion 8

There exists a state function S which is defined as

$$dS \equiv \frac{dQ_{rev}}{T} \qquad (16-16)$$

Conclusion 6 said that for a reversible process,

$$\frac{|\widehat{Q}_H|}{|\widehat{Q}_C|} = \frac{T_H}{T_C} \tag{6 - 14}$$

from which we see that

$$\frac{|\widehat{Q}_H|}{T_H} = \frac{|\widehat{Q}_C|}{T_C} \tag{16 - 17}$$

and, in fact, using signed values rather than absolute values,

$$\frac{\widehat{Q}_H}{T_H} + \frac{\widehat{Q}_C}{T_C} = 0 \tag{16 - 18}$$

That is, if an element of the working fluid traverses through a complete (revesible) cycle during which $\widehat{Q}_H$ was transferred at T_H and $\widehat{Q}_C$ was transferred at T_C, then at the end of a complete cycle there was no net transfer of $\widehat{Q}\,/\,T$.

Consider now a differential scale process, that is, one operating between hot and cold reservoirs which are differentially close together (with respect to temperature) and that has differential amounts of heat transfer. If it is a reversible process, then we may write the $\widehat{Q}\,/\,T$ value at a given temperature as $d\widehat{Q}_{\text{rev}}\,/\,T$. Then, in traversing a complete cycle we may write the sum of such individual terms will be zero.

Based on this last result it is apparent that there is a state function which can be defined in terms of the differential heat transferred divided by temperature. This new function is called entropy and defined according to equation 16-16.

16.3.10 Conclusion 9

> For any isolated system, the only processes which can occur are those which result in an increase in entropy.

This conclusion is a statement of the Second Law of Thermodynamics. An alternate statement is that, considering the system and surroundings together, the combined entropy change must be greater than or equal to zero. This result can be understood by referring again to Figure 16-5. Recall that in this figure the Carnot engine provides work to the Carnot refrigerator and that there is no net transfer of heat between the cold and hot reservoirs so long as the work associated with both the engine and the refrigerator have the same absolute value. In the previous discussion we considered a heat engine which might be operating more efficiently than the Carnot engine and were led to a contradiction of the experimental observation that heat cannot be transferred from a cold temperature to a high temperature without a net performance of work. This implied that the heat engine cannot be more efficient than the Carnot engine.

We consider now a heat engine which is *less efficient* than a Carnot engine so that to deliver the required work to the Carnot refrigerator requires that *more heat* be extracted from the high temperature reservoir than for the Carnot cycle. Considering the two processes together then, there is a net transfer of heat from the high temperature reservoir to the processes. On the low temperature side, again because we are considering a heat engine which is less efficient than the Carnot engine, the amount of heat that must be rejected to the low temperature must be greater than that for the Carnot engine; there is a net rejection of heat from the two processes together to the low temperature reservoir. Overall then, considering the two processes together as a system, there is a net transfer of heat from high temperature to low temperature, certainly a possible outcome.

We now must consider the entropy change of the system plus surroundings which arises from this process. We previously showed that for a Carnot engine operating in a complete cycle, that there was no change in entropy of the system plus surroundings brought about by this process. For the less efficient heat engine, however, we may write the entropy gain of the surroundings at the high temperature reservoir as

$$(\Delta \widehat{S}_{\text{sur}})_{T_H} = -\frac{(\widehat{Q}_H)_{\text{Car}} + \epsilon}{T_H} \tag{16-19}$$

and that at the low temperature reservoir as

$$(\Delta \widehat{S}_{\text{sur}})_{T_C} = \frac{(\widehat{Q}_C)_{\text{Car}} + \epsilon}{T_C} \tag{16-20}$$

Here we have written the heat transfer terms as the sum of the equivalent Carnot heat transfer plus the extra amount (ϵ) which exists for this less efficient process. Now the entropy change for the engine itself, because we are considering it for one complete cycle, must be zero. Consequently, the total entropy change is given by

$$\Delta \widehat{S}_{\text{tot}} = -\frac{\epsilon}{T_H} + \frac{\epsilon}{T_C} \tag{16-21}$$

and because $T_H > T_C$ it must be true that this total change is positive i.e.

$$\Delta \widehat{S}_{\text{tot}} > 0 \tag{16-22}$$

We have then as a final result for the (closed) system plus surroundings that either the entropy change is zero (for a reversible process) or it is greater than zero (for a less efficient i.e. irreversible, process). Now from the above conclusions, entropy is an extensive state property and therefore we can count it in the same way that we have counted other extensive properties. However, as a result of conclusion 9, the entropy of the system plus surroundings may increase as a result of irreversibilities so that entropy is not a conserved property. Our accounting equation then for entropy must make allowances for the generation of entropy in the system and surroundings. Accordingly, we have as statements for the system and surroundings those given by equations 8-6 and 8-7.

16.4 AVAILABILITY

As discussed previously in Chapter 8 and the above paragraphs of this chapter, experience convinces us that there are constraints placed upon processes which limit what can actually occur. Heat cannot be transferred spontaneously from a cold object to a hotter one. Work cannot be produced from heat with 100 percent efficiency. A single-phase mixture of multiple species will not spontaneously unmix, etc. These experimental observations are embodied in the concept of entropy and the second law of thermodynamics.

With this understanding that processes are constrained by the second law, it makes sense to ask ourselves "what are the practical implications of these limits?" For example, if we have a steady-state flow process with the objective to produce work to perform a task, what is the *maximum* amount of work that we can expect to get from such a process? Or, if we have a closed system containing mass at a given state and we change the state of that mass to another state through heat transfer and work, what is the maximum useful work that we can extract from such a process? Questions like these sometimes are addressed using the concept of *availability*. Availability is the maximum amount of work which is theoretically obtainable from material in a given state without violating the second law of thermodynamics, i.e. without the entropy change of the system plus surroundings being negative.

16.4.1 Steady-State Flow Processes

Consider for example a steady-state flow process with one entering and leaving flow stream. Mass entering and leaving the system exchanges internal energy, enthalpy, kinetic energy, potential energy, entropy, etc. with the surroundings. Also heat and work may exchange between the system and surroundings. For such a process the conservation of total energy, in enthalpy form, says

$$0 = (\widehat{H} + \widehat{KE} + \widehat{PE})_1 - (\widehat{H} + \widehat{KE} + \widehat{PE})_2 + \widehat{Q} + \widehat{W}_{\text{shaft}} \tag{7 − 61}$$

The work term here is all forms of work except that associated with the flow of the fluid at the system boundaries due to pressure which you recall, is included in the enthalpy terms. Now this result says nothing whatsoever about constraints on the process. It just says that whatever is happening with respect to changes in the state of the material between inlet and outlet streams and with respect to heat transfer and work, this result must be satisfied. The second law of thermodynamics goes a step further and says that in addition, the entropy of the system and surroundings together cannot decrease, i.e.,

$$\left(\frac{dS_{\text{sys}}}{dt}\right) + \left(\frac{dS_{\text{sur}}}{dt}\right) \geq 0 \tag{8 − 2}$$

Now, because the process is operating at steady state, the system entropy does not change with time (by definition of steady state). Consequently, the second law constraint becomes

$$\left(\frac{dS_{\text{sur}}}{dt}\right) \geq 0 \tag{8 − 2}$$

Now we are interested, as stated above, in the most work (i.e. the availability) which can be obtained from such a process. Any irreversibilities which occur in the system or surroundings will necessarily result in a reduced amount of work. Consequently, we are interested in the work associated with a reversible process for which the entropy of the change of the universe is zero, i.e.

$$\left(\frac{dS_{\text{sur}}}{dt}\right) = 0 \tag{8 − 2}$$

Now the entropy of the surroundings in this case is affected 1) by the entropy entering and leaving the system due to the exchange of mass between the system and surroundings and 2) by the transfer of heat between the system and surroundings. Adding these two effects to calculate the change in the entropy of the surroundings and equating this change to zero for a reversible process gives

$$0 = -(\dot{m}\widehat{S})_1 + (\dot{m}\widehat{S})_2 - \left(\frac{\dot{Q}}{T_o}\right) \tag{16 − 23}$$

Here we assumed that the temperature of the surroundings is uniform at T_o and that Q is written with respect to the system so that if Q results in the transfer of energy from the system to the surroundings, then Q is negative. This result then says that

$$\widehat{Q} = T_o(\widehat{S}_2 - \widehat{S}_1) \tag{16 − 24}$$

Note that if the material of the process is going to a state of lower entropy ($\widehat{S}_2 < \widehat{S}_1$) then heat must be rejected to the surroundings in order to not violate the second law. Q then is a minimum amount of heat which must be rejected to the surroundings. Any less than this and the second law will be violated. Any more than this, however, will subtract from the work which might be obtained from the process, in accordance with the conservation of total energy. Consequently we can use this result in Equation 7-61 to obtain the maximum work available in accordance with the second law:

$$
\begin{aligned}
|\widehat{W}_{\text{shaft}}|\text{max} &= -\widehat{W}_{\text{shaft}} \\
&= (\widehat{H}_1 - \widehat{H}_2) + (\widehat{KE}_1 - \widehat{KE}_2) + (\widehat{PE}_1 - \widehat{PE}_2) - T_o(\widehat{S}_1 - \widehat{S}_2)
\end{aligned}
\tag{16-25}
$$

To summarize, if we know the entering and exiting states of the process fluid and the surroundings' temperature, then it is a simple matter to calculate the maximum work which can possibly be obtained from such a process. This we will refer to as the availability for this steady-state flow process.

16.4.2 Closed Systems

The availability associated with a closed system in a given state can be obtained by a completely analogous reasoning process for a closed system. The statement of energy conservation gives

$$
(\widehat{U} + \widehat{KE} + \widehat{PE})_{\text{end}} - (\widehat{U} + \widehat{KE} + \widehat{PE})_{\text{beg}} = \widehat{Q} + \widehat{W}
\tag{16-26}
$$

The energy of the system changes because of direct exchange with the surroundings of heat and work where the work term includes any work associated with the volume of the system changing size. Now, as in the preceding discussion, this equation says nothing whatsoever about limits on the process; for that we must consult the second law. For this situation, and in contrast to the steady-state process, the entropy of the system may change as well as that of the surroundings. Consequently, the second law statement for a closed, reversible, process is

$$
\begin{aligned}
\Delta \widehat{S}_{\text{sys}} + \Delta \widehat{S}_{\text{sur}} &= 0 \\
\widehat{S}_{\text{sys,end}} - \widehat{S}_{\text{sys,beg}} &= \Delta \widehat{S}_{\text{sys}} \\
\widehat{S}_{\text{sur,end}} - \widehat{S}_{\text{sur,beg}} &= \left(\frac{\widehat{Q}}{T_o} \right) \\
&= \Delta \widehat{S}_{\text{sur}}
\end{aligned}
\tag{16-27}
$$

Again, Q is written with respect to the system so that if heat is transferred from the system to the surroundings, i.e. if heat is rejected to the surroundings, then $\widehat{Q}$ is less than zero. The total entropy change of a system plus surroundings then is given by

$$
\widehat{S}_{\text{sys,end}} - \widehat{S}_{\text{sys,beg}} - \left(\frac{\widehat{Q}}{T_o} \right) = 0
\tag{16-28}
$$

from which we have

$$
\widehat{Q} = T_o(\widehat{S}_{\text{sys,end}} - \widehat{S}_{\text{sys,beg}})
\tag{16-29}
$$

From equations 16-28 and 16-31, the work can now be written as

$$
\widehat{W} = (\widehat{U} + \widehat{KE} + \widehat{PE})_{\text{end}} - (\widehat{U} + \widehat{KE} + \widehat{PE})_{\text{beg}} - T_o(\widehat{S}_{\text{end}} - \widehat{S}_{\text{beg}})
\tag{16-30}
$$

Now this is not quite the maximum useful work which can be obtained from this process. Any work done against the surroundings pressure is not useful. Consequently, we subtract this work to obtain the maximum useful work:

$$\widehat{W}_{\max} = (\widehat{U} + \widehat{KE} + \widehat{PE})_{\text{end}} - (\widehat{U} + \widehat{KE} + \widehat{PE})_{\text{beg}}$$
$$- T_o(\widehat{S}_{\text{end}} - \widehat{S}_{\text{beg}}) + P_o(\widehat{V}_{\text{end}} - \widehat{V}_{\text{beg}}) \qquad (16-31)$$

Note that the sign of this total work will be negative when energy is transferred from the system to surroundings and that an expansion of the system from the beginning to the end ($\widehat{V}_{\text{beg}} < \widehat{V}_{\text{end}}$) will be positive and therefore will decrease this energy transfer. If the end state is at P_o and T_o (the ambient conditions, a "dead" reference state) then the portion of the maximum useful work which excludes the kinetic energy and potential energy, is termed the availability:

$$\boxed{\Phi = (\widehat{U}_{\text{beg}} - \widehat{U}_o) - T_o(\widehat{S}_{\text{beg}} - \widehat{S}_o) + P_o(\widehat{V}_{\text{beg}} - \widehat{V}_o)} \qquad (16-32)$$

Physically this represents the maximum useful work which can be obtained from the thermal energy of the material in its initial state. Additional work of course may be obtained because of changes in kinetic and potential energy.

16.5 IRREVERSIBILITY

In the above situations the maximum useful work was calculated for a reversible process for which the entropy change of the universe was zero. This was an ideal limiting case. In practice, of course, the entropy change will be positive and consequently the useful work will be reduced accordingly. In accordance with the first law in each situation, as the work is reduced, heat transfer must be increased. *For the steady-state process* such an increase in heat transfer will produce a consequent increase in the entropy of the surroundings and hence of the universe. The net result is that the decrease in work is simply the ambient temperature T_o times the total increase in the entropy of the universe. In fact, this same result is obtained for the closed system also and leads to the definition of a quantity called irreversibility:

$$\boxed{I = T_o(S_{\text{uni,end}} - S_{\text{uni,beg}})} \qquad (16-33)$$

Irreversibility is the amount (or rate) of work which is not obtained because of irreversibilities which occur in the process. This irreversibility is also referred to as lost work in that it represents work which is not obtained from the process but which would have been obtained if irreversibilities had been avoided.

16.6 REVIEW

1) A number of physical observations provide convincing evidence that there are a number of processes which cannot occur even though they would not be in conflict with conservation laws. Two observations related to the power cycles are that

    (a) No apparatus can convert heat completely into work with no other effect on the system and surroundings (i.e. without rejecting heat to a lower temperature) and

(b) The transfer of heat from cold to hot as a sole result of a process is not possible.

2) A Carnot heat engine has the property that, at least conceptually, it can be made to operate in the reverse direction to transfer heat from a low temperature to a high temperature without violating the observation stated in 1(b). Hence, this process requires work from the surroundings. Such a Carnot engine, which is reversible, is the most efficient process for converting heat to work and for transferring heat from a low temperature to a high temperature.

4) The efficiency of a Carnot engine or refrigerator depends solely upon the temperatures of the hot and cold reservoirs between which it is transferring heat. The efficiency is not dependent upon the method for achieving the engine or upon the working fluid used in the engine.

5) Temperature is a property of matter which has a well-defined basis in Carnot engine efficiency. The temperature scale which is defined by the ideal gas law is completely compatible with these implications of Carnot engines, that is, the absolute temperature scales which we frequently use (Kelvin or Rankine) are valid thermodynamic temperatures.

6) Arising out of the discussion of Carnot engines, heat transfer, and temperature is the observation that a state function exists which is written in terms of reversible heat transfer and temperature. This state function is called entropy and defined by the relation

$$d\widehat{S} \equiv \frac{d\widehat{Q}_{\text{rev}}}{T}$$

7) The total entropy change associated with any process, meaning the entropy change of the system plus the entropy change of the surroundings which results from this process, must satisfy the relation

$$d\widehat{S}_{\text{Tot}} \geq 0$$

This result is known as the Second Law of Thermodynamics.

16.7 NOTATION

The following notation is used in this chapter.

Scalar Variables and Descriptions		Dimensions
H	Enthalpy	[energy]
KE	kinetic energy	[energy]
m	total mass	[mass]
$\dot{m}$	total mass flow rate	[mass / time]
P	pressure	[force / length^2]
PE	potential energy	[energy]
Q	heat	[energy]
$\dot{Q}$	heat flow rate	[energy / time]
S	Entropy	[energy / temperature]
t	time	[time]

T	temperature	[temperature]
U	Internal energy	[energy]
V	volume	[length^3]
W	work	[energy]
$\dot{W}$	rate of work or power	[energy / time]
ϵ	extra heat transfer over Carnot heat transfer	[energy]
η	efficiency	
Φ	available energy	[energy]

Subscripts

beg	the beginning of the time period
Car	Carnot
end	the end of the time period
in	input or entering
irr	designating an irrevesible process
max	the maximum
out	output or leaving
rev	designating a to revesible process
shaft	work associated with rotating equipment
sur	corresponding to the surrounding
sys	corresponding to the system
th	thermal
uni	corresponding to the entire universe

Superscripts

$\hat{}$ property per mass

16.8 SUGGESTED READING

Abbott, M. M., and H. C. Van Ness, *Theory and Problems of Thermodynamics*, 2nd edition, McGraw-Hill, New York, 1989

Çengel, Y. A., and M. A. Boles, *Thermodynamics, An Engineering Approach*, McGraw-Hill New York 1989

Halliday, D. and R. Resnick, *Physics for Students of Science and Engineering, Part I*, John Wiley & Sons, 1960

Hougen, O. A., K. M. Watson, and R. A. Ragatz, *Chemical Process Principles, Part II Thermodynamics*, 2nd edition, John Wiley & Sons, New York, 1959

Sandler, S. I., *Chemical and Engineering Thermodynamics*, John Wiley & Sons New York 1989

Smith, J. M., and H. C. Van Ness, *Introduction to Chemical Engineering Thermodynamics*, 4th edition, McGraw-Hill, New York, 1987

Van Wylen, G. J., and R. E. Sonntag, *Fundamentals of Classical Thermodynamics*, 3rd edition, John Wiley & Sons, New York, 1986

QUESTIONS

1) What is a heat engine?

2) What is a refrigerator and how does it differ from a heat engine? How is it similar?

3) What is the objective of a heat engine?

4) What is the objective of a refrigerator?

5) What constraint is placed upon the operation of a heat engine by the Second Law of Thermodynamics (this constraint is really a physical observation which led to the establishment of the Second Law of Thermodynamics)?

6) What is a Carnot heat engine?

7) How is the efficiency of a heat engine defined?

8) How is the thermal efficiency of a Carnot heat engine calculated?

9) What is meant by the term thermodynamic temperature scale?

10) What can happen to the entropy of an isolated system?

11) A tank contains one gallon of water which is not at uniform temperature; it is hotter at the top than it is at the bottom. Assuming that the tank is perfectly insulated and that no work or mass exchanges with the surroundings, what will happen with time to the temperature of the water? What kind of process does the water undergo over time with respect to entropy and the second law of thermodynamics?

12) True or False: The entropy of a system can never decrease. Explain your answer.

13) What role does the second law of thermodynamics play in our engineering analysis that is different from the other laws with which we deal?

SCALES

1) For superheated steam in an initial state (state 1) of 14.7 psia and 400°F and a final state (state 2) of 600 psia and 1,000°F, what is the change in entropy (per lb_m) from state 1 to state 2?

2) For water in an initial state of liquid water at 100 psia and 328°F and a final state of 400 psia and 800°F, what is the change in entropy (per lb_m) from state 1 to state 2?

3) For 5 lb-moles of air (in the ideal gas state) in an initial state of 50°F and a final state of 200°F, what is its change in entropy?

4) For 5 kg of water in an initial state of 1 atm and 5°C and a final state of 90°C, what is the change in entropy from state 1 to state 2?

5) For 3 moles of ammonia in an initial state of 75 kPa and 0°C and a final state of 400 kPa and 80°C, what is the change in entropy from state 1 to state 2?

6) For 3 moles of ammonia in the ideal gas state at an initial temperature of 0°C and a final temperature of 80°C, what is the change in entropy from state 1 to state 2?

7) Compare the results from exercises 5 and 6.

8) For hydrogen chloride (HCl) in the ideal gas state at an initial temperature of 20°C and a final temperature of 180°C, what is the change in entropy (per mole) from state 1 to state 2?

9) If γ for a particular ideal gas is 1.4 and its initial state is defined by $P_1 = 1$ atm, $T_1 = 300$ K, then what is its temperature after an adiabatic, reversible compression to 3 atm?

10) Saturated water at 200°C enters a process at a mass flow rate of 0.5 kg/s. At what rate does entropy enter the process due to this flow?

11) A rifle fires a bullet. If, when the bullet starts from rest, there is a gas charge of 10^{-6} m^3 at a pressure of 10,000 psi, then what is the pressure of the gas just before the bullet leaves the rifle, assuming that the gas expansion is a reversible and adiabatic and that γ for the gas mixture is 1.4? The barrel volume is 4.6×10^{-5} m^3. What gases comprise the mixture?

12) What is the Carnot efficiency for a heat engine operating between 200°C and 50°C?

13) What is the Carnot efficiency for a heat engine operating between 400°F and 80°F?

14) In a closed, reversible process, 30 kJ of heat are transferred to the process system which is at a constant temperature of 200 K. What is the change in entropy of the system?

PROBLEMS

1) A power cycle is operating between a high temperature reservoir of 500 °F and a low temperature reservoir of 212 °F. Calculate the Carnot efficiency of this power cycle.

2) Calculate of the Carnot thermal efficiency of a power cycle operating between a high temperature reservoir of 600 °C and a low temperature reservoir of 100 °C.

3) Calculate the Carnot efficiency of a power cycle operating between a high temperature reservoir of 800 K and a low temperature reservoir of 0 K.

4) What can you conclude about η_{th} as $T_C \rightarrow 0$? as $T_H \rightarrow \infty$?

5) A defined region of space is at constant temperature 75 °F. No mass enters the system and no work is done. During some time period, 35 Btu of heat is transferred to the system. Calculate the entropy entering the system due to the heat transfer. If the process is reversible, calculate the change in entropy of the system.

6) A system is at constant temperature 100 °C. The surroundings is at constant temperature 75 °C. No mass crosses the system boundary and no work is done. 50 kJ of energy is transferred as heat from the system to the surroundings. Calculate the entropy leaving the system due to the heat transfer. Calculate the entropy entering the surroundings due to the heat transfer. Calculate the change in entropy of the universe due to this heat transfer.

7) A simple power cycle is proposed where $\widehat{Q}_H = 1200$ Btu / lb$_m$ and $\widehat{W}_{net} = -700$ Btu / lb$_m$. The high temperature reservoir is at 212 °F and the cold temperature reservoir is 32 °F. Use the conservation of energy to determine $\widehat{Q}_C$. Is this process thermodynamically possible and why? If this process were a Carnot cycle, determine the thermal efficiency. For a Carnot cycle determine $|\widehat{W}_{net}|$.

8) It is claimed that a power cycle operates between a high temperature reservoir of 300 °C and a cold temperature reservoir of 50 °C. The heat transferred at the high temperature is $\widehat{Q}_H = 700$ kJ / kg, and the heat transferred at the cold temperature is $\widehat{Q}_C = -200$ kJ / kg. Use the conservation of energy to determine $\widehat{W}_{net}$. Is this process thermodynamically possible and why? If this process were a Carnot cycle, determine the thermal efficiency. For a Carnot cycle determine $|\widehat{W}_{net}|$.

9) A steady-state process is using air to convert thermal and mechanical energy to work. The entering and exiting conditions are given. You may treat air as an ideal gas at these conditions.

$$\widehat{KE}_{in} = 5 \text{ Btu / lb}_m \qquad \widehat{KE}_{out} = 3 \text{ Btu / lb}_m$$

$$\widehat{PE}_{in} = 10 \text{ Btu / lb}_m \qquad \widehat{PE}_{out} = 4 \text{ Btu / lb}_m$$

$$T_{in} = 400°F \qquad T_{out} = 200°F$$

$$P_{in} = 5 \text{ atm} \qquad P_{out} = 1 \text{ atm}$$

The temperature of the surroundings is 75 °F. The air specific heat capacity is $\widehat{C}_P = 7R / (2(MW))$. Determine the maximum available shaft work in Btu / lb$_m$.

10) Steam is used in a steady-state process to convert thermal energy to mechanical energy. The inlet temperature and pressure of the superheated steam is 1600 psi, 1200 °F. The outlet temperature of the saturated vapor is 212 °F. Neglect the changes in the kinetic and potential energy from the inlet to the outlet conditions. The temperature of the surroundings is 75 °F. Determine the maximum available shaft work in Btu / lb$_m$.

11) To convert thermal energy to mechanical energy, steam is used in a steady-state process . The inlet temperature and pressure of the superheated steam is 15000 kPa, 600 °C. The outlet pressure of the saturated vapor is 101 kPa. Neglect the changes in the kinetic and potential energy from the inlet to the outlet conditions. The temperature of the surrounding is 25 °C. Determine the maximum available shaft work in Btu / lb$_m$.

12) One kg of nitrogen is used in a closed system to convert thermal energy to work. The nitrogen has a constant volume specific heat capacity of $\widehat{C}_V = 5R / (2(MW))$. The gas obeys the ideal gas law. The initial condition is $T_{beg} = 200$ °C, $P_{beg} = 20$ atm. The reference conditions are $T_o = 25$ °C, $P_o = 1$ atm. Calculate the maximum useful work (kJ).

13) One lb$_m$ of helium is used in a closed system to convert thermal energy to work. The helium has a constant volume specific heat capacity of $\widehat{C}_V = 3R / (2(MW))$. The gas obeys the ideal gas law. The initial condition is $T_{beg} = 500$ °F, $P_{beg} = 400$ psi. The reference conditions are $T_o = 75$ °F, $P_o = 14.7$ psi. Calculate the maximum useful work (Btu).

14) A heat engine takes heat (Q_H) from a high temperature reservoir (in the surroundings), produces work (to the surroundings $W < 0$), and rejects heat (Q_L) to a low temperature reservoir (also the surroundings). The process uses a working fluid in a cyclic process so that, considering the process as a whole, there is no flow of mass into or out of the process.

   a) Represent this information in a sketch which shows the two temperature reservoirs, the process, and the pertinent exchanges between the process and the surroundings. State any assumptions which you would like to make about the process.

The process does not operate at steady state, but rather is cyclic. Nevertheless, there is a time period for which the working fluid of the process (system - say steam, for example) undergoes changes which return it (at the end of the time period) to its initial state (at the beginning of the time period). In this sense, the process is like a steady-state process which has no exchange of mass between the system and surroundings. During this time period $\widehat{Q}_H = 3500$ (kJ / kg) and $\widehat{Q}_C = -1500$ (kJ / kg).

   b) What was the value of $\widehat{W}_{shaft}$ during the same time period?

   c) If the high temperature reservoir is at 600 K and the low temperature reservoir is at 300 K, is this a viable process?

   d) What is the minimum amount of heat which can be rejected to the low temperature reservoir for the same value of $\widehat{Q}_H$? Why?

CHAPTER
SEVENTEEN
POWER AND REFRIGERATION CYCLES

17.1 INTRODUCTION

In previous chapters we presented a foundation for thermodynamics which consists of several elements. First, conservation of mass and energy principles provide a need for calculating energy functions such as internal energy and enthalpy. Second, heat capacities and equations-of-state, are used to calculate these changes in the energy functions for single-phase, compositionally homogeneous materials. When phase changes or chemical reactions are involved, heats of phase transition and reaction are also required. Third, the second law of thermodynamics places constraints on processes such as that heat cannot be converted totally to work and that heat will not transfer from a low temperature to a higher temperature spontaneously. We are now ready to use this foundation to understand and to analyze power generation and refrigeration cycles.

As will be seen, power and refrigeration cycles have a lot in common in terms of their analysis even though their objectives are quite opposite. The objective of power cycles is to transfer heat from a high temperature reservoir and extract from that heat usable work whereas the objective of a refrigeration cycle is to use work to remove heat from a low temperature reservoir (provide refrigeration). In the analysis, both kinds of cycles will be compared to a corresponding Carnot cycle. The Carnot cycle, as discussed earlier, is a special kind of reversible process in that it is the most efficient process for performing the desired function. The Carnot cycle then will serve as a benchmark of performance for the power and refrigeration cycles.

Each section on power cycles then will consist of a discussion on the corresponding Carnot cycle which will then be followed by discussions of various actual cycles which are compared to the ideal Carnot cycle.

17.2 POWER CYCLES

As mentioned above, the objective of a power cycle or heat engine is to produce work from heat transferred from a high temperature reservoir. Now according to the second law of thermodynamics we cannot do this with 100 percent efficiency; the heat transferred cannot be converted totally into work. Consequently, of the heat which is absorbed at the high temperature reservoir (T_H) some is converted to work (W) and some must be rejected as heat at a lower temperature (T_C). Such a heat engine is depicted in Figure 17-1a.

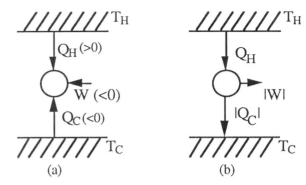

Figure 17-1: Power cycle

Note that both the heat transfer at the high temperature and at the low temperature as well as the work are considered to be positive when energy is added to the system (the heat engine). Consequently, for this process Q_H is positive, W is negative and Q_C also is negative. This figure is drawn in an equivalent alternative way in Figure 17-1b where the *magnitudes* of all the heat transfer and work are shown and the direction of energy flow adjusted accordingly.

17.2.1 Carnot Cycles

Now the most efficient process for converting heat to work is the Carnot cycle. Such a process is reversible with no entropy generated by either heat transfer (heat transferred at each reservoir is over a zero temperature difference) or by work (no mechanical irreversibilities, i.e. there is no conversion of mechanical energy to internal energy due to viscous dissipation). The Carnot cycle proceeds in the following way.

1) The working fluid of the process undergoes a reversible adiabatic compression which raises its temperature from that of the low temperature reservoir to that of the high temperature reservoir. This is an isentropic process.

2) With the fluid now at the temperature of the high temperature reservoir, heat is transferred isothermally to the fluid in the amount Q_H.

3) The fluid is now allowed to undergo a reversible adiabatic expansion during which the temperature drops from that of the high temperature reservoir to that of the low temperature reservoir (T_C). This is another isentropic process.

4) Finally, with the fluid at T_C, heat is rejected to the low temperature reservoir in the amount Q_C. This is a second isothermal process.

A particularly useful way of observing this process is on a temperature entropy diagram. Such a diagram is shown in Figure 17-2 for a Carnot cycle with each path numbered according to the steps above. Along any path of this cycle, because the process is reversible, the heat transfer is given by:

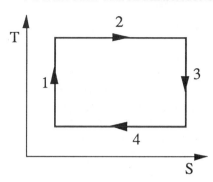

Figure 17-2: Carnot TS diagram

$$Q_{\text{rev}} = \int_{\text{path}} T dS \qquad (17-1)$$

Accordingly then, the heat transfer along path 1 is zero (adiabatic process); along path 2 it is represented by the area under this path projected onto the entropy axis; along path 3 it is zero again; and along path 4 the heat transfer is the area under that path projected onto the entropy axis. Accordingly then, from the first law (conservation of total energy for this complete cycle, during which the fluid goes through changes but returns to its original state and hence during which the change in state of the system is zero) the area inside the rectangle representing the process is the net work done by the process. This is the objective of the process: to take Q_H, the area under the high temperature isotherm and produce work, the area inside the cyclic rectangle. The area underneath the rectangle is Q_C, the amount of energy which could not be converted to work due to second law constraints. From the previous chapter, the thermodynamic efficiency of this process is given by

$$\boxed{\eta_{\text{th}} = 1 - \frac{T_C}{T_H}} \qquad (17-2)$$

The efficiency is not determined by the nature of the working fluid, but rather by the temperatures of the two reservoirs. By way of a practical calculation note that if the ambient temperature T_C is approximately 300 K and the high temperature T_H is approximately 600 K, then the thermal efficiency is about 50%. Thus 50% is about the best we can do for heat engines for normal processes. In actuality we find that real processes which involve irreversibilities in both heat transfer and work have at best about 35% efficiency.

Consider now this same Carnot process on a P-V diagram (Figure 17-3). This figure is shown for an ideal gas and the area inside the P-V lens for the process represents the work $(-\int P dV)$ because in this case the process is reversible (of course if the process were not reversible we could not calculate the actual work in this way).

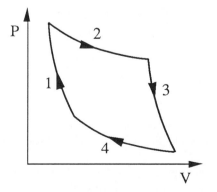

Figure 17-3: Carnot cycle
PV diagram

Now because the Carnot cycle is the most efficient process for converting heat to work, we would like to be able to design our real processes as close to Carnot efficient as possible. However, if we use a gas as a working fluid there are two major problems. First, with a gas it is difficult to achieve a constant-temperature expansion and compression. Second, it is very difficult to achieve processes which are close enough to reversible to provide adequate efficiency. Processes which use gases lose considerable work to thermal energy through friction. There are two notable exceptions to this, however, where gases can be used in a cycle with good efficiency. These are the Stirling engine and the Erickson cycle. We will not consider these further here, but they may, especially the Stirling engine, be important processes in the future as technical difficulties with materials and design are overcome.

17.2.2 Vapor Compression Cycles: The Rankine Cycle

A vapor compression cycle takes advantage of the fact that during a change of phase a fluid changes volume greatly (for a liquid to vapor transition) while at the same time necessarily maintaining constant temperature and pressure. In this way, control of the process at constant temperature for at least a major part of the cycle is guaranteed and automatic.

Such a process is depicted on a T-S diagram in Figure 17-4. We will begin the cycle with the working fluid as a subcooled liquid at the low temperature of the process. Then the ideal (mechanically reversible) process proceeds in the following way.

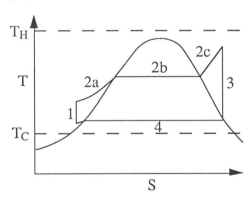

Figure 17-4: Rankine cycle

1) The subcooled liquid is compressed adiabatically and reversibly (in a pump) to a high pressure, shown as path 1 in Figure 17-4. This process also raises the temperature of the fluid somewhat. It is an adiabatic process, and therefore isentropic, when done reversibly.

2) The subcooled liquid (liquid outside the two-phase envelope) is now passed through a boiler at constant pressure where heat is added to achieve the following:

   a) Heat the liquid at constant pressure to a saturated liquid.
   b) Vaporize the saturated liquid to saturated steam (at constant temperature) and
   c) Superheat the steam at constant pressure.

3) Expand the steam adiabatically and reversibly (i.e. constant entropy) to the system low pressure in a turbine thereby producing work (the objective of the cycle) and saturated steam/water near 100% steam conditions.

4) The low pressure steam/water (up to about 10% water) is now passed through a condenser, again at constant pressure, wherein heat is transferred from the steam to condense it to the original starting condition.

Each of these four paths is shown in Figure 17-4. A schematic of the process with each path associated with the equipment is shown in Figure 17-5. Note also that in Figure 17-4 the temperatures of the hot and cold reservoirs are indicated. The high temperature reservoir is necessarily greater than the highest temperature achieved by the superheated steam and the low temperature reservoir is, necessarily, lower than the temperature in the condenser. In a thermally reversible process these temperature differences between the working fluid and the reservoirs would be zero but this would also require an infinite amount of heat transfer area or an infinitely slow circulation rate of the working fluid in the exchangers, hardly a practical idea.

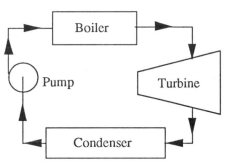

Figure 17-5: Power cycle

We consider now the heat and work of this Rankine cycle leading to a comparison to the Carnot-cycle heat and work. Considering the flow-through boiler as the system, the conservation of total energy says that

$$H_{\text{in}} - H_{\text{out}} + Q_H = 0 \qquad (17-3)$$

from which, because of the UHAG relationship for enthalpy, we see that

$$Q_H = \int_{\text{Boiler}} T\,dS \qquad (17-4)$$

Consequently, for the boiler at constant pressure the heat transferred in the boiler is the area under the T-S path which is followed by the fluid in traversing the boiler. This is true whether the boiler is a reversible process or not, so long as it is constant pressure (or nearly so). (If constant pressure is not an acceptable approximation, then revert to calculating Q_H as the increase in enthalpy through the boiler.) Consequently, we can represent the area under the boiler path as the heat transferred to the fluid in the boiler (Figure 17-6).

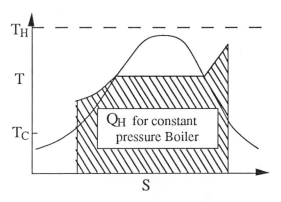

Figure 17-6: Rankine cycle
heat transfer in the boiler

Now along the turbine, if it is truly adiabatic, the heat transfer is zero. Through the condenser, following the same reasoning as for the boiler, we again have that the heat transferred is the area under the condenser path shown in Figure 17-7.

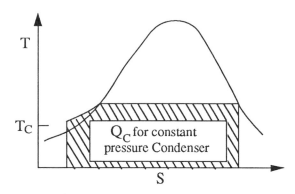

Figure 17-7: Rankine cycle
heat transfer in the condenser

Finally, for an adiabatic pump process, the heat transferred is zero. Considering energy conversion for the entire Rankine cycle, then (during which the fluid in traversing the entire cycle is returned to its original state and therefore has zero change in its energy), gives that the net work of the process is the difference of these two areas (Figure 17-8). For a mechanically reversible process this net work is simply the area inside the cyclic process.

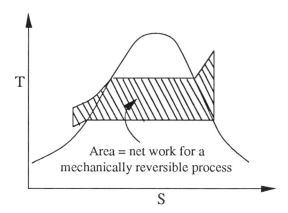

Figure 17-8: Rankine cycle
net work

If there are mechanical irreversibilities associated with the pump in the turbine then these processes, instead of occurring at constant entropy, cause an increase in entropy and the paths followed may be similar to that shown in Figure 17-9. In this case, extra heat must be removed in the condenser to account for the entropy production in the turbine and less heat is added in the boiler because of the entropy production in the pump.

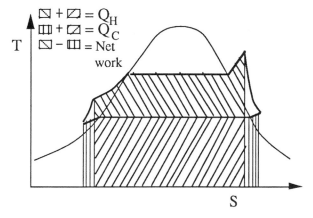

Figure 17-9: Rankine cycle
with mechanical irreversbilities

This complicates the picture of subtracting the two heat transfers to get work; the area inside the cycle is no longer the work but we can see how we could calculate the work from the Q_C and Q_H areas. The picture is still very useful.

We will now compare this Rankine cycle (the mechanically reversible version) to the Carnot cycle and then perform the same comparison for a mechanically irreversible Rankine cycle. To perform the comparison, consider a Carnot cycle which transfers the same amount of heat at the high temperature. The question then will be how much work will be extracted from this heat for the Rankine cycle compared to the Carnot cycle. We know that the Carnot cycle is the most efficient and therefore it will provide the most work. The question is how much less work will the Rankine cycle, operating on the same input heat (Q_H) deliver.

The comparison is shown in Figure 17-10. A Carnot cycle using the same amount of heat will have a constant temperature process at the high temperature reservoir such that the area under this isothermal path is equal to Q_H. Accordingly, a rectangular area is shown which is the same as the area under the boiler path of the ideal Rankine cycle.

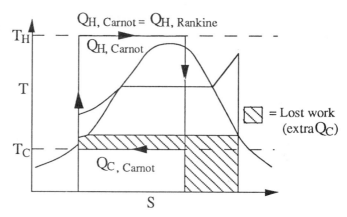

Figure 17-10: Carnot cycle
with reversible Rankine cycle

Note that T_H for the Carnot process is the same as for the Rankine cycle and must be higher than the maximum temperature achieved by the superheated steam at the exit of the boiler. This rectangle is positioned in Figure 17-10 so that, as a matter of convenience, its constant entropy compression cycle passes through the state of the water entering the boiler in the ideal Rankine cycle. The equal-area condition then says that the isentropic expansion for this Carnot cycle occurs at a lower entropy than it does in the Rankine cycle. Now once Q_H and T_H are established for the Rankine cycle and T_C is established (it is given), then Q_C for the Carnot cycle is also established as the area under the isothermal compression portion of the cycle at T_C.

So, for this Carnot cycle superimposed on the Rankine cycle, we see that the work inside the Carnot loop compares to the work of the Rankine loop and we can see how Q_C for the Carnot cycle compares to Q_C for the Rankine cycle. Because the Carnot cycle is most efficient for this fixed amount of heat transferred at the high temperature, the amount of heat rejected at T_C for this Carnot cycle must be a minimum and that for the Rankine cycle must be greater than for the Carnot cycle. We see this in the figure and in fact the amount by which Q_C exceeds ($Q_{C,Car}$) is the amount of energy that would have been converted to work were the Rankine cycle as efficient as the Carnot cycle. Because this amount of energy was not converted to work it may be considered to be lost work; to the extent the work of a real process is less than that of the corresponding Carnot process, work is lost.

We can do a similar comparison for a mechanically irreversible Rankine cycle and this is shown in Figure 17-11. In this case the Carnot cycle Q_H must again equal the area under the boiler portion of the Rankine cycle path (neglecting pressure drop). Again this Q_H rectangle, along with T_C establishes Q_C for the Carnot cycle. In this case we can see that the heat actually rejected at the low temperature exceeds the Carnot Q_C by an even greater amount due to the additional area which results from the turbine irreversibilities, and due to the pump irreversibilities.

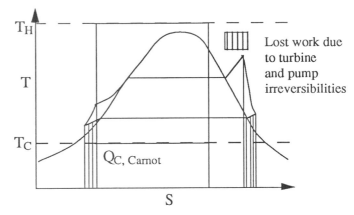

Figure 17-11: Carnot cycle with mechanically irreversible Rankine cycle

The lost work is even greater and by comparing this figure to lost work for the ideal Rankine cycle, we can establish the portions of the condenser heat transfer (lost work) for this mechanically irreversible process which arise from turbine and pump irreversibilities. The remaining part of the excess is the same as it was for the mechanically reversible process under the same conditions and arises because of the fact that the temperature differences between the process and the reservoirs on both the high temperature and low temperature sides are finite. In fact, on the high temperature side this difference is considerable over much of the boiler process. This is true because in order to superheat the steam, T_H must exceed the final superheated value which itself is in excess of the fluid temperature over most of the boiler path. The heat transfer irreversibilities account for a considerable reduction of inefficiency of the Rankine cycle compared to the Carnot efficiency.

This comparison of the Rankine cycle to the Carnot cycle which operates between the same high and low temperature reservoirs and with the same Q_H input provides a clear rationale for efforts which have been made to increase the efficiency of the Rankine cycle. Any process parameter change which results in a closer approach of the average fluid temperature through the boiler increases the area of the Carnot cycle loop relative to the Rankine cycle loop thereby moving its constant entropy expansion step closer to the Rankine cycle expansion. This also increases the amount of Carnot Q_C proportionally and thereby decreases the difference between the Rankine cycle Q_C and the Carnot cycle Q_C; there is less lost work in such a process. Likewise, any parameter change which results in a closer approach of the Rankine cycle condenser temperature to T_C reduces the difference between the Rankine Q_C and the Carnot Q_C and therefore results in less lost work. Such considerations are the motivation for a number of process modifications outlined below:

1) Increase the operating pressure of the boiler. This raises the temperature at which vaporization occurs as shown in Figure 17-12 by the dashed boiler curve. In this case the same maximum temperature limit for the superheated steam exists so that maximum entropy of the steam for the Rankine cycle is also reduced. The corresponding Carnot cycle maximum entropy is increased as is Q_C and the amount of lost work is reduced.

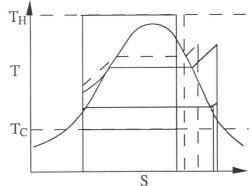

Figure 17-12: Modified Rankine cycle

2) Reduce the pressure of the condenser thereby moving the temperature of the condenser closer to T_C. This results in a reduced Q_C which is then closer to the Carnot cycle Q_C and therefore results in less lost work.

3) Allow the superheated steam to approach closer to T_H thereby raising the average boiler temperature. In order to achieve this a boiler with greater heat transfer area would be required and hence a greater capital cost. The net effect however is to achieve a greater heat transfer and therefore a shift in the corresponding Carnot cycle isentropic expansion path to a higher entropy, resulting in relatively less lost work.

4) After expanding through a high pressure turbine, reheat the steam. Note more than one reheat cycle is possible. This change is shown in Figure 17-13 and results in a higher average temperature of the boiler process and therefore improved efficiency. It also again increases the capital cost because of the additional turbine(s) and heat exchangers. This modification is especially useful in combination with a higher boiler pressure which then virtually necessitates a reheat cycle or cycles in order to produce a low pressure steam which does not contain excessive moisture (the steam must be carried far enough beyond the phase envelope so that when it is expanded to low pressure it does not penetrate too far into the two-phase region. Excessive moisture in the turbine causes poor efficiency and excessive wear.).

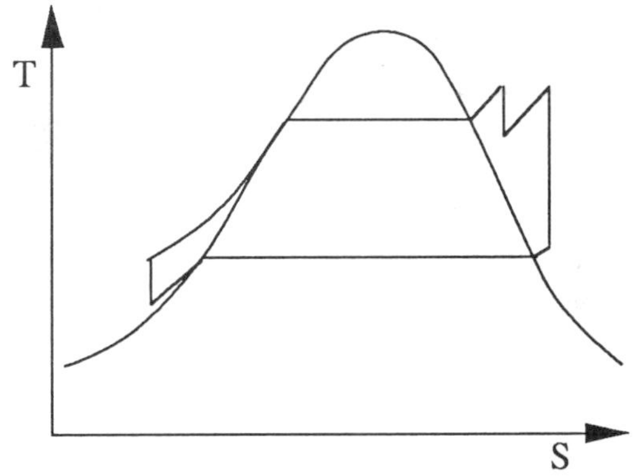

Figure 17-13: Rankine cycle
with reheater

Other possibilities exist. One method is to use feedwater heaters whose purpose is to raise the average temperature in the boiler by raising the temperature of the feed water to the boiler. These will not be discussed further here but discussions may be readily found in recent thermodynamics textbooks. Another method of improving the approach of the high temperature side of the process to the high temperature reservoir is to use a dual vapor compression cycle that involves the use of two different fluids in separate cycles (Figure 17-14).

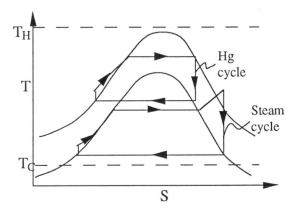

Figure 17-14: Rankine dual
vapor compression cycles

The heat of the topping cycle is rejected to the high temperature side of the bottoming cycle. The net effect of this combined process is to reduce the temperature differences between the high temperature reservoir and the fluid cycle thereby improving the thermodynamic efficiency. In practice, mercury has been used in the topping cycle and steam in the bottoming cycle and efficiencies of 40 to 42% were achieved. This process is little (if at all) used today because of its cost relative to the cost of fossil fuels (cost of T_H) and the inherent safety hazards associated with using mercury. The concept is still valid however and other binary pairs are possible, the main point being that one of the pairs should have a very high critical temperature, well above the high temperature reservoir.

17.2.3 Quantitative Calculations of Heat and Work

While the above analysis provides a very useful picture of the relations between the amounts of heat transferred, the useful work, and the lost work (relative to the corresponding Carnot cycle) it is not the way quantitative calculations are made. Considering the boiler as a system at steady-state, no work occurs and we may neglect changes in kinetic and potential energy between the entering and leaving fluid. Hence, for the boiler, we have:

$$\widehat{H}_{out} - \widehat{H}_{in} = \widehat{Q}_H$$

(17 − 5)

where $\widehat{Q}_H$ is the heat transferred per unit mass. In terms of total enthalpies and heat transfer, this result is

$$\widehat{H}_{out}\dot{m} - \widehat{H}_{in}\dot{m} = \dot{Q}_H$$

(17 − 6)

Similarly, for the condenser

$$\widehat{H}_{out} - \widehat{H}_{in} = \widehat{Q}_C$$

(17 − 7)

Calculations of these changes in enthalpy to determine Q_H or Q_C can be made using the relations for enthalpy in terms of heat capacities, equations-of-state, and latent heats of vaporization or, more simply, they can be determined using tabulated values of properties for the working fluid (e.g., steam tables) in which case someone has already made the calculations for us. For computer calculations, the values would normally be calculated as needed using the heat capacity, EOS, and latent heat data rather than retrieved from stored data.

Equations for the turbine and pump are similar except that work appears instead of heat. For the turbine, we may assume steady-state, adiabatic, and negligible potential energy and kinetic energy effects so that

$$\widehat{H}_{out} - \widehat{H}_{in} = \widehat{W}_{turb}$$

(17 − 8)

Similarly, for the pump

$$\widehat{H}_{out} - \widehat{H}_{in} = \widehat{W}_{pump}$$

(17 − 9)

These results for the work hold whether the processes are isentropic or not. Note that the turbine work will be negative while the pump work will be positive (for the convention that work is positive when energy is added to the system).

17.2.4 Calculations of Isentropic Work and Adjustments Using Efficiency

Note also that the changes in enthalpy presented above can be expressed in terms of the UHAG relation for enthalpy:

$$d\widehat{H} = T d\widehat{S} + \widehat{V} dP \qquad (15-6)$$

which at constant entropy is

$$d\widehat{H} = \widehat{V} dP$$

Now if the fluid (liquid) in the pump is essentially constant density ($\widehat{V}^{-1}$), we can calculate the (isentropic) pump work as

$$\boxed{\widehat{W}_{\text{pump,isen}} = \widehat{V}(P_{\text{out}} - P_{\text{in}})} \qquad (17-10)$$

and, in fact, this is the normal way to calculate the pump work. If there are pump irreversibilities, the work is calculated this way and adjusted (increased) according to this reduced efficiency:

$$\boxed{\widehat{W}_{\text{pump}} = \frac{\widehat{W}_{\text{pump,isen}}}{\eta}} \qquad (17-11)$$

where η is the mechanical efficiency of the pump. For the turbine, we could do a similar calculation, assuming that we know the fluid density as a function of pressure (e.g., for isentropic expansion of an ideal gas $P\widehat{V}^{\gamma}$ = constant). However, the change in enthalpy usually is obtained directly from tabulated thermodynamic data for the working fluid. The change for the isentropic process is calculated first to give the isentropic turbine work and then this is adjusted according to the turbine mechanical efficiency:

$$\boxed{\widehat{W}_{\text{turbine}} = \widehat{W}_{\text{turbine,isen}}\eta} \qquad (17-12)$$

Note that in this case, the isentropic work is multiplied by the efficiency whereas for the pump the work is divided by the efficiency. For the turbine, an efficiency less than 100% means that less than the isentropic work is obtained from the process whereas for the pump, an efficiency of less than 100% means that more than the isentropic work must be supplied to the process.

Example 17-1. Steam is used as a working fluid in a power plant. The high pressure steam is 1600 psi and the low pressure exhaust from the turbine is 14.7 psi. The boiler heats the steam to a maximum temperature of 1200 °F. The exhaust from the turbine is saturated vapor at 14.7 psi. The water leaving the condenser is saturated at 14.7 psi. The entire cycle is operating at steady state.

    a) Calculate the work per unit mass required to pump the saturated water from atmospheric pressure to 1600 psi. The pump operates adiabatically and reversibly.

    b) Calculate the heat transfer per unit mass to convert the water to steam and increase the temperature to 1200 °F.

    c) Calculate the work per unit mass from the expansion through the turbine. The turbine operates adiabatically.

d) Calculate the heat transfer per unit mass to condense the saturated steam to saturated liquid.

e) Calculate the thermal efficiency for this cycle.

SOLUTION: The Rankine cycle for this plant is shown in Figure 17-4. First we will go to the steam table given in Table C-22 and gather the data.

State 1: Saturated Liquid $T = 212°F$ $P = 14.7$ psi	$\widehat{V}_1 = 0.0167$ ft^3 / lb$_m$ $\widehat{H}_1 = 180.17$ Btu / lb$_m$ $\widehat{S}_1 = 0.3121$ Btu / lb$_m$ R	State 2: Compressed Liquid $T = ?$ °F $P = 1600$ psi	$\widehat{V}_2 = 0.0167$ ft^3 / lb$_m$ $\widehat{H}_2 = ?$ Btu / lb$_m$ $\widehat{S}_2 = ?$ Btu / lb$_m$ R
State 3: Superheated Vapor $T = 1200°F$ $P = 1600$ psi	$\widehat{V}_3 = 0.5915$ ft^3 / lb$_m$ $\widehat{H}_3 = 1605.6$ Btu / lb$_m$ $\widehat{S}_3 = 1.6678$ Btu / lb$_m$ R	State 4: Saturated Vapor $T = 212°F$ $P = 14.7$ psi	$\widehat{V}_4 = 26.8$ ft^3 / lb$_m$ $\widehat{H}_4 = 1150.5$ Btu / lb$_m$ $\widehat{S}_4 = 1.7568$ Btu / lb$_m$ R

The work per unit mass to compress the water from atmospheric pressure to 1600 psi is calculated from equation (17-8). This equation assumes that the density of the liquid is constant and that the pump is isentropic.

$$\widehat{W}_{\text{pump}} = \int \widehat{V} dP = \widehat{V}(P_{\text{high}} - P_{\text{low}})$$

$$= 0.0167(1600 - 14.7) \text{ (ft}^3 \text{ lb}_f / \text{lb}_m \text{ in}^2)$$

$$= \frac{26.47 \text{ ft}^3 \text{ lb}_f}{\text{lb}_m \text{ in}^2} \left| \frac{144 \text{ in}^2}{1 \text{ ft}^2} \right| \frac{1 \text{ Btu}}{778 \text{ lb}_f \text{ ft}} = 4.9 \text{ Btu} / \text{lb}_m$$

Next the specifc enthalpy of the compressed liquid at State 2 is calculated considering steady-state adiabatic operation of the pump.

$$\widehat{H}_1 - \widehat{H}_2 + \widehat{W}_{\text{pump}} = 0$$

Rearranging

$$\widehat{H}_2 = \widehat{H}_1 + \widehat{W}_{\text{pump}}$$

$$= 180.16 + 4.9 = 185.1 \text{ Btu} / \text{lb}_m$$

Now that all of the enthalpy values of the steam at each condition are known, the heat transfer and work can be calculated. The heat transfer from the boiler is calculated from the entering and exiting enthalpy values. No work is done in the boiler.

$$\widehat{H}_2 - \widehat{H}_3 + \widehat{Q}_H = 0$$

Rearranging

$$\widehat{Q}_H = \widehat{H}_3 - \widehat{H}_2$$

$$= 1605.6 - 185.06 = 1420.5 \text{ Btu} / \text{lb}_m$$

The work per unit mass done on the turbine is calculated from the entering and exiting enthalpy values. The turbine operates adiabatically.

$$\widehat{H}_3 - \widehat{H}_4 + \widehat{W}_{\text{turb}} = 0$$

The exhaust from the turbine is saturated vapor. Note that the value of the work is negative since energy is leaving the system or the fluid. The work is:

$$\widehat{W}_{turb} = \widehat{H}_4 - \widehat{H}_3$$

$$= 1150.5 - 1605.6 = -455.1 \text{ Btu / lb}_m$$

Also note, from the entropy values, that the turbine is not operating reversibly. Finally, the heat transfer per unit mass to condense the saturated vapor is calculated from the entering the exiting conditions of the condenser.

$$\widehat{H}_4 - \widehat{H}_1 + \widehat{Q}_C = 0$$

Rearranging and evaluating gives:

$$\widehat{Q}_C = \widehat{H}_1 - \widehat{H}_4$$

$$= 180.17 - 1150.5 = -970.3 \text{ Btu / lb}_m$$

The thermal efficiency of the cycle is the absolute value of the net work divided by the heat transfer from the high temperature reservoir

$$|\widehat{W}_{net}| = |\widehat{W}_{turb} + \widehat{W}_{pump}|$$

$$= |-455.1 + 4.9| = 450.2 \text{ (Btu / lb}_m)$$

The thermal efficiency is:

$$\eta_{th} = \frac{|\widehat{W}_{net}|}{Q_H}$$

$$= \frac{450.2}{1420.5} = 0.32 \quad \text{or} \quad 32\%$$

Only one-third of the energy transferred to the steam at the high temperature is available to produce useful work.

Example 17-2 You are a young engineer at the Central Power company outside Watsonville. The city uses energy at the rate of 100,000 kW during a summer day. The Central Power plant uses steam as the working fluid; however, the mass flow rate can be varied. The steam leaving the boiler and entering the turbine is at 1600 psi and 1200 °F. The exhaust from the turbine is at a pressure of 14.7 psia. The water leaving the condenser is saturated liquid at 14.7 psi. The turbine operates reversibly and adiabatically. Calculate the thermal efficiency for this power cycle. Determine the mass flow rate of steam required to meet the city's need. The plant is coal fired. If coal has an average energy value of 12,000 Btu / lb$_m$, determine the rate at which coal is burned in the furnace.

SOLUTION: Many of the same conditions from problem 17-1 exist in this problem. The only major difference is the adiabatic, reversible operation of the turbine. This means that the entropy of the steam leaving the turbine is equal to the entropy of the steam entering the turbine. This also means that the exhaust is a mixture of both saturated liquid and vapor (see discussion below). The data from Table C-22 and example 17-1 are given below.

State 1: Saturated Liquid $T = 212°F$ $P = 14.7$ psi	$\widehat{V}_1 = 0.0167$ ft^3 / lb$_m$ $\widehat{H}_1 = 180.17$ Btu / lb$_m$ $\widehat{S}_1 = 0.3121$ Btu / lb$_m$ R	State 2: Compressed Liquid $T = ?$ °F $P = 1600$ psi	$\widehat{V}_2 = 0.0167$ ft^3 / lb$_m$ $\widehat{H}_2 = 185.1$ Btu / lb$_m$ $\widehat{S}_2 = ?$ Btu / lb$_m$ R
State 3: Superheated Vapor $T = 1200°F$ $P = 1600$ psi	$\widehat{V}_3 = 0.5915$ ft^3 / lb$_m$ $\widehat{H}_3 = 1605.6$ Btu / lb$_m$ $\widehat{S}_3 = 1.6678$ Btu / lb$_m$ R	State 4: Liquid/Vapor $T = ?°F$ $P = 14.7$ psi	$\widehat{V}_4 = ?$ ft^3 / lb$_m$ $\widehat{H}_4 = ?$ Btu / lb$_m$ $\widehat{S}_4 = 1.6678$ Btu / lb$_m$ R

We now consider the state of the water leaving the turbine in more detail. Because the operation is steady-state and isentropic, we know that the entropy values entering and leaving the turbine must be the same.

$$\widehat{S}_3 - \widehat{S}_4 = 0$$

Then, because the entropy of saturated vapor at 14.7 psi is greater than that of the entering steam, it must be true that the water leaving the turbine is a mixture of saturated liquid and saturated vapor. So

$$\widehat{S}_4 = x_{\text{vap}}\widehat{S}_{4,\text{vap}} + x_{\text{liq}}\widehat{S}_{4,\text{liq}}$$

where x_{vap} is the fraction of the steam that is vapor, and x_{liq} is the fraction of the steam that is liquid. The percentage of the steam in the liquid and vapor phase is unknown. Since there are only two phases present, liquid and vapor, the following constraint equation is true:

$$1 = x_{\text{vap}} + x_{\text{liq}}$$

or

$$x_{\text{liq}} = 1 - x_{\text{vap}}$$

substituting and rearranging gives:

$$x_{\text{vap}} = \frac{\widehat{S}_3 - \widehat{S}_{4,\text{liq}}}{\widehat{S}_{4,\text{vap}} - \widehat{S}_{4,\text{liq}}}$$

The vapor and liquid specific entropies are:

$$\widehat{S}_{4,\text{vap}} = 1.7568 \text{ Btu / lb}_\text{m} \text{ R} \qquad \text{and} \qquad \widehat{S}_{4,\text{liq}} = 0.3121 \text{ Btu / lb}_\text{m} \text{ R}$$

Substituting and evaluting gives:

$$x_{\text{vap}} = \frac{1.6678 - 0.3121}{1.7568 - 0.3121} = 0.938$$

The exhaust is 93.8 % saturated vapor and 6.2 % saturated liquid. Now the specific enthalpy of the turbine exhaust is:

$$\widehat{H}_4 = x_{\text{vap}}\widehat{H}_{4,\text{vap}} + x_{\text{liq}}\widehat{H}_{4,\text{liq}}$$

Substituting in the known values gives:

$$\widehat{H}_4 = 0.938(1150.5) + 0.062(180.2) = 1090.4 \text{ Btu / lb}_\text{m}$$

The work of the turbine is calculated

$$\widehat{W}_{\text{turb}} = \widehat{H}_4 - \widehat{H}_3$$
$$= 1090.4 - 1605.6 = -515.2 \text{ Btu / lb}_\text{m}$$

The heat transfer at the high temperature is calculated in Example 17-1, $Q_H = 1420.5$ (Btu / lb$_\text{m}$). The thermal efficiency for this the power cycle operating with an isentropic expansion is:

$$\eta_{\text{th}} = \frac{|\widehat{W}_{\text{net}}|}{\widehat{Q}_H} = \frac{|-515.2 + 4.9|}{|1420.5|} = 0.359$$

Only 36 % of the heat transferred at the high temperature is converted to useful work. The massflow rate of the steam is calculated.

$$\dot{m} |\widehat{W}_{net}| = 100,000 \text{ kW}$$

Rearranging and substituting gives:

$$\dot{m} = \frac{100,000}{|\widehat{W}_{net}|} = \frac{100,000 \text{ kW}}{|510.3 \text{ Btu}|} \left| \frac{1 \text{ lb}_m}{1 \text{ kW hr}} \right| \frac{3413 \text{ Btu}}{|3600 \text{ s}|} \frac{1 \text{ hr}}{}$$

$$= 186 \text{ lb}_m / \text{s}$$

We will assume that all of the heating value in the coal is completely transferred to the steam. In reality this is not the case. The rate at which coal is burned is calculated:

$$\dot{m}_{coal}(\text{Coal heating value}) = \widehat{Q}_H \dot{m}_{steam}$$

Rearranging and substituting gives:

$$\dot{m}_{coal} = \frac{1420.5(186)}{12000} = 22 \text{ (lb}_m \text{ coal} / \text{s)}$$

Example 17-3. Now consider that the turbine and pumps in the power cycle have efficiencies less than 100 %. In example 17-2 we assumed that the turbines and pumps were 100 % efficient. The real turbine has an efficiency of 0.85 and the real pump has an efficiency of 0.80. Determine the thermal efficiency of the cycle given in example 17-2, the mass flow rate of steam, and the mass flow rate of coal.

SOLUTION: The pump work is corrected using the given efficiency:

$$\widehat{W}_{pump,real} = \frac{\widehat{W}_{pump,ideal}}{\eta_{pump}}$$

Substituting and evaluating gives:

$$\widehat{W}_{pump,real} = \frac{4.9}{0.80} = 6.1 \text{ (Btu} / \text{lb}_m)$$

The turbine work is corrected using the given efficiency:

$$\widehat{W}_{turb,real} = \eta_{turb} \widehat{W}_{pump,ideal}$$

Substuting and evaluating gives:

$$\widehat{W}_{turb,real} = 0.85(-515.2) = -437.9 \text{ (Btu} / \text{lb}_m)$$

Now the absolute value of the real net work is calculated:

$$|\widehat{W}_{net}| = |-437.9 + 6.1| = 431.8 \text{ (Btu} / \text{lb}_m)$$

The thermal efficiency is:

$$\eta_{th} = \frac{431.8}{1420.5} = 0.304$$

The steam mass flow rate is calculated:

$$\dot{m} = \frac{100,000}{|\widehat{W}_{\text{net}}|} = \frac{100,000 \text{ kW}}{} \left| \frac{1 \text{ lb}_m}{431.8 \text{ Btu}} \right| \frac{3413 \text{ Btu}}{1 \text{ kW hr}} \left| \frac{1 \text{ hr}}{3600 \text{ s}} \right.$$

$$= 219.6 \; (\text{lb}_m / \text{s})$$

The coal mass flow rate is:

$$\dot{m}_{\text{coal}} = \frac{1420.5(219.6)}{12000} = 26 \; (\text{lb}_m\text{coal} / \text{s})$$

17.2.5 Internal Combustion Gas Power Cycles

Gas power cycles such as the Stirling engine and the vapor compression cycle such as the Rankine cycle rely upon the external generation of a high temperature (combustion of fossil fuels, e.g.) and then the transfer of heat from this high temperature reservoir through a heat exchanger to the process. This transfer of heat from outside the process through a heat exchanger causes a reduction in efficiency because the heat must transfer through a heat exchanger and therefore requires a temperature difference between the reservoir and the process. This temperature difference introduces an inherent thermodynamic irreversibility and consequently a reduced efficiency.

Through the use of internal combustion engines heat can be generated from within the process by means of the conversion of internal (chemical) energy to internal (thermal) energy, thereby mimicking the transfer of heat but without a heat transfer apparatus required. This has both the advantage of very efficient "heat transfer" and minimal equipment or capital costs. This is an important advantage of internal combustion engines. However because of its design, the working fluid must be continually refreshed with the combustible mixture. The fuel and oxidant must replace the exhaust gases in one part of the cycle. One further advantage of the internal combustion engine apart from the reduced equipment required is that, because heat exchangers are not required, higher temperatures can be sustained in a combustion chamber. Therefore the higher temperatures result in an inherently higher thermodynamic efficiency.

The simpler mechanical design (no heat exchangers) of the internal combustion engine is counteracted to some extent by a more complicated process for analysis. Obviously, because the gases must be replaced after each combustion, this cannot be a closed, cyclic process. Furthermore, because of the combustion, the number of moles of gas in the process and the heat capacity of the gas mixture changes. Finally the process cannot be separated into a sequence of well-defined and easily analyzed paths. Hence an accurate analysis of internal combustion processes is quite involved.

However, these complications do not preclude a reasonable approximate analysis. In doing so, results are obtained which agree reasonably well with experimental measurements and therefore suggest that the primary physical phenomena are represented in spite of the simplifications. The simplified analysis replaces the real internal combustion cycle with one where the fluid is circulated in a closed process and heat is suddenly added at the time of combustion and suddenly removed at the time that the exhaust gases are allowed to leave the combustion chamber.

The Otto Cycle. To clarify this further we consider first the standard four cycle internal combustion automobile cycle as depicted in Figure 17-15. We begin the cycle at the point that the exhaust gases have been swept out of the combustion chamber and so the piston has been displaced to its maximum swept position, leaving the combustion chamber at its minimum volume. At this point the intake valve is opened and as the piston recedes, a mixture of air and fuel is drawn into the cylinder at essentially constant pressure.

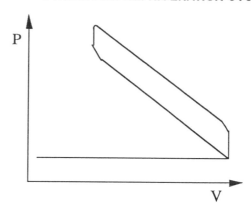

Figure 17-15: Otto Cycle

When this stroke is finished, so that the cylinder volume is a maximum, the intake valve closes and the piston begins its travel in the opposite direction, thereby compressing the gas until finally it reaches the minimum volume again. Near this minimum combustion chamber volume (near TDC, top dead center) a spark ignites the combustible air/fuel mixture and rapid combustion releases "heat" (converts chemical bond internal energy to thermal internal energy) and the temperature of the gases rises dramatically. As mentioned above, the number of moles also increases due to the combustion stoichiometry. At top dead center, the maximum pressure is achieved and the piston begins its travel again towards larger volumes and sweeps out a path of increasing volume and decreasing pressure until it is almost again to the maximum volume. At this point the exhaust valve opens, allowing the gases to leave the chamber and the pressure to drop suddenly to near atmospheric. The piston then again reverses direction and sweeps out a volume to top dead center with the exhaust valve open thereby sweeping the combustible gases out of the cylinder (except for those remaining in the clearance volume) and completing the cycle before beginning another repeat sequence.

Now as mentioned above, this is a fairly complicated process but it can be replaced by a simpler process which can be analyzed quite easily, the air standard Otto cycle. This cycle replaces the compression and expansion steps by adiabatic reversible steps and replaces the nearly constant volume heat transfer steps (the combustion step and hot gas expulsion when the intake valve is first opened) with true constant-volume processes undergoing an increase or decrease in pressure due to heat transfer. This idealized process is depicted in Figure 17-16.

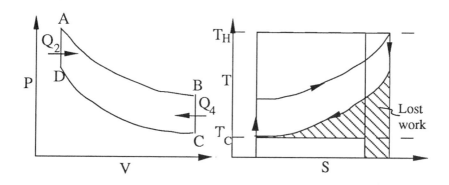

Figure 17-16: Otto cycle PV and TS diagrams

Because this process is reversible, the work is the area inside the P-V loop and can be readily calculated. The work associated with each of the isentropic processes is simply the area under the P-V curve, i.e. $-\int P dV$. The difference of these two areas, i.e. the area inside the loop, is the net work for the entire cycle. We can even calculate this work directly by doing these integrals and recognizing that for an isentropic process of an ideal gas PV^γ is constant for a given value of the entropy from which:

$$\int P dV = \frac{K}{1-\gamma}(V_2^{(1-\gamma)} - V_1^{(1-\gamma)}) \tag{17 - 13}$$

where K is the value of PV^γ.

Alternatively, we can use the first law to calculate the work in terms of the two heat transfer steps and then from those calculate the efficiency of the process. This latter approach is more direct and simple mathematically. From the first law for one cycle of this process we can write

$$Q_2 + Q_4 + W = 0 \tag{17 - 14}$$

Hence the thermal efficiency of the process (defined to be the work produced divided by the heat input) is given by

$$\begin{aligned} \eta_{\text{th}} &= \frac{|W|}{Q_2} \\ &= \frac{Q_2 + Q_4}{Q_2} \end{aligned} \tag{17 - 15}$$

Now the heat transfer steps are constant volume processes that involve zero work so that from the first law we have

$$Q_2 = U_A - U_D \tag{17 - 16}$$

and for an ideal gas at constant volume the change in internal energy is an integral of heat capacity over temperature so that (assuming C_V is temperature independent)

$$Q_2 = C_V(T_A - T_D) \tag{17 - 17}$$

Likewise, for the other constant volume process we have

$$Q_4 = C_V(T_C - T_B) \tag{17 - 18}$$

from which the thermal efficiency of this cycle is given by

$$\begin{aligned} \eta_{\text{th}} &= \frac{C_V[(T_A - T_D) + (T_C - T_B)]}{C_V(T_A - T_D)} \\ &= 1 + \left(\frac{(T_C - T_B)}{T_A - T_D}\right) \\ &= 1 + \left(\frac{T_C - T_B}{T_A - T_D}\right) \end{aligned} \tag{17 - 19}$$

Now, through the ideal gas law this ratio of temperature differences can be converted to a ratio involving pressures and volumes according to

$$\eta_{\text{th}} = 1 + \frac{(P_C V_C - P_B V_B)}{(P_A V_A - P_D V_D)} \tag{17 - 20}$$

which can be further simplified by recognizing that V_C equals V_B and V_A equals V_D. Accordingly

$$
\begin{aligned}
\eta_{th} &= 1 + \frac{V_C(P_C - P_B)}{V_D(P_A - P_D)} \\
&= 1 - \frac{V_C(1 - P_B / P_C)P_C}{V_D(1 - P_A / P_D)P_D}
\end{aligned}
$$

(17 – 21)

Now furthermore because paths 1 and 3 are isentropic, for an ideal gas we have that

$$ P_C V_C^{\gamma} = P_D V_D^{\gamma} $$

(17 – 22)

and

$$ P_B V_B^{\gamma} = P_A V_A^{\gamma} $$

(17 – 23)

and from the ratio of these two equations we obtain (again recognizing that V_C equals V_B and V_D equals V_A)

$$ \left(\frac{P_C}{P_B}\right) = \left(\frac{P_A}{P_D}\right) $$

(17 – 24)

so that the efficiency now can be written as

$$
\begin{aligned}
\eta_{th} &= 1 - \frac{V_C}{V_D}\left(\frac{V_D^{\gamma}}{V_C^{\gamma}}\right) \\
&= 1 - \left(\frac{V_D}{V_C}\right)^{\gamma-1}
\end{aligned}
$$

(17 – 25)

or

$$ \boxed{\text{Otto Cycle:} \qquad \eta_{th} = 1 - \left(\frac{1}{r}\right)^{\gamma-1}} $$

(17 – 26)

In this last equation r is the compression ratio, the ratio of V_C to V_D. This result says that the larger the compression ratio, the larger the efficiency of this cycle and is in agreement with experimental observations. It also says that efficiency is independent of the amount of combustion, i.e. of the "heat transfer." Of course the absolute amount of work done depends directly on the amount of heat available.

Example 17-4. Air is used in an Otto cycle to convert thermal energy into mechanical energy. Assume that the air obeys the ideal gas law. The molar heat capacity is $\widetilde{C}_P = 7R / 2$ and $\widetilde{C}_V = 5R / 2$. The cycle is shown in Figure 17-16. The volumes at states A, B, C, and D, are given:

$$ V_A = V_D = 10 \text{ in}^3 $$
$$ V_B = V_C = 40 \text{ in}^3 $$

Calculate the thermal efficiency of this power cycle.

SOLUTION: From the volume data, we can calculate the compression ratio r.

$$r = \frac{V_C}{V_D}$$

Substituting in the values gives:

$$r = \frac{40}{10} = 4$$

Now the ratio of the heat capacities is calculated:

$$\gamma = \frac{\widetilde{C}_P}{\widetilde{C}_V} = \frac{7R/2}{5R/2} = 1.4$$

The Otto cycle thermal efficiency is

$$\eta_{th} = 1 - \left(\frac{1}{r}\right)^{\gamma-1}$$

Substituting in the values gives:

$$\eta_{th} = 1 - \left(\frac{1}{4}\right)^{1.4-1} = 0.426$$

The thermal efficiency of the Otto cycle is 0.426.

___

The Diesel Cycle. From the previous discussion it is evident that the higher the compression ratio, the higher the efficiency for such an internal combustion process. However, a practical problem encountered in going to higher compression ratios is that the higher pressures produce higher temperatures, leading to spontaneous preignition (dieseling). A diesel engine cycle (Figure 17-17) circumvents this problem by allowing the compression to take place without fuel and then on the expansion step the fuel is injected and, due to the high temperatures which exist in the combustion chamber as it is injected, it ignites.

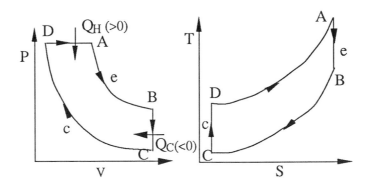

Figure 17-17: Diesel cycle PV and TS diagrams

The injection process occurs relatively slowly compared to the expansion and therefore this part of the process is more nearly at constant pressure than at constant volume. The main effect of this kind of process is that higher compression ratios are achieved and therefore higher efficiency than can be achieved in the Otto cycle. Note that in this cycle the compression ratios along the two isentropic paths are different due to the fact that V_A does not equal V_D whereas V_B equals V_C. The ratios are r_c and r_e for the compression expansion steps respectively. An analysis of this process gives an expression for the efficiency as

$$\text{Diesel Cycle:} \qquad \eta_{\text{th}} = 1 - \frac{1}{\gamma}\left(\frac{(1/r_e)^\gamma - (1/r_c)^\gamma}{(1/r_e) - (1/r_c)}\right) \qquad (17-27)$$

Example 17-5. Air is used in a Diesel cycle to convert thermal energy into mechanical energy. Assume that the air obeys the ideal gas law. The molar heat capacity is $\widetilde{C}_P = 7R/2$ and $\widetilde{C}_V = 5R/2$. The cycle is shown in Figure 17-17. The volumes at states A, B, C, and D, are given:

$$V_A = 20 \text{ in}^3$$
$$V_B = V_C = 40 \text{ in}^3$$
$$V_D = 10 \text{ in}^3$$

Calculate the thermal efficiency of this power cycle.

SOLUTION: From the volume data, the two compression ratios are calculated:

$$r_e = \frac{V_B}{V_A} = \frac{40}{20} = 2$$

$$r_c = \frac{V_C}{V_D} = \frac{40}{10} = 4$$

The ratio of the heat capacities is 1.4. Substituting everything into the Diesel cycle thermal efficiency equation gives:

$$\eta_{\text{th}} = 1 - \frac{1}{\gamma}\left(\frac{(1/r_e)^\gamma - (1/r_c)^\gamma}{(1/r_e) - (1/r_c)}\right)$$

$$= 1 - \frac{1}{1.4}\left(\frac{(1/2)^{1.4} - (1/4)^{1.4}}{(1/2) - (1/4)}\right)$$

$$= 1 - (1/1.4)\left(\frac{0.379 - 0.144}{0.5 - 0.25}\right)$$

$$= 0.329$$

The Diesel cycle has a thermal efficiency of 0.329.

Other Internal Combustion Power Plants. Three other significant internal combustion processes are the combustion gas turbine, jet engines, and rocket engines. The combustion gas turbine takes advantage of both internal combustion efficiencies and turbine efficiencies; turbines are inherently more efficient than reciprocating compressors because the motion is continuous in one direction and frictional dissipation can be reduced.

Jet and rocket engines perform still differently in that in these engines the energy due to combustion, rather than being converted to a reciprocating or rotating shaft work, is converted to kinetic energy of gases. This kinetic energy provides impulse through the conservation of linear momentum. Further consideration of these devices is deferred to other references.

17.3 REFRIGERATION CYCLES

The objective of the refrigeration cycle is to remove heat from a low temperature reservoir and reject it at a high temperature reservoir. The second law, however, says that we cannot do this without performing work to accomplish it. Accordingly, our discussion of refrigeration cycles, in addition to considering the physical aspects of the process, will address the amount of work that is required to remove a given amount of heat from the low temperature reservoir, that is the amount of work required to achieve the stated objective. In fact, the ratio of the amount of heat removed from the low temperature reservoir to the amount of work required is defined as the *coefficient of performance* for a refrigeration cycle. The higher this ratio the less the amount of work required in order to carry out the heat removal.

As was true for power cycles, refrigeration cycles can be carried out using gases. However, the performance of these cycles is generally not very good and therefore they find only a limited application. Therefore we will not discuss further such cycles as the air refrigeration cycle. We will, however, discuss the vapor compression cycle, which follows a process similar to the vapor compression power cycle, and we will discuss absorption refrigeration cycles, a process which does not have a parallel in power cycles. As with the power cycle discussion we will now move to a discussion of a Carnot refrigeration cycle.

17.3.1 The Carnot Refrigeration Cycle

As for the power cycles, the Carnot refrigeration cycle is the most efficient refrigeration process. This process consists again of two isothermal steps and two isentropic steps:

1) A reversible adiabatic compression of the working fluid from a low temperature to a high temperature.

2) An isothermal rejection of heat to a high temperature reservoir from the working fluid which is at the same temperature as this reservoir.

3) An adiabatic reversible expansion of the working fluid from the high temperature to the low temperature.

4) An isothermal absorption of heat from the low temperature reservoir to the working fluid at the same temperature of this reservoir.

This process is depicted in Figure 17-18 on a temperature entropy diagram. Again as with the power cycle, the heat transferred during each of the isothermal steps is the area under that path and the resulting work of the entire process is the area inside the process rectangle. The coefficient of performance for this process is given by:

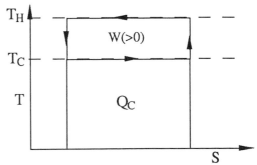

Figure 17-18: Carnot refrigeration cycle

$$
\boxed{
\begin{aligned}
\text{COP} &= \left(\frac{Q_C}{W}\right) \\[2mm]
&= \left(\frac{T_C}{T_H - T_C}\right)
\end{aligned}
}
\qquad (17-28)
$$

Because of the rectangular geometry of each of the areas we can say that Q_C is proportional to T_C and W is proportional to $(T_H - T_C)$ which results in the rightmost equality of the previous equation. As was true for the power cycle, the coefficient of performance for a Carnot refrigerator can be calculated entirely in terms of the hot and cold reservoir temperatures. In this case notice that the coefficient of performance increases as T_C increases and T_H decreases.

17.3.2 Vapor Compression Refrigeration

The ideal (mechanically reversible) vapor compression refrigeration cycle takes place largely within the two-phase envelope of the working fluid. Starting with a saturated vapor at a low temperature it consists of

1) An adiabatic reversible compression of the vapor to a superheated vapor.

2) Condensation of the superheated vapor to saturated liquid by removal of heat at the high temperature reservoir.

3) The adiabatic reversible expansion of the working fluid to the temperature of the low-temperature reservoir. This places it inside the two-phase envelope as, primarily, liquid.

4) The isothermal vaporization of the working fluid by absorbing heat from the low temperature reservoir.

The process is shown in Figure 17-19. Note that the paths of this process look very similar to the power cycle but the direction in which the cycle is traversed is reversed. Also note that the high-temperature and low-temperature reservoir temperatures lie within the bounds of the high and low temperatures of the process, that is, the high temperature reservoir is lower than the high temperature side of the process so as to be able to remove heat on this high temperature side and to condense the fluid. Likewise, the low temperature reservoir temperature is greater than the low temperature side of the working fluid so as to be able to vaporize the fluid in the evaporator.

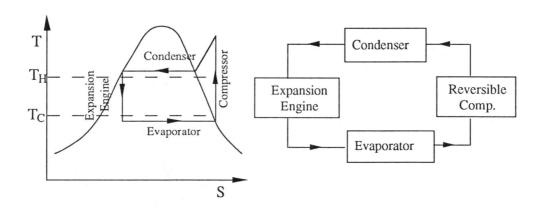

Figure 17-19: Ideal vapor compression refrigeration

This process just described is an ideal process with isentropic compression and expansion. A non-ideal process is shown in Figure 17-20 where the entropy increases in both the compression and expansion portions of the process. In the real (non-ideal) process an expansion valve, a very small opening restricting flow, is used for the gas expansion as a very simple and therefore inexpensive piece of equipment.

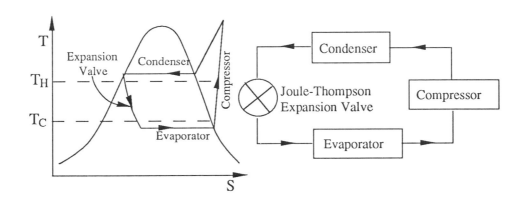

Figure 17-20: Actual vapor compression refrigeration cycle

In either the ideal process or the non-ideal process (actual process) the condenser and evaporator (refrigerator) are essentially constant-pressure devices. Therefore the heat transferred in each of these parts of the process again is equal to the area under the T-S curve representing that part of the cycle. For the mechanically reversible cycle the area inside the process loop is the work required to achieve the given refrigeration.

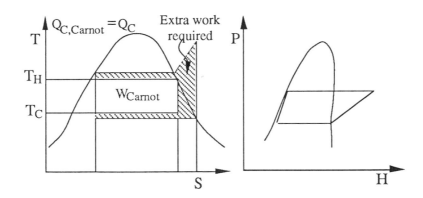

Figure 17-21: Ideal vapor compression refrigeration and Carnot cycle

This work is compared to that required for an equivalent Carnot cycle (equivalent in terms of delivering the same amount of cooling) in Figure 17-21. Again, for the Carnot cycle we begin by constructing an area which contains the same amount of heat transfer in the refrigerator. Now because T_C is actually greater than the low temperature of the process we will achieve this same area with a rectangle which is narrower than that for the real process. Then once the width of the rectangle is established by Q_C and T_C, the value of T_H establishes Q_H for this Carnot cycle and therefore the work required to extract Q_C. From the figure it is evident that the work of the actual cycle exceeds that of the Carnot cycle. In a process which is not mechanically reversible (Figure 17-22) we see a similar comparison. Again the width of the Carnot cycle rectangle which gives the same Q_C is less than that for the actual process and mechanical irreversibilities further reduce the coefficient of performance for a refrigeration cycle.

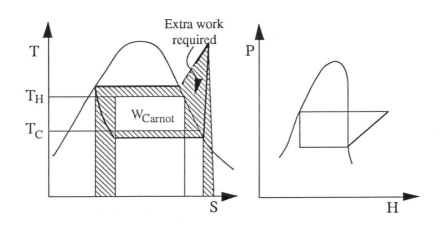

Figure 17-22: Actual vapor compression refrigeration and Carnot cycle

The expansion of the gas through the expansion valve takes advantage of the Joule-Thompson effect for a gas. That is, as it expands through this valve to a lower pressure, it undergoes cooling. This change in temperature with pressure at constant enthalpy (the first law says that this process with zero work and zero heat transfer at steady state and with no kinetic energy or potential energy changes is a constant-enthalpy process) is defined to be the Joule-Thompson coefficient (μ_J)

$$\mu_J = \left(\frac{\partial T}{\partial P}\right)_H \qquad\qquad (17-29)$$

If this coefficient is positive, then the gas cools upon expansion, i.e. it cools in going to a lower pressure. The cooling of the refrigerant upon expansion through this Joule-Thompson valve is obviously crucial to the operation of the cycle.

Example 17-6. A refrigeration cycle that uses ammonia as the working fluid is proposed. The high pressure side operates at 140 psi, and the low pressure side operates at 30 psi. In the expansion, the system contains an isentropic turbine as the expansion engine. The compressor also operates isentropically. Figure 17-19 shows a schematic of the process. Calculate the coefficient of performance (COP) for the cycle.

SOLUTION: Table C-24 contains thermal properties of saturated and superheated ammonia. A partial listing is given below.

State 1: Saturated Vapor $T = -0.6°F$ $P = 30$ psi	$\widehat{V}_1 = 9.236$ ft^3 / lb$_m$ $\widehat{H}_1 = 611.6$ Btu / lb$_m$ $\widehat{S}_1 = 1.3364$ Btu / lb$_m$ R	State 2: Superheated Vapor $T = ?$ °F $P = 140$ psi	$\widehat{V}_2 = ?$ ft^3 / lb$_m$ $\widehat{H}_2 = ?$ Btu / lb$_m$ $\widehat{S}_2 = 1.3364$ Btu / lb$_m$ R
State 3: Saturated Liquid $T = 74.8°F$ $P = 140$ psi	$\widehat{V}_3 = 0.02649$ ft^3 / lb$_m$ $\widehat{H}_3 = 126.0$ Btu / lb$_m$ $\widehat{S}_3 = 0.2638$ Btu / lb$_m$ R	State 4: Liquid/Vapor mixture $T = -0.6°F$ $P = 30$ psi	$\widehat{V}_4 = ?$ ft^3 / lb$_m$ $\widehat{H}_4 = ?$ Btu / lb$_m$ $\widehat{S}_4 = 0.2638$ Btu / lb$_m$ R

The first thing to do is to calculate the specific enthalpy of the ammonia at state 2. The compression is assumed to be isentropic. By knowning the specific entropy and the pressure, we can determine the specific enthalpy by linear interpolation of the data in the table.

$$x = \frac{1.3364 - 1.305}{1.342 - 1.305} = 0.838$$

Now applying this to the enthalpy gives:

$$\widehat{H}_2 = 0.838(709.9 - 686.0) + 686.0 = 706 \ (\text{Btu / lb}_m)$$

The next thing to do is to calculate the enthalpy of state 4. Here, the expansion is constant entropy so we must determine the enthalpy of the vapor/liquid mixture. As was done in example 17-2, we will first calculate the fraction that is liquid and vapor from the entropy.

$$x_{vap} = \frac{\widehat{S}_3 - \widehat{S}_{4,liq}}{\widehat{S}_{4,vap} - \widehat{S}_{4,liq}}$$

Substituting the known values gives:

$$x_{vap} = \frac{0.2638 - 0.0962}{1.3364 - 0.0962} = 0.135$$

The mass fraction of vapor is 0.135 and the mass fraction of liquid is 0.865. The enthalpy of the mixture is calculated:

$$\widehat{H}_4 = x_{vap}\widehat{H}_{4,vap} + x_{liq}\widehat{H}_{4,liq}$$
$$= 0.135(611.6) + 0.865(42.3)$$
$$= 119.1 \ \text{Btu / lb}_m$$

The work of compression is calculated:

$$\widehat{W}_{comp} = \widehat{H}_2 - \widehat{H} - 1$$
$$= 706 - 611.6 = 94.4 \ \text{Btu / lb}_m$$

The work of the turbine is calculated:

$$\widehat{W}_{turp} = \widehat{H}_4 - \widehat{H}_3$$
$$= 119.1 - 126.0 = -6.9 \ \text{Btu / lb}_m$$

The heat transfer at the cold temperature is calculated:

$$\widehat{Q}_C = H_1 - H_4$$
$$= 611.6 - 119.1 = 492.5 \ \text{Btu / lb}_m$$

The COP is calculated:

$$\text{COP} = \frac{Q_C}{|W_{\text{net}}|}$$

$$= \frac{492.5}{|94.4 - 6.9|} = 5.63$$

This refrigeration cycle has a coefficient of performace of 5.63. Note that COP values are greater than unity.

Example 17-7. For the refrigeration system in Example 17-6, the turbine for the expansion has been replaced by a throttling valve. Calculate the COP for this refrigeration cycle.

SOLUTION: The calculated data is given in the table below. The new value for the heat transferred at the cold temperature is calculated.

State 1:	$\widehat{V}_1 = 9.236$ ft^3 / lb$_m$	State 2:	$\widehat{V}_2 =$? ft^3 / lb$_m$
Saturated Vapor	$\widehat{H}_1 = 611.6$ Btu / lb$_m$	Superheated Vapor	$\widehat{H}_2 = 706$ Btu / lb$_m$
$T = -0.6°$F	$\widehat{S}_1 = 1.3364$ Btu / lb$_m$ R	$T =$? °F	$\widehat{S}_2 = 1.3364$ Btu / lb$_m$ R
$P = 30$ psi		$P = 140$ psi	
State 3:	$\widehat{V}_3 = 0.02649$ ft^3 / lb$_m$	State 4:	$\widehat{V}_4 =$? ft^3 / lb$_m$
Saturated Liquid	$\widehat{H}_3 = 126.0$ Btu / lb$_m$	Liquid/Vapor mixture	$\widehat{H}_4 = 126.0$ Btu / lb$_m$
$T = 74.8°$F	$\widehat{S}_3 = 0.2638$ Btu / lb$_m$ R	$T =$-0.6°F	$\widehat{S}_4 =$? Btu / lb$_m$ R
$P = 140$ psi		$P = 30$ psi	

$$\widehat{Q}_C = H_1 - H_4$$

$$= 611.6 - 126.0 = 485.6 \text{ Btu / lb}_m$$

The only work done on the fluid is during the compression step. The turbine has been replaced by an expansion valve and hence $\widehat{H}_4 = \widehat{H}_3$. The COP is:

$$\text{COP} = \frac{Q_C}{|W_{\text{net}}|}$$

$$= \frac{485.6}{94.4} = 5.14$$

This refrigeration cycle has a coefficient of performace of 5.14.

17.3.3 Heat Pump

A heat pump is actually a refrigeration cycle but operated with a different objective. Here the purpose is to provide heat to the high temperature reservoir by extracting it from the low temperature reservoir. Again work is required to do this and the process looks exactly the same as the refrigeration cycle but the objective addresses the high temperature heat transfer rather than low temperature heat transfer. Obviously the analysis and the description of the cycle are identical. It's as though if we were using a heat pump to heat our house during the winter, we are treating the outdoors as a refrigerator and making it colder by performing work and then rejecting the heat into our house to make it warmer. During the summer the heat pump can be reversed so that now the house becomes the refrigerator and the outdoors the high temperature reservoir. In either case the cycle path and direction looks the same on the T-S diagram.

17.3.4 Absorption Refrigeration

In an absorption refrigeration cycle the vapor compression part of the cycle is replaced by a pump which pumps a liquid to a high pressure. In order to achieve the compression of a vapor from a low pressure to a high pressure by this pumping of a liquid we must first absorb the vapor at low pressure into the liquid and then desorb it at high pressure to produce the high pressure gas.

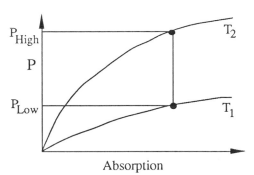

This absorption and desorption is accomplished by means of heat transfer and is best explained in terms of the vapor liquid equilibrium for the refrigerant/absorber binary system. Such an equilibrium is shown in Figure 17-23. On the low temperature side if the absorber is maintained at a low enough temperature then it can absorb an appreciable amount of refrigerant at the low pressure of the refrigerator side.

Figure 17-23: Absorption refrigeration

Then the absorber containing the refrigerant is pumped to a higher pressure where heat is now added to the binary mixture thereby shifting the equilibrium in favor of a higher pressure, even when a smaller amount of a refrigerant is absorbed, thereby driving off the refrigerant from the absorbent. Consequently, the absorber at the low temperature side is cooled and that at the high temperature side is heated. An external heat source such as a gas flame is used to heat the absorber on the high temperature side. Thus we have the seemingly paradoxical situation of achieving refrigeration by using a hot gas flame.

If a third component is added to the absorber (hydrogen, e.g.) then use of the pump can be eliminated. In this case the total pressure of the system is the same in the absorber and the regenerator. The partial pressures of the refrigerant and the hydrogen, however, are different because of the different temperatures. In this way the refrigerant is still forced to cycle around the loop and refrigeration is achieved. Now we really seem to have a paradox in that the only source of energy for the process is a gas flame and that is used to bring about cooling!

17.4 REVIEW

1) Cyclic processes which take a working fluid through a series of expansion, cooling, compression and heating cycles can be used to produce power from heat and to use work to remove heat at a low temperature in a refrigeration process.

2) A Carnot cycle which consists of isothermal heating and cooling steps and isentropic expansion and compression steps is the most efficient cyclic power cycle and refrigeration cycle. As such it serves as the benchmark to which to compare the real processes.

3) The vapor compression power cycle is the most common actual power cycle and is called the Rankine cycle. Power cycle are evaluated by their efficiency which is the ratio of the amount of work produced to the amount of heat used to produce the work. Thermal efficiencies are less than unity.

4) Power cycles involving a gas (the working fluid exists as a gas throughout the power cycle) exist but are not in common use. A gas power cycle which has some inherent advantages and may see more use in the future is the Stirling cycle.

5) A considerable cause of reduced efficiency (relative to the corresponding Carnot cycle) in the Rankine process is the fact that the process high and low temperatures are significantly different from reservoir temperatures. Efforts to improve the Rankine cycle have focused on reducing this difference between the process and reservoir temperatures.

6) The vapor compression refrigeration cycle is the most commonly used refrigeration process. Refrigeration processes are evaluated according to their coefficient of performance which is the ratio of the amount of refrigeration to the amount of work required to achieve that refrigeration. Coefficients of performance are geater than unity.

17.5 NOTATION

The following notation was used in this chapter.

Scalar Variables and Descriptions		Dimensions
C_V	constant volume heat capacity	[energy / temperature]
C_P	constant pressure heat capacity	[energy / temperature]
COP	coefficient of performance	
H	enthalpy	[energy]
m	total mass	[mass]
$\dot{m}$	total mass flow rate	[mass / time]
P	pressure	[force / length^2]
PE	potential energy	[energy]
Q	heat	[energy]
$\dot{Q}$	heat flow rate	[energy / time]
r	compression ratio	
S	entropy	[energy / temperature]
t	time	[time]
T	temperature	[temperature]
U	internal energy	[energy]
V	volume	[length^3]
W	work	[energy]
$\dot{W}$	rate of work or power	[energy / time]
γ	ratio of C_V and C_P	
η	efficiency	
μ_J	Joule-Thompson coefficient	[temperature / pressure]

Subscripts	
C	designating the cold reservoir
c	designating the compression
e	designating the expansion

Car	Carnot
H	designating the hot reservoir
in	input or entering
out	output or leaving
rev	designating a to revesible process
shaft	work associated with rotating equipment
sur	corresponding to the surrounding
sys	corresponding to the system
th	thermal

Superscripts

property per mass

17.6 SUGGESTED READING

Abbott, M. M., and H. C. Van Ness, *Theory and Problems of Thermodynamics*, 2nd edition, McGraw-Hill, New York, 1989

Çengel, Y. A., and M. A. Boles, *Thermodynamics, An Engineering Approach*, McGraw-Hill New York 1989

Hougen, O. A., K. M. Watson, and R. A. Ragatz, *Chemical Process Principles, Part II Thermodynamics*, 2nd edition, John Wiley & Sons, New York, 1959

Sandler, S. I., *Chemical and Engineering Thermodynamics*, John Wiley & Sons New York 1989

Smith, J. M., and H. C. Van Ness, *Introduction to Chemical Engineering Thermodynamics*, 4th edition, McGraw-Hill, New York, 1987

Van Wylen, G. J., and R. E. Sonntag, *Fundamentals of Classical Thermodynamics*, 3rd edition, John Wiley & Sons, New York, 1986

QUESTIONS

1) What is a power cycle? What sequence of steps might a reversible power cycle follow to be an example Carnot cycle?

2) What is the thermal efficiency of a power cycle? How is it defined and how is it calculated for a Carnot cycle?

3) Why is a gas a difficult working fluid for a power cycle?

4) What is the entropy change for a closed system which is undergoing a reversible adiabatic process? How would you answer this question if the system is not closed?

5) From a practical viewpoint, how might the heat transferred across the boiler (or across the condenser) be calculated for a power cycle? Also, how would the pump and turbine work be calculated?

6) Define what is meant by the mechanical efficiency of a pump in a power cycle. Also define what is meant by the mechanical efficiency of the turbine. Explain the differences between these two definitions of efficiency.

7) Describe the Rankine cycle in terms of the steps which occur throughout the cycle and also in terms of a path followed on a temperature-entropy diagram. On your T-S diagram, show the phase envelope for the two-phase region.

8) For what kind of process or processes does the area under a path on a temperature-entropy diagram represent heat transfer?

9) Give an example of how the work of a power cycle can be depicted on a temperature-entropy diagram for:

    a) a Carnot cycle

    b) a mechanically reversible Rankine cycle (but with heat transfer irreversibilities due to differences between the high and low temperatures and the hot and cold reservoirs) and

    c) a Rankine cycle with reversibilities in the pump and turbine.

10) List two ways in which the standard vapor compression (Rankine) cycle can be adjusted or changed in its operation so as to improve its efficiency.

11) Sketch the automobile power cycle on a pressure volume diagram and label the sequence of steps which occurs throughout a complete cycle with respect to the operation of the engine. Sketch the air standard Otto cycle on a P-V diagram and label each step with respect to the process that occurs. In what ways is this air standard cycle similar to the automobile cycle and in what ways is it different?

12) Sketch the air standard Otto cycle on a temperature-entropy diagram and identify each step with respect to those indicated in the diagrams of the previous question.

13) Sketch the diesel cycle on a pressure-volume diagram and explain why the picture is different from the air standard Otto cycle.

14) What is the objective of a refrigeration cycle?

15) The coefficient of performance is a parameter which characterizes the effectiveness of a refrigeration cycle. How is this parameter defined and how is it calculated for a reversible refrigeration cycle (a Carnot refrigerator)?

16) What steps might be used to make a Carnot refrigeration cycle?

17) Explain the operation of a vapor compression refrigeration cycle and compare the diagram on a temperature-entropy diagram with that for a Carnot cycle. In a practical or real refrigeration cycle, a Joule-Thompson valve is used in place of an expansion engine. How do these two pieces of equipment compare in terms of what they do? What is an expansion engine able to do that the expansion valve cannot? Compare the temperature-entropy diagram for a refrigeration cycle that uses a Joule-Thompson valve with one that uses an expansion engine.

18) What is the definition of a Joule-Thompson coefficient and why is it important to a refrigeration cycle?

19) What is a heat pump and how does it compare to a refrigeration cycle?

20) What is an example of a gas absorption refrigeration cycle?

PROBLEMS

1) A power cycle uses air to convert thermal energy to mechanical energy. The high temperature reservoir is 600 °F and the low temperature reservoir is 75 °F. Calculate the Carnot efficiency for this cycle. If the working fluid is changed from air to neon and the low and high temperature reservoirs remain at the same temperature, calculate the Carnot efficiency.

2) A power cycle uses steam to convert thermal energy to mechanical energy. The superheated steam entering the turbine is 10,000 kPa and 600 °C. The exhaust from the turbine is saturated vapor at 101.3 kPa. The liquid leaving the condenser is saturated at 101.3 kPa. The turbine and pump operate adiabatically and you may assume that the pump is reversible.

a) Calculate the $\widehat{W}_{pump}$ to pump the fluid from 101.3 kPa to 10,000 kPa.

b) Calculate the $\widehat{Q}_H$ for the boiler operation.

c) Calculate the $\widehat{W}_{turb}$ for the expansion from 10,000 kPa to 101.3 kPa

d) Calculate the $\widehat{Q}_C$ for the condenser.

e) Calculate the thermal efficiency for the power cycle.

3) A power cycle using steam at 600 °C and 10,000 kPa converts thermal energy to mechanical energy. The exhaust pressure from the turbine is 101.3 kPa. The condenser produces saturated liquid at 101.3 kPa. The turbine operates abiatically and reversibly (isentropic). The pump is adiabatic and reversible.

a) Calculate the $\widehat{W}_{pump}$ to pump the fluid from 101.3 kPa to 10,000 kPa.

b) Calculate the $\widehat{Q}_H$ for the boiler operation.

c) Calculate the $\widehat{W}_{turb}$ for the expansion from 10,000 kPa to 101.3 kPa

d) Calculate the $\widehat{Q}_C$ for the condenser.

e) Calculate the thermal efficiency for the power cycle.

4) Steam at 10,000 kPa and 600 °C is expanded through a turbine to an exhaust pressure of 101.3 kPa. The efficiency of the turbine in 0.9. The condenser produces saturated liquid at 101.3 kPa. The efficiency on the pump is 0.85. Both the pump and the turbine are adiabatic.

a) Calculate the $\widehat{W}_{pump}$ to pump the fluid from 101.3 kPa to 10,000 kPa.

b) Calculate the $\widehat{Q}_H$ for the boiler operation.

c) Calculate the $\widehat{W}_{turb}$ for the expansion from 10,000 kPa to 101.3 kPa

d) Calculate the $\widehat{Q}_C$ for the condenser.

e) Calculate the thermal efficiency for the power cycle.

5) Steam at 1000 psi and 1200 °F is expanded through an adiabatic reversible turbine to an exhaust pressure of 14.7 psi. The condenser produces saturated liquid. All the pumps and turbines are mechanically reversible. Calculate the thermal efficiency of the cycle. If the power requirement is 50,000 kW, determine the steam mass flow rate.

6) Steam at 800 psi and 1200 °F is expanded through an adiabatic reversible turbine to an exhaust pressure of 14.7 psi. The condenser produces saturated liquid. All the pumps and turbines are mechanically reversible. Calculate the thermal efficiency of the cycle. If the power requirement is 50,000 kW, determine the steam mass flow rate.

7) Steam at 600 psi and 1200 °F is expanded through an adiabatic reversible turbine to an exhaust pressure of 14.7 psi. The condenser produces saturated liquid. All the pumps and turbines are mechanically reversible. Calculate the thermal efficiency of the cycle. If the power requirement is 50,000 kW, determine the steam mass flow rate.

8) Steam at 400 psi and 1200 °F is expanded through an adiabatic reversible turbine to an exhaust pressure of 14.7 psi. The condenser produces saturated liquid. All the pumps and turbines are mechanically reversible. Calculate the thermal efficiency of the cycle. If the power requirement is 50,000 kW, determine the steam mass flow rate.

9) In problems 5, 6, 7, and 8 the only condition that changed was the operating pressure of the steam in the boiler. Make a graph of the thermal efficiency of the cycle versus the high temperature steam pressure. Also, make a graph of the steam mass flowrate versus high temperature steam pressure.

10) Helium, a monotonic gas, is used in an Otto cycle. The molar heat capacities of the monotonic gas are, $\widetilde{C}_P = 5R / 2$, and $\widetilde{C}_P = 3R / 2$. The range of the volumes of the cycle is 30 cm³ and 10 cm³. Calculate the thermal efficiency for this cycle.

11) Nitrogen, a diatomic molecule, is used in a Otto cycle. The molar heat capacities of the diatomic gas are, $\widetilde{C}_P = 7R / 2$, and $\widetilde{C}_V = 5R / 2$. The range of the volumes of the cycle is 30 cm^3 and 10 cm^3. Calculate the thermal efficiency for this cycle.

12) Helium, a monontonic gas is used in a Diesel cycle. The compression ratios for the cycle are given $r_e = 3$, and $r_c = 6$. The molar heat capacities of the gas are $\widetilde{C}_P = 5R / 2$, and $\widetilde{C}_V = 3R / 2$. Calculate the thermal efficiency of the cycle.

13) Nitrogen, a diatomic gas is used in a Diesel cycle. The compression ratios for the cycle are given $r_e = 3$, and $r_c = 6$. The molar heat capacities of the gas are $\widetilde{C}_P = 7R / 2$, and $\widetilde{C}_V = 5R / 2$. Calculate the thermal efficiency of the cycle.

14) A Carnot refrigerator is operating between a high temperature reservoir of 200 °F and a low temerature reservoir of 0 °F. Calculate the COP for this refrigeration cycle.

15) A refrigeration cycle is using ammonia as the working fluid. The fluid entering the compressor is saturated vapor at 200 kPa. The ammonia is compressed, adiabatically and reversibly, to 800 kPa. The refrigerator uses a reversible adiabatic engine in the expansion. Calculate the coefficient of performance. If the heat from the cold temperature reservoir is removed at 3 kJ / s, determine the mass flow rate of ammonia and the power required to run the compressor. The discharge from the condenser is saturated liquid.

16) A refrigeration cycle is using ammonia as the working fluid. The fluid entering the compressor is saturated vapor at 200 kPa. The ammonia is compressed, adiabatically and reversibly, to 800 kPa. The expansion is through a throttling valve. Calculate the coefficient of performance. If the heat from the cold temperature reservoir is removed at 3 kJ / s, determine the mass flow rate of ammonia and the power required to run the compressor. The discharge from the condenser is saturated liquid.

17) Consider a three-step cycle involving one mole of an ideal gas as follows:

   a) a reversible, isothermal compression from 300 K, 1 bar to 2 bar.
   b) a reversible, isobaric expansion
   c) a reversible, isochoric compression to the initial state (300K, 1 bar)

Sketch this cycle on a P versus V diagram.

For each step of the cycle, determine the changes in: temperature, pressure, volume, internal energy, enthalpy, entropy, and the heat and work.

18) You are a young, enlightened engineer in the year 1879 and have been assigned the task of designing a steam locomotive for pulling a train from Durango (elev. 6520 ft) to Silverton (elev. 9300 ft), Colorado, a distance of 46 miles. The engine and tender, fully loaded, are 286,600 lb$_m$ and 10 cars at 15,000 lb$_m$ each total 150,000 for a total train mass of 436,600 lb$_m$. Coal with a heating value of 11,920 BTU/lbm will be available to power the engine using a steam piston engine. Water, available along the way at storage tanks filled from local streams, is to be used in a "cycle" which takes the water from ambient conditions (45 °F, 1 atm) through a boiler (where it reaches entering engine conditions of 190 psia, 380 °F), through the engine (after which it is saturated vapor at 300 °F), and finally to the surroundings. This is not a true cycle in that the water is not actually returned to the beginning state and reused; fresh water must be supplied in order to "complete" the cycle. (This is also true of our internal combustion engines which use air as the working fluid.) The temperature of the combustion gases in the boiler is 3800 °F and the surroundings temperature on a typical summer day is 75 °F. Your initial effort will be a "back of the envelope" calculation.

   a) Estimate what you think is a reasonable overall efficiency for such an engine. In making your estimate, you may want to consider: (i) the Carnot efficiency for operating between the combustion gas temperature and the surroundings temperature, (ii) the Carnot efficiency based upon the working fluid temperatures entering and exiting the pistons, and (iii) any reasonable adjustments for mechanical inefficiencies.

   b) Based on your answer to (a), estimate the amount of coal necessary for a one-way trip from Durango to Silverton.

c) How much water will you need to use as the working fluid for a one-way trip and how many times do you estimate the train will need to stop for water? State any assumptions.

19) In parts (a) and (b), two different gases are put through the same series of changes in state in a steady state flow process. The initial state is 20 °F and 14.7 (psia). The four steps are:

i) A reversible adiabatic compression to 100 (psia) followed by:
ii) A reversible constant pressure temperature change of the material to 240 °F followed by:
iii) A reversible isothermal expansion to 14.7 (psia) followed by:
iv) A reversible constant pressure temperature change to 20 °F.

In parts (a) and (b), calculate the heat transfer per unit mass and the work per unit mass for the entire four-step process above for the gas indicated. Also, for each change in state calculate, ΔT, ΔP, ΔV and ΔU, ΔH, ΔS.

a) One pound mass per second of nitrogen gas with constant pressure and constant volume heat capacities:

$$\widehat{C}_P = \frac{7}{2}\frac{R}{(MW)}$$

$$\widehat{C}_V = \frac{5}{2}\frac{R}{(MW)}$$

This gas behaves as an ideal gas.

b) One pound mass per second of ammonia (non-ideal) gas. The thermal data for ammonia are located in Appendix C.

20) A heat pump for home heating is a refrigeration process operated so as to refrigerate the outdoors and heat the house. Show how you might design a system that would use, in a single machine, the same heat exchangers, compressor, and expansion valve for both air conditioning during the summer and heating during the winter, i.e., sketch a process schematic showing the necessary plumbing, valves, etc. Be sure to include the path followed by the refrigerant for each mode of operation. Also, sketch a $T - S$ diagram for each mode of operation. You may assume an adiabatic, reversible compressor.

Also, to compare air conditioning to heat pump operation, calculate the Carnot coefficients of performance for typical ambient (outdoors) and house temperatures in the southern U.S. climates and make a comparison for northern U.S. climates and summarize your four COP number, along with you assumed temperatures. What conclusions might you draw about the effects of the ambient temperature on the practicality of using a heat pump. (Hint: Consider a T-S diagram and the effect of changing T_C on the COP.)

VII

PERSPECTIVES AGAIN

The purpose of this book has not been, (believe it or not) primarily, to teach you rigid body mechanics, or fluid mechanics, or electrical circuits, or thermodynamics. There are more complete references for doing that. The purpose instead has been to focus on a single theme or concept and to convey to you the importance of this concept to engineering and science. In this sense, all engineers do the same thing and in the same way, we count things.

The conservation and accounting equations which we have discussed throughout this course are summarized in Table 18-1. At this point you should be viewing each equation as far more than a mathematical statement relating abstract symbols. Rather, you should be viewing each one automatically in the context of this counting process. A physical understanding of these equations allows you to approach a problem and its solution from a different viewpoint than otherwise would be possible and allows you to apply your physical understanding and intuition of the situation to your solution approach.

This is not the end of your exposure to conservation and accounting concepts. Apart from more detail and application of the material covered in this text to more advanced problems and even to peripheral areas, you will have exposure to applications of the same conservation and accounting concepts to differential-sized volumes (i.e. points in a continuum) rather than the macroscopic systems which we have considered. In fact, parallelling the approach and result which we have presented in this text, is another set of equations dealing with continua which describe exactly the same phenomena in exactly the same way but for a differential-sized system. Table 18-2 summarizes equations which correspond to those given in Table 18-1 but for a differential element. These equations look considerably

Table 18-1: Conservation (C) and Accounting (A) Equations

Extensive Property	Type	Course 1: Macroscopic
Total Mass	C*	$\left(\dfrac{dm_{sys}}{dt}\right) = \sum \dot{m}_{in/out}$
Species Mass	A	$\left(\dfrac{d(m_i)_{sys}}{dt}\right) = \sum (\dot{m}_i)_{in/out} + \sum (\dot{m}_i)_{gen/con}$
Net Charge	C	$\left(\dfrac{dq_{sys}}{dt}\right) = \sum i_{in/out}$
Linear Momentum	C	$\left(\dfrac{d\mathbf{p}_{sys}}{dt}\right) = \sum (\dot{m}\mathbf{v})_{in/out} + \sum \mathbf{f}_{ext}$
Angular Momentum	C	$\left(\dfrac{d\mathbf{l}_{sys}}{dt}\right) = \sum (\mathbf{r} \times \mathbf{v}\dot{m})_{in/out} + \sum (\mathbf{r} \times \mathbf{f})_{ext}$
Total Energy	C*	$\left(\dfrac{dE_{sys}}{dt}\right) = \sum \left(\dot{m}(\widehat{H} + \widehat{PE} + \widehat{KE}) \right)_{in/out} + \sum \dot{Q} + \sum \dot{W}_{non-flow}$
Steady State Mechanical Energy	A	$0 = \dot{m}(\widehat{PE} + \widehat{KE})_{in} - \dot{m}(\widehat{PE} + \widehat{KE})_{out} + \dot{m}\displaystyle\int_{P_{out}}^{P_{in}} \dfrac{dP}{\rho} + \sum \dot{W}_{shaft} - \sum \dot{F}$
Steady State Thermal Energy	A	$0 = \dot{m}(\widehat{H}_{in} - \widehat{H}_{out}) - \dot{m}\displaystyle\int_{P_{out}}^{P_{in}} \dfrac{dP}{\rho} + \sum \dot{Q} + \sum \dot{F}$
Entropy	A	$\dfrac{d}{dt}\left(S_{sys} + S_{sur} \right)_{in/out} \geq 0$
Entropy-System	A	$\left(\dfrac{dS_{sys}}{dt}\right) = \sum (\dot{m}\widehat{S})_{in/out} + \sum \left(\dfrac{\dot{Q}}{T}\right)_{sys} + \sum (\dot{S}_{gen(non-Q)})_{sys}$
Entropy-Surrounding	A	$\left(\dfrac{dS_{sur}}{dt}\right) = -\sum (\dot{m}\widehat{S})_{in/out} - \sum \left(\dfrac{\dot{Q}}{T}\right)_{sur} + \sum (\dot{S}_{gen(non-Q)})_{sur}$

* Neglecting "conversions" between mass and energy.

more formidable because they are partial differential equations rather than ordinary differential equations and the notation is more unfamiliar to you. However, you should realize that the time derivative still represents, physically, an accumulation. Furthermore, your familiarity with the divergence theorem of mathematics should convince you that appearance of $-\nabla \cdot (\ )$ in each of these equations represents a net input or output from the differential system being considered. With these two conceptual keys, you are in a position to view these equations in exactly the same way with respect to conservation and accounting as you do those in Table 18-1. Further expansion on this theme must await a later course.

Finally, this effort to present a unified picture and focus on conservation and accounting for all engineers is motivated by more than an academic perspective. We live in a world where, more and more, the boundaries between disciplines are becoming fuzzy and where projects are big enough to be approached from a multidisciplinary viewpoint, by teams of individuals having a wide spectrum of education and experiences. These facts motivate a need for a common understanding and perspective amongst the various disciplines and the various phenomena with which we deal. It is the tenet of this course that each of us in our own discipline can best understand what those in other disciplines are doing and how their work interrelates with our efforts by employing a conservation viewpoint. Conservation and accounting are the basis for the structure of engineering.

Table 18-2: Continuum Conservation (C) and Accounting (A) Equations

Extensive Property	Type	Course 4: Continuum
Total Mass	C*	$\left(\dfrac{\partial \rho}{\partial t}\right) = -\nabla \cdot (\rho \mathbf{v})$
Species Mass	A	$\left(\dfrac{\partial \rho_i}{\partial t}\right) = -\nabla \cdot (\rho \mathbf{v})_i + r_i$
Charge	C	$\left(\dfrac{\partial \rho}{\partial t}\right) = -\nabla \cdot \mathbf{J}$
Linear Momentum	C	$\left(\dfrac{\partial (\rho \mathbf{v})}{\partial t}\right) = -\nabla \cdot (\rho \mathbf{v}\mathbf{v}) - \nabla P + \nabla \cdot \mathbf{S} + \rho \mathbf{g}$
Angular Momentum	C	$\mathbf{T} = \mathbf{T}^T$
Total Energy	C*	$\left(\dfrac{\partial (\rho \widehat{E})}{\partial t}\right) = -\nabla \cdot (\rho \widehat{E} \mathbf{v}) - \nabla \cdot \mathbf{q} - \nabla \cdot (P\mathbf{v}) + \nabla \cdot (\mathbf{S} \cdot \mathbf{v}) + \rho (\mathbf{v} \cdot \mathbf{g})$
Mechanical Energy	A	$\left(\dfrac{\partial (\rho \widehat{K})}{\partial t}\right) = -\nabla \cdot (\rho \widehat{K} \mathbf{v}) - \mathbf{v} \cdot \nabla P + \mathbf{v} \cdot (\nabla \cdot \mathbf{S}) + \rho (\mathbf{v} \cdot \mathbf{g})$
Thermal Energy	A	$\left(\dfrac{\partial (\rho \widehat{U})}{\partial t}\right) = -\nabla \cdot (\rho \widehat{U} \mathbf{v}) - \nabla \cdot \mathbf{q} - P(\nabla \cdot \mathbf{v}) + \mathrm{tr}(\mathbf{S} \cdot \nabla \mathbf{v})$
Entropy	A	$\left(\dfrac{\partial (\rho \widehat{S})}{\partial t}\right) \geq -\nabla \cdot (\rho \widehat{S} \mathbf{v}) - \nabla \cdot \left(\dfrac{\mathbf{q}}{T}\right)$ $-\mathbf{q} \cdot T \geq 0; \ \mathrm{tr}(\mathbf{S} \cdot \nabla \mathbf{v}) \geq 0$

* Neglecting "conversions" between mass and energy.

$\widehat{K} = \widehat{PE} + \widehat{KE}$

A.1 INTRODUCTION

In addition to the concepts of a system, extensive property, and conservation principles, the engineer needs other tools to adequately understand and solve problems. The conservation and accounting concept produce the mathematical equations that the engineer must solve. However, unlike purely mathematical equations where we are solving for unknown numbers, the equations generated from the conservation or accounting concepts represent physical quantities such as mass, forces, energy, length, and electrical current. That is, the variables in the equations are measured in units of some quantity or another.

A dimension is a fundamental property that can be measured such as time, mass, length, temperature, or charge. In order to make the measurement, a unit of measurement must be defined. For example, length can be measured in units of feet, meters, yards, microns, etc. The numerical value represents the number of the chosen units required for the measurement. The value and the units are both required to provide a meaningful measure. Some examples are listed below.

$$5 \text{ miles} \qquad 3.2 \text{ lb}_m \qquad 0.01 \text{ s} \qquad 32\ ^oF$$

Physical properties are also calculated by multiplying or dividing different dimensions such as volume (length cubed), current (charge divided by time), and density (mass divided by volume). Examples of these dimensions are given.

$$30 \text{ cm}^3 \qquad 32.2 \text{ ft} / \text{s}^2 \qquad 1 \text{ g} / \text{mL} \qquad 25.5 \text{ N} / \text{m}^2$$

When variables that have specific units attached to them appear in equations, the quantities can be added and subtracted only if the units or dimensions are the same. This was the exact same principle you learned in algebra.

$$9 \text{ grams} + 1 \text{ gram} = 10 \text{ grams}$$
$$9x + 1x = 10x$$

but

$$5 \text{ s} - 3 \text{ ft} =? \quad \text{(meaningless)}$$
$$5y - 3z =? \quad \text{(meaningless)}$$

However, multiplication and division of units in equations is always allowed provided that certain axioms of algebra such as division by zero are not violated.

$$32 \text{ ft} / \text{s}^2 \times 2 \text{ lb}_\text{m} = 64 \text{ lb}_\text{m} \text{ ft} / \text{s}^2$$
$$20 \text{ miles} \times 2 \text{ hr}^{-1} = 40 \text{ miles} / \text{hr}$$
$$1.5 \text{ in} \times 2 \text{ in} = 3 \text{ in}^2$$
$$5 \text{ C} \times 0.5 \text{ s}^{-1} = 2.5 \text{ C} / \text{s}$$

Also, some properties combine in such a way that the units cancel out, giving dimensionless numbers. Furthermore, these dimensionless numbers are very important in engineering fields and are used extensively to perform many calculations. Some simple examples are shown.

$$3 \text{ s} \times \left(\frac{1}{10 \text{ s}} \right) = \frac{3}{10} \quad \text{(dimensionless)}$$
$$\left(\frac{9 \text{ lb}_\text{m}}{\text{ft}^3} \right) \times \left(\frac{0.001 \text{ ft}^3}{\text{lb}_\text{m}} \right) = 0.009 \quad \text{(dimensionless)}$$

Units and dimensions are some of the most important concepts you should understand. Units and dimensions allow engineers and scientists to quantify physical properties. These dimensions represent different magnitudes. As you know, there is a big difference between 2 pounds mass and 2 tons. Therefore, in all the equations that are developed, including all the units will help you better understand what you are trying to solve for.

As mentioned, quantities that can be measured can be expressed in several different ways. In freshmen Chemistry you used terms such as grams to quantify, mass and liters to describe volume. However, volumes of gasoline are usually measured in gallons (in the U.S.) and metals such as aluminum and steel in pounds mass or tons. A simple example is the representation of linear velocity (length/time). All the following are valid expressions for linear velocity.

$$\text{miles} / \text{hour} \quad \text{ft} / \text{s} \quad \text{cm} / \text{min} \quad \text{meter} / \text{year} \quad \text{meters} / \text{hour}$$

The magnitude of the numerical value of the linear velocity is dependent on the units of measure used.

Equivalent measures using different units of measure and dimensions can be used to form useful ratios called conversion factors. Examples of some equivalent measures are given along with their corresponding conversion factors. Remember, there are literally thousands of equivalent measures, and they can be found in most handbooks.

$$1 \text{ mile} = 5,280 \text{ ft} \qquad 5,280 \ \frac{\text{ft}}{\text{mile}}$$

$$1 \text{ min} = 60 \text{ s} \qquad 60 \ \frac{\text{s}}{\text{min}}$$

$$1 \text{ hour} = 60 \text{ min} \qquad 60 \ \frac{\text{min}}{\text{hr}}$$

$$1 \text{ lb}_m = 454 \text{ g} \qquad 454 \ \frac{\text{g}}{\text{lb}_m}$$

$$1 \text{ ft}^3 = 7.48 \text{ gal} \qquad 7.48 \ \frac{\text{gal}}{\text{ft}^3}$$

$$1 \text{ mL} = 1 \text{ cm}^3 \qquad 1 \ \frac{\text{cm}^3}{\text{mL}}$$

$$1 \text{ yd} = 3 \text{ ft} \qquad 3 \ \frac{\text{ft}}{\text{yd}}$$

$$1 \text{ in} = 2.54 \text{ cm} \qquad 2.54 \ \frac{\text{cm}}{\text{in}}$$

Conversion factors provide a way that we can convert from one set of units to another. Conversion of a quantity expressed in one set of units to another set of units involves multiplying or dividing by a conversion factor. Therefore, it is important to carry all the units when solving problems so you know what all of the variables represent. An example is to determine how many feet are equivalent to 5 yards.

$$5 \text{ yds} \times \frac{3 \text{ ft}}{1 \text{ yd}} = 15 \text{ ft}$$

Notice how the units cancel when the conversion factor is used properly to give the answer as 15 feet. If we had made a mistake and did the conversion:

$$5 \ (\text{yds}) \frac{1 \text{ yd}}{3 \text{ ft}} = \left(\frac{5 \text{ yd}^2}{3 \text{ ft}} \right)$$

the units clearly tell us that an error has been made.

Example A-1. Determine the linear velocity of 65 mph in terms of cm/s.

SOLUTION: The conversion factors needed to solve this problem are listed in the following equation.

$$\frac{65 \text{ miles}}{\text{hr}} \left| \frac{1 \text{ hr}}{60 \text{ min}} \right| \frac{1 \text{ min}}{60 \text{ s}} \left| \frac{5280 \text{ ft}}{1 \text{ mile}} \right| \frac{12 \text{ in}}{1 \text{ ft}} \left| \frac{2.54 \text{ cm}}{1 \text{ in}} \right| = 2,906 \text{ cm} / \text{s}$$

The solution is 2,906 cm/s is equivalent to 65 mph.

As an engineer your will be exposed to a variety of different units and systems of units. For example, the scientist or researcher may give you data in milliliters (volume), Atmospheres (pressure), and Molarity (concentration). However, the production line understands gallons (volume), pounds per square inch (pressure), and weight fraction (concentration). The engineer must be able to convert the data from one source to another in a useful and understandable form for all parties involved. In this way, the engineer acts as a communication link.

There are three (3) major systems of units: the SI or "Systeme International d'Unites", the CGS system, and the American Engineering system. The SI system of units was formulated in 1960 and has the base units of meter (m) for length, kilogram (kg) for mass, and the second (s) for time. The CGS system, used most often in the scientific community, has the base units of grams (g) for mass, centimeter (cm) for length, and second (s) for time. These systems use measurements that you know as the metric system. Because they are a part of the metric system, prefixes are used to represent powers of ten. Common prefixes include:

$$\begin{array}{lll} \text{Nano} & (\eta) = & 10^{-9} \\ \text{Micro} & (\mu) = & 10^{-6} \\ \text{Milli} & (\text{m}) = & 10^{-3} \\ \text{Centi} & (\text{c}) = & 10^{-2} \\ \text{Kilo} & (\text{k}) = & 10^{3} \\ \text{Mega} & (\text{M}) = & 10^{6} \end{array}$$

The American Engineering System base units are the foot (ft) for length, pound-mass (lb_m) for mass, and the second (s) for time. The American system is very difficult to make conversions in because, unlike the metric system, the conversions are not multiples of ten. For example,

$$1 \text{ yd} = 3 \text{ ft} = 36 \text{ in}$$

compared with

$$1 \text{ m} = 100 \text{ cm} = 1000 \text{ m}$$

A.2 FORCE AND WEIGHT

Force and weight have the same dimensions and represent the same quantity. The dimensions for force and weight are mass times length divided by time squared. In the SI system, the Newton (N) is used to describe force. The definition is as follows:

$$1 \text{ N} \equiv 1 \text{ kg m} / \text{s}^2$$

The CGS system uses the erg to describe force and weight.

$$1 \text{ erg} \equiv 1 \text{ g cm} / \text{s}^2$$

The derived unit of force in the American Engineering system is defined as that required to accelerate 1 pound mass accelerating at $32.174 \text{ ft} / \text{s}^2$ (g at mean sea level):

$$1 \text{ lb}_f \equiv 32.174 \text{ lb}_m \text{ ft} / \text{s}^2$$

A special conversion factor to convert the force from a defined unit such as lb_f to lb_m ft / s^2 is sometimes called g_c.

$$g_c \equiv 32.174 \, \frac{\text{lb}_m \text{ ft}}{\text{lb}_f \text{ s}^2}$$

g_c is simply a conversion factor used to convert the units of lb_m ft $/ \, s^2$ to lb_f. Its value is a *constant* independent of the actual value of g. We will simply use g_c in this way; when we need to convert lb_m ft $/ \, s^2$ to lb_f, we will use the factor $32.174 \, \dfrac{lb_m \, ft}{s^2 \, lb_f}$ just as we will use $1 \, \dfrac{kg \, m}{s^2 \, N}$ to convert between kg m $/ \, s^2$ and N.

Example A-2. What is the force in lb_f if 2 lb_m is accelerating at 10 ft $/ \, s^2$.

SOLUTION:

$$F = ma$$

$$= 2 \, lb_m \; 10 \; ft \, / \, s^2$$

$$= 20 \, \frac{lb_m \, ft}{s^2} \left| \frac{1 \; lb_f}{32.174 \; lb_m \; ft \, / \, s^2} \right. = 0.62 \; lb_f$$

Dimensional homogeneity means that all additive terms on both sides of the equation must have the same units for the equation to be valid.

$$s \; (cm) = s_O \; (cm) + v \; (cm/s)t \; (s)$$

is dimensionally homogeneous.

$$s \; (cm) = s_O \; (cm) + v \; (cm/s)$$

is not dimensionally homogeneous and is not valid. Furthermore, transcendental functions (sin, log, exp...), exponents (3 in y^3) and the arguments of transcendental functions (θ in $\cos(\theta)$) must be dimensionless quantities. As an example, $e^{3 \; ft}$, $\sin(2 \; lb_m)$ and $\ln(5 \; s)$ make no sense.

Even though the conversions are more difficult in the American system than the other two systems, the American system is used and understood by millions of people, and to effectively communicate, you also must learn to operate in the American system. Furthermore, the physical quantities and measurements are probably more familiar to you in the American system than the SI and CGS systems.

A list has been provided of different systems and the base units including temperature, charge, and light intensity which we have not discussed. Also, a fairly exhaustive list of equivalent units is also listed. These lists should not be committed to memory but should be used as a reference source. Hopefully, through continued use in problem solving, many of the conversion factors will become memorized.

Table A-1: Base units for the different systems

Quantity	SI	CGS	American Eng.
Length	Meter (m)	Centimeter (cm)	Foot (ft)
Mass	Kilogram (kg)	Gram (g)	Pound (lb_m)
Time	Second (s)	Second (s)	Second (s)
Temperature	Kelvin (K)	Kelvin (K)	Rankine ($^{\circ}R$)
Electric Current	Ampere (A)	Ampere (A)	Ampere (A)

Another system of units is based on force instead of mass. The FPS system gives:

Table A-2: Base units for FPS system

Length	Foot (ft)
Time	Second (s)
Force	Pound force (lb_f)
Temperature	Rankin (R)
Electric Current	Ampere (A)

The mass unit is the slug and it is the amount of mass that accelerates at 1 ft / s^2 when subjected to sea level gravitational force per mass: 1 slug = 32.174 lb_m. Note also that 1 slug accelerating at 32.174 ft / s^2 has a net force exerted on it of 32.174 lb_f.

A.3 IDEAL GAS CONSTANT VALUES

Table A-3: Values for the Gas Constant[a,b]

R		R	
	= 1.987 Btu / lbmole °R		= 1.987 cal / mol K
	= 10.73 ft^3 psi / lbmol °R		= 82.05 cm^3 atm / mol K
	= 1545 ft lb$_f$ / lbmol °R		= 8.314 × 10^3 N m / kgmol K
	= 0.730 ft^3 atm / lbmol °R		= 8.314 J / mol K
			= 8.314 × 10^{7}dyne cm / mol K
			= 8.314 m^3 Pa / mol K
			= 83.14 cm^3 bar / mol K
			= 8.314 kJ / kgmol K

[a] The SI "mol" is 1 gmole, i.e., in SI, 1 mol of water is 18 g.
[b] 1 kgmol = 1,000 mol.

A.4 GREEK ALPHABET

An essential part of engineering and science notation is the Greek alphabet. Table A-4 presents both upper and lower case letters.

Table A-4: The Greek Alphabet

Letter	Upper Case	Lower Case
Alpha	A	α
Beta	B	β
Gamma	Γ	γ
Delta	Δ	δ
Epsilon	E	ϵ
Zeta	Z	ζ
Eta	H	η
Theta	Θ	θ
Iota	I	ι
Kappa	K	κ
Lambda	Λ	λ
Mu	M	μ
Nu	N	ν
Xi	Ξ	ξ
Omicron	O	o
Pi	Π	π
Rho	R	ρ
Sigma	Σ	σ
Tau	T	τ
Upsilon	Υ	υ
Phi	Φ	ϕ
Chi	X	χ
Psi	Ψ	ψ
Omega	Ω	ω

Table A-5: Conversion factors

Quantity		Equivalent values
Mass	1 lb_m:	16 oz / lb_m
		2000 lb_m / ton
		0.4536 kg / lb_m
		0.03108 slug / lb_m
	1 kg:	1000 g / kg
		1000 kg / metric ton
		2.205 lb_m / kg
		0.06853 slug / kg
Charge	1 Coulomb (C):	6.24×10^{18} electron charges / C
	1 C/s:	1 A/(C/s)
	1 farad (f):	1 (C/V)/f
Length	1 ft:	12 in / ft
		3 ft / yd
		0.3048 m / ft
	1 m:	100 cm / m
		1000 mm / m
		10^{10} Angstroms / m
		3.2808 ft / m
Volume	1 ft^3:	7.48 gal / ft^3
		1728 in^3 / ft^3
		0.02832 m^3 / ft^3
	1 m^3:	1000 L / m^3
		10^6 cm^3 / m^3
		10^6 mL / m^3
		35.31 ft^3 / m^3
		264.17gal / m^3
Force	1 lb_f:	32.174 lb_m ft / s^2 / lb_f
		4.448 N / lb_f
	1 N:	1 kg m / s^2 / N
		10^5 dynes / N
		10^5cm g / s^2 / N
		0.225 lb_f / N

Pressure	1 atm:	14.696 psi / atm
		29.921 in Hg @ 32 °F / atm
		33.9 ft H_2O @ 39.2 °F / atm
		14.696 lb_f / in² / atm
	1 atm:	101.3 kPa / atm
		1.013×10^5 N / m² / atm
		1.013×10^5 Pa / atm
		760 mm Hg @ 0 °C / atm
		760 torr / atm
		1.013 bars / atm
		10.333 m H_2O @ 4 °C / atm
Temperature	°F:	°R = °F + 459.67
		1 F° = 1.8 C° (temperature difference)
	°K:	K = °C + 273.15
Energy	1 Btu:	778.2 ft lb_f / Btu
		1055 J / Btu
		252 cal / Btu
		2554.5 Btu / hp hr
		3412.2 Btu / kW hr
		5.4 psi ft³ / Btu
	1 J:	1N m / J
		10^7 ergs / J
		10^7 dyne cm / J
		9.48×10^{-4} Btu / J
		1 W s / J
		1 cal / 4.184 J
Power	1 hp:	550 ft lb_f / s / hp
		745.7 W / hp
		0.707 Btu / s / hp
	1 W:	1 J / s / W
		1 kW / 1000 W
		1.341×10^{-3} hp / W
Viscosity	1 cp:	2.419 lb_m / ft h / cp
		1 Pa s / 1000 cp
		1 lb_m / ft s / 1488.2 cp
	1 Pa s:	1 N s / m² / Pa s
		1000 cp / Pa s

B.1 INTRODUCTION

In the study of motion and forces in solids and fluids, vectors, their mathematics and their operations, are essential to problem statements and to problem solving. Scalar quantities are represented by a single number, their magnitude, whereas vectors are quantities which have directions in space, as well as a magnitude. Having a well-defined and well-ordered mathematical method for quantifying vector quantities and for calculating and representing the physical phenomena which involve vectors is essential to the problem statements and problem solutions with which we are confronted in the areas of both solid and fluid mechanics.

To describe vectors, we define a three dimensional space by using a coordinate system. The selection of the coordinate system may be done at our convenience according to the problem we are trying to solve. In each case, however, a coordinate system is described or represented by coordinate directions and by its base vectors. By defining three base vectors, each defining a different direction, all other vectors can be described as linear combinations of these base vectors. Any three non-coplanar vectors can be used as base vectors. However, almost always (always in this course) we will define *orthogonal* vectors (mutually perpendicular) that form a right-handed system.

Some other terminology is used to describe the various types of coordinate systems and base vectors. Base vectors are *cartesian* if they are of unit length, that is, if they have been normalized to unit length. If they are *both orthogonal and cartesian*, then they are *orthonormal*. Third, base vectors are *rectangular cartesians* if they are orthonormal and everywhere the same (in both magnitude and direction). In general the base vectors for a coordinate system need not be independent of position. For example, the base vectors for a cylindrical or spherical system vary with position (e.g., in a cylindrical system the radius vector changes in direction as the angular coordinate varies and while holding the z coordinate constant). Fourth, we will refer to a coordinate system as *curvilinear* if it is not rectangular cartesian.

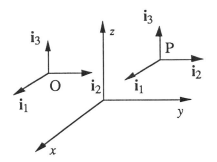

These definitions and concepts can be illustrated with the three common coordinate systems with which we ordinarily deal: rectangular cartesian, cylindrical, and spherical. Figure B-1 depicts a rectangular cartesian coordinate system. Here we have three mutually perpendicular axes, the 1, 2, and 3 directions (x, y, z directions) and no matter where we are in the coordinate system those directions remain the same.

Figure B-1: Rectangular cartesian
coordinate system

Therefore, at any point in the coordinate system, we can define a base vector as being a vector pointing in one of those three directions and of unit length. We will indicate these vectors as i_1, i_2, and i_3 (it is common usage also to indicate these for the rectangular cartesian system as i, j, k). These three base vectors then can be used as a basis for representing any vector of the vector space as a linear combination of these three. For example, a vector v we could write as:

$$v = v_1 i_1 + v_2 i_2 + v_3 i_3$$

$$= \sum_{m=1}^{3} v_m i_m$$

These vectors are shown in the figure at point O. If we move to another point in space, say point P, and we depict base vectors at that point, then for this rectangular cartesian system, by definition, they are the same as they are at point O; each vector has the same magnitude (unity) and direction as its counterpart at point O. The three vector components, v_1, v_2, and v_3 are used to form the linear combination of the base vectors i_1, i_2, and i_3 to represent v as shown in Figure B-2.

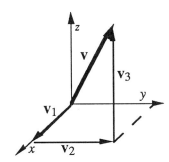

Figure B-2: Linear combination
of base vectors

This representation says nothing whatsoever about where this vector is acting and in fact the "origin" of a vector is not defined for a vector; only its magnitude and direction are defined. Consequently any vector which operates at any point or exists at any arbitrary point of our coordinate system can be represented in this way. If a vector acting at the origin is of the same magnitude and direction as another vector acting at point Q, for example, then these vectors will have exactly the same representation, in terms of the linear combination of the base vectors, for this rectangular cartesian system as presented in Figure B-3.

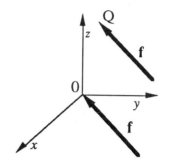

Figure B-3: Magnitude and direction

Consider now a cylindrical coordinate system shown in Figure B-4. For comparison the cylindrical directions and dimensions are superimposed upon a rectangular system, a normal procedure for representing a curvilinear coordinate system. In this case the 1 direction for the cylindrical system is the distance from the z axis of the rectangular coordinate system (measured perpendicular to the z axis), the 2 direction is an angular displacement away from the x direction in the rectangular coordinate system and toward the y direction, and the 3 direction is the same as the z dimension in the rectangular cartesian system.

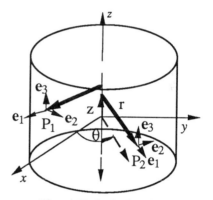

Figure B-4: Cylindrical coordinate system

A point in space then is defined in this cylindrical system by the coordinates r, θ, and z. The base vectors at this point in space are defined to be unit vectors at that point in the 1, 2, and 3 directions so that the 1 direction base vector points along the radial direction. We will call that vector e_1. The 2-direction base vector is perpendicular to the 1 and points in the 2 direction, which is tangent to a circle centered at the z axis and of radius r. e_1 and e_2 are mutually perpendicular. The 3-direction base vector is perpendicular to the plane of these first two and points in the z direction.

The base vectors are indicated in Figure B-4 at point P1. At point P2 we have a different set of base vectors. Base vector e_1 is in a direction of the radial dimension at point P2. This vector e_2 is perpendicular to that one again tangent to a circle centered at the z direction axis, but because the point P2 is at a different place from P1, this tangent direction is rotated relative to that at point P1. The third base vector e_3 is the only one that is the same as the base vector at point P1; for this coordinate system e_3 is not a function of position whereas e_1 and e_2 are. For this system we can represent the r, θ, and z coordinates as functions of the rectangular cartesian x, y, and z coordinates by the following transformations:

$$r = \sqrt{x^2 + y^2}$$

$$\theta = \tan^{-1}\left(\frac{y}{x}\right)$$

$$z = z$$

and

$$e_1 = \cos(\theta)i_1 + \sin(\theta)i_2 + (0)i_3$$
$$e_2 = -\sin(\theta)i_1 + \cos(\theta)i_2 + (0)i_3$$
$$e_3 = (0)i_1 + (0)i_2 + (1)i_3$$

A vector described at a point in space using this cylindrical curvilinear system will be represented as a linear combination of the base vectors at that point. Consequently, a different set of base vectors will be used at point P1 to describe a vector acting at point P1 from the set used at point P2 for a vector acting at point P2. This means that if a vector quantity such as a force, velocity. or acceleration is a function of position, then we must take into account not just its change in magnitude and direction but also the change of its representation due to the dependence of its base vectors on position. The magnitude and direction of two vectors acting at different points in space may be the same, but in a curvilinear coordinate system their components and base vectors will be different. This is a mathematical artifact of a curvilinear coordinate system. We will not be confronted with this complication in this course. However, it is one of which you should be aware and which you will see more of when working with spatial derivatives of vectors in Course Four of this sequence and in mathematical applications involving vectors. Spatial derivatives of vectors require recognizing the spatial dependence, not just of the vector components, but also of the base vectors.

Finally, note that the base vectors in this coordinate system at each point in space form a right handed system as they did for the rectangular cartesian system. That is, if we rotate base vector e_1 into base vector e_2 through its 90° angle with the fingers of our right hand, then our thumb is pointing in the 3-direction, the direction of base vector e_3.

Figure B-5 shows a similar figure for a spherical coordinate system. For this system the 1-direction is along a radius from the origin. The 2-direction is tangent to an arc measured down from the z rectangular coordinate system direction. The 3-direction is tangent to a circle centered at the z axis and forming a right hand system with the 1 and 2 vectors. We refer to these dimensions as r, θ, and ϕ and they are the spherical coordinates.

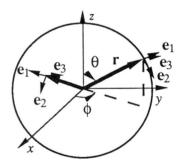

Figure B-5: Spherical
coordinate system

The base vectors e_1, e_2, and e_3 for this system are also functions of position as they were for the cylindrical system. In this case, in fact, all three base vectors vary with position. Geometrically we can relate the r, θ, and ϕ coordinates to the rectangular x, y, and z coordinates by the following transformations:

$$r = \sqrt{x^2 + y^2 + z^2}$$

$$\theta = \cos^{-1}\left(\frac{\sqrt{x^2 + y^2 + z^2}}{z}\right)$$

$$\phi = \tan^{-1}\left(\frac{y}{x}\right)$$

and

$$e_1 = \left(\sin(\theta)\cos(\phi) \right)i_1 + \left(\sin(\theta)\sin(\phi) \right)i_2 + \cos(\theta)i_3$$

$$e_2 = \left(\cos(\theta)\cos(\phi) \right)i_1 + \left(\cos(\theta)\sin(\phi) \right)i_2 - \sin(\theta)i_3$$

$$e_3 = -\sin(\phi)i_1 + \cos(\phi)i_2 + (0)i_3$$

As for the cylindrical system, base vectors for this spherical system at two different points in space will be different set of base vectors. Each set will be orthonormal and right handed but they will be rotated in space with respect to each other.

The right-handed constraint is a matter of convention and for each of these coordinate systems. We could, if we so wished, define a left-handed system. However, whatever we do we must be consistent; otherwise, we would encounter ridiculous property characteristics such as negative density. Consequently the universally accepted convention is to use right handed coordinate systems. Other coordinate systems could be defined (such as parabolic, ellipsoidal confocal, ellipsoidal, etc.). In each case there is a set of transformations for calculating that coordinate system's coordinates in terms of a rectangular cartesian coordinate system coordinates. Also, in each case base vectors are defined which are orthonormal and right handed. The utility of these coordinate systems is for special geometries where defining one coordinate system or another simplifies the problem in some way either by reducing the number of equations required to describe the problem or by simplifying the statement of boundary conditions for the problem. We will not pursue such applications any further.

B.2 OPERATIONS WITH VECTORS

B.2.1 The Dot Product of Two Vectors

The dot product of two vectors is a scalar quantity which is the product of the magnitude of one vector times the magnitude of the component of the other vector which is in the direction of the first vector. In other words, if the length of the projection of the second vector onto the first vector is multiplied by the length of the first vector then the resulting product is the dot product. This dot product is a scalar product having magnitude only with no associated direction. Note also that it doesn't matter which vector is called the first and which is the second. The projection of either one onto the other and multiplied by the length of the other one will give the same scalar product. This is depicted in Figure B-6.

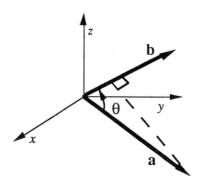

Figure B-6: Dot product of two vectors

Mathematically this scalar product is represented as:

$$\mathbf{a} \cdot \mathbf{b} = |\mathbf{a}||\mathbf{b}|\cos(\theta)$$

If one of the vectors is a base vector then the scalar product is simply the magnitude of the projection of the other vector onto that base vector or in other words, the component of the vector which is acting in the direction of the base vector. For example if mass is moving with velocity $\mathbf{v}$ through a cross-sectional area having normal vector $\mathbf{n}$, then the component of the velocity which is actually moving mass across the area is the component which is in the direction perpendicular to the area, i.e., which is in the direction of the normal vector. Consequently we can calculate the velocity at which mass is moving across the area as the dot product of the velocity vector with the area normal vector. The component of the velocity which lies within the cross-sectional area, i.e., which is tangent to the area, does nothing to convey mass across the area.

The components of a vector in a given coordinate system can be written in terms of this scalar or dot product as well. The dot product of a vector with a base vector in a given coordinate direction, by definition of the dot product, gives the magnitude of the component of the vector in that direction. In fact, this gives us an alternate representation of a vector in terms of its magnitude and the direction that it makes in space represented as an angle with respect to each of the coordinate directions as shown in Figure B-7.

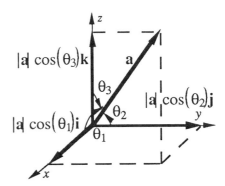

Figure B-7: Coordinate directions

Consequently, then, the component in each of the three directions can be calculated as the dot product of the vector with the corresponding base vectors.

$$\mathbf{a} = (\mathbf{a} \cdot \mathbf{i})\mathbf{i} + (\mathbf{a} \cdot \mathbf{j})\mathbf{j} + (\mathbf{a} \cdot \mathbf{k})\mathbf{k}$$

or:

$$\mathbf{a} = |\mathbf{a}|\cos(\theta_1)\mathbf{i} + |\mathbf{a}|\cos(\theta_2)\mathbf{j} + |\mathbf{a}|\cos(\theta_3)\mathbf{k}$$

Also, if the components of two vectors for a given coordinate system are known (represented in terms of the same three base vectors for a curvilinear coordinate system) then the dot product of these two vectors is easily calculated in terms of these vector components:

$$\mathbf{a} \cdot \mathbf{b} = a_1 b_1 + a_2 b_2 + a_3 b_3$$

Other calculations involving dot products can also be made easily. First, if the magnitudes of both vectors and their relative directions (the angle between them) are known, then the dot product may be calculated directly from the definition . Second, if each vector is known in terms of its components in a given coordinate system and in terms of the same base vectors (i.e., both vectors are written at the same point in space so that their base vectors are the same for a curvilinear coordinate system) then the angle between the two vectors (or at least the cosine of the angle between them) can be determined by using the definition of the dot product:

$$\cos(\theta) = \left(\frac{\mathbf{a} \cdot \mathbf{b}}{|\mathbf{a}||\mathbf{b}|}\right)$$

Finally, if a vector is known in terms of its components in its coordinate system, then its magnitude can be determined in terms of these components simply by taking the dot product with itself, for in that case the angle between the two vectors is zero and the cosine unity so that the scalar product is simply the square of the magnitude of the vector:

$$\mathbf{a} \cdot \mathbf{a} = a_1^2 + a_2^2 + a_3^2$$

B.2.2 The Addition of Two Vectors

Two vectors may be added simply by adding their corresponding components (provided the set of base vectors is the same for both vectors, a constraint for curvilinear and coordinate systems). Figure B-8 depicts the addition of two vectors.

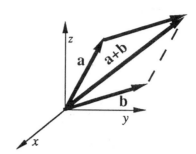

Figure B-8: Addition of two vectors

Thus we have:

$$\mathbf{a} + \mathbf{b} = (a_1 + b_1)\mathbf{i}_1 + (a_2 + b_2)\mathbf{i}_2 + (a_3 + b_3)\mathbf{i}_3$$

B.2.2 Multiplication of a Vector by a Scalar

The product of a scalar with a vector is defined to be a vector in the same direction as the original vector but with magnitude equal to the product of the scalar and the magnitude of the original vector. Therefore we can express the new vector in terms of components of the old vector simply by writing each new component as the scalar times the old component, i.e.

$$\alpha\mathbf{a} = \alpha a_1\mathbf{e}_1 + \alpha a_2\mathbf{e}_2 + \alpha a_3\mathbf{e}_3$$

B.2.3 The Cross Product of Two Vectors

The cross product of two vectors (**a** and **b**) is defined to be a third vector which is normal to the plane of the first two in such a way that if **a**, **b**, and the third vector form a right handed system shown in Figure B-9. The magnitude of this third vector is equal to the product of the magnitudes of the two vectors and the sine of the angle between the two vectors.

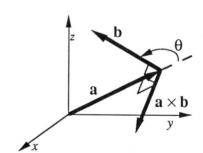

Figure B-9: Cross product of two vectors

If, the components of the two vectors are known in terms of the same set of base vectors, then the cross product is easily calculated from the determinant of an array containing the base vectors and the components of the two vectors in the following way.

$$\mathbf{a} \times \mathbf{b} = \begin{vmatrix} \mathbf{i}_1 & \mathbf{i}_2 & \mathbf{i}_3 \\ a_1 & a_2 & a_3 \\ b_1 & b_2 & b_3 \end{vmatrix}$$
$$= (a_2 b_3 - b_2 a_3)\mathbf{i}_1 - (-a_3 b_1 + b_3 a_1)\mathbf{i}_2 + (a_1 b_2 - b_1 a_2)\mathbf{i}_3$$

Calculating the cross product in this way gives a vector with the properties as given in the definition above and gives it in a convenient calculational form, that is, in terms of its components and base vectors. Note that

$$\mathbf{a} \times \mathbf{b} = -\mathbf{b} \times \mathbf{a}$$

and that in the determinant the components of $\mathbf{a}$ are in the second row and those of $\mathbf{b}$ in the third.

B.2.4 The Triple Scalar Product

The triple scalar product is a scalar product of three vectors. It is formed by taking the dot product of one vector with a cross product of the other two. Note, that even though the order of the two vectors does not matter when forming a dot product, it certainly does matter for the cross product and hence, it also matters for the triple scalar product. Specifically, given three vectors, $\mathbf{a}$, $\mathbf{b}$, and $\mathbf{c}$, although there are twelve products which can be formed (corresponding to the order of the vectors and the position of the cross and dot operations, there are only two different *values* for the triple scalar products which can be formed:

$$\mathbf{a} \cdot (\mathbf{b} \times \mathbf{c}) = (\mathbf{a} \times \mathbf{b}) \cdot \mathbf{c} = \mathbf{b} \cdot (\mathbf{c} \times \mathbf{a}) = (\mathbf{b} \times \mathbf{c}) \cdot \mathbf{a} = \mathbf{c} \cdot (\mathbf{a} \times \mathbf{b}) = (\mathbf{c} \times \mathbf{a}) \cdot \mathbf{b}$$

and

$$\mathbf{a} \cdot (\mathbf{c} \times \mathbf{b}) = (\mathbf{a} \times \mathbf{c}) \cdot \mathbf{b} = \mathbf{b} \cdot (\mathbf{a} \times \mathbf{c}) = (\mathbf{b} \times \mathbf{a}) \cdot \mathbf{c} = \mathbf{c} \cdot (\mathbf{b} \times \mathbf{a}) = (\mathbf{c} \times \mathbf{b}) \cdot \mathbf{a}$$

The first set is the negative of the second. Hence, for example, $\mathbf{a} \cdot (\mathbf{b} \times \mathbf{c}) = -\mathbf{a} \cdot (\mathbf{c} \times \mathbf{b})$. Note that each of the products in the first set is an "even" permutation of the three vectors while each of those in the second set is an "odd" permutation. The odd permutation is obtained from an even by interchanging two adjacent vectors. Note that if the order of two vectors in a cross product is interchanged, then the negative result is obtained; this is the reason that the odd permutation products are the negative of the even ones. Also, note that the dot and cross can be interchanged without changing the value of the product.

Geometrically, the triple scalar product corresponds to the *volume* of the parallelepiped formed by the three vectors placed with their tails together, thereby forming three adjacent edges of the solid.

B.3 POSITION VECTORS

As an example of the representation of vectors in the three main coordinate systems, consider a vector which describes the position of a point in space with respect to the origin of the coordinate system. This is a vector of magnitude representing the distance from the origin to the point of interest and the direction along the line from the origin to that point. In a rectangular cartesian system that position vector is easily represented as a linear combination of the coordinates of that point with the base vectors, i.e.

$$\mathbf{r} = (x_1)\mathbf{i}_1 + (y_1)\mathbf{i}_2 + (z_1)\mathbf{i}_3$$

In a cylindrical coordinate system the representation is a little bit trickier, however, and the position vector there is represented as

$$\mathbf{r} = (r)\mathbf{e}_1 + (z_1)\mathbf{e}_3$$

where r is the cylindrical system radial coordinate, and $\mathbf{e}_1$ is the radial base vector, which is a function through θ. Likewise, in a spherical system the position vector is not represented explicitly in terms of all three coordinates but instead very simply as

$$\mathbf{r} = (r)\mathbf{e}_1$$

where r is the spherical system radial coordinate, and $\mathbf{e}_1$, the radial coordinate base vector, is a function of position through θ, and ϕ. In each case, the base vectors are those at the point in space which is of interest.

B.3.1 Dependence of Coordinate System Base Vectors on Position

Although for a rectangular cartesian coordinate system the base vectors are independent of position, this is not necessarily the case for all coordinate systems. A curvilinear coordinate system is one for which the base vectors are functions of position. coordinate systems are curvilinear. In the previous section two curvilinear systems, cylindrical and spherical, were defined and the position vector for a point in space was represented in terms of the base vectors which exist at that point in space; the position vectors for two different points in space are written in terms of different base vectors, in general. However, nothing has been said yet about precisely how those base vectors change with position. In this section we will address this question of the dependence of the coordinate system base vectors on position.

First, we must consider some mathematics of a function of multiple independent variables. For example, if a function f may depend upon several independent variables, say x, y, and z, i.e. we may write

$$f = f(x, y, z)$$

The variables x, y, and z are independent variables and the function f is a recipe for taking values for these independent variables and calculating a number. In general, if either x, y, or z is changed then the recipe will calculate a different number for f. For example, we may think of the pressure exerted by a gas in a container as a function of the number of moles of that gas, the temperature of that gas, and the volume of the container:

$$P = f(n, T, V)$$

If we change the number of moles in a given container (fixed volume) while keeping the temperature constant, the pressure must change accordingly. Likewise, if we change the temperature (again for a fixed volume) and number of moles, the pressure will change. Finally, if the size of the container is changed for a given number of moles and temperature, then again the pressure will change. The simplest such relation that we have is known as the ideal gas law, that is:

$$P = \left(\frac{nRT}{V} \right)$$

where R is the gas constant.

A useful concept having to do with functions of variables is that of calculating the change in the dependent variable or property which arises from changes in the various independent variables, either individually or all changes taken together. For example, if we have a function of a single independent variable, say

$$f = f(x)$$

then we can calculate changes in the value of f which arise from changes in the value of x according to the equation

$$df = \left(\frac{df}{dx} \right) dx$$

Here if we have some small change in x, then the change in f can be calculated from that change in x knowing the derivative of f with respect to x. From calculus you are familiar with the concept of the derivative of a function having the interpretation as the slope of a tangent to the curve. Consequently, for small changes in x, we can calculate changes in f in accordance with their tangent slope.

For a function of multiple independent variables

$$f = f(x, y, z)$$

we have a similar relation except that we may treat each independent variable separately and look at derivatives of the function with respect to each of these variables while holding the other variables constant. Each partial derivative then contributes an amount to the total change in f in accordance with the change in that independent variable and the value of the partial derivative. Accordingly we may write that the total differential (change) in a function f as defined above is given by:

$$df = \left(\frac{\partial f}{\partial x}\right)_{y,z} dx + \left(\frac{\partial f}{\partial y}\right)_{x,z} dy + \left(\frac{\partial f}{\partial z}\right)_{x,y} dx$$

Again each partial derivative of the function represents changes in f as the indicated independent variable is changed but while holding the other two independent variables constant, i.e. treating them exactly like other constants in the functional relationship while performing this differentiation. The partial derivatives contain information about the function f; each one tells how sensitive (i.e. how susceptible to change) f is to changes in a specific independent variable. A total change in f then is calculated by summing each of these separate changes due to each of the independent variables.

Cylindrical Coordinate System. We are now prepared to look at changes in the base vectors for a curvilinear coordinate system as position changes. In a curvilinear coordinate system each base vector can be written as a function of the position in space. For example, for the cylindrical system the base vectors are $\mathbf{e}_r$, $\mathbf{e}_\theta$, and $\mathbf{e}_z$ as shown in the figure. As position is changed, the base vectors change, not in length (they are always unit vectors) but rather in direction. Accordingly, we may think of each base vector as a function of the r, θ and z position coordinates:

$$\mathbf{e}_r = \mathbf{e}_r(r, \theta, z)$$
$$\mathbf{e}_\theta = \mathbf{e}_\theta(r, \theta, z)$$
$$\mathbf{e}_z = \mathbf{e}_z(r, \theta, z)$$

In accordance with the preceding discussion then, in order to establish the total differential or the total change of any one position vector which arises from a change in position (change in each coordinate), we consider the change of that base vector due to changes in each of the position coordinates taken separately and then add these changes together. For example we can write:

$$d\mathbf{e}_r = \left(\frac{\partial \mathbf{e}_r}{\partial r}\right)_{\theta,z} dr + \left(\frac{\partial \mathbf{e}_r}{\partial \theta}\right)_{r,z} d\theta + \left(\frac{\partial \mathbf{e}_r}{\partial z}\right)_{r,\theta} dz$$

Again each of the partial derivatives represents a change in the position vector with respect to one of the coordinates holding the other two coordinates constant.

Each of these partial derivatives can be determined fairly easily by considering the position vectors and their geometrical relationships. For example, the change in the $\mathbf{e}_r$ base vector due to the change in the radial coordinate holding θ and z constant must be zero because at constant θ and z there is no change in the direction of the $\mathbf{e}_r$ base vector and it is always unit length. Therefore we write

$$\left(\frac{\partial \mathbf{e}_r}{\partial r}\right)_{\theta,z} = 0$$

Likewise, $\mathbf{e}_r$ does not change with the z coordinate holding r and θ constant. Its direction is unchanged by moving along the z axis. Accordingly,

$$\left(\frac{\partial \mathbf{e}_r}{\partial z}\right)_{r,\theta} = 0$$

However, $\mathbf{e}_r$ does change as the angle θ is changed holding r and z constant. In Figure B-10 it is evident that $(\mathbf{e}_r)_{\theta+d\theta} - (\mathbf{e}_r)_\theta = d\theta \mathbf{e}_\theta$ from which we can see that $\left(\dfrac{\partial \mathbf{e}_r}{\partial \theta}\right)_{r,z} = \mathbf{e}_\theta$. Now putting these three results together we obtain a result for the total differential change in $\mathbf{e}_r$ which arises from differential changes in the r, θ, and z coordinates: $d\mathbf{e}_r = \mathbf{e}_\theta \, d\theta$.

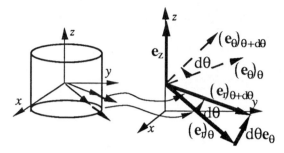

Figure B-10: Cylindrical
coordinate systems

Similarly, we can consider $\mathbf{e}_\theta$ and $\mathbf{e}_z$. Obviously, changing r only does not affect the base vector $\mathbf{e}_\theta$. Likewise, changing z only does not affect $\mathbf{e}_\theta$. However, changing θ does change $\mathbf{e}_\theta$ according to:

$$\left(\frac{\partial \mathbf{e}_\theta}{\partial \theta}\right)_{r,z} = -\mathbf{e}_r$$

Accordingly, we can write

$$d\mathbf{e}_\theta = -\mathbf{e}_r \, d\theta$$

Finally, the base vector $\mathbf{e}_z$ is independent of r, θ, and z at all points in space so that:

$$d\mathbf{e}_z = 0$$

Relations of these sorts are necessary for describing changes in vectors in curvilinear coordinate systems. For example, we can write that the velocity of a component is equal to the time rate of change of the position vector. In a cylindrical coordinate system the position vector is written:

$$\mathbf{r} = r\mathbf{e}_r + z\mathbf{e}_z$$

Consequently, the velocity vector which is the time derivative of this position vector is written

$$\begin{aligned}
\mathbf{v} &= \left(\frac{d\mathbf{r}}{dt}\right) \\
&= \frac{d}{dt}\left(r\mathbf{e}_r + z\mathbf{e}_z\right) \\
&= \left(\frac{dr}{dt}\right)\mathbf{e}_r + r\left(\frac{d\mathbf{e}_r}{dt}\right) + \left(\frac{dz}{dt}\right)\mathbf{e}_z + z\left(\frac{d\mathbf{e}_z}{dt}\right)
\end{aligned}$$

Then from the above relationships for the total differentials of the various base vectors, it is clear that this can be written:

$$\mathbf{v} = \left(\frac{dr}{dt}\right)\mathbf{e}_r + r\left(\frac{d\theta}{dt}\right)\mathbf{e}_\theta + \left(\frac{dz}{dt}\right)\mathbf{e}_z$$

from which we can write the components of the velocity vector as:

$$v_r = \left(\frac{dr}{dt}\right)$$

$$v_\theta = r\left(\frac{d\theta}{dt}\right)$$

$$v_z = \left(\frac{dz}{dt}\right)$$

The r, θ, z components of $\left(\dfrac{d\mathbf{e}_r}{dt}\right)$, $\left(\dfrac{d\mathbf{e}_\theta}{dt}\right)$, and $\left(\dfrac{d\mathbf{e}_z}{dt}\right)$ are summarized in Table B-1 with the dot notation. Note the skew symmetry of this table.

Table B-1: Cylindrical Components

	r-comp	θ-comp	z-comp
$\left(\dfrac{d\mathbf{e}_r}{dt}\right)$	0	$\dot{\theta}$	0
$\left(\dfrac{d\mathbf{e}_\theta}{dt}\right)$	$-\dot{\theta}$	0	0
$\left(\dfrac{d\mathbf{e}_z}{dt}\right)$	0	0	0

Spherical Coordinate System. Results can be obtained in a similar way for the spherical coordinate system where the position vector for the base vectors are functions of the r, θ, and z position coordinates. Accordingly, we may write

$$\mathbf{e}_r = \mathbf{e}_r(r, \theta, \phi)$$
$$\mathbf{e}_\theta = \mathbf{e}_\theta(r, \theta, \phi)$$
$$\mathbf{e}_\phi = \mathbf{e}_\phi(r, \theta, \phi)$$

Again, each partial derivative can be considered in turn and through the geometrical relationships can be evaluated. Accordingly, we have the non-zero partial derivatives are:

$$\left(\frac{\partial \mathbf{e}_r}{\partial \theta}\right)_{r,\phi} = \mathbf{e}_\theta$$

$$\left(\frac{\partial \mathbf{e}_r}{\partial \phi}\right)_{r,\theta} = \sin(\theta)\mathbf{e}_\phi$$

$$\left(\frac{\partial \mathbf{e}_\theta}{\partial \theta}\right)_{r,\phi} = -\mathbf{e}_r$$

$$\left(\frac{\partial \mathbf{e}_\theta}{\partial \phi}\right)_{r,\theta} = \cos(\theta)\mathbf{e}_\phi$$

$$\left(\frac{\partial \mathbf{e}_\phi}{\partial \phi}\right)_{r,\theta} = -\sin(\theta)\mathbf{e}_r - \cos(\theta)\mathbf{e}_\theta$$

The r, θ and ϕ components, then, of $\left(\dfrac{d\mathbf{e}_r}{dt}\right)$, $\left(\dfrac{d\mathbf{e}_\theta}{dt}\right)$, and $\left(\dfrac{d\mathbf{e}_\phi}{dt}\right)$ are given in Table B-2. Again, the table is skew symmetric.

Table B-2: Spherical Components

	r-comp	θ-comp	ϕ-comp
$\left(\dfrac{d\mathbf{e}_r}{dt}\right)$	0	$\dot\theta$	$\sin(\theta)\dot\phi$
$\left(\dfrac{d\mathbf{e}_\theta}{dt}\right)$	$-\dot\theta$	0	$\cos(\theta)\dot\phi$
$\left(\dfrac{d\mathbf{e}_\phi}{dt}\right)$	$-\sin(\theta)\dot\phi$	$-\cos(\theta)\dot\phi$	0

As an example, we can again look at the position vector which in spherical coordinates is given by:

$$\mathbf{r} = r\mathbf{e}_r$$

Then the velocity is equal to:

$$\mathbf{v} = \left(\frac{d\mathbf{r}}{dt}\right)$$

$$= \left(\frac{dr}{dt}\right)\mathbf{e}_r + r\left(\frac{d\mathbf{e}_r}{dt}\right)$$

$$= \left(\frac{dr}{dt}\right)\mathbf{e}_r + r\left(\frac{d\theta}{dt}\right)\mathbf{e}_\theta + r\sin(\theta)\left(\frac{d\theta}{dt}\right)\mathbf{e}_\phi$$

from which we can write that the r, θ and ϕ components of the velocity vector are:

$$v_r = \left(\frac{dr}{dt}\right)$$

$$v_\theta = r\left(\frac{d\theta}{dt}\right)$$

$$v_\phi = r\sin(\theta)\left(\frac{d\phi}{dt}\right)$$

Other vectors such as the acceleration vector obtained by differentiating this velocity vector are discussed in the section on rigid body kinematics in Chapter 10.

Coordinate System Defined at a Point Along a Particle's Path. In some cases it is convenient to set up a coordinate system along a particle's path and orient it in such a way that one of the axes is tangent to that path. This leads to two parameters which can be used to characterize a curve in space. These parameters are the radius of curvature and the torsion. Each of these parameters may change along the length of the curve but have a definite value at any particular point of that curve.

To define such a coordinate system we imagine a curve in space (Figure B-11). This curve can be thought to represent the path of a particle's trajectory or the path that a particle would take in moving through space. If we associate specific times to points along this curve then from the curve we can calculate the velocity of the particle by differentiating this position vector with respect to time. The tangent to the curve then is the velocity vector at any particular point in space. At a point along the curve where we desire to set up a coordinate system, it is convenient to select one of the major axes as tangent to this curve, i.e. in the direction of the particle velocity.

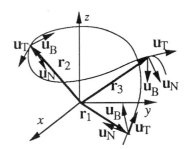

Figure B-11: Coordinate system defined along a path

A suitable (unit) base vector along this curve then would be the velocity divided by the speed of the particle, i.e.

$$\mathbf{u}_T = \left(\frac{\mathbf{v}}{|\mathbf{v}|}\right) = \left(\frac{\mathbf{v}}{v}\right)$$

Note that $\mathbf{u}_T$ points in the direction of $\mathbf{v}$. This base vector, being a unit vector, when dotted with itself gives

$$\mathbf{u}_T \cdot \mathbf{u}_T = 1$$

The derivative with respect to time of this dot product is necessarily zero (is a constant) from which it is evident that the time rate of change of this unit vector is normal to the tangent unit vector. Mathematically,

$$\frac{d(\mathbf{u}_T \cdot \mathbf{u}_T)}{dt} = 0$$

so that

$$\mathbf{u}_T \cdot \left(\frac{d\mathbf{u}_T}{dt} \right) = 0$$

and hence

$$\mathbf{u}_T \perp \left(\frac{d\mathbf{u}_T}{dt} \right)$$

Evidently the time derivative of this tangent vector can serve as the direction of another axis in our coordinate system which will be normal to the unit tangent axis. Another way to view this normal vector to the curve is by differentiating the unit tangent in terms of the velocity and the speed of the particle along the path. Because we can write

$$\mathbf{u}_T = \frac{\mathbf{v}}{v}$$

the derivative of $\mathbf{u}_T$ is

$$\left(\frac{d\mathbf{u}_T}{dt} \right) = \left(\frac{\mathbf{a}}{v} \right) - \left(\frac{\mathbf{v}}{v^2} \left(\frac{dv}{dt} \right) \right)$$

and we see that this normal vector to the path is composed of the sum of two vectors, one being in the direction of the velocity and the other being in the direction of the acceleration. The two together give a vector that is normal to the curve; the acceleration as a general rule is not normal to the curve.

In talking about the motion of a particle along a path, we have described the position and velocity and other such quantities as functions of time. This is a natural association to make. However, mathematically it is actually more convenient and a better physical representation if we use the length along the path as an independent variable rather than time. Obviously the two are related and if we have the path described as a function of time then for "normal" curves we can also relate the time to arc length and describe the position as a function of length along the curve just as well as we can of time. With this kind of description we can then talk about the tangent vector, the normal vector, and the cross product of these two using these derivatives with respect to arc length rather than the derivatives with respect to time. The derivative of the position vector with respect to arc length then is related to the velocity according to:

$$\left(\frac{d\mathbf{r}}{ds} \right) = \frac{\left(\dfrac{d\mathbf{r}}{dt} \right)}{\left(\dfrac{ds}{dt} \right)} = \left(\frac{\mathbf{v}}{v} \right)$$

which gives the same unit tangent vector that we had before in terms of velocity. In this sense the arc length is the natural parameter to use for describing the curve; the derivative of the position vector with respect to arc length is automatically a unit tangent vector and needs no further normalization as does the velocity vector (the derivative of the position vector with respect to time).

Likewise, from the expression of the unit tangent vector as a function of arc length, we can calculate the normal vector to the curve. It turns out that the magnitude of this normal vector is the radius of curvature, ρ, of the curve at the point of consideration. Consequently, by normalizing by this radius of curvature we obtain a unit vector which is perpendicular to the unit tangent vector:

$$\frac{d\mathbf{u}_T}{ds} = \frac{\mathbf{u}_N}{\rho}$$

$$\left|\frac{d\mathbf{u}_T}{ds}\right| = \left|\frac{d}{ds}\left(\frac{d\mathbf{r}}{ds}\right)\right|$$

$$= \frac{1}{\rho}$$

Note that $\mathbf{u}_N$ points to the "inside" of the path. The cross product of these two unit vectors gives another unit vector $\mathbf{u}_B$, the binormal, which serves as the third base vector for the coordinate system established at the point on the curve:

$$\mathbf{u}_B = \mathbf{u}_T \times \mathbf{u}_N$$

So, $\mathbf{u}_B$ is normal to both $\mathbf{u}_T$ and $\mathbf{u}_N$. ($\mathbf{u}_B \cdot \mathbf{u}_N = 0$ and $\mathbf{u}_B \cdot \mathbf{u}_T = 0$) and $\mathbf{u}_T$, $\mathbf{u}_N$, and $\mathbf{u}_B$ form the right-handed coordinate system.

We now would like to know how $\mathbf{u}_T$, $\mathbf{u}_N$, and $\mathbf{u}_B$ vary with position (say, in terms of the arc length). Now, we already know that $d\mathbf{u}_T / ds = \mathbf{u}_N / \rho$. Looking next at $\mathbf{u}_B$, because it is a unit vector,

$$\mathbf{u}_B \cdot \mathbf{u}_B = 1$$

and hence

$$\frac{d\mathbf{u}_B}{ds} \cdot \mathbf{u}_B = 0$$

Consequently, $d\mathbf{u}_B / ds$ lies in a plane perpendicular to $\mathbf{u}_B$ (i.e., in a plane defined by $\mathbf{u}_T$ and $\mathbf{u}_N$), but the question is, in which direction? Because $\mathbf{u}_B \cdot \mathbf{u}_T = 0$,

$$\frac{d}{ds}(\mathbf{u}_B \cdot \mathbf{u}_T) = 0 \qquad \text{or} \qquad \frac{d\mathbf{u}_B}{ds} \cdot \mathbf{u}_T = -\mathbf{u}_B \cdot \frac{d\mathbf{u}_T}{ds}$$

But, from before, $d\mathbf{u}_T / ds = \mathbf{u}_N / \rho$ so that

$$\frac{d\mathbf{u}_B}{ds} \cdot \mathbf{u}_T = -\mathbf{u}_B \cdot \frac{\mathbf{u}_N}{\rho} = 0$$

So that $d\mathbf{u}_B / ds$ is normal to $\mathbf{u}_T$. Hence, because $d\mathbf{u}_B / ds$ is normal to both $\mathbf{u}_B$, and $\mathbf{u}_T$ is must be in the direction of $\mathbf{u}_N$. It is conventional to let $|d\mathbf{u}_B / ds| = -1 / \sigma$ so that

$$\frac{d\mathbf{u}_B}{ds} = -\frac{\mathbf{u}_N}{\sigma}$$

The scalar $1 / \sigma$ is called the torsion.

Also, we can now determine $d\mathbf{u}_N / ds$. Because $\mathbf{u}_N = \mathbf{u}_B \times \mathbf{u}_T$, and from the previous results for $d\mathbf{u}_B / ds$ and $d\mathbf{u}_T / ds$, we have that

$$\frac{d\mathbf{u}_N}{ds} = \frac{d}{ds}(\mathbf{u}_B \times \mathbf{u}_T) = \frac{d\mathbf{u}_B}{ds} \times \mathbf{u}_T + \mathbf{u}_B \times \frac{d\mathbf{u}_T}{ds}$$

$$= -\frac{\mathbf{u}_N}{\sigma} \times \mathbf{u}_T + \mathbf{u}_B \times \frac{\mathbf{u}_N}{\rho}$$

$$= \frac{\mathbf{u}_B}{\sigma} - \frac{\mathbf{u}_T}{\rho}$$

We now have the dependence of all three path base vectors on position (as defined by path length, s) which are summarized in Table B-3. These three relations are called the Serret-Frenet equations and are defined by the path. Note, as before, the skew symmetry of this table.

Table B-3: Path Coordinates[a,b]

	T-comp	N-comp	B-comp
$\dfrac{d\mathbf{u}_T}{ds}$	0	$\dfrac{1}{\rho}$	0
$\dfrac{d\mathbf{u}_N}{ds}$	$-\dfrac{1}{\rho}$	0	$\dfrac{1}{\sigma}$
$\dfrac{d\mathbf{u}_B}{ds}$	0	$-\dfrac{1}{\sigma}$	0

[a] ρ = radius of curvature; $1/\sigma$ = torsion; ρ and $1/\sigma$ are functions of position along the path, i.e., $\rho = \rho(s)$; $\sigma = \sigma(s)$.

[b] s = arc length; $ds/dt = v$ (speed), so that, $\dfrac{d\mathbf{u}_T}{dt} = v\dfrac{d\mathbf{u}_T}{ds} = \dfrac{v}{\rho}\mathbf{u}_N$

B.3.2 Time Derivatives of Position Vectors Using Position-Independent Base Vectors

The velocity vector of a particle is the time rate of change of the position vector. That is given the position vector $\mathbf{r}$ with the position being a function of time and the base vectors independent of the position:

$$\mathbf{r} = x(t)\mathbf{i}_1 + y(t)\mathbf{i}_2 + z(t)\mathbf{i}_3$$

the velocity ($\mathbf{v}$) of the particle from a preselected base point is:

$$\mathbf{v} = \left(\frac{d\mathbf{r}}{dt}\right) = \left(\frac{dx}{dt}\right)\mathbf{i}_1 + \left(\frac{dy}{dt}\right)\mathbf{i}_2 + \left(\frac{dz}{dt}\right)\mathbf{i}_3$$

The acceleration vector $\mathbf{a}$ of the motion of the particle is defined as the second derivative of the position vector with respect to time from a preselected base point.

$$\mathbf{a} = \left(\frac{d\mathbf{v}}{dt}\right) = \left(\frac{d^2\mathbf{r}}{dt^2}\right) = \left(\frac{d^2x}{dt^2}\right)\mathbf{i}_1 + \left(\frac{d^2y}{dt^2}\right)\mathbf{i}_2 + \left(\frac{d^2z}{dt^2}\right)\mathbf{i}_3$$

Finally, the time derivative of the products of two vectors $\mathbf{a}(t)$ and $\mathbf{b}(t)$ are given:

Dot Product
$$\left(\frac{d(\mathbf{a} \cdot \mathbf{b})}{dt}\right) = \left(\frac{d\mathbf{a}}{dt}\right) \cdot \mathbf{b} + \mathbf{a} \cdot \left(\frac{d\mathbf{b}}{dt}\right)$$

Cross Product
$$\left(\frac{d(\mathbf{a} \times \mathbf{b})}{dt}\right) = \left(\frac{d\mathbf{a}}{dt}\right) \times \mathbf{b} + \mathbf{a} \times \left(\frac{d\mathbf{b}}{dt}\right)$$

B.4 VECTOR IDENTITIES

Vector operations involving the dot product and cross product are useful in describing physical phenomena in 3-dimensions. A brief list of mathematical laws and indentities is provided for you as a basis for understanding vector manipulations.

$$(\mathbf{a} \cdot k\mathbf{b}) = k(\mathbf{a} \cdot \mathbf{b}) \qquad\qquad (\mathbf{a} \cdot \mathbf{b}) = (\mathbf{b} \cdot \mathbf{a})$$
$$(\mathbf{a} \cdot \mathbf{b})\mathbf{c} \neq \mathbf{a}(\mathbf{b} \cdot \mathbf{c}) \qquad\qquad \mathbf{a} \cdot (\mathbf{b} + \mathbf{c}) = (\mathbf{a} \cdot \mathbf{b}) + (\mathbf{a} \cdot \mathbf{c})$$
$$(\mathbf{a} \times k\mathbf{b}) = k(\mathbf{a} \times \mathbf{b}) \qquad\qquad (\mathbf{a} \times \mathbf{b}) = -(\mathbf{b} \times \mathbf{a})$$
$$\mathbf{a} \times (\mathbf{b} \times \mathbf{c}) \neq (\mathbf{a} \times \mathbf{b}) \times \mathbf{c} \qquad\qquad (\mathbf{a} + \mathbf{b}) \times \mathbf{c} = (\mathbf{a} \times \mathbf{b}) + (\mathbf{b} \times \mathbf{c})$$
$$\mathbf{a} \times (\mathbf{b} \times \mathbf{c}) = (\mathbf{a} \cdot \mathbf{c})\mathbf{b} - (\mathbf{a} \cdot \mathbf{b})\mathbf{c} \qquad\qquad \mathbf{a} \cdot (\mathbf{b} \times \mathbf{c}) = \mathbf{b} \cdot (\mathbf{c} \times \mathbf{a})$$
$$\mathbf{a} \cdot (\mathbf{b} \times \mathbf{c}) = (\mathbf{a} \times \mathbf{b}) \cdot \mathbf{c} \qquad\qquad (\mathbf{a} \times \mathbf{b}) \cdot (\mathbf{c} \times \mathbf{d}) = (\mathbf{a} \cdot \mathbf{c})(\mathbf{b} \cdot \mathbf{d}) - (\mathbf{a} \cdot \mathbf{d})(\mathbf{b} \cdot \mathbf{d})$$

B.5 DYADIC PRODUCT OF TWO VECTORS

The dyadic product of two vectors is analogous to the matrix multiplication of two vectors and gives a (3×3) matrix called a second order tensor:

$$\mathbf{ab} = \begin{bmatrix} a_1\mathbf{i} \\ + \quad a_2\mathbf{j} \\ + \quad a_3\mathbf{k} \end{bmatrix} [\, b_1\mathbf{i} \quad + \quad b_2\mathbf{j} \quad + \quad b_3\mathbf{k}\,]$$

$$= \begin{bmatrix} a_1 b_1 \mathbf{ii} & + & a_1 b_2 \mathbf{ij} & + & a_1 b_3 \mathbf{ik} \\ + \quad a_2 b_1 \mathbf{ji} & + & a_2 b_2 \mathbf{jj} & + & a_2 b_3 \mathbf{jk} \\ + \quad a_3 b_1 \mathbf{ki} & + & a_3 b_2 \mathbf{kj} & + & a_3 b_3 \mathbf{kk} \end{bmatrix}$$

The dyadic product of the position vector $\mathbf{r}$ in a cartesian coordinate system is:

$$\mathbf{rr} = \begin{bmatrix} x\mathbf{i} \\ + \quad y\mathbf{j} \\ + \quad z\mathbf{k} \end{bmatrix} [\, x\mathbf{i} \quad + \quad y\mathbf{j} \quad + \quad z\mathbf{k}\,]$$

$$= \begin{bmatrix} x^2\mathbf{ii} & + & xy\mathbf{ij} & + & xz\mathbf{ik} \\ + \quad yx\mathbf{ji} & + & y^2\mathbf{jj} & + & yz\mathbf{jk} \\ + \quad zx\mathbf{ki} & + & zy\mathbf{kj} & + & z^2\mathbf{kk} \end{bmatrix}$$

B.6 IDENTITY TENSOR

The identity tensor is defined as a (3×3) matrix with ones down the main diagonal and zeros every place else:

$$\mathbf{I} \equiv \begin{bmatrix} 1\mathbf{ii} & + & 0\mathbf{ij} & + & 0\mathbf{ik} \\ + & 0\mathbf{ji} & + & 1\mathbf{jj} & + & 0\mathbf{jk} \\ + & 0\mathbf{ki} & + & 0\mathbf{kj} & + & 1\mathbf{kk} \end{bmatrix}$$

A term of the moment of inertia tensor using the identity matrix is calculated. Given the position vector:

$$\mathbf{r} = x\mathbf{i} + y\mathbf{j} + z\mathbf{k}$$

The dot product of the position vector with itself is:

$$\mathbf{r} \cdot \mathbf{r} = r^2$$
$$= x^2 + y^2 + z^2$$

Using the identity tensor:

$$r^2\mathbf{I} = \begin{bmatrix} (x^2+y^2+z^2)\mathbf{ii} & + & 0\mathbf{ij} & + & 0\mathbf{ik} \\ + & 0\mathbf{ji} & + & (x^2+y^2+z^2)\mathbf{jj} & + & 0\mathbf{jk} \\ + & 0\mathbf{ki} & + & 0\mathbf{kj} & + & (x^2+y^2+z^2)\mathbf{kk} \end{bmatrix}$$

B.7 THE DOT PRODUCT OF A TENSOR AND A VECTOR

The dot product of a tensor and a vector arises in the calculation of angular momentum. The dot product of a tensor and a vector is the projection of the tensor in the direction of the vector. For the given tensor $\mathbf{T}$ and vector $\mathbf{v}$ the dot product is:

$$\mathbf{T} \cdot \mathbf{v} = \begin{bmatrix} T_{11}\mathbf{ii} & + & T_{12}\mathbf{ij} & + & T_{13}\mathbf{ik} \\ + & T_{21}\mathbf{ji} & + & T_{22}\mathbf{jj} & + & T_{23}\mathbf{jk} \\ + & T_{31}\mathbf{ki} & + & T_{32}\mathbf{kj} & + & T_{33}\mathbf{kk} \end{bmatrix} \cdot \begin{bmatrix} v_1\mathbf{i} \\ + & v_2\mathbf{j} \\ + & v_3\mathbf{k} \end{bmatrix}$$

$$= \begin{bmatrix} (T_{11}v_1 + T_{12}v_2 + T_{13}v_3)\mathbf{i} \\ + & (T_{21}v_1 + T_{22}v_2 + T_{23}v_3)\mathbf{j} \\ + & (T_{31}v_1 + T_{32}v_2 + T_{33}v_3)\mathbf{k} \end{bmatrix}$$

To calculate the scalar moment of inertia, the pre dot and post dot vector manipulations are required:

$$\mathbf{v} \cdot \mathbf{T} \cdot \mathbf{v} = [v_1\mathbf{i} + v_2\mathbf{j} + v_3\mathbf{k}] \cdot \begin{bmatrix} T_{11}\mathbf{ii} & + & T_{12}\mathbf{ij} & + & T_{13}\mathbf{ik} \\ + & T_{21}\mathbf{ji} & + & T_{22}\mathbf{jj} & + & T_{23}\mathbf{jk} \\ + & T_{31}\mathbf{ki} & + & T_{32}\mathbf{kj} & + & T_{33}\mathbf{kk} \end{bmatrix} \cdot \begin{bmatrix} v_1\mathbf{i} \\ + & v_2\mathbf{j} \\ + & v_3\mathbf{k} \end{bmatrix}$$

$$= \left\{ \begin{array}{ccccc} T_{11}v_1^2 & + & T_{21}v_2v_1 & + & T_{31}v_3v_1 \\ + & T_{12}v_1v_2 & + & T_{22}v_2^2 & + & T_{32}v_3v_2 \\ + & T_{13}v_1v_3 & + & T_{23}v_2v_3 & + & T_{33}v_3^2 \end{array} \right\}$$

The last terms in the equation is a sum of scalars and therefore is a scalar too.

B.8 REVIEW

Vectors are mathematical representations of physical quantities such as forces, momentum, velocity etc. having both magnitude and direction. To describe vectors, three dimensional space is defined using a coordinate system. The base vectors establish the coordinate system and all other vectors can be written as linear combinations of those base vectors. The three most common coordinate systems are, rectangular cartesian, cylindrical, and spherical.

The mathematical operations of vector quantities differ from the mathematical operations of scalars. The dot product and cross product of two vectors are two different types of vector multiplication. The dot product of two vectors yields a scalar quantity while the cross product gives a vector quantity perpendicular to the original two vectors. The addition of two vectors is accomplished by the addition of the corresponding components. Finally, position vectors and the time derivatives of position vectors are used to describe velocity, acceleration and motion in 3-dimensions.

Identities for vector manipulations and definitions of a dyadic product, the identity tensor and operations with the identity tensor are given. Also the dot product of a tensor and a vector are explained.

APPENDIX
C
ENGINEERING DATA

This appendix contains a variety of engineering data providing the necessary information to solve the problems in this book. A list of references is provided for more extensive data. The order of the data presentation in this appendix parallels the structure of the text as a whole.

For more information:

Handbook of Chemistry and Physics, Chemical Rubber Company Press

International Critical Tables of Numerical Data, Physics, Chemistry and Technology, McGraw Hill

Mechanical Engineers' Handbook, John Wiley and Sons

Perry's Chemical Engineers' Handbook, McGraw Hill

Standard Handbook for Civil Engineers, McGraw Hill

Standard Handbook for Electrical Engineers, McGraw Hill

Table C-1: Density and Viscosity
of Liquid Water at 1 atm (1.01 bar)

T °C	ρ 10^3 kg / m^3	ρ lb$_m$ / ft^3	μ mPa s	μ cp
0	1.000	62.4	1.79	1.79
5	1.000	62.4	1.52	1.52
10	1.000	62.4	1.31	1.31
15	0.999	62.3	1.14	1.14
20	0.998	62.2	1.00	1.00
25	0.997	62.2	0.890	0.890
30	0.996	62.1	0.798	0.798
35	0.994	62.0	0.719	0.719
40	0.992	61.9	0.653	0.653
45	0.990	61.7	0.596	0.596
50	0.988	61.6	0.547	0.547
55	0.986	61.5	0.504	0.504
60	0.983	61.3	0.466	0.466
65	0.980	61.1	0.434	0.434
70	0.978	61.0	0.404	0.404
75	0.975	60.8	0.378	0.378
80	0.972	60.6	0.355	0.355
85	0.969	60.4	0.344	0.334
90	0.965	60.2	0.315	0.315
95	0.962	60.0	0.298	0.298
100	0.958	59.8	0.282	0.282

1 poise = 1 g / cm s = 0.1 kg / m s =
100 cp = 6.720 × 10^{-2} lb$_m$ / ft s

Table C-2: Density and Viscosity
of Mercury at 1 atm

T °C	ρ 10^3 kg / m^3	ρ lb$_m$ / ft^3	μ mPa s	μ cp
0	13.60	848.6	1.68	1.68
5	13.58	847.4		
10	13.57	846.8	1.62	1.62
15	13.56	846.1		
20	13.55	845.5	1.55	1.55
25	13.53	844.7		
30	13.52	843.6	1.50	1.50
35	13.51	843.0		
40	13.50	842.4	1.45	1.45
50	13.47	840.5	1.41	1.41
60	13.45	839.3	1.37	1.37
70	13.42	837.4	1.33	1.33
80	13.40	836.2	1.30	1.30
90	13.38	834.9	1.27	1.27
100	13.35	833.0	1.24	1.24

Table C-3: Density and Viscosity
of Dry Air at 1 atm

T °F	ρ 10^3 kg / m^3	ρ lb$_m$ / ft^3	μ mPa s	μ cp
0	1.4×10^{-3}	0.086	0.016	0.016
32	1.3×10^{-3}	0.081	0.017	0.017
50	1.3×10^{-3}	0.078	0.018	0.018
100	1.1×10^{-3}	0.071	0.019	0.019
150	1.0×10^{-3}	0.065	0.020	0.020
200	9.6×10^{-4}	0.060	0.022	0.022
250	9.0×10^{-4}	0.056	0.023	0.023
300	8.3×10^{-4}	0.052	0.024	0.024
350	7.9×10^{-4}	0.049	0.025	0.025
400	7.4×10^{-4}	0.046	0.026	0.026
450	7.1×10^{-4}	0.044	0.027	0.027
500	6.6×10^{-4}	0.041	0.028	0.028

Table C-4: Density and Viscosity of Various Liquids at 1 atm

Name	T °C	ρ 10^3 kg / m^3	ρ lb$_m$ / ft^3	μ mPa s	μ cp
Acetone	20	0.79	49.4		
Ethanol	20	0.79	49.4	1.20	1.20
Methanol	0	0.81	50.5	0.82	0.82
Benzene	0	0.90	56.1	0.91	0.91
Carbon disul.	0	1.29	80.7	0.44	0.44
Carbon tet.	20	1.60	99.6	0.97	0.97
Chloroform	20	1.49	93.0	0.58	0.58
Diethylether	0	0.74	45.9	0.28	0.28
Gasoline		0.66-0.69	41.0-43.0		
Sea water	15	1.02	63.99		

Table C-5: Density of Various Solids at T = 25 °C

Name	ρ 10^3 kg / m^3	ρ lb$_m$ / ft^3
Aluminum hard drawn	2.55-2.80	159.2-179.8
Asbestos	2.0-2.8	125-175
Beeswax	0.96-0.97	60-61
Bone	1.7-2.0	106-125
Butter	0.86-0.87	53-54
Chalk	1.9-2.8	118-175
Copper cast rolled	8.8-8.95	565.2-574.9
Cork	0.22-0.26	14-16
Diamond	3.01-3.52	188-220
Glass	2.4-2.8	150-175
Gold cast hammered	19.25-19.35	1,236-1,243
Ivory	1.83-1.92	114-120
Lead	11.34	728.4
Nickel	8.9	571.6
Quartz	2.65	165
Silver cast hammered	10.4-10.6	668.0-680.8
Starch	1.53	95
Steel, Carbon	7.85	490
Sugar	1.59	99

Table C-6: Centers of mass for 2-dimensional objects

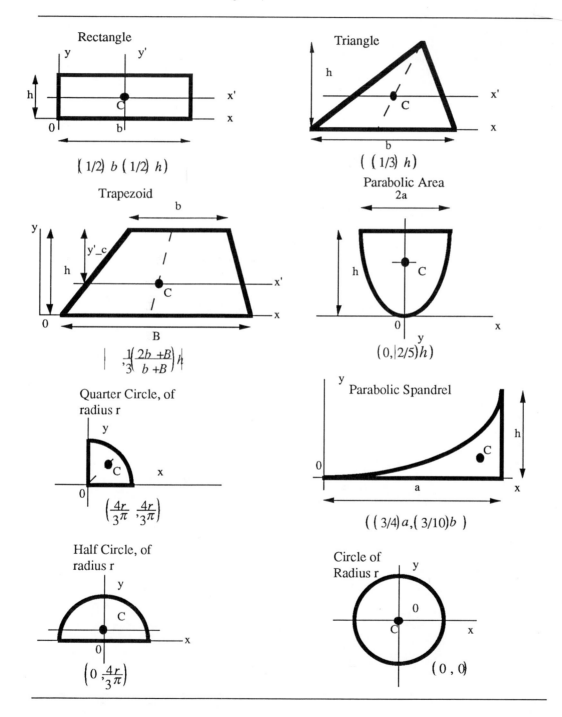

Rectangle

$\left(\tfrac{1}{2}\, b \;\; \tfrac{1}{2}\, h\right)$

Triangle

$\left(\;\; \tfrac{1}{3}\, h\right)$

Trapezoid

$\tfrac{1}{3}\left(\dfrac{2b + B}{b + B}\right) h$

Parabolic Area

$\left(0,\; \tfrac{2}{5} h\right)$

Quarter Circle, of radius r

$\left(\dfrac{4r}{3\pi}\; ,\; \dfrac{4r}{3\pi}\right)$

Parabolic Spandrel

$\left(\;\tfrac{3}{4}\, a,\; \tfrac{3}{10}\, b\;\right)$

Half Circle, of radius r

$\left(0\; ,\; \dfrac{4r}{3\pi}\right)$

Circle of Radius r

$\left(0\; ,\; 0\right)$

Table C-7: Moments of inertia about coordinate axis for 3-dimensional objects

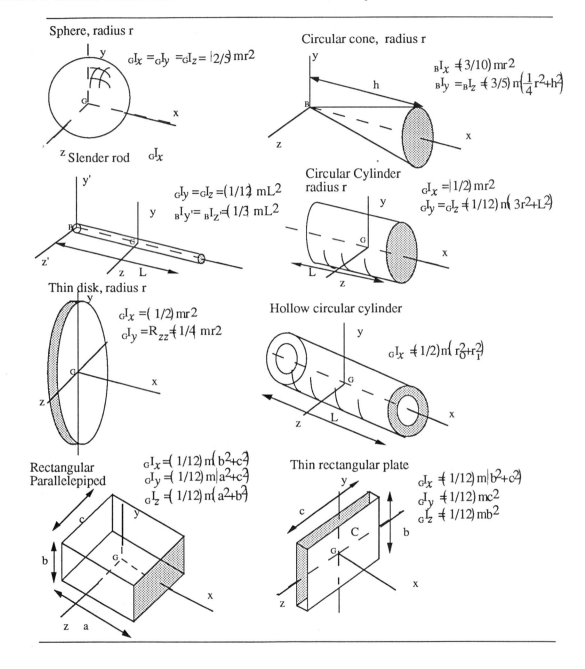

Sphere, radius r

$$_Gl_x = {_G}l_y = {_G}l_z = (2/3) mr^2$$

Circular cone, radius r

$$_Bl_x = (3/10) mr^2$$
$$_Bl_y = {_B}l_z = (3/5) m\left(\frac{1}{4} r^2 + h^2\right)$$

Slender rod $_Gl_x$

$$_Gl_y = {_G}l_z = (1/12) mL^2$$
$$_Bl_{y'} = {_B}l_{z'} = (1/3) mL^2$$

Circular Cylinder
radius r

$$_Gl_x = (1/2) mr^2$$
$$_Gl_y = {_G}l_z = (1/12) m(3r^2 + L^2)$$

Thin disk, radius r

$$_Gl_x = (1/2) mr^2$$
$$_Gl_y = R_{zz} = (1/4) mr^2$$

Hollow circular cylinder

$$_Gl_x = (1/2) m\left(r_o^2 + r_i^2\right)$$

Rectangular
Parallelepiped

$$_Gl_x = (1/12) m(b^2 + c^2)$$
$$_Gl_y = (1/12) m(a^2 + c^2)$$
$$_Gl_z = (1/12) m(a^2 + b^2)$$

Thin rectangular plate

$$_Gl_x = (1/12) m(b^2 + c^2)$$
$$_Gl_y = (1/12) mc^2$$
$$_Gl_z = (1/12) mb^2$$

Table C-8: Static and Kinetic Coefficients of Friction

Material	Condition	T °C	μ_s	μ_k
Glass on glass	clean		0.9-1.0	
Glass on metal	clean		0.5-0.7	
Diamond on diamond	clean		0.1	
Rubber on solids	clean		1-4	
Wood on wood	clean and dry		0.25-0.5	
Brake material/iron	clean		0.4	
	wet		0.2	
Steel on Steel	clean		0.58	
	light oil	100	0.19	
Al on Al	air		1.9	
Cu on Cu	air		1.6	
	H_2 or N_2		4.0	
Gold on gold	air		2.8	
	H_2 or N_2		4.0	
Ice on ice	clean	0	0.5-0.15	0.02 †
	clean	-12	0.3	0.035 †
Ski wax on ice	wet	0	0.1	
	dry	-10	0.2	0.18 ‡
Al on ice	wet	0	0.4	
	dry	-10	0.38	

† 4 m/s
‡ 0.1 m/s
(Reprinted with permission from "Handbook of Chemistry and Physics" 64th ed., Chemical Rubber Publishing Company, 1983, Pg. F-17 through F-19)

Table C-9: Support Reactions for 2-dimensional Rigid Bodies

Connection	Reaction	Number of unknowns
1) Light string		One
2) Massless link		One
3) Smooth pin		Two
4) Smooth surface		One
5) Roller		One
6) Fixed support		Three

Table C-10: Support Reactions for 3-dimensional objects

	Connection	Reactions	Number of unknowns
1) Light cable		f	One
2) Smooth surface		f	One
3) Ball and socket		f_y, f_x, f_z	Three
4) Smooth bearing		$(r \times f)_y$, f_y, f_z, $(r \times f)_z$	Four
5) Smooth pin		f_y, $(r \times f)_y$, f_x, $(r \times f)_x$, f_z	Five
6) Hinge		$(r \times f)_y$, f_y, f_x, $(r \times f)_z$, f_z	Five
7) Fixed support		$(r \times f)_y$, f_y, $(r \times f)_z$, f_x, f_z, $(r \times f)_x$	Six

Table C-11: Surface Tensions of Various Liquids

Material	In contact	T °C	σ dyne / cm	σ N / m	σ lb$_f$ / ft
Water	air	18	73.05	0.07305	4.93×10^{-3}
Ethanol	air	0	24.05	0.02405	1.62×10^{-3}
Methanol	air	0	24.49	0.02449	1.65×10^{-3}
	air	20	22.61	0.02261	1.52×10^{-3}
Benzene	air	10	30.22	0.03022	2.04×10^{-3}
	air	20	28.85	0.02885	1.95×10^{-3}
Acetone	air	0	26.21	0.02621	1.77×10^{-3}
	air	20	23.70	0.02370	1.60×10^{-3}
Nitrogen	vapor	−183	6.6	0.0066	4.45×10^{-4}
	vapor	−193	8.27	0.00827	5.58×10^{-4}
Mercury	vac	20	472	0.472	3.18×10^{-2}

(Reprinted with permission from "Handbook of Chemistry and Physics" 64th ed., Chemical Rubber Publishing Company, 1983, Pg. F-21 to F-36)

Table C-12: Steel-pipe Dimensions

Nominal pipe size in	OD in	Schedule No.	ID in	Flow area in^2
1/8	0.405	40†	0.269	0.058
		80‡	0.215	0.036
1/4	0.540	40	0.364	0.104
		80	0.302	0.072
3/8	0.675	40	0.493	0.192
		80	0.423	0.141
1/2	0.840	40	0.622	0.304
		80	0.546	0.235
3/4	1.05	40	0.824	0.534
		80	0.742	0.432
1	1.32	40	1.049	0.864
		80	0.957	0.718
1 1/4	1.66	40	1.380	1.50
		80	1.278	1.28
1 1/2	1.90	40	1.610	2.04
		80	1.500	1.76
2	2.38	40	2.067	3.35
		80	1.939	2.95
2 1/2	2.88	40	2.469	4.79
		80	2.323	4.23
3	3.50	40	3.068	7.38
		80	2.900	6.61
4	4.50	40	4.026	12.7
		80	3.826	11.5
6	6.625	40	6.065	28.9
		80	5.761	26.1
8	8.625	40	7.981	50.0
		80	7.625	45.7
10	10.75	40	10.02	78.8
		60	9.75	74.6
12	12.75	30	12.09	115
16	16.0	30	15.25	183
20	20.0	20	19.25	291
24	24.0	20	23.25	425

† Schedule 40 designates former "standard wall" pipe.

‡ Schedule 80 designates former "extra-strong wall" pipe.

(Adapted from Perry, R.H. and Green D., "Perry's Chemical Engineers' Handbook," 6th. ed., McGraw-Hill, 1984, Pg. 6-42 to 6-44)

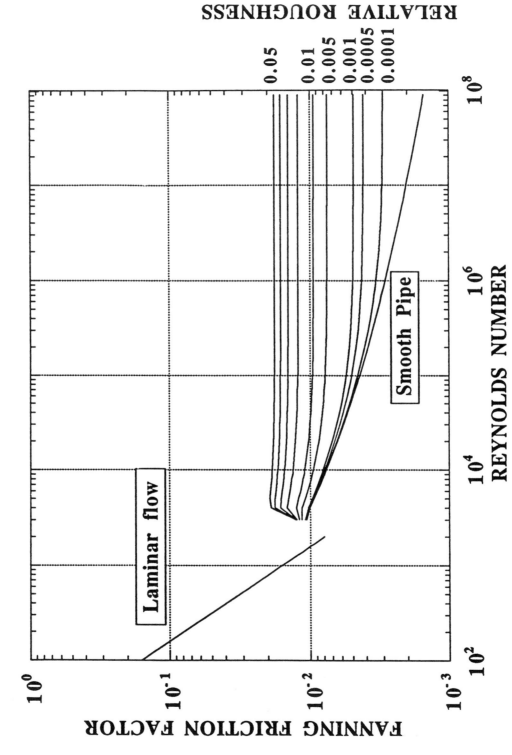

Figure C-1: Fanning friction factors calculated from the Churchill Equation [Based on L.F. Moody, *Trans. ASME*, 66:671-684 (1944)]

Table C-13: Dimensionless Equivalent Lengths
for Fittings and Valves

Description	(l_{eq} / D) per fitting
45 ° elbows	10
90 ° elbows, std. radius	32
90 ° elbows, medium radius	26
90 ° elbows, long sweep	20
90 ° square elbows	60
180 ° close return bends	75
180 ° medium-radius return bends	50
Tee (used as elbow, entering run)	60
Tee (used as elbow, entering branch)	90
Couplings	Negligible
Unions	Negligible
Gate valves, open	7
Globe valves, open	300
Angle valves, open	170
Water meters, disk	400
Water meters, piston	600
Water meters, impulse wheel	300

(Adapted from Peters, M.S. and K.D. Timmerhaus, "Plant Design and Economics for Chemical Engineers," 3rd. ed., McGraw-Hill, 1980, Pg. 514 and 515)

Table C-14: Roughness Values for Various Materials

Material	Roughness, ϵ in
Drawn tubing	0.00006
Commercial steel	0.0018
Wrought iron	0.0018
Asphalted cast iron	0.0048
Galvanized iron	0.0060
Cast iron	0.0102

(Adapted from Moody, L.F. *Trans*. *ASME*, **66**:671-684 (1944).)

Table C-15: Ideal-Gas State Heat Capacities of Selected Organics
for values of T from 273.15 K to 1500 K
for the equation $\widetilde{C}_p(T) / R = A + BT + CT^2 + DT^{-2}$

Chemical		A	B	C	D
Paraffins:					
Methane	CH_4	1.702	9.081×10^{-3}	-21.64×10^{-7}	–
Ethane	C_2H_6	1.131	19.225×10^{-3}	-55.61×10^{-7}	–
Propane	C_3H_8	1.213	28.785×10^{-3}	-88.24×10^{-7}	–
n-Butane	C_4H_{10}	2.281	36.371×10^{-3}	-111.80×10^{-7}	–
n-Pentane	C_5H_{12}	2.974	44.514×10^{-3}	-137.84×10^{-7}	–
n-Hexane	C_6H_{14}	3.763	52.553×10^{-3}	-163.42×10^{-7}	–
n-Heptane	C_7H_{16}	4.557	60.570×10^{-3}	-188.87×10^{-7}	–
n-Octane	C_8H_{18}	5.348	68.595×10^{-3}	-214.36×10^{-7}	–
1-Alkenes:					
Ethylene	C_2H_4	1.424	14.394×10^{-3}	-43.92×10^{-7}	–
Propylene	C_3H_6	1.637	22.706×10^{-3}	-69.15×10^{-7}	–
1-Butene	C_4H_8	2.583	31.082×10^{-3}	-97.24×10^{-7}	–
1-Pentene	C_5H_{10}	3.611	38.669×10^{-3}	-120.80×10^{-7}	–
1-Hexene	C_6H_{12}	4.333	46.850×10^{-3}	-147.05×10^{-7}	–
1-Heptene	C_7H_{14}	5.108	54.919×10^{-3}	-172.74×10^{-7}	–
Miscellaneous organics:					
Acetylene	C_2H_2	6.132	1.952×10^{-3}	–	-1.299×10^5
Acetaldehyde	C_2H_4O	3.735	14.609×10^{-3}	-44.00×10^{-7}	–
Ethanol	C_2H_6O	3.518	20.001×10^{-3}	-60.02×10^{-7}	–
1,3-Butadiene	C_4H_6	2.734	26.786×10^{-3}	-88.82×10^{-7}	–
Benzene	C_6H_6	-0.206	39.064×10^{-3}	-133.01×10^{-7}	–
Cyclohexane	C_6H_{12}	-3.876	63.249×10^{-3}	-209.28×10^{-7}	–
Toluene	C_7H_8	0.290	47.052×10^{-3}	-157.16×10^{-7}	–
Styrene	C_8H_8	2.050	50.192×10^{-3}	-166.62×10^{-7}	–
Ethylbenzene	C_8H_{10}	1.124	55.380×10^{-3}	-184.76×10^{-7}	–

(Selected from H.M. Spencer, *Ind. Eng. Chem.*, **40**:2152, 1948)

Table C-16: Ideal-Gas State Heat Capacities of Selected Inorganics
for values of T from 273.15 K to T_{max} K
for the equation $\widetilde{C}_p(T) / R = A + BT + DT^{-2}$

Chemical		A	B	D	T_{max}
Air		3.355	0.575×10^{-3}	-0.016×10^5	2,000
Ammonia	NH_3	3.578	3.020×10^{-3}	-0.19×10^5	1,800
Bromine	Br_2	4.493	0.056×10^{-3}	-0.154×10^5	3,000
Carbon monoxide	CO	3.376	0.557×10^{-3}	-0.031×10^5	2,000
Carbon dioxide	CO_2	5.457	1.045×10^{-3}	-1.157×10^5	2,000
Carbon disulfide	CS_2	6.266	0.805×10^{-3}	-0.906×10^5	1,800
Chlorine	Cl_2	4.442	0.089×10^{-3}	-0.344×10^5	3,000
Hydrogen	H_2	3.249	0.422×10^{-3}	0.083×10^5	3,000
Hydrogen sulfide	H_2S	3.931	1.490×10^{-3}	-0.232×10^5	2,300
Hydrogen chloride	HCl	3.156	0.624×10^{-3}	0.151×10^5	2,000
Hydrogen cyanide	HCN	4.736	1.359×10^{-3}	-0.725×10^5	2,500
Nitrogen	N_2	3.280	0.593×10^{-3}	0.040×10^5	2,000
Dinitrogen oxide	N_2O	5.328	1.214×10^{-3}	-0.928×10^5	2,000
Nitric oxide	NO	3.387	0.629×10^{-3}	0.014×10^5	2,000
Nitrogen dioxide	NO_2	4.982	1.195×10^{-3}	-0.792×10^5	2,000
Oxygen	O_2	3.639	0.506×10^{-3}	-0.227×10^5	2,000
Sulfur dioxide	SO_2	5.699	0.801×10^{-3}	-1.015×10^5	2,000
Sulfur trioxide	SO_3	8.060	1.056×10^{-3}	-2.028×10^5	2,000
Water	H_2O	3.470	1.450×10^{-3}	0.121×10^5	2,000

(Selected from K.K. Kelley, *U.S. Bur. Mines Bull.*, 584, 1960;
L.B. Pankratz, *U.S. Bur. Mines Bull.*, 672, 1982)

Table C-17: Liquid Heat Capacities
for values of T_{min} K to T_{max} K
for the equation $\widetilde{C}_p(T) / R = A + BT + CT^2 + ET^3$

Chemical	A	B	C	E	Temp Range K T_{min} - T_{max}
Ammonia	-16.45	266.08×10^{-3}	948.82×10^{-6}	1177.25×10^{-9}	196.15 - 373.15
Benzene	-38.14	606.88×10^{-3}	-1715.14×10^{-6}	1730.76×10^{-9}	279.15 - 523.15
1,3-Butadiene	10.29	28.51×10^{-3}	-156.56×10^{-6}	373.41×10^{-9}	164.15 - 393.15
Carbon tet.	0.95	159.5×10^{-3}	-545.63×10^{-6}	667.31×10^{-9}	250.15 - 533.15
Cyclohexane	-54.28	566.06×10^{-3}	-1483.8×10^{-6}	1364.21×10^{-9}	280.15 - 533.15
Ethanol	-8.1	221.3×10^{-3}	-876.48×10^{-6}	1263.8×10^{-9}	160.15 - 453.15
Ethylene oxide	6.27	51.97×10^{-3}	258.2×10^{-6}	444.87×10^{-9}	161.15 - 453.15
Methanol	13.50	-52.03×10^{-3}	133.60×10^{-6}	-2.72×10^{-9}	371.15 - 493.15
Sulfur trioxide	-160.16	1616.91×10^{-3}	-4690.5×10^{-6}	4750.9×10^{-9}	290.15 - 493.15
Toluene	-6.76	212.24×10^{-3}	-623.2×10^{-6}	659.79×10^{-9}	178.15 - 583.15
Water	6.11	25.59×10^{-3}	-75.83×10^{-6}	77.92×10^{-9}	273.15 - 623.15

(From J.W. Miller, Jr., G.R. Schorr, and C.L. Yaws, *Chem. Eng.*, **83** (23): 129, 1976)

Table C-18: Solid Heat Capacities
for values of 298.15 K to T_{max} K
for the equation $\widetilde{C}_p(T) / R = A + BT + DT^{-2}$

Chemical	A	B	D	T_{max}
CaO	6.104	0.443×10^{-3}	-1.047×10^5	2,000
CaCO$_3$	12.572	2.637×10^{-3}	-3.120×10^5	1,200
Ca(OH)$_2$	9.597	5.435×10^{-3}	–	700
CaC$_2$	8.254	1.429×10^{-3}	-1.042×10^5	720
CaCl$_2$	8.646	1.530×10^{-3}	-0.302×10^5	1,055
C(graphite)	1.771	0.771×10^{-3}	-0.867×10^5	2,000
Cu	2.677	0.815×10^{-3}	0.035×10^5	1,357
CuO	5.780	0.973×10^{-3}	-0.874×10^5	1,400
Fe(α)	-0.111	6.111×10^{-3}	1.150×10^5	1,043
Fe$_2$O$_3$	11.812	9.697×10^{-3}	-1.976×10^5	960
Fe$_3$O$_4$	9.594	27.112×10^{-3}	0.409×10^5	850
FeS	2.612	13.286×10^{-3}	–	411
I$_2$	6.481	1.502×10^{-3}	–	386
NH$_4$Cl	5.939	16.105×10^{-3}	–	457
Na	1.988	4.688×10^{-3}	–	371
NaCl	5.526	1.963×10^{-3}	–	1,073
NaOH	0.121	16.316×10^{-3}	1.948×10^5	566
NaHCO$_3$	5.128	18.148×10^{-3}	–	400
S(rhombic)	4.114	-1.728×10^{-3}	-0.783×10^5	368
SiO$_2$(quartz)	4.871	5.365×10^{-3}	-1.001×10^5	847

(Selected from K.K. Kelley, *U.S. Bur. Mines Bull.*, 584, 1960;
L.B. Pankratz, *U.S. Bur. Mines Bull.*, 672, 1982)

Table C-19: Critical Values for Various Chemicals

Chemical	T_c K	P_c atm	$\widetilde{V}_c$ cm^3 / gmol	Z_c	ω
Acetic acid	594.4	57.1	171.	0.200	0.454
Acetone	508.1	46.4	209.	0.232	0.309
Acetylene	308.3	60.6	113.	0.271	0.184
Ammonia	405.6	111.3	72.5	0.242	0.250
Argon	150.8	48.1	74.9	0.291	−0.004
Benzene	562.1	48.3	259.	0.271	0.212
Bromine	584.0	102.0	127.	0.270	0.132
n-Butane	425.2	37.5	255.	0.274	0.193
Isobutane	408.1	36.0	263.	0.283	0.176
Carbon dioxide	304.2	72.8	94.0	0.274	0.225
Carbon monoxide	132.9	34.5	93.1	0.295	0.049
Carbon disulfide	552.	78.0	170.	0.293	0.115
Carbon tetrachloride	556.4	45.0	276.0	0.272	0.194
Chlorine	417.0	76.0	124.	0.275	0.073
Chloroform	536.4	54.0	239	0.293	0.216
Cyclohexane	553.4	40.2	308.	0.273	0.213
Diethyl ether	466.7	35.9	280.	0.262	0.281
Ethane	305.4	48.2	148.	0.285	0.098
Ethanol	516.2	63.0	167.	0.248	0.635
Ethylene	282.4	49.7	129.	0.276	0.085
Hydrogen	33.2	12.8	65	0.305	−0.220
Hydrogen chloride	324.6	82.0	81.0	0.249	0.12
Hydrogen cyanide	456.8	53.2	139	0.197	0.407
Hydrogen sulfide	373.2	88.2	98.5	0.284	0.100
Methane	190.6	45.4	99.0	0.288	0.008
Methanol	512.6	79.9	118.	0.224	0.559
Methyl chloride	416.3	65.9	139.	0.268	0.156
Methyl ethyl ketone	535.6	41.0	267.	0.249	0.329
Neon	44.4	27.2	41.7	0.311	0.00
Nitric oxide (NO)	180.	64.	58.	0.250	0.507
Nitrogen	126.2	33.5	89.5	0.290	0.040
Nitrous oxide (N$_2$O)	309.6	71.5	97.4	0.274	0.160
Oxygen	154.6	49.8	73.4	0.288	0.021
n-Pentane	469.6	33.3	304.	0.262	0.251
Isopentane	433.8	31.6	303.	0.269	0.197
Propane	369.8	41.9	203.	0.281	0.152
Propylene	365.0	45.6	181.	0.275	0.148
Sulphur dioxide	430.8	77.8	122.	0.268	0.251
Sulphur trioxide	491.0	81.	130.	0.26	0.41
Toluene	591.7	40.6	316.	0.264	0.257
Water	647.3	217.6	56.0	0.229	0.344

(From R.C. Reid, J.M. Prausnitz, T.K. Sherwood, "The Properties of Gases and Liquids" 3rd. ed., McGraw-Hill, 1977, Pg. 629 to 677)

Table C-20: Standard Enthalpy of Formation at 298 K and 1 bar for Selected Organic Compounds

Chemical		State[a]	$\Delta \widetilde{H}^o_{298}$ J / gmol[b]
Paraffins:			
Methane	CH_4	g	$-74,520$
Ethane	C_2H_6	g	$-83,820$
Propane	C_3H_8	g	$-104,680$
n-Butane	C_4H_{10}	g	$-125,790$
n-Pentane	C_5H_{12}	g	$-146,760$
n-Hexane	C_6H_{14}	g	$-166,920$
n-Heptane	C_7H_{16}	g	$-187,780$
n-Octane	C_8H_{18}	g	$-208,750$
1-Alkenes:			
Ethylene	C_2H_4	g	$52,500$
Propylene	C_3H_6	g	$20,000$
1-Butene	C_4H_8	g	-540
1-Pentene	C_5H_{10}	g	$-21,300$
1-Hexene	C_6H_{12}	g	$-42,000$
1-Heptene	C_7H_{14}	g	$-62,800$
Miscellaneous organics:			
Formaldehyde	CH_2O	g	$-108,570$
Methanol	CH_4O	g	$-200,660$
Methanol	CH_4O	l	$-238,660$
Acetylene	C_2H_2	g	$228,200$
Acetaldehyde	C_2H_4O	g	$-166,190$
Ethylene oxide	C_2H_4O	g	$-52,630$
Acetic acid	$C_2H_4O_2$	l	$-484,500$
1,3-Butadiene	C_4H_6	g	$109,240$
Ethanol	C_2H_6O	g	$-235,100$
Ethanol	C_2H_6O	l	$-277,690$
Benzene	C_6H_6	g	$82,930$
Benzene	C_6H_6	l	$49,080$
Cyclohexane	C_6H_{12}	g	$-123,140$
Cyclohexane	C_6H_{12}	l	$-156,230$
Toluene	C_7H_8	g	$50,170$
Toluene	C_7H_8	l	$12,180$
Styrene	C_8H_8	g	$148,000$
Ethylbenzene	C_8H_{10}	g	$29,920$

[a] g: gas, l: liquid
[b] With respect to the substance formed
(Selected from "TRC Thermodynamic Tables-Hydrocarbons," Thermodynamics
Research Center, Texas A & M Univ. System, College Station, Texas;
"The NBS Tables of Chemical Thermodynamic Properties," J. Physical and
Chemical Reference Data, vol. 11, supp. 2, 1982)

Table C-21: Standard Enthalpy of Formation at 298 K and 1 bar
for Selected Inorganic Compounds

Chemical		State[a]	$\Delta \widetilde{H}^o_{298}$ J / gmol[b]
Ammonia	NH_3	g	$-46,110$
Calcium carbide	CaC_2	s	$-59,800$
Calcium carbonate	$CaCO_3$	s	$-1,206,920$
Calcium chloride	$CaCl_2$	s	$-795,800$
Calcium hydroxide	$Ca(OH)_2$	s	$-986,090$
Calcium oxide	CaO	s	$-635,090$
Carbon dioxide	CO_2	g	$-393,509$
Carbon monoxide	CO	g	$-110,525$
Hydrochloric acid	HCl	g	$-92,307$
Hydrogen cyanide	HCN	g	$135,100$
Hydrogen sulfide	H_2S	g	$-20,630$
Iron oxide	FeO	s	$-272,000$
Iron oxide (hematite)	Fe_2O_3	s	$-824,200$
Iron oxide (magnetite)	Fe_3O_4	s	$-1,118,400$
Iron sulfide (pyrite)	FeS_2	s	$-178,200$
Lithium chloride	$LiCl$	s	$-408,610$
Nitric acid	HNO_3	l	$-174,100$
Nitrogen oxides	NO	g	$90,250$
	NO_2	g	$33,180$
	N_2O	g	$82,050$
	N_2O_4	g	$9,160$
Sodium carbonate	Na_2CO_3	s	$-1,130,680$
Sodium chloride	$NaCl$	s	$-411,153$
Sodium hydroxide	$NaOH$	s	$-425,609$
Sulfur dioxide	SO_2	g	$-296,830$
Sulfur trioxide	SO_3	g	$-395,720$
Sulfur trioxide	SO_3	l	$-441,040$
Sulfuric acid	H_2SO_4	l	$-813,989$
Water	H_2O	g	$-241,818$
Water	H_2O	l	$-285,830$

[a] g: gas, l: liquid, s: solid
[b] With respect to the substance formed
(Selected from "The NBS Tables of Chemical Thermodynamic Properties,"
J. Physical and Chemical Reference Data, vol. 11, supp. 2, 1982)

Table C-22: Thermal Properties of Saturated and Superheated Steam (English Units)

P psi	T^{sat} °F		Saturated Water	Saturated Steam	Temperature °F 400	600	800	1000	1200	1400
14.7	212	$\hat{V}$	0.0167	26.799	34.67	42.86	51.00	59.13	67.25	75.36
		$\hat{U}$	180.12	1077.6	1145.7	1218.7	1294.5	1373.7	1456.5	1543.0
		$\hat{H}$	180.17	1150.5	1239.9	1335.2	1433.2	1534.5	1639.4	1747.9
		$\hat{S}$	0.3121	1.7568	1.8743	1.9739	2.0585	2.1332	2.2005	2.2621
100	328	$\hat{V}$	0.01774	4.431	4.935	6.216	7.443	8.655	9.860	11.060
		$\hat{U}$	298.21	1105.2	1136.0	1214.5	1291.9	1371.8	1455.1	1542.0
		$\hat{H}$	298.54	1187.2	1227.4	1329.6	1429.7	1532.0	1637.6	1746.7
		$\hat{S}$	0.4743	1.6187	1.6516	1.7586	1.8451	1.9205	1.9883	2.0502
200	382	$\hat{V}$	0.01839	2.2873	2.3598	3.0583	3.6915	4.3077	4.9165	5.521
		$\hat{U}$	354.82	1113.7	1122.8	1209.4	1288.9	1369.7	1453.5	1540.9
		$\hat{H}$	355.51	1198.3	1210.1	1322.6	1425.5	1529.1	1635.4	1745.3
		$\hat{S}$	0.5438	1.5454	1.5593	1.6773	1.7663	1.8426	1.9109	1.9732
400	445	$\hat{V}$	0.01934	1.1610	–	1.4763	1.8151	2.1339	2.4450	2.752
		$\hat{U}$	422.74	1118.7	–	1198.2	1282.7	1365.3	1450.2	1538.7
		$\hat{H}$	424.17	1204.6	–	1307.4	1417.0	1523.3	1631.2	1742.4
		$\hat{S}$	0.6217	1.4847	–	1.5901	1.6850	1.7632	1.8325	1.8956
600	486	$\hat{V}$	0.02013	0.7697	–	0.9456	1.1892	1.4093	1.6211	1.8289
		$\hat{U}$	469.46	1118.2	–	1185.3	1276.2	1360.9	1447.0	1536.5
		$\hat{H}$	471.70	1203.7	–	1290.3	1408.3	1517.4	1627.0	1739.5
		$\hat{S}$	0.6723	1.4461	–	1.5329	1.6351	1.7155	1.7859	1.8497

P (psia)	T (°F)									
800	518	$\hat{V}$	0.02087	0.5690	—	0.6774	0.8759	1.0470	1.2093	1.3674
		$\hat{U}$	506.7	1115.2	—	1170.8	1269.5	1356.4	1443.7	1534.2
		$\hat{H}$	509.81	1199.4	—	1271.1	1399.1	1511.4	1622.7	1736.6
		$\hat{S}$	0.7111	1.4163	—	1.4869	1.5980	1.6807	1.522	1.8167
1000	544	$\hat{V}$	0.02159	0.4460	—	0.5137	0.6875	0.8295	0.9622	1.0905
		$\hat{U}$	538.6	1110.4	—	1154.3	1262.4	1351.9	1440.4	1531.9
		$\hat{H}$	542.55	1192.9	—	1249.3	1389.6	1505.4	1618.4	1733.7
		$\hat{S}$	0.7434	1.3910	—	1.4457	1.5677	1.6530	1.7256	1.7909
1200	567	$\hat{V}$	0.02232	0.3624	—	0.4016	0.5615	0.6845	0.7974	0.9055
		$\hat{U}$	566.9	1104.3	—	1135.0	1255.0	1347.4	1437.1	1528.2
		$\hat{H}$	571.85	1184.8	—	1224.2	1379.7	1499.4	1614.2	1729.4
		$\hat{S}$	0.7714	1.3683	—	1.4061	1.5415	1.6298	1.7035	1.7694
1400	587	$\hat{V}$	0.02307	0.3018	—	0.3176	0.4712	0.5809	0.6798	0.7737
		$\hat{U}$	592.9	1097.1	—	1111.8	1247.3	1342.7	1433.8	1525.7
		$\hat{H}$	598.83	1175.3	—	1194.1	1369.3	1493.2	1609.9	1726.3
		$\hat{S}$	0.7966	1.3474	—	1.3652	1.5182	1.6096	1.6845	1.7508
1600	605	$\hat{V}$	0.02387	0.2555	—	—	0.4032	0.5031	0.5915	0.6748
		$\hat{U}$	617.1	1088.9	—	—	1239.1	1338.0	1430.4	1523.3
		$\hat{H}$	624.2	1164.5	—	—	1358.5	1486.9	1605.6	1723.2
		$\hat{S}$	0.8199	1.3274	—	—	1.4968	1.5916	1.6678	1.7347
1800	621	$\hat{V}$	0.02472	0.2186	—	—	0.3500	0.4426	0.5229	0.5980
		$\hat{U}$	640.3	1079.5	—	—	1230.6	1333.2	1427.1	1520.8
		$\hat{H}$	648.49	1152.3	—	—	1374.2	1480.6	1601.2	1720.1
		$\hat{S}$	0.8417	1.3079	—	—	1.4768	1.5753	1.6528	1.7204

Units: $\hat{V}$ [=] ft³/lb$_m$; $\hat{U}$ [=] Btu/lb$_m$; $\hat{H}$ [=] Btu/lb$_m$; $\hat{S}$ [=] Btu/lb$_m$ R

Reference State is 32 °F and 0.0886 psia where $\hat{H}_{Liq}^{Sat} = 0.000$ and $\hat{S}_{Liq}^{Sat} = 0.000$

(Adapted from the 1967 ASME STEAM TABLES; by The American Society of Mechanical Engineers)

Table C-23: Thermal Properties of Saturated and Superheated Steam (SI Units)

P kPa	T^{sat} °C		Saturated Water	Saturated Steam	Temperature °C					
					200	300	400	500	600	800
101.3	100	$\hat{V}$	0.001044	1.673	2.1438	2.6042	3.0619	3.5187	3.9750	4.952
		$\hat{U}$	418.96	2506.5	2658.1	2810.6	2968.0	3131.6	3302.0	3663.5
		$\hat{H}$	419.06	2676.0	2875.3	3074.4	3278.2	3488.1	3704.8	4158.6
		$\hat{S}$	1.3069	7.3554	7.8288	8.2105	8.5381	8.8287	9.0922	9.5652
200	120	$\hat{V}$	0.001061	0.885	1.0804	1.3162	1.5492	1.7812	2.0129	2.475
		$\hat{U}$	504.49	2529.2	2654.4	2808.8	2966.9	3130.8	3301.4	3663.1
		$\hat{H}$	504.70	2706.3	2870.5	3072.1	3276.7	3487.0	3704.0	4158.2
		$\hat{S}$	1.5301	7.1268	7.5072	7.8937	8.2226	8.5139	8.7776	9.2449
400	144	$\hat{V}$	0.001084	0.462	0.5343	0.6548	0.7725	0.8892	1.0054	1.2372
		$\hat{U}$	604.24	2552.7	2646.7	2805.3	2964.6	3129.2	3300.2	3662.4
		$\hat{H}$	604.67	2737.6	2860.4	3067.2	3273.6	3484.9	3702.3	4157.3
		$\hat{S}$	1.7764	6.8943	7.1708	7.5675	7.8994	8.1919	8.4563	8.9244
600	159	$\hat{V}$	0.001101	0.315	0.3520	0.4344	0.5136	0.5918	0.6696	0.8245
		$\hat{U}$	669.76	2566.2	2638.5	2801.6	2962.4	3127.6	3298.9	3661.8
		$\hat{H}$	670.42	2755.5	2849.7	3062.3	3270.6	3482.7	3700.7	4156.5
		$\hat{S}$	1.9308	6.7575	6.9662	7.3740	7.7090	8.0027	8.2678	8.7367
800	170	$\hat{V}$	0.001115	0.240	0.2608	0.3241	0.3842	0.4432	0.5017	0.6181
		$\hat{U}$	720.04	2757.3	2629.9	2797.9	2960.2	3125.9	3297.7	3661.1
		$\hat{H}$	720.94	2767.5	2838.6	3057.3	3267.5	3480.5	3699.1	4155.6
		$\hat{S}$	2.0457	6.6596	6.8148	7.2348	7.5729	7.8678	8.1336	8.6033

(kPa)	(°C)									
1000	180	$\hat{V}$	0.001127	0.194	0.2059	0.2580	0.3065	0.3540	0.4010	0.4943
		$\hat{U}$	761.48	2581.9	2620.9	2794.2	2957.9	3124.3	3296.4	3660.4
		$\hat{H}$	762.60	2776.2	2826.8	3052.1	3264.4	3478.3	3697.4	4154.7
		$\hat{S}$	2.1382	6.5828	6.6922	7.1251	7.4665	7.7627	8.0292	8.4996
2000	212	$\hat{V}$	0.001177	0.0995	—	0.1255	0.1511	0.1755	0.1995	0.2467
		$\hat{U}$	906.24	2598.2	—	2774.0	2946.4	3116.2	3290.2	3657.0
		$\hat{H}$	908.59	2797.2	—	3025.0	3248.7	3467.3	3689.2	4150.3
		$\hat{S}$	2.4469	6.3366	—	6.7696	7.1296	7.4323	7.7022	8.1765
5000	264	$\hat{V}$	0.001286	0.0394	—	0.0453	0.0578	0.0685	0.0786	0.09811
		$\hat{U}$	1148.0	2597.0	—	2699.0	2909.3	3091.2	3271.4	3646.6
		$\hat{H}$	1154.5	2794.2	—	2925.5	3198.3	3433.7	3664.5	4137.1
		$\hat{S}$	2.9206	5.9735	—	6.2105	6.6508	6.9770	7.2578	7.7440
10000	311	$\hat{V}$	0.001453	0.0180	—	—	0.0264	0.0328	0.0383	0.04859
		$\hat{U}$	1393.5	2547.3	—	—	2835.8	3047.0	3239.5	3628.9
		$\hat{H}$	1408.0	2727.7	—	—	3099.9	3374.6	3622.7	4114.8
		$\hat{S}$	3.3605	5.6198	—	—	6.2182	6.5994	6.9013	7.4077
15000	342	$\hat{V}$	0.001658	0.0103	—	—	0.0156	0.0208	0.0249	0.02385
		$\hat{U}$	1586.1	2459.9	—	—	2744.2	2998.7	3206.5	3592.7
		$\hat{H}$	1611.0	2615.0	—	—	2979.1	3310.6	3579.8	4069.7
		$\hat{S}$	3.6859	5.3178	—	—	5.8876	6.3487	6.6764	7.0544
20000	366	$\hat{V}$	0.002037	0.00587	—	—	0.00995	0.0148	0.0182	0.02385
		$\hat{U}$	1785.7	2300.8	—	—	2621.6	2945.7	3172.3	3592.7
		$\hat{H}$	1826.5	2418.4	—	—	2820.5	3241.1	3535.5	4069.7
		$\hat{S}$	4.0149	4.9412	—	—	5.5585	6.1456	6.5043	7.0544

Units: $\hat{V}$ [=] m^3/kg; $\hat{U}$ [=] kJ/kg; $\hat{H}$ [=] kJ/kg; $\hat{S}$ [=] kJ/kg K

Reference State is 0.01 °C and 0.611 kPa where $\hat{H}_{\mathrm{Liq}}^{\mathrm{Sat}} = 0.000$ and $\hat{S}_{\mathrm{Liq}}^{\mathrm{Sat}} = 0.000$

(Adapted from the 1967 ASME STEAM TABLES; by The American Society of Mechanical Engineers)

Table C-24: Thermal Properties of Saturated and Superheated Ammonia (English Units)

P psi	T^{sat} °F		Saturated Liquid	Saturated Vapor	Temperature °F						
					0	40	80	120	160	200	240
10	−41	$\hat{V}$	0.0232	25.81	28.58	31.20	33.78	36.35	38.90	41.45	—
		$\hat{H}$	−1.4	597.1	618.9	639.3	659.7	680.3	701.1	722.2	—
		$\hat{S}$	−0.0034	1.4276	1.477	1.520	1.559	1.596	1.631	1.664	—
20	−17	$\hat{V}$	0.02376	13.5	14.09	15.45	16.78	18.08	19.37	20.66	—
		$\hat{H}$	25.0	606.2	615.5	637.0	658.0	678.9	700.0	721.2	—
		$\hat{S}$	0.0578	1.3700	1.391	1.436	1.476	1.513	1.549	1.582	—
30	−0.6	$\hat{V}$	0.02417	9.236	9.25	10.2	11.10	11.99	12.87	13.73	—
		$\hat{H}$	42.3	611.6	661.9	634.6	656.2	677.5	698.8	720.3	—
		$\hat{S}$	0.0962	1.3364	1.337	1.385	1.426	1.464	1.500	1.533	—
50	21.7	$\hat{V}$	0.02479	5.710	—	5.988	6.564	7.117	7.655	8.185	—
		$\hat{H}$	66.5	618.2	—	629.5	652.6	674.7	696.6	718.5	—
		$\hat{S}$	0.1475	1.1464	—	1.317	1.361	1.401	1.437	1.472	—
70	37.7	$\hat{V}$	0.02526	4.151	—	4.177	4.615	5.025	5.420	5.807	6.187
		$\hat{H}$	84.2	622.4	—	623.9	648.7	671.8	694.3	716.6	738.9
		$\hat{S}$	0.1835	1.2658	—	1.269	1.317	1.358	1.395	1.430	1.463
100	56.0	$\hat{V}$	0.02584	2.952	—	—	3.149	3.454	3.743	4.021	4.294
		$\hat{H}$	104.7	626.5	—	—	642.6	667.3	690.8	713.7	736.5
		$\hat{S}$	0.2237	1.2356	—	—	1.266	1.310	1.349	1.385	1.419

P	T										
140	74.8	$\hat{V}$	0.02649	2.132	—	—	2.166	2.404	2.622	2.830	3.030
		$\hat{H}$	126.0	629.9	—	—	633.8	661.1	686.0	709.9	733.3
		$\hat{S}$	0.2638	1.2068	—	—	1.214	1.263	1.305	1.342	1.376
180	89.8	$\hat{V}$	0.02706	1.667	—	—	—	1.818	1.999	2.167	2.328
		$\hat{H}$	143.3	632.0	—	—	—	654.4	681.0	705.9	730.1
		$\hat{S}$	0.2954	1.1850	—	—	—	1.225	1.269	1.308	1.344
220	102.4	$\hat{V}$	0.02758	1.367	—	—	—	1.443	1.601	1.745	1.881
		$\hat{H}$	158.0	633.2	—	—	—	647.3	675.8	701.9	726.8
		$\hat{S}$	0.3216	1.1671	—	—	—	1.192	1.239	1.280	1.317
240	108.1	$\hat{V}$	0.02782	1.253	—	—	—	1.302	1.452	1.587	1.741
		$\hat{H}$	164.7	633.6	—	—	—	643.5	673.1	699.8	725.1
		$\hat{S}$	0.3332	1.1592	—	—	—	1.176	1.226	1.268	1.305
260	113.4	$\hat{V}$	0.02829	1.155	—	—	—	1.182	1.326	1.453	1.572
		$\hat{H}$	171.1	633.9	—	—	—	639.5	670.4	697.7	723.4
		$\hat{S}$	0.3441	0.1518	—	—	—	1.162	1.213	1.256	1.294

Units: $\hat{V}$ [=] ft^3/lb$_m$; $\hat{U}$ [=] Btu/lb$_m$; $\hat{H}$ [=] Btu/lb$_m$; $\hat{S}$ [=] Btu/lb$_m$; $\hat{S}$ [=] Btu/lb$_m$ R

Reference State is -40 °F and 10.41 psia where $\hat{H}_{Liq}^{Sat} = 0.000$ and $\hat{S}_{Liq}^{Sat} = 0.000$

(From *U.S. Natl. Bur. Stand. Circ.* 142 (1923))

Table C-25: Thermal Properties of Saturated and Superheated Ammonia (SI Units)

P kPa	T^{sat} °C		Saturated Liquid	Saturated Vapor	−20	0	20	40	60	80	100
							Temperature °C				
75	−39.2	$\hat{V}$	0.001451	1.5434	1.6233	1.7591	1.8932	2.0261	2.1584	2.2903	—
		$\hat{H}$	2.168	1390.7	1433.0	1476.1	1518.9	1561.8	1605.1	1648.9	—
		$\hat{S}$	0.008402	5.9558	6.1190	6.2828	6.4339	6.5756	6.7096	6.8373	—
150	−25.2	$\hat{V}$	0.001488	0.7958	0.7984	0.8697	0.9388	1.0068	1.0740	1.1408	1.2072
		$\hat{H}$	65.260	1412.2	1424.1	1469.8	1514.1	1558.0	1602.0	1646.3	1691.1
		$\hat{S}$	0.2704	5.7111	5.7526	5.9266	6.0831	6.2280	6.3641	6.4933	6.6167
200	−18.9	$\hat{V}$	0.001506	0.6026	—	0.6471	0.7001	0.7519	0.8029	0.8533	0.9035
		$\hat{H}$	94.456	1421.4	—	1465.5	1510.9	1555.5	1599.9	1644.6	1689.6
		$\hat{S}$	0.3870	5.6090	—	5.7737	5.9342	6.0813	6.2189	6.3491	6.4732
300	−9.2	$\hat{V}$	0.001535	0.4276	—	0.4243	0.4613	0.4968	0.5316	0.5658	0.5997
		$\hat{H}$	136.48	1433.0	—	1456.3	1504.2	1550.3	1595.7	1641.1	1686.7
		$\hat{S}$	0.5481	5.0569	—	5.5493	5.7186	5.8707	6.0114	6.1437	6.2693
400	−1.9	$\hat{V}$	0.001559	0.3173	—	0.3125	0.3417	0.3692	0.3959	0.4220	0.4478
		$\hat{H}$	171.24	1441.9	—	1446.5	1497.2	1544.9	1591.5	1637.6	1683.7
		$\hat{S}$	0.6782	5.0008	—	5.3803	5.5597	5.7173	5.8613	5.9957	6.1228
500	4.1	$\hat{V}$	0.00158	0.2528	—	—	0.2698	0.2926	0.3144	0.3357	0.3565
		$\hat{H}$	199.91	1448.5	—	—	1489.9	1539.5	1587.1	1634.0	1680.7
		$\hat{S}$	0.7826	5.2889	—	—	5.4314	5.5950	5.7425	5.8793	6.0079

600	9.3	$\widehat{V}$	0.00160	0.2165	—	0.2217	0.2414	0.2600	0.2781	0.2957
		$\widehat{H}$	223.0	1453.1	—	1482.4	1533.8	1582.7	1630.4	1667.7
		$\widehat{S}$	0.8640	5.2277	—	5.3222	5.4923	5.6436	5.7826	5.9129
800	17.9	$\widehat{V}$	0.00163	0.1637	—	0.1615	0.1773	0.1920	0.2060	0.2196
		$\widehat{H}$	263.46	1460.4	—	1446.3	1522.2	1573.7	1623.1	1671.6
		$\widehat{S}$	1.0041	5.1247	—	5.1387	5.3232	5.4827	5.6268	5.7603
1000	24.9	$\widehat{V}$	0.001658	0.1294	—	—	0.1388	0.1511	0.1627	0.1739
		$\widehat{H}$	298.22	1465.8	—	—	1510.0	1564.4	1615.6	1665.4
		$\widehat{S}$	1.1214	5.0410	—	—	5.1840	5.3525	5.5021	5.6392
1200	31.0	$\widehat{V}$	0.001684	0.1083	—	—	0.1129	0.1238	0.1338	0.1434
		$\widehat{H}$	327.41	1469.3	—	—	1497.1	1554.7	1608.0	1659.2
		$\widehat{S}$	1.2174	4.9747	—	—	5.0629	5.2416	5.3970	5.5379
1600	41.0	$\widehat{V}$	0.001731	0.081007	—	—	—	0.0895	0.0977	0.1053
		$\widehat{H}$	377.00	1473.4	—	—	—	1534.4	1592.3	1646.4
		$\widehat{S}$	1.3760	4.8663	—	—	—	5.0543	5.3722	5.3722

Units: $\widehat{V}$ [=] m³/kg; $\widehat{U}$ [=] kJ/kg; $\widehat{H}$ [=] kJ/kg; $\widehat{S}$ [=] kJ/kg K

Reference State is -40 °C and 71.8 kPa where $\widehat{H}_{\text{Liq}}^{\text{Sat}} = 0.000$ and $\widehat{S}_{\text{Liq}}^{\text{Sat}} = 0.000$

(From *U.S. Natl. Bur. Stand. Circ.* 142 (1923))

Index

A

absolute temperature, 543, 546, 547, 549, 594
absorption refrigeration, 638, 676
accelertion, 314
accounting, definition of, 8
    and extensive property, 8
    and generation and consumption, 10
    and system, 8
    and time period, 9
    versus conservation, 13
accounting equations,
    and algebraic equations, 22-23
    and ordinary differential equations, 23
    and partial differential equations, 23
accounting equations summary, 22, 252-253
accumulation, 10
adiabatic, 185
air conditioner (refrigeration cycle), 671-675
angular momentum,
    and center of mass frame, 151, 155
    conservation of, 135-174, 157, 155-165
    and a constant direction axis, 145
    definition of, 136
    exchange due to forces, 136-137
    exchange due to mass exchange, 136
    and an inertial frame, 155-156
    and moment of inertia, 151
    and parallel axis theorem, 157-160
    for a particle, 142-145
    per unit mass ($r \times v$), 136
    and planes of symmetry, 163
    rate equation, 137, 155-160
    for a rigid body, 147-155
    and rigid body statics, 140-142
    and steady-state, 140
    summary of rigid body forms, 159
    of the system, 142-155
    versus linear momentum, 165-166
angular velocity, 144
anode, 84
Antoine equation for vapor pressure, 606
availability, 640-643
    for closed systems, 642-643
    for steady-state flow processes, 641

B

balance equation, 19
base vectors, 697-701

basis, 9
battery, 84
beginning of time period, 10
Benedict-Webb-Rubin equation of state, 578
Bernoulli's equation, 199, 449-450
buoyancy, 404-407
buoyant force, 404
bypass stream, 52

C

capacitor, 86-87
Carnot cycle, 633, 650-652
Carnot efficiency, 637
cathode, 84
center of mass, 110
centroid, 269
charge accounting rate equation, 89-90
charge accumulation on a plate, 91
charge, conservation of, 79-104, 81
    accounting of, 79
    Kirchhoff's current law, 94
chemical reactions and mass conservation, 59-66
Churchill equation for friction factor, 437
Clapeyron equation, 605
Clausius-Clapeyron equation, 605
closed system, 190
coefficient of performance, 671
compressibility factors, 596
    and equations for H, S, 596
concentrated load, 267
conservation and accounting equations summary table, 22, 252-253
conservation, definition of, 13
    and extensive property, 13-14
    and system, 13-14
    and time period, 14
    versus accounting, 13
    versus balance, 19
conserved property, 13
constitutive relationships, 20
consumption, 10
contact angle, 410-412
corresponding states correlations, 579
critical point, 573
cross product, 136, 703-704
cubic equations of state, 576-578
current, 90-91, 477
current divider circuit, 487
curvilinear coordinate systems, 317-321
cycles, 630-633
cylindrical coordinate system, 699-700, 706-708

and kinematics, 317-318

D

Daniell cell, 84-85
Darcy friction factor, 451
degrees of freedom for Gibbs's phase rule, 563
density, 55-56
    tables, 718-720
design, 20-21
diesel cycle, 668-669
dimensions, 694-695
distillation, 15-16, 49-50
distributed forces, 262
    friction, 264-267
    gravity, 263-264
    loads, 267-270
dot product, 701-702
    and work, 186
drag force, 130, 339-344, 378
dynamics,
    of fluids, 423
    and particle kinematics, 314-327
    of particles, 338-352
    of rigid bodies, 352-372
    and rigid body kinematics, 327-338

E

efficiency,
    Carnot, 638
    thermal, 634
electrical energy per charge (voltage), 203
electric field, 86, 203
end of time period, 10
energy,
    conservation of, 175-230
    conversion, 197
    electrical, 183-185, 203-218
    exchange of due to heat, 185
    exchange of due to mass exchange, 177-183
    exchange of due to work, 186
    forms of, 177-180
    heat, 185
    internal, 182-183, 540, 546-557
    kinetic, 177-180
    mechanical energy accounting, 197-200
    potential, 180-182
    thermal energy accounting, 201-203
    work, 186-187
energy analysis of dynamics, 347-352, 364-372
engineering and analysis picture, 20
enthalpy, 188-189, 201, 537-559, 595-599
    calculation of, 547, 580, 599

form of energy conservation, 188-189
    and heats of formation, 610
    and heats of reaction, 609-610
    and heats of vaporization, 605
    for ideal gases, 585
    for non-ideal gases, 594-599
    for solids and liquids, 599
    and thermal energy accounting, 201
    as a thermodynamic property, 537-559
    residual, 595
entropy, 231
    accounting of, 234-242
    addition due to heat transfer, 233, 236
    and availability, 640-643
    calculations of, 549, 586, 599
    exchange with mass exchange, 233, 235
    and free expansion of a gas, 240-241
    generation, 234, 237
    and heat engines, 632, 633-639, 651-670
    irreversibility, 643
    and refrigeration cycles, 633, 670-673
    residual, 596
    and the second law of thermodynamics, 231-250
    as a state property of matter, 233
entropy generation,
    and chemical reaction equilibrium,
    definition, 630
    and free expansion of a gas, 240-241
    and mixing of two unlike gases, 630
    and temperature equilibration, 243, 630
    and vapor liquid equilibrium, 243
    and viscous dissipation, 630
entropy generation due to heat transfer, 234, 237
equation of state, 538, 572-585
equations of state ($P - \hat{V} - T$ relations)
    for gases, 574-579
    for liquids, 579
equations summary, 252-253
equivalent force, 267-277
equivalent resistance, 480-488
Erickson cycle, 652
extensive property, 8

F

Fanning friction factor, 436-437, 451, 728
first-order response, 519-527
flash distillation, 15-16, 49
flow measurement, 452-459
flow work, 188-190
fluid statics, 385-422
forces and moving fluids, 459-467

on pipe bends, 459-460
    and rockets, 460
frame indifference, 124
frames, 286
friction, 264-267
friction coefficient,
    kinetic, 265
    static, 265
friction factor, 436-437, 451, 728
friction losses, 198, 435
    in pipe fittings, 440-441, 729
    for straight pipe, 436-437
fulcrum, 295
fusion (melting), 572

G

generation, 9-10
Gibb's free energy, 538, 545-546, 559-561
    and reference states, 558-559
Gibb's phase rule, 563-566
Greek alphabet, 693

H

heat,
    and conservation of energy, 187
    definition, 185-86
    and the second law of thermodynamics, 233-237,
        629-643
    sign convention, 185
heat capacity at constant pressure($\widehat{C}_P$), 547-570
    and enthalpy, 547-548, 580, 595
    and entropy, 549, 586, 595-596
    and internal energy, 549
    related to $\widehat{C}_V$, 549, 571, 586
heat capacity at constant volume($\widehat{C}_V$), 548, 566
    and entropy, 549, 585
    and internal energy, 547-548, 585
    related to $\widehat{C}_P$, 549, 570-571, 586
heat engine, 632, 650-670
heat of evaporization, 605
heat pump, 676
heats of formation, 612
heats of reaction, 609
Helmholtz free energy, 538, 545-546, 558-560

I

ideal gas constant, R, 594, 692
ideal gas equation, 574
ideal gas temperature, 594
inductor, 203, 210-211, 514-515, 519-520, 527-529
inertia,
    and angular momentum, 155-159

for a particle, 145
    relative to the center of mass, 151
    for a rigid body, 148
input, 10
intensive property, 537, 540, 541, 563
interfacial tension, 410-413
internal combustion cycles, 664-670
internal energy, 182-183, 537-538, 540-541, 543, 545-
        549, 558-559, 564-567
    calculations of, 547, 580, 585
irreversibility, 643
irreversible, 234, 236
isentrope, isentropic, 549
isobar, isobaric, 549
isochore, isochoric, 549
isolated systems, 191
isometric, 549
isotherm, isothermal, 549

J

jet engines, 670
Joule-Thompson coefficient, 673

K

kinematics, 314
    interdependent motion, 333-334
    of particles, 314-327
    relative motion, 332-333
    of rigid bodies, 327-338
kinetic energy, 177-180
Kirchhkoff's current law, 94, 478
Kirchhkoff's voltage law, 204-205, 478

L

laminar flow, 426
lever and fulcrum, 295
linear momentum, conservation of, 16-19, 105-133
    and center of mass, 109-110
    exchange due to forces, 106
    exchange due to mass exchange, 106
    and Newton's laws of motion, 114
    for a particle, 258
    for a rigid body, 109-110, 258, 327
    and rigid body dynamics, 313-375
    and rigid body statics, 257-311
    and velocity of the center of mass, 109

M

machines, 294
magnetic fields, 203-204, 210-214, 515
mass composition, 40-42
mass, conservation of, 14, 35-78, 38-39
    accounting, 36
    bypass stream, 52

and the chemical elements, 38
and chemical reactions, 38, 59-66
and closed systems, 67
density, 55
multi-component accounting, 38, 44
and multiple inlets, 48
and multiple unit processes, 50
and open systems, 67
rate equations, 42-45
recycle stream, 52
and steady state, 48
mass flow rate, 42, 426
and angular momentum exchange, 136
and density, 55
and energy exchange, 187
and entropy, 235
and linear momentum exchange, 107
and moving system boundary, 461
mass fraction, 40-42
maximum power transfer, 502
Maxwell relations, 567-571
mechanical energy, 197, 347-348, 424-425
mechanical energy conversion, 197
mesh analysis, 497, 517
method of joints, 278
method of sections, 278
model, 20
modeling foundation, 20-21
molar composition, 41-42
molar volume, 574
mole fraction, 41
moment, 136
moment arm, 136
moment of inertia, 144, 722
and angular momentum, 144
and the parallel axis theorem, 157
momentum analysis of dynamics, 338-347, 352-364
momentum in transit (forces), 106
momentum of a rigid body, 108-110
momentum per mass (velocity), 106
money and accounting, 10-11
multi-component rate equations, 44
multiple unit process, 50
multi-stage process, 50

N

negative charge, 79-90
Newton's laws of motion, 114
nodal analysis, 493, 517
non-conserved property, 14
non-Newtonian fluids, pipe flow, 449

n-type material, 88
nuclear reactions and charge conservation, 82-84

O

Ohms law, 206, 478
orifice meter, 453
Otto cycle, 665-667
output, 9-10

P

parallel axis theorem, 157
particle statics, 259-262
particle, versus rigid body, 257
path coordinate system, 710-713
and kinematics, 320-321
path function, 540-545
path property, 540-545
Peng-Robinson equation of state, 577
phase transitions, 603
pin connections, and structures, 277, 724-725
pipe flow, 433-459
and the extended Bernouilli equation, 433-434
and flow measurement, 452-459
and friction factor, 436, 437, 451
and friction losses, 435-437
laminar flow, 426, 434, 436
and non-circular conduits, 450-451
and non-Newtonian fluids, 449
and pipe fittings, 440-441, 729
and pipe networks, 449
Reynolds number, 435
turbulent flow, 426, 434, 436
and types of problems, 441-442
Pitot tube, 454
p-n diode, 88
position vector, 109-110, 704
cyclindrical coordinates, 704
rectangular cartesian coordinates, 704
spherical coordinates, 704
positive charge, 79-90
potential energy, 180-182
power, 187
and electrical energy, 480
power cycle, 632, 650-670
Carnot cycles, 650-652
diesel cycle, 668-669
heat and work calculations, 658-659
internal combustion cycles, 664-670
jet engine, 670
Otto cycle, 665-668
Rankine cycle, 652-658
rocket engines, 670

pressure, 119-122, 386-388
    and buoyancy, 404-407
    dependence on position, 386-388
    forces on a submerged area, 392-404
    and interfacial tension, 410-413
    and pressure vessels, 407-410
pressure energy, 425
pressure vessels, 407-410
psia, 118
psig, 118
p-type material, 88
pulleys, 296-298

R

Rackett equation, 579
radial velocity component, 144
Rankine cycle, 652-658
rate equation, 23
    for angular momentum conservation, 137
    for charge conservation, 89-90
    for energy conservation, 187
    for linear momentum conservation, 107
    for mass conservation, 42-43
    for the second law of thermodynamics, 233
rectangular, cartesian coordinate system, 698
    and kinematics, 315-316
recycle stream, 52
Redlich-Kwong equation of state, 576
reference frame,
    and angular momentum, 155-160
    and linear momentum, 124
reference states for thermodynamic functions, 558-559
refrigeration cycle, 633, 670-676
    absorption refrigeration, 676
    Carnot cycle, 670-671
    and the heat pump, 675
    vapor compression cycle, 671-675
Reidel equation for vapor pressure, 606
relative roughness, 436, 728
residual enthalpy, 595
residual entropy, 595
resistor, 206-207, 477-505, 514
resistors in parallel, 484
resistors in series, 480
reversible, 236
Reynold's number, 435-437, 728
rigid body, versus particle, 257
rigid body statics, 262
rocket engines, 670
rotational kinetic energy, 179

S

second-order response, 527-529
semi-conductor and charge conservation, 87-88
Soave-Redlich-Kwong equation of state, 577
source transformation, 489
space truss, 277, 281-286
specific volume, 190
    related to molar volume, 574
spherical coordinate system, 708-710
    and kinematics, 319-320
state function, 540-541
state property, 540-541
statics,
    fluid, 385-422
    rigid body, 257-311
steady state, 48
Stirling cycle, 652
sublimation, 572
summary of equations, 252-253
supercritical fluid, 572
superposition principle, 491
supports for structures, 277, 724-725
system, 8

T

Tait equation, 580
temperature,
    and Carnot cycle, 636-637
    ideal gas absolute temperature, 594, 637-638
    and thermodynamic efficiency, 638
    and the zeroth law, 634
thermal energy, 201
thermodynamic efficiency, 635, 638
thermodynamic temperature, 636-637
Thevenin equivalent circuit, 500-502
    Norton form, 502
thrust and rockets, 460
time period, 9
    differential time period, 23
    finite time period, 23
torque, 136-137, 157
torque about an axis, 157
translational kinetic energy, 179
transverse velocity component, 144
triple point, 572
truss, 277-286

U

UHAG relations, 545-546
units of measure, 687-695
unit vector, 698

V

Van der Waals equation, 576

vaporization, 572
vapor pressure, 605-606
vectors, 697-716
velocity, 314
    average, 426-427
    profile, 426
velocity correction factor,
    for Bernoulli's equation, 429
    for momentum, 429
Venturi meter, 452-453
virial equations of state, 575-576
viscosity,
    and the Reynolds number, 435
    tables, 718-720
voltage, 203-204, 477
voltage divider circuit, 482
volume expansivity, 599
volumetric properties, 572-580
virtual work, 294

W

Watson equation, 605
work, 186
"work-energy" theorem, 366